ENANTIOSELECTIVE CHEMICAL SYNTHESIS

Methods, Logic and Practice

Direct Book Publishing, LLC is a cooperative enterprise that provides superior compensation for authors and sells books at very affordable prices. For more information, write to the following e-mail address: directbookpublishing@gmail.com. To order books, visit http://directbookpublishing.com/.

ENANTIOSELECTIVE CHEMICAL SYNTHESIS

Methods, Logic and Practice

E.J. Corey and László Kürti

Department of Chemistry
and
Chemical Biology
Harvard University

DIRECT BOOK
PUBLISHING

Published by *Direct Book Publishing, LLC*, Dallas, Texas.

Cover and *Direct Book Publishing* Logo design: László Kürti

Library of Congress Catalog-in-Publication Data is available.

ISBN: 978-0-615-39515-9 (hardcover)

Printed in the United States of America.

PREFACE

The prodigious growth of knowledge in the chemical sciences of the 20th century continues in the 21st. Accretive and gradual, more like a growing ocean wave than a consequence of dramatic or revolutionary changes, the advance is particularly significant in synthetic chemistry because of its centrality and value to society. For more than a century the rate of progress in synthetic chemistry has been such that every 15 years, or so, problems could be solved that were too difficult for scientists in the preceding period. The subject matter of this book provides abundant evidence of the increasing power of chemistry in recent years. It is a remarkable story considering that even in the mid-20th century the ability to control the absolute stereochemical course of chemical reactions seemed beyond reach. Chemists have been keenly aware for more than 100 years of the awesome catalytic power of nature's catalysts, the enzymes, to produce complex organic molecules with complete stereocontrol, including absolute configuration. Yet, it is only now that centamolecular catalysts (molecular weights in the hundreds) are being created by chemists that begin to rival the much larger and more complex enzymes of biochemistry.

In Part I of this book we have tried to present clearly, comprehensively and concisely the most useful enantioselective processes available to synthetic chemists. Part II provides an extensive discussion of the most logical ways to apply these new enantioselective methods to the planning of syntheses of stereochemically complex molecules. This hitherto neglected area is essential for the advancement of enantioselective synthesis to a more rational and powerful level. Finally, Part III describes in detail many reaction sequences which have been used successfully for the construction of a wide variety of complex target molecules.

Use of Colors in This Book

The substrates, reactants and reagents of the chemical transformations that appear in Part I and II of this book are shown in three different colors: blue, red and occasionally magenta. The ligand or catalyst (or chiral controller) is always shown in green as illustrated in Figure 1. The newly formed bonds are shown in black and the coloring of the product reflects the origin of the subunits that have been incorporated as a result of the transformation.

All the multi-step reaction sequences displayed in Part III utilize the color red to show newly formed bonds. Figure 2 illustrates the ring-opening of aziridine **15** to product **16,** with the new C-O bond shown in red. The red

C-N bond in the precursor **15** had been formed in the previous step.

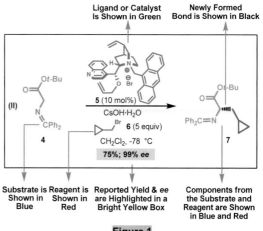

Figure 1

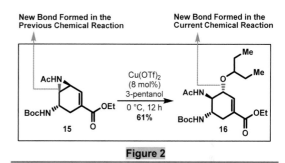

Figure 2

Our goal in writing this book has been to make accessible to researchers and students of synthetic chemistry the vast and still-expanding field of enantioselective synthesis in the hope that this fascinating area of chemistry can more readily be mastered.

Many scientific colleagues have perused pre-publication versions of this book. We have been encouraged by their very positive comments. It would be gratifying if this book empowers students and colleagues all over the world and accelerates progress in synthetic chemistry.

E.J. Corey & László Kürti

Harvard University, Cambridge, MA
August 2010

CONTENTS

PART I.

TOOLS FOR ENANTIOSELECTIVE SYNTHESIS

PART II.

PLANNING ENANTIOSELECTIVE SYNTHESES

PART III.

ENANTIOSELECTIVE MULTI-STEP SYNTHESIS: EXAMPLES

PART I.

TOOLS FOR
ENANTIOSELECTIVE SYNTHESIS

TOOLS FOR ENANTIOSELECTIVE SYNTHESIS

Introduction

During the last three decades the field of enantioselective methodology for chemical synthesis has been transformed by a large number of important discoveries and ideas. Those advances have markedly enhanced the ability of chemists to assemble complex organic structures with multiple stereocenters. There are now a very large number of reactions for generating new stereocenters by addition to the sp^2 carbon of C=C, C=O, C=N and Michael acceptors among others. These reactions can lead to C-C, C-O, C-N or C-heteroatom bond formation. They may involve the use of chiral reagents or catalysts, either with or without metal centers. The result, in addition to the generation of new stereocenters, may be chain-extension, functional group introduction or addition, ring-formation, rearrangement or appendage attachment. For all these reasons, the field of multistep assembly of complex organic (i.e., carbogenic = member of the family of carbon compounds) molecular assembly has been both expanded and enriched.

One consequence of the intense activity in this area by many research scientists has been a dramatic increase in the possibilities for applying synthetic chemistry to human needs and well-being. However, a less favorable byproduct of recent progress is the ever-growing task for synthetic chemists to identify or select the most effective methods for solving a synthetic chemical problem. The first part of this book is meant to alleviate this burden on chemists by providing an extensive summary of the most useful new tools for enantioselective synthesis.

Topics are presented in a standardized way with generous use of structural formulas and examples. The organization is both informal and intuitive, so as to facilitate locating the needed information or comparing alternative tools for a particular synthetic task. The referencing is both recent and extensive to allow easy access to the vast supporting literature. Particular attention has been given to the inclusion of the most useful review articles on each subject. An attempt has been made to present each topic clearly and concisely. Another objective has been to serve the needs of both the intermediate student of synthetic chemistry and the experienced researcher.

Standard Layout of Sections in Part I

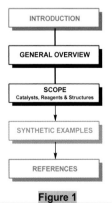

Figure 1

The block diagram in Figure 1 depicts the standard layout of sections in Part I of this book. Figure 2 illustrates a set of reactions as they usually appear in the subsequent sections of Part I. Each chemical transformation is labeled with a roman numeral. The catalyst or ligand loading is shown above the arrow and the name or structure of the reagent or reactant appears below along with the reaction conditions. The reported yield and enantiomeric excess are highlighted in bright yellow boxes. If appropriate, the diastereomeric ratio is displayed under the structure of the major product. The structures and identifying numbers of the catalysts and ligands appear in the lower part of the scheme. Finally, the appropriate references are indicated for each reaction entry.

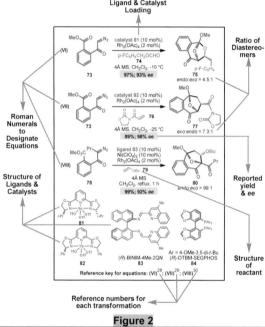

Figure 2

References are given in a standard format throughout the book, as exemplified in Figure 3. Each reference contains a complete list of the authors, full title and full journal citation.

References

28. (a) Suga, H., Inoue, K., Inoue, S. & Kakehi, A. Highly Enantioselective 1,3-Dipolar Cycloaddition Reactions of 2-Benzopyrylium-4-olate Catalyzed by Chiral Lewis Acids. *J. Am. Chem. Soc.* **2002**, 124, 14836-14837; (b) Suga, H., Inoue, K., Inoue, S., Kakehi, A. & Shiro, M. Chiral 2,6-Bis(oxazolinyl)pyridine-Rare Earth Metal Complexes as Catalysts for Highly Enantioselective 1,3-Dipolar Cycloaddition Reactions of 2-Benzopyrylium-4-olates. *J. Org. Chem.* **2005**, 70, 47-56.
29. Suga, H., Suzuki, T., Inoue, K. & Kakehi, A. Asymmetric cycloaddition reactions between 2-benzopyrylium-4-olates and 3-(2-alkenoyl)-2-oxazolidinones in the presence of 2,6-bis(oxazolinyl)pyridine-lanthanoid complexes. *Tetrahedron* **2006**, 62, 9218-9225.
30. Suga, H., Ishimoto, D., Higuchi, S., Ohtsuka, M., Arikawa, T., Tsuchida, T., Kakehi, A. & Baba, T. Dipole-LUMO/Dipolarophile-HOMO Controlled Asymmetric Cycloadditions of Carbonyl Ylides Catalyzed by Chiral Lewis Acids. *Org. Lett.* **2007**, 9, 4359-4362.

Figure 3

TOOLS FOR ENANTIOSELECTIVE SYNTHESIS
CONTENTS

TOOLS FOR ENANTIOSELECTIVE SYNTHESIS
CONTENTS

TOOLS FOR ENANTIOSELECTIVE SYNTHESIS
CONTENTS

1.6 Cyclization

ENANTIOSELECTIVE ADDITION OF H_2 TO C=C

Background: Enantioselective Homogeneous Hydrogenation

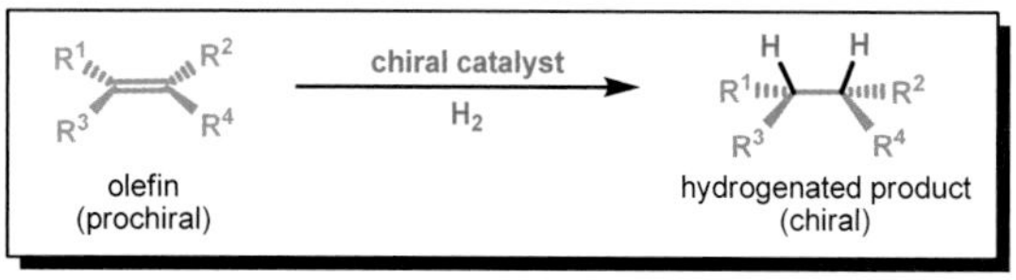

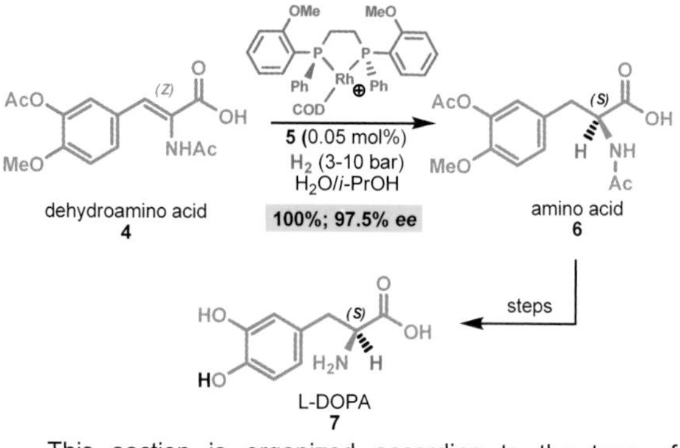

The discovery by G. Wilkinson that the solution of the rhodium complex **1**, catalyzes the reduction of unhindered alkenes by H_2 gas at 1 atmosphere pressure[1] paved the way for W.S. Knowles[2] and co-workers to show that the use of the chiral monodentate phosphorous ligand (*S*)-CAMP (**2**) resulted in good enantioselectivity in the Wilkinson hydrogenation of dehydroamino acids. Studies by H. Kagan[3] revealed that C_2-symmetric bidentate phosphine ligands such as **3** gave comparable results.

Wilkinson's cat. (1966)
1

(*S*)-CAMP (1972)
2

(*S*,*S*)-DIOP (1971)
3

An even more effective bidentate phosphine ligand-Rh complex, (*S*,*S*)-DIPAMP-Rh (**5**), that was developed by Knowles was used in the commercial production at Monsanto of L-DOPA **7**, an anti-Parkinson's drug.[4]

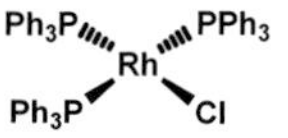

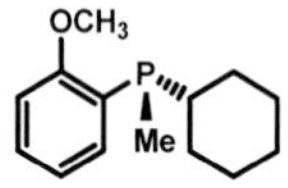

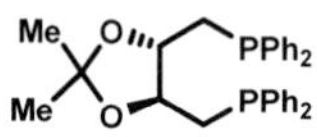

This section is organized according to the type of unsaturated substrate and the transition metal (Rh, Ru, Ir, etc.) used for reduction by H_2.

Most Common Substrate Types

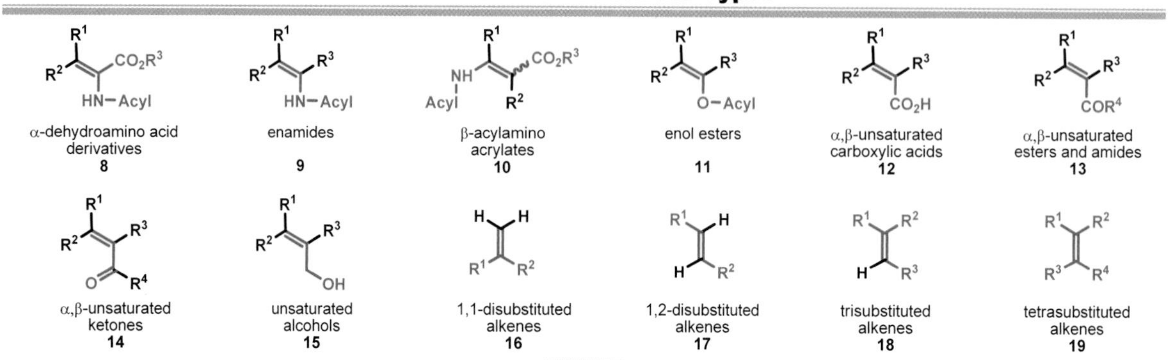

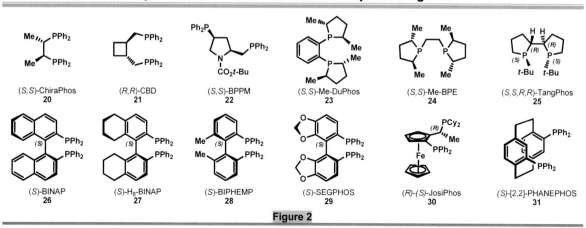

Figure 1

Representative Chiral Bidentate Phosphorus Ligands[5]

Figure 2

ENANTIOSELECTIVE ADDITION OF H₂ TO C=C

Chiral Monodentate Phosphorus, P,N and Non-Phosphorus Ligands[6]

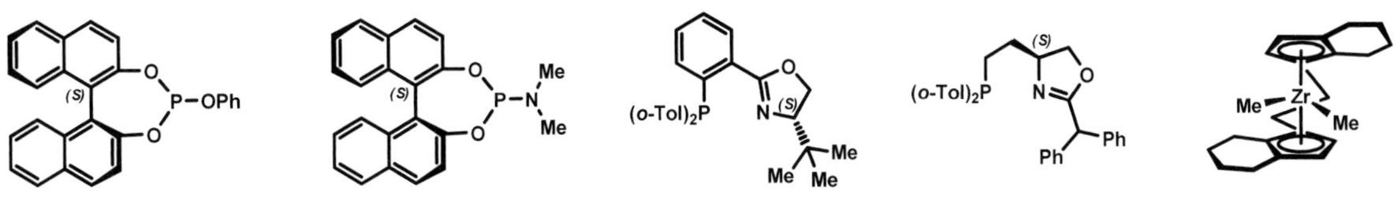

| monophosphite 32 | (S)-MonoPhos 33 | (S)-PHOX 34 | (S)-JM-Phos 35 | zirconocene 36 |

Figure 3

Comments on Figures 1-3

(1) Most of the successful catalytic enantioselective hydrogenations have utilized functionalized olefinic substrates (Figure 1, **8-15**) that have one or more heteroatom-containing groups in proximity to the double bond which can coordinate with the catalytic metal at C=C.[7] The enantioselective hydrogenation of unfunctionalized olefins (**16-19**) has proven to be much more challenging, due to less constraining monodentate coordination of these substrates to the metal.[8]

(2) Many chiral bidentate ligands have been developed for metal-catalyzed enantioselective hydrogenation of C=C since no one is universally applicable.[9a,9b,5c,9c]

(3) Other transition metals besides rhodium,[10] especially ruthenium (Ru)[11] and iridium (Ir),[12] have become important since the existing Rh-based catalysts are not effective for the enantioselective hydrogenation of all substrate types (Figure 1, **10-19**).[8b]

(4) The bidentate phosphorus ligands shown in Figure 2 represent a sampling of the main structural types.[5a,5c]

(5) Ligands **20** and **21** are C_2-symmetric, in common with Kagan's DIOP ligand (**3**). Me-DuPhos (**23**) and Me-BPE (**24**) contain C_2-symmetric phosphacycles separated by aromatic and aliphatic two-carbon spacers, respectively. TangPhos (**25**) is unique because it has chirality both on carbon and phosphorus. (S)-BINAP (**26**) and (S)-H₈–BINAP (**27**) are both axially chiral (atropisomeric) and C_2-symmetric. Two other important atropisomeric ligands are BIPHEMP (**28**) and SEGPHOS (**29**). JosiPhos (**30**) is one of many ferrocene-based bisphosphine ligands while PHANEPHOS (**31**) is based on a [2,2]-paracyclophane.

(6) Recently, monophosphine ligands (Figure 3, **32** and **33**) have been found to be effective in Rh-catalyzed asymmetric hydrogenations.[13a,13b,6b,13c] P,N ligands,[6c] such as **34** and **35**, have found use in the enantioselective iridium (Ir)-catalyzed hydrogenation of unfunctionalized alkenes.[8a] The zirconocene catalyst **36** was shown to catalyze the highly enantioselective hydrogenation of tetrasubstituted alkenes.[8b]

Enantioselective Hydrogenation of α- and β-Dehydroamino Acid Derivatives[5a,14]

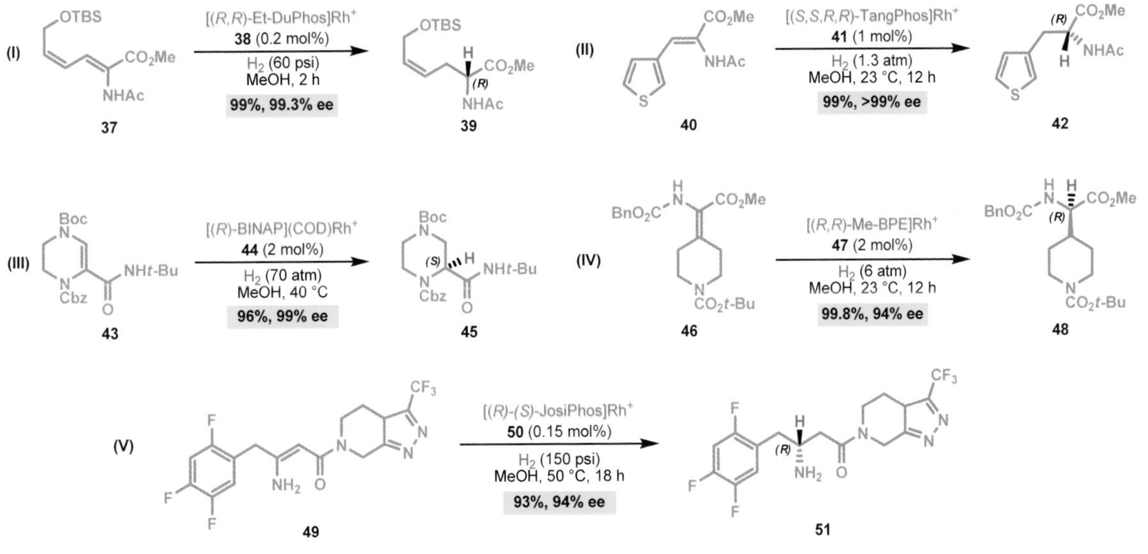

Reference Key for Equations: (I)[15]; (II)[16]; (III)[17]
Scheme 1
Reference Key for Equations: (IV)[18]; (V)[19]

ENANTIOSELECTIVE ADDITION OF H$_2$ TO C=C

Enantioselective Hydrogenation of α- and β-Dehydroamino Acids

Comments on Scheme 1

(1) The hydrogenations summarized in Scheme 1 share several features of the reaction pathway, including:

    (a) bidentate coordination of the π-electrons of the C=C bond undergoing reduction and a lone pair of the nearby carbonyl oxygen with the metal center;

    (b) metal insertion into H$_2$ (oxidative addition) to form a dihydride intermediate;

    (c) addition to C=C to form new C-H and C-Rh-H bonds and;

    (d) reductive elimination of the C-Rh-H subunit to form the second C-H bond.

There is probably some variation with differing reactants in the fine details of kinetics and pre-transition state assembly geometry.[20]

(2) Tetrahydropyrazine **43** gave piperazine **45**, a key chiral intermediate in the preparation of HIV protease inhibitor Indinavir sulfate, in 96% yield and 99% ee upon hydrogenation in the presence of the cationic (R)-BINAP-Rh complex **44**, which was selected as a result of extensive screening of several other Rh-catalysts.[17]

(3) The synthesis of (R)-N-Boc-4-piperidinylglycine was achieved in high enantiomeric excess by the Rh-catalyzed hydrogenation of tetrasubstituted olefin **46**. The ligand (R,R)-Me-BPE (ent-**24**, Figure 2) is more effective than either DuPhos (**23**) or DIPAMP (**5**).[18]

(4) The JosiPhos-Rh complex (**50**) has been applied to the industrial preparation of sitagliptin (Januvia™).[21,19]

Enantioselective Hydrogenation of Enamides[5a,5c,5e]

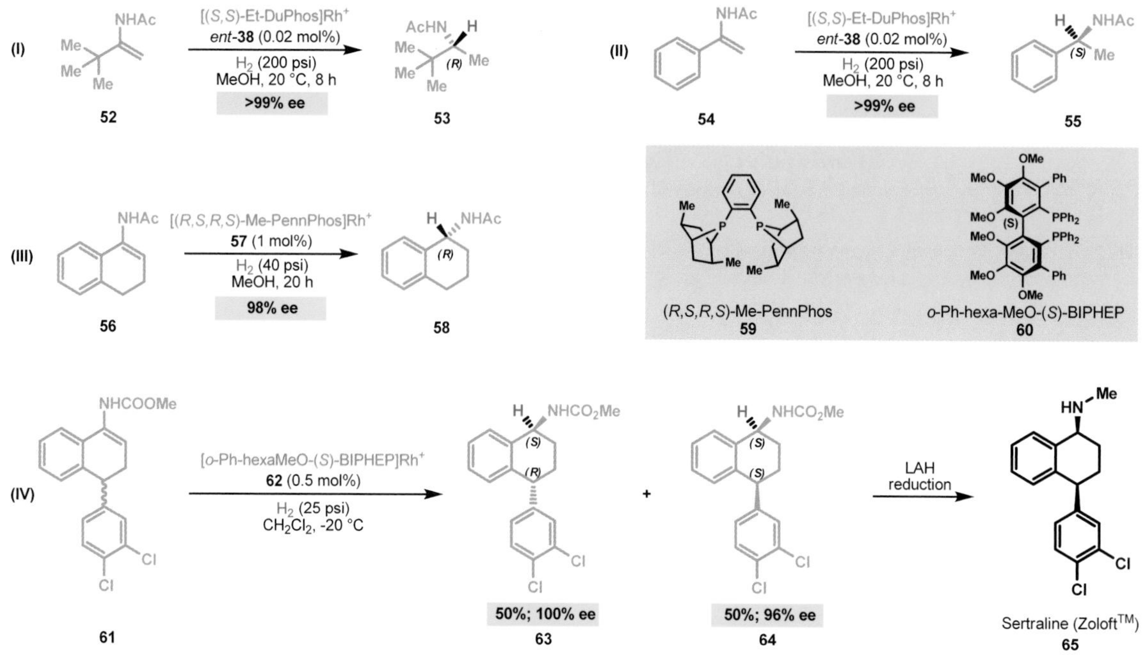

Scheme 2

Reference Key for Equations: (I)[22]; (II)[22] Reference Key for Equations: (III)[23]; (IV)[24]

Description of Scheme 2

(1) The hydrogenations of a bulky alkyl enamide (e.g., **52**→**53**) and an aryl-substituted enamide (e.g., **54**→**55**) with the catalyst (S,S)-Et-DuPHOS-Rh (ent-**38**) occur with opposite π-facial selectivity.[22]

(2) The highly enantioselective reduction of cyclic enamides (e.g., **56**→**58**) is possible using the Rh-(R,S,R,S)-Me-PennPhos catalyst (**57**).[23]

(3) Structurally complex cyclic encarbamates are hydrogenated with high enantioselectivity in the presence of Rh-complex **62** which contains the ortho-substituted BIPHEP ligand **60**. Racemic substrate **61** afforded both the cis and trans products (**63** and **64**) with high enantiomeric excess.[24] Reduction of carbamate **64** gave rise to the antidepressant sertraline (**65**, Zoloft™).

ENANTIOSELECTIVE ADDITION OF H$_2$ TO C=C

Enantioselective Hydrogenation of Enamides (continued)

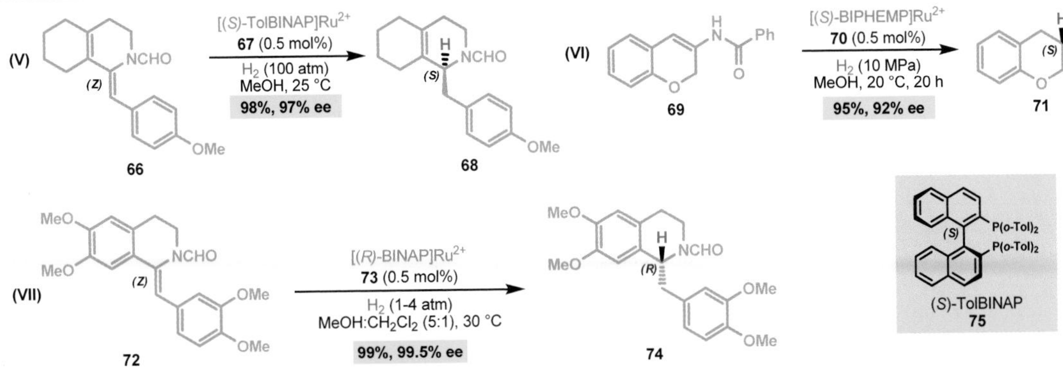

Scheme 3

Reference Key for Equations: (V)[25]; (VI)[26]

Reference Key for Equation: (VII)[25]

Comments on Scheme 3

(1) The cyclic (Z)-enamides **66** and **72** have been hydrogenated with high enantioselectivity using the Ru complexes **67** (from **75**) and **73**. These were shown to be superior to the corresponding Rh complexes. The enantioselective hydrogenation of **72** demonstrates a general asymmetric pathway to isoquinoline alkaloids.[25]

(2) The highly enantioselective conversion of the trisubstituted cyclic enamide **69** (derived from 3-chromanone) to the 3-aminochroman **71** has been achieved using the Ru-(S)-BIPHEMP catalyst **70**.[26]

Enantioselective Hydrogenation of Enol Esters[5a,5c,5e]

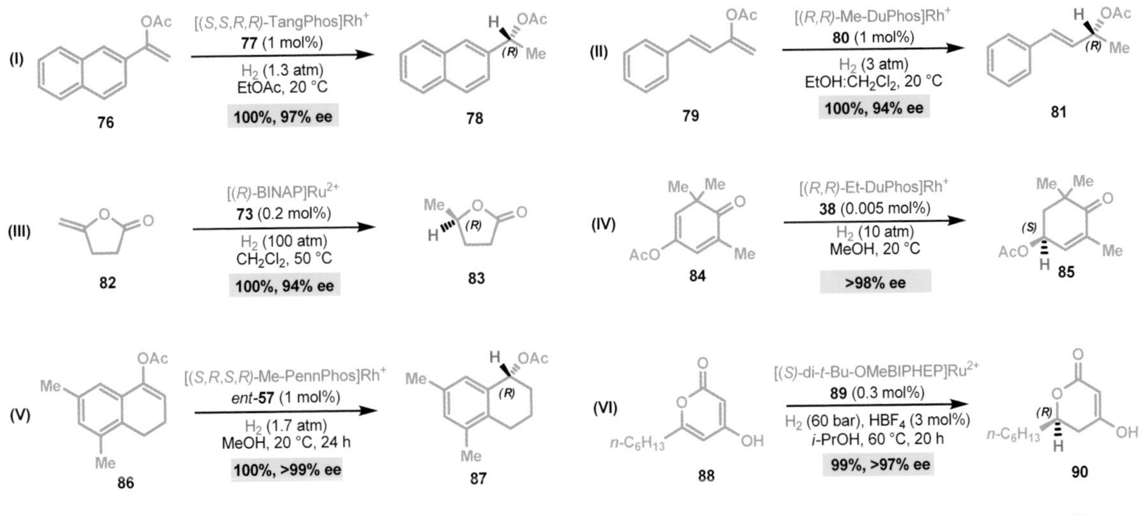

Scheme 4

Reference Key for Equations: (I)[27]; (II)[28]; (III)[29]

Reference Key for Equations: (IV)[5c]; (V)[30]; (VI)[31]

Comments on Scheme 4

(1) The highly enantioselective catalytic hydrogenation of enol esters is more challenging than that of dehydroamino acids (see Scheme 1), perhaps because of weaker coordination of the carbonyl oxygen atom to the metal center.[5e]

(2) Both ruthenium and rhodium bisphosphine complexes (e.g., **77**, **80**, **73**) have been applied to the reduction of acyclic[28,27] and cyclic[30,5c] enol esters generally under mild conditions, although relatively high H$_2$ pressure is required for certain cyclic substrates (e.g., **82**, **88**).[29,31]

(3) The reduction of cyclic enol acetate **84** provided a route to the carothenoid zeaxhantin.[5c]

ENANTIOSELECTIVE ADDITION OF H_2 TO C=C

Enantioselective Hydrogenation of α,β-Unsaturated Carboxylic Acids and Derivatives[5a,5e]

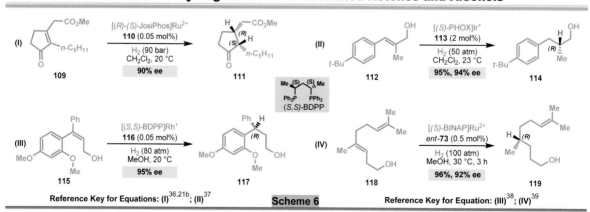

Reference Key for Equations: (I)[32]; (II)[33]; (III)[27]   **Scheme 5**   Reference Key for Equations: (IV)[34]; (V)[13c]; (VI)[35]

Comments on Scheme 5

(1) Ruthenium-bisphosphine complexes function well in the enantioselective hydrogenation of di-, tri- and tetrasubstituted α,β-unsaturated acids (e.g, **91** and **94**).[33,32] The transformation **91→93** has been used in the large-scale production of (S)-ibuprofen. The hydrogenation of the tetrasubstituted carboxylic acid **94** affords **96**, a key building block for the calcium antagonist Mibefradil.[33]

(2) The monodentate phosphoramidite ligand **107**, in combination with ruthenium(II), is an effective catalyst (**102**) in the hydrogenation of **101** to **103**, a key

intermediate in the synthesis of the antihypertensive drug Aliskiren.[6b,13c]

(3) Hydrogenation of itaconic acid derivative **97** with Rh(I)-complex **77** furnishes succinate **98** in high yield and enantioselectivity.[27] The reduction of E/Z olefin mixture **99** affords nearly enantiopure dimethyl succinate derivative **100**.[34]

(4) The FerroTane-Ru(II) complex **105** was used in the kg-scale enantioselective synthesis of 2-substituted succinate **106**, an intermediate for the synthesis of the MMP-3 inhibitor UK-370,106, from the cyclohexylamine salt of itaconate **104**.[35,5c]

Enantioselective Hydrogenation of Unsaturated Ketones and Alcohols[5a,5e]

Reference Key for Equations: (I)[36,21b]; (II)[37]   **Scheme 6**   Reference Key for Equation: (III)[38]; (IV)[39]

10

ENANTIOSELECTIVE ADDITION OF H$_2$ TO C=C

Enantioselective Hydrogenation of Unsaturated Ketones and Alcohols

Comments on Scheme 6

(1) The fragrance (+)-cis-methyl dihydro-jasmonate (**111**) is produced from the unsaturated ketone α,β-enone **109**. The Ru-JosiPhos catalyst (**110**) was the most enantioselective for this challenging hydrogenation of the tetrasubstituted C=C bond in **109**.[36,21b]

(2) The trisubstituted allylic alcohol **115** was hydrogenated in the presence of the Rh-BDPP complex **116** to provide **117**, an intermediate that can

be used to prepare several members of the dalbergione family of quinone natural products.[38]

(3) Trisubstituted allylic alcohols (e.g, **112**) can also be hydrogenated with the Ir-PHOX catalyst **113** which is made from the N,P-ligand (S)-PHOX (**34**), Figure 3.[37]

(4) The homoallylic alcohol homogeraniol (**118**) undergoes hydrogenation with high position- and enantioselectivity in the presence of the Ru-BINAP complex *ent*-**73**.[39]

Enantioselective Hydrogenation of Monodentate Olefins[8,40]

Reference Key for Equations: (I)[41]; (II)[42]; (III)[41]; (IV)[43]; (V)[44]   **Scheme 7**   Reference Key for Equation: (VI)[44]; (VII)[45]; (VIII)[46]; (IX)[43]

ENANTIOSELECTIVE ADDITION OF H₂ TO C=C

Enantioselective Hydrogenation of Monodentate Olefins

Comments on Scheme 7

(1) The enantioselective hydrogenation of olefins lacking other coordinating groups has been more problematic than that of substrates bearing donor groups near the olefinic linkage. The chiral Rh and Ru catalysts which are effective for bidentate substrates (i.e., ones with C=C and a nearby functional group) are much less so for monodentate olefins. In contrast, chiral cationic Ir(I) complexes can catalyze highly enantioselective hydrogenation of monodentate olefins.[8,40]

(2) The success of these cationic Ir(I) complexes in the enantioselective hydrogenation of monodentate

olefins is partly due to the use of the weakly coordinating counter-ion BARF⁻ (**151**) and also to the stronger π-coordinating ability of Ir versus Rh.

BARF⁻ = tetra(3,5-bistrifluoromethyl)phenyl borate **151**

(3) Substrates with di-, tri- and tetrasubstituted double bonds which are in conjugation with aromatic rings (e.g., **120**[41], **123**[42], **132**[44], **134**[44] and **136**[45]) can be hydrogenated with high enantioselectivity. The Ir(I) catalysts **145**, **146** and **149** (with chiral bidentate P,N- or C,N-type ligands) have been especially useful. The chiral zirconocene catalyst **36** (Figure 3) has also been successfully used (e.g, **132**→**133**[44] and **134**→**135**[44]).

(4) The highly enantioselective hydrogenation of 1,3-dienes such as **139** (with trisubstituted double bonds that are in conjugation with an aromatic ring) has been achieved using the novel C,N-type ligand **150** in combination with Ir(I).[46] With other types of ligands the hydrogenation of conjugated dienes leads to mixtures of products.[8a]

(5) Purely alkyl-substituted olefin substrates (e.g., **129** and **142**) can be hydrogenated with very high enantioselectivity in the presence of the P,N-type pyridine-phosphinite-Ir complexes **130** and **143**.[43]

Appendix: π-Face-Selective Hydrogenation Directed by a Functional Group

In just the same way that chiral cationic Rh or Ir catalysts can direct π-face-selective hydrogenation of an achiral olefin to a chiral product, π-face-selective hydrogenation can be achieved under direction of a coordinating functional group attached to a stereocenter in proximity to C=C by an achiral Rh⁺ or Ir⁺ catalyst.[47] This process is illustrated by two examples which are shown at right.[48]

Functional group-directed addition to C=C, especially in allylic and homoallylic substrates, provides a major approach to stereoselective synthesis using reactions such as epoxidation, methylenation and cycloaddition.[49]

152 → [(NDB)(DIPHOS-4)]Rh⁺BF₄⁻ (**153**, 1.8 mol%), H₂ (950 psi), THF, 25 °C, **93%** → **154**

155 → [(COD)(py)(PCy₃)]Ir⁺PF₆⁻ (**153**, 20 mol%), H₂ (1 atm), CH₂Cl₂, 23 °C, **> 90%** → **156**

References

1. Osborn, J. A., Jardine, F. H., Young, J. F. & Wilkinson, G. Preparation and properties of tris(triphenylphosphine)halorhodium(I) and some reactions thereof including catalytic homogeneous hydrogenation of olefins and acetylenes and their derivatives. *J. Chem. Soc., A* **1966**, 1711-1732.

2. Knowles, W. S., Sabacky, M. J. & Vineyard, B. D. Catalytic asymmetric hydrogenation. *J. Chem. Soc., Chem. Commun.* **1972**, 10-11.

3. Kagan, H. B. & Dang, T. P. Asymmetric synthesis of hydratropic acid and amino acids by homogeneous catalytic hydrogenation. *J. Chem. Soc. D.* **1971**, 481.

4. (a) Knowles, W. S. Asymmetric hydrogenations (Nobel Lecture). *Angew. Chem., Int. Ed.* **2002**, 41, 1998-2007; (b) Knowles, W. S. Asymmetric hydrogenations - The Monsanto L-dopa process in *Asymmetric Catalysis on Industrial Scale* 23-38 (**2004**).

5. (a) Tang, W. & Zhang, X. New Chiral Phosphorus Ligands for Enantioselective Hydrogenation. *Chem. Rev. (Washington, DC, U. S.)* **2003**, 103, 3029-3069; (b) Shimizu, H., Nagasaki, I. & Saito, T. Recent advances in biaryl-type bisphosphine ligands. *Tetrahedron* **2005**, 61, 5405-5432; (c) Cobley, C. J. & Moran, P. H. Enantioselective hydrogenation: phospholane ligands in *Handbook of Homogeneous Hydrogenation* 773-831 (**2007**); (d) Blaser, H.-U., Lotz, M. & Spindler, F.

Enantioselective hydrogenation of alkenes with ferrocene-based ligands in *Handbook of Homogeneous Hydrogenation* 833-851 (**2007**); (e) Chi, Y., Tang, W. & Zhang, X. The other bisphosphine ligands for enantioselective alkene hydrogenation in *Handbook of Homogeneous Hydrogenation* 853-882 (**2007**).

6. (a) Kok, S. H. L., Au-Yeung, T. T. L., Cheung, H. Y., Lam, W. S., Chan, S. S. & Chan, A. S. C. Bidentate ligands containing a heteroatom-phosphorus bond in *Handbook of Homogeneous Hydrogenation* 883-993 (**2007**); (b) van den Berg, M., Feringa, B. L. & Minnaard, A. J. Enantioselective alkene hydrogenation: monodentate ligands in *Handbook of Homogeneous Hydrogenation* 995-1027 (**2007**); (c) Pfaltz, A. & Bell, S. P,N and non-phosphorus ligands in *Handbook of Homogeneous Hydrogenation* 1029-1048 (**2007**).

7. Ager, D. Enantioselective alkene hydrogenation: Introduction and historic overview in *Handbook of Homogeneous Hydrogenation* 745-772 (**2007**).

8. (a) Cui, X. & Burgess, K. Catalytic homogeneous asymmetric hydrogenations of largely unfunctionalized alkenes. *Chem. Rev.* **2005**, 105, 3272-3296; (b) Pfaltz, A. & Bell, S. Enantioselective hydrogenation of unfunctionalized alkenes in *Handbook of Homogeneous Hydrogenation* 1049-1072 (**2007**).

9. (a) Xie, J.-H., Zhu, S.-F., Fu, Y., Hu, A.-G. & Zhou, Q.-L. New chiral phosphorus ligands with spirobiindane backbone for asymmetric hydrogenations. *Pure Appl. Chem.* **2005**, 77, 2121-2132; (b) Ohkuma, T., Kitamura, M. & Noyori, R. Ligand design for catalytic asymmetric reduction in *New Frontiers in Asymmetric Catalysis* 1-32 (**2007**); (c) de Vries, J. G. & Lefort, L. High-throughput experimentation and ligand libraries in *Handbook of Homogeneous Hydrogenation* 1245-1278 (**2007**).

10. Chi, Y., Tang, W. & Zhang, X. Rhodium-catalyzed asymmetric hydrogenation in *Modern Rhodium-Catalyzed Organic Reactions* 1-31 (**2005**).

11. Morris, R. H. Ruthenium and Osmium in *Handbook of Homogeneous Hydrogenation* 45-70 (**2007**).

12. (a) Kaellstroem, K., Munslow, I. & Andersson, P. G. Ir-catalyzed asymmetric hydrogenation: ligands, substrates and mechanism. *Chem.-- Eur. J.* **2006**, 12, 3194-3200; (b) Crabtree, R. H. Iridium in *Handbook of Homogeneous Hydrogenation* 31-44 (**2007**); (c) Roseblade, S. J. & Pfaltz, A. Iridium-Catalyzed Asymmetric Hydrogenation of Olefins. *Acc. Chem. Res.* **2007**, 40, 1402-1411; (d) Church, T. L. & Andersson, P. G. Iridium catalysts for the asymmetric hydrogenation of olefins with nontraditional functional substituents. *Coord. Chem. Rev.* **2008**, 252, 513-531; (e) Blaser, H.-U. Application of iridium catalysts in the fine chemicals industry in *Iridium Complexes in Organic Synthesis* 1-14 (**2009**).

13. (a) van den Berg, M., Minnaard, A. J., Haak, R. M., Leeman, M., Schudde, E. P., Meetsma, A., Feringa, B. L., de Vries, A. H. M., Maljaars, C. E. P., Willans, C. E., Hyett, D., Boogers, J. A. F., Henderickx, H. J. W. & de Vries, J. G. Monodentate phosphoramidites: A breakthrough in rhodium-catalyzed asymmetric hydrogenation of olefins. *Adv. Synth. Catal.* **2003**, 345, 308-323; (b) de Vries, J. G. Asymmetric olefin hydrogenation using monodentate BINOL- and bisphenol-based ligands: phosphonites, phosphites, and phosphoramidites in *Handbook of Chiral Chemicals (2nd Edition)* 269-286 (**2006**); (c) Minnaard, A. J., Feringa, B. L., Lefort, L. & de Vries, J. G. Asymmetric Hydrogenation Using Monodentate Phosphoramidite Ligands. *Acc. Chem. Res.* **2007**, 40, 1267-1277.

14. Najera, C. & Sansano, J. M. Catalytic Asymmetric Synthesis of α-Amino Acids. *Chem. Rev.* **2007**, 107, 4584-4671.

15. Burk, M. J., Allen, J. G. & Kiesman, W. F. Highly Regio- and Enantioselective Catalytic Hydrogenation of Enamides in Conjugated Diene Systems: Synthesis and Application of γ,δ-Unsaturated Amino Acids. *J. Am. Chem. Soc.* **1998**, 120, 657-663.

16. Tang, W. & Zhang, X. A chiral 1,2-bisphospholane ligand with a novel structural motif: applications in highly enantioselective Rh-catalyzed hydrogenations. *Angew. Chem., Int. Ed.* **2002**, 41, 1612-1614.

17. Rossen, K., Weissman, S. A., Sager, J., Reamer, R. A., Askin, D., Volante, R. P. & Reider, P. J. Asymmetric hydrogenation of tetrahydropyrazines: synthesis of (S)-piperazine-2-tert-butylcarboxamide, an intermediate in the preparation of the HIV protease inhibitor indinavir. *Tetrahedron Lett.* **1995**, 36, 6419-6422.

18. Shieh, W.-C., Xue, S., Reel, N., Wu, R., Fitt, J. & Repic, O. An enantioselective synthesis of (R)-4-piperidinylglycine. *Tetrahedron: Asymmetry* **2001**, 12, 2421-2425.

19. Shultz, C. S. & Krska, S. W. Unlocking the Potential of Asymmetric Hydrogenation at Merck. *Acc. Chem. Res.* **2007**, 40, 1320-1326.

20. (a) Pettinari, C., Marchetti, F. & Martini, D. Metal complexes as hydrogenation catalysts in *Comprehensive Coordination Chemistry II* 75-139 (**2004**); (b) Gridnev, I. D. & Imamoto, T. On the Mechanism of Stereoselection in Rh-Catalyzed Asymmetric Hydrogenation: A General Approach for Predicting the Sense of Enantioselectivity. *Acc. Chem. Res.* **2004**, 37, 633-644; (c) Drexler, H. J., Preetz, A., Schmidt, T. & Heller, D. Kinetics of homogeneous hydrogenations: measurement and interpretation in *Handbook of Homogeneous Hydrogenation* 257-293 (**2007**); (d) Brown, J. M. Mechanism of enantioselective hydrogenation in *Handbook of Homogeneous Hydrogenation* 1073-1103 (**2007**).

21. (a) Oro, L. A., Carmona, D. & Fraile, J. M. Hydrogenation reactions in *Metal-Catalysis in Industrial Organic Processes* 79-113 (**2006**); (b) Blaser, H.-U., Spindler, F. & Thommen, M. Industrial applications in *Handbook of Homogeneous Hydrogenation* 1279-1324 (**2007**).

22. Burk, M. J., Casy, G. & Johnson, N. B. A Three-Step Procedure for Asymmetric Catalytic Reductive Amination of Ketones. *J. Org. Chem.* **1998**, 63, 6084-6085.

23. Zhang, Z., Zhu, G., Jiang, Q., Xiao, D. & Zhang, X. Highly Enantioselective Hydrogenation of Cyclic Enamides Catalyzed by a Rh-PennPhos Catalyst. *J. Org. Chem.* **1999**, 64, 1774-1775.

24. Tang, W., Chi, Y. & Zhang, X. An ortho-Substituted BIPHEP Ligand and Its Applications in Rh-Catalyzed Hydrogenation of Cyclic Enamides. *Org. Lett.* **2002**, 4, 1695-1698.

25. Kitamura, M., Hsiao, Y., Ohta, M., Tsukamoto, M., Ohta, T., Takaya, H. & Noyori, R. General asymmetric synthesis of isoquinoline alkaloids. Enantioselective hydrogenation of enamides catalyzed by BINAP-ruthenium(II) complexes. *J. Org. Chem.* **1994**, 59, 297-310.

26. Renaud, J. L., Dupau, P., Hay, A. E., Guingouain, M., Dixneuf, P. H. & Bruneau, C. Ruthenium-catalysed enantioselective hydrogenation of trisubstituted enamides derived from 2-tetralone and 3-chromanone: Influence of substitution on the amide arm and the aromatic ring. *Adv. Synth. Catal.* **2003**, 345, 230-238.

27. Tang, W., Liu, D. & Zhang, X. Asymmetric Hydrogenation of Itaconic Acid and Enol Acetate Derivatives with the Rh-TangPhos Catalyst. *Org. Lett.* **2003**, 5, 205-207.

28. Boaz, N. W. Asymmetric hydrogenation of acyclic enol esters. *Tetrahedron Lett.* **1998**, 39, 5505-5508.

29. Ohta, T., Miyake, T., Seido, N., Kumobayashi, H. & Takaya, H. Asymmetric Hydrogenation of Olefins with Aprotic Oxygen Functionalities Catalyzed by BINAP-Ru(II) Complexes. *J. Org. Chem.* **1995**, 60, 357-363.

30. Jiang, Q., Xiao, D., Zhang, Z., Chao, P. & Zhang, X. Highly enantioselective hydrogenation of cyclic enol acetates catalyzed by a Rh-PennPhos complex. *Angew. Chem., Int. Ed.* **1999**, 38, 516-518.

31. Fehr, M. J., Consiglio, G., Scalone, M. & Schmid, R. Asymmetric Hydrogenation of Substituted 2-Pyrones. *J. Org. Chem.* **1999**, 64, 5768-5776.

32. Kumobayashi, H., Miura, T., Sayo, N., Saito, T. & Zhang, X. Recent advances of BINAP chemistry in the industrial aspects. *Synlett* **2001**, 1055-1064.

33. Crameri, Y., Foricher, J., Scalone, M. & Schmid, R. Practical synthesis of (S)-2-(4-fluorophenyl)-3-methylbutanoic acid, key building block for the calcium antagonist Mibefradil. *Tetrahedron: Asymmetry* **1997**, 8, 3617-3623.

34. Burk, M. J., Bienewald, F., Harris, M. & Zanotti-Gerosa, A. Practical access to 2-alkylsuccinates through asymmetric catalytic hydrogenation of Stobbe-derived itaconates. *Angew. Chem., Int. Ed.* **1998**, 37, 1931-1933.

35. Ashcroft, C. P., Challenger, S., Derrick, A. M., Storey, R. & Thomson, N. M. Asymmetric Synthesis of an MMP-3 Inhibitor Incorporating a 2-Alkyl Succinate Motif. *Org. Process Res. Dev.* **2003**, 7, 362-368.

36. Genet, J.-P. Asymmetric Catalytic Hydrogenation. Design of New Ru Catalysts and Chiral Ligands: From Laboratory to Industrial Applications. *Acc. Chem. Res.* **2003**, 36, 908-918.

37. Lightfoot, A., Schnider, P. & Pfaltz, A. Enantioselective hydrogenation of olefins with iridium-phosphanodihydrooxazole catalysts. *Angew. Chem., Int. Ed.* **1998**, 37, 2897-2899.

38. Bissel, P., Nazih, A., Sablong, R. & Lepoittevin, J.-P. Stereoselective Synthesis of (R)- and (S)-4-Methoxydalbergione via Asymmetric Catalytic Hydrogenation. *Org. Lett.* **1999**, 1, 1283-1285.

39. Ohta, T., Takaya, H., Kitamura, M., Nagai, K. & Noyori, R. Asymmetric hydrogenation of unsaturated carboxylic acids catalyzed by BINAP-ruthenium(II) complexes. *J. Org. Chem.* **1987**, 52, 3174-3176.

40. Diesen, J. S. & Andersson, P. G. Hydrogenation of unfunctionalized alkenes in *Modern Reduction Methods* 39-64 (**2008**).

41. Cozzi, P. G., Zimmermann, N., Hilgraf, R., Schaffner, S. & Pfaltz, A. Chiral phosphinopyrrolyl-oxazolines: A new class of easily prepared, modular P,N-ligands. *Adv. Synth. Catal.* **2001**, 343, 450-454.

42. McIntyre, S., Hoermann, E., Menges, F., Smidt, S. P. & Pfaltz, A. Iridium-catalyzed enantioselective hydrogenation of terminal alkenes. *Adv. Synth. Catal.* **2005**, 347, 282-288.

43. Bell, S., Wuestenberg, B., Kaiser, S., Menges, F., Netscher, T. & Pfaltz, A. Asymmetric Hydrogenation of Unfunctionalized, Purely Alkyl-Substituted Olefins. *Science (Washington, DC, U. S.)* **2006**, 311, 642-644.

44. Troutman, M. V., Appella, D. H. & Buchwald, S. L. Asymmetric hydrogenation of unfunctionalized tetrasubstituted olefins with a cationic zirconocene catalyst. *J. Am. Chem. Soc.* **1999**, 121, 4916-4917.

45. Engman, M., Cheruku, P., Tolstoy, P., Bergquist, J., Voelker, S. F. & Andersson, P. G. Highly selective iridium-catalyzed asymmetric hydrogenation of trifluoromethyl olefins: a new route to trifluoromethyl-bearing stereocenters. *Adv. Synth. Catal.* **2009**, 351, 375-378.

46. (a) Cui, X. & Burgess, K. Iridium-Mediated Asymmetric Hydrogenation of 2,3-Diphenylbutadiene: A Revealing Kinetic Study. *J. Am. Chem. Soc.* **2003**, 125, 14212-14213; (b) Cui, X., Ogle, J. W. & Burgess, K. Stereoselective hydrogenations of aryl-substituted dienes. *Chem. Commun. (Cambridge, U. K.)* **2005**, 672-674.

47. (a) For a seminal publication, see: Crabtree, R. H. & Davis, M. W. Occurrence and origin of a pronounced directing effect of a hydroxyl group in hydrogenation with [Ir(cod)P(C₆H₁₁)₃(py)]PF₆. *Organometallics* **1983**, 2, 681-682; (b) For a recent review, see: Yamagishi, T. Diastereoselective hydrogenation in *Handbook of Homogeneous Hydrogenation* 631-712 (**2007**).

48. (a) Stork, G. & Kahne, D. E. Stereocontrol in homogeneous catalytic hydrogenation via hydroxyl group coordination. *J. Am. Chem. Soc.* **1983**, 105, 1072-1073; (b) Corey, E. J., Desai, M. C. & Engler, T. A. Total synthesis of (±)-retigeranic acid. *J. Am. Chem. Soc.* **1985**, 107, 4339-4341.

49. Hoveyda, A. H., Evans, D. A. & Fu, G. C. Substrate-directable chemical reactions. *Chem. Rev.* **1993**, 93, 1307-1370.

ENANTIOSELECTIVE α,β-REDUCTION OF C=C-C=O

Enantioselective Reduction of α,β-Unsaturated Carbonyl Compounds

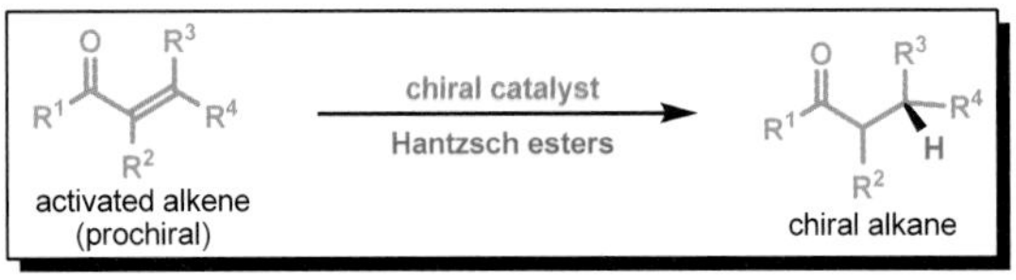

activated alkene (prochiral) → chiral alkane

1,4-Dihydropyridines in the Hantzsch ester series (e.g., **3** or **4**), in combination with chiral catalysts such as **5-8** (Figure 1), are capable of reducing the C=C bond in α,β-unsaturated aldehydes or ketones with high enantio- and position selectivity.[1]

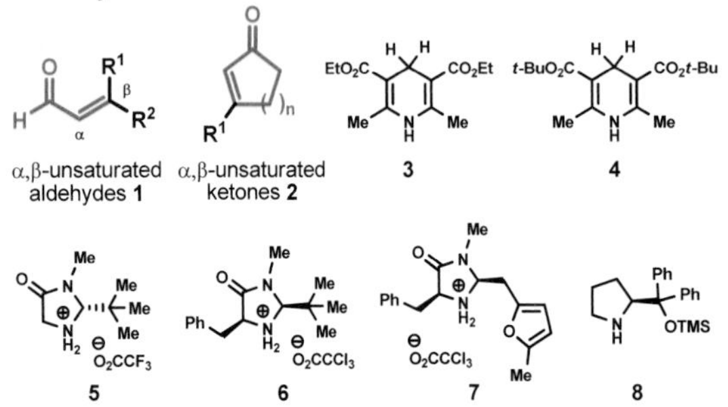

Examples of this process are shown in Schemes 2 and 3. Current methodology is not applicable to α-substituted enals and enones.

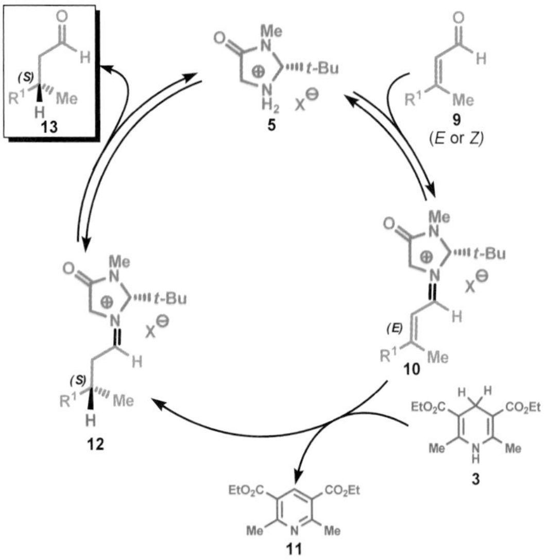

α,β-unsaturated aldehydes **1**   α,β-unsaturated ketones **2**

3   **4**

5   **6**   **7**   **8**

Figure 1

A possible pathway for the enantioselective Hantzsch ester mediated conjugate reduction via chiral iminium intermediate is illustrated for the imidazolidinone catalyst **5** in Scheme 1.

Scheme 1

Enantioselective Conjugate Reduction by 1,4-Dihydropyridines via Iminium Intermediates

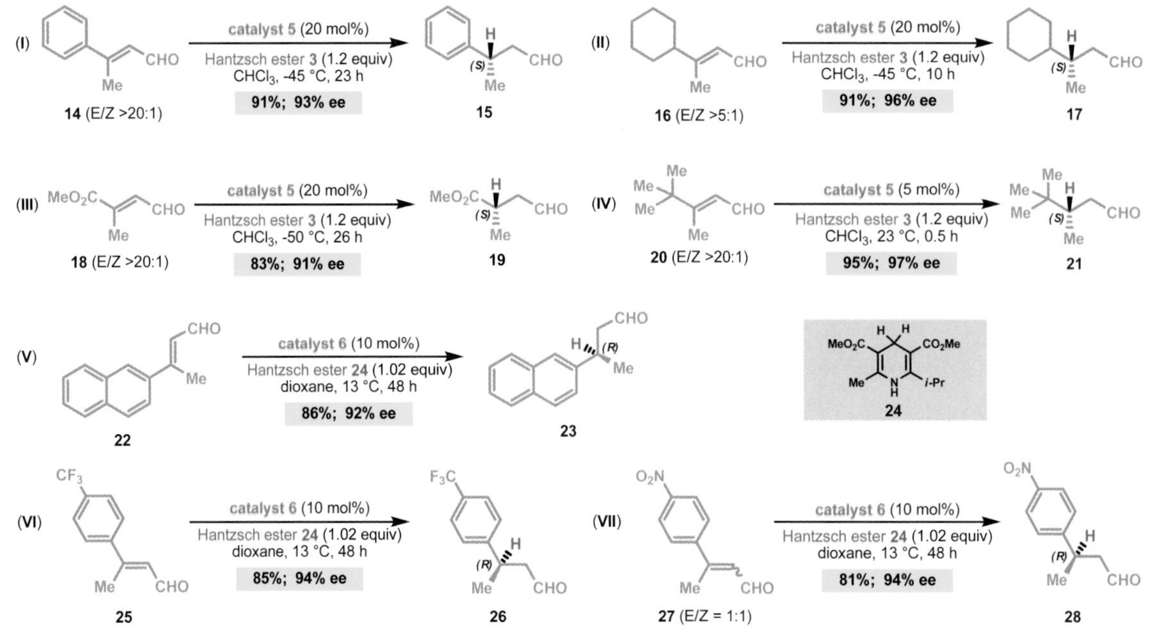

(I) **14** (E/Z >20:1) — catalyst 5 (20 mol%), Hantzsch ester 3 (1.2 equiv), CHCl₃, -45 °C, 23 h — 91%; 93% ee — (S) **15**

(II) **16** (E/Z >5:1) — catalyst 5 (20 mol%), Hantzsch ester 3 (1.2 equiv), CHCl₃, -45 °C, 10 h — 91%; 96% ee — (S) **17**

(III) **18** (E/Z >20:1) — catalyst 5 (20 mol%), Hantzsch ester 3 (1.2 equiv), CHCl₃, -50 °C, 26 h — 83%; 91% ee — (S) **19**

(IV) **20** (E/Z >20:1) — catalyst 5 (5 mol%), Hantzsch ester 3 (1.2 equiv), CHCl₃, 23 °C, 0.5 h — 95%; 97% ee — (S) **21**

(V) **22** — catalyst 6 (10 mol%), Hantzsch ester 24 (1.02 equiv), dioxane, 13 °C, 48 h — 86%; 92% ee — (R) **23** — **24**

(VI) **25** — catalyst 6 (10 mol%), Hantzsch ester 24 (1.02 equiv), dioxane, 13 °C, 48 h — 85%; 94% ee — (R) **26**

(VII) **27** (E/Z = 1:1) — catalyst 6 (10 mol%), Hantzsch ester 24 (1.02 equiv), dioxane, 13 °C, 48 h — 81%; 94% ee — (R) **28**

Reference Key for Equations: (I-III)[2]   **Scheme 2**   Reference Key for Equations: (IV)[2]; (V-VII)[3]

ENANTIOSELECTIVE α,β-REDUCTION OF C=C-C=O

Proton-Accelerated Enantioselective Conjugate Reduction of Activated Alkenes

Citral (29) → Citronellal (30)
catalyst 39 (20 mol%)
Hantzsch ester 24 (1.1 equiv)
THF, 23 °C, 24 h
71%; 90% ee

(IX) 31 → 32
catalyst 39 (20 mol%)
Hantzsch ester 24 (1.1 equiv)
dioxane, 23 °C, 24 h
84%; 98% ee

(X) 33 → 34
catalyst 7 (20 mol%)
Hantzsch ester 4 (1.1 equiv)
Et₂O, 0 °C, 6 h
81%; 96% ee

(XI) 35 → 36
catalyst 7 (20 mol%)
Hantzsch ester 4 (1.1 equiv)
Et₂O, 0 °C, 40 h
66%; 98% ee

(XII) 37 → 38
catalyst 7 (20 mol%)
Hantzsch ester 4 (1.1 equiv)
Et₂O, 0 °C, 9 h
70%; 92% ee

Ar = 2,4,6-triisopropyl-phenyl 39

(R)-DTBM-SEGPHOS 40

Reference Key for Equations: (VIII-IX)[4]; (X)[5]   Scheme 3   Reference Key for Equations: (XI-XII)[5]

Enantioselective conjugate reduction of α,β-unsaturated carbonyl compounds can also be effected by chiral bisphosphine Cu(I) hydride complexes generated from CuO*t*-Bu, bisphosphine and polymethylhydrosiloxane (PMHS) in toluene.[6] Exclusion of air is crucial to the success of these reactions. A detailed procedure which illustrates this method has been published[6b] (**41→42**, 65 g scale) and is shown at right.

41 → 42
catalyst 40 (0.0004 mol%)
CuCl (1 mol%), NaO*t*-Bu (1 mol%)
PMHS (2 equiv)
toluene, -35 °C, 72 h
88%; 98.5% ee

References

1. (a) Yang Jung, W., Hechavarria Fonseca Maria, T., Vignola, N. & List, B. Metal-free, organocatalytic asymmetric transfer hydrogenation of α,β-unsaturated aldehydes. *Angew. Chem. Int. Ed. Engl.* **2004**, 44, 108-110; (b) Rosen, J. Enantioselective organocatalytic reduction of α,β-unsaturated aldehydes. *Chemtracts* **2005**, 18, 65-71; (c) Adolfsson, H. Organocatalytic hydride transfers: a new concept in asymmetric hydrogenations. *Angew. Chem., Int. Ed.* **2005**, 44, 3340-3342; (d) Lelais, G. & MacMillan, D. W. C. Iminium catalysis. *Enantioselective Organocatalysis* **2007**, 95-120; (e) Lelais, G. & MacMillan, D. W. C. History and perspective of chiral organic catalysts in *New Frontiers in Asymmetric Catalysis* 313-358 (**2007**); (f) Kagan, H. B. Organocatalytic enantioselective reduction of olefins, ketones, and imines in *Enantioselective Organocatalysis* 391-401 (**2007**); (g) You, S.-L. Recent developments in asymmetric transfer hydrogenation with Hantzsch esters: a biomimetic approach. *Chemistry--An Asian Journal* **2007**, 2, 820-827; (h) Ouellet, S. G., Walji, A. M. & Macmillan, D. W. C. Enantioselective Organocatalytic Transfer Hydrogenation Reactions using Hantzsch Esters. *Acc. Chem. Res.* **2007**, 40, 1327-1339; (i) Connon, S. J. Asymmetric organocatalytic reductions mediated by dihydropyridines. *Org. Biomol. Chem.* **2007**, 5, 3407-3417; (j) Pan, S. C. & List, B. New concepts for organocatalysis. *Ernst Schering Foundation Symposium Proceedings* **2008**, 1-43; (k) Saini, A., Kumar, S. & Sandhu, J. S. Hantzsch reaction: recent advances in Hantzsch 1,4-dihydropyridines. *J. Sci. Ind. Res.* **2008**, 67, 95-111; (l) Adolfsson, H. Alkene and imino reductions by organocatalysis in *Modern Reduction*

Methods 341-361 (**2008**); (m) Melchiorre, P., Marigo, M., Carlone, A. & Bartoli, G. Asymmetric aminocatalysis gold rush in organic chemistry. *Angew. Chem., Int. Ed.* **2008**, 47, 6138-6171.

2. Ouellet, S. G., Tuttle, J. B. & MacMillan, D. W. C. Enantioselective Organocatalytic Hydride Reduction. *J. Am. Chem. Soc.* **2005**, 127, 32-33.

3. Yang, J. W., Hechavarria Fonseca, M. T., Vignola, N. & List, B. Metal-free, organocatalytic asymmetric transfer hydrogenation of α,β-unsaturated aldehydes. *Angew. Chem., Int. Ed.* **2005**, 44, 108-110.

4. Mayer, S. & List, B. Asymmetric counteranion-directed catalysis. *Angew. Chem., Int. Ed.* **2006**, 45, 4193-4195.

5. Tuttle, J. B., Ouellet, S. G. & MacMillan, D. W. C. Organocatalytic Transfer Hydrogenation of Cyclic Enones. *J. Am. Chem. Soc.* **2006**, 128, 12662-12663.

6. (a) Hughes, G., Kimura, M. & Buchwald, S. L. Catalytic Enantioselective Conjugate Reduction of Lactones and Lactams. *J. Am. Chem. Soc.* **2003**, 125, 11253-11258; (b) Lipshutz, B. H., Servesko, J. M., Petersen, T. B., Papa, P. P. & Lover, A. A. Asymmetric 1,4-Reductions of Hindered β-Substituted Cycloalkenones Using Catalytic SEGPHOS-Ligated CuH. *Org. Lett.* **2004**, 6, 1273-1275; (c) Deutsch, C., Krause, N. & Lipshutz, B. H. CuH-Catalyzed Reactions. *Chem. Rev. (Washington, DC, U. S.)* **2008**, 108, 2916-2927; (d) Lipshutz, B. H. Rediscovering organocopper chemistry through copper hydride. It's all about the ligand. *Synlett* **2009**, 509-524.

ENANTIOSELECTIVE α,β-REDUCTION OF C=C-C=O

"CuH" Catalyzed Enantioselective Reduction of α,β-Unsaturated Carbonyl Compounds

Scheme 1

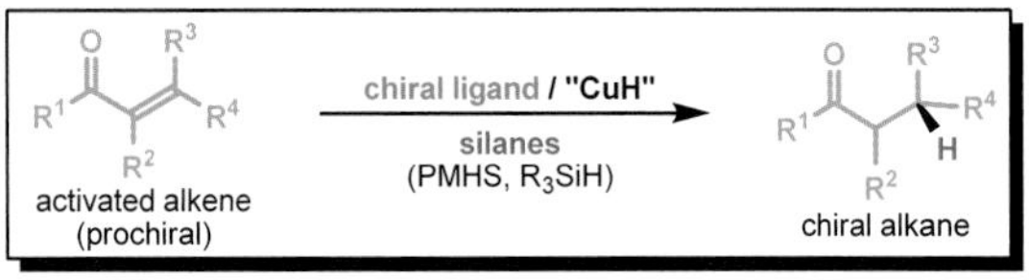

In the late 1980's Stryker developed the phosphine-stabilized hexameric copper-hydride [(PPh$_3$)CuH]$_6$, "Stryker's reagent", for the effective 1,4-reduction of α,β-unsaturated carbonyl compounds.[1] Subsequently, enantioselective versions of this transformation have been developed using catalytic amounts of "Cu-H"/chiral bisphosphine complexes that are formed in situ.[2] Some of the successful chiral bisphosphine ligands are shown in Figure 1. Silanes such as polymethylhydrosiloxane (PMHS) and trialkylsilanes, were found to be the most effective stoichiometric reducing agents.

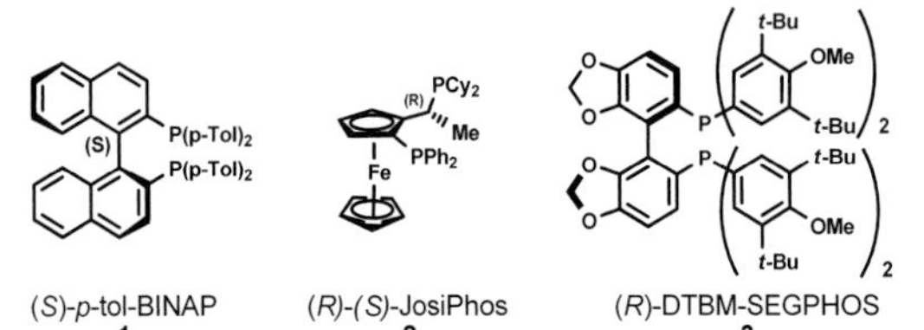

Figure 1

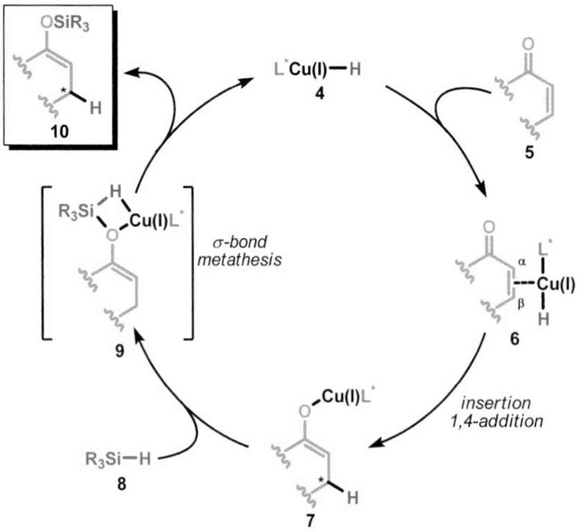

A reasonable catalytic cycle has been proposed (Scheme 1).[2] The copper hydride/chiral ligand complex 4 coordinates with 5 to give the π-complex 6 that undergoes an enantioselective insertion of hydride at the β-carbon.

"CuH" Catalyzed Enantioselective Reductions

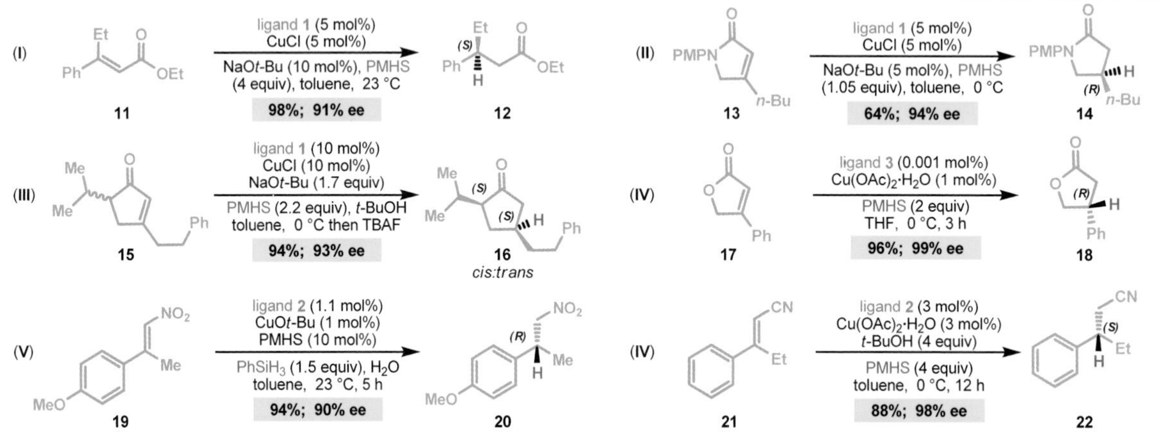

Scheme 2

Reference Key for Equations: (I-II)[3]; (III)[4]

Reference Key for Equations: (IV)[5]; (V)[6]; (VI)[7]

References

1. Mahoney, W. S., Brestensky, D. M. & Stryker, J. M. Selective hydride-mediated conjugate reduction of α,β-unsaturated carbonyl compounds using [(Ph$_3$P)CuH]$_6$. *J. Am. Chem. Soc.* **1988**, 110, 291-293.
2. (a) Rendler, S. & Oestreich, M. Polishing a diamond in the rough: "Cu-H" catalysis with silanes. *Angew. Chem., Int. Ed.* **2007**, 46, 498-504; (b) Deutsch, C., Krause, N. & Lipshutz, B. H. CuH-Catalyzed Reactions. *Chem. Rev.* **2008**, 108, 2916-2927; (c) Lipshutz, B. H. Rediscovering organocopper chemistry through copper hydride. It's all about the ligand. *Synlett* **2009**, 509-524.
3. Appella, D. H., Moritani, Y., Shintani, R., Ferreira, E. M. & Buchwald, S. L. Asymmetric conjugate reduction of α,β-unsaturated esters using a chiral phosphine-copper catalyst. *J. Am. Chem. Soc.* **1999**, 121, 9473-9474.
4. Jurkauskas, V. & Buchwald, S. L. Dynamic Kinetic Resolution via Asymmetric Conjugate Reduction: Enantio- and Diastereoselective Synthesis of 2,4-Dialkyl Cyclopentanones. *J. Am. Chem. Soc.* **2002**, 124, 2892-2893.
5. Hughes, G., Kimura, M. & Buchwald, S. L. Catalytic Enantioselective Conjugate Reduction of Lactones and Lactams. *J. Am. Chem. Soc.* **2003**, 125, 11253-11258.
6. Czekelius, C. & Carreira, E. M. Catalytic enantioselective conjugate reduction of β,β-disubstituted nitroalkenes. *Angew. Chem., Int. Ed.* **2003**, 42, 4793-4795.
7. Lee, D., Yang, Y. & Yun, J. Copper hydride-catalyzed enantioselective conjugate reduction of unsaturated nitriles. *Synthesis* **2007**, 2233-2235.

ENANTIOSELECTIVE ADDITION OF H₂ TO C=O

Background: Highly Enantioselective Homogeneous Hydrogenation of Ketones

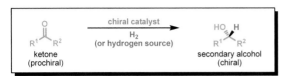

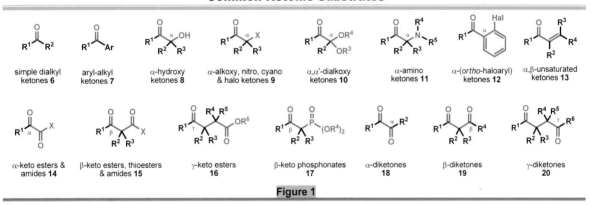

The first transition metal-catalyzed highly enantioselective homogeneous hydrogenation of functionalized ketones to form enantiopure secondary alcohols was reported by R. Noyori et al. in the late 1980s.[1] Using either (R)- or (S)-BINAP-Ru(II) complexes (2 or 3), β-keto esters such as 1 were converted to the corresponding (R)- or (S)-β-hydroxy esters (4 or 5) in high yield and enantiomeric purity. Catalysts 2 and 3 were also highly effective for the enantioselective hydrogenation of a wide variety of α- and β-functionalized ketones (e.g., α-amino ketones and β-hydroxy ketones).[2] Later it was discovered that Ru-catalysts having both chiral bisphosphine and chiral diamine ligands catalyzed the enantioselective reduction of unfunctionalized ketones (e.g., dialkyl ketones, aryl-alkyl ketones).[3]

It was also demonstrated that hydrogen donors, such as small ethanol, 2-propanol and triethylammonium formate, can replace hydrogen gas (H₂) as the hydrogen source in the enantioselective reduction of ketones, a process called *asymmetric transfer hydrogenation*.[4] A significant advantage of this version is that it does not require the use of pressurized reaction vessels.

Common Ketonic Substrates

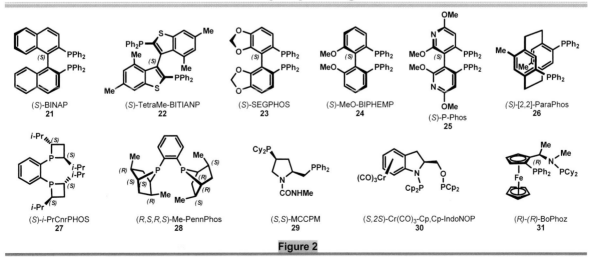

Figure 1

Some Bisphosphine Ligands

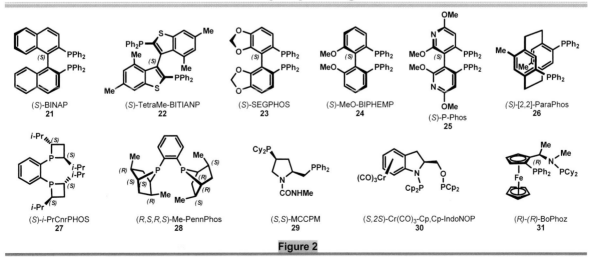

Figure 2

ENANTIOSELECTIVE ADDITION OF H₂ TO C=O

Chiral 1,2- and 1,4-Diamine Ancillary Ligands

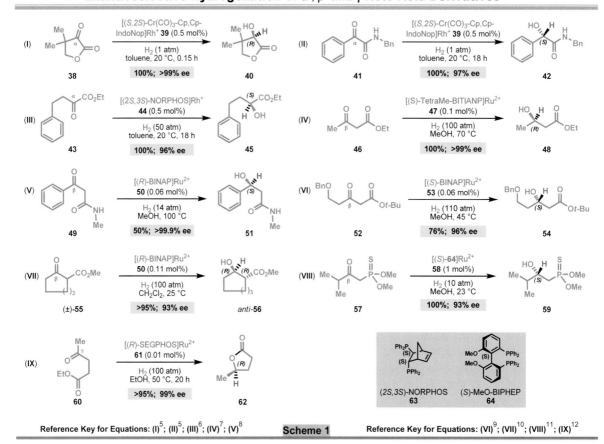

(S,S)-DPEN **32**   (S,S)-DACH **33**   (S)-DAIPEN **34**   (R)-DM-DABN **35**   (R)-IPHAN **36**

Figure 3

Comments on Figures 1-3

(1) Enantioselective carbonyl hydrogenation has been applied to 1,2-, 1,3- and 1,4-dicarbonyl compounds (**14-20**, Figure 1), including α-, β- and γ-diketones (**18-20**) as well as α-, β- and γ-keto acid derivatives (**14-17**). These bifunctional substrates are thought to undergo reduction via chelated transition states.[3k] Unfunctionalized prochiral ketones (e.g., **6** and **7**) and α-functionalized ketones (**8-12**) have also been useful substrates.[3c,3h,3j]

(2) The bisphosphines **21-25** (Figure 2) are atropisomeric and C₂-symmetric, while the ligands **26-31** belong to completely different structural classes. Variation of these ligands in combination with the reaction conditions allows substantial control of the stereoselectivity of ketone hydrogenation.[3e,3i]

(3) The simultaneous chelation of chiral 1,2- and 1,4-diamine ligands **32-36** (Figure 3) to Ru(II) gives rise to complexes (e.g., **37**) that are highly effective catalysts for the enantioselective hydrogenation of simple ketones. With these catalysts the chemoselective hydrogenation of C=O in the presence of C=C, as well as other sensitive functional groups, can be achieved.[3e,4e]

37

Enantioselective Hydrogenation of α-, β- and γ-Keto Acid Derivatives[3g,3k,3l]

ENANTIOSELECTIVE ADDITION OF H₂ TO C=O

Enantioselective Hydrogenation of α-, β- and γ-Keto Acid Derivatives

Comments on Scheme 1

(1) The enantioselective hydrogenation of α-, β- and γ-keto esters leads to synthetically versatile chiral hydroxyl ester building blocks. Both Ru and Rh-bisphosphine complexes are effective for these transformations.

(2) The Rh(I)-complex **39**, prepared using the aminophosphine-phosphinite diphosphine ligand **30** (Figure 2),[13] catalyzes the highly enantioselective hydrogenation of cyclic α-keto ester **38** and α-keto amide **41**.[5]

(3) α-Hydroxy ester **45**, an important intermediate for the synthesis of the angiotensin-converting enzyme (ACE) inhibitor benazepril, is formed by the hydrogenation of α-keto ester **43** in the presence of the norbornyl bisphosphine ligand (2S,3S)-NORPHOS (**63**).[6]

(4) The Ru(II)-complexes of atropisomeric bisphosphines such as **21** and **22** (Figure 2) have been found useful for the enantioselective hydrogenation of cyclic and acyclic β-keto esters (e.g., **55**,[10] **46**,[7] **49**[8] and **52**[9]).

(a) The reduction of cyclic racemic β-keto ester **55**[10] proceeds via dynamic kinetic resolution (in situ racemization of the labile α-stereocenter) to afford the chiral anti-product **56** in high yield and enantioselectivity.

(b) The enantioselective reduction of the benzoyl-acetamide **49** in the presence of Ru(II)-(R)-BINAP complex **50** provides **51**, an intermediate en route to fluoxetine.[8]

(c) The reduction **52→54** proceeds with high enantioselectivity presumably due to sufficient steric differentiation between the two oxygen atoms during the coordination event.[9]

(5) The β-keto thiophosphonate **57** undergoes enantioselective hydrogenation under conditions similar to those used for β-keto esters.[11]

(6) The γ-keto ester **60** was reduced to the corresponding γ-lactone **62** more efficiently with the Ru(II)-SEGPHOS complex **61** than with various Ru(II)-BINAP complexes[12,14].

Mechanism: Catalytic Cycle of the Ru(II)-Catalyzed Hydrogenation of β-Keto Esters[3k]

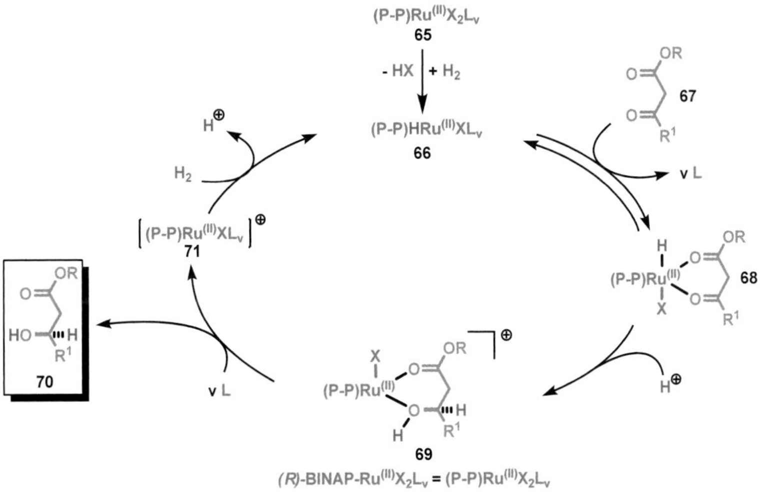

Scheme 2

A proposed catalytic cycle for the enantioselective hydrogenation of β-keto esters is outlined in Scheme 2.[3k] The individual steps in the catalytic cycle are:

(a) Precatalyst **65** reacts with hydrogen to afford the Ru(II)-hydride complex **66**, the active catalytic species.

(b) The substrate **67** coordinates reversibly with **66** to give the chelate **68**.

(c) Protonation of the carbonyl oxygen in **68** affords cationic complex in which the increased electrophilicity of the keto group induces a facile hydride transfer from Ru to carbon to give complex **69**.

(d) The release of product **70** from complex **69** is replaced with molecules of the solvent (L) and reaction of **71** with H₂ regenerates the Ru(II)-hydride complex **66**.

19

ENANTIOSELECTIVE ADDITION OF H₂ TO C=O

Enantioselective Hydrogenation of α-, β- and γ-Functionalized Ketones

(I)

72 → **74**

[(S,S)-MCCPM]Rh⁺ **73** (0.001 mol%)
H₂ (20 atm), EtOH, NEt₃, 50 °C, 20 h
100%; 96% ee

(II)

75 → **77**

[(S)-**95**][(R,R)-dpen]Ru²⁺ **76** (0.2 mol%)
H₂ (80 bar), i-PrOH, t-BuOK, 50 °C
98%; 97% ee

(III)

78 → **80**

trans-[(S)-**96**][(S)-daipen]Ru²⁺ **79** (0.01 mol%)
H₂ (8 atm), i-PrOH, t-BuOK, 25 °C
96%; 97.5% ee

(IV)

81 → **83**

[(S,R)-**97**]Rh⁺ **82** (0.02 mol%)
H₂ (50 bar), i-PrOH, K₂CO₃, 50 °C
92%; 99% ee

(V)

84 → **85**

trans-[(S)-**96**][(S)-daipen]Ru²⁺ **79** (0.01 mol%)
H₂ (8 atm), i-PrOH, t-BuOK, 25 °C, 32 h
97%; 99% ee

(VI)

86 → **88**

[(R)-BINAP]Ru²⁺ **50** (0.11 mol%)
H₂ (93 atm), MeOH, 32 °C
100%; 92% ee

(VII)

89 → **90**

[(R)-BINAP]Ru²⁺ **50** (0.11 mol%)
H₂ (70 atm), EtOH, 32 °C
100%; 98% ee

(VIII)

91 → **92**

(S,S)-Ru(II)-catalyst **98** (0.5 mol%)
HCO₂H/NEt₃ (6.0:2.4), DMF (1 M), 30 °C, 16 h
67%; 95% ee

(IX)

93 → **94**

(R,R)-Rh(I)-catalyst **99** (0.1 mol%)
HCO₂H/NEt₃, EtOAc (1 M), 25 °C
93%; 98% ee

(S)-MeO-BIPHEP3 **95**

(S)-BINAP derivative **96**

(S꜀,R_P)-DuanPhos **97**

RuCl[(S,S)-Tsdpen][η⁶-arene] **98**

RuCl[(R,R)-Tsdpen][Cp*] **99**

Reference Key for Equations: (I)[15]; (II)[16]; (III)[17]; (IV)[18]; (V)[17] **Scheme 3** Reference Key for Equations: (VI)[1]; (VII)[1]; (VIII)[19]; (IX)[20]

Comments on Scheme 3

(1) The acyclic and cyclic α-amino ketones **72** and **75** are efficiently hydrogenated in the presence of chiral Rh(I)-bisphosphine complexes (e.g., **73**) or Ru(II)-bisphosphine/1,2-diamine complexes (e.g., **76**).[15-16] The β-amino-α-phenylethanol derivative **74** was utilized for the synthesis of (S)-levamisole,[15] an imidazothiazole antibiotic, while the cyclic amino alcohol **77** is an intermediate en route to the synthesis of NMDA 2B receptor antagonist Ro 67-8867.[16]

(2) The enantioselective hydrogenation of β- and γ-amino ketones (e.g., **78**,[17] **81**[18] and **84**[17]) is possible with either Ru(II)- and Rh(I) catalysts (e.g., **79** and **82**).

(3) The products **80** and **83** are valuable chiral building blocks for the production of a number of antidepressants. The 1,4-amino alcohol **85** is a potent antipsychotic agent, BMS1801100.[17]

(4) α-Functionalized aromatic ketones **91**[19] and **93**[20] have been efficiently reduced under transfer hydrogenation conditions in the presence of the well-defined Ru- and Rh-complexes **98** and **99**. Optically active alcohols **92** and **94** can be further transformed into a variety of useful compounds.

ENANTIOSELECTIVE ADDITION OF H_2 TO C=O

Enantioselective Hydrogenation of Diketones and Olefinic Ketones[21,3e,3k,4e]

(I) 100 → [(S)-BINAP]Ru²⁺ **101** (0.11 mol%), H_2 (20 atm), MeOH, 23 °C → 102 **26%; 100% ee** + 74% meso product

(II) 103 → [Walphos-**120**]Ru²⁺ **104** (0.1 mol%), H_2 (100 atm), MeOH, 80 °C → 105 **98%; 97% ee**

(III) 106 → [(S,S)-Ru(II)-catalyst **121** (0.001 mol%), H_2 (80 atm), K_2CO_3, i-PrOH, 28 °C, 43 h → 107 **100%; 97% ee**

(IV) 108 → [(S,S)]Ru(II)-catalyst **121** (0.2 mol%), H_2 (10 atm), K_2CO_3, i-PrOH, 30 °C → 109 **>96%; 94% ee**

(V) 110 → [(S,S)]Ru(II)-catalyst **121** (0.2 mol%), H_2 (10 atm), K_2CO_3, i-PrOH, 30 °C → 111 **>96%; 94% ee**

(VI) 112 → [(S,S)]Ru(II)-catalyst **121** (0.2 mol%), H_2 (10 atm), K_2CO_3, i-PrOH, 30 °C → 113 **>96%; 99% ee**

(VII) 114 → [(S)-BINAP][(S,S)-DPEN]Ru(II) **115** (0.4 mol%), H_2 (8 atm), KOH, i-PrOH, 28 °C, 16 h → 116 **97%; 98% ee**

(VIII) 117 → [(±)-**122**][(S,S)-DPEN]Ru(II) **118** (0.01 mol%), H_2 (10 atm), KOH, i-PrOH:toluene, 0 °C, 6 h → 119 **100%; 95% ee**

Walphos **120**
[(S)-XylBINAP/(S)-DAIPEN]Ru(II) **121**
(S)-TolBINAP **122**

Reference Key for Equations (I)[1]; (II)[22]; (III)[23]; (IV)[3e] **Scheme 4** Reference Key for Equations: (V)[3e]; (VI)[24]; (VII)[25]; (VIII)[26]

Comments on Scheme 4

(1) Enantioselective hydrogenation of the 1,2-diketone **100** afforded the optically active 1,2-diol **102** with high enantioselectivity but in only modest yield.[1] On the other hand, the 1,3-diketone **103** gave rise to the 1,3-anti diol **105** in very high yield and enantiomeric purity using the ferrocene-based catalyst Walphos **120**.[22]

(2) The highly enantioselective hydrogenation of olefinic ketones such as **106**,[23] **108**,[3e] **110**,[3e] **112**, **114** and **117** became possible during the mid-1990s when Ru-bisphosphine/diamine complexes emerged as catalysts.[24,3e] In these hydrogenations with Ru-bisphosphine/diamine catalysts such as **121** the olefinic subunit is not reduced even though the diamine-free Ru-BINAP complexes are highly effective for the hydrogenation of C=C bonds.

(3) The catalysts used in the reduction of olefinic ketones require the presence of a base (e.g., K_2CO_3, KOH) to prevent the formation of byproducts.

(4) Cyclic enones **114**[25] and **117**[26] were effectively hydrogenated using Ru(II)-BINAP/DPEN (**115**) and Ru(II)-TolBINAP/DPEN (**118**) catalysts, respectively. The transformation **117**→**119** is an example of using a racemic bisphopshine ligand (e.g., (±)-**122**) complexed with Ru(II), along with a nonracemic 1,2-diamine ligand such as (S,S)-DPEN (**32**, Figure 3) to effect high enantioselectivity – a process called *asymmetric (enantioselective) activation*.[26,3k] The catalytic species is **123** rather than the diastereomer **124**.[25-26]

[(R)-TolBINAP/(S,S)-DPEN]Ru(II) **123** (more active)

[(S)-TolBINAP/(S,S)-DPEN]Ru(II) **124** (less active)

ENANTIOSELECTIVE ADDITION OF H$_2$ TO C=O

Enantioselective Hydrogenation of Simple Ketones[3b,3g,3k,4e]

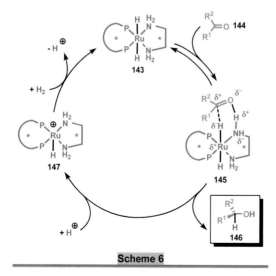

(I) $\quad$ **125** $\xrightarrow[\substack{H_2 \text{ (30 atm)} \\ 2,6\text{-lutidine, KBr} \\ MeOH, 23\ ^\circ C, 96\ h}]{\substack{[Me\text{-PennPhos-28}]Rh^+ \\ \textbf{126}\ (1\ mol\%)}}$ **127** $\quad$ **51%; 94% ee**

(II) $\quad$ **128** $\xrightarrow[\substack{H_2 \text{ (8 atm), } t\text{-BuOK} \\ i\text{-PrOH, 28}\ ^\circ C, 12\ h}]{\substack{[(S,S)]Ru(II)\text{-catalyst} \\ \textbf{121}\ (0.011\ mol\%)}}$ **129** $\quad$ **96%; 95% ee**

(III) $\quad$ **130** $\xrightarrow[\substack{H_2 \text{ (4 atm), KOH} \\ i\text{-PrOH, 28}\ ^\circ C, 11\ h}]{\substack{[(S)\text{-BINAP}][(R,R)\text{-DPEN}]Ru^{2+} \\ \textbf{131}\ (0.05\ mol\%)}}$ **132** $\quad$ **93% ee**

(IV) $\quad$ **133** $\xrightarrow[\substack{H_2 \text{ (10 atm), } t\text{-BuOK} \\ i\text{-PrOH, 80}\ ^\circ C, 1\ h}]{\substack{[(S,SS)]Ru(II)\text{-catalyst} \\ \textbf{124}\ (0.00025\ mol\%)}}$ **134** $\quad$ **93%; 91% ee**

(V) $\quad$ **135** $\xrightarrow[\substack{H_2 \text{ (8 atm), } t\text{-BuOK} \\ i\text{-PrOH, 28}\ ^\circ C, 16\ h}]{\substack{[(S,S)]Ru(II)\text{-catalyst} \\ \textbf{121}\ (0.001\ mol\%)}}$ **136** $\quad$ **99%; 99% ee**

(VI) $\quad$ **137** $\xrightarrow[\substack{H_2 \text{ (8 atm), } t\text{-BuOK} \\ i\text{-PrOH, 28}\ ^\circ C, 4\ h}]{\substack{[(S,S)]Ru(II)\text{-catalyst} \\ \textbf{121}\ (0.01\ mol\%)}}$ **138** $\quad$ **100%; 96% ee**

(VII) $\quad$ **139** $\xrightarrow[\substack{H_2 \text{ (8 atm), } t\text{-BuOK} \\ i\text{-PrOH, 30}\ ^\circ C, 16\ h}]{\substack{[(S,S)]Ru(II)\text{-catalyst} \\ \textbf{121}\ (0.05\ mol\%)}}$ **140** $\quad$ **>96%; 100% ee**

(VIII) $\quad$ **141** $\xrightarrow[\substack{H_2 \text{ (8 atm), } t\text{-BuOK} \\ i\text{-PrOH, 28}\ ^\circ C, 15\ h}]{\substack{[(S,S)]Ru(II)\text{-catalyst} \\ \textbf{121}\ (0.05\ mol\%)}}$ **142** $\quad$ **>99%; 98% ee**

Reference Key for Equations (I)[27]; (II)[23]; (III)[28]; (IV)[3a] $\qquad$ **Scheme 5** $\qquad$ Reference Key for Equations: (V)[23]; (VI)[23]; (VII)[29]; (VIII)[30]

Comments on Scheme 5

(1) The challenging hydrogenation of simple (unfunctionalized) ketones that cannot chelate to metals is thought to involve the catalytic cycle shown in Scheme 6.[31,3k]

Scheme 6

(2) The structure **145** represents a proposed pre-transition state assembly involving hydrogen bonding of coordinated N-H to the ketone oxygen and simultaneous hydride transfer from Ru to carbon.[31,3k]

(3) The highly enantioselective hydrogenation of aliphatic (alkyl-alkyl) ketones, such as **125**,[27] **128**[23] and **130**,[28] with two very similar alkyl groups, requires the use of catalysts such as Rh(I)-Me-PennPhos (e.g., **126**) and Ru(II)-bisphosphine/1,2-diamine (e.g., **121** and **131**).[3e]

(4) The hydrogenation of acetophenone and acetylnaphthalene derivatives (e.g., **133**,[3a] **135**,[23] **137**[23]) as well as heteroaromatic ketones (e.g., **139**[29]) proceeds with very high yield and enantioselectivity with Ru(II)-bisphosphine/1,2-diamine catalysts.

(5) Although *ortho*-substituted benzophenones (e.g., **141**[30]) undergo efficient hydrogenation, *meta*- and *para*-substituted benzophenones do not, as might be expected from the steric similarity of the two aryl groups.

References

1. Kitamura, M., Ohkuma, T., Inoue, S., Sayo, N., Kumobayashi, H., Akutagawa, S., Ohta, T., Takaya, H. & Noyori, R. Homogeneous asymmetric hydrogenation of functionalized ketones. *J. Am. Chem. Soc.* **1988**, 110, 629-631.
2. Noyori, R., Ohkuma, T., Kitamura, M., Takaya, H., Sayo, N., Kumobayashi, H. & Akutagawa, S. Asymmetric hydrogenation of β-keto carboxylic esters. A practical, purely chemical access to β-hydroxy esters in high enantiomeric purity. *J. Am. Chem. Soc.* **1987**, 109, 5856-5858.

3. (a) Doucet, H., Ohkuma, T., Murata, K., Yokozawa, T., Kozawa, M., Katayama, E., England, A. F., Ikariya, T. & Noyori, R. *Trans*-[RuCl₂(phosphine)₂(1,2-diamine)] and chiral trans-[RuCl₂(diphosphine)(1,2-diamine)]: shelf-stable precatalysts for the rapid, productive, and stereoselective hydrogenation of ketones. *Angew. Chem., Int. Ed.* **1998**, 37, 1703-1707; (b) Fehring, V. & Selke, R. Highly enantioselective complex-catalyzed reduction of ketones - now with purely aliphatic derivatives too. *Angew. Chem., Int. Ed.* **1998**, 37, 1827-1830; (c) Ohkuma, T. & Noyori, R. Hydrogenation of carbonyl groups in *Comprehensive Asymmetric Catalysis I-III* 199-246 (**1999**); (d) Palmer, M. J. & Wills, M. Asymmetric transfer hydrogenation of C:O and C:N bonds. *Tetrahedron: Asymmetry* **1999**, 10, 2045-2061; (e) Noyori, R. & Okhuma, T. Asymmetric catalysis by architectural and functional molecular engineering: practical chemo- and stereoselective hydrogenation of ketones. *Angew. Chem., Int. Ed.* **2001**, 40, 40-73; (f) Noyori, R. Asymmetric catalysis: Science and opportunities (nobel lecture 2001). *Adv. Synth. Catal.* **2003**, 345, 15-32; (g) Tang, W. & Zhang, X. New Chiral Phosphorus Ligands for Enantioselective Hydrogenation. *Chem. Rev. (Washington, DC, U. S.)* **2003**, 103, 3029-3069; (h) Ohkuma, T. & Noyori, R. Hydrogenation of carbonyl groups. *Compr. Asymmetric Catal., Suppl.* **2004**, 1, 1-41; (i) Zanotti-Gerosa, A., Herns, W., Groarke, M. & Hancock, F. Ruthenium-catalysed asymmetric reduction of ketones. Diphosphine ligands in hydrogenations for pharmaceutical synthesis. *Platinum Met. Rev.* **2005**, 49, 158-165; (j) Kwong, F. Y., Qiu, L., Lam, W. H. & Chan, A. S. C. Recent developments in asymmetric hydrogenation of C=O motif compounds. *Advances in Organic Synthesis* **2005**, 1, 261-299; (k) Ohkuma, T. & Noyori, R. Enantioselective ketone and β-keto ester hydrogenations (including mechanisms) in *Handbook of Homogeneous Hydrogenation* 1105-1163 (**2007**); (l) Mortreux, A. & Karim, A. Rhodium-catalyzed enantioselective hydrogenation of functionalized ketones in *Handbook of Homogeneous Hydrogenation* 1165-1192 (**2007**); (m) Klingler, F. D. Asymmetric Hydrogenation of Prochiral Amino Ketones to Amino Alcohols for Pharmaceutical Use. *Acc. Chem. Res.* **2007**, 40, 1367-1376; (n) Somanathan, R., Cortez, N. A., Parra-Hake, M., Chavez, D. & Aguirre, G. Immobilized chiral metal catalysts for enantioselective hydrogenation of ketones. *Mini-Rev. Org. Chem.* **2008**, 5, 313-322.
4. (a) Fujii, A., Hashiguchi, S., Uematsu, N., Ikariya, T. & Noyori, R. Ruthenium(II)-Catalyzed Asymmetric Transfer Hydrogenation of Ketones Using a Formic Acid-Triethylamine Mixture. *J. Am. Chem. Soc.* **1996**, 118, 2521-2522; (b) Morris, D. J. & Wills, M. Weighing is optional: highly active and practical catalysts for asymmetric ketone transfer hydrogenation. *PharmaChem* **2006**, 5, 8-11; (c) Noyori, R., Ohkuma, T., Sandoval, C. A. & Muniz, K. Asymmetric hydrogenation through metal-ligand bifunctional catalysis in *Asymmetric Synthesis* 321-325 (**2007**); (d) Blacker, A. J. Enantioselective transfer hydrogenation in *Handbook of Homogeneous Hydrogenation* 1215-1244 (**2007**); (e) Ikariya, T. & Blacker, A. J. Asymmetric Transfer Hydrogenation of Ketones with Bifunctional Transition Metal-Based Molecular Catalysts. *Acc. Chem. Res.* **2007**, 40, 1300-1308; (f) Wu, X. & Xiao, J. Aqueous-phase asymmetric transfer hydrogenation of ketones - a greener approach to chiral alcohols. *Chem. Commun. (Cambridge, U. K.)* **2007**, 2449-2466.
5. Pasquier, C., Naili, S., Pelinski, L., Brocard, J., Mortreux, A. & Agbossou, F. Synthesis and application in enantioselective hydrogenation of new free and chromium complexed aminophosphine-phosphinite ligands. *Tetrahedron: Asymmetry* **1998**, 9, 193-196.
6. Takahashi, H., Sakuraba, S., Takeda, H. & Achiwa, K. Asymmetric reactions catalyzed by chiral metal complexes. 41. Highly efficient asymmetric hydrogenation of amino ketone derivatives leading to practical syntheses of (S)-propranolol and related compounds. *J. Am. Chem. Soc.* **1990**, 112, 5876-5878.
7. Benincori, T., Brenna, E., Sannicolo, F., Trimarco, L., Antognazza, P., Cesarotti, E., Demartin, F. & Pilati, T. New Class of Chiral Diphosphine Ligands for Highly Efficient Transition Metal-Catalyzed Stereoselective Reactions: The Bis(diphenylphosphino) Five-membered Biheteroaryls. *J. Org. Chem.* **1996**, 61, 6244-6251.
8. Huang, H.-L., Liu, L. T., Chen, S.-F. & Ku, H. The synthesis of a chiral fluoxetine intermediate by catalytic enantioselective hydrogenation of benzoylacetamide. *Tetrahedron: Asymmetry* **1998**, 9, 1637-1640.
9. Rychnovsky, S. D. & Hoye, R. C. Convergent Synthesis of the Polyene Macrolide (-)-Roxaticin. *J. Am. Chem. Soc.* **1994**, 116, 1753-1765.
10. Genet, J. P., Pfister, X., Ratovelomanana-Vidal, V., Pinel, C. & Laffitte, J. A. Dynamic kinetic resolution of cyclic β²-keto esters with preformed or in-situ prepared chiral diphosphine-ruthenium(II) catalysts. *Tetrahedron Lett.* **1994**, 35, 4559-4562.
11. Gautier, I., Ratavelomanana-Vidal, V., Savignac, P. & Genet, J.-P. Asymmetric hydrogenation of β-ketophosphonates and β-ketothiophosphonates with chiral Ru(II) catalysts. *Tetrahedron Lett.* **1996**, 37, 7721-7724.
12. Saito, T., Yokozawa, T., Ishizaki, T., Moroi, T., Sayo, N., Miura, T. & Kumobayashi, H. New chiral diphosphine ligands designed to have a narrow dihedral angle in the biaryl backbone. *Adv. Synth. Catal.* **2001**, 343, 264-267.
13. Agbossou, F., Carpentier, J.-F., Hapiot, F., Suisse, I. & Mortreux, A. The aminophosphine-phosphinites and related ligands: synthesis, coordination chemistry and enantioselective catalysis. *Coord. Chem. Rev.* **1998**, 178-180, 1615-1645.
14. Kumobayashi, H., Miura, T., Sayo, N., Saito, T. & Zhang, X. Recent advances of BINAP chemistry in the industrial aspects. *Synlett* **2001**, 1055-1064.
15. Takeda, H., Tachinami, T., Aburatani, M., Takahashi, H., Morimoto, T. & Achiwa, K. Asymmetric reactions catalyzed by chiral metal complexes. XXVII. Efficient asymmetric hydrogenation of α-aminoacetophenone derivatives leading to practical synthesis of (S)-(-)-Levamisole. *Tetrahedron Lett.* **1989**, 30, 363-366.
16. Scalone, M. & Waldmeier, P. Efficient Enantioselective Synthesis of the NMDA 2B Receptor Antagonist Ro 67-8867. *Org. Process Res. Dev.* **2003**, 7, 418-425.
17. Ohkuma, T., Ishii, D., Takeno, H. & Noyori, R. Asymmetric Hydrogenation of Amino Ketones Using Chiral RuCl2(diphosphine)(1,2-diamine) Complexes. *J. Am. Chem. Soc.* **2000**, 122, 6510-6511.
18. Liu, D., Gao, W., Wang, C. & Zhang, X. Practical synthesis of enantiopure γ-amino alcohols by rhodium-catalyzed asymmetric hydrogenation of β-secondary-amino ketones. *Angew. Chem., Int. Ed.* **2005**, 44, 1687-1689.
19. Watanabe, M., Murata, K. & Ikariya, T. Practical Synthesis of Optically Active Amino Alcohols via Asymmetric Transfer Hydrogenation of Functionalized Aromatic Ketones. *J. Org. Chem.* **2002**, 67, 1712-1715.
20. Hamada, T., Torii, T., Izawa, K., Noyori, R. & Ikariya, T. Practical Synthesis of Optically Active Styrene Oxides via Reductive Transformation of 2-Chloroacetophenones with Chiral Rhodium Catalysts. *Org. Lett.* **2002**, 4, 4373-4376.
21. Ohkuma, T. & Noyori, R. Carbonyl hydrogenations in *Transition Metals for Organic Synthesis* 25-69 (**1998**).
22. Sturm, T., Weissensteiner, W. & Spindler, F. A novel class of ferrocenyl-aryl-based diphosphine ligands for Rh- and Ru-catalysed enantioselective hydrogenation. *Adv. Synth. Catal.* **2003**, 345, 160-164.
23. Ohkuma, T., Koizumi, M., Doucet, H., Pham, T., Kozawa, M., Murata, K., Katayama, E., Yokozawa, T., Ikariya, T. & Noyori, R. Asymmetric Hydrogenation of Alkenyl, Cyclopropyl, and Aryl Ketones. RuCl2(xylbinap)(1,2-diamine) as a Precatalyst Exhibiting a Wide Scope. *J. Am. Chem. Soc.* **1998**, 120, 13529-13530.
24. Ohkuma, T., Ooka, H., Ikariya, T. & Noyori, R. Preferential hydrogenation of aldehydes and ketones. *J. Am. Chem. Soc.* **1995**, 117, 10417-10418.
25. Ohkuma, T., Ikehira, H., Ikariya, T. & Noyori, R. Asymmetric hydrogenation of cyclic α,β-unsaturated ketones to chiral allylic alcohols. *Synlett* **1997**, 467-468.
26. Ohkuma, T., Doucet, H., Pham, T., Mikami, K., Korenaga, T., Terada, M. & Noyori, R. Asymmetric Activation of Racemic Ruthenium(II) Complexes for Enantioselective Hydrogenation. *J. Am. Chem. Soc.* **1998**, 120, 1086-1087.
27. Jiang, Q., Jiang, Y., Xiao, D., Cao, P. & Zhang, X. Highly enantioselective hydrogenation of simple ketones catalyzed by a Rh-PennPhos complex. *Angew. Chem., Int. Ed.* **1998**, 37, 1100-1103.
28. Ohkuma, T., Ooka, H., Yamakawa, M., Ikariya, T. & Noyori, R. Stereoselective Hydrogenation of Simple Ketones Catalyzed by Ruthenium(II) Complexes. *J. Org. Chem.* **1996**, 61, 4872-4873.
29. Ohkuma, T., Koizumi, M., Yoshida, M. & Noyori, R. General Asymmetric Hydrogenation of Heteroaromatic Ketones. *Org. Lett.* **2000**, 2, 1749-1751.
30. Ohkuma, T., Koizumi, M., Ikehira, H., Yokozawa, T. & Noyori, R. Selective Hydrogenation of Benzophenones to Benzhydrols. Asymmetric Synthesis of Unsymmetrical Diarylmethanols. *Org. Lett.* **2000**, 2, 659-662.
31. Sandoval, C. A., Ohkuma, T., Muniz, K. & Noyori, R. Mechanism of Asymmetric Hydrogenation of Ketones Catalyzed by BINAP/1,2-Diamine-Ruthenium(II) Complexes. *J. Am. Chem. Soc.* **2003**, 125, 13490-13503.

ENANTIOSELECTIVE REDUCTION OF C=O

Background: Catalytic Enantioselective Reduction of Prochiral Ketones by Boranes

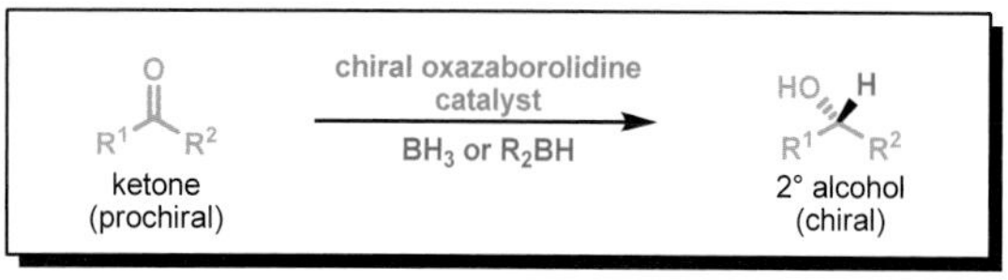

Chiral B-substituted oxazaborolidines such as **1**, derived from diphenylprolinol (**2**, 1,1-diphenylpyrrolidinomethanol), are very useful as catalysts for the enantioselective synthesis of chiral secondary alcohols from ketones using various boranes as the stoichiometric reductant.[1]

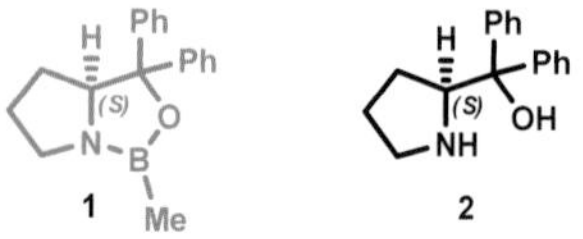

The reduction of acetophenone (**3**) by the borane-tetrahydrofuran complex (BH$_3$·THF, 0.6 equiv) in the presence of 10 mol% of **1** in THF at 2 °C to form (R)-1-phenylethanol (**4**) in 99% yield and 96.5% enantiomeric purity represents a simple and typical example of the general method.[2]

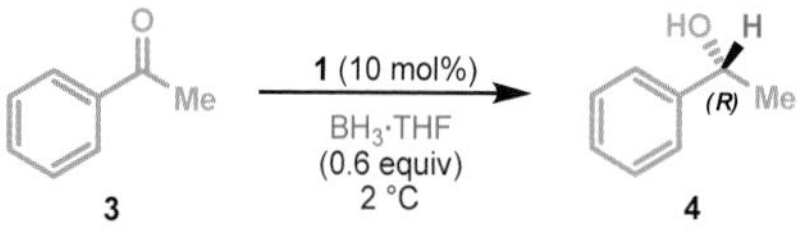

The scope of the enantioselective reduction is broad.[1b] The mechanistic details of the reaction pathway are sufficiently clear that the absolute stereochemical course of the reaction is generally predictable. That pathway is illustrated for the case of acetophenone in Scheme 1.[2a]

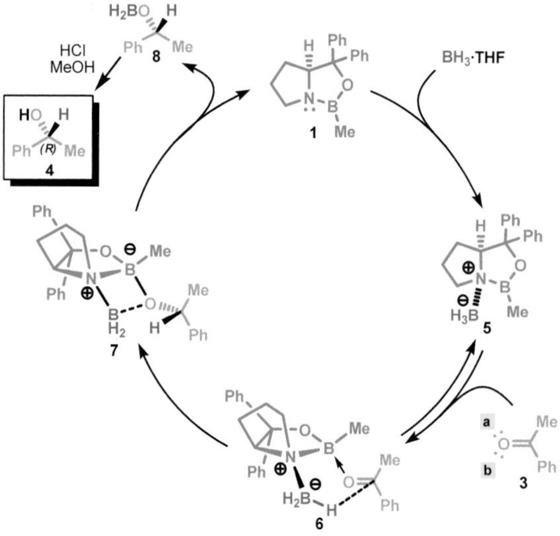

Scheme 1

The oxazaborolidine **1** is very weakly basic at nitrogen, but nonetheless can coordinate with BH$_3$ to form complex **5** (Scheme 1) having a *cis* fusion of the 5,5-ring system, as shown by isolation as a crystalline compound and structure

determination by X-ray diffraction analysis. The Lewis acidity at boron is greatly enhanced in **5** and, in consequence, it readily coordinates with the ketonic substrate **3** to give ternary complex **6**. The complexation selectively occurs at the sterically more accessible lone pair at the oxygen of **3** (lone pair **a**) and leads to a structure in which rotation about the B-O bond is restricted by the steric bulk of the nearby phenyl group of the catalyst. Complex **6** is doubly activated for migration of H from boron to carbon, which occurs with π-facial selectivity so as to lead to the secondary alcohol (R)-**4**.

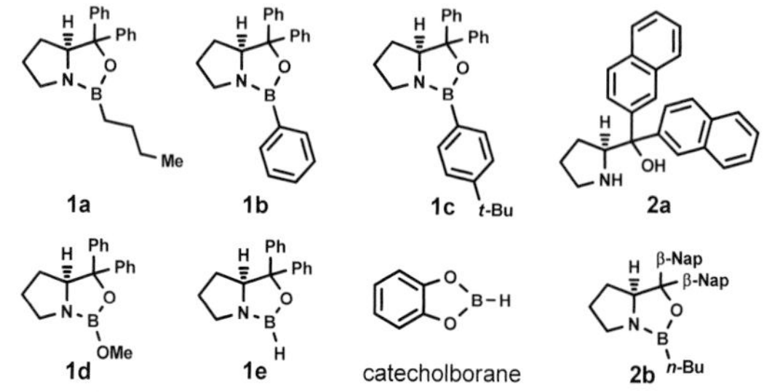

Figure 1

Several structural variants of catalyst **1** have been employed successfully (Figure 1). The methyl group at boron can be replaced by longer chain alkyl (e.g., n-Bu), aryl (e.g., phenyl or 4-t-butylphenyl) or even methoxy. The phenyl groups of **2** may be replaced by β-naphthyl substituents at carbon. In addition to BH$_3$·THF other boranes, especially catecholborane and BH$_3$·Me$_2$S, have been employed as stoichiometric reductants. The B-n-Bu analog of **1** (**1a**, Figure 1), which is very soluble in toluene, has often been used for reductions with cathecolborane in that solvent.

The chiral oxazaborolidine-catalyzed reduction of prochiral ketones is generally highly enantioselective with substrates in which the two substituents attached to the carbonyl carbon differ in steric bulk or if there are other factors favoring complexation by boron at one of the two lone pairs at the carbonyl oxygen. Hundreds of ketonic substrates have been successfully reduced with high enantioselectivity to form the corresponding secondary alcohol with the absolute configuration predicted from a pathway analogous to that depicted for acetophenone in Scheme 1.[1b]

Both (R)- and (S)-diphenylprolinols are commercially available and relatively inexpensive. In addition these ligands can generally be recovered from reaction mixtures for reuse. The tricyclic oxazaborolidines **9a** (R=Me) and **9b** (R=n-Bu) are also readily prepared and can be used interchangeably with the (R)-enantiomer of **1**. Aldehydes are deuterated to form the corresponding ^{2}H-labeled primary alcohols.[3]

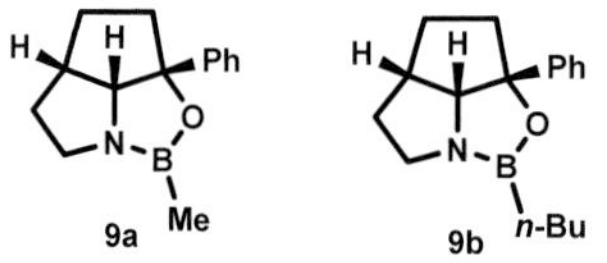

The scope and utility of the enantioselective reduction of ketones using **1** as catalyst is apparent from the examples shown in Schemes 2 and 3. They key reduction step and the final target structure are shown.

ENANTIOSELECTIVE REDUCTION OF C=O

Enantioselective Reduction of Prochiral Ketones by the CBS Method

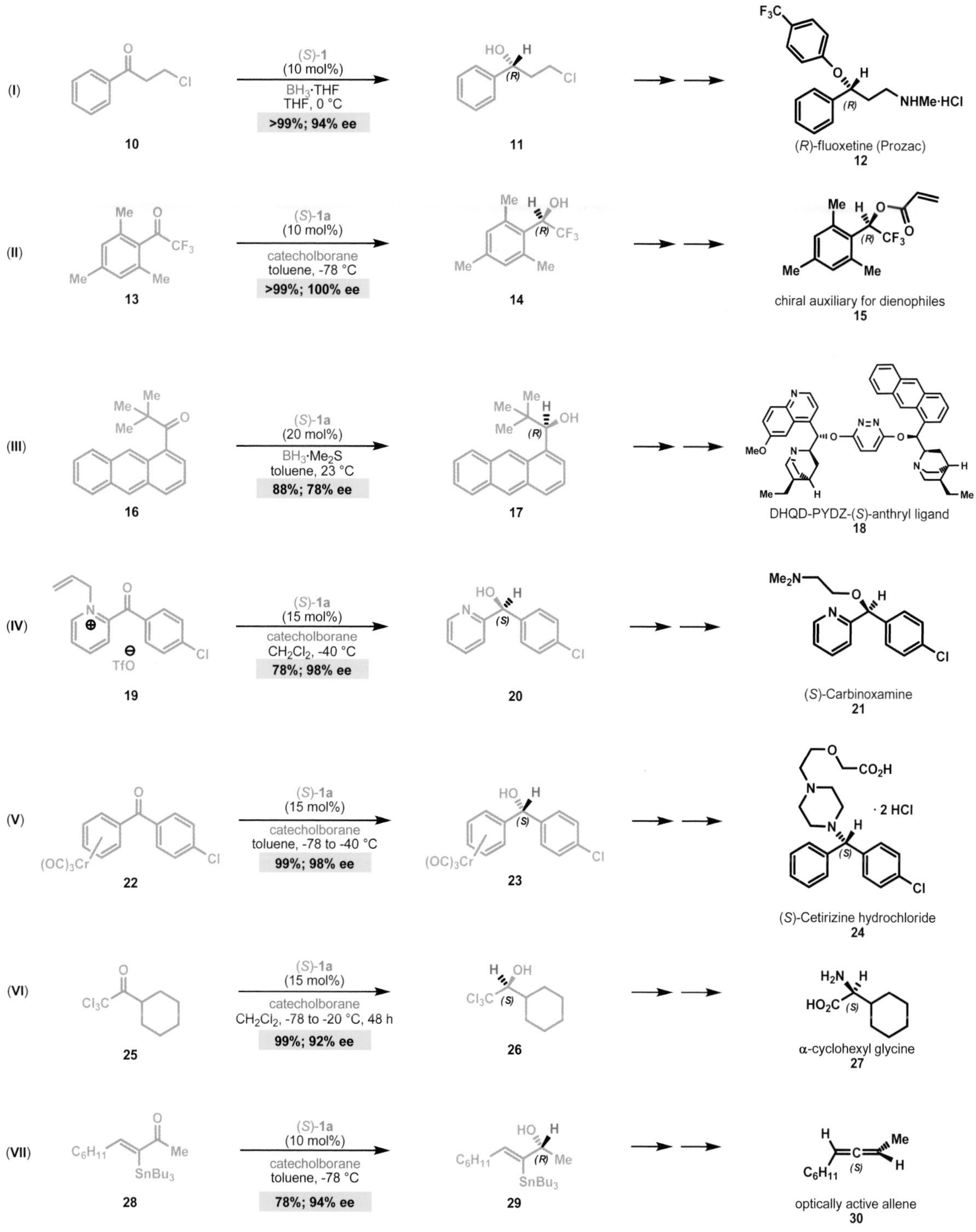

(I)

10 → (S)-1 (10 mol%), BH₃·THF, THF, 0 °C, >99%; 94% ee → 11 → → (R)-fluoxetine (Prozac) 12

(II)

13 → (S)-1a (10 mol%), catecholborane, toluene, -78 °C, >99%; 100% ee → 14 → → chiral auxiliary for dienophiles 15

(III)

16 → (S)-1a (20 mol%), BH₃·Me₂S, toluene, 23 °C, 88%; 78% ee → 17 → → DHQD-PYDZ-(S)-anthryl ligand 18

(IV)

19 → (S)-1a (15 mol%), catecholborane, CH₂Cl₂, -40 °C, 78%; 98% ee → 20 → → (S)-Carbinoxamine 21

(V)

22 → (S)-1a (15 mol%), catecholborane, toluene, -78 to -40 °C, 99%; 98% ee → 23 → → (S)-Cetirizine hydrochloride 24

(VI)

25 → (S)-1a (15 mol%), catecholborane, CH₂Cl₂, -78 to -20 °C, 48 h, 99%; 92% ee → 26 → → α-cyclohexyl glycine 27

(VII)

28 → (S)-1a (10 mol%), catecholborane, toluene, -78 °C, 78%; 94% ee → 29 → → optically active allene 30

Reference Key for Equations: (I)[4]; (II)[5]; (III)[6]; (IV)[7]   **Scheme 2**   Reference Key for Equations: (V)[8]; (VI)[9]; (VII)[10]

ENANTIOSELECTIVE REDUCTION OF C=O

Enantioselective Reduction of Prochiral Ketones by the CBS Method (Continued)

(VIII) **31** → (S)-1 (10 mol%), $BH_3 \cdot THF$, THF, 10 °C, **88%; 93% ee** → **32** → → ginkgolide B **33**

(IX) **34** → (S)-1a (20 mol%), catecholborane, $PhNEt_2$ toluene, -50 °C, **86%; 92% ee** → **35** → → (+)-estrone **36**

(X) **37** → (S)-1e (10 mol%), catecholborane toluene, -78 °C, **93%; >96% ee** → **38** → → (−)-morphine **39**

(XI) (R)-40a ⇌ *in equilibrium* (S)-40b → (S)-1 (10 mol%), $BH_3 \cdot THF$ THF; 30 °C, 3 h, **94%; 97% ee**, *kinetic resolution* → **41** → single atropisomer **42**

Reference Key for Equations: (VIII)[11]; (IX)[12]   **Scheme 3**   Reference Key for Equations: (X)[13]; (XI)[14]

References

1. (a) Itsuno, S. Enantioselective reduction of ketones. *Organic Reactions (Hoboken, NJ, United States)* **1998**, 52, 395-576; (b) Corey, E. J. & Helal, C. J. Reduction of carbonyl compounds with chiral oxazaborolidine catalysts: A new paradigm for enantioselective catalysis and a powerful new synthetic method. *Angew. Chem., Int. Ed.* **1998**, 37, 1986-2012; (c) Cho, B. T. Oxazaborolidines as asymmetric inducers for the reduction of ketones and ketimines. *Boronic Acids* **2005**, 411-439; (d) Cho, B. T. Recent advances in the synthetic applications of the oxazaborolidine-mediated asymmetric reduction. *Tetrahedron* **2006**, 62, 7621-7643; (e) Stemmler, R. T. CBS oxazaborolidines, versatile catalysts for asymmetric synthesis. *Synlett* **2007**, 997-998; (f) Arai, N. & Ohkuma, T. Carbonyl hydroboration in *Modern Reduction Methods* 159-181 (**2008**).
2. (a) Corey, E. J., Bakshi, R. K. & Shibata, S. Highly enantioselective borane reduction of ketones catalyzed by chiral oxazaborolidines. Mechanism and synthetic implications. *J. Am. Chem. Soc.* **1987**, 109, 5551-5553; (b) Corey, E. J., Bakshi, R. K., Shibata, S., Chen, C. P. & Singh, V. K. A stable and easily prepared catalyst for the enantioselective reduction of ketones. Applications to multistep syntheses. *J. Am. Chem. Soc.* **1987**, 109, 7925-7926.
3. Corey, E. J. & Link, J. O. A new chiral catalyst for the enantioselective synthesis of secondary alcohols and deuterated primary alcohols by carbonyl reduction. *Tetrahedron Lett.* **1989**, 30, 6275-6278.
4. Corey, E. J. & Reichard, G. A. Enantioselective and practical syntheses of R- and S-fluoxetines. *Tetrahedron Lett.* **1989**, 30, 5207-5210.
5. Corey, E. J., Cheng, X. M. & Cimprich, K. A. 1-Mesityl-2,2,2-trifluoroethanol, an outstanding new chiral controller for catalyzed Diels-Alder reactions. *Tetrahedron Lett.* **1991**, 32, 6839-6842.
6. Corey, E. J., Noe, M. C. & Grogan, M. J. A mechanistically designed mono-cinchona alkaloid is an excellent catalyst for the enantioselective dihydroxylation of olefins. *Tetrahedron Lett.* **1994**, 35, 6427-6430.
7. Corey, E. J. & Helal, C. J. Asymmetric synthesis of (S)-carbinoxamine. New aspects of oxazaborolidine-catalyzed enantioselective carbonyl reduction. *Tetrahedron Lett.* **1996**, 37, 5675-5678.
8. Corey, E. J. & Helal, C. J. Catalytic enantioselective synthesis of the second generation histamine antagonist cetirizine hydrochloride. *Tetrahedron Lett.* **1996**, 37, 4837-4840.
9. (a) Corey, E. J. & Link, J. O. A general, catalytic, and enantioselective synthesis of α-amino acids. *J. Am. Chem. Soc.* **1992**, 114, 1906-1908; (b) Corey, E. J., Link, J. O. & Bakshi, R. K. A mechanistic and structural analysis of the basis for high enantioselectivity in the oxazaborolidine-catalyzed reduction of trihalomethyl ketones by catecholborane. *Tetrahedron Lett.* **1992**, 33, 7107-7110.
10. Konoike, T. & Araki, Y. Concise allene synthesis from propargylic alcohols by hydrostannation and deoxystannylation: a new route to chiral allenes. *Tetrahedron Lett.* **1992**, 33, 5093-5096.
11. Corey, E. J. & Gavai, A. V. Enantioselective route to a key intermediate in the total synthesis of ginkgolide B. *Tetrahedron Lett.* **1988**, 29, 3201-3204.
12. Yeung, Y.-Y., Chein, R.-J. & Corey, E. J. Conversion of Torgov's Synthesis of Estrone into a Highly Enantioselective and Efficient Process. *J. Am. Chem. Soc.* **2007**, 129, 10346-10347.
13. Hong, C. Y., Kado, N. & Overman, L. E. Asymmetric synthesis of either enantiomer of opium alkaloids and morphinans. Total synthesis of (-)- and (+)-dihydrocodeinone and (-)- and (+)-morphine. *J. Am. Chem. Soc.* **1993**, 115, 11028-11029.
14. Bringmann, G. & Vitt, D. Stereoselective Ring-Opening Reaction of Axially Prostereogenic Biaryl Lactones with Chiral Oxazaborolidines: An AM1 Study of the Complete Mechanistic Course. *J. Org. Chem.* **1995**, 60, 7674-7681.

ENANTIOSELECTIVE REDUCTION OF C=O

Enantioselective Hydrosilylation of Ketones

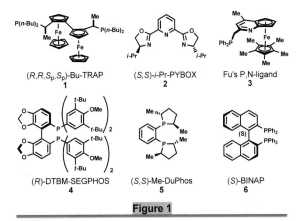

ketone (prochiral) → chiral catalyst, silanes (e.g., HSiCl₃, Ph₂SiH₂, PMHS) → 2° alcohol (chiral)

The discovery in the 1970s[1] that the reduction of ketones by hydrosilylation could be effected using various silanes and Wilkinson's catalyst [RhCl(PPh₃)₃], paved the way for enantioselective hydrosilylation of ketones with chiral complexes of Rh, Ru or Ir.[2] A selection of chiral ligands for complexation and some complexes with these metals are shown in Figures 1 and 2. The reactive species in these processes appears to be one derived by insertion of the complexed metal into the Si-H bond, forming intermediates of the type $R_3Si[M]H(L_n)$.[1,3,2e]

(R,R,Sₚ,Sₚ)-Bu-TRAP
1

(S,S)-i-Pr-PYBOX
2

Fu's P,N-ligand
3

(R)-DTBM-SEGPHOS
4

(S,S)-Me-DuPhos
5

(S)-BINAP
6

Figure 1

(S,S)-(EBTHI)TiF₂
7

Imamoto's catalyst
8

Evans' P,S-catalyst
9

Oxazolinylcarbene-Rh
10

TADDOL-based Rh catalyst
11

Figure 2

A wide variety of prochiral ketones have been reduced enantioselectively using the hydrosilylation method (see examples in Schemes 1 and 2).

The enantioselective hydrosilylation of sterically hindered ketones using an iron(II)-Me-DuPhos complex (**44→45**, Scheme 2) has been described.[4] However, it has not been established that the iron acetate used in these reactions was copper-free.[5] Two examples of copper-bisphosphine-catalyzed hydrosilylation of ketones (**48→49** and **50→51**) are shown in Scheme 2.[6]

Enantioselective hydrosilylation of ketones has also been reported for compounds of Ti, Zr, Cu and Zn.[2e]

Enantioselective Hydrosilyation of Structurally Diverse Ketones

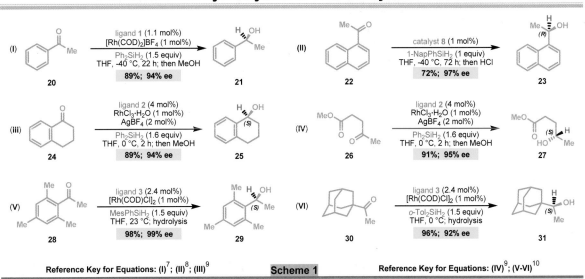

(I) 20 — ligand 1 (1.1 mol%), [Rh(COD)₂]BF₄ (1 mol%), Ph₂SiH₂ (1.5 equiv), THF, -40 °C, 22 h; then MeOH — **89%; 94% ee** → 21

(II) 22 — catalyst 8 (1 mol%), 1-NapPhSiH₂ (1 equiv), THF, -40 °C, 72 h; then HCl — **72%; 97% ee** → 23 (R)

(III) 24 — ligand 2 (4 mol%), RhCl₃·H₂O (1 mol%), AgBF₄ (2 mol%), Ph₂SiH₂ (1.6 equiv), THF, 0 °C, 2 h; then MeOH — **89%; 94% ee** → 25 (S)

(IV) 26 — ligand 2 (4 mol%), RhCl₃·H₂O (1 mol%), AgBF₄ (2 mol%), Ph₂SiH₂ (1.6 equiv), THF, 0 °C, 2 h; then MeOH — **91%; 95% ee** → 27 (S)

(V) 28 — ligand 3 (2.4 mol%), [Rh(COD)Cl]₂ (1 mol%), MesPhSiH₂ (1.5 equiv), THF, 23 °C; hydrolysis — **98%; 99% ee** → 29 (S)

(VI) 30 — ligand 3 (2.4 mol%), [Rh(COD)Cl]₂ (1 mol%), o-Tol₂SiH₂ (1.5 equiv), THF, 0 °C; hydrolysis — **96%; 92% ee** → 31 (S)

Reference Key for Equations: (I)[7]; (II)[8]; (III)[9]   **Scheme 1**   Reference Key for Equations: (IV)[9]; (V-VI)[10]

ENANTIOSELECTIVE REDUCTION OF C=O

Enantioselective Hydrosilylation of Structurally Diverse Ketones

(VII) **32** → ligand 3 (2.4 mol%) [Rh(COD)Cl]$_2$ (1 mol%) / o-Tol$_2$SiH$_2$ (1.5 equiv) THF, 0 °C; hydrolysis → **96%; 92% ee** → **33** (S)

(VIII) **34** → catalyst 9 (1 mol%) / 1-NapPhSiH$_2$ (1.5 equiv) THF, -20 °C; then HCl → **95%; 94% ee** → **35** (S)

(IX) **36** → catalyst 9 (1 mol%) / 1-NapPhSiH$_2$ (1.5 equiv) THF, -20 °C; then HCl → **91%; 92% ee** → **37** (S)

(X) **38** → catalyst 7 (1 mol%) pyrrolidine/MeOH / 1-NapPhSiH$_2$ (1.5 equiv) THF, 60 °C, 10 h → **80%; >98% ee** → **39** (S)

(XI) **40** → catalyst 7 (2 mol%) pyrrolidine/MeOH / 1-NapPhSiH$_2$ (1.5 equiv) THF, 60 °C, 5 h → **87%; 96% ee** → **41** (S)

(XII) **42** → catalyst 7 (2 mol%) pyrrolidine/MeOH / 1-NapPhSiH$_2$ (1.5 equiv) THF, 60 °C, 4 h → **90%; 84% ee** → **43** (S)

(XIII) **44** → ligand 5 (10 mol%) Fe(OAc)$_2$ (5 mol%) / PMHS (4 equiv) THF, 65 °C, 24 h → **78%; 99% ee** → **45** (R)

(XIV) **46** → catalyst 10 (1 mol%) AgBF$_4$ (1.2 mol%) / Ph$_2$SiH$_2$ (1.1 equiv) CH$_2$Cl$_2$, -60 °C, 10 h → **87%; 96% ee** → **47** (S)

(XV) **48** → ligand 4 (0.4 mol%) CuCl (5 mol%) NaO-tBu (5 mol%) / PMHS (2 equiv) toluene, 0 °C, 5 h → **98%; 93% ee** → **49** (R)

(XVI) **50** → ligand 4 (0.4 mol%) CuCl (5 mol%) NaO-tBu (5 mol%) / PMHS (2 equiv) toluene, -30 °C, 20 h → **98%; 93% ee** → **51** (S)

Reference Key: (VII)[10]; (VIII-IX)[11]; (X-XI)[12] **Scheme 2** Reference Key: (XII)[12]; (XIII)[4]; (XIV)[13]; (XV-XVI)[6]

References

1. Ojima, I., Nihonyanagi, M. & Nagai, Y. Stereoselective reduction of ketones with hydrosilane-rhodium(I) complex combinations. *Bull. Chem. Soc. Jap.* **1972**, 45, 3722.
2. (a) Nishiyama, H. Hydrosilylation of carbonyl and imino groups. *Compr. Asymmetric Catal. I-III* **1999**, 1, 267-287; (b) Carpentier, J.-F. & Bette, V. Chemo- and enantioselective hydrosilylation of carbonyl and imino groups. An emphasis on non-traditional catalyst systems. *Curr. Org. Chem.* **2002**, 6, 913-936; (c) Nishiyama, H. Hydrosilylations of carbonyl and imine compounds. *Transition Metals for Organic Synthesis (2nd Edition)* **2004**, 2, 182-191; (d) Riant, O., Mostefai, N. & Courmarcel, J. Recent advances in the asymmetric hydrosilylation of ketones, imines, and electrophilic double bonds. *Synthesis* **2004**, 2943-2958; (e) Rendler, S. & Oestreich, M. Diverse modes of silane activation for the hydrosilylation of carbonyl compounds in *Modern Reduction Methods* 183-207 (**2008**).
3. Ojima, I., Kogure, T., Kumagai, M., Horiuchi, S. & Sato, T. Reduction of carbonyl compounds via hydrosilylation. II. Asymmetric reduction of ketones via hydrosilylation catalyzed by a rhodium(I) complex with chiral phosphine ligands. *J. Organomet. Chem.* **1976**, 122, 83-97.
4. Shaikh, N. S., Enthaler, S., Junge, K. & Beller, M. Iron-catalyzed enantioselective hydrosilylation of ketones. *Angew. Chem., Int. Ed.* **2008**, 47, 2497-2501.
5. For a recent correspondence on this topic, see: S.L. Buchwald and C. Bolm. On the Role of Metal Contaminants in Catalyses with FeCl$_3$. *Angew. Chem. Intl. Ed. Eng.* **2009**, 48, 5586-5587.
6. Lee, C.-T. & Lipshutz, B. H. Nonracemic Diarylmethanols From CuH-Catalyzed Hydrosilylation of Diaryl Ketones. *Org. Lett.* **2008**, 10, 4187-4190.
7. Kuwano, R., Uemura, T., Saitoh, M. & Ito, Y. A trans-chelating bisphosphine possessing only planar chirality and its application to catalytic asymmetric reactions. *Tetrahedron Asymmetry* **2004**, 15, 2263-2271.
8. Imamoto, T., Itoh, T., Yamanoi, Y., Narui, R. & Yoshida, K. Highly enantioselective hydrosilylation of simple ketones catalyzed by rhodium complexes of P-chiral diphosphine ligands bearing tert-butylmethylphosphino groups. *Tetrahedron Asymmetry* **2006**, 17, 560-565.
9. Nishiyama, H., Yamaguchi, S., Kondo, M. & Itoh, K. Electronic substituent effect of nitrogen ligands in catalytic asymmetric hydrosilylation of ketones: chiral 4-substituted bis(oxazolinyl)pyridines. *J. Org. Chem.* **1992**, 57, 4306-4309.
10. Tao, B. & Fu, G. C. Application of a new family of P,N ligands to the highly enantioselective hydrosilylation of aryl alkyl and dialkyl ketones. *Angew. Chem., Int. Ed.* **2002**, 41, 3892-3894.
11. Evans, D. A., Michael, F. E., Tedrow, J. S. & Campos, K. R. Application of Chiral Mixed Phosphorus/Sulfur Ligands to Enantioselective Rhodium-Catalyzed Dehydroamino Acid Hydrogenation and Ketone Hydrosilylation Processes. *J. Am. Chem. Soc.* **2003**, 125, 3534-3543.
12. Yun, J. & Buchwald, S. L. Titanocene-Catalyzed Asymmetric Ketone Hydrosilylation: The Effect of Catalyst Activation Protocol and Additives on the Reaction Rate and Enantioselectivity. *J. Am. Chem. Soc.* **1999**, 121, 5640-5644.
13. Gade, L. H., Cesar, V. & Bellemin-Laponnaz, S. A modular assembly of chiral oxazolinylcarbene-rhodium complexes: Efficient phosphane-free catalysts for the asymmetric hydrosilylation of dialkyl ketones. *Angew. Chem., Int. Ed.* **2004**, 43, 1014-1017.

ENANTIOSELECTIVE ADDITION OF H₂ TO C=N

Background: Highly Enantioselective Homogeneous Hydrogenation of Imines

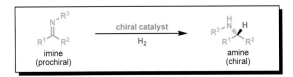

imine (prochiral) → chiral catalyst / H₂ → amine (chiral)

The enantioselective catalytic hydrogenation of imines to chiral amines provides a route to many useful intermediates, pharmaceuticals and agrochemicals.[1] There are fewer effective catalysts for this transformation than for the enantioselective reduction of C=O because of the following problems: (a) stronger coordination of the transition metal to the unshared electron than to the C=N π-bond resulting in low catalytic activity/turnover; (b) frequent problems in preparing pure (E) and (Z) imines; (c) hydrolytic instability of the imine substrate and (d) requirement for H₂ at elevated pressure.

An early and modestly enantioselective catalytic reduction of an imine by hydrosilylation, using a chiral Rh(I) complex was reported by H. Kagan in 1973.[2]

Subsequent to Kagan's report, a large number of Ru-, Ir- and Rh-chiral bisphosphine complexes have been examined for imine hydrogenation. Various additives (e.g., TBAI, KI, I₂) have been found to influence enantioselectivity.[1b]

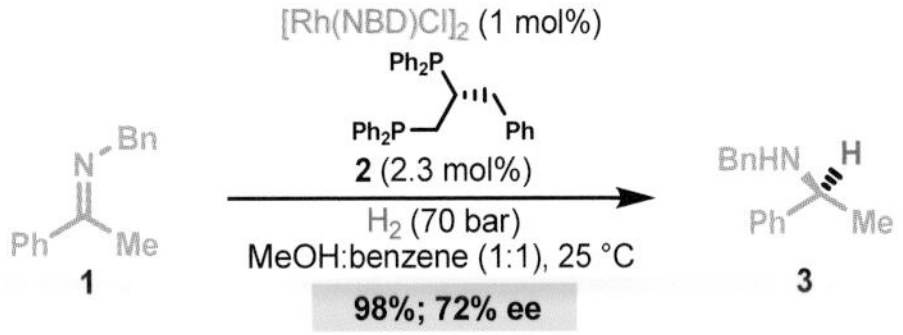

[Rh(NBD)Cl]₂ (1 mol%)
2 (2.3 mol%)
H₂ (70 bar)
MeOH:benzene (1:1), 25 °C

98%; 72% ee

An effective catalytic homogeneous hydrogenation of imines (e.g., 1→3) was demonstrated by Markó et al.[3] However, it was not until the late 1990s that highly effective chiral catalysts emerged for this transformation using either hydrogen gas or various hydrogen donors such as 2-propanol or triethylammonium formate.[1k,1n,1p]

Common Imine Substrates

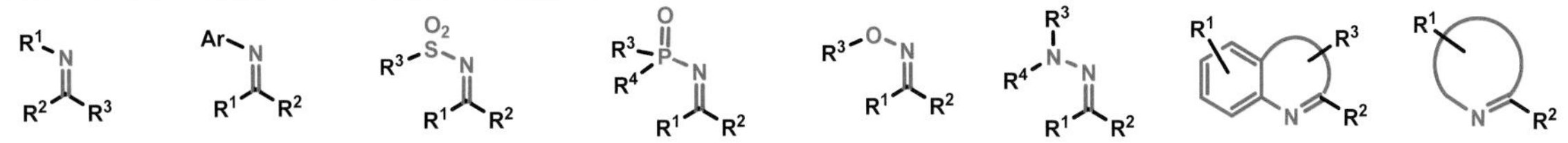

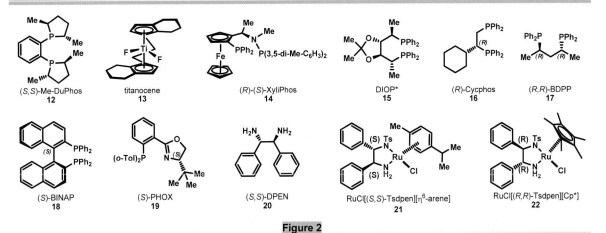

| acyclic N-alkylimines 4 | acyclic N-arylimines 5 | acyclic N-sulfonylimines 6 | acyclic N-phosphynylimines 7 | acyclic oximes 8 | acyclic hydrazones 9 | cyclic N-arylimines 10 | cyclic N-alkylimines 11 |

Figure 1

Miscellaneous Ligands and Catalysts

(S,S)-Me-DuPhos **12** | titanocene **13** | (R)-(S)-XyliPhos **14** | DIOP* **15** | (R)-Cycphos **16** | (R,R)-BDPP **17**

(S)-BINAP **18** | (S)-PHOX **19** | (S,S)-DPEN **20** | RuCl[(S,S)-Tsdpen][η⁶-arene] **21** | RuCl[(R,R)-Tsdpen][Cp*] **22**

Figure 2

Comments on Figures 1 and 2

(6) Figure 1 illustrates the various types of imines that have been reduced enantioselectively; each structural class requires a particular catalyst system.

(7) Various chiral ligands, e.g., bisphosphines, diamines, P,N-ligands and metallocenes (Figure 2) in combination with Ir, Rh and Ru complexes have been applied to the enantioselective hydrogenation of imines.[1m,1o]

ENANTIOSELECTIVE ADDITION OF H₂ TO C=N

Enantioselective Hydrogenation of *N*-Arylimines[1f,1m]

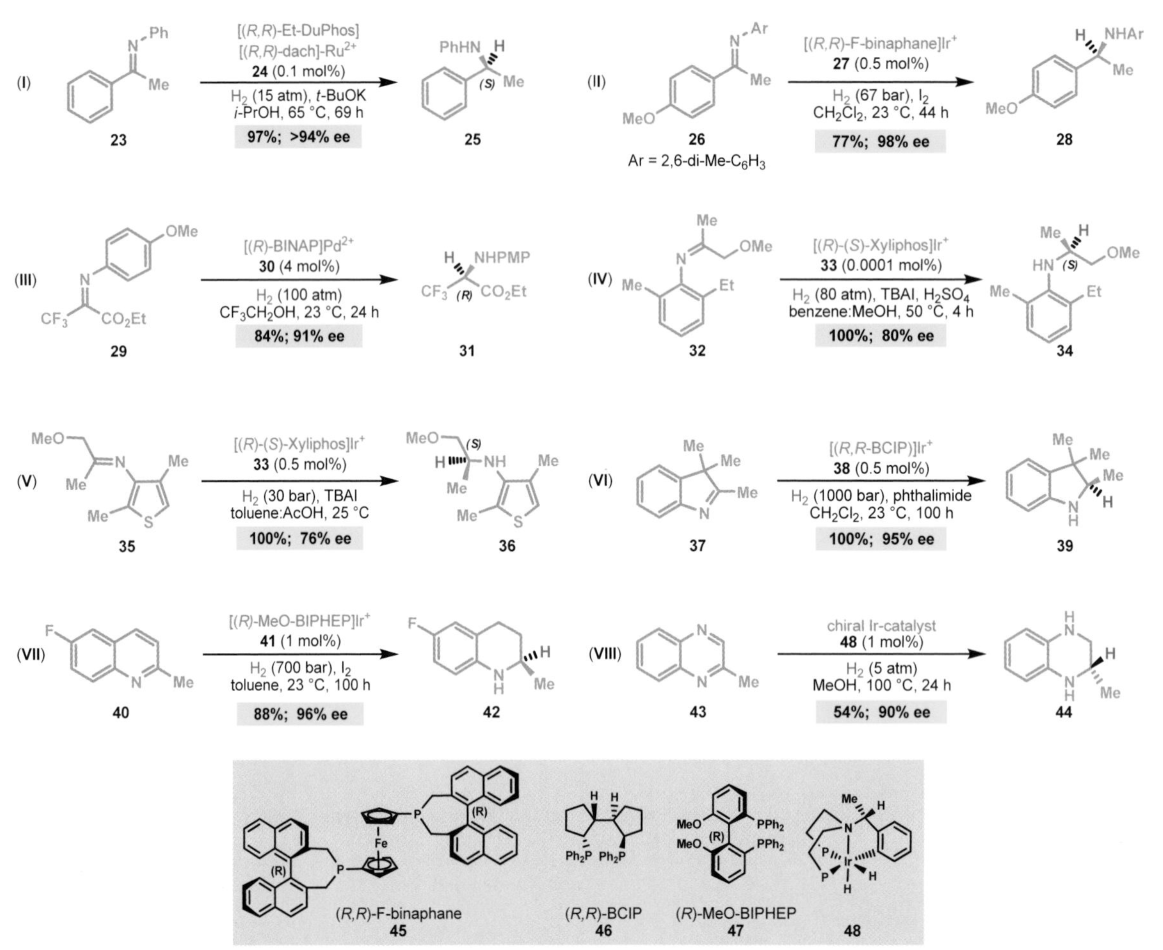

Scheme 1

Reference Key for Equations: (I)[4]; (II)[5]; (III)[6]; (IV)[7]   Reference Key for Equations: (V)[8]; (VI)[9]; (VII)[10]; (VIII)[11]

Comments on Scheme 1

(1) The acetophenone imine derivatives **23**[4] and **26**[5] are hydrogenated enantioselectively using chiral Ru-bisphosphine/diamine complexes, for example Ru complex **24**, or the chiral Ir-complex **27** (which contains the ferrocene-based ligand (*R*,*R*)-F-binaphane **45**). In the case of **26**→**28**, added iodine enhances enantioselectivity.

(2) The α-fluorinated iminoester **29**[5] is hydrogenated efficiently in the presence of the Pd-BINAP complex **30**, allowing the synthesis of enantioenriched β-fluorinated α-amino esters such as **31**. Enantioselectivity depends significantly on solvent (30% ee with EtOH vs. 91% ee with 2,2,2-trifluoroethanol).[5]

(3) The transformation **32**→**34** is the key step in a large scale (10^4 tons/year) industrial synthesis of the herbicide (*S*)-metolachlor.[7] The Ir-complex **33** is prepared *in situ* by using the ferrocenyl diphosphine ligand (*R*)-(*S*)-xyliphos (**14**, Figure 2). The resulting catalyst is extraordinarily active and therefore a substrate/catalyst ratio of $>10^6$ can be used. The use of the additives tetrabutylammonium iodide and sulfuric acid is essential in this process.

(4) The enantioselective hydrogenation of cyclic aromatic and heteroaromatic imines (e.g., **37**[9], **40**[10] and **43**[11]) has been challenging partly due to the high stability of the substrates and catalyst deactivation/poisoning by the basic nitrogen atoms. The emergence of chiral Ir-bisphosphine complexes such as **38**, **41** and **48** has allowed the preparation of saturated chiral heterocyclic compounds (e.g., **39**, **42** and **44**) in high enantiomeric purity.

ENANTIOSELECTIVE ADDITION OF H₂ TO C=N

Enantioselective Hydrogenation of *N*-Alkylimines[1f,1m,1p]

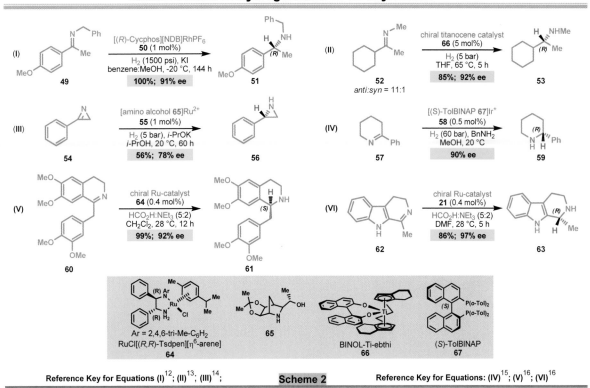

Reference Key for Equations (I)[12]; (II)[13]; (III)[14];   **Scheme 2**   Reference Key for Equations: (IV)[15]; (V)[16]; (VI)[16]

Enantioselective Hydrogenation of C=N-X : Hydrazones, Oximes & Sulfonyl-imines[1f,1m]

Reference Key for Equations (I)[17]; (II)[18]   **Scheme 3**   Reference Key for Equation: (III)[19]

Comments on Schemes 2 and 3

(1) The cyclic imines **54**[14], **57**[15], **60**[16], **62**[16] cannot undergo *syn/anti* isomerization and consequently the enantioselectivity of hydrogenation is higher than for acyclic imines.

(2) Complex six-membered cyclic imines **60**[16] and **62**[16] afford chiral tetrahydroisoquinolines in high yield and enantiomeric purity under transfer hydrogenation conditions. These products are intermediates for the synthesis of certain isoquinoline alkaloids.

(3) The *N*-heteroatom-substituted (activated) imine derivatives **68**[17], **72**[18] and **75**[19] can also be hydrogenated enantioselectively in high yield (Scheme 3) at lower pressure and temperature than non-activated imine substrates.

ENANTIOSELECTIVE ADDITION OF H$_2$ TO C=N

Mechanism of Enantioselective Imine Hydrogenation[1m]

A proposed pathway for the hydrogenation **26→28** (Scheme 1) is shown at right. The active Ir(III) species **83** is formed from the Ir(I) complex **80** by the oxidative addition of the additive I$_2$ (see below).[5]

References

1. (a) Blaser, H.-U. & Spindler, F. Hydrogenation of imino groups. *Compr. Asymmetric Catal. I-III* **1999**, 1, 247-265; (b) Kobayashi, S. & Ishitani, H. Catalytic Enantioselective Addition to Imines. *Chem. Rev.* **1999**, 99, 1069-1094; (c) Palmer, M. J. & Wills, M. Asymmetric transfer hydrogenation of C:O and C:N bonds. *Tetrahedron: Asymmetry* **1999**, 10, 2045-2061; (d) Brunel, J. M. Recent advances in asymmetric reduction of imines and imine derivatives. *Rec. Res. Dev. Org. Chem.* **2003**, 7, 155-190; (e) Lennon, I. C. & Moran, P. H. Asymmetric hydrogenation of pharmaceutically interesting substrates. *Curr. Opin. Drug Discovery Dev.* **2003**, 6, 855-875; (f) Tang, W. & Zhang, X. New Chiral Phosphorus Ligands for Enantioselective Hydrogenation. *Chem. Rev.* **2003**, 103, 3029-3069; (g) Spindler, F. & Blaser, H.-U. Enantioselective reduction of C=N bonds and enamines with hydrogen in *Transition Metals for Organic Synthesis (2nd Edition)* 113-123 (**2004**); (h) Tararov, V. I. & Boerner, A. Approaching highly enantioselective reductive amination. *Synlett* **2005**, 203-211; (i) Vassylyev, O., Panarello, A. & Khinast, J. G. Enantioselective hydrogenations with chiral titanocenes. *Molecules* **2005**, 10, 587-619; (j) Coperet, C. Hydrogenation with early transition metal, lanthanide and actinide complexes in *Handbook of Homogeneous Hydrogenation* 111-151 (**2007**); (k) Roszkowski, P. & Czarnocki, Z. Selected recent developments in the enantioselective reduction of imines by asymmetric transfer hydrogenation. *Mini-Rev. Org. Chem.* **2007**, 4, 190-200; (l) Clarke, M. L. & Roff, G. J. Homogeneous hydrogenation of aldehydes, ketones, imines and carboxylic acid derivatives: chemoselectivity and catalytic activity in *Handbook of Homogeneous Hydrogenation* 413-454 (**2007**); (m) Spindler, F. & Blaser, H.-U. Enantioselective hydrogenation of C=N functions and enamines in *Handbook of Homogeneous Hydrogenation* 1193-1214 (**2007**); (n) Ouellet, S. G., Walji, A. M. & Macmillan, D. W. C. Enantioselective Organocatalytic Transfer Hydrogenation Reactions using Hantzsch Esters. *Acc. Chem. Res.* **2007**, 40, 1327-1339; (o) Claver, C. & Fernandez, E. Imine hydrogenation in *Modern Reduction Methods* 237-269 (**2008**); (p) Wills, M. Imino reductions by transfer hydrogenation in *Modern Reduction Methods* 271-296 (**2008**).
2. Langlois, N., Dang Tuan, P. & Kagan, H. B. Asymmetric synthesis of amines by hydrosilylation of imines catalyzed by a chiral complex of rhodium. *Tetrahedron Lett.* **1973**, 4865-4868.
3. Vastag, S., Bakos, J., Toros, S., Takach, N. E., King, R. B., Heil, B. & Marko, L. Rhodium phosphine complexes as homogeneous catalysts. 14. Asymmetric hydrogenation of a Schiff base of acetophenone - effect of phosphine and catalyst structure on enantioselectivity. *J. Mol. Catal.* **1984**, 22, 283-287.
4. Cobley, C. J. & Henschke, J. P. Enantioselective hydrogenation of imines using a diverse library of ruthenium dichloride(diphosphine)(diamine) precatalysts. *Adv. Synth. Catal.* **2003**, 345, 195-201.
5. Xiao, D. & Zhang, X. Highly enantioselective hydrogenation of acyclic imines catalyzed by Ir-f-binaphane complexes. *Angew. Chem., Int. Ed.* **2001**, 40, 3425-3428.

6. Abe, H., Amii, H. & Uneyama, K. Pd-catalyzed asymmetric hydrogenation of alpha-fluorinated iminoesters in fluorinated alcohol: a new and catalytic enantioselective synthesis of fluoro alpha-amino acid derivatives. *Org. Lett.* **2001**, 3, 313-315.
7. Blaser, H.-U., Malan, C., Pugin, B., Spindler, F., Steiner, H. & Studer, M. Selective hydrogenation for fine chemicals: Recent trends and new developments. *Adv. Synth. Catal.* **2003**, 345, 103-151.
8. Blaser, H. U., Buser, H. P., Hausel, R., Jalett, H. P. & Spindler, F. Tunable ferrocenyl diphosphine ligands for the Ir-catalyzed enantioselective hydrogenation of N-aryl imines. *J. Organomet. Chem.* **2001**, 621, 34-38.
9. Zhu, G. & Zhang, X. Additive effects in Ir-BICP catalyzed asymmetric hydrogenation of imines. *Tetrahedron: Asymmetry* **1998**, 9, 2415-2418.
10. Wang, W.-B., Lu, S.-M., Yang, P.-Y., Han, X.-W. & Zhou, Y.-G. Highly Enantioselective Iridium-Catalyzed Hydrogenation of Heteroaromatic Compounds, Quinolines. *J. Am. Chem. Soc.* **2003**, 125, 10536-10537.
11. Bianchini, C., Barbaro, P., Scapacci, G., Farnetti, E. & Graziani, M. Enantioselective Hydrogenation of 2-Methylquinoxaline to (-)-(2S)-2-Methyl-1,2,3,4-tetrahydroquinoxaline by Iridium Catalysis. *Organometallics* **1998**, 17, 3308-3310.
12. Becalski, A. G., Cullen, W. R., Fryzuk, M. D., James, B. R., Kang, G. J. & Rettig, S. J. Catalytic asymmetric hydrogenation of imines. Use of rhodium(I)/phosphine complexes and characterization of rhodium(I)/imine complexes. *Inorg. Chem.* **1991**, 30, 5002-5008.
13. Willoughby, C. A. & Buchwald, S. L. Catalytic Asymmetric Hydrogenation of Imines with a Chiral Titanocene Catalyst: Scope and Limitations. *J. Am. Chem. Soc.* **1994**, 116, 8952-8965.
14. Roth, P., Andersson, P. G. & Somfai, P. Asymmetric reduction of azirines; a new route to chiral aziridines. *Chem. Commun. (Cambridge, U. K.)* **2002**, 1752-1753.
15. Tani, K., Onouchi, J.-i., Yamagata, T. & Kataoka, Y. Iridium(I)-catalyzed asymmetric hydrogenation of prochiral imines; protic amines as catalyst improvers. *Chem. Lett.* **1995**, 955-956.
16. Uematsu, N., Fujii, A., Hashiguchi, S., Ikariya, T. & Noyori, R. Asymmetric Transfer Hydrogenation of Imines. *J. Am. Chem. Soc.* **1996**, 118, 4916-4917.
17. Burk, M. J. & Feaster, J. E. Enantioselective hydrogenation of the C:N group: a catalytic asymmetric reductive amination procedure. *J. Am. Chem. Soc.* **1992**, 114, 6266-6267.
18. Xie, Y., Mi, A., Jiang, Y. & Liu, H. Enantioselective hydrogenation of ketone oxime catalyzed by Ir-DPAMPP complex. *Synth. Commun.* **2001**, 31, 2767-2771.
19. Cobley, C. J., Foucher, E., Lecouve, J.-P., Lennon, I. C., Ramsden, J. A. & Thominot, G. The synthesis of S 18986, a chiral AMPA receptor modulator, via catalytic asymmetric hydrogenation. *Tetrahedron Asymmetry* **2003**, 14, 3431-3433.

ENANTIOSELECTIVE REDUCTION OF C=N

Introduction: Enantioselective Hydrosilylation of Imines

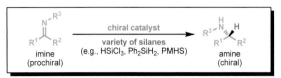

A modestly enantioselective, catalytic hydrosilylation of an imine (**1**→**3**) was reported by H. Kagan in 1973 using the Rh(I)/(R,R)-DIOP complex **2** (Scheme 1).[1]

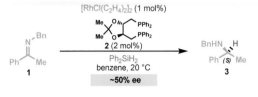

Scheme 1

The catalytic enantioselective hydrosilylation of imines, first developed during the 1990s,[2] is operationally much simpler than catalytic enantioselective hydrogenation (see page 29) since it generally occurs at atmospheric pressure and ambient temperature. The range of substrates for this transformation is broad (Figure 1).

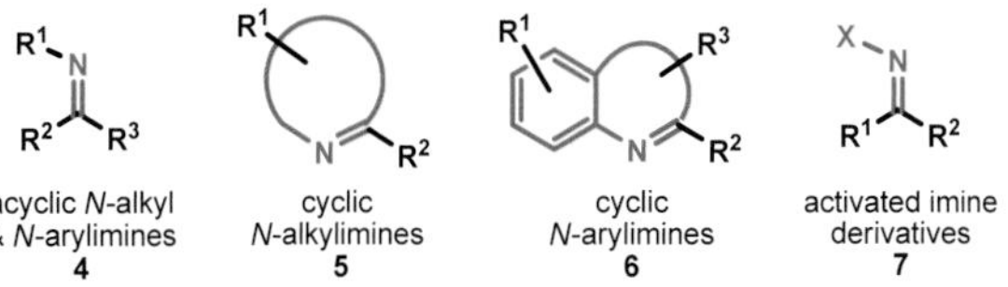

R[1] N R[2] R[3]
acyclic N-alkyl & N-arylimines **4**

R[1] N R[2] (cyclic)
cyclic N-alkylimines **5**

R[1] R[3] N R[2]
cyclic N-arylimines **6**

X N R[1] R[2]
activated imine derivatives **7**

Figure 1

Useful catalysts for imine hydrosilylation include **8** and complexes of Ru-, Rh-, Ir- and Zn with the ligands **9-11** (shown in Figure 2).[2g]

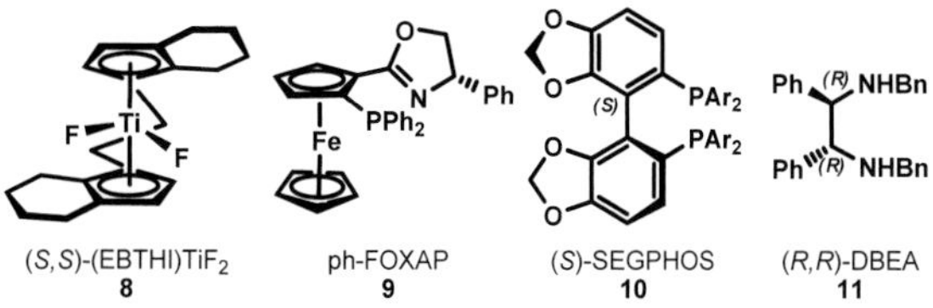

(S,S)-(EBTHI)TiF$_2$ **8**

ph-FOXAP **9**

(S)-SEGPHOS **10**
Ar = 4-OMe-3,5-di-t-Bu

(R,R)-DBEA **11**

Figure 2

Additives such as amines have also been found to widen the substrate scope or improve enantioselectivity of imine hydrosilylation.

The hydrosilylation catalyst **8** (Figure 2), developed by S. Buchwald et al.,[2a] initially was useful only with cyclic imines. However, the use of a primary amine additive allowed broader application.[3]

A mechanism has been proposed for imine hydrosilylation using (S,S)-(EBTHI)-TiF$_2$ (**8**) in which a catalytically active Ti(III)-hydride reacts with the imine in the enantiomeric product-determining step.[3] The precise nature of the pre-transition state assembly and the reason for acceleration of the reduction by a primary amine are uncertain.

Another system for imine hydrosilylation involves the use of trichlorosilane together with chiral bidentate Lewis bases[4] such as **18** and **19** (Figure 3);[2g] see p. 170 for pathway.

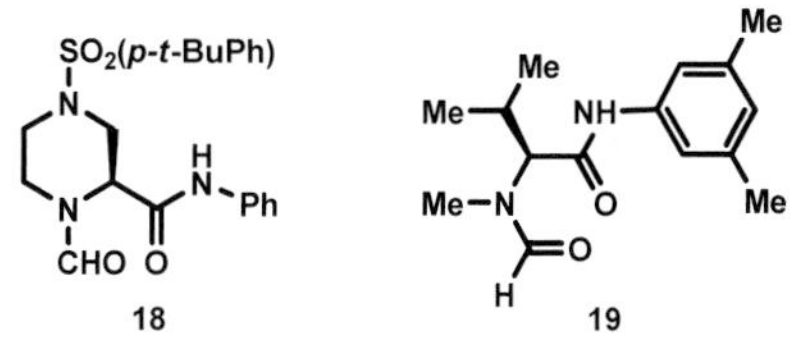

SO$_2$(p-t-BuPh) ... **18**

... **19**

Figure 3

Enantioselective Hydrosilylation of Structurally Diverse Imines

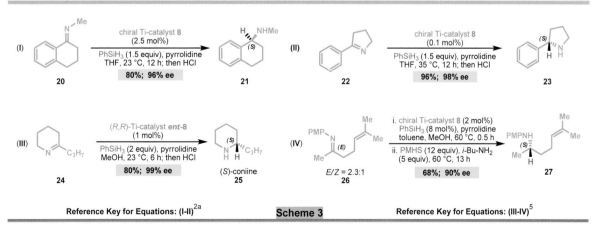

(I)
20 → chiral Ti-catalyst **8** (2.5 mol%), PhSiH$_3$ (1.5 equiv), pyrrolidine THF, 23 °C, 12 h; then HCl → 80%; 96% ee → (S) **21**

(II)
22 → chiral Ti-catalyst **8** (0.1 mol%), PhSiH$_3$ (1.5 equiv), pyrrolidine THF, 35 °C, 12 h; then HCl → 96%; 98% ee → (S) **23**

(III)
24 → (R,R)-Ti-catalyst ent-**8** (1 mol%), PhSiH$_3$ (2 equiv), pyrrolidine MeOH, 23 °C, 6 h; then HCl → 80%; 99% ee → (S)-coniine **25**

(IV)
26 E/Z = 2.3:1 → i. chiral Ti-catalyst **8** (2 mol%) PhSiH$_3$ (8 mol%), pyrrolidine toluene, MeOH, 60 °C, 0.5 h; ii. PMHS (12 equiv), i-Bu-NH$_2$ (5 equiv), 60 °C, 13 h → 68%; 90% ee → (S) **27**

Reference Key for Equations: (I-II)[2a]

Scheme 3

Reference Key for Equations: (III-IV)[5]

ENANTIOSELECTIVE REDUCTION OF C=N

Enantioselective Hydrosilylation of Structurally Diverse Imines

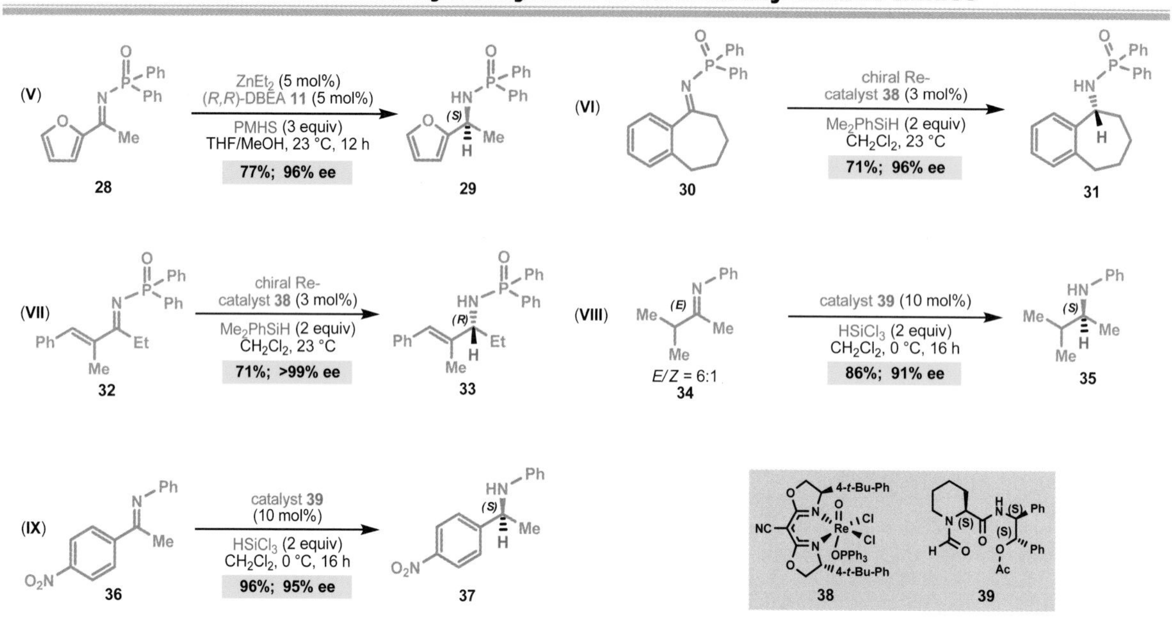

(V)

28 → 29
ZnEt$_2$ (5 mol%)
(R,R)-DBEA 11 (5 mol%)
PMHS (3 equiv)
THF/MeOH, 23 °C, 12 h
77%; 96% ee

(VI)

30 → 31
chiral Re-catalyst 38 (3 mol%)
Me$_2$PhSiH (2 equiv)
CH$_2$Cl$_2$, 23 °C
71%; 96% ee

(VII)

32 → 33
chiral Re-catalyst 38 (3 mol%)
Me$_2$PhSiH (2 equiv)
CH$_2$Cl$_2$, 23 °C
71%; >99% ee

(VIII)

34 → 35
E/Z = 6:1
catalyst 39 (10 mol%)
HSiCl$_3$ (2 equiv)
CH$_2$Cl$_2$, 0 °C, 16 h
86%; 91% ee

(IX)

36 → 37
catalyst 39 (10 mol%)
HSiCl$_3$ (2 equiv)
CH$_2$Cl$_2$, 0 °C, 16 h
96%; 95% ee

38   39

Reference Key for Equations: (V)[6]; (VI)[7]

Scheme 4

Reference Key for Equations: (VII)[7]; (VIII-IX)[8,9]

Comments on Schemes 3 and 4

(1) The catalyst **8** can be used for the enantioselective hydrosilylation of both acyclic and cyclic imines (**20**, **22** and **24**) at low catalyst loadings.[2a,5]

(2) It is noteworthy that the imine **26** was reduced with high enantioselectivity using catalyst **8** in combination with the inexpensive polymethylhydrosiloxane (PMHS) as stoichiometric reducing agent to form **27** in 90% enantiomeric excess.[5]

(3) The transformation **28**→**29** proceeded with excellent enantioselectivity using the Et$_2$Zn/chiral diamine **11** catalyst system and PMHS.[6]

(4) The reduction of the α,β-unsaturated imine **32** proceeds with complete chemoselectivity. The chiral Re-oxo complex **38** was shown to catalyze the non-oxidative enantioselective reduction of structurally diverse imines. The stability of this complex in air makes it possible to conduct this reaction under "open-flask" conditions.[7]

References

1. Langlois, N., Dang Tuan, P. & Kagan, H. B. Asymmetric synthesis of amines by hydrosilylation of imines catalyzed by a chiral complex of rhodium. *Tetrahedron Lett.* **1973**, 4865-4868.
2. (a) Verdaguer, X., Lange, U. E. W., Reding, M. T. & Buchwald, S. L. Highly Enantioselective Imine Hydrosilylation Using (S,S)-Ethylenebis(η^5-tetrahydroindenyl)titanium Difluoride. *J. Am. Chem. Soc.* **1996**, 118, 6784-6785; (b) Kobayashi, S. & Ishitani, H. Catalytic Enantioselective Addition to Imines. *Chem. Rev.* **1999**, 99, 1069-1094; (c) Nishiyama, H. & Itoh, K. Asymmetric Hydrosilylation and Related reactions in *Catal. Asymmetric Synth. (2nd Edition)* 111-143 (**2000**); (d) Carpentier, J.-F. & Bette, V. Chemo- and enantioselective hydrosilylation of carbonyl and imino groups. An emphasis on non-traditional catalyst systems. *Curr. Org. Chem.* **2002**, 6, 913-936; (e) Nishiyama, H. Hydrosilylations of carbonyl and imine compounds. *Transition Metals for Organic Synthesis (2nd Edition)* **2004**, 2, 182-191; (f) Riant, O., Mostefai, N. & Courmarcel, J. Recent advances in the asymmetric hydrosilylation of ketones, imines, and electrophilic double bonds. *Synthesis* **2004**, 2943-2958; (g) Riant, O. Hydrosilylation of imines in *Modern Reduction Methods* 321-337 (**2008**).
3. Verdaguer, X., Lange, U. E. W. & Buchwald, S. L. Amine additives greatly expand the scope of asymmetric hydrosilylation of imines. *Angew. Chem., Int. Ed.* **1998**, 37, 1103-1107.
4. Wang, Z., Cheng, M., Wu, P., Wei, S. & Sun, J. L-Piperazine-2-carboxylic Acid Derived N-Formamide as a Highly Enantioselective Lewis Basic Catalyst for Hydrosilylation of N-Aryl Imines with an Unprecedented Substrate Profile. *Org. Lett.* **2006**, 8, 3045-3048.
5. Reding, M. T. & Buchwald, S. L. Short Enantioselective Total Syntheses of the Piperidine Alkaloids (S)-Coniine and (2R,6R)-trans-Solenopsin A via Catalytic Asymmetric Imine Hydrosilylation. *J. Org. Chem.* **1998**, 63, 6344-6347.
6. Park, B.-M., Mun, S. & Yun, J. Zinc-catalyzed enantioselective hydrosilylation of imines. *Adv. Synth. Catal.* **2006**, 348, 1029-1032.
7. Nolin, K. A., Ahn, R. W. & Toste, F. D. Enantioselective Reduction of Imines Catalyzed by a Rhenium(V)-Oxo Complex. *J. Am. Chem. Soc.* **2005**, 127, 12462-12463.
8. Wang, Z., Ye, X., Wei, S., Wu, P., Zhang, A. & Sun, J. A Highly Enantioselective Lewis Basic Organocatalyst for Reduction of N-Aryl Imines with Unprecedented Substrate Spectrum. *Org. Lett.* **2006**, 8, 999-1001.
9. Malkov, A.V., Vrankova, K., Stroncius, S. & Kocovsky, P. Asymmetric Reduction of Imines with Tricholorosilane, Catalyzed by Sigamide, an Amino Acid-Derived Formamide: Scope and Limitations. *J. Org. Chem.* **2009**, 74, 5839-5849.

ENANTIOSELECTIVE REDUCTION OF C=N

Enantioselective Reduction of Imines Using 1,4-Dihydropyridines

imine (prochiral) → chiral catalyst / Hantzsch esters → amine (chiral)

1,4-Dihydropyridines, such as Hantzsch esters (**1**), are capable of reducing imines that have been activated by *N*-protonation. Such reductions are directly analogous to enzymatic reductions of C=N and C=O involving the 1,4-dihydropyridine NADH (**2**) as hydride source.[1] An enantioselective version of the Hantzsch ester-promoted reduction of C=N was reported in 2005 using a chiral proton donor.[2]

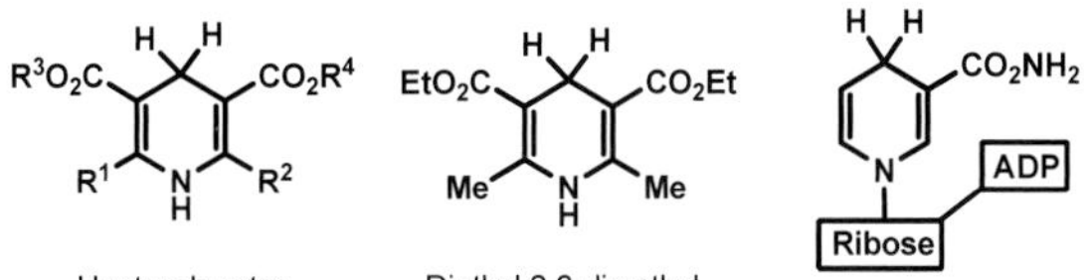

Figure 1

Hantzsch ester (general formula) **1**

Diethyl 2,6-dimethyl-1,4-dihydropyridine-3,5-dicarboxylate **1a**

NADH **2**

This reduction is operable with a broad range of imines as substrates and also for imines that are generated *in situ* from a primary amine and a prochiral ketone. Symmetrical Hantzsch esters such as **1a** are readily available in one step from β-keto esters, formaldehyde and ammonia (Hantzsch dihydropyridine synthesis).[3]

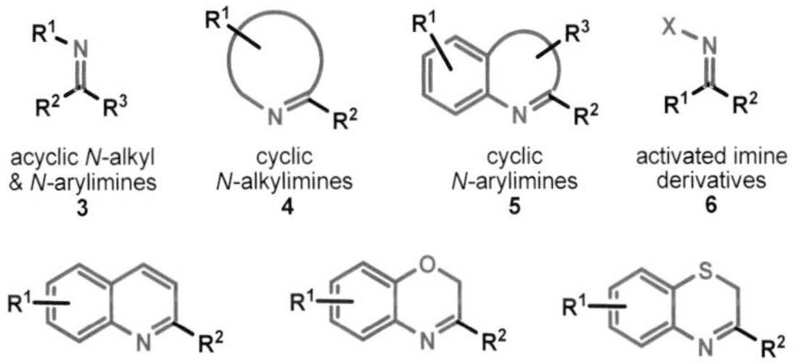

Figure 2

acyclic *N*-alkyl & *N*-arylimines **3**

cyclic *N*-alkylimines **4**

cyclic *N*-arylimines **5**

activated imine derivatives **6**

substituted quinolines **7**

substituted benzoxazines **8**

substituted benzothiazines **9**

Some of the chiral phosphoric acid derivatives that are effective as catalysts in enantioselective imine reductions are shown in Figures 3 and 4.[1]

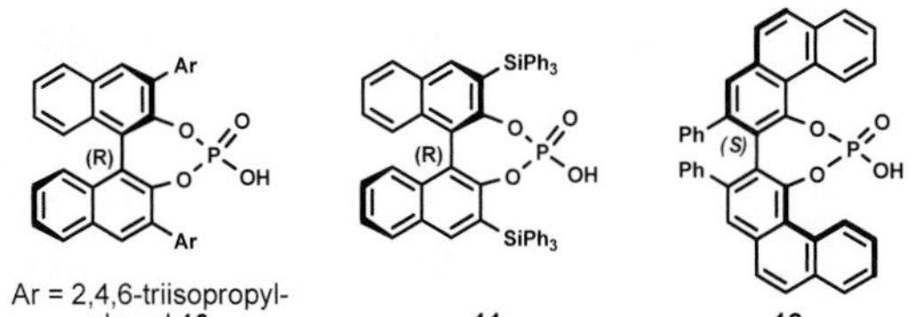

Figure 3

Ar = 2,4,6-triisopropyl-phenyl **10**

11

12

Figure 4

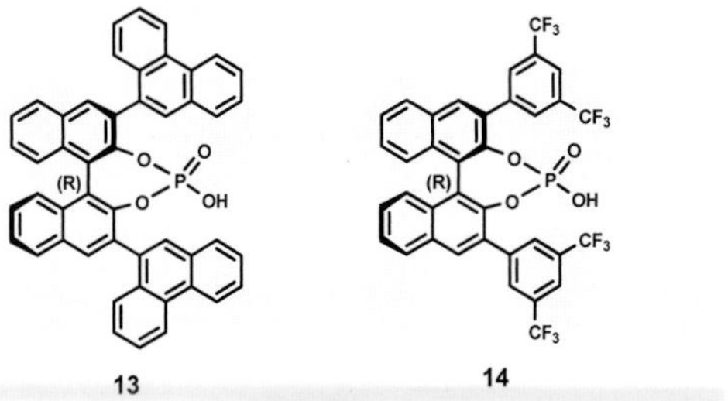

13   **14**

A reasonable mechanism of the enantioselective Hantzsch ester mediated hydride transfer to imines catalyzed by a chiral phosphoric acid is shown in Scheme 1:

(a) the chiral phosphoric acid **10** protonates the imine **15** in toluene solution to form a contact ion pair (**A**);

(b) the resulting iminium ion pair is now activated for an enantioselective nucleophilic attack by a hydrogen of the Hantzsch ester **1**;

(c) driving force for the hydride-transfer derives from the aromatization of **1** to **16**;

(d) dissociation of the ion pair **B** affords the amine **17** and regenerates the chiral phosphoric acid **10**.

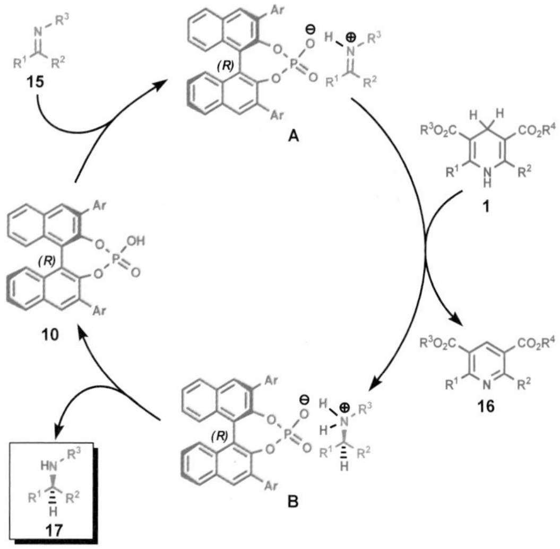

Scheme 1

The current methodology is not ideal since Hantzsch esters must be used as stoichoiometric reductants which can cause difficulties in the purification of product. In addition, the Hantzsch ester **1a** is not inexpensive and as much as 20 mol% of chiral proton donor may be required for optimum results.

The enantioselective reduction of prochiral ketones by Hantzsch esters has also been described,[4] but this method appears to be less practical than the other approaches to carbonyl reductions described on pages 24-26 and 27-28.

ENANTIOSELECTIVE REDUCTION OF C=N

Enantioselective Reduction of Imines Using 1,4-Dihydropyridines

(I) **18** → catalyst **10** (1 mol%), Hantzsch ester **1a** (1.4 equiv), toluene, 35 °C, 60 h → **19** 80%; 90% ee

(II) **20** → catalyst **10** (1 mol%), Hantzsch ester **1a** (1.4 equiv), toluene, 35 °C, 60 h → **21** 91%; 93% ee

(III) **22** → catalyst (S)-**12** (5 mol%), Hantzsch ester **1a** (1.4 equiv), toluene, 50 °C, 19 h → **23** 93%; 96% ee

(IV) **24** → catalyst (S)-**12** (5 mol%), Hantzsch ester **1a** (1.4 equiv), toluene, 50 °C, 21 h → **25** 88%; 99% ee

(V) **26** → catalyst (S)-**13** (5 mol%), Hantzsch ester **1a** (4 equiv), benzene, 60 °C → **27** 91% ee

(VI) **28** → catalyst (S)-**13** (2 mol%), Hantzsch ester **1a** (2.4 equiv), benzene, 60 °C, 12 h → **29** 93%; 91% ee

(VII) **30** → catalyst (S)-**13** (0.1 mol%), Hantzsch ester **1a** (1.25 equiv), CHCl₃, 23 °C → **31** 95%; 98% ee

(VIII) PhCOMe + **32** → catalyst (S)-**11** (10 mol%), Hantzsch ester **1a**, 5 Å MS, benzene, 40-50 °C, 24-96 h → **33** 70%; 91% ee

Reference Key for Equations: (I)²ᵇ; (II)²ᵇ; (III)⁵; (IV)⁵   **Scheme 2**   Reference Key for Equations: (V)⁶; (VI)⁷; (VII)⁸; (VIII)⁹

Comments on Scheme 2

(1) Reduction of benzoxazines and benzothiazines is noteworthy because the transformation (e.g., **30**→**31**) proceeds in excellent yield and enantioselectivity at very low catalyst loadings.[8]

(2) The enantioselective coupling of acetophenone with **32** to form **33** using **11** as catalyst exemplifies the reduction of an *in situ* generated imine.[9]

References

1. (a) Kagan, H. B. Organocatalytic enantioselective reduction of olefins, ketones, and imines in *Enantioselective Organocatalysis* 391-401 (2007); (b) You, S.-L. Recent developments in asymmetric transfer hydrogenation with Hantzsch esters: a biomimetic approach. *Chemistry--An Asian Journal* 2007, 2, 820-827; (c) Ouellet, S. G., Walji, A. M. & Macmillan, D. W. C. Enantioselective Organocatalytic Transfer Hydrogenation Reactions using Hantzsch Esters. *Acc. Chem. Res.* 2007, 40, 1327-1339; (d) Connon, S. J. Asymmetric organocatalytic reductions mediated by dihydropyridines. *Org. Biomol. Chem.* 2007, 5, 3407-3417; (e) Adair, G., Mukherjee, S. & List, B. TRIP-a powerful Bronsted acid catalyst for asymmetric synthesis. *Aldrichimica Acta* 2008, 41, 31-39; (f) Saini, A., Kumar, S. & Sandhu, J. S. Hantzsch reaction: recent advances in Hantzsch 1,4-dihydropyridines. *J. Sci. Ind. Res.* 2008, 67, 95-111; (g) Adolfsson, H. Alkene and imino reductions by organocatalysis in *Modern Reduction Methods* 341-361 (2008).

2. (a) Rueping, M., Sugiono, E., Azap, C., Theissmann, T. & Bolte, M. Enantioselective Bronsted Acid Catalyzed Transfer Hydrogenation: Organocatalytic Reduction of Imines. *Org. Lett.* 2005, 7, 3781-3783; (b) Hoffmann, S., Seayad, A. M. & List, B. A powerful Bronsted acid catalyst for the organocatalytic asymmetric transfer hydrogenation of imines. *Angew. Chem., Int. Ed.* 2005, 44, 7424-7427.

3. Kürti, L. & Czakó, B. Hantzsch dihydropyridine synthesis in *Strategic Applications of Named Reactions in Organic Synthesis* 194-195 (Academic Press/Elsevier, San Diego, 2005).

4. Yang, J. W. & List, B. Catalytic Asymmetric Transfer Hydrogenation of α-Keto esters with Hantzsch Esters. *Org. Lett.* 2006, 8, 5653-5655.

5. Li, G., Liang, Y. & Antilla, J. C. A Vaulted Biaryl Phosphoric Acid-Catalyzed Reduction of α-Imino Esters: The Highly Enantioselective Preparation of α-Amino Esters. *J. Am. Chem. Soc.* 2007, 129, 5830-5831.

6. Rueping, M. & Antonchick, A. P. Organocatalytic enantioselective reduction of pyridines. *Angew. Chem., Int. Ed.* 2007, 46, 4562-4565.

7. Rueping, M., Antonchick, A. P. & Theissmann, T. A highly enantioselective Bronsted acid catalyzed cascade reaction: organocatalytic transfer hydrogenation of quinolines and their application in the synthesis of alkaloids. *Angew. Chem., Int. Ed.* 2006, 45, 3683-3686.

8. Rueping, M., Antonchik, A. P. & Theissmann, T. Remarkably low catalyst loading in Bronsted acid catalyzed transfer hydrogenations: enantioselective reduction of benzoxazines, benzothiazines, and benzoxazinones. *Angew. Chem., Int. Ed.* 2006, 45, 6751-6755.

9. Storer, R. I., Carrera, D. E., Ni, Y. & MacMillan, D. W. C. Enantioselective Organocatalytic Reductive Amination. *J. Am. Chem. Soc.* 2006, 128, 84-86.

ENANTIOSELECTIVE REDUCTION OF C=N

Introduction: Enantioselective Reduction of Oxime Ethers Catalyzed by Oxazaborolidines

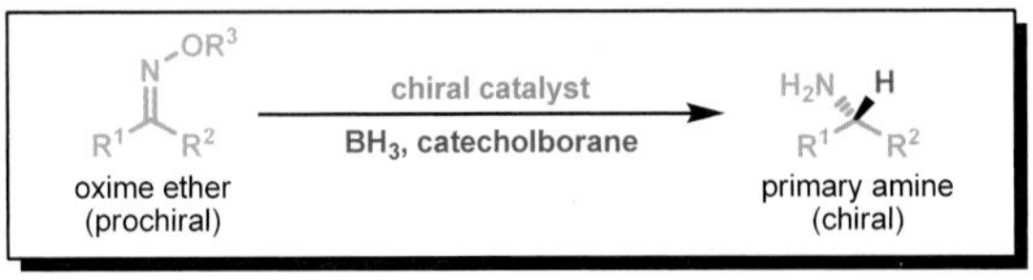

oxime ether (prochiral) → primary amine (chiral)

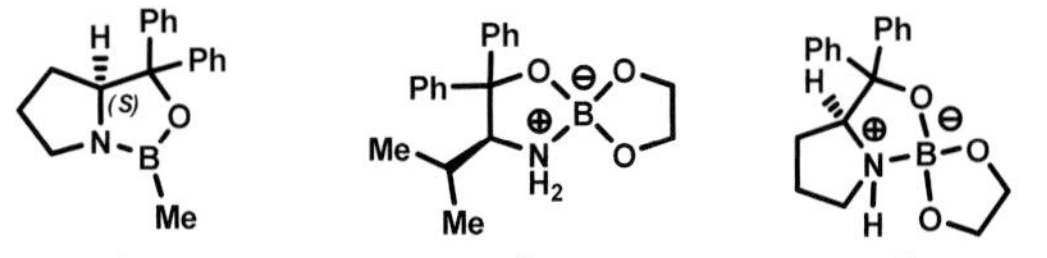

chiral catalyst, BH$_3$, catecholborane

Figure 1

Chiral oxazaborolidines (e.g., **1**) catalyze the enantioselective reduction of ketones using boranes as stoichiometric reductants and catalyst loading in the range of 0.1-10 mol% (see pages 164, 208, 233).[1] Although oxazaborolidines catalyze the enantioselective reduction of C=N bonds, stoichiometric quantities are generally required because of tight product binding to the catalyst.[2] Recently, catalytic systems have been developed utilizing spiroborate esters such as **2** and **3** (Figure 1) that are effective at 10 mol% when oxime ethers are used as substrates.[3]

The pathway for these reductions appears to be analogous to that for the oxazaborolidine-mediated/catalyzed reduction of prochiral ketones, the active species being the type of B-alkoxy oxazaborolidine **4** shown on the left. The role of the B-alkoxy substituent is to diminish product inhibition because of electron-donation to boron.

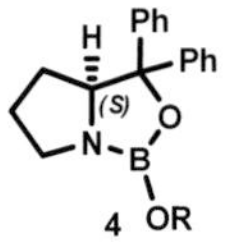

CBS Reduction of Structurally Diverse Oxime Ethers

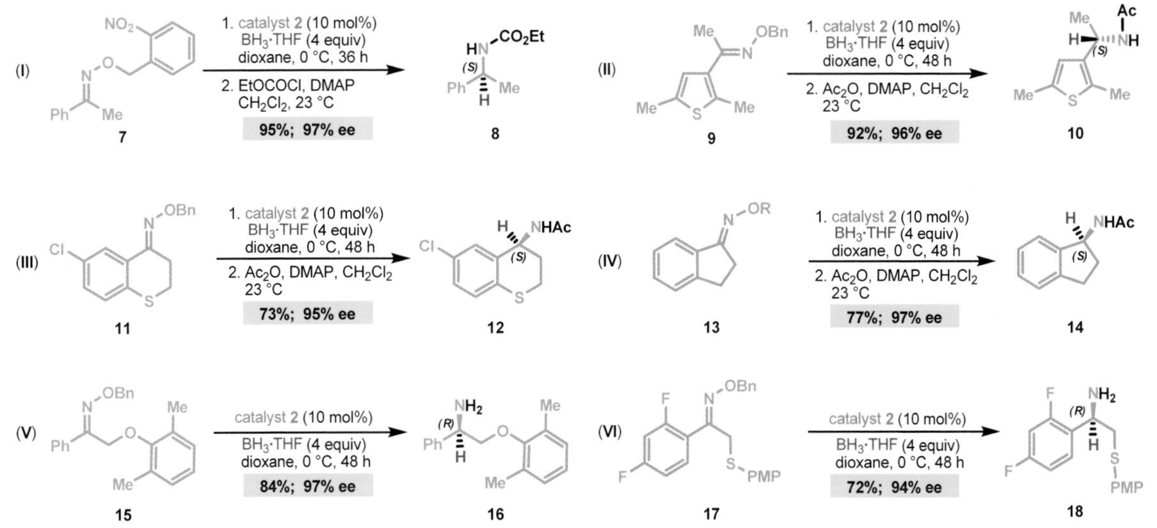

Scheme 1

(I) 7 → 1. catalyst 2 (10 mol%), BH$_3$·THF (4 equiv), dioxane, 0 °C, 36 h; 2. EtOCOCl, DMAP, CH$_2$Cl$_2$, 23 °C → 8; 95%; 97% ee

(II) 9 → 1. catalyst 2 (10 mol%), BH$_3$·THF (4 equiv), dioxane, 0 °C, 48 h; 2. Ac$_2$O, DMAP, CH$_2$Cl$_2$, 23 °C → 10; 92%; 96% ee

(III) 11 → 1. catalyst 2 (10 mol%), BH$_3$·THF (4 equiv), dioxane, 0 °C, 48 h; 2. Ac$_2$O, DMAP, CH$_2$Cl$_2$, 23 °C → 12; 73%; 95% ee

(IV) 13 → 1. catalyst 2 (10 mol%), BH$_3$·THF (4 equiv), dioxane, 0 °C, 48 h; 2. Ac$_2$O, DMAP, CH$_2$Cl$_2$, 23 °C → 14; 77%; 97% ee

(V) 15 → catalyst 2 (10 mol%), BH$_3$·THF (4 equiv), dioxane, 0 °C, 48 h → 16; 84%; 97% ee

(VI) 17 → catalyst 2 (10 mol%), BH$_3$·THF (4 equiv), dioxane, 0 °C, 48 h → 18; 72%; 94% ee

Reference Key for Equations: (I-III)[3d]   Reference Key for Equations: (IV)[3b]; (V-VI)[4]

References

1. Corey, E. J. & Helal, C. J. Reduction of carbonyl compounds with chiral oxazaborolidine catalysts: A new paradigm for enantioselective catalysis and a powerful new synthetic method. *Angew. Chem., Int. Ed.* **1998**, 37, 1986-2012.
2. Cho, B. T. Recent advances in the synthetic applications of the oxazaborolidine-mediated asymmetric reduction. *Tetrahedron* **2006**, 62, 7621-7643.
3. (a) Chu, Y., Shan, Z., Liu, D. & Sun, N. Asymmetric Reduction of Oxime Ethers Promoted by Chiral Spiroborate Esters with an O$_3$BN Framework. *J. Org. Chem.* **2006**, 71, 3998-4001; (b) Huang, X., Ortiz-Marciales, M., Huang, K., Stepanenko, V., Merced, F. G., Ayala, A. M., Correa, W. & De Jesus, M. Asymmetric Synthesis of Primary Amines via the Spiroborate-Catalyzed Borane Reduction of Oxime Ethers. *Org. Lett.* **2007**, 9, 1793-1795; (c) Stepanenko, V., De Jesus, M., Correa, W., Guzman, I.,

Vazquez, C., Ortiz, L. & Ortiz-Marciales, M. Spiroborate esters in the borane-mediated asymmetric synthesis of pyridyl and related heterocyclic alcohols. *Tetrahedron Asymmetry* **2007**, 18, 2738-2745; (d) Huang, K., Merced, F. G., Ortiz-Marciales, M., Melendez, H. J., Correa, W. & De Jesus, M. Highly Enantioselective Borane Reduction of Heteroaryl and Heterocyclic Ketoxime Ethers Catalyzed by Novel Spiroborate Ester Derived from Diphenylvalinol: Application to the Synthesis of Nicotine Analogues. *J. Org. Chem.* **2008**, 73, 4017-4026.
4. Huang, K., Ortiz-Marciales, M., Stepanenko, V., De Jesus, M. & Correa, W. A Practical and Efficient Route for the Highly Enantioselective Synthesis of Mexiletine Analogues and Novel β-Thiophenoxy and Pyridyl Ethers. *J. Org. Chem.* **2008**, 73, 6928-6931.

ENANTIOSELECTIVE EPOXIDATION OF C=C

Katsuki-Sharpless Epoxidation

allylic alcohol → epoxy alcohol

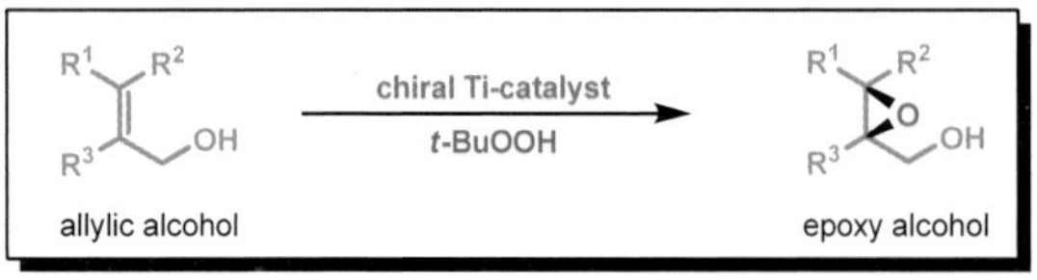

In 1980, T. Katsuki and K.B. Sharpless made the surprising discovery that allylic alcohols undergo enantioselective epoxidation in the presence of Ti(Oi-Pr)$_4$, diethyl tartrate (1) or diisopropyl tartrate (2) and *tert*-butylhydroperoxide (3) (Figure 1).[1] Later, they reported that the inclusion of 4 Å molecular sieves allows the use of *substoichiomeric* amounts of Ti(Oi-Pr)$_4$ and tartrate ester when dry *t*-BuOOH is used. This process (Katsuki-Sharpless epoxidation) is now one of the most widely used catalytic asymmetric oxidations in organic synthesis.[2] The 2,3-epoxy alcohols produced by it are valuable chiral building blocks and have been used extensively in the preparation of pharmaceutical intermediates and in the synthesis of many complex natural products.[3]

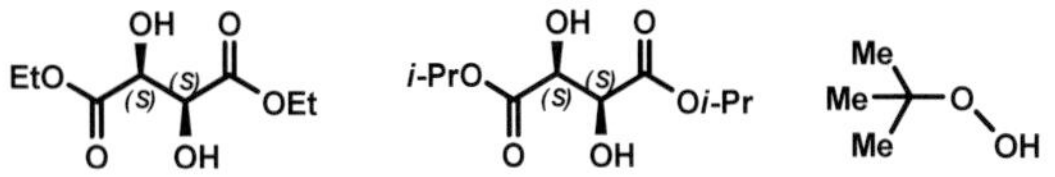

Allylic alcohols are the most favorable substrates and presence of the hydroxyl group is essential. The reaction proceeds enantioselectively with virtually any substitution pattern on C=C. The stereochemistry is reagent-controlled and either enantiomeric 2,3-epoxy alcohol (4 or 5) can be obtained by use of the appropriate tartrate ester (S,S or R,R). The enantiofacial selectivity of the reaction may be predicted using the diagram shown in Figure 2.

The Katsuki-Sharpless epoxidation has also been applied to the kinetic resolution of racemic secondary allylic alcohols and to the desymmetrization of meso bis-allylic alcohols.[4] Labile epoxides have been utilized for further transformation without isolation.[5] There are also examples of successful enantioselective epoxidation with homoallylic, bis-homoallylic and tris-homoallylic alcohols.[6]

The detailed mechanism remains uncertain despite experimental[7] and theoretical[8] studies. Two pathways have been proposed, one via the dimeric structure (B, Figure 3) and the other involving a zwitterionic titanium species (A, Figure 3).[8a]

In contrast to the Katsuki-Sharpless epoxidation, which must be carried out under anhydrous conditions because the reaction is inhibited by water, there is a water-tolerant catalytic system for the enantioselective epoxidation of allylic alcohols based on vanadium complexes of type C.[9]

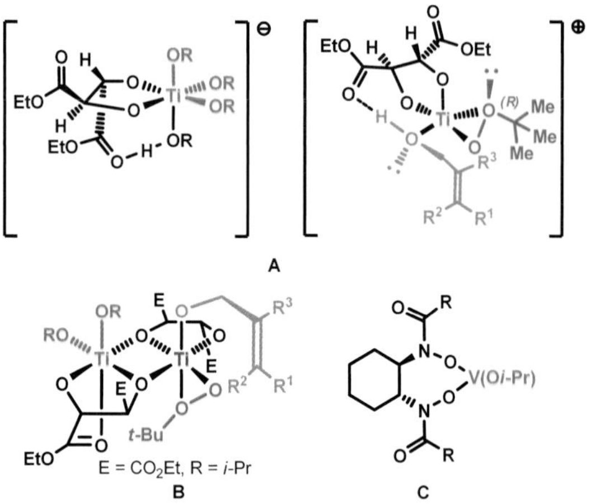

(S,S)-diethyl tartrate (DET) 1   (S,S)-diisopropyl tartrate (DIPT) 2   TBHP 3

Figure 1

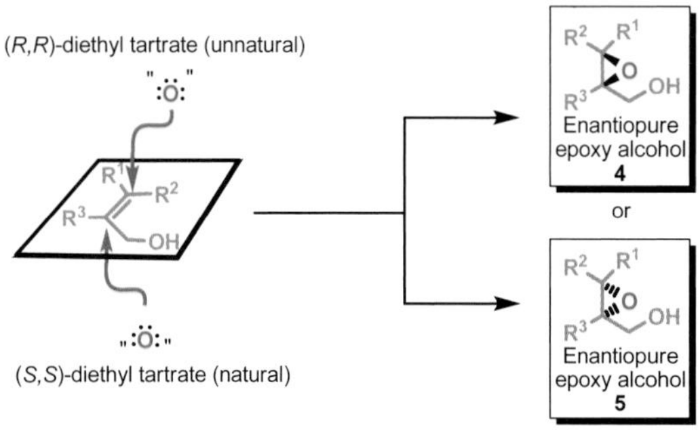

(R,R)-diethyl tartrate (unnatural)

Enantiopure epoxy alcohol 4

or

Enantiopure epoxy alcohol 5

(S,S)-diethyl tartrate (natural)

Figure 2

A

E = CO$_2$Et, R = *i*-Pr
B

C

Figure 3

Asymmetric Epoxidation of Structurally Diverse Allylic Alcohols

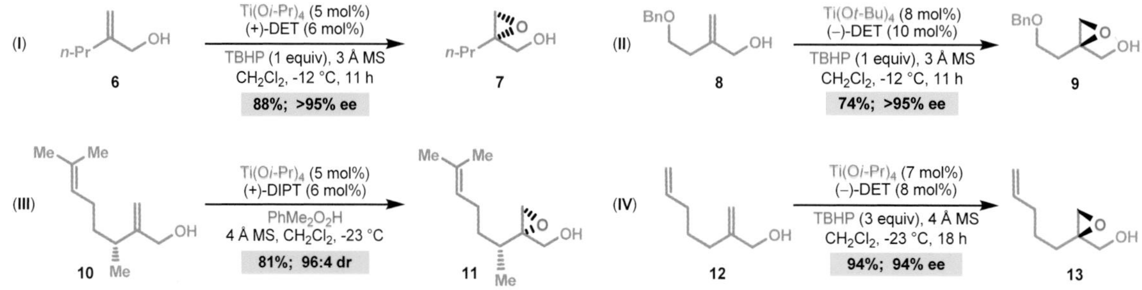

(I) 6 → Ti(Oi-Pr)$_4$ (5 mol%) (+)-DET (6 mol%); TBHP (1 equiv), 3 Å MS, CH$_2$Cl$_2$, -12 °C, 11 h → 7   **88%; >95% ee**

(II) 8 → Ti(Ot-Bu)$_4$ (8 mol%) (–)-DET (10 mol%); TBHP (1 equiv), 3 Å MS, CH$_2$Cl$_2$, -12 °C, 11 h → 9   **74%; >95% ee**

(III) 10 → Ti(Oi-Pr)$_4$ (5 mol%) (+)-DIPT (6 mol%); PhMe$_2$O$_2$H, 4 Å MS, CH$_2$Cl$_2$, -23 °C → 11   **81%; 96:4 dr**

(IV) 12 → Ti(Oi-Pr)$_4$ (7 mol%) (–)-DET (8 mol%); TBHP (3 equiv), 4 Å MS, CH$_2$Cl$_2$, -23 °C, 18 h → 13   **94%; 94% ee**

Reference Key for Equations: (I)[10]; (II)[11]   **Scheme 1**   Reference Key for Equations: (III)[12]; (IV)[13]

ENANTIOSELECTIVE EPOXIDATION OF C=C

Asymmetric Epoxidation of Structurally Diverse Allylic Alcohols

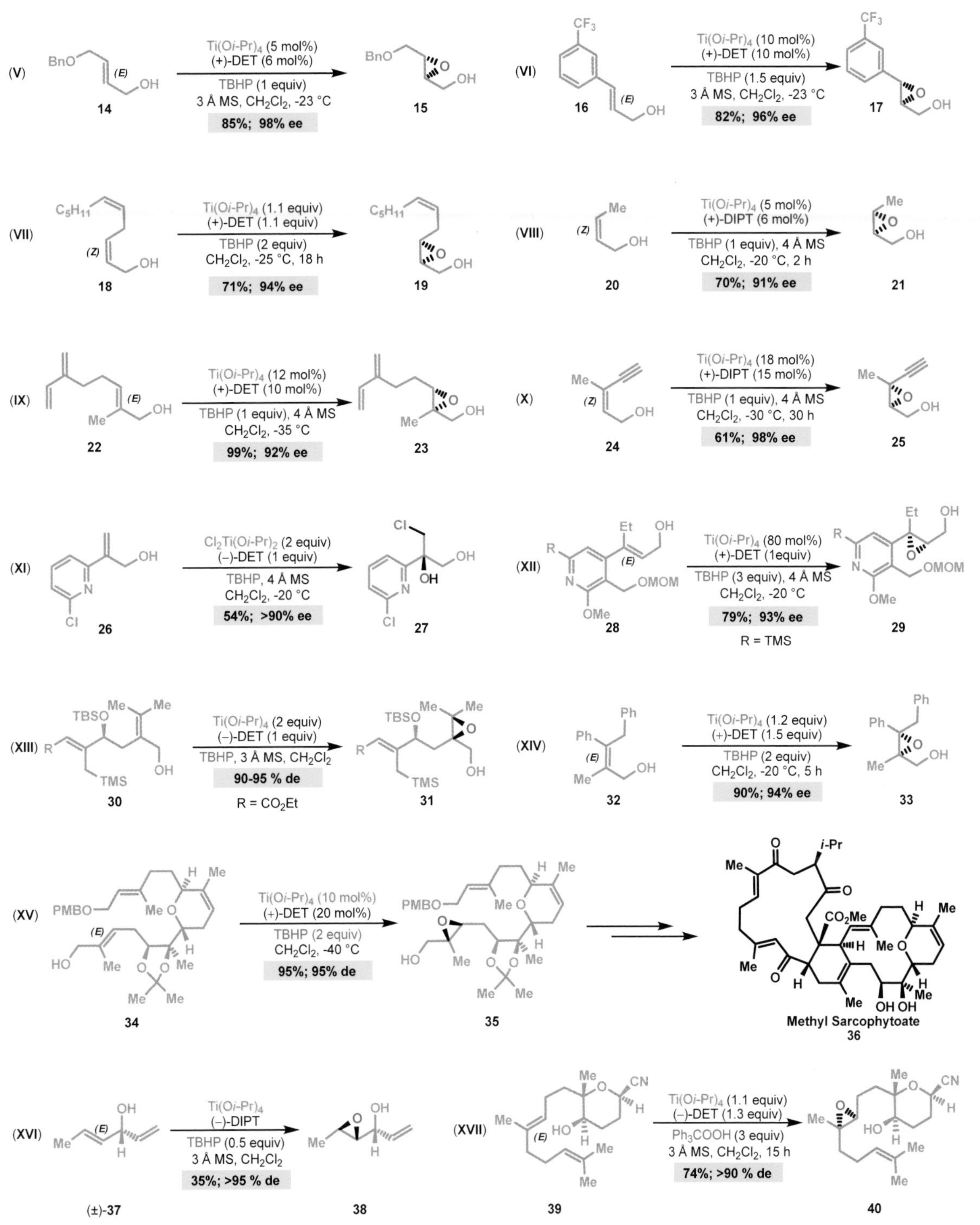

(V)
BnO ... (E) ... OH
14
Ti(Oi-Pr)₄ (5 mol%)
(+)-DET (6 mol%)
TBHP (1 equiv)
3 Å MS, CH₂Cl₂, -23 °C
BnO ... O ... OH
15
85%; 98% ee

(VI)
CF₃ ... (E) ... OH
16
Ti(Oi-Pr)₄ (10 mol%)
(+)-DET (10 mol%)
TBHP (1.5 equiv)
3 Å MS, CH₂Cl₂, -23 °C
CF₃ ... O ... OH
17
82%; 96% ee

(VII)
C₅H₁₁ ... (Z) ... OH
18
Ti(Oi-Pr)₄ (1.1 equiv)
(+)-DET (1.1 equiv)
TBHP (2 equiv)
CH₂Cl₂, -25 °C, 18 h
C₅H₁₁ ... O ... OH
19
71%; 94% ee

(VIII)
(Z) ... Me ... OH
20
Ti(Oi-Pr)₄ (5 mol%)
(+)-DIPT (6 mol%)
TBHP (1 equiv), 4 Å MS
CH₂Cl₂, -20 °C, 2 h
Me ... O ... OH
21
70%; 91% ee

(IX)
Me (E) ... OH
22
Ti(Oi-Pr)₄ (12 mol%)
(+)-DET (10 mol%)
TBHP (1 equiv), 4 Å MS
CH₂Cl₂, -35 °C
Me ... O ... OH
23
99%; 92% ee

(X)
Me (Z) ... OH
24
Ti(Oi-Pr)₄ (18 mol%)
(+)-DIPT (15 mol%)
TBHP (1 equiv), 4 Å MS
CH₂Cl₂, -30 °C, 30 h
Me ... O ... OH
25
61%; 98% ee

(XI)
... OH, N, Cl
26
Cl₂Ti(Oi-Pr)₂ (2 equiv)
(–)-DET (1 equiv)
TBHP, 4 Å MS
CH₂Cl₂, -20 °C
Cl ... N ... OH ... OH, Cl
27
54%; >90% ee

(XII)
R ... Et ... OH (E) ... OMOM, N, OMe
28
R = TMS
Ti(Oi-Pr)₄ (80 mol%)
(+)-DET (1equiv)
TBHP (3 equiv), 4 Å MS
CH₂Cl₂, -20 °C
R ... Et ... OH ... O ... OMOM, N, OMe
29
79%; 93% ee

(XIII)
TBSO, Me, Me
R ... TMS ... OH
30
R = CO₂Et
Ti(Oi-Pr)₄ (2 equiv)
(–)-DET (1 equiv)
TBHP, 3 Å MS, CH₂Cl₂
TBSO, Me, Me ... O
R ... TMS ... OH
31
90-95 % de

(XIV)
Ph ... Ph (E) ... Me ... OH
32
Ti(Oi-Pr)₄ (1.2 equiv)
(+)-DET (1.5 equiv)
TBHP (2 equiv)
CH₂Cl₂, -20 °C, 5 h
Ph ... Ph ... O, Me ... OH
33
90%; 94% ee

(XV)
PMBO ... H, Me, Me O ... (E) ... HO, Me, O O, Me, Me, Me
34
Ti(Oi-Pr)₄ (10 mol%)
(+)-DET (20 mol%)
TBHP (2 equiv)
CH₂Cl₂, -40 °C
95%; 95% de
PMBO ... O ... H, Me, Me O ... HO, Me, O O, Me, Me, Me
35
→
Me, O, i-Pr, O, CO₂Me, H, Me, H, Me, O, Me, H, H, Me, O, Me, OH OH
Methyl Sarcophytoate
36

(XVI)
Me (E) ... OH ... H
(±)-37
Ti(Oi-Pr)₄
(–)-DIPT
TBHP (0.5 equiv)
3 Å MS, CH₂Cl₂
35%; >95 % de
Me ... O ... OH ... H
38

(XVII)
Me ... Me, O, CN ... HO, H, H ... (E), Me, Me
39
Ti(Oi-Pr)₄ (1.1 equiv)
(–)-DET (1.3 equiv)
Ph₃COOH (3 equiv)
3 Å MS, CH₂Cl₂, 15 h
74%; >90 % de
Me ... O ... Me, O, CN ... HO, H, H ... Me, Me
40

Reference Key for Equations: (V)[14]; (VI)[15]; (VII)[16]; (VIII)[10]; (IX)[17]; (X)[18]

Reference Key for Equations: (XI)[19]; (XII)[20]; (XIII)[21]; (XIV)[22] (XV)[23]; (XVI)[3b]; (XVII)[6]

Scheme 2

39

ENANTIOSELECTIVE EPOXIDATION OF C=C

Chemistry of 2,3-Epoxy Alcohols and Derivatives

(XVIII) — **41** → SnCl$_4$, CH$_2$Cl$_2$, 20 °C, 24 h → **42**

(XIX) — **43** → Ti(Oi-Pr)$_4$, CH$_2$Cl$_2$, 23 °C, **50%** → **44**

(XX) — **45** → K$_2$CO$_3$, MeOH, 24 h, **71-79%** → **46**

(XXI) — **47** (99% ee) → Cr(III)-tetra-Ph porphyrin-triflate (1 mol%), ClCH$_2$CH$_2$Cl, 83 °C, **96%** → **48** (99% ee)

Reference Key for Equations: (XVIII)[3d]; (XIX)[24]   **Scheme 1**   Reference Key for Equations: (XX)[25]; (XXI)[26]

References

1. Katsuki, T. & Sharpless, K. B. The first practical method for asymmetric epoxidation. *J. Am. Chem. Soc.* **1980**, 102, 5974-5976.
2. (a) Katsuki, T. & Martin, V. Asymmetric epoxidation of allylic alcohols: The Katsuki-Sharpless epoxidation reaction. *Organic Reactions (Hoboken, NJ, United States)* **1996**, 48, 1-299; (b) Johnson, R. A. & Sharpless, K. B. Catalytic asymmetric epoxidation of allylic alcohols. *Catal. Asymmetric Synth. (2nd Ed.)* **2000**, 231-280; (c) Sharpless, K. B. Searching for new reactivity (Nobel Lecture). *Angew. Chem., Int. Ed.* **2002**, 41, 2024-2032; (d) Xia, Q. H., Ge, H. Q., Ye, C. P., Liu, Z. M. & Su, K. X. Advances in Homogeneous and Heterogeneous Catalytic Asymmetric Epoxidation. *Chem. Rev. (Washington, DC, U. S.)* **2005**, 105, 1603-1662; (e) Adolfsson, H. & Balan, D. Metal-catalyzed synthesis of epoxides. *Aziridines and Epoxides in Organic Synthesis* **2006**, 185-228; (f) Ramon, D. J. & Yus, M. In the arena of enantioselective synthesis, titanium complexes wear the laurel wreath. *Chem. Rev.* **2006**, 106, 2126-2208; (g) Hussain, M. M. & Walsh, P. J. Tandem Reactions for Streamlining Synthesis: Enantio- and Diastereoselective One-Pot Generation of Functionalized Epoxy Alcohols. *Acc. Chem. Res.* **2008**, 41, 883-893.
3. (a) Behrens, C. H. & Sharpless, K. B. New transformations of 2,3-epoxy alcohols and related derivatives. Easy routes to homochiral substances. *Aldrichimica Acta* **1983**, 16, 67-80; (b) Sharpless, K. B., Behrens, C. H., Katsuki, T., Lee, A. W. M., Martin, V. S., Takatani, M., Viti, S. M., Walker, F. J. & Woodard, S. S. Stereo and regioselective openings of chiral 2,3-epoxy alcohols. Versatile routes to optically pure natural products and drugs. Unusual kinetic resolutions. *Pure Appl. Chem.* **1983**, 55, 589-604; (c) Hanson, R. M. The synthetic methodology of nonracemic glycidol and related 2,3-epoxy alcohols. *Chem. Rev.* **1991**, 91, 437-476; (d) Pena, P. C. A. & Roberts, S. M. The chemistry of epoxy alcohols. *Curr. Org. Chem.* **2003**, 7, 555-571.
4. Martin, V. S., Woodard, S. S., Katsuki, T., Yamada, Y., Ikeda, M. & Sharpless, K. B. Kinetic resolution of racemic allylic alcohols by enantioselective epoxidation. A route to substances of absolute enantiomeric purity? *J. Am. Chem. Soc.* **1981**, 103, 6237-6240.
5. (a) Lu, L. D. L., Johnson, R. A., Finn, M. G. & Sharpless, K. B. Two new asymmetric epoxidation catalysts. Unusual stoichiometry and inverse enantiofacial selection. *J. Org. Chem.* **1984**, 49, 728-731; (b) Klunder, J. M., Ko, S. Y. & Sharpless, K. B. Asymmetric epoxidation of allyl alcohol: efficient routes to homochiral β-adrenergic blocking agents. *J. Org. Chem.* **1986**, 51, 3710-3712.
6. Corey, E. J. & Ha, D. C. Total synthesis of venustatriol. *Tetrahedron Lett.* **1988**, 29, 3171-3174.
7. (a) Sharpless, K. B., Woodard, S. S. & Finn, M. G. On the mechanism of titanium-tartrate-catalyzed asymmetric epoxidation. *Pure Appl. Chem.* **1983**, 55, 1823-1836; (b) Woodard, S. S., Finn, M. G. & Sharpless, K. B. Mechanism of asymmetric epoxidation. 1. Kinetics. *J. Am. Chem. Soc.* **1991**, 113, 106-113; (c) Finn, M. G. & Sharpless, K. B. Mechanism of asymmetric epoxidation. 2. Catalyst structure. *J. Am. Chem. Soc.* **1991**, 113, 113-126; (d) Potvin, P. G. & Bianchet, S. The nature of the Katsuki-Sharpless asymmetric epoxidation catalyst. *J. Org. Chem.* **1992**, 57, 6629-6635.
8. (a) Corey, E. J. On the origin of enantioselectivity in the Katsuki-Sharpless epoxidation procedure. *J. Org. Chem.* **1990**, 55, 1693-1694; (b) Cui, M., Adam, W., Shen, J. H., Luo, X. M., Tan, X. J., Chen, K. X., Ji, R. Y. & Jiang, H. L. A Density-Functional Study of the Mechanism for the Diastereoselective Epoxidation of Chiral Allylic

Alcohols by the Titanium Peroxy Complexes. *J. Org. Chem.* **2002**, 67, 1427-1435.
9. Zhang, W., Basak, A., Kosugi, Y., Hoshino, Y. & Yamamoto, H. Enantioselective epoxidation of allylic alcohols by a chiral complex of vanadium: An effective controller system and a rational mechanistic model. *Angew. Chem., Int. Ed.* **2005**, 44, 4389-4391.
10. Gao, Y., Klunder, J. M., Hanson, R. M., Masamune, H., Ko, S. Y. & Sharpless, K. B. Catalytic asymmetric epoxidation and kinetic resolution: modified procedures including in situ derivatization. *J. Am. Chem. Soc.* **1987**, 109, 5765-5780.
11. Tanner, D. & Somfai, P. Asymmetric synthesis of (R)-(+)-3,3,4-trimethyl-4-(hydroxymethyl)-1,3-dioxolane of high enantiomeric purity. *Tetrahedron* **1986**, 42, 5985-5990.
12. White, J. D., Amedio, J. C., Jr., Gut, S. & Jayasinghe, L. Synthesis of the macrolactone alkaloid (+)-usaramine via necic acid coupling to a pyrrolizidine borane. *J. Org. Chem.* **1989**, 54, 4268-4270.
13. Schomaker, J. M. & Borhan, B. Total synthesis of (+)-tanikolide via oxidative lactonization. *Org. Biomol. Chem.* **2004**, 2, 621-624.
14. Katsuki, T., Lee, A. W. M., Ma, P., Martin, V. S., Masamune, S., Sharpless, K. B., Tuddenham, D. & Walker, F. J. Synthesis of saccharides and related polyhydroxylated natural products. 1. Simple alditols. *J. Org. Chem.* **1982**, 47, 1373-1378.
15. Goument, B., Duhamel, L. & Mauge, R. Asymmetric syntheses of (S)-fenfluramine using Sharpless epoxidation methods. *Tetrahedron* **1994**, 50, 171-188.
16. Mills, L. S. & North, P. C. A short synthesis of a key leukotriene B$_4$ synthon. *Tetrahedron Lett.* **1983**, 24, 409-410.
17. Yuasa, H., Makado, G. & Fukuyama, Y. Determination of the absolute configuration of vibsanin F by asymmetric synthesis via π-allylpalladium complex. *Tetrahedron Lett.* **2003**, 44, 6235-6239.
18. Ghosh, A. K. & Lei, H. An enantioselective synthesis of the core unit of the non-nucleoside reverse transcriptase inhibitor taurospongin A. *Tetrahedron Asymmetry* **2003**, 14, 629-634.
19. Klunder, J. M., Caron, M., Uchiyama, M. & Sharpless, K. B. Chlorohydroxylation of olefins with peroxides and titanium tetrachloride. *J. Org. Chem.* **1985**, 50, 912-915.
20. Gabarda, A. E., Du, W., Isarno, T., Tangirala, R. S. & Curran, D. P. Asymmetric total synthesis of (20R)-homocamptothecin, substituted homocamptothecins and homosilatecans. *Tetrahedron* **2002**, 58, 6329-6341.
21. Pettersson, L., Frejd, T. & Magnusson, G. An enantiospecific synthesis of a taxol A-ring building unit. *Tetrahedron Lett.* **1987**, 28, 2753-2756.
22. Erickson, T. J. Asymmetric synthesis of Darvon alcohol. *J. Org. Chem.* **1986**, 51, 934-935.
23. Ichige, T., Okano, Y., Kanoh, N. & Nakata, M. Total Synthesis of Methyl Sarcophytoate, a Marine Natural Biscembranoid. *J. Org. Chem.* **2009**, 74, 230-243.
24. Morgans, D. J., Jr., Sharpless, K. B. & Traynor, S. G. Epoxy alcohol rearrangements: hydroxyl-mediated delivery of Lewis acid promoters. *J. Am. Chem. Soc.* **1981**, 103, 462-464.
25. Hanson, R. M. Epoxide migration (Payne rearrangement) and related reactions. *Org. React.* **2002**, 60, 1-156.
26. Takanami, T. & Suda, K. Metalloporphyrins and Phthalocyanines as Efficient Lewis Acid Catalysts with a Unique Reaction-Field. *Journal of Synthetic Organic Chemistry - Japan* **2009**, 67, 595-605.

ENANTIOSELECTIVE EPOXIDATION OF C=C

Jacobsen (salen)Mn(III)-Catalyzed Oxidation of Unfunctionalized Olefins

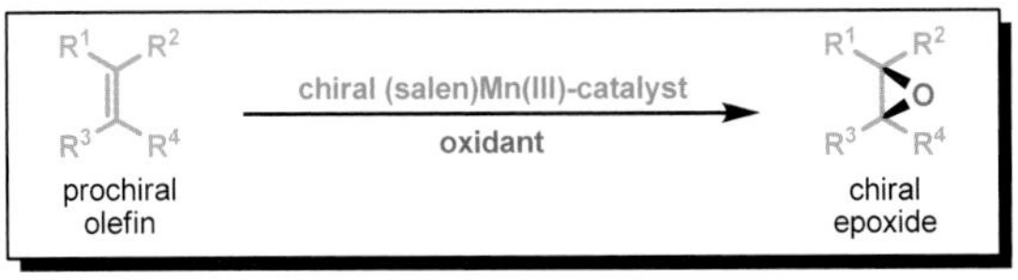

prochiral olefin → chiral epoxide

chiral (salen)Mn(III)-catalyst / oxidant

Kochi's report[1] in 1986 that the (salen)Mn(III) complex[2] **1** catalyzes the efficient epoxidation of olefins by iodosyl benzene (PhIO) led to Jacobsen's finding in 1990[3] that this reaction is fairly enantioselective (33-93% ee) if the ethylenediamine part of the ligand is replaced by (S,S)- or (R,R)-1,2-diphenyldiaminoethane, *and* if the positions *ortho* to the phenolic hydroxyl carry a bulky group, as in the (R,R)-complex **2** (Figure 1).[4] Further improvements included the use of the related catalyst **3**, and NaOCl as the terminal oxidant.[5]

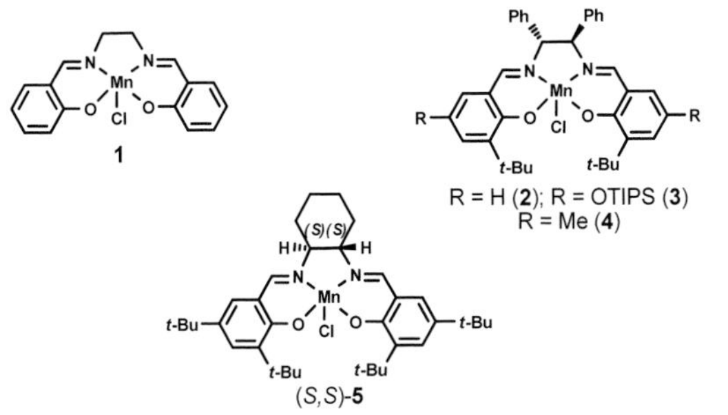

The Jacobsen epoxidation has been applied to the enantioselective epoxidation of mono-, di-, tri- and tetrasubstituted olefins and to unsymmetrical (Z)-disubstituted olefins with good results.[8] However, the epoxidation of (E)-disubstituted alkenes is usually only poorly enantioselective.

The Jacobsen epoxidation takes place with higher yield and enantioselectivity if the olefinic bond is conjugated with a π-system. The rate of the epoxidation, the yield and the enantioselectivity can be affected by the use of various additives, for example pyridine N-oxide, which may imply that a hexacoordinate oxomanganese(V) species is the effective epoxidation reagent, possibly with C_2-symmetric, canted, non-planar six-membered chelate rings.[8] The geometric details of the pre-transition state assembly and the basis of enantioselectivity have been analyzed (see page 167 for pathway).[9] Twelve examples of the Jacobsen epoxidation are shown in Schemes 1 and 2. Entries **VII-X** in Scheme 2 illustrate the formation of a *trans* epoxide from a *cis*-double bond. These cases show that the Jacobsen epoxidation can occur by a 2-step pathway with intervening C-C bond rotation. The 2-step process is also favored by the use of certain chiral quaternary ammonium halides.[11c]

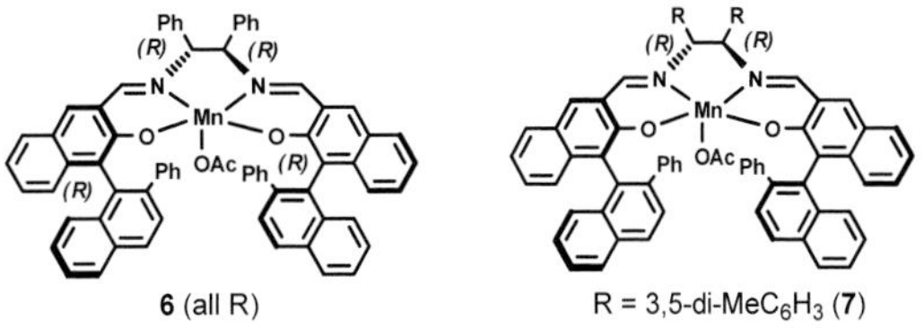

1

Ph Ph

R = H (**2**); R = OTIPS (**3**)
R = Me (**4**)

(S,S)-**5**

Figure 1

6 (all R)

R = 3,5-di-MeC$_6$H$_3$ (**7**)

Figure 2

A similar catalyst system was developed by Katsuki and coworkers (Figure 2).[6] The most widely used catalysts are **3** and **5** which are commercially available (Figure 1).[7]

Asymmetric Epoxidation of Structurally Diverse Unfunctionalized Alkenes

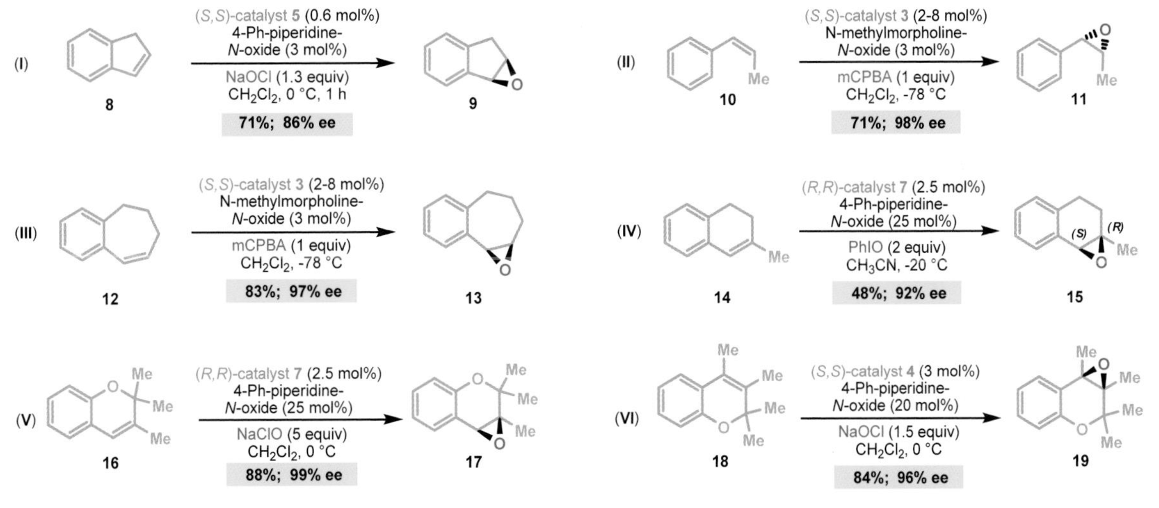

(I) **8** → **9**

(S,S)-catalyst **5** (0.6 mol%)
4-Ph-piperidine-N-oxide (3 mol%)
NaOCl (1.3 equiv)
CH$_2$Cl$_2$, 0 °C, 1 h
71%; 86% ee

(II) **10** → **11**

(S,S)-catalyst **3** (2-8 mol%)
N-methylmorpholine-N-oxide (3 mol%)
mCPBA (1 equiv)
CH$_2$Cl$_2$, -78 °C
71%; 98% ee

(III) **12** → **13**

(S,S)-catalyst **3** (2-8 mol%)
N-methylmorpholine-N-oxide (3 mol%)
mCPBA (1 equiv)
CH$_2$Cl$_2$, -78 °C
83%; 97% ee

(IV) **14** → **15**

(R,R)-catalyst **7** (2.5 mol%)
4-Ph-piperidine-N-oxide (25 mol%)
PhIO (2 equiv)
CH$_3$CN, -20 °C
48%; 92% ee

(V) **16** → **17**

(R,R)-catalyst **7** (2.5 mol%)
4-Ph-piperidine-N-oxide (25 mol%)
NaClO (5 equiv)
CH$_2$Cl$_2$, 0 °C
88%; 99% ee

(VI) **18** → **19**

(S,S)-catalyst **4** (3 mol%)
4-Ph-piperidine-N-oxide (20 mol%)
NaOCl (1.5 equiv)
CH$_2$Cl$_2$, 0 °C
84%; 96% ee

Reference Key for Equations: (I)[10]; (II-III)[10a]   **Scheme 1**   Reference Key for Equations: (IV-V)[11]; (VI)[12]

ENANTIOSELECTIVE EPOXIDATION OF C=C

Asymmetric Epoxidation of Structurally Diverse Unfunctionalized Alkenes

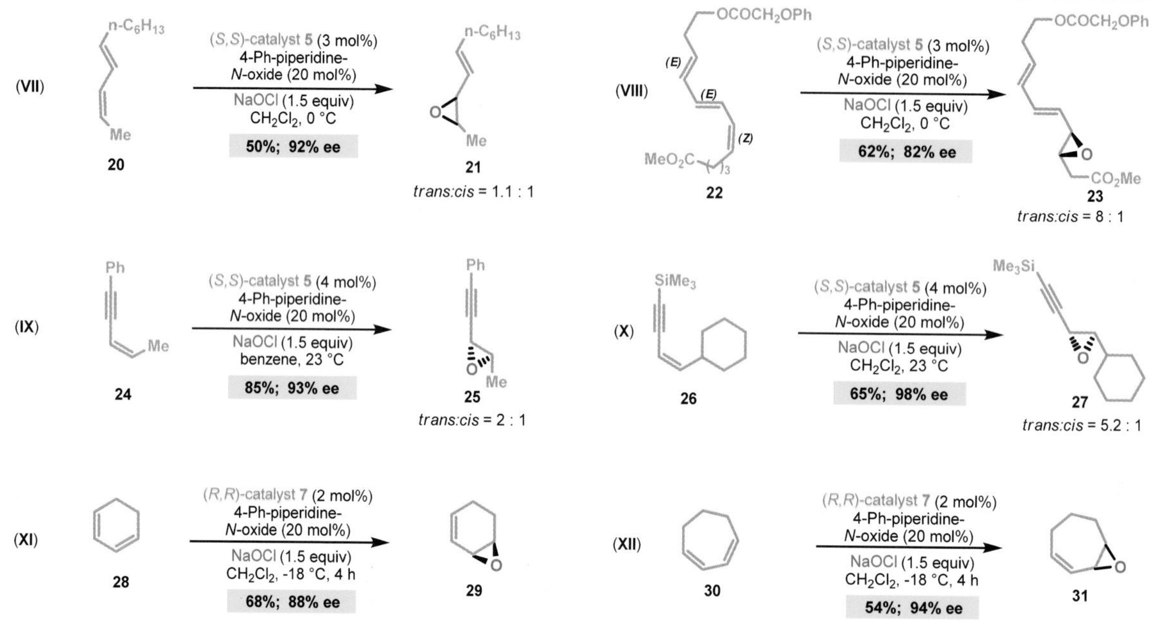

(VII)

$n\text{-}C_6H_{13}$

20

(S,S)-catalyst **5** (3 mol%)
4-Ph-piperidine-
N-oxide (20 mol%)
NaOCl (1.5 equiv)
CH₂Cl₂, 0 °C
50%; 92% ee

$n\text{-}C_6H_{13}$

21

trans:cis = 1.1 : 1

(VIII)

(E)

(E)

(Z)

MeO₂C

OCOCH₂OPh

22

(S,S)-catalyst **5** (3 mol%)
4-Ph-piperidine-
N-oxide (20 mol%)
NaOCl (1.5 equiv)
CH₂Cl₂, 0 °C
62%; 82% ee

OCOCH₂OPh

O
CO₂Me

23

trans:cis = 8 : 1

(IX)

Ph

Me

24

(S,S)-catalyst **5** (4 mol%)
4-Ph-piperidine-
N-oxide (20 mol%)
NaOCl (1.5 equiv)
benzene, 23 °C
85%; 93% ee

Ph

O
Me

25

trans:cis = 2 : 1

(X)

SiMe₃

26

(S,S)-catalyst **5** (4 mol%)
4-Ph-piperidine-
N-oxide (20 mol%)
NaOCl (1.5 equiv)
CH₂Cl₂, 23 °C
65%; 98% ee

Me₃Si

O

27

trans:cis = 5.2 : 1

(XI)

28

(R,R)-catalyst **7** (2 mol%)
4-Ph-piperidine-
N-oxide (20 mol%)
NaOCl (1.5 equiv)
CH₂Cl₂, -18 °C, 4 h
68%; 88% ee

O

29

(XII)

30

(R,R)-catalyst **7** (2 mol%)
4-Ph-piperidine-
N-oxide (20 mol%)
NaOCl (1.5 equiv)
CH₂Cl₂, -18 °C, 4 h
54%; 94% ee

O

31

Reference Key for Equations: (VII-VIII)[13]; (IX)[14]

Scheme 2

Reference Key for Equations: (X);[14] (XI-XII)[15]

References

1. Srinivasan, K., Michaud, P. & Kochi, J. K. Epoxidation of olefins with cationic (salen)manganese(III) complexes. The modulation of catalytic activity by substituents. *J. Am. Chem. Soc.* **1986**, 108, 2309-2320.
2. The name "salen" refers to ligands that feature the *N,N*-ethylenebis(salicylideneamidato) core.
3. Zhang, W., Loebach, J. L., Wilson, S. R. & Jacobsen, E. N. Enantioselective epoxidation of unfunctionalized olefins catalyzed by salen manganese complexes. *J. Am. Chem. Soc.* **1990**, 112, 2801-2803.
4. For selected reviews on salen-metal complexes, see: (a) Baleizao, C. & Garcia, H. Chiral Salen Complexes: An Overview to Recoverable and Reusable Homogeneous and Heterogeneous Catalysts. *Chem. Rev.* **2006**, 106, 3987-4043; (b) Caputo, C. A. & Jones, N. D. Developments in asymmetric catalysis by metal complexes of chiral chelating nitrogen-donor ligands. *Dalton Trans.* **2007**, 4627-4640.
5. (a) Zhang, W. & Jacobsen, E. N. Asymmetric olefin epoxidation with sodium hypochlorite catalyzed by easily prepared chiral manganese(III) salen complexes. *J. Org. Chem.* **1991**, 56, 2296-2298; (b) Lee, N. H., Muci, A. R. & Jacobsen, E. N. Enantiomerically pure epoxychromans via asymmetric catalysis. *Tetrahedron Lett.* **1991**, 32, 5055-5058; (c) Jacobsen, E. N., Zhang, W., Muci, A. R., Ecker, J. R. & Deng, L. Highly enantioselective epoxidation catalysts derived from 1,2-diaminocyclohexane. *J. Am. Chem. Soc.* **1991**, 113, 7063-7064.
6. (a) Irie, R., Noda, K., Ito, Y., Matsumoto, N. & Katsuki, T. Catalytic asymmetric epoxidation of unfunctionalized olefins. *Tetrahedron Lett.* **1990**, 31, 7345-7348; (b) Irie, R., Noda, K., Ito, Y. & Katsuki, T. Enantioselective epoxidation of unfunctionalized olefins using chiral (salen)manganese(III) complexes. *Tetrahedron Lett.* **1991**, 32, 1055-1058.
7. Larrow, J. F., Jacobsen, E. N., Gao, Y., Hong, Y., Nie, X. & Zepp, C. M. A Practical Method for the Large-Scale Preparation of [N,N'-Bis(3,5-di-tertbutylsalicylidene)-1,2-cyclohexanediaminato(2-)]manganese(III) chloride, a Highly Enantioselective Epoxidation Catalyst. *J. Org. Chem.* **1994**, 59, 1939-1942.
8. For selected reviews on the Jacobsen epoxidation, see: (a) Larrow, J. F. & Jacobsen, E. N. Asymmetric processes catalyzed by chiral (salen)metal complexes. *Top. Organomet. Chem.* **2004**, 6, 123-152;

(b) Muniz-Fernandez, K. & Bolm, C. Manganese-catalyzed epoxidations. *Transition Metals for Organic Synthesis (2nd Edition)* **2004**, 2, 344-356; (c) McGarrigle, E. M. & Gilheany, D. G. Chromium- and Manganese-salen Promoted Epoxidation of Alkenes. *Chem. Rev. (Washington, DC, U. S.)* **2005**, 105, 1563-1602; (d) Armstrong, A. Oxidation reactions in *Enantioselective Organocatalysis* 403-424 **(2007)**.
9. (a) Linker, T. The Jacobsen-Katsuki epoxidation and its controversial mechanism. *Angew. Chem., Int. Ed. Engl.* **1997**, 36, 2060-2062; (b) Scheurer, A., Maid, H., Hampel, F., Saalfrank, R. W., Toupet, L., Mosset, P., Puchta, R. & van Eikema Hommes, N. J. R. Influence of the conformation of salen complexes on the stereochemistry of the asymmetric epoxidation of olefins. *Eur. J. Org. Chem.* **2005**, 2566-2574; (c) Kürti, L., Blewett, M. & Corey, E. J. Origin of Enantioselectivity in the Jacobsen Epoxidation of Olefins. *Org. Lett.* **2009**, 11, 4592-4595.
10. (a) Palucki, M., McCormick, G. J. & Jacobsen, E. N. Low temperature asymmetric epoxidation of unfunctionalized olefins catalyzed by (salen)Mn(III) complexes. *Tetrahedron Lett.* **1995**, 36, 5457-5460; (b) Larrow, J. F., Roberts, E., Verhoeven, T. R., Ryan, K. M., Senanayake, C. H., Reider, P. J. & Jacobsen, E. N. (1S,2R)-1-aminoindan-2-ol (1H-inden-2-ol, 1-amino-2,3-dihydro-(1S-cis)-). *Org. Synth.* **1999**, 76, 46-56.
11. Fukuda, T., Irie, R. & Katsuki, T. Mn-salen catalyzed asymmetric epoxidation of conjugated trisubstituted olefins. *Synlett* **1995**, 197-198.
12. Brandes, B. D. & Jacobsen, E. N. Enantioselective catalytic epoxidation of tetrasubstituted olefins. *Tetrahedron Lett.* **1995**, 36, 5123-5126.
13. Chang, S., Lee, N. H. & Jacobsen, E. N. Regio- and enantioselective catalytic epoxidation of conjugated polyenes. Formal synthesis of LTA₄ methyl ester. *J. Org. Chem.* **1993**, 58, 6939-6941.
14. Lee, N. H. & Jacobsen, E. N. Enantioselective epoxidation of conjugated dienes and enynes. Trans-Epoxides and cis-olefins. *Tetrahedron Lett.* **1991**, 32, 6533-6536.
15. Mikane, D., Hamada, T., Irie, R. & Katsuki, T. Mn-salen catalyzed asymmetric epoxidation of 1,3-cycloalkadienes and dialkylsubstituted Z-olefins. *Synlett* **1995**, 827-828.

ENANTIOSELECTIVE EPOXIDATION OF C=C

Epoxidation of Unfunctionalized Olefins by Chiral Dioxiranes (Shi Epoxidation)

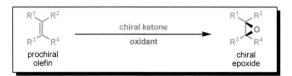

prochiral olefin → chiral ketone / oxidant → chiral epoxide

Dioxiranes such as **1** and **2** can be generated by reaction of potassium peroxomonosulfate ($KHSO_5$) or its equivalent Oxone ($2KHSO_5 \cdot KHSO_4$) with acetone or methyl trifluoromethyl ketone, respectively. The reaction of **1** or the more reactive **2** with olefins produces epoxides in high yield (Scheme 1).[1]

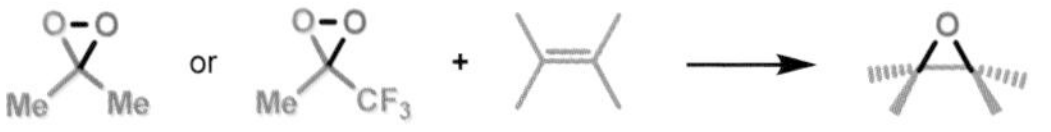

Scheme 1

Various chiral dioxiranes have been prepared and examined as reagents for enantioselective epoxidation of double bonds. Of these the most widely used are the dioxiranes derived from ketones **5** and *ent*-**5** introduced by Y. Shi and coworkers (Figure 1).[2] Catalyst **5** can be prepared[3] from D-fructose and *ent*-**5** can be prepared from L-sorbose via L-fructose.[4]

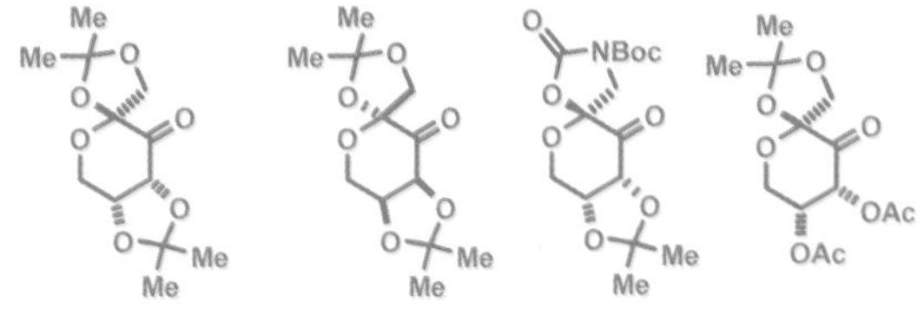

5
Shi catalyst derived from D-fructose

ent-**5**
Shi catalyst derived from L-fructose

6

7

Figure 1

The stereocenters in ketones **5** and *ent*-**5** provide a dissymmetric steric environment that discriminates among various approach orientations of the olefin to the corresponding dioxirane. The transfer of oxygen from the dioxirane to the olefin is generally considered to occur by backside approach to the O-O bond, with the olefinic π-plane being orthogonal to the dioxirane ring plane ("spiro" transition

state geometry, Figure 2). The Shi epoxidation is generally carried out at pH 7-10 to minimize decomposition of the dioxirane by other pathways, e.g. Baeyer-Villiger rearrangement to lactone, a troublesome side reaction that generally necessitates the use of 10 or more mol% of catalyst.[5]

The mechanism of the Shi asymmetric epoxidation has been studied extensively.[4,6] The catalytic cycle is depicted in Scheme 2.

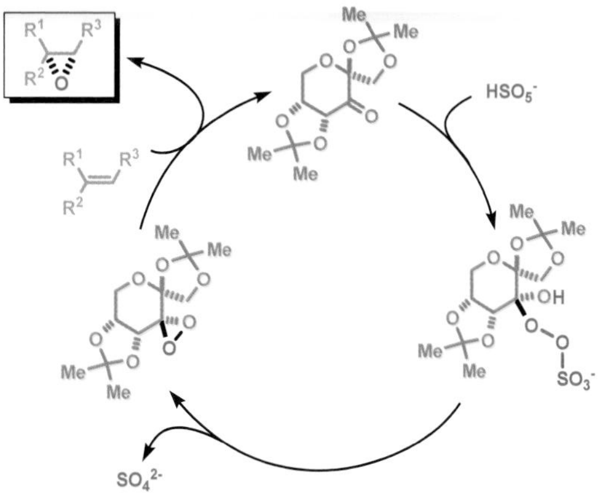

Scheme 2

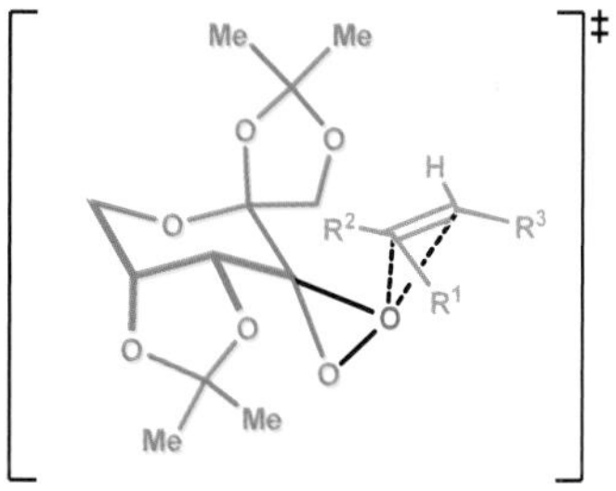

Figure 2

The scope of the Shi epoxidation is illustrated by twelve examples shown in Schemes 3 and 4.

Asymmetric Epoxidation of Structurally Diverse Unfunctionalized Alkenes

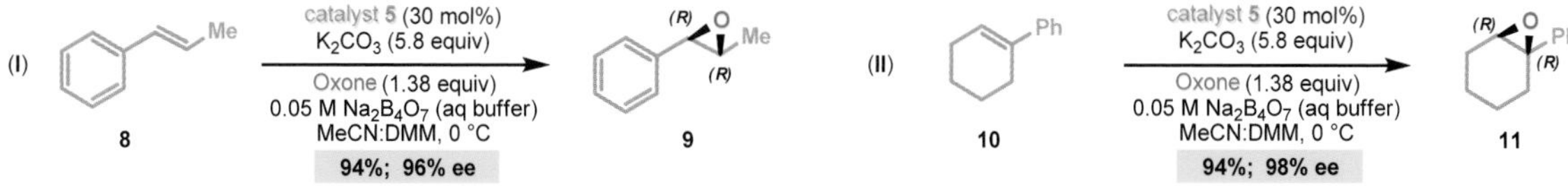

(I) **8** → catalyst **5** (30 mol%) / K_2CO_3 (5.8 equiv) / Oxone (1.38 equiv) / 0.05 M $Na_2B_4O_7$ (aq buffer) / MeCN:DMM, 0 °C → **9**
94%; 96% ee

(II) **10** → catalyst **5** (30 mol%) / K_2CO_3 (5.8 equiv) / Oxone (1.38 equiv) / 0.05 M $Na_2B_4O_7$ (aq buffer) / MeCN:DMM, 0 °C → **11**
94%; 98% ee

Reference Key for Equations: (I)[7]

Scheme 3

Reference Key for Equations: (II)[7]

ENANTIOSELECTIVE EPOXIDATION OF C=C

Asymmetric Epoxidation of Structurally Diverse Unfunctionalized Alkenes

(III) **12** — catalyst 5 (30 mol%), K_2CO_3 (5.8 equiv), Oxone (1.38 equiv), 0.05 M $Na_2B_4O_7$ (aq buffer), MeCN:DMM, 0 °C → **13** — 83%; 95% ee

(IV) **14** — catalyst 5 (30 mol%), K_2CO_3 (5.8 equiv), Oxone (1.38 equiv), 0.05 M $Na_2B_4O_7$ (aq buffer), MeCN:DMM, 0 °C → **15** — 97%; 87% ee

(V) **16** — catalyst 5 (30 mol%), K_2CO_3 (5.8 equiv), Oxone (1.38 equiv), 0.05 M $Na_2B_4O_7$ (aq buffer), MeCN:DMM, 0 °C → **17** — 89%; 94% ee

(VI) **18** — catalyst *ent*-5 (30 mol%), K_2CO_3 (5.8 equiv), Oxone (1.38 equiv), 0.05 M $Na_2B_4O_7$ (aq buffer), MeCN:DMM, 0 °C → **19** — 41%; 97% ee

(VII) **20** — catalyst 5 (65 mol%), K_2CO_3 (5.8 equiv), Oxone (1.38 equiv), 0.05 M $Na_2B_4O_7$ (aq buffer), MeCN:DMM, 0 °C → **21** — 74%; 94% ee

(VIII) **22** — catalyst 5 (30 mol%), K_2CO_3 (5.8 equiv), Oxone (1.38 equiv), 0.05 M $Na_2B_4O_7$ (aq buffer), MeCN:DMM, 0 °C → **23** — 71%; 93% ee

(IX) **24** — catalyst 5 (20 mol%), K_2CO_3 (5.8 equiv), Oxone (1.12 equiv), 0.05 M $Na_2B_4O_7$ (aq buffer), MeCN:DMM, 0 °C → **25** — 77%; 97% ee

(X) **26** — catalyst 5 (20 mol%), K_2CO_3 (5.8 equiv), Oxone (1.12 equiv), 0.05 M $Na_2B_4O_7$ (aq buffer), MeCN:DMM, 0 °C → **27** — 82%; 95% ee

(XI) **28** — catalyst 5 (30 mol%), K_2CO_3 (5.8 equiv), Oxone (1.38 equiv), 0.05 M $Na_2B_4O_7$ (aq buffer), MeCN:DMM, 0 °C → **29** — 78%; 93% ee

(XII) **30** — catalyst 7 (30 mol%), $NaHCO_3$ (15.5 equiv), Oxone (5 equiv), 0.05 M $Na_2B_4O_7$ (aq buffer), MeCN:DMM, 0 °C → **31** — 73%; 96% ee

Reference Key for Equations: (III-VI)[7]; (VII-VIII)[8] **Scheme 4** Reference Key for Equations: (IX-X)[9];(XI)[10]; (XII)[11]

References

1. For a seminal publication, see: Curci, R., Fiorentino, M. & Serio, M. R. Asymmetric epoxidation of unfunctionalized alkenes by dioxirane intermediates generated from potassium peroxomonosulfate and chiral ketones. *J. Chem. Soc., Chem. Commun.* **1984**, 155-156.

2. Tu, Y., Wang, Z.-X. & Shi, Y. An Efficient Asymmetric Epoxidation Method for *trans*-Olefins Mediated by a Fructose-Derived Ketone. *J. Am. Chem. Soc.* **1996**, 118, 9806-9807.

3. Tu, Y., Frohn, M., Wang, Z.-X. & Shi, Y. Synthesis of 1,2:4,5-di-o-isopropylidene-D-erythro-2,3-hexodiulo-2,6-pyranose. A highly enantioselective ketone catalyst for epoxidation. *Org. Synth.* **2003**, 80, 1-8.

4. For recent reviews on the Shi Asymmetric Epoxidation and other related organocatalytic epoxidations of olefins, see: (a) Wong, O. A., Shi, Y. Organocatalytic Oxidation. Asymmetric Epoxidation of Olefins Catalyzed by Chiral Ketones and Iminium Salts. *Chem. Rev.* **2008**, 108, 3958-3987; (b) Shi, Y. in *Handbook of Chiral Chemicals* (2nd Edition) 147-163 (**2006**); (c) Kürti, L., Czakó, B. Shi Asymmetric Epoxidation in *Strategic Applications of Named Reactions in Organic Synthesis* **2005**, pp 410-411 (Academic Press/Elsevier, San Diego); (d) Shi, Y. Organocatalytic Asymmetric Epoxidation of Olefins by Chiral Ketones. *Acc. Chem. Res.* **2004**, 37, 488-496; (e) Yang, D. Ketone-Catalyzed Asymmetric Epoxidation Reactions. *Acc. Chem. Res.* **2004**, 37, 497-505.

5. (a) At high pH values (above pH 11) the Oxone ($KHSO_5$) undergoes rapid decomposition, while at lower pH values the ketone catalyst undergoes Baeyer-Villiger oxidation. At the optimum pH level (pH ~7-10), the nucleophilicity of the Oxone toward the ketone catalyst is enhanced and the overall efficiency of the epoxidation process is thus

improved; (b) Tian, H., She, X. & Shi, Y. Electronic probing of ketone catalysts for asymmetric epoxidation. Search for more robust catalysts. *Org. Lett.* **2001**, 3, 715-718; (c) Tian, H., She, X., Yu, H., Shu, L. & Shi, Y. Designing New Chiral Ketone Catalysts. Asymmetric Epoxidation of *cis*-Olefins and Terminal Olefins. *J. Org. Chem.* **2002**, 67, 2435-2446.

6. Singleton, D. A. & Wang, Z. Isotope Effects and the Nature of Enantioselectivity in the Shi Epoxidation. The Importance of Asynchronicity. *J. Am. Chem. Soc.* **2005**, 127, 6679-6685.

7. Wang, Z.-X., Tu, Y., Frohn, M., Zhang, J.-R. & Shi, Y. An Efficient Catalytic Asymmetric Epoxidation Method. *J. Am. Chem. Soc.* **1997**, 119, 11224-11235.

8. Warren, J. D. & Shi, Y. Chiral Ketone-Catalyzed Asymmetric Epoxidation of 2,2-Disubstituted Vinylsilanes. *J. Org. Chem.* **1999**, 64, 7675-7677.

9. Frohn, M., Dalkiewicz, M., Tu, Y., Wang, Z.-X. & Shi, Y. Highly Regio- and Enantioselective Monoepoxidation of Conjugated Dienes. *J. Org. Chem.* **1998**, 63, 2948-2953.

10. Wang, Z.-X., Cao, G.-A. & Shi, Y. Chiral Ketone Catalyzed Highly Chemo- and Enantioselective Epoxidation of Conjugated Enynes. *J. Org. Chem.* **1999**, 64, 7646-7650.

11. Wu, X.-Y., She, X. & Shi, Y. Highly Enantioselective Epoxidation of α,β-Unsaturated Esters by Chiral Dioxirane. *J. Am. Chem. Soc.* **2002**, 124, 8792-8793.

ENANTIOSELECTIVE EPOXIDATION OF C=C−C=O

Nucleophilic Epoxidation of α,β-Unsaturated Carbonyl Compounds

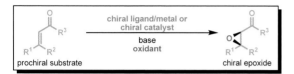

prochiral substrate → chiral epoxide

chiral ligand/metal or chiral catalyst

base
oxidant

The nucleophilic epoxidation[1] of α,β-unsaturated carbonyl compounds[2] by NaOCl, hydrogen peroxide (H$_2$O$_2$) or hydroperoxides such as t-BuOOH was first rendered enantioselective by the use of poly-(S)-alanine as an additive for the substrate benzalacetophenone (1→2, Scheme 1).[3]

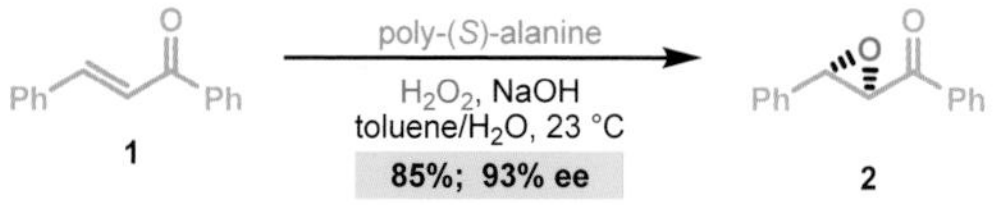

poly-(S)-alanine

H$_2$O$_2$, NaOH
toluene/H$_2$O, 23 °C

85%; 93% ee

1 → 2

Scheme 1

Later chiral quaternary ammonium salts were found to direct the enantioselective epoxidation of various aryl vinyl ketones under phase transfer conditions.[4] The cinchona alkaloid quaternary ammonium bromide[5] (5, Figure 1), for example, accelerates and directs the face-selective epoxidation of 1 by KOCl to ent-2.[4,6] A rational explanation of the preferred pre-transition state assembly for this reaction has been presented.[6]

Enantioselective epoxidation of some α,β-unsaturated cinnamaldehydes has been described using chiral pyrrolidinomethanols of type 6 with t-BuOOH or H$_2$O$_2$ as oxidant.[2] These reactions proceed via the iminium ion 7

formed by the addition of the chiral pyrrolidine to the aldehyde formyl group.[7] In addition to the previously mentioned approaches, a catalyst formed from (R) or (S)-BINOL, La(Oi-Pr)$_3$ and Ph$_3$P=O or Ph$_3$As=O promotes the enantioselective α,β-epoxidation of vinyl aryl ketones and certain α,β-unsaturated acid derivatives.[8] A selection of illustrative examples of the enantioselective epoxidation of α,β-unsaturated carbonyl compounds is shown in Schemes 2 and 3. Some of these examples illustrate less common methods that have been used for enantioselective epoxidations.

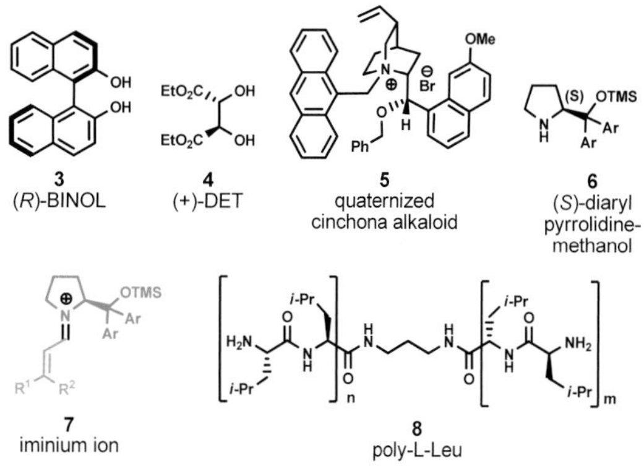

3
(R)-BINOL

4
(+)-DET

5
quaternized
cinchona alkaloid

6
(S)-diaryl
pyrrolidine-
methanol

7
iminium ion

8
poly-L-Leu

Figure 1

Asymmetric Epoxidation of Structurally Diverse α,β-Unsaturated Carbonyl Compounds

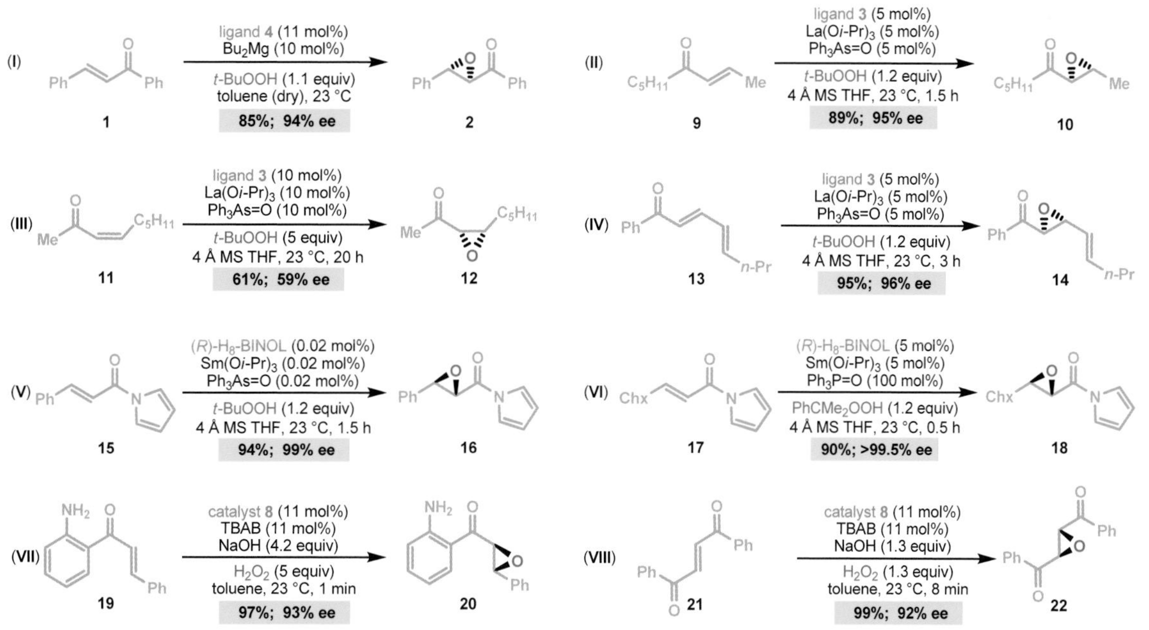

(I) 1 → 2
ligand 4 (11 mol%)
Bu$_2$Mg (10 mol%)
t-BuOOH (1.1 equiv)
toluene (dry), 23 °C
85%; 94% ee

(II) 9 → 10
ligand 3 (5 mol%)
La(Oi-Pr)$_3$ (5 mol%)
Ph$_3$As=O (5 mol%)
t-BuOOH (1.2 equiv)
4 Å MS THF, 23 °C, 1.5 h
89%; 95% ee

(III) 11 → 12
ligand 3 (10 mol%)
La(Oi-Pr)$_3$ (10 mol%)
Ph$_3$As=O (10 mol%)
t-BuOOH (5 equiv)
4 Å MS THF, 23 °C, 20 h
61%; 59% ee

(IV) 13 → 14
ligand 3 (5 mol%)
La(Oi-Pr)$_3$ (5 mol%)
Ph$_3$As=O (5 mol%)
t-BuOOH (1.2 equiv)
4 Å MS THF, 23 °C, 3 h
95%; 96% ee

(V) 15 → 16
(R)-H$_8$-BINOL (0.02 mol%)
Sm(Oi-Pr)$_3$ (0.02 mol%)
Ph$_3$As=O (0.02 mol%)
t-BuOOH (1.2 equiv)
4 Å MS THF, 23 °C, 1.5 h
94%; 99% ee

(VI) 17 → 18
(R)-H$_8$-BINOL (5 mol%)
Sm(Oi-Pr)$_3$ (5 mol%)
Ph$_3$P=O (100 mol%)
PhCMe$_2$OOH (1.2 equiv)
4 Å MS THF, 23 °C, 0.5 h
90%; >99.5% ee

(VII) 19 → 20
catalyst 8 (11 mol%)
TBAB (11 mol%)
NaOH (4.2 equiv)
H$_2$O$_2$ (5 equiv)
toluene, 23 °C, 1 min
97%; 93% ee

(VIII) 21 → 22
catalyst 8 (11 mol%)
TBAB (11 mol%)
NaOH (1.3 equiv)
H$_2$O$_2$ (1.3 equiv)
toluene, 23 °C, 8 min
99%; 92% ee

Reference Key for Equations: (I)[9]; (II-IV)[8b]   ### Scheme 2   Reference Key for Equations: (V-VI)[8c]; (VII-VIII)[10]

ENANTIOSELECTIVE EPOXIDATION OF C=C–C=O

Asymmetric Epoxidation of Structurally Diverse α,β-Unsaturated Carbonyl Compounds

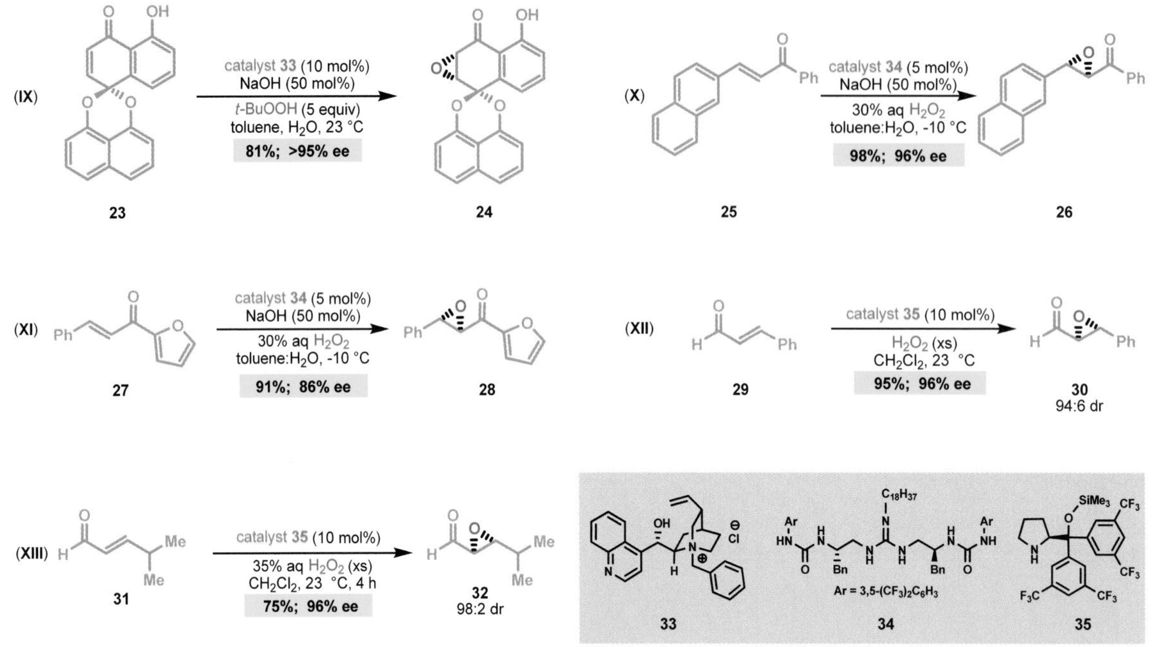

(IX)
23 → **24**

catalyst **33** (10 mol%)
NaOH (50 mol%)
t-BuOOH (5 equiv)
toluene, H$_2$O, 23 °C
81%; >95% ee

(X)
25 → **26**

catalyst **34** (5 mol%)
NaOH (50 mol%)
30% aq H$_2$O$_2$
toluene:H$_2$O, -10 °C
98%; 96% ee

(XI)
27 → **28**

catalyst **34** (5 mol%)
NaOH (50 mol%)
30% aq H$_2$O$_2$
toluene:H$_2$O, -10 °C
91%; 86% ee

(XII)
29 → **30**
94:6 dr

catalyst **35** (10 mol%)
H$_2$O$_2$ (xs)
CH$_2$Cl$_2$, 23 °C
95%; 96% ee

(XIII)
31 → **32**
98:2 dr

catalyst **35** (10 mol%)
35% aq H$_2$O$_2$ (xs)
CH$_2$Cl$_2$, 23 °C, 4 h
75%; 96% ee

33   **34**   **35**

Reference Key for Equations: (IX)[11]; (X-XI)[12]

Scheme 3

Reference Key for Equations: (XII-XIII)[7]

References

1. For a seminal publication, see: Julia, S., Masana, J. & Vega, J. C. Synthetic enzyme: highly stereoselective epoxidation of chalcone in the three-phase system toluene-water-poly-(S)-alanine. *Angew. Chem.* **1980**, 92, 968-969.
2. For recent reviews on the enantioselective epoxidation of α,β-unsaturated carbonyl compounds, see: (a) Lattanzi, A. Advances in asymmetric epoxidation of α,β-unsaturated carbonyl compounds: the organocatalytic approach. *Curr. Org. Synth.* **2008**, 5, 117-133; (b) Armstrong, A. Oxidation reactions in *Enantioselective Organocatalysis* 403-424 (**2007**).
3. For selected reviews that discuss the Juliá-Colonna epoxidation, see: (a) Colby Davie, E. A., Mennen, S. M., Xu, Y. & Miller, S. J. Asymmetric Catalysis Mediated by Synthetic Peptides. *Chem. Rev.* **2007**, 107, 5759-5812; (b) Jarvo, E. R. & Miller, S. J. Amino acids and peptides as asymmetric organocatalysts. *Tetrahedron* **2002**, 58, 2481-2495; (c) Lauret, C. & Roberts, S. M. Asymmetric epoxidation of α,β-unsaturated ketones catalyzed by poly(amino acids). *Aldrichimica Acta* **2002**, 35, 47-51.
4. For recent reviews on asymmetric phase-transfer catalysis, see: (a) Maruoka, K. Practical Aspects of Recent Asymmetric Phase-Transfer Catalysis. *Org. Process Res. Dev.* **2008**, 12, 679-697; (b) Maruoka, K. Binaphthyl- and biphenyl-modified chiral phase-transfer catalysts for asymmetric synthesis. *Asymmetric Phase Transfer Catalysis* **2008**, 71-113; (c) Hashimoto, T. & Maruoka, K. Recent Development and Application of Chiral Phase-Transfer Catalysts. *Chem. Rev.* **2007**, 107, 5656-5682; (d) Albanese, D. New applications of phase transfer catalysis in organic synthesis. *Mini-Rev. Org. Chem.* **2006**, 3, 195-217; (e) Kacprzak, K. & Gawronski, J. Cinchona alkaloids and their derivatives: versatile catalysts and ligands in asymmetric synthesis. *Synthesis* **2001**, 961-998.
5. Corey, E. J., Xu, F. & Noe, M. C. A Rational Approach to Catalytic Enantioselective Enolate Alkylation Using a Structurally Rigidified and Defined Chiral Quaternary Ammonium Salt under Phase Transfer Conditions. *J. Am. Chem. Soc.* **1997**, 119, 12414-12415.
6. Corey, E. J. & Zhang, F.-Y. Mechanism and conditions for highly enantioselective epoxidation of α,β-enones using charge-accelerated

catalysis by a rigid quaternary ammonium salt. *Org. Lett.* **1999**, 1, 1287-1290.
7. Marigo, M., Franzen, J., Poulsen, T. B., Zhuang, W. & Jorgensen, K. A. Asymmetric Organocatalytic Epoxidation of α,β-Unsaturated Aldehydes with Hydrogen Peroxide. *J. Am. Chem. Soc.* **2005**, 127, 6964-6965.
8. (a) For a seminal publication, see: Daikai, K., Kamaura, M. & Junji, I. Remarkable ligand effect on the enantioselectivity of the chiral lanthanum complex-catalyzed asymmetric epoxidation of enones. *Tetrahedron Lett.* **1998**, 39, 7321-7322. For a review, see: Inanaga, J., Furuno, H. & Hayano, T. Asymmetric Catalysis and Amplification with Chiral Lanthanide Complexes. *Chem. Rev.* **2002**, 102, 2211-2225; (b) Nemoto, T., Ohshima, T., Yamaguchi, K. & Shibasaki, M. Catalytic Asymmetric Epoxidation of Enones Using La-BINOL-Triphenylarsine Oxide Complex: Structural Determination of the Asymmetric Catalyst. *J. Am. Chem. Soc.* **2001**, 123, 2725-2732; (c) Matsunaga, S., Qin, H., Sugita, M., Okada, S., Kinoshita, T., Yamagiwa, N. & Shibasaki, M. Catalytic asymmetric epoxidation of α,β-unsaturated N-acyl pyrroles as monodentate and activated ester equivalent acceptors. *Tetrahedron* **2006**, 62, 6630-6639.
9. Elston, C. L., Jackson, R. F. W., Macdonald, S. J. F. & Murray, P. J. Asymmetric epoxidation of chalcones with chirally modified lithium and magnesium tert-butyl peroxides. *Angew. Chem., Int. Ed. Engl.* **1997**, 36, 410-412.
10. Gerlach, A. & Geller, T. Scale-up studies for the asymmetric Julia-Colonna epoxidation reaction. *Adv. Synth. Catal.* **2004**, 346, 1247-1249.
11. Barrett, A. G. M., Blaney, F., Campbell, A. D., Hamprecht, D., Meyer, T., White, A. J. P., Witty, D. & Williams, D. J. Unified Route to the Palmarumycin and Preussomerin Natural Products. Enantioselective Synthesis of (-)-Preussomerin G. *J. Org. Chem.* **2002**, 67, 2735-2750.
12. Tanaka, S. & Nagasawa, K. Guanidine-urea bifunctional organocatalyst for asymmetric epoxidation of 1,3-diarylenones with hydrogen peroxide. *Synlett* **2009**, 667-670.

ENANTIOSELECTIVE DIHYDROXYLATION OF C=C

Sharpless Asymmetric Dihydroxylation of Unfunctionalized Olefins

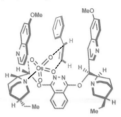

prochiral olefin → chiral 1,2-diol

The suprafacial (*cis*) dihydroxylation of olefins by OsO_4, which is strongly catalyzed by amines, can be converted into an enantioselective form simply by the use of certain chiral 1,2-diamines or various chinchona alkaloid derivatives.[1] Although the 1,2-diamine-catalyzed versions can be highly selective, they suffer from the disadvantage of being stoichiometric in both ligand and the expensive oxidant OsO_4.[2] The optimization of cinchona alkaloid derivatives as ligands for OsO_4 by Sharpless and coworkers made possible the use of substoichiometric quantities of ligand and OsO_4 for highly enantioselective dihydroxylation.[3] The *bis*-dihydro quinidine derivative in (DHQD)$_2$PHAL (**2**) and the corresponding bis-dihydro quinine (DHQ)$_2$PHAL (**1**) are most frequently used and commercially available. A whole range of related catalysts have been devised by modification of the linker group (e.g., **4** and **5** in Figure 1).[1]

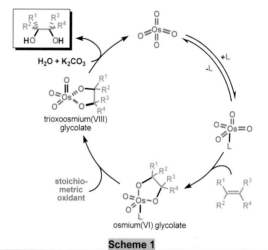

1 (DHQ)$_2$PHAL

2 (DHQD)$_2$PHAL

3

4 (DHQD)$_2$PYDZ

5 (DHQD)$_2$PYR

Figure 1

The remarkable enantioselectivity of the Sharpless dihydroxylation is currently best explained by a model which is illustrated for the case of styrene and **1** by the pre-transition state assembly shown in Figure 2.[4]

In this model the substrate styrene is held in a U-shaped binding region of the catalyst-OsO_4 complex, while one axial and one equatorial oxygen of the OsO_4 subunit attack C=C to form a cyclic osmate ester by [3+2]-cycloaddition (Scheme 1). The mechanistic model of the Sharpless dihydroxylation correctly predicts the absolute configuration of the major dihydroxylation product and also provides insights with regard to modifying a substrate structure so as to improve enantioselectivity.[5] Although initially controversial, the transition state model depicted in Figure 2 is strongly supported by a wide range of experimental data.[1,4]

The Sharpless dihydroxylation is generally carried out with $K_3Fe(CN)_6$ as the stoichiometric oxidant, K_2CO_3 as base to cleave the intermediate osmate ester and ca. 1 mol% of

OsO_4 or its equivalent $K_2OsO_2(OH)_4$ in *t*-BuOH-H_2O as medium (generally biphasic).

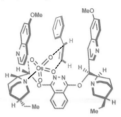

Figure 2

Scheme 1

osmium(VI) glycolate

trioxoosmium(VIII) glycolate

Ligands **1** and **2** are also commercially available as pre-formulated solids (AD-mix-α and AD-mix-β).The empirical model shown in Figure 3 can be used to predict the absolute configuration of the expected product (R_S = small substituent, R_M = medium substituent and R_L = large substituent).

The main limitations of the Sharpless asymmetric dihydroxylation arise with (Z)-1,2-disubstituted olefins having substituents of similar size and with sterically congested tetrasubstitued olefins.

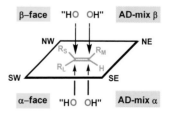

β–face   "HO OH"   AD-mix β

α–face   "HO OH"   AD-mix α

AD-mix α: (DHQ)$_2$PHAL + $K_2OsO_2(OH)_4$ + $K_3Fe(CN)_6$

AD-mix β: (DHQD)$_2$PHAL + $K_2OsO_2(OH)_4$ + $K_3Fe(CN)_6$

Figure 3

ENANTIOSELECTIVE DIHYDROXYLATION OF C=C

Asymmetric Dihydroxylation of Structurally Diverse Alkenes

Reference Key for Equations: (I)[6]; (II)[5b]; (III)[7]; (IV)[8]   **Scheme 4**   **Reference Key for Equations:** (V)[9]; (VI)[10]; (VII)[11]; (VIII)[12]

References

1. For recent reviews on the enantioselective dihydroxylation of alkenes, see: (a) Noe, M. C., Letavic, M. A. & Snow, S. L. Asymmetric dihydroxylation of alkenes. *Organic Reactions* **2005**, 66, 109-625; (b) Zaitsev, A. B. & Adolfsson, H. Recent developments in asymmetric dihydroxylations. *Synthesis* **2006**, 1725-1756; (c) Kolb, H. C. & Sharpless, K. B. Asymmetric dihydroxylation. *Transition Metals for Organic Synthesis (2nd Edition)* **2004**, 2, 275-298; (d) Kolb, H. C., VanNieuwenhze, M. S. & Sharpless, K. B. Catalytic Asymmetric Dihydroxylation. *Chem. Rev.* **1994**, 94, 2483-2547.

2. (a) Tomioka, K., Nakajima, M. & Koga, K. Enantioface differentiation in cis dihydroxylation of carbon-carbon double bonds by osmium tetroxide using a chiral diamine with D2 symmetry. *J. Am. Chem. Soc.* **1987**, 109, 6213-6215; (b) Corey, E. J., Jardine, P. D., Virgil, S., Yuen, P. W. & Connell, R. D. Enantioselective vicinal hydroxylation of terminal and (E)-1,2-disubstituted olefins by a chiral complex of osmium tetroxide. An effective controller system and a rational mechanistic model. *J. Am. Chem. Soc.* **1989**, 111, 9243-9244.

3. (a) Jacobsen, E. N., Marko, I., Mungall, W. S., Schroeder, G. & Sharpless, K. B. Asymmetric dihydroxylation via ligand-accelerated catalysis. *J. Am. Chem. Soc.* **1988**, 110, 1968-1970; (b) Sharpless, K. B., Amberg, W., Bennani, Y. L., Crispino, G. A., Hartung, J., Jeong, K. S., Kwong, H. L., Morikawa, K., Wang, Z. M. & et al. The osmium-catalyzed asymmetric dihydroxylation: a new ligand class and a process improvement. *J. Org. Chem.* **1992**, 57, 2768-2771.

4. (a) Corey, E. J. & Noe, M. C. Rigid and highly enantioselective catalyst for the dihydroxylation of olefins using osmium tetraoxide clarifies the origin of enantiospecificity. *J. Am. Chem. Soc.* **1993**, 115, 12579-12580; (b) Corey, E. J. & Noe, M. C. A Critical Analysis of the Mechanistic Basis of Enantioselectivity in the Bis-Cinchona Alkaloid Catalyzed Dihydroxylation of Olefins. *J. Am. Chem. Soc.* **1996**, 118, 11038-11053.

5. (a) Corey, E. J., Noe, M. C. & Grogan, M. J. A mechanistically designed mono-cinchona alkaloid is an excellent catalyst for the enantioselective dihydroxylation of olefins. *Tetrahedron Lett.* **1994**, 35, 6427-6430; (b) Corey, E. J., Guzman-Perez, A. & Noe, M. C. The application of a mechanistic model leads to the extension of the Sharpless asymmetric dihydroxylation to allylic 4-methoxybenzoates and conformationally related amine and homoallylic alcohol derivatives. *J. Am. Chem. Soc.* **1995**, 117, 10805-10816; (c) Corey, E. J. & Zhang, J. Highly Effective Transition Structure Designed Catalyst for the Enantio- and Position-Selective Dihydroxylation of Polyisoprenoids. *Org. Lett.* **2001**, 3, 3211-3214.

6. Hanessian, S., Meffre, P., Girard, M., Beaudoin, S., Sanceau, J. Y. & Bennani, Y. Asymmetric dihydroxylation of olefins with a simple chiral ligand. *J. Org. Chem.* **1993**, 58, 1991-1993.

7. Wang, Z. M. & Sharpless, K. B. Asymmetric dihydroxylation of α-substituted styrene derivatives. *Synlett* **1993**, 603-604.

8. Krysan, D. J. A dramatic reversal of facial selectivity in the Sharpless asymmetric dihydroxylation of a sterically hindered 3-methylidene-benzofuran. *Tetrahedron Lett.* **1996**, 37, 1375-1376.

9. Tietze, L. F. & Goerlitzer, J. Preparation of chiral building blocks for a highly convergent vitamin E synthesis. Systematic investigations on the enantioselectivity of the Sharpless bishydroxylation. *Synthesis* **1998**, 873-878.

10. Shao, H. & Goodman, M. An Enantiomeric Synthesis of allo-Threonines and β-Hydroxyvalines. *J. Org. Chem.* **1996**, 61, 2582-2583.

11. Bassindale, A. R., Taylor, P. G. & Xu, Y. Asymmetric dihydroxylation of vinyl- and allyl-silanes. *J. Chem. Soc., Perkin Trans. 1* **1994**, 1061-1067.

12. Xie, L., Crimmins, M. T. & Lee, K.-H. Asymmetric synthesis of 3',4'-Di-O-(-)-camphanoyl-(+)-cis-khellactone (DCK), a potent anti-HIV agent. *Tetrahedron Lett.* **1995**, 36, 4529-4532.

ENANTIOSELECTIVE AZIRIDINATION OF C=C

Catalytic Asymmetric Aziridination of Alkenes

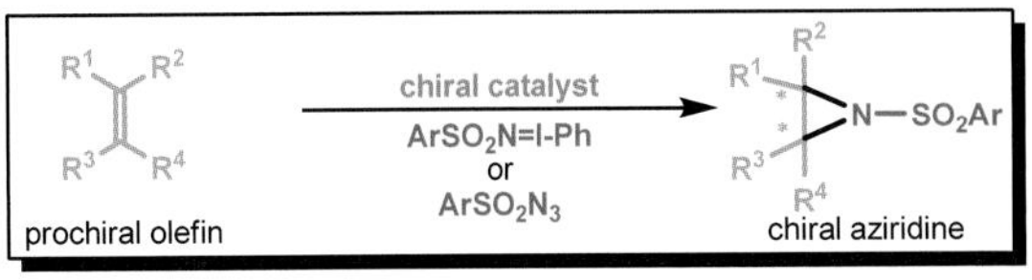

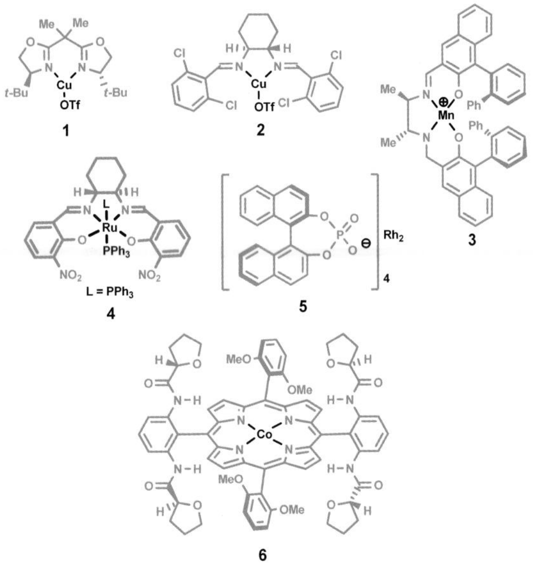

The catalytic aziridination of olefins[1] with fair to good enantioselectivities has been demonstrated with nitrene sources such as TsN=IPh, TsN3 and catalysts **1-6** (Figure 1).[2] The chiral catalysts which have been most effective are Mn(III)-complex **3**,[2b] Ru(II)complex **4**[2f] and Co(II)-catalyst **6**.[2g]

The reactive species in these processes is thought to be the corresponding metal-nitrene complex, $L_nM=NSO_2Ar$. Representative examples of enantioselective aziridination of olefins are shown in Scheme 1.

Although most of the catalyst development has been carried out with $ArSO_2$ = Ts, other useful N-substituents which are more readily removed are p-$NO_2C_6H_4SO_2$ and $Me_3SiCH_2CH_2SO_2$. The range of suitable olefinic substrates remains quite limited.

Chiral aziridines can also be synthesized by other methods[3] including: (1) addition of ethyl diazoacetate to imines catalyzed by either chiral Cu-bisoxazoline complexes[4] or chiral Lewis acids[5] and (2) by reaction of chiral sulfonium ylides with imines.[6]

Figure 1

Catalytic Asymmetric Aziridination of Alkenes

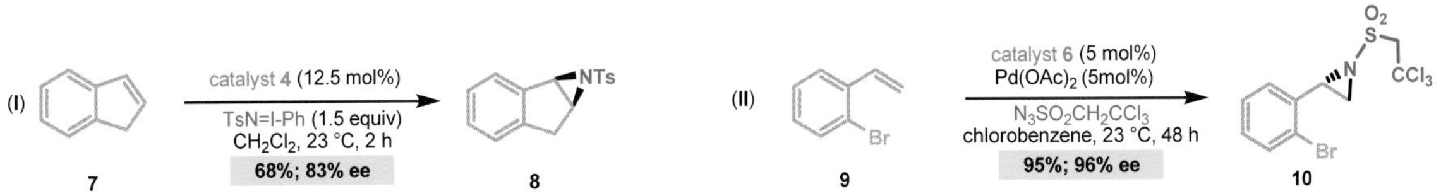

Scheme 1

References

1. For recent reviews on enantioselective aziridination, see: (a) Sweeney, J. B. Synthesis of aziridines in *Aziridines and Epoxides in Organic Synthesis* 117-144 (2006); (b) Mueller, P. & Fruit, C. Enantioselective Catalytic Aziridinations and Asymmetric Nitrene Insertions into CH Bonds. *Chem. Rev.* 2003, 103, 2905-2919; (c) Faul, M. M. & Evans, D. A. Asymmetric aziridination. *Asymmetric Oxid. React.* 2001, 115-128.
2. (a) Pirrung, M. C. & Zhang, J. Asymmetric dipolar cycloaddition reactions of diazo compounds mediated by a binaphtholphosphate rhodium catalyst. *Tetrahedron Lett.* 1992, 33, 5987-5990; (b) Noda, K., Hosoya, N., Irie, R., Ito, Y. & Katsuki, T. Asymmetric aziridination by using optically active (salen)manganese(III) complexes. *Synlett* 1993, 469-471; (c) Li, Z., Conser, K. R. & Jacobsen, E. N. Asymmetric alkene aziridination with readily available chiral diimine-based catalysts. *J. Am. Chem. Soc.* 1993, 115, 5326-5327; (d) Evans, D. A., Faul, M. M., Bilodeau, M. T., Anderson, B. A. & Barnes, D. M. Bis(oxazoline)-copper complexes as chiral catalysts for the enantioselective aziridination of olefins. *J. Am. Chem. Soc.* 1993, 115, 5328-5329; (e) Bennani, Y. L. & Hanessian, S. trans-1,2-Diaminocyclohexane Derivatives As Chiral Reagents, Scaffolds, and Ligands for Catalysis: Applications in Asymmetric Synthesis and Molecular Recognition. *Chem. Rev.* 1997, 97, 3161-3195; (f) Liang, J.-L., Yu, X.-Q. & Che, C.-M. Amidation of silyl enol ethers and cholesteryl acetates with chiral ruthenium(ii) Schiff-base catalysts: catalytic and enantioselective studies. *Chem. Commun. (Cambridge, U. K.)* 2002, 124-125; (g) Subbarayan, V., Ruppel, J. V., Zhu, S., Perman, J. A. & Zhang, X. P. Highly asymmetric cobalt-catalyzed aziridination of alkenes with trichloroethoxysulfonyl azide (TcesN3). *Chem. Commun. (Cambridge, U. K.)* 2009, 4266-4268.
3. For a review, see: Aggarwal, V. K., Badine, M. D. & Moorthie, V. A. Asymmetric Synthesis of Epoxides and Aziridines from Aldehydes and Imines in *Aziridines and Epoxides in Organic Synthesis* 1-35 (2006).
4. Hansen, K. B., Finney, N. S. & Jacobsen, E. N. Carbenoid transfer to imines: a new asymmetric catalytic synthesis of aziridines. *Angew. Chem., Int. Ed. Engl.* 1995, 34, 676-678.
5. Antilla, J. C. & Wulff, W. D. Catalytic asymmetric aziridination with arylborate catalysts derived from VAPOL and VANOL ligands. *Angew. Chem., Int. Ed.* 2000, 39, 4518-4521.
6. Aggarwal, V. K., Ferrara, M., O'Brien, C. J., Thompson, A., Jones, R. V. H. & Fieldhouse, R. Scope and limitations in sulfur ylide mediated catalytic asymmetric aziridination of imines: use of phenyldiazomethane, diazo esters and diazoacetamides. *J. Chem. Soc., Perkin Trans. 1* 2001, 1635-1643.

ENANTIOSELECTIVE ADDITION OF H–Y TO C=C

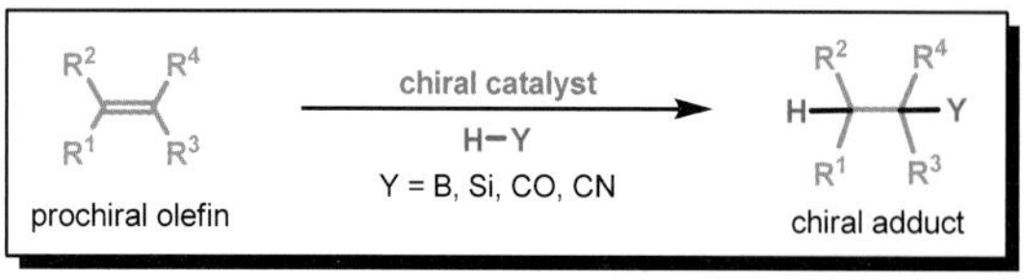

prochiral olefin → chiral adduct (Y = B, Si, CO, CN)

I. Addition of H-B to C=C

Since H.C. Brown's early work on the π-face-selective addition of the chiral borane diisopinocampheylborane (Ipc$_2$BH, **1**) to olefins[1] (for example **2**→**3**, Scheme 1), the area of enantioselective hydroboration has progressed to catalytic versions involving an achiral borane (e.g., catecholborane) and a chiral Rh(I)-phosphine complex.[2]

1. Ipc$_2$BH (**1**)
2. H$_2$O$_2$, OH$^-$
>90%; 89% ee

Scheme 1

The following examples using this approach, which seems to be limited to fairly simple olefins, indicate the current state of this methodology (Scheme 2).

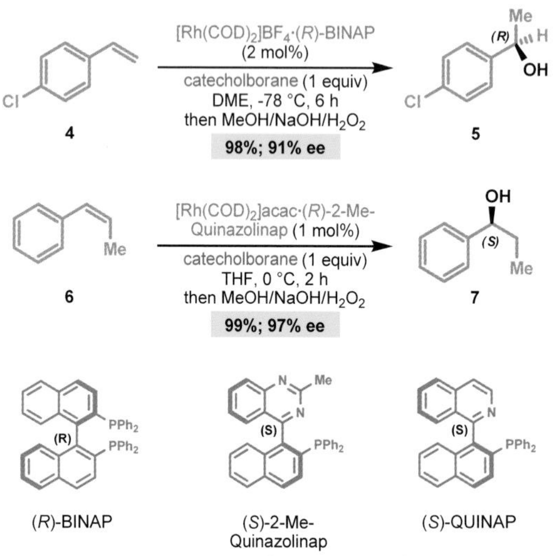

[Rh(COD)$_2$]BF$_4$·(R)-BINAP (2 mol%)
catecholborane (1 equiv)
DME, -78 °C, 6 h
then MeOH/NaOH/H$_2$O$_2$
98%; 91% ee

[Rh(COD)$_2$]acac·(R)-2-Me-Quinazolinap (1 mol%)
catecholborane (1 equiv)
THF, 0 °C, 2 h
then MeOH/NaOH/H$_2$O$_2$
99%; 97% ee

(R)-BINAP   (S)-2-Me-Quinazolinap   (S)-QUINAP

Scheme 2

Enantioselective hydroboration of C=C also provides a pathway for the synthesis of chiral amines through reactions which replace B by N. A representative example of this transformation (**8**→**9**) is shown in Scheme 3.[3]

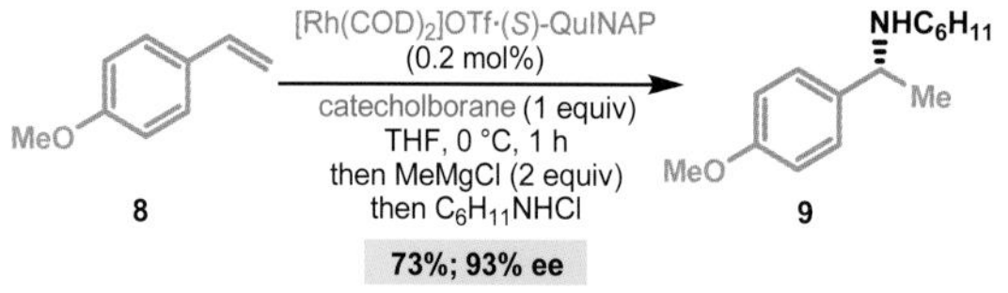

[Rh(COD)$_2$]OTf·(S)-QuiNAP (0.2 mol%)
catecholborane (1 equiv)
THF, 0 °C, 1 h
then MeMgCl (2 equiv)
then C$_6$H$_{11}$NHCl
73%; 93% ee

Scheme 3

II. Addition of H-Si to C=C

Enantioselective hydrosilylation of C=C has been realized with just a limited range of olefins – principally reactive substrates such as norbornene, cyclopentadiene and various styrenes.[4] Three examples involving Pd-catalysis are shown In Scheme 4.[5]

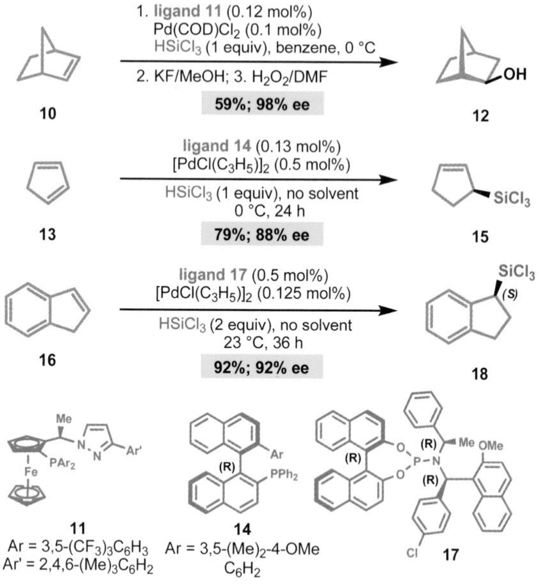

1. ligand **11** (0.12 mol%)
Pd(COD)Cl$_2$ (0.1 mol%)
HSiCl$_3$ (1 equiv), benzene, 0 °C
2. KF/MeOH; 3. H$_2$O$_2$/DMF
59%; 98% ee

ligand **14** (0.13 mol%)
[PdCl(C$_3$H$_5$)]$_2$ (0.5 mol%)
HSiCl$_3$ (1 equiv), no solvent
0 °C, 24 h
79%; 88% ee

ligand **17** (0.5 mol%)
[PdCl(C$_3$H$_5$)]$_2$ (0.125 mol%)
HSiCl$_3$ (2 equiv), no solvent
23 °C, 36 h
92%; 92% ee

11
Ar = 3,5-(CF$_3$)$_3$C$_6$H$_3$
Ar' = 2,4,6-(Me)$_3$C$_6$H$_2$

14 Ar = 3,5-(Me)$_2$-4-OMe C$_6$H$_2$

17

Scheme 4

III. Addition of H/CHO to C=C

The most efficient enantioselective hydroformylation[6] processes have involved the use of H$_2$ and CO (at elevated pressures) with terminal olefins and chiral Rh(I)-bis-phosphine complexes as catalysts. Chiral Rh(I)-bis-phosphite complexes have also been studied, but generally show diminished enantioselectivity. Two exceptionally enantioselective hydroformylation reactions are shown in Scheme 5.[7]

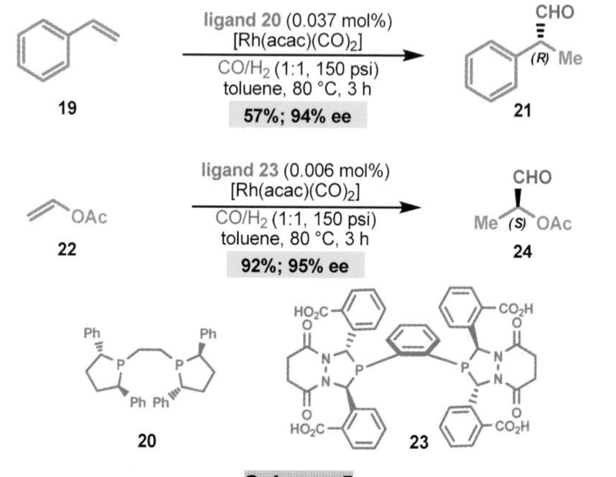

ligand **20** (0.037 mol%)
[Rh(acac)(CO)$_2$]
CO/H$_2$ (1:1, 150 psi)
toluene, 80 °C, 3 h
57%; 94% ee

ligand **23** (0.006 mol%)
[Rh(acac)(CO)$_2$]
CO/H$_2$ (1:1, 150 psi)
toluene, 80 °C, 3 h
92%; 95% ee

20   **23**

Scheme 5

ENANTIOSELECTIVE ADDITION OF H–Y TO C=C

IV. Addition of HCN to C=C

Enantioselective hydrocyanation[8] provides another pathway for adding a single carbon to C=C. In its present form this process has been only moderately enantioselective and also limited to conjugated vinyl groups. Three reprsentative examples are shown in Scheme 6.[9]

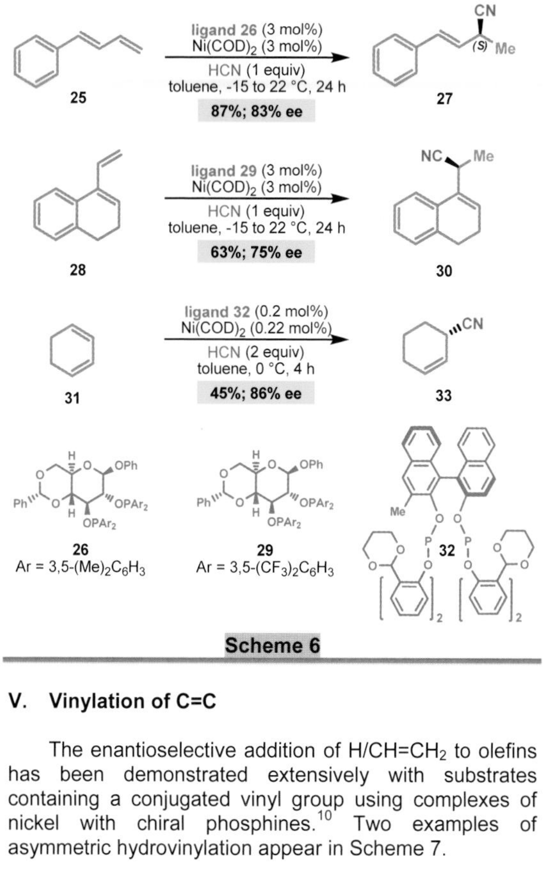

Scheme 6

V. Vinylation of C=C

The enantioselective addition of H/CH=CH$_2$ to olefins has been demonstrated extensively with substrates containing a conjugated vinyl group using complexes of nickel with chiral phosphines.[10] Two examples of asymmetric hydrovinylation appear in Scheme 7.

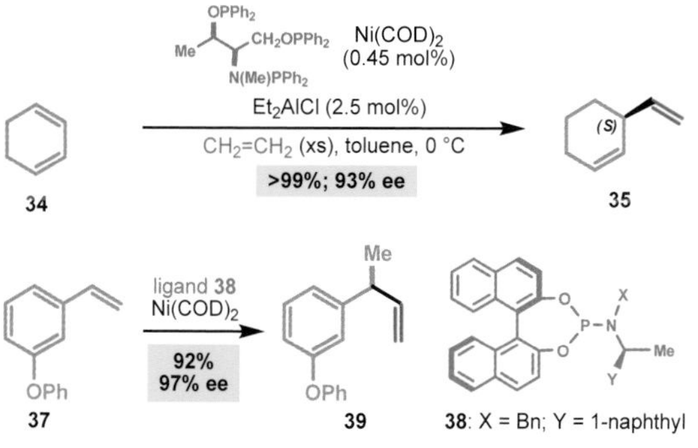

Scheme 7

References

1. For a seminal paper on olefin hydroboration/oxidation, see: Brown, H. C. & Zweifel, G. Hydroboration as a convenient procedure for the

asymmetric synthesis of alcohols of high optical purity. *J. Am. Chem. Soc.* **1961**, 83, 486-487; for a review, see: Brown, H. C., Jadhav, P. K. & Mandal, A. K. Asymmetric syntheses via chiral organoborane reagents. *Tetrahedron* **1981**, 37, 3547-3587.

2. For recent reviews on the catalytic enantioselective hydroboration of olefins, see: (a) Coyne, A. G. & Guiry, P. J. The development and application of rhodium-catalyzed hydroboration of alkenes in *Modern Reduction Methods* 65-86 (**2008**); (b) 1. Carroll, A.-M., O'Sullivan, T. P. & Guiry, P. J. The development of enantioselective Rhodium-catalyzed hydroboration of olefins. *Adv. Synth. Catal.* **2005**, 347, 609-631; (c) Brown, J. M. Rhodium-catalyzed hydroborations and related reactions. *Modern Rhodium-Catalyzed Organic Reactions* **2005**, 33-53; (d) Vogels, C. M. & Westcott, S. A. Recent advances in organic synthesis using transition metal-catalyzed hydroborations. *Curr. Org. Chem.* **2005**, 9, 687-699; (e) Hayashi, T. Hydroboration of carbon-carbon double bonds. *Compr. Asymmetric Catal. I-III*, 1, 351-364 (**1999**).

3. Fernandez, E., Maeda, K., Hooper, M. W. & Brown, J. M. Catalytic asymmetric hydroboration/amination and alkylamination with rhodium complexes of 1,1'-(2-diarylphosphino-1-naphthyl)isoquinoline. *Chem.-Eur. J.* **2000**, 6, 1840-1846.

4. For reviews on the hydrosilylation of olefins, see: (a) Mayes, P. A. & Perlmutter, P. Alkene reduction: hydrosilylation in *Modern Reduction Methods* 87-105 (**2008**); (b) Gibson, S. E. & Rudd, M. The role of secondary interactions in the asymmetric palladium-catalyzed hydrosilylation of olefins with monophosphine ligands. *Adv. Synth. Catal.* **2007**, 349, 781-795; (c) Nishiyama, H. & Itoh, K. Asymmetric hydrosilylation and related reactions in *Catal. Asymmetric Synth. (2nd Ed.)* 111-143 (**2000**); (d) Hayashi, T. Hydrosilylation of carbon-carbon double bonds. *Compr. Asymmetric Catal. I-III*, 1, 319-333 (**1999**).

5. (a) Pioda, G. & Togni, A. Highly enantioselective palladium-catalyzed hydrosilylation of norbornene with trichlorosilane using ferrocenyl ligands. *Tetrahedron Asymmetry* **1998**, 9, 3903-3910; (b) Hayashi, T., Han, J. W., Takeda, A., Tang, J., Nohmi, K., Mukaide, K., Tsuji, H. & Uozumi, Y. Modification of chiral monodentate phosphine ligands (MOP) for palladium-catalyzed asymmetric hydrosilylation of cyclic 1,3-dienes. *Adv. Synth. Catal.* **2001**, 343, 279-283; (c) Li, X., Song, J., Xu, D. & Kong, L. Asymmetric hydrosilylation of styrenes by use of new chiral phosphoramidites. *Synthesis* **2008**, 925-931.

6. For selected reviews on the enantioselective hydroformylation of olefins, see: (a) Claver, C., Godard, C., Ruiz, A., Pamies, O. & Dieguez, M. Enantioselective carbonylation reactions in *Modern Carbonylation Methods* 65-92 (**2008**); (b) Nozaki, K. Activation of small molecules (CO, HCN, RNC, and CO$_2$) in *New Frontiers in Asymmetric Catalysis* 101-127 (**2007**); (c) Klosin, J. & Landis, C. R. Ligands for Practical Rhodium-Catalyzed Asymmetric Hydroformylation. *Acc. Chem. Res.* **2007**, 40, 1251-1259; (d) Claver, C., Dieguez, M., Pamies, O. & Castillon, S. Asymmetric hydroformylation. *Top. Organomet. Chem.* **2006**, 18, 35-64; (e) Dieguez, M., Pamies, O. & Claver, C. Recent advances in Rh-catalyzed asymmetric hydroformylation using phosphite ligands. *Tetrahedron Asymmetry* **2004**, 15, 2113-2122; (f) Breit, B. Synthetic aspects of stereoselective hydroformylation. *Acc. Chem. Res.* **2003**, 36, 264-275.

7. Axtell, A. T., Cobley, C. J., Klosin, J., Whiteker, G. T., Zanotti-Gerosa, A. & Abboud, K. A. Highly regio- and enantioselective asymmetric hydroformylation of olefins mediated by 2,5-disubstituted phospholane ligands. *Angew. Chem., Int. Ed.* **2005**, 44, 5834-5838.

8. For selected reviews on the enantioselective hydrocyanation of olefins, see: (a) Nozaki, K. Activation of small molecules (CO, HCN, RNC, and CO$_2$) in *New Frontiers in Asymmetric Catalysis* 101-127 (**2007**); (b) Wilting, J. & Vogt, D. Asymmetric hydrocyanation of alkenes in *Handbook of C-H Transformations* 87-96 (**2005**); (c) RajanBabu, T. V. & Casalnuovo, A. L. Hydrocyanation of carbon-carbon double bonds. *Compr. Asymmetric Catal. I-III*, 1, 367-378 (**1999**).

9. (a) Saha, B. & RajanBabu, T. V. Nickel(0)-Catalyzed Asymmetric Hydrocyanation of 1,3-Dienes. *Org. Lett.* **2006**, 8, 4657-4659; (b) Wilting, J., Janssen, M., Mueller, C., Lutz, M., Spek, A. L. & Vogt, D. Binaphthol-based diphosphite ligands in asymmetric nickel-catalyzed hydrocyanation of styrene and 1,3-cyclohexadiene: influence of steric properties. *Adv. Synth. Catal.* **2007**, 349, 350-356.

10. For reviews on the enantioselective hydrovinylation of olefins, see: (a) RajanBabu, T. V. In pursuit of an ideal carbon-carbon bond-forming reaction: Development and applications of the hydrovinylation of olefins. *Synlett* **2009**, 853-885; (b) RajanBabu, T. V. Asymmetric Hydrovinylation Reaction. *Chem. Rev.* **2003**, 103, 2845-2860; (c) Goossen, L. J. Asymmetric hydrovinylation: new perspectives through use of modular ligand systems. *Angew. Chem., Int. Ed.* **2002**, 41, 3775-3778.

NUCLEOPHILIC ADDIT. OF HETEROATOMS TO C=C

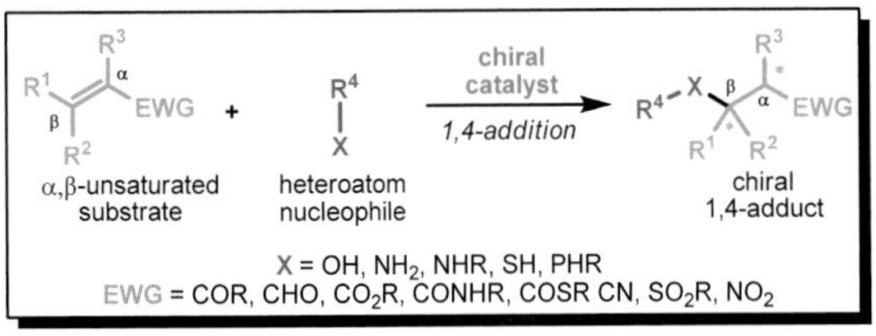

α,β-unsaturated substrate + heteroatom nucleophile X → chiral 1,4-adduct

X = OH, NH$_2$, NHR, SH, PHR
EWG = COR, CHO, CO$_2$R, CONHR, COSR CN, SO$_2$R, NO$_2$

Introduction

The 1,4-addition of heteroatoms (N, O, S, P) to activated alkenes (aza-Michael reaction, oxa-Michael reaction, etc.) leads to synthetically valuable β-oxy-, β-amino- as well as other β-hetero-substituted carbonyl or nitro compounds.[1] The β-oxygenated carbonyl compounds can also be obtained using the aldol reaction, therefore the oxa-Michael reaction is a useful synthetic alternative to the aldol reaction. Enantioselective variants of the hetero-Michael reaction used to rely mainly on chiral controller groups or chiral reagents (e.g., lithium amides) to achieve asymmetric induction. However, highly enantioselective versions utilizing small organic molecules as catalysts have since been developed. The first catalytic and highly enantioselective examples of aza-Michael and oxa-Michael addition were reported by S. Miller[2] and T. Ishikawa[3], respectively (see Examples I & II in Scheme 1).

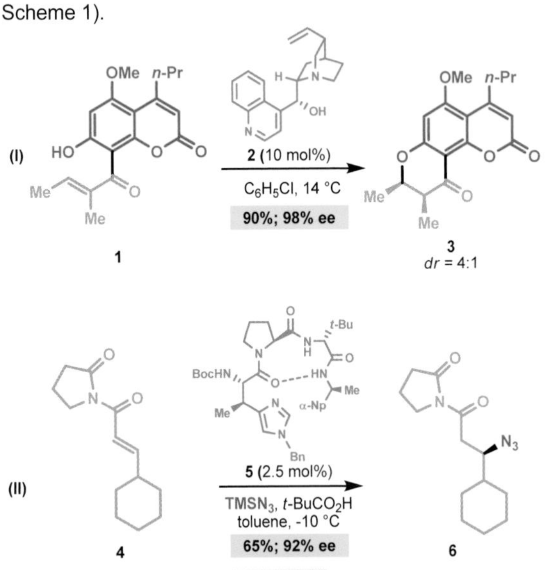

(I)

2 (10 mol%)
C$_6$H$_5$Cl, 14 °C
90%; 98% ee

(II)

5 (2.5 mol%)
TMSN$_3$, t-BuCO$_2$H
toluene, -10 °C
65%; 92% ee

Scheme 1

I. Catalytic Enantioselective Conjugate Addition of Nitrogen Nucleophiles (Aza-Michael Addition)

A selection of efficient catalytic enantioselective aza-Michael reactions utilizing a wide variety of nitrogen nucleophiles (e.g., hydroxylamine derivatives, imides, azides, benzotriazoles, tetrazoles, imidazoles, anilines) and Michael acceptors are shown in Scheme 3

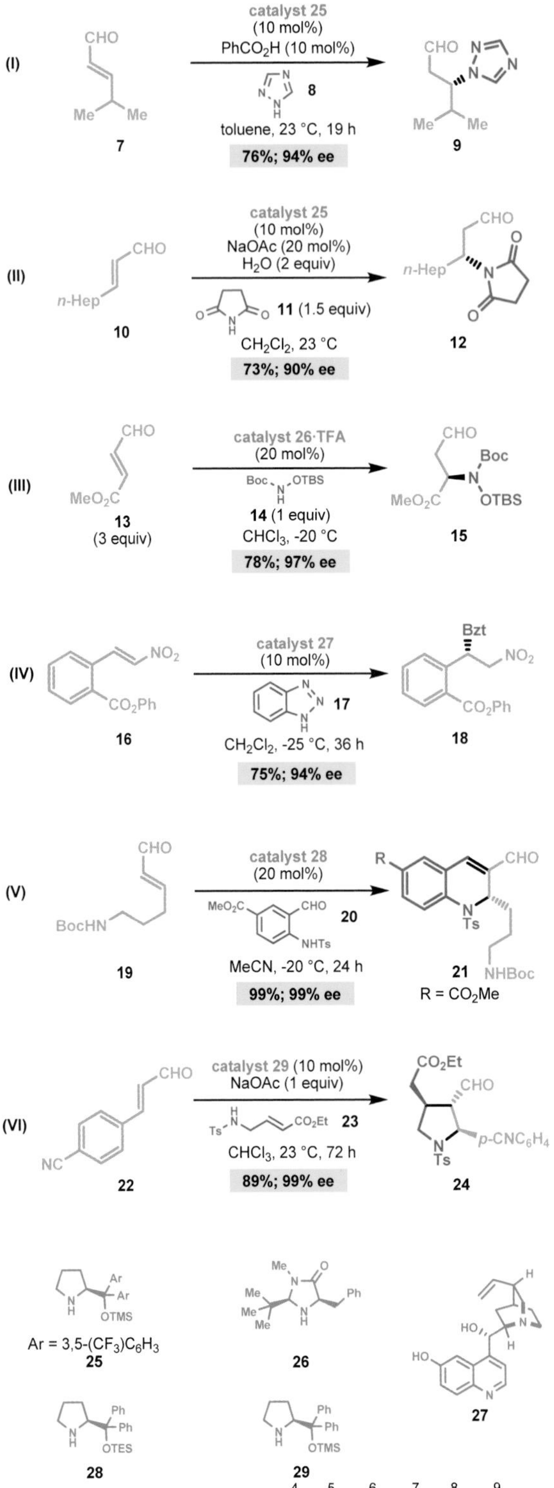

(I)

catalyst 25 (10 mol%)
PhCO$_2$H (10 mol%)
8
toluene, 23 °C, 19 h
76%; 94% ee

(II)

catalyst 25 (10 mol%)
NaOAc (20 mol%)
H$_2$O (2 equiv)
11 (1.5 equiv)
CH$_2$Cl$_2$, 23 °C
73%; 90% ee

(III)

catalyst 26·TFA (20 mol%)
14 (1 equiv)
CHCl$_3$, -20 °C
78%; 97% ee

(IV)

catalyst 27 (10 mol%)
17
CH$_2$Cl$_2$, -25 °C, 36 h
75%; 94% ee

(V)

catalyst 28 (20 mol%)
20
MeCN, -20 °C, 24 h
99%; 99% ee

(VI)

catalyst 29 (10 mol%)
NaOAc (1 equiv)
23
CHCl$_3$, 23 °C, 72 h
89%; 99% ee

Ar = 3,5-(CF$_3$)C$_6$H$_3$
25

26

27

28

29

Reference key for equations: (I)[4]; (II)[5]; (III)[6]; (IV)[7]; (V)[8]; (VI)[9]

Scheme 3

NUCLEOPHILIC ADDIT. OF HETEROATOMS TO C=C

II. Catalytic Enantioselective Conjugate Addition of Oxygen, Sulfur and Phosphorous Nucleophiles (Oxa-, Sulfa- and Phospha-Michael Addition)

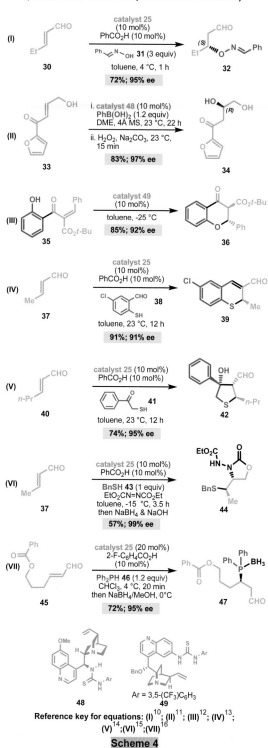

Reference key for equations: (I)[10]; (II)[11]; (III)[12]; (IV)[13]; (V)[14];(VI)[15];(VII)[16]

Scheme 4

Examples III-VI in Scheme 4 highlight the power of the oxa- and sulfa-Michael addition reaction to create five- and six membered heterocycles using both enals and enones as Michael acceptors. Especially noteworthy is the multicomponent reaction (37→44, Example VI) in which highly functionalized chiral oxazolidinones are fomed in high yield and with high enantiomeric purity.

References

1. (a) Enders, D., Saint-Dizier, A., Lannou, M.-I. & Lenzen, A. The phospha-Michael addition in organic synthesis. *Eur. J. Org. Chem.* **2005**, 29-49; (b) Vicario, J. L., Badia, D., Carrillo, L., Etxebarria, J., Reyes, E. & Ruiz, N. The asymmetric aza-Michael reaction. A review. *Org. Prep. Proced. Int.* **2005**, 37, 513-538; (c) Enders, D., Luettgen, K. & Narine, A. A. Asymmetric sulfa-Michael additions. *Synthesis* **2007**, 959-980; (d) Nising, C. F. & Braese, S. The oxa-Michael reaction: from recent developments to applications in natural product synthesis. *Chem. Soc. Rev.* **2008**, 37, 1218-1228; (e) Krishna, P. R., Sreeshailam, A. & Srinivas, R. Recent advances and applications in asymmetric aza-Michael addition chemistry. *Tetrahedron* **2009**, 65, 9657-9672; (f) Enders, D., Wang, C. & Liebich, J. X. Organocatalytic Asymmetric Aza-Michael Additions. *Chem.--Eur. J.* **2009**, 15, 11058-11076.
2. Guerin, D. J. & Miller, S. J. Asymmetric Azidation-Cycloaddition with Open-Chain Peptide-Based Catalysts. A Sequential Enantioselective Route to Triazoles. *J. Am. Chem. Soc.* **2002**, 124, 2134-2136.
3. Tanaka, T., Kumamoto, T. & Ishikawa, T. Solvent effects on stereoselectivity in 2,3-dimethyl-4-chromanone cyclization by quinine-catalyzed asymmetric intramolecular oxo-Michael addition. *Tetrahedron Asymmetry* **2000**, 11, 4633-4637.
4. Diner, P., Nielsen, M., Marigo, M. & Joergensen, K. A. Enantioselective organocatalytic conjugate addition of N Heterocycles to α,β-unsaturated aldehydes. *Angew. Chem., Int. Ed.* **2007**, 46, 1983-1987.
5. Jiang, H., Nielsen, J. B., Nielsen, M. & Joergensen, K. A. Organocatalyzed asymmetric β-amination and multicomponent syn-selective diamination of α,β-unsaturated aldehydes. *Chem.--Eur. J.* **2007**, 13, 9068-9075.
6. Chen, Y. K., Yoshida, M. & MacMillan, D. W. C. Enantioselective Organocatalytic Amine Conjugate Addition. *J. Am. Chem. Soc.* **2006**, 128, 9328-9329.
7. Wang, J., Li, H., Zu, L. & Wang, W. Enantioselective Organocatalytic Michael Addition Reactions between *N*-Heterocycles and Nitroolefins. *Org. Lett.* **2006**, 8, 1391-1394.
8. Yoshitomi, Y., Arai, H., Makino, K. & Hamada, Y. Enantioselective synthesis of martinelline chiral amine and its diastereomer using asymmetric tandem Michael-aldol reaction. *Tetrahedron* **2008**, 64, 11568-11579.
9. Li, H., Zu, L., Xie, H., Wang, J. & Wang, W. Highly enantio- and diastereoselective organocatalytic cascade aza-Michael-Michael reactions: a direct method for the synthesis of trisubstituted chiral pyrrolidines. *Chem. Commun. (Cambridge, U. K.)* **2008**, 5636-5638.
10. Bertelsen, S., Diner, P., Johansen, R. L. & Jorgensen, K. A. Asymmetric Organocatalytic β-Hydroxylation of α,β-Unsaturated Aldehydes. *J. Am. Chem. Soc.* **2007**, 129, 1536-1537.
11. Li, D. R., Murugan, A. & Falck, J. R. Enantioselective, Organocatalytic Oxy-Michael Addition to γ/δ-Hydroxy-α,β-enones: Boronate-Amine Complexes as Chiral Hydroxide Synthons. *J. Am. Chem. Soc.* **2008**, 130, 46-48.
12. Biddle, M. M., Lin, M. & Scheidt, K. A. Catalytic Enantioselective Synthesis of Flavanones and Chromanones. *J. Am. Chem. Soc.* **2007**, 129, 3830-3831.
13. Wang, W., Li, H., Wang, J. & Zu, L. Enantioselective Organocatalytic Tandem Michael-Aldol Reactions: One-Pot Synthesis of Chiral Thiochromenes. *J. Am. Chem. Soc.* **2006**, 128, 10354-10355.
14. Brandau, S., Maerten, E. & Jorgensen, K. A. Asymmetric synthesis of highly functionalized tetrahydrothiophenes by organocatalytic domino reactions. *J. Am. Chem. Soc.* **2006**, 128, 14986-14991.
15. Marigo, M., Schulte, T., Franzen, J. & Jorgensen, K. A. Asymmetric multicomponent domino reactions and highly enantioselective conjugated addition of thiols to α,β-unsaturated aldehydes. *J. Am. Chem. Soc.* **2005**, 127, 15710-15711.
16. Ibrahem, I., Rios, R., Vesely, J., Hammar, P., Eriksson, L., Himo, F. & Cordova, A. Enantioselective organocatalytic hydrophosphination of α,β-unsaturated aldehydes. *Angew. Chem., Int. Ed.* **2007**, 46, 4507-4510.

α-FUNCTIONALIZATION OF CARBONYL COMPOUNDS

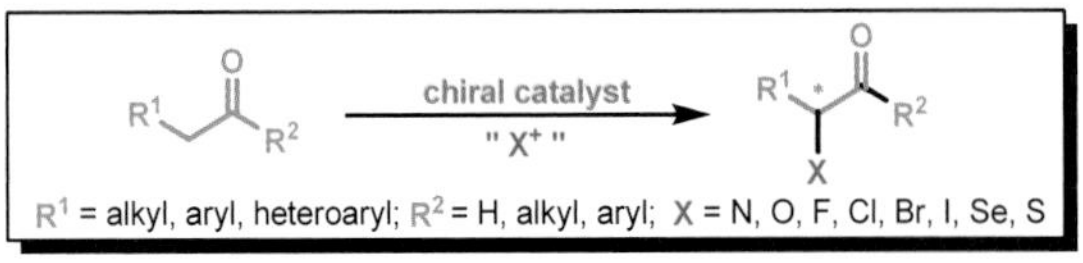

$R^1 \underset{R^2}{\overset{O}{\diagup}}$ →(chiral catalyst, "X⁺")→ $R^1 \overset{O}{\underset{X}{\diagup}} R^2$

R¹ = alkyl, aryl, heteroaryl; R² = H, alkyl, aryl; X = N, O, F, Cl, Br, I, Se, S

Introduction

The enantioselective replacement of hydrogen alpha to carbonyl by oxygen, nitrogen, halogen or sulfur has been demonstrated using as catalyst a variety of chiral bases, most commonly secondary amines which can generate chiral enamine intermediates.[1] A variety of such α-C-H replacements are exemplified in Schemes 1-4.

I. Catalytic Enantioselective α-Amination of Aldehydes, Ketones and Carboxylic Acid Derivatives

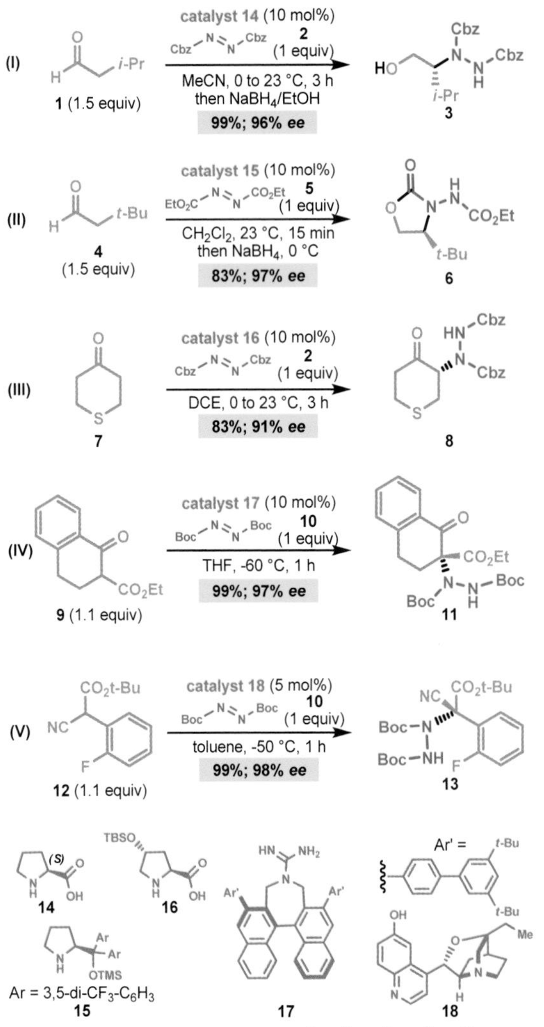

(I) **1** (1.5 equiv) — catalyst 14 (10 mol%), 2 (1 equiv) Cbz-N=N-Cbz, MeCN, 0 to 23 °C, 3 h then NaBH₄/EtOH — **99%; 96% ee** — **3**

(II) **4** (1.5 equiv) — catalyst 15 (10 mol%), 5 (1 equiv) EtO₂C-N=N-CO₂Et, CH₂Cl₂, 23 °C, 15 min then NaBH₄, 0 °C — **83%; 97% ee** — **6**

(III) **7** — catalyst 16 (10 mol%), 2 (1 equiv) Cbz-N=N-Cbz, DCE, 0 to 23 °C, 3 h — **83%; 91% ee** — **8**

(IV) **9** (1.1 equiv) — catalyst 17 (10 mol%), 10 (1 equiv) Boc-N=N-Boc, THF, -60 °C, 1 h — **99%; 97% ee** — **11**

(V) **12** (1.1 equiv) — catalyst 18 (5 mol%), 10 (1 equiv) Boc-N=N-Boc, toluene, -50 °C, 1 h — **99%; 98% ee** — **13**

14 (TBSO, (S)); 16; 15 (Ar = 3,5-di-CF₃-C₆H₃); 17; 18 (Ar' = t-Bu)

Reference key for equations: (I)²; (II)³; (III)⁴; (IV)⁵; (V)⁶

Scheme 1

II. Catalytic Enantioselective α-Oxygenation of Aldehydes and Ketones

(I) **19** (1.2 equiv) — catalyst 14 (10 mol%), Ph-N=O **20** (1 equiv), DMSO, 23 °C, 10 min then NaBH₄/EtOH — **80%; 99% ee** — **21**

(II) **22** (2 equiv) — catalyst 16 (10 mol%), Ph-N=O **20** (1 equiv, slow addit.), DMF, 0 °C, 2 h — **45%; 99% ee** — **23**

(III) **24** (1 equiv) — catalyst 31 (20 mol%), 25 (1.5 equiv) Bz-O-O-Bz, THF, 23 °C, 20 h — **72%; 94% ee** — **26**

(IV) **27** — catalyst 32 (20 mol%) tetraphenylporphyrine (TPP, 1 mol%), O₂, UV light, DMSO, 23 °C, 30 min — **67%; 72% ee** — **28**

(singlet oxygen incorporation)

(IV) **29** — catalyst 33 (20 mol%) tetraphenylporphyrine (TPP, 1 mol%), O₂, UV light, CHCl₃, 0 °C, 6 h then NaBH₄, MeOH — **68%; 98% ee** — **30**

(singlet oxygen incorporation)

31; 32; 33; 34 (· DCA)

Reference key for equations: (I)⁷; (II)⁸; (III)⁹; (IV)¹⁰; (V)¹¹

Scheme 2

III. Catalytic Enantioselective α-Halogenation of Aldehydes and Ketones[1a-c,1h,1j,1k]

(I) **35** (1.5 equiv) — catalyst 15 (1 mol%), PhO₂S-N(F)-SO₂Ph **36** (1 equiv), MTBE, 23 °C, 2 h then NaBH₄/MeOH — **95%; 96% ee** — **37**

(II) **38** (1.5 equiv) — catalyst 34 (20 mol%), PhO₂S-N(F)-SO₂Ph **36** (1 equiv), THF: i-PrOH (9:1), 10 °C, 10 h then NaBH₄ — **85%; 98% ee** — **39**

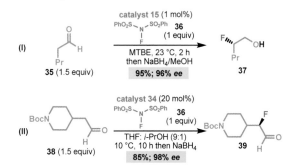

Reference key for equations: (I)¹²; (II)¹³

Scheme 3

α-FUNCTIONALIZATION OF CARBONYL COMPOUNDS

Catalytic Enantioselective α-Halogenation of Aldehydes and Ketones (Continued)

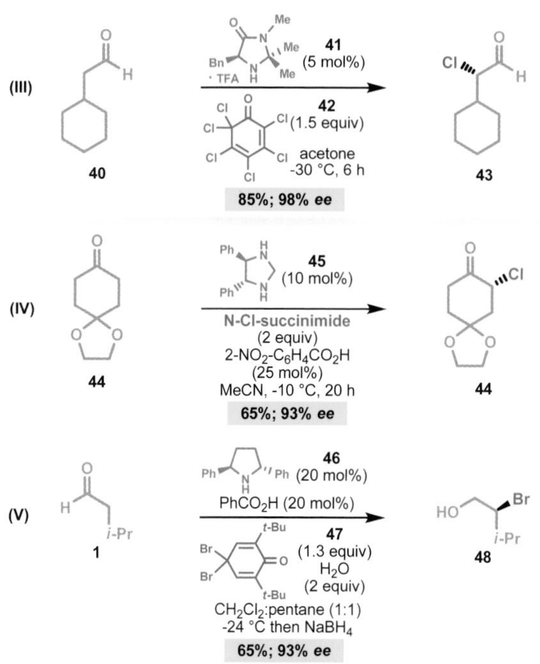

(III)

85%; 98% ee

(IV)

65%; 93% ee

(V)

65%; 93% ee

Reference key for equations: (I)[14]; (II)[15]; (III)[16]

Scheme 3

IV. Catalytic Enantioselective α-Sulfenylation of Aldehydes and β-Dicarbonyl Compounds

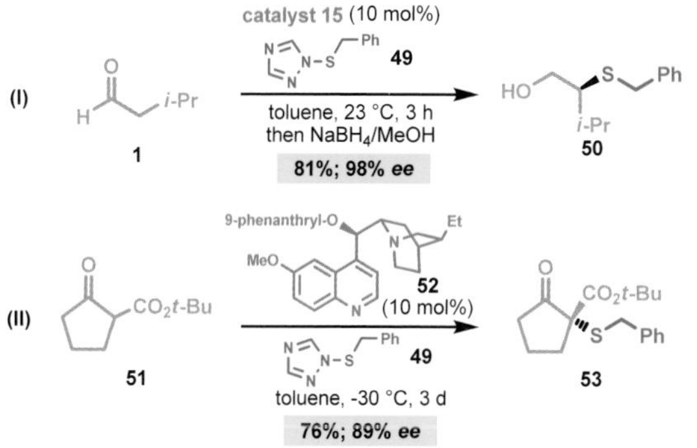

(I)

catalyst 15 (10 mol%)

toluene, 23 °C, 3 h
then NaBH₄/MeOH

81%; 98% ee

(II)

toluene, -30 °C, 3 d

76%; 89% ee

Reference key for equations: (I)[17]; (II)[18]

Scheme 4

References

1. (a) Ma, J.-A. & Cahard, D. Asymmetric Fluorination, Trifluoromethylation, and Perfluoroalkylation Reactions. *Chem. Rev. (Washington, DC, U. S.)* **2004**, 104, 6119-6146; (b) France, S., Weatherwax, A. & Lectka, T. Recent developments in catalytic, asymmetric α-halogenation: a new frontier in asymmetric catalysis. *Eur. J. Org. Chem.* **2005**, 475-479; (c) Oestreich, M. Strategies for catalytic asymmetric electrophilic α-halogenation of carbonyl

compounds. *Angew. Chem., Int. Ed.* **2005**, 44, 2324-2327; (d) Guillena, G. & Ramon, D. J. Enantioselective α-heterofunctionalization of carbonyl compounds: organocatalysis is the simplest approach. *Tetrahedron Asymmetry* **2006**, 17, 1465-1492; (e) Marigo, M. & Jorgensen, K. A. Organocatalytic direct asymmetric α-heteroatom functionalization of aldehydes and ketones. *Chem. Commun. (Cambridge, U. K.)* **2006**, 2001-2011; (f) Bobbio, C. & Gouverneur, V. Catalytic asymmetric fluorinations. *Org. Biomol. Chem.* **2006**, 4, 2065-2075; (g) Marigo, M. & Joergensen, K. A. α-Heteroatom functionalization in *Enantioselective Organocatalysis* 56-76 (**2007**); (h) Shibata, N., Ishimaru, T., Nakamura, S. & Toru, T. New approaches to enantioselective fluorination: Cinchona alkaloids combinations and chiral ligands/metal complexes. *J. Fluorine Chem.* **2007**, 128, 469-483; (i) Yamamoto, H. & Kawasaki, M. Nitroso and azo compounds in modern organic synthesis: late blooming but very rich. *Bull. Chem. Soc. Jpn.* **2007**, 80, 595-607; (j) Ueda, M., Kano, T. & Maruoka, K. Organocatalyzed direct asymmetric α-halogenation of carbonyl compounds. *Org. Biomol. Chem.* **2009**, 7, 2005-2012; (k) Kang, Y. K. & Kim, D. Y. Recent advances in catalytic enantioselective fluorination of active methines. *Curr. Org. Chem.* **2010**, 14, 917-927; (l) Vilaivan, T. & Bhanthumnavin, W. Organocatalyzed asymmetric α-oxidation, α-aminoxylation and α-amination of carbonyl compounds. *Molecules* **2010**, 15, 917-958.

2. List, B. Direct Catalytic Asymmetric α-Amination of Aldehydes. *J. Am. Chem. Soc.* **2002**, 124, 5656-5657.

3. Franzen, J., Marigo, M., Fielenbach, D., Wabnitz, T. C., Kjrsgaard, A. & Jorgensen, K. A. A General Organocatalyst for Direct α-Functionalization of Aldehydes: Stereoselective C-C, C-N, C-F, C-Br, and C-S Bond-Forming Reactions. *J. Am. Chem. Soc.* **2005**, 127, 18296-18304.

4. Hayashi, Y., Aratake, S., Imai, Y., Hibino, K., Chen, Q.-Y., Yamaguchi, J. & Uchimaru, T. Direct asymmetric α-amination of cyclic ketones catalyzed by siloxyproline. *Chemistry--An Asian Journal* **2008**, 3, 225-232.

5. Terada, M., Nakano, M. & Ube, H. Axially Chiral Guanidine as Highly Active and Enantioselective Catalyst for Electrophilic Amination of Unsymmetrically Substituted 1,3-Dicarbonyl Compounds. *J. Am. Chem. Soc.* **2006**, 128, 16044-16045.

6. Saaby, S., Bella, M. & Jorgensen, K. A. Asymmetric construction of quaternary stereocenters by direct organocatalytic amination of α-substituted α-cyanoacetates and β-dicarbonyl compounds. *J. Am. Chem. Soc.* **2004**, 126, 8120-8121.

7. Zhong, G. A facile and rapid route to highly enantiopure 1,2-diols by novel catalytic asymmetric α-aminoxylation of aldehydes. *Angew. Chem., Int. Ed.* **2003**, 42, 4247-4250.

8. Hayashi, Y., Yamaguchi, J., Hibino, K., Sumiya, T., Urushima, T., Shoji, M., Hashizume, D. & Koshino, H. A highly active 4-siloxyproline catalyst for asymmetric synthesis. *Adv. Synth. Catal.* **2004**, 346, 1435-1439.

9. Gotoh, H. & Hayashi, Y. Diphenylprolinol silyl ether as a catalyst in an asymmetric, catalytic and direct α-benzoyloxylation of aldehydes. *Chem. Commun. (Cambridge, U. K.)* **2009**, 3083-3085.

10. Sunden, H., Engqvist, M., Casas, J., Ibrahem, I. & Cordova, A. Direct amino acid-catalyzed asymmetric α-oxidation of ketones with molecular oxygen. *Angew. Chem., Int. Ed.* **2004**, 43, 6532-6535.

11. Ibrahem, I., Zhao, G.-L., Sunden, H. & Cordova, A. A route to 1,2-diols by enantioselective organocatalytic α-oxidation with molecular oxygen. *Tetrahedron Lett.* **2006**, 47, 4659-4663.

12. Marigo, M., Fielenbach, D., Braunton, A., Kjoersgaard, A. & Jorgensen, K. A. Enantioselective formation of stereogenic carbon-fluorine centers by a simple catalytic method. *Angew. Chem., Int. Ed.* **2005**, 44, 3703-3706.

13. Beeson, T. D. & MacMillan, D. W. C. Enantioselective Organocatalytic α-Fluorination of Aldehydes. *J. Am. Chem. Soc.* **2005**, 127, 8826-8828.

14. Brochu, M. P., Brown, S. P. & MacMillan, D. W. C. Direct and Enantioselective Organocatalytic α-Chlorination of Aldehydes. *J. Am. Chem. Soc.* **2004**, 126, 4108-4109.

15. Marigo, M., Bachmann, S., Halland, N., Braunton, A. & Jorgensen, K. A. Highly enantioselective direct organocatalytic α-chlorination of ketones. *Angew. Chem., Int. Ed.* **2004**, 43, 5507-5510.

16. Bertelsen, S., Halland, N., Bachmann, S., Marigo, M., Braunton, A. & Jorgensen, K. A. Organocatalytic asymmetric α-bromination of aldehydes and ketones. *Chem. Commun. (Cambridge, U. K.)* **2005**, 4821-4823.

17. Marigo, M., Wabnitz, T. C., Fielenbach, D. & Jorgensen, K. A. Enantioselective organo-catalyzed α sulfenylation of aldehydes. *Angew. Chem., Int. Ed.* **2005**, 44, 794-797.

18. Sobhani, S., Fielenbach, D., Marigo, M., Wabnitz, T. C. & Jorgensen, K. A. Direct organocatalytic asymmetric α-sulfenylation of activated C-H bonds in lactones, lactams, and β-dicarbonyl compounds. *Chem.--Eur. J.* **2005**, 11, 5689-5694.

ENANTIOSELECTIVE α-ALKYLATION - PART I.

Introduction

Since the early 1970s a large number of removable chiral controller groups (X$_c$, chiral auxiliaries) have been developed for the synthesis of enantiomerically pure compounds utilizing various C-C bond-forming diastereoselective reactions, including the alkylation of enolates.[1] Figure 1 shows the structure of a number of chiral controllers (Evans[2], Meyers[3], Enders[4], Oppolzer[5], Schöllkopf[6] and Myers[7]) that have been widely applied. The most useful controllers share the following properties: (1) synthetically accessible; (2) provide high stereocontrol; (3) allow simple purification of principal reaction product; (4) readily removable without compromising stereocenters or functionality and (5) afford high yields of C-C coupling products.

Meyers

Evans

Enders

Myers

Schöllkopf

Oppolzer

The examples in Scheme 1 illustrate the use of various commonly utilized oxazolidinone controllers, perhaps the most widely used. In these examples the sense of diastereoselectivity can be rationalized in terms of a metal chelated oxazolidinone enolate which undergoes alkylation at the sterically less shielded face of the resulting bicyclic intermediate. Scheme 2 documents a variety of examples of stereocontrolled C-C bond formation using the camphorsultam controller of W. Oppolzer. The α-alkylation of enolates derived from *N*-acyl pseudoephedrine derivatives by a wide variety of alkyl halides is illustrated in Scheme 3. It is important to note that the enolates derived from *N*-acyl pseudoephedrine derivatives are significantly more nucleophilic than the enolates derived from *N*-acylated Evans and Oppolzer chiral controllers. Examples I-III in Scheme 3 demonstrate that the stereoselective construction of quaternary all-carbon stereocenters is possible using pseudoephedrine as the chiral controller group. The α-alkylation of carboxylic acid enolates utilizing miscellaneous chiral controller groups are shown in Scheme 4. In all of these cases the preferred alkylating species is a primary iodide. Less reactive halides (e.g., 2°) are rarely used.

I. Diastereoselective α-Alkylation of *N*-Acyl Oxazolidinone-Derived Enolates:

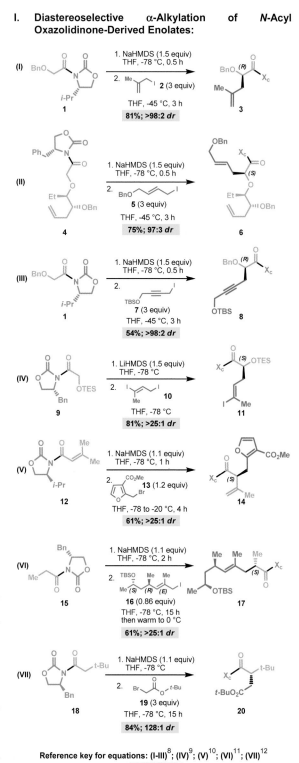

Reference key for equations: (I-III)[8]; (IV)[9]; (V)[10]; (VI)[11]; (VII)[12]

Scheme 1

ENANTIOSELECTIVE α-ALKYLATION - PART I.

II. Diastereoselective α-Alkylation of N-Acyl Camphorsultam-Derived Enolates:

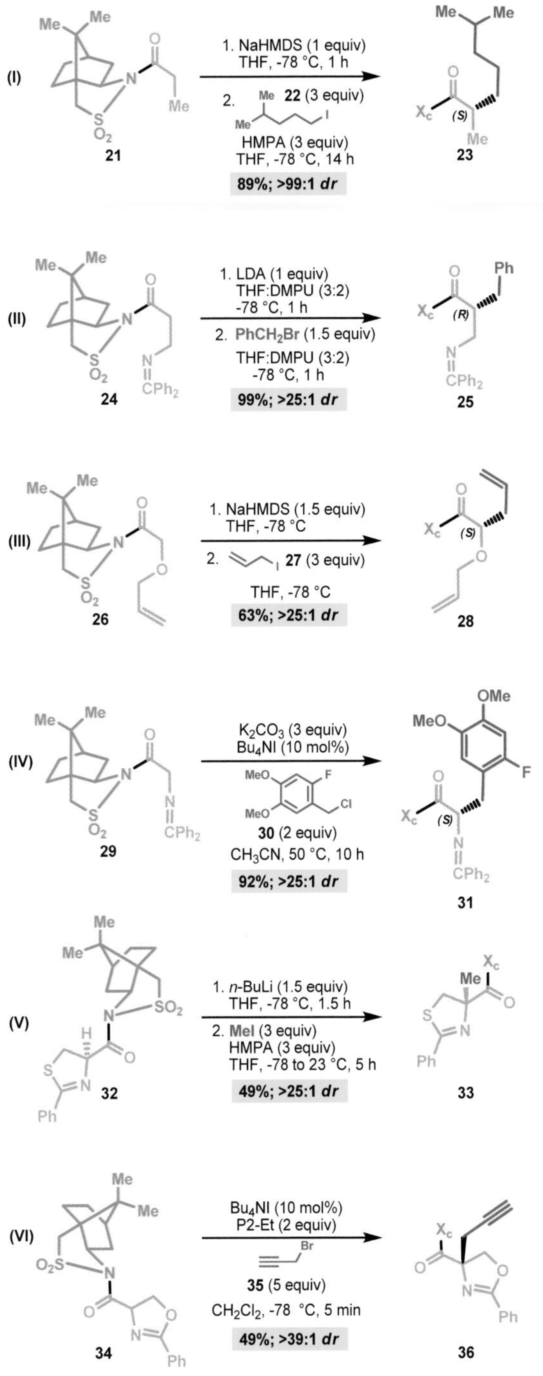

(I)

21

1. NaHMDS (1 equiv) THF, -78 °C, 1 h
2. **22** (3 equiv)
   HMPA (3 equiv) THF, -78 °C, 14 h

23

89%; >99:1 dr

(II)

24

1. LDA (1 equiv) THF:DMPU (3:2) -78 °C, 1 h
2. PhCH₂Br (1.5 equiv) THF:DMPU (3:2) -78 °C, 1 h

25

99%; >25:1 dr

(III)

26

1. NaHMDS (1.5 equiv) THF, -78 °C
2. **27** (3 equiv)
   THF, -78 °C

28

63%; >25:1 dr

(IV)

29

K₂CO₃ (3 equiv) Bu₄NI (10 mol%)

30 (2 equiv) CH₃CN, 50 °C, 10 h

31

92%; >25:1 dr

(V)

32

1. n-BuLi (1.5 equiv) THF, -78 °C, 1.5 h
2. MeI (3 equiv) HMPA (3 equiv) THF, -78 to 23 °C, 5 h

33

49%; >25:1 dr

(VI)

34

Bu₄NI (10 mol%) P2-Et (2 equiv)

35 (5 equiv) CH₂Cl₂, -78 °C, 5 min

36

49%; >39:1 dr

Reference key for equations: (I)[5]; (II)[13]; (III)[14]; (IV)[15]; (V)[16];(VI)[17]

Scheme 2

III. Diastereoselective α-Alkylation of N-Acyl Pseudoephedrine-Derived Enolates:

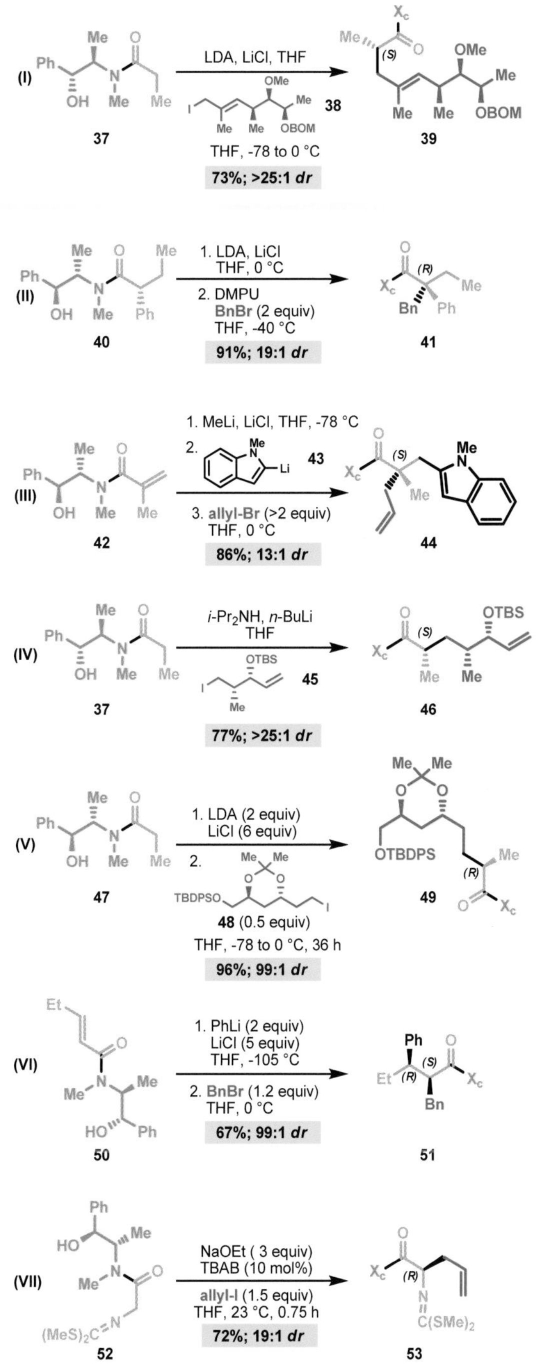

(I)

37

LDA, LiCl, THF

38

THF, -78 to 0 °C

39

73%; >25:1 dr

(II)

40

1. LDA, LiCl THF, 0 °C
2. DMPU
   BnBr (2 equiv) THF, -40 °C

41

91%; 19:1 dr

(III)

42

1. MeLi, LiCl, THF, -78 °C
2. **43**
3. allyl-Br (>2 equiv) THF, 0 °C

44

86%; 13:1 dr

(IV)

37

i-Pr₂NH, n-BuLi THF

45

46

77%; >25:1 dr

(V)

47

1. LDA (2 equiv) LiCl (6 equiv)
2. **48** (0.5 equiv)
   THF, -78 to 0 °C, 36 h

49

96%; 99:1 dr

(VI)

50

1. PhLi (2 equiv) LiCl (5 equiv) THF, -105 °C
2. BnBr (1.2 equiv) THF, 0 °C

51

67%; 99:1 dr

(VII)

52

NaOEt (3 equiv) TBAB (10 mol%)

allyl-I (1.5 equiv) THF, 23 °C, 0.75 h

53

72%; 19:1 dr

Reference key for equations: (I)[18]; (II-III)[19]; (IV)[20]; (V)[21];(VI)[22];(VII)[23]

Scheme 3

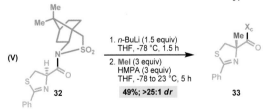
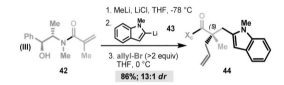
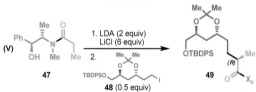

ENANTIOSELECTIVE α-ALKYLATION - PART I.

IV. Diastereoselective α-Alkylation of Carboxylic Acid Enolates Using Removable Chiral Controller Groups (Miscellaneous Controller Groups):

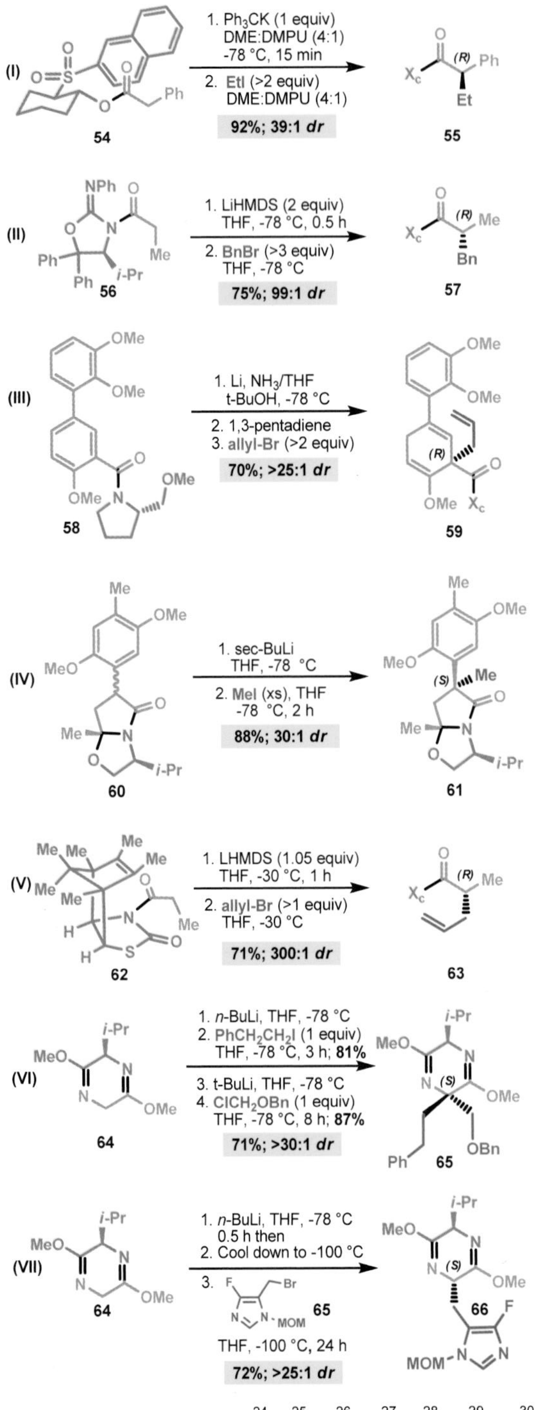

(I)
54
1. Ph₃CK (1 equiv) DME:DMPU (4:1) -78 °C, 15 min
2. EtI (>2 equiv) DME:DMPU (4:1)
92%; 39:1 dr
55 — X_c (R) Ph Et

(II)
56
1. LiHMDS (2 equiv) THF, -78 °C, 0.5 h
2. BnBr (>3 equiv) THF, -78 °C
75%; 99:1 dr
57 — X_c (R) Me Bn

(III)
58
1. Li, NH₃/THF t-BuOH, -78 °C
2. 1,3-pentadiene
3. allyl-Br (>2 equiv)
70%; >25:1 dr
59

(IV)
60
1. sec-BuLi THF, -78 °C
2. MeI (xs), THF -78 °C, 2 h
88%; 30:1 dr
61

(V)
62
1. LHMDS (1.05 equiv) THF, -30 °C, 1 h
2. allyl-Br (>1 equiv) THF, -30 °C
71%; 300:1 dr
63

(VI)
64
1. n-BuLi, THF, -78 °C
2. PhCH₂CH₂I (1 equiv) THF, -78 °C, 3 h; **81%**
3. t-BuLi, THF, -78 °C
4. ClCH₂OBn (1 equiv) THF, -78 °C, 8 h; **87%**
71%; >30:1 dr
65

(VII)
64
1. n-BuLi, THF, -78 °C 0.5 h then
2. Cool down to -100 °C
3. 65
THF, -100 °C, 24 h
72%; >25:1 dr
66

Reference keys for equations: (I)[24];(II)[25];(III)[26];(IV)[27];(V)[28];(VI)[29];(VII)[30]

Scheme 4

V. Diastereoselective α-Alkylation of Ketone Enolates Using Removable Chiral Controller Groups (Enders SAMP and RAMP Hydrazones):

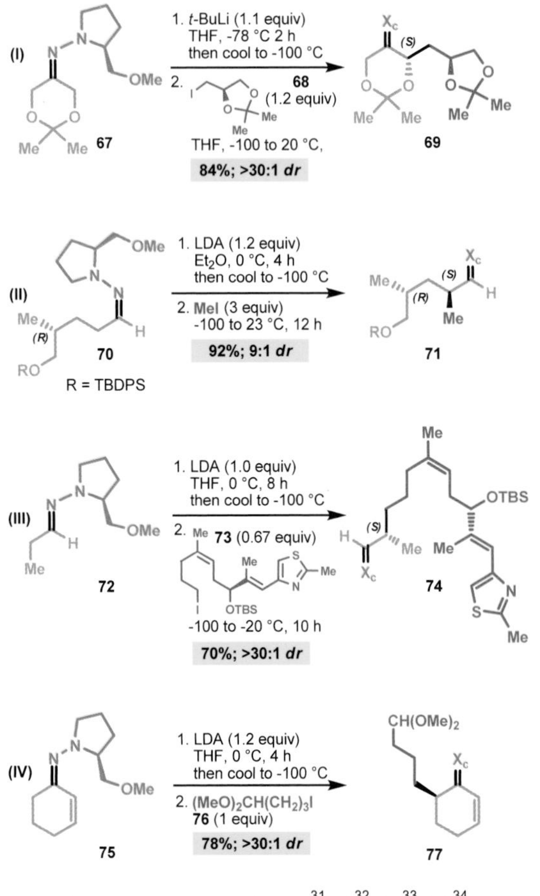

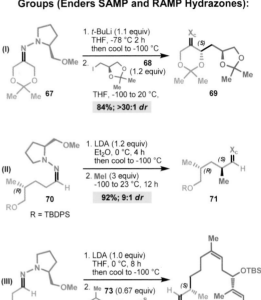

(I)
67
1. t-BuLi (1.1 equiv) THF, -78 °C 2 h then cool to -100 °C
2. 68 (1.2 equiv)
THF, -100 °C to 20 °C,
84%; >30:1 dr
69

(II)
70
R = TBDPS
1. LDA (1.2 equiv) Et₂O, 0 °C, 4 h then cool to -100 °C
2. MeI (3 equiv) -100 to 23 °C, 12 h
92%; 9:1 dr
71

(III)
72
1. LDA (1.0 equiv) THF, 0 °C, 8 h then cool to -100 °C
2. 73 (0.67 equiv)
-100 to -20 °C, 10 h
70%; >30:1 dr
74

(IV)
75
1. LDA (1.2 equiv) THF, 0 °C, 4 h then cool to -100 °C
2. (MeO)₂CH(CH₂)₃I 76 (1 equiv)
78%; >30:1 dr
77

Reference key for equations: (I)[31]; (II)[32]; (III)[33]; (IV)[34]

Scheme 5

VI. Diastereoselective α-Alkylation of Chiral Enolates

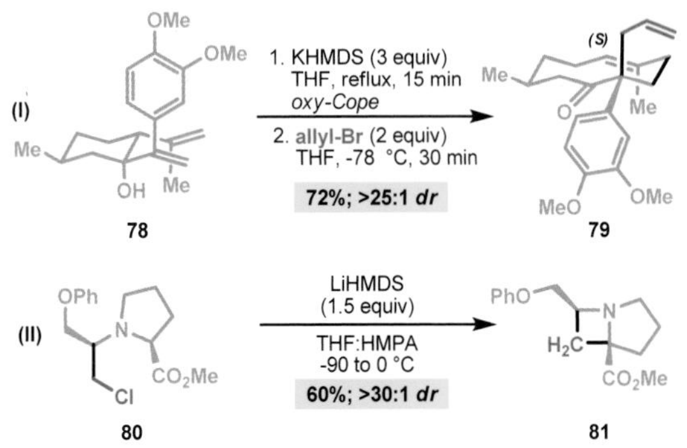

(I)
78
1. KHMDS (3 equiv) THF, reflux, 15 min oxy-Cope
2. allyl-Br (2 equiv) THF, -78 °C, 30 min
72%; >25:1 dr
79

(II)
80
LiHMDS (1.5 equiv)
THF:HMPA -90 to 0 °C
60%; >30:1 dr
81

Reference key for equations: (I)[35]; (II)[36]

Scheme 6

ENANTIOSELECTIVE α-ALKYLATION - PART I.

Diastereoselective α-Alkylation of Chiral Enolates (Continued):

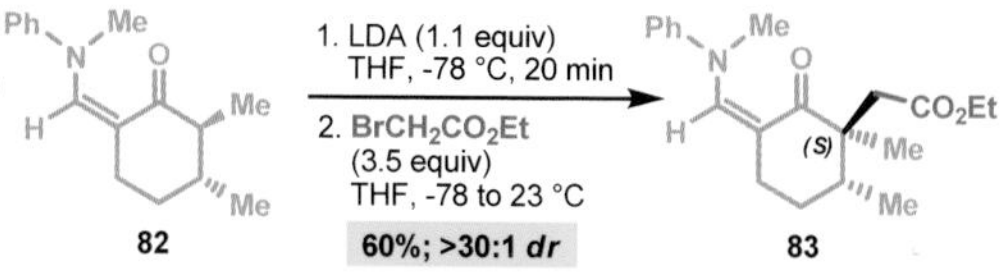

Reference key for equation: (III)[37]

Scheme 7

References

1. (a) Oppolzer, W. Camphor derivatives as chiral auxiliaries in asymmetric synthesis. *Tetrahedron* **1987**, 43, 1969-2004; (b) Ager, D. J., Prakash, I. & Schaad, D. R. 1,2-Amino Alcohols and Their Heterocyclic Derivatives as Chiral Auxiliaries in Asymmetric Synthesis. *Chem. Rev.* **1996**, 96, 835-875; (c) Reiser, O. Oppolzer sultams. *Org. Synth. Highlights IV* **2000**, 11-16; (d) Job, A., Janeck, C. F., Bettray, W., Peters, R. & Enders, D. The SAMP/RAMP-hydrazone methodology in asymmetric synthesis. *Tetrahedron* **2002**, 58, 2253-2329; (e) Gnas, Y. & Glorius, F. Chiral auxiliaries - principles and recent applications. *Synthesis* **2006**, 1899-1930; (f) Evans, D. A., Helmchen, G., Rueping, M. & Wolfgang, J. Chiral auxiliaries in asymmetric synthesis. *Asymmetric Synth.* **2007**, 3-9; (g) Lee, A. W. M., Chan, W. H., Zhang, S.-J. & Zhang, H.-K. Non camphor based sultam chiral auxiliaries. *Curr. Org. Chem.* **2007**, 11, 213-228; (h) Zappia, G., Cancelliere, G., Gacs-Baitz, E., Delle Monache, G., Misiti, D., Nevola, L. & Botta, B. Oxazolidin-2-one ring, a popular framework in synthetic organic chemistry part 2 [1]. Applications and modifications. *Curr. Org. Synth.* **2007**, 4, 238-307; (i) Lazny, R. & Nodzewska, A. N,N-Dialkylhydrazones in Organic Synthesis. From Simple N,N-Dimethylhydrazones to Supported Chiral Auxiliaries. *Chem. Rev. (Washington, DC, U. S.)* **2010**, ACS ASAP.
2. Evans, D. A., Ennis, M. D. & Mathre, D. J. Asymmetric alkylation reactions of chiral imide enolates. A practical approach to the enantioselective synthesis of α-substituted carboxylic acid derivatives. *J. Am. Chem. Soc.* **1982**, 104, 1737-1739.
3. (a) Meyers, A. I. Asymmetric carbon-carbon bond formation from chiral oxazolines. *Acc. Chem. Res.* **1978**, 11, 375-381; (b) Meyers, A. I. Chiral Oxazolines and Their Legacy in Asymmetric Carbon-Carbon Bond-Forming Reactions. *J. Org. Chem.* **2005**, 70, 6137-6151; (c) Meyers, A. I. The use of chiral oxazolines in early C-C bond forming reactions. *Asymmetric Synth.* **2007**, 37-39.
4. Enders, D. & Eichenauer, H. Asymmetric synthesis of α-substituted ketones by metalation and alkylation of chiral hydrazones. *Angew. Chem.* **1976**, 88, 579-581.
5. Oppolzer, W., Moretti, R. & Thomi, S. Asymmetric alkylation of N-acylsultams: a general route to enantiomerically pure, crystalline C(α,α)-disubstituted carboxylic acid derivatives. *Tetrahedron Lett.* **1989**, 30, 5603-5606.
6. Schoellkopf, U. Enantioselective synthesis of non-proteinogenic amino acids via metalated bis-lactim ethers of 2,5-diketopiperazines. *Tetrahedron* **1983**, 39, 2085-2091.
7. Myers, A. G., Yang, B. H., Chen, H. & Gleason, J. L. Use of Pseudoephedrine as a Practical Chiral Auxiliary for Asymmetric Synthesis. *J. Am. Chem. Soc.* **1994**, 116, 9361-9362.
8. Crimmins, M. T., Emmitte, K. A. & Katz, J. D. Diastereoselective Alkylations of Oxazolidinone Glycolates: A Useful Extension of the Evans Asymmetric Alkylation. *Org. Lett.* **2000**, 2, 2165-2167.
9. Chappell, M. D., Stachel, S. J., Lee, C. B. & Danishefsky, S. J. En Route to a Plant Scale Synthesis of the Promising Antitumor Agent 12,13-Desoxyepothilone B. *Org. Lett.* **2000**, 2, 1633-1636.
10. Cases, M., Gonzalez-Lopez de Turiso, F., Hadjisoteriou, M. S. & Pattenden, G. Synthetic studies towards furanocembrane diterpenes. A total synthesis of bis-deoxylophotoxin. *Org. Biomol. Chem.* **2005**, 3, 2786-2804.
11. Wattanasereekul, S. & Maier, M. E. Synthesis of the 8-hydroxy acid of jasplakinolide. *Adv. Synth. Catal.* **2004**, 346, 855-861.
12. Evans, D. A., Wu, L. D., Wiener, J. J. M., Johnson, J. S., Ripin, D. H. B. & Tedrow, J. S. A General Method for the Synthesis of Enantiomerically Pure β-Substituted, β-Amino Acids through α-Substituted Succinic Acid Derivatives. *J. Org. Chem.* **1999**, 64, 6411-6417.
13. Ponsinet, R., Chassaing, G., Vaissermann, J. & Lavielle, S. Diastereoselective synthesis of β^2-amino acids. *Eur. J. Org. Chem.* **2000**, 83-90.
14. Schmidt, B. & Wildemann, H. Synthesis of enantiomerically pure divinyl- and diallylcarbinols. *J. Chem. Soc., Perkin Trans. 1* **2002**, 1050-1060.
15. Deng, W.-P., Wong, K. A. & Kirk, K. L. Convenient syntheses of 2-, 5- and 6-fluoro- and 2,6-difluoro-L-DOPA. *Tetrahedron Asymmetry* **2002**, 13, 1135-1140.
16. Singh, S., Rao Samala, J. & Pennington Michael, W. Efficient asymmetric synthesis of (S)- and (R)-N-Fmoc-S-trityl-alpha-methylcysteine using camphorsultam as a chiral auxiliary. *J. Org. Chem.* **2004**, 69, 4551-4554.
17. Lee, J., Lee, Y.-I., Kang, M. J., Lee, Y.-J., Jeong, B.-S., Lee, J.-H., Kim, M.-J., Choi, J.-y., Ku, J.-M., Park, H.-g. & Jew, S.-s. Enantioselective Synthetic Method for α-Alkylserine via Phase-Transfer Catalytic Alkylation of 2-Phenyl-2-oxazoline-4-carbonylcamphorsultam. *J. Org. Chem.* **2005**, 70, 4158-4161.
18. Murray, T. J. & Forsyth, C. J. Total Synthesis of GEX1A. *Org. Lett.* **2008**, 10, 3429-3431.
19. Kummer, D. A., Chain, W. J., Morales, M. R., Quiroga, O. & Myers, A. G. Stereocontrolled Alkylative Construction of Quaternary Carbon Centers. *J. Am. Chem. Soc.* **2008**, 130, 13231-13233.
20. Prasad Narasimhulu, C., Iqbal, J., Mukkanti, K. & Das, P. Studies towards the total synthesis of narbonolide: stereoselective preparation of the C1-C10 fragment. *Tetrahedron Lett.* **2008**, 49, 3185-3188.
21. Su, Y., Xu, Y., Han, J., Zheng, J., Qi, J., Jiang, T., Pan, X. & She, X. Total Synthesis of (-)-Bitungolide F. *J. Org. Chem.* **2009**, 74, 2743-2749.
22. Reyes, E., Vicario, J. L., Carrillo, L., Badia, D., Iza, A. & Uria, U. Tandem Asymmetric Conjugate Addition/α-Alkylation Using (S,S)-(+)-Pseudoephedrine as Chiral Auxiliary. *Org. Lett.* **2006**, 8, 2535-2538.
23. Guillena, G. & Najera, C. The imine (+)-pseudoephedrine glycinamide: a useful reagent for the asymmetric synthesis of (R)-α-amino acids. *Tetrahedron Asymmetry* **2001**, 12, 181-183.
24. Sarakinos, G. & Corey, E. J. A Practical New Chiral Controller for Asymmetric Diels-Alder and Alkylation Reactions. *Org. Lett.* **1999**, 1, 1741-1744.
25. Le, T. N., Nguyen, Q. P. B., Kim, J. N. & Kim, T. H. 5,5-Dimethyl-2-phenylamino-2-oxazoline as an effective chiral auxiliary for asymmetric alkylations. *Tetrahedron Lett.* **2007**, 48, 7834-7837.
26. Malachowski, W. P., Paul, T. & Phounsavath, S. The Enantioselective Synthesis of (-)-Lycoramine with the Birch-Cope Sequence. *J. Org. Chem.* **2007**, 72, 6792-6796.
27. Kuwahara, S. & Saito, M. Enantioselective total synthesis of enokipodins A-D. *Tetrahedron Lett.* **2004**, 45, 5047-5049.
28. Hoshimoto, S., Matsunaga, H. & Kunieda, T. Sterically constrained "roofed" 2-thiazolidinones as excellent chiral auxiliaries. *Chem. Pharm. Bull.* **2000**, 48, 1541-1544.
29. Yiotakis, A., Magriotis, P. A. & Vassiliou, S. A simple synthesis of the metabotropic receptor ligand (2S)-α-(hydroxymethyl)-glutamic acid and its Fmoc protected derivatives. *Tetrahedron Asymmetry* **2007**, 18, 873-877.
30. Hajduch, J., Cramer, J. C. & Kirk, K. L. An enantioselective synthesis of (S)-4-fluorohistidine. *J. Fluorine Chem.* **2008**, 129, 807-810.
31. Andre, C., Bolte, J. & Demuynck, C. Syntheses of 4-deoxy-D-fructose and enzymic affinity study. *Tetrahedron Asymmetry* **1998**, 9, 3737-3739.
32. Chandrasekhar, S., Yaragorla, S. R., Sreelakshmi, L. & Reddy, C. R. Formal total synthesis of (-)-spongidepsin. *Tetrahedron* **2008**, 64, 5174-5183.
33. Nicolaou, K. C., Ninkovic, S., Sarabia, F., Vourloumis, D., He, Y., Vallberg, H., Finlay, M. R. V. & Yang, Z. Total Syntheses of Epothilones A and B via a Macrolactonization-Based Strategy. *J. Am. Chem. Soc.* **1997**, 119, 7974-7991.
34. Jung, M. Current developments in the chemistry of artemisinin and related compounds. *Curr. Med. Chem.* **1994**, 1, 35-49.
35. Arns, S., Lebrun, M.-E., Grise, C. M., Denissova, I. & Barriault, L. Diastereoselective Construction of Quaternary Carbons Directed via Macrocyclic Ring Conformation: Formal Synthesis of (-)-Mesembrine. *J. Org. Chem.* **2007**, 72, 9314-9322.
36. Sivaprakasam, M., Couty, F., David, O., Marrot, J., Sridhar, R., Srinivas, B. & Rao, K. R. A straightforward synthesis of enantiopure bicyclic azetidines. *Eur. J. Org. Chem.* **2007**, 5734-5739.
37. Pena-Lopez, M., Martinez, M. M., Sarandeses, L. A. & Sestelo, J. P. Total synthesis of (+)-neomarinone. *Chem.--Eur. J.* **2009**, 15, 910-916.

ENANTIOSELECTIVE α-ALKYLATION - PART II.

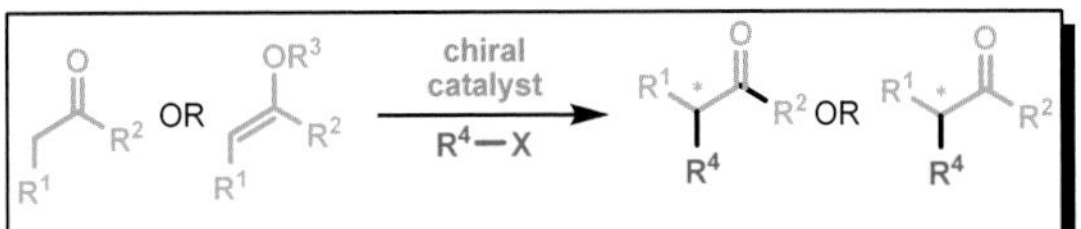

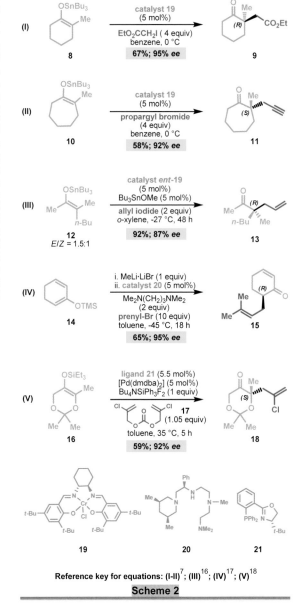

R^1 = alkyl, aryl, heteroaryl; R^2 = alkyl, aryl or H; R^3 = SnR_3, SiR_3, Li, allyloxycarbonyl; R^4 = allyl, aryl, alkyl; X = Cl, Br, I, OTf

Introduction

The enantioselective alkylation of enolates has been under active development since the 1970s. The use of chiral controller groups (chiral auxiliaries) dominated the field until recently (see pages 56-59), since high yields and consistently high levels of stereoselectivity can often be achieved. Controller-free *catalytic* enantioselective α-alkylation[1] of carbonyl compounds, generating tertiary or quaternary carbon stereocenters[2], is a more recent development. Discussed in this section are a number of methods for catalytic enantioselective α-arylation[3] and α-vinylation[4] of carbonyl compounds as well as alkylation. A number of different approaches have been used to effect the highly enantioselective catalytic α-alkylation and α-arylation of enolates, including: (a) enantioselective deprotonation of symmetrical ketones using a chiral lithium amide as base;[5,1b] (b) memory of chirality that is based on the dynamic chirality of enolates;[6] (c) alkylation of tributylstannyl enolates using chiral Cr-salen complexes;[7] (d) alkylation of achiral lithium enolates in the presence of chiral tetraamine ligands;[8] (e) *in situ* alkylation of enolates generated by use of a chiral phase-transfer catalyst;[9] (f) intramolecular chiral 2°-amine catalyzed α-alkylation of aldehydes to form 3-, 5- and 6-membered rings;[10] (g) intermolecular alkylation of aldehydes via transient chiral enamine derivatives;[11] (h) alkylation of racemic α-bromocarbonyl compounds using the Hiyama and Kumada reactions;[12] (i) alkylation of enamides catalyzed by chiral protic acids;[13] (j) Pd-catalyzed allylation of enol carbonates;[14] and (k) Pd-catalyzed α-arylation of carbonyl compounds.[15,3]

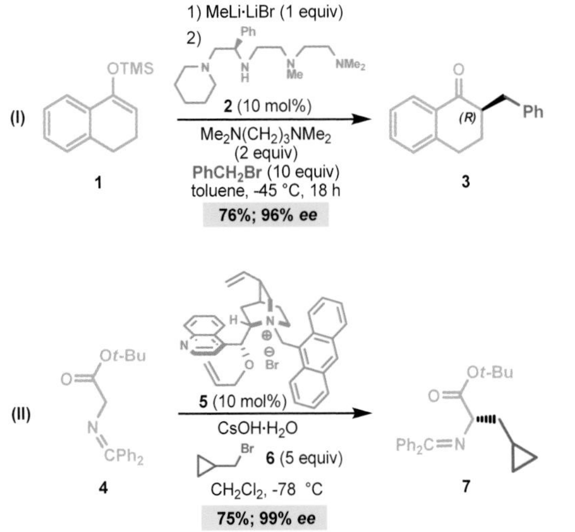

The first example of catalytic and highly enantioselective α-benzylation of achiral lithium enolates, generated from silyl enol ethers, was reported by Koga and co-workers[8] (Example I, Scheme 1) who utilized a chiral tetraamine as catalyst. The highly effective α-alkylation of *in situ*-generated enolates with a chiral quaternary ammonium salt as a phase-transfer catalyst is especially useful for the synthesis of α-amino acids and also mechanistically explained (Example II, Scheme 1); see pages 167-168.[9c]

I. Enantioselective α-Alkylation of Preformed Enolates and Enol Ethers

Scheme 1

Scheme 2

Reference key for equations: (I-II)[7]; (III)[16]; (IV)[17]; (V)[18]

ENANTIOSELECTIVE α-ALKYLATION - PART II.

Enantioselective α-Alkylation of Preformed Enolates and Enol Ethers (Continued)

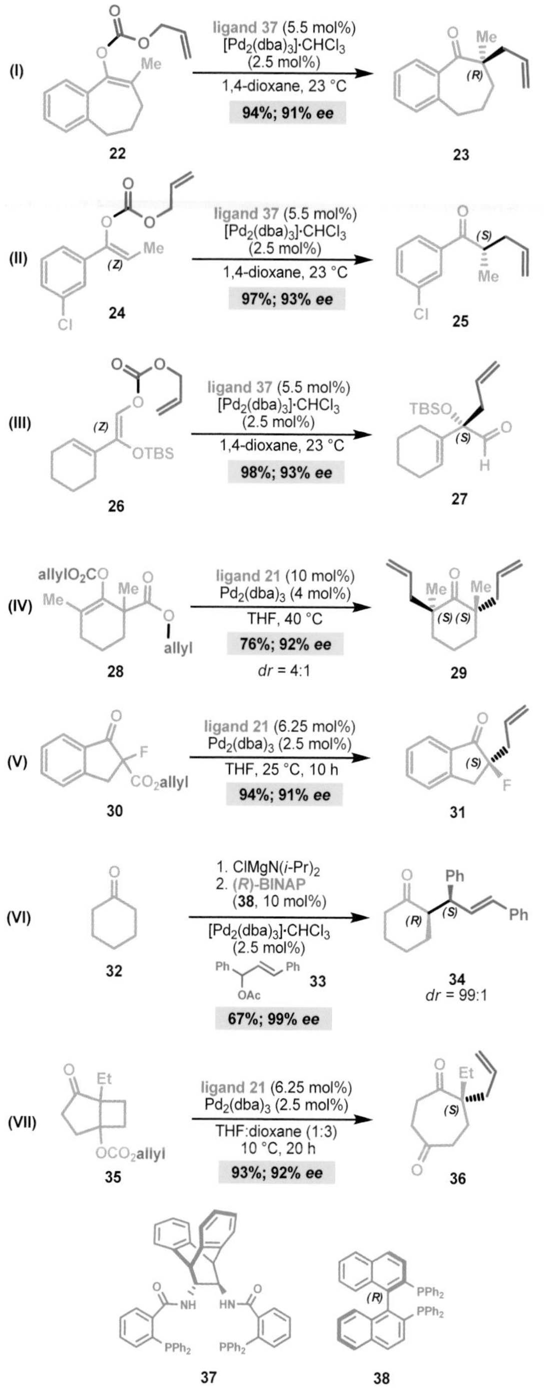

II. Direct Enantioselective Intra- and Intermolecular α-Alkylation of Aldehydes

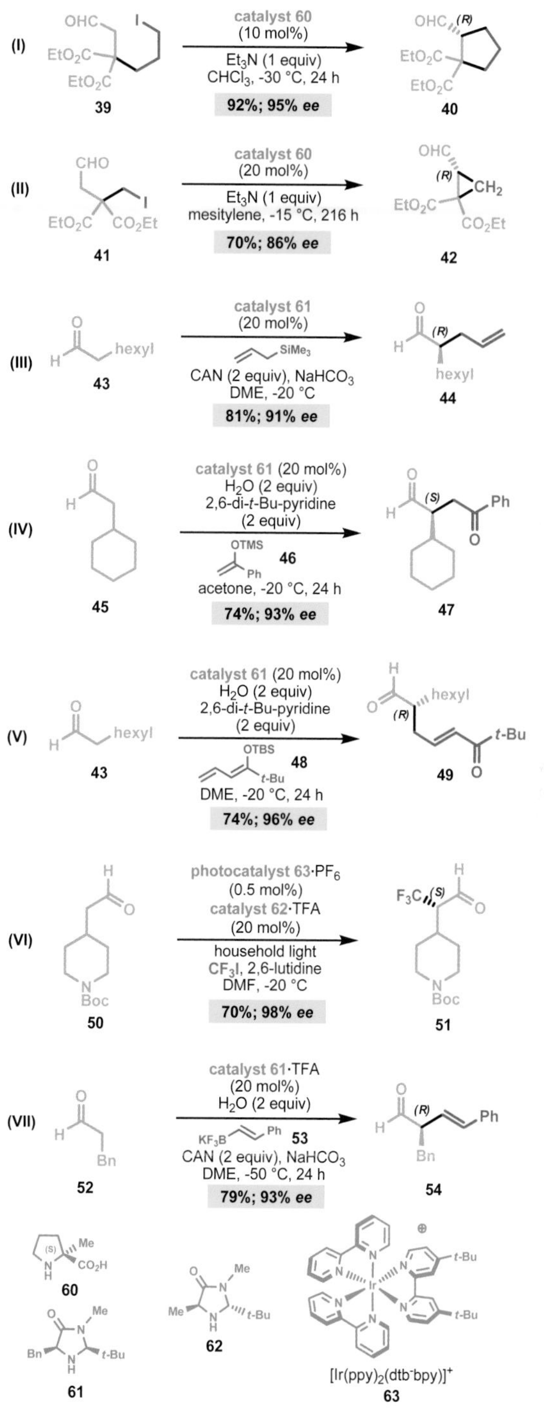

Reference key for equations: (I)[14b];(II)[19];(III)[20];(IV)[21];(V)[22];(VI)[23];(VII)[24]

Scheme 3

Reference key for equations: (I-II)[10]; (III)[11a]; (IV-V)[25]; (VI)[26]; (VII)[4]

Scheme 4

ENANTIOSELECTIVE α-ALKYLATION - PART II.

III. Direct Enantioselective Intra- and Intermolecular α-Arylation of Carbonyl Compounds (Aldehydes, Ketones and Carboxylic Acid Derivatives)

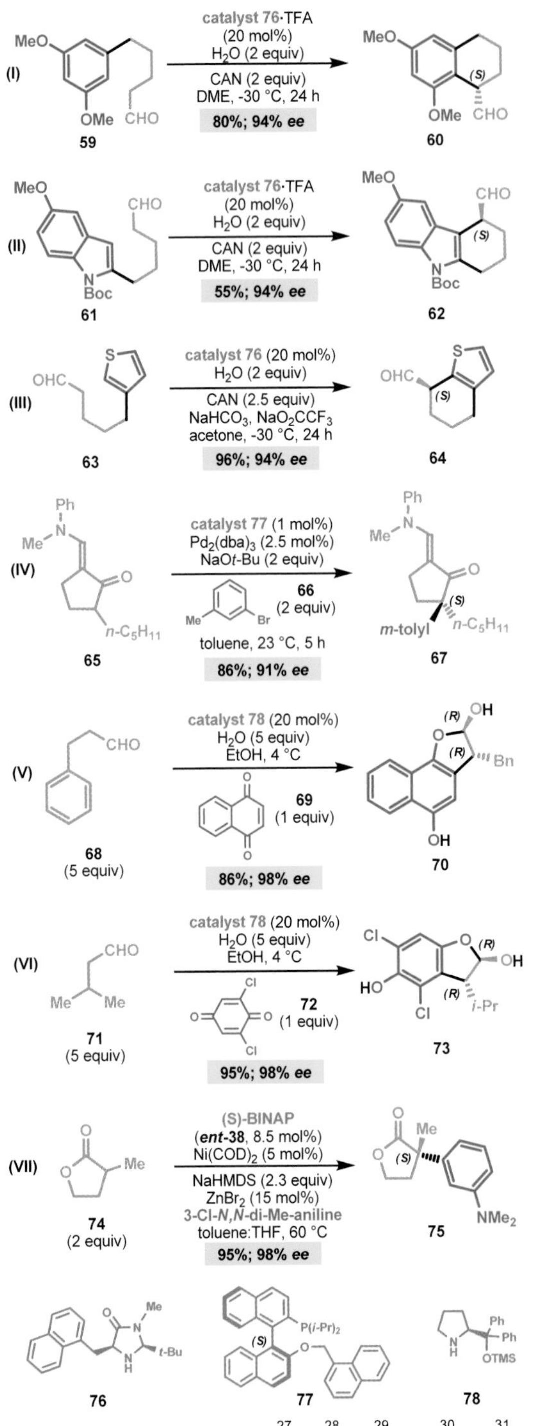

Direct Enantioselective Intra- and Intermolecular α-Arylation of Carbonyl Compounds (Continued)

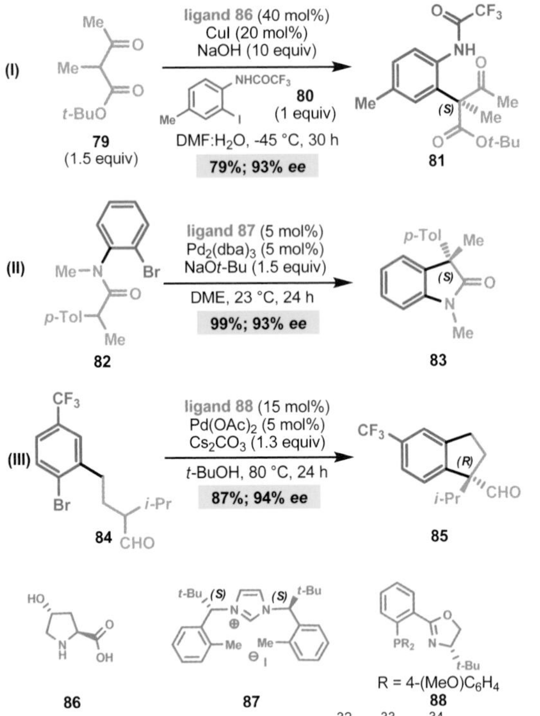

(I) catalyst 76·TFA (20 mol%) H₂O (2 equiv) CAN (2 equiv) DME, -30 °C, 24 h — **80%; 94% ee** — 59 → 60

(II) catalyst 76·TFA (20 mol%) H₂O (2 equiv) CAN (2 equiv) DME, -30 °C, 24 h — **55%; 94% ee** — 61 → 62

(III) catalyst 76 (20 mol%) H₂O (2 equiv) CAN (2.5 equiv) NaHCO₃, NaO₂CCF₃ acetone, -30 °C, 24 h — **96%; 94% ee** — 63 → 64

(IV) catalyst 77 (1 mol%) Pd₂(dba)₃ (2.5 mol%) NaOt-Bu (2 equiv) 66 (2 equiv) toluene, 23 °C, 5 h — **86%; 91% ee** — 65 → 67

(V) catalyst 78 (20 mol%) H₂O (5 equiv) EtOH, 4 °C 69 (1 equiv) — **86%; 98% ee** — 68 (5 equiv) → 70

(VI) catalyst 78 (20 mol%) H₂O (5 equiv) EtOH, 4 °C 72 (1 equiv) — **95%; 98% ee** — 71 (5 equiv) → 73

(VII) (S)-BINAP (ent-38, 8.5 mol%) Ni(COD)₂ (5 mol%) NaHMDS (2.3 equiv) ZnBr₂ (15 mol%) 3-Cl-N,N-di-Me-aniline toluene:THF, 60 °C — **95%; 98% ee** — 74 (2 equiv) → 75

Reference key for equations: (I-II)[27]; (III)[28]; (IV)[29]; (V-VI)[30]; (VII)[31]

Scheme 5

(I) ligand 86 (40 mol%) CuI (20 mol%) NaOH (10 equiv) 80 (1 equiv) DMF:H₂O, -45 °C, 30 h — **79%; 93% ee** — 79 (1.5 equiv) → 81

(II) ligand 87 (5 mol%) Pd₂(dba)₃ (5 mol%) NaOt-Bu (1.5 equiv) DME, 23 °C, 24 h — **99%; 93% ee** — 82 → 83

(III) ligand 88 (15 mol%) Pd(OAc)₂ (5 mol%) Cs₂CO₃ (1.3 equiv) t-BuOH, 80 °C, 24 h — **87%; 94% ee** — 84 → 85

R = 4-(MeO)C₆H₄ 88

Reference key for equations: (I)[32]; (II)[33]; (III)[34]

Scheme 6

IV. Ni-Mediated Enantioselective α-Alkylation and α-Arylation of α-Bromo Carbonyl Compounds

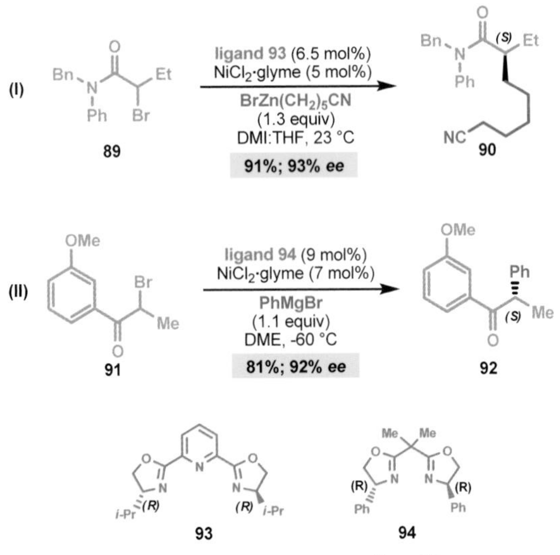

(I) ligand 93 (6.5 mol%) NiCl₂·glyme (5 mol%) BrZn(CH₂)₅CN (1.3 equiv) DMI:THF, 23 °C — **91%; 93% ee** — 89 → 90

(II) ligand 94 (9 mol%) NiCl₂·glyme (7 mol%) PhMgBr (1.1 equiv) DME, -60 °C — **81%; 92% ee** — 91 → 92

Reference key for equations: (I)[35]; (II)[12b]

Scheme 7

ENANTIOSELECTIVE α-ALKYLATION - PART II.

Enantioselective C$_\alpha$-C Bond Formation Between an Enamide and a Chiral Anion-Cation Pair

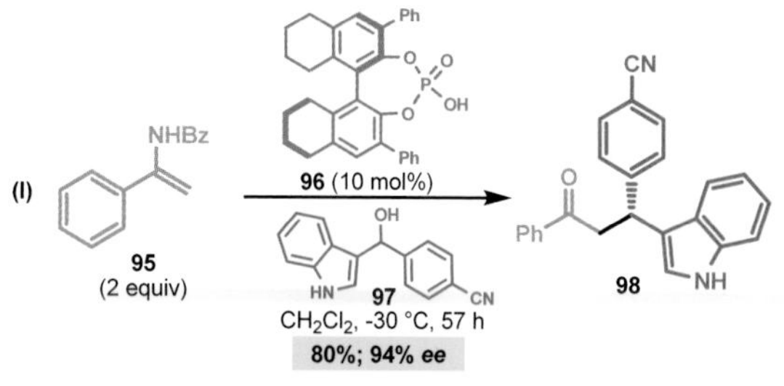

Reference key for equation: (I)[13]

Scheme 7

References

1. (a) Hughes, D. L. Alkylation of enolates. *Compr. Asymmetric Catal. I-III* **1999**, 3, 1273-1294; (b) Wu, G. G. & Huang, M. Organolithium in asymmetric processes. *Top. Organomet. Chem.* **2004**, 6, 1-35; (c) Birsa, M. L. Carbanions and electrophilic aliphatic substitution. *Organic Reaction Mechanisms* **2008**, 249-276.
2. (a) Christoffers, J. & Baro, A. Stereoselective construction of quaternary stereocenters. *Adv. Synth. Catal.* **2005**, 347, 1473-1482; (b) Trost, B. M. & Jiang, C. Catalytic enantioselective construction of all-carbon quaternary stereocenters. *Synthesis* **2006**, 369-396; (c) Cozzi, P. G., Hilgraf, R. & Zimmermann, N. Enantioselective catalytic formation of quaternary stereogenic centers. *Eur. J. Org. Chem.* **2007**, 5969-5994.
3. Burtoloso, A. C. B. Catalytic enantioselective α-arylation of carbonyl compounds. *Synlett* **2009**, 320-327.
4. Kim, H. & MacMillan, D. W. C. Enantioselective Organo-SOMO Catalysis: The α-Vinylation of Aldehydes. *J. Am. Chem. Soc.* **2008**, 130, 398-399.
5. O'Brien, P. Recent advances in asymmetric synthesis using chiral lithium amide bases. *J. Chem. Soc., Perkin Trans. 1* **1998**, 1439-1458.
6. Kawabata, T. & Fuji, K. Memory of chirality: asymmetric induction based on the dynamic chirality of enolates. *Top. Stereochem.* **2003**, 23, 175-205.
7. Doyle, A. G. & Jacobsen, E. N. Enantioselective alkylations of tributyltin enolates catalyzed by Cr(salen)Cl: Access to enantiomerically enriched all-carbon quaternary centers. *J. Am. Chem. Soc.* **2005**, 127, 62-63.
8. Imai, M., Hagihara, A., Kawasaki, H., Manabe, K. & Koga, K. Catalytic Asymmetric Benzylation of Achiral Lithium Enolates Using a Chiral Ligand for Lithium in the Presence of an Achiral Ligand. *J. Am. Chem. Soc.* **1994**, 116, 8829-8830.
9. (a) Dolling, U. H., Davis, P. & Grabowski, E. J. J. Efficient catalytic asymmetric alkylations. 1. Enantioselective synthesis of (+)-indacrinone via chiral phase-transfer catalysis. *J. Am. Chem. Soc.* **1984**, 106, 446-447; (b) O'Donnell, M. J., Bennett, W. D. & Wu, S. The stereoselective synthesis of α-amino acids by phase-transfer catalysis. *J. Am. Chem. Soc.* **1989**, 111, 2353-2355; (c) Corey, E. J., Xu, F. & Noe, M. C. A Rational Approach to Catalytic Enantioselective Enolate Alkylation Using a Structurally Rigidified and Defined Chiral Quaternary Ammonium Salt under Phase Transfer Conditions. *J. Am. Chem. Soc.* **1997**, 119, 12414-12415.
10. Vignola, N. & List, B. Catalytic asymmetric intramolecular α-alkylation of aldehydes. *J. Am. Chem. Soc.* **2004**, 126, 450-451.
11. (a) Beeson, T. D., Mastracchio, A., Hong, J.-B., Ashton, K. & MacMillan, D. W. C. Enantioselective Organocatalysis Using SOMO Activation. *Science (Washington, DC, U. S.)* **2007**, 316, 582-585; (b) Nicewicz, D. A. & MacMillan, D. W. C. Merging Photoredox Catalysis with Organocatalysis: The Direct Asymmetric Alkylation of Aldehydes. *Science (Washington, DC, U. S.)* **2008**, 322, 77-80; (c) Melchiorre, P. Light in aminocatalysis: the asymmetric intermolecular α-alkylation of aldehydes. *Angew. Chem., Int. Ed.* **2009**, 48, 1360-1363.
12. (a) Dai, X., Strotman, N. A. & Fu, G. C. Catalytic Asymmetric Hiyama Cross-Couplings of Racemic α-Bromo Esters. *J. Am. Chem. Soc.* **2008**, 130, 3302-3303; (b) Lou, S. & Fu, G. C. Nickel/Bis(oxazoline)-Catalyzed Asymmetric Kumada Reactions of Alkyl Electrophiles:

13. Guo, Q.-X., Peng, Y.-G., Zhang, J.-W., Song, L., Feng, Z. & Gong, L.-Z. Highly Enantioselective Alkylation Reaction of Enamides by Bronsted-Acid Catalysis. *Org. Lett.* **2009**, 11, 4620-4623.
14. (a) Behenna, D. C. & Stoltz, B. M. The enantioselective Tsuji allylation. *J. Am. Chem. Soc.* **2004**, 126, 15044-15045; (b) Trost, B. M. & Xu, J. Regio- and Enantioselective Pd-Catalyzed Allylic Alkylation of Ketones through Allyl Enol Carbonates. *J. Am. Chem. Soc.* **2005**, 127, 2846-2847; (c) Mohr, J. T. & Stoltz, B. M. Enantioselective Tsuji allylations. *Chemistry--An Asian Journal* **2007**, 2, 1476-1491.
15. Aahman, J., Wolfe, J. P., Troutman, M. V., Palucki, M. & Buchwald, S. L. Asymmetric Arylation of Ketone Enolates. *J. Am. Chem. Soc.* **1998**, 120, 1918-1919.
16. Doyle, A. G. & Jacobsen, E. N. Enantioselective alkylation of acyclic α,α-disubstituted tributyltin enolates catalyzed by a {Cr(salen)} complex. *Angew. Chem., Int. Ed.* **2007**, 46, 3701-3705.
17. Kuramochi, A., Usuda, H., Yamatsugu, K., Kanai, M. & Shibasaki, M. Total Synthesis of (±)-Garsubellin A. *J. Am. Chem. Soc.* **2005**, 127, 14200-14201.
18. Enquist, J. A., Jr. & Stoltz, B. M. The total synthesis of (-)-cyanthiwigin F by means of double catalytic enantioselective alkylation. *Nature (London, U. K.)* **2008**, 453, 1228-1231.
19. Trost, B. M. & Xu, J. Palladium-Catalyzed Asymmetric Allylic α-Alkylation of Acyclic Ketones. *J. Am. Chem. Soc.* **2005**, 127, 17180-17181.
20. Trost, B. M., Xu, J. & Reichle, M. Enantioselective Synthesis of α-Tertiary Hydroxyaldehydes by Palladium-Catalyzed Asymmetric Allylic Alkylation of Enolates. *J. Am. Chem. Soc.* **2007**, 129, 282-283.
21. Mohr, J. T., Behenna, D. C., Harned, A. M. & Stoltz, B. M. Deracemization of quaternary stereocenters by Pd-catalyzed enantioconvergent decarboxylative allylation of racemic β-keto esters. *Angew. Chem., Int. Ed.* **2005**, 44, 6924-6927.
22. Nakamura, M., Hajra, A., Endo, K. & Nakamura, E. Synthesis of chiral α-fluoro ketones through catalytic enantioselective decarboxylation. *Angew. Chem., Int. Ed.* **2005**, 44, 7248-7251.
23. Braun, M., Laicher, F. & Meier, T. Diastereoselective and enantioselective palladium-catalyzed allylic substitution with nonstabilized ketone enolates. *Angew. Chem., Int. Ed.* **2000**, 39, 3494-3497.
24. Schulz, S. R. & Blechert, S. Palladium-catalyzed synthesis of substituted cycloheptane-1,4-diones by an asymmetric ring-expanding allylation (AREA). *Angew. Chem., Int. Ed.* **2007**, 46, 3966-3970.
25. Jang, H.-Y., Hong, J.-B. & MacMillan, D. W. C. Enantioselective Organocatalytic Singly Occupied Molecular Orbital Activation: The Enantioselective α-Enolation of Aldehydes. *J. Am. Chem. Soc.* **2007**, 129, 7004-7005.
26. Nagib, D. A., Scott, M. E. & MacMillan, D. W. C. Enantioselective α-Trifluoromethylation of Aldehydes via Photoredox Organocatalysis. *J. Am. Chem. Soc.* **2009**, 131, 10875-10877.
27. Nicolaou, K. C., Reingruber, R., Sarlah, D. & Brase, S. Enantioselective Intramolecular Friedel-Crafts-Type α-Arylation of Aldehydes. *J. Am. Chem. Soc.* **2009**, 131, 2086-2087.
28. Conrad, J. C., Kong, J., Laforteza, B. N. & MacMillan, D. W. C. Enantioselective α-Arylation of Aldehydes via Organo-SOMO Catalysis. An Ortho-Selective Arylation Reaction Based on an Open-Shell Pathway. *J. Am. Chem. Soc.* **2009**, 131, 11640-11641.
29. Hamada, T., Chieffi, A., Ahman, J. & Buchwald, S. L. An Improved Catalyst for the Asymmetric Arylation of Ketone Enolates. *J. Am. Chem. Soc.* **2002**, 124, 1261-1268.
30. Aleman, J., Cabrera, S., Maerten, E., Overgaard, J. & Joergensen, K. A. Asymmetric organocatalytic α-arylation of aldehydes. *Angew. Chem., Int. Ed.* **2007**, 46, 5520-5523, S5520/5521-S5520/5519.
31. Spielvogel, D. J. & Buchwald, S. L. Nickel-BINAP Catalyzed Enantioselective α-Arylation of α-Substituted γ-Butyrolactones. *J. Am. Chem. Soc.* **2002**, 124, 3500-3501.
32. Xie, X., Chen, Y. & Ma, D. Enantioselective Arylation of 2-Methylacetoacetates Catalyzed by CuI/trans-4-Hydroxy-L-proline at Low Reaction Temperatures. *J. Am. Chem. Soc.* **2006**, 128, 16050-16051.
33. Kuendig, E. P., Seidel, T. M., Jia, Y.-x. & Bernardinelli, G. Bulky chiral carbene ligands and their application in the palladium-catalyzed asymmetric intramolecular α-arylation of amides. *Angew. Chem., Int. Ed.* **2007**, 46, 8484-8487.
34. Garcia-Fortanet, J. & Buchwald, S. L. Asymmetric palladium-catalyzed intramolecular α-arylation of aldehydes. *Angew. Chem., Int. Ed.* **2008**, 47, 8108-8111.
35. Fischer, C. & Fu, G. C. Asymmetric nickel-catalyzed Negishi cross-couplings of secondary α-bromo amides with organozinc reagents. *J. Am. Chem. Soc.* **2005**, 127, 4594-4595.

STEREOCONTROLLED ALDOL REACTION

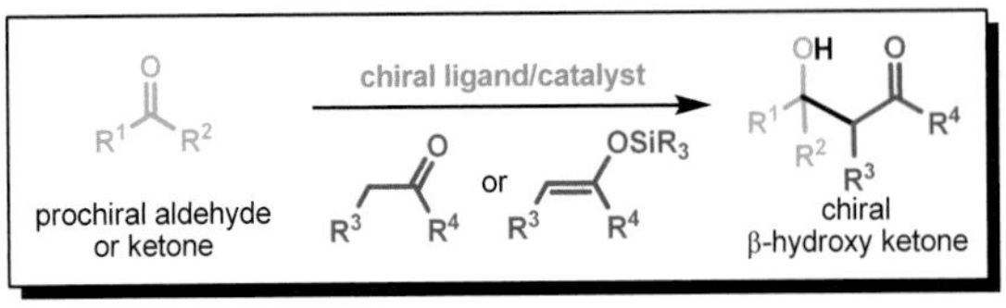

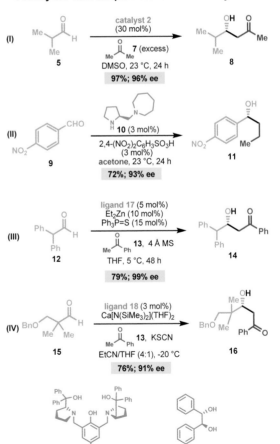

Introduction

Biosynthetic C-C bond formation commonly involves the aldol condensation of an aldehyde or ketone with a nucleophilic enolate equivalent, a fact that is evident from the structures of many naturally occurring molecules. The enolate equivalent may be derived from an aldehyde, ketone, ester or other carboxylic acid derivative. *Enzyme-catalyzed aldol reactions generally take place with complete control of configuration at the carbons that become linked. Either one or two* new stereocenters can be created. In synthetic chemistry, the control of stereochemistry at the newly created stereocenters has been a *major* challenge, and only in the past few decades have chiral controller groups and chiral reagents or catalysts been developed. The problem was simplified by findings in the 1960s that aldol reactions in which two new stereocenters are created could be performed with *diastereocontrol* under conditions favoring a six-membered cyclic (i.e., highly organized) transition state as shown in Scheme 1.

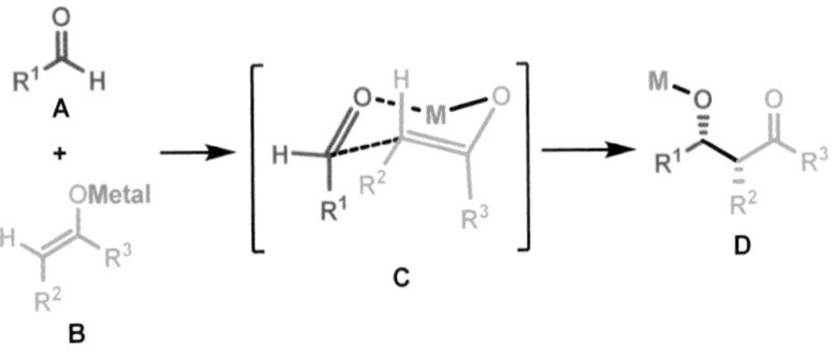

Scheme 1

Further developments included:

(1) the use of temporary, i.e., removable, chiral controller groups ("chiral auxiliaries")[1] as R[3];

(2) the use of chiral ligands on the metal;[2]

(3) a modified version of the aldol process (Mukaiyama aldol[3]), involving the use of an enol ether instead of an enolate with activation of the electrophilic carbonyl partner by complexation with a chiral Lewis acid catalyst;

(4) the use of catalytic amounts of a chiral amine which can form a reactive iminium ion intermediate with a carbonyl substrate.[4]

Eder, Wichert and Sauer[5] and later Hajos and Parrish[6] reported the (S)-proline-catalyzed enantioselective intramolecular aldol reaction (**1**→**4** via **3**[7], Scheme 2; patents filed on March 20, 1970 and January 20, 1971, respectively).[5-6,8] More recently the scope of this transformation has been expanded to include intermolecular aldol reactions involving a wide range of aldehydes and ketones. This section provides examples of (1) direct catalytic enantioselective aldol reactions between aldehydes and ketones; (2) aldol reactions with enolates or enolate

equivalents with a chiral ligand attached to the metal and (3) aldol reactions controlled by chiral auxiliaries.

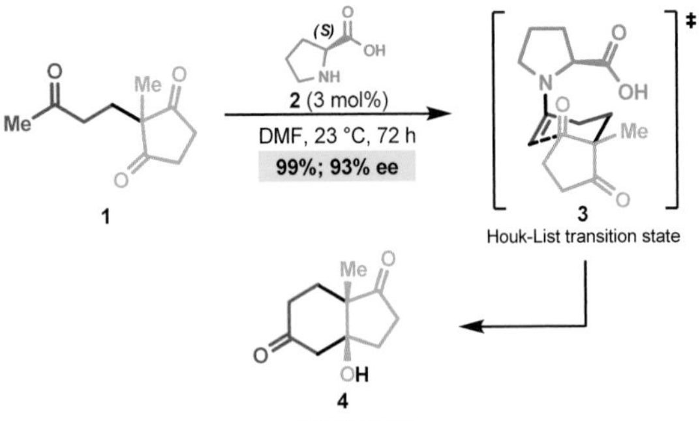

Scheme 2

I. Catalytic Enantioselective Aldol Reaction of Simple Aldehydes with Simple Ketones and Aldehydes

Reference key for equations: (I)[9]; (II)[10]; (III)[11]; (IV)[12]

Scheme 3

STEREOCONTROLLED ALDOL REACTION

II. Catalytic Enantio- and Diastereoselective Aldol Reaction of Aldehydes and Ketones

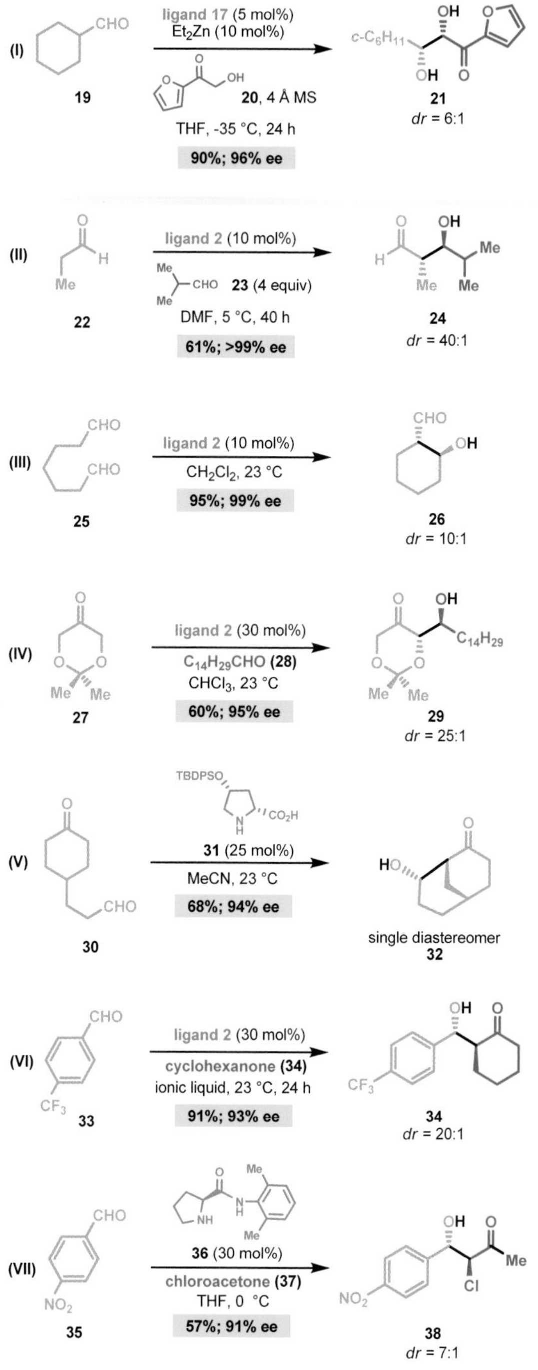

(I)

19

ligand 17 (5 mol%)
Et$_2$Zn (10 mol%)

20, 4 Å MS

THF, -35 °C, 24 h

90%; 96% ee

c-C$_6$H$_{11}$

21
dr = 6:1

(II)

22

ligand 2 (10 mol%)

23 (4 equiv)

DMF, 5 °C, 40 h

61%; >99% ee

24
dr = 40:1

(III)

25

ligand 2 (10 mol%)

CH$_2$Cl$_2$, 23 °C

95%; 99% ee

26
dr = 10:1

(IV)

27

ligand 2 (30 mol%)

C$_{14}$H$_{29}$CHO (28)

CHCl$_3$, 23 °C

60%; 95% ee

29
dr = 25:1

(V)

30

31 (25 mol%)

MeCN, 23 °C

68%; 94% ee

single diastereomer
32

(VI)

33

ligand 2 (30 mol%)

cyclohexanone (34)

ionic liquid, 23 °C, 24 h

91%; 93% ee

34
dr = 20:1

(VII)

35

36 (30 mol%)

chloroacetone (37)

THF, 0 °C

57%; 91% ee

38
dr = 7:1

Reference key for equations: (I)[13]; (II)[14]; (III)[15]; (IV)[16];
(V)[17]; (VI)[18]; (VII)[19]

Scheme 4

III. Catalytic Enantioselective Aldol Reaction Using Silyl Enol Ethers as Nucleophiles

Various metal-catalyzed (e.g., B, Cu, Sn, Ti and Zr) enantioselective Mukaiyama aldol reactions are shown in Scheme 5.

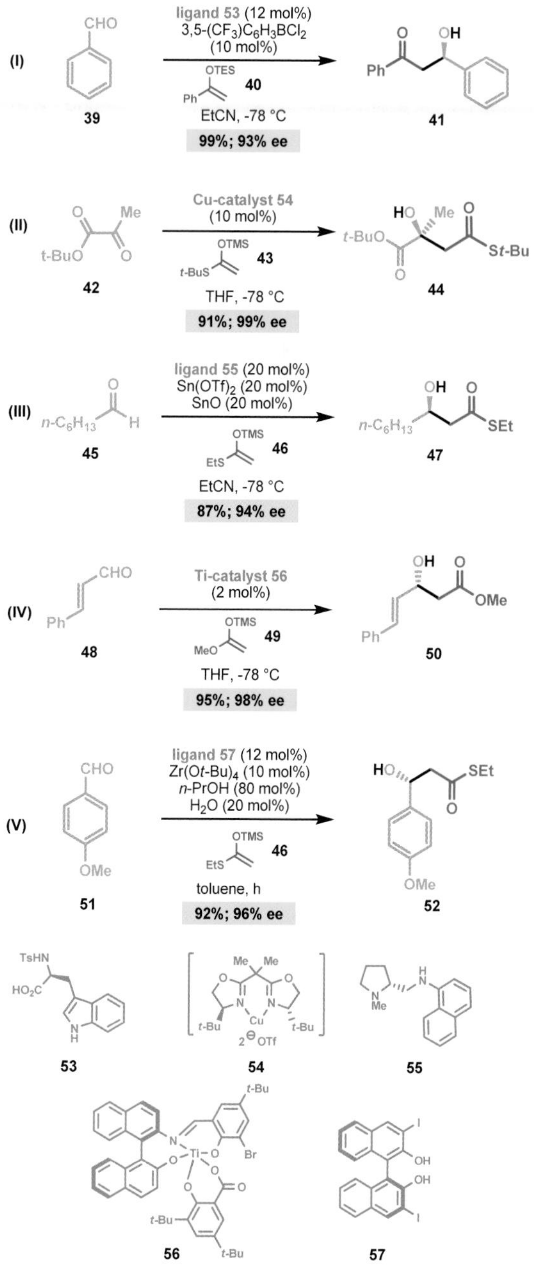

(I)

39

ligand 53 (12 mol%)
3,5-(CF$_3$)C$_6$H$_3$BCl$_2$
(10 mol%)

40

EtCN, -78 °C

99%; 93% ee

41

(II)

42

Cu-catalyst 54
(10 mol%)

43

THF, -78 °C

91%; 99% ee

44

(III)

45

ligand 55 (20 mol%)
Sn(OTf)$_2$ (20 mol%)
SnO (20 mol%)

46

EtCN, -78 °C

87%; 94% ee

47

(IV)

48

Ti-catalyst 56
(2 mol%)

49

THF, -78 °C

95%; 98% ee

50

(V)

51

ligand 57 (12 mol%)
Zr(Ot-Bu)$_4$ (10 mol%)
n-PrOH (80 mol%)
H$_2$O (20 mol%)

46

toluene, h

92%; 96% ee

52

53

54

55

56

57

Reference key for equations: (I)[20]; (II)[21]; (III)[22];
(IV)[23]; (V)[24]

Scheme 5

STEREOCONTROLLED ALDOL REACTION

IV. Catalytic Enantio- and Diastereoselective Aldol Reaction Using Enolsilanes as Nucleophiles

The reaction of an enolate with a chlorosilane such as Me₃SiCl reliably forms the corresponding silyl enol ether (vinyloxysilane) very cleanly. Although silyl ethers are generally insufficiently nucleophilic to react rapidly with aldehydes or ketones at room temperature or below, the reaction can be promoted by Lewis acids that activate the aldehyde or ketone partner by complexation. This process, extensively studied by Mukaiyama, provides a valuable extension of the aldol coupling reaction because it can provide control of stereochemistry.

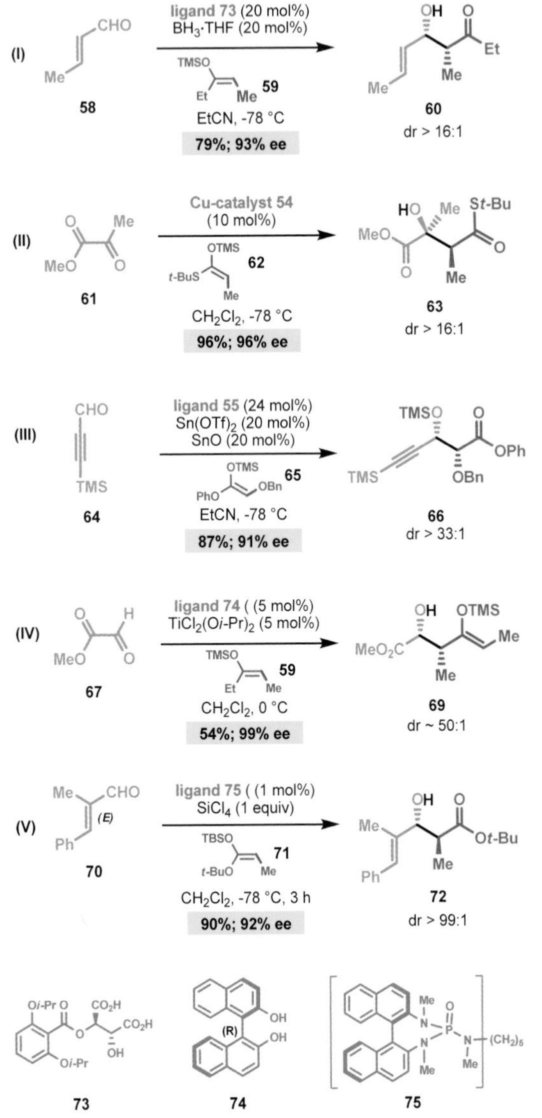

Reference key for equations: (I)[25]; (II)[26]; (III)[27]; (IV)[28]; (V)[29]

Scheme 6

The use of chiral Lewis acid complexes can provide control of absolute as well as relative configuration at two new stereocenters and lead to a very powerful synthetic construction. Vinyloxy boranes are also very useful enolate equivalents, especially with recoverable chiral attachments to boron (see examples in next section).

V. Enantio- and Diastereoselective Aldol Reaction Using Chiral Controllers: An Introduction

A common form of the stereocontrolled aldol reaction utilizes a chiral enolate partner formed by attachment of a carboxylic acid to a chiral controller/auxiliary group.[30] Because this approach frequently provides good stereocontrol and yield along with predictable product configuration, ease of purification and recoverability of the controller, it can be advantageous despite the need for stoichiometric amounts of the chiral auxiliary. Four chiral auxiliaries for aldol reactions with acetate derivatives are shown in Figure 1.

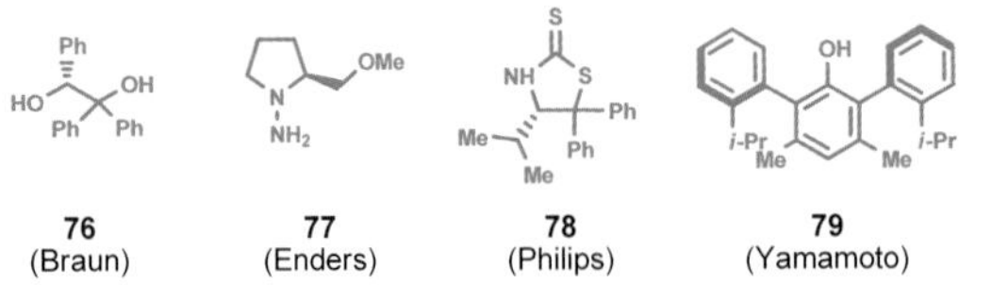

Figure 1

Some chiral auxiliaries (see Figure 2) allow propionate aldol reactions with excellent *syn* or *anti* stereochemistry and their use is illustrated in Scheme 9.

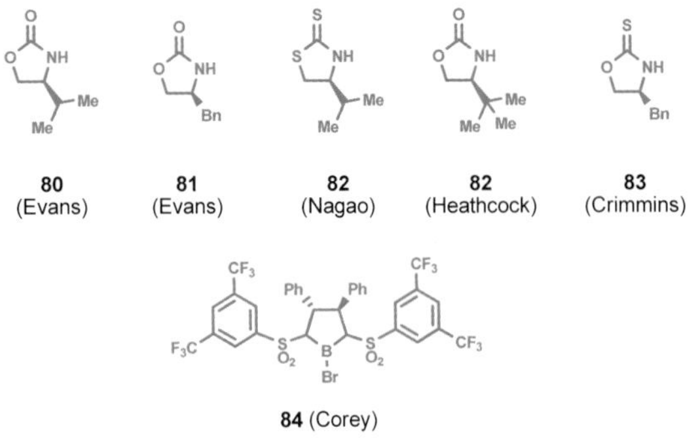

Figure 2

In the case of chiral *N*-propionyl-2-oxazolidinones (or thiones) a (*Z*)-enolate structure can lead to different *syn* aldol products (bicoordinate *syn* aldol as well as the tricoordinate *syn* aldol, see Figure 3) depending on whether the pathway involves pre-transition state assembly **E** or **F**.

STEREOCONTROLLED ALDOL REACTION

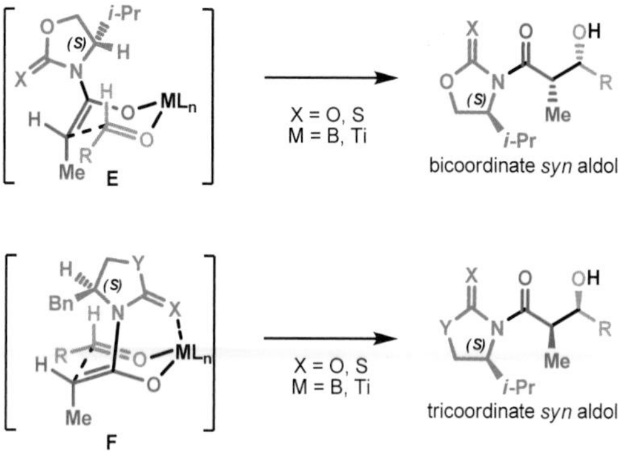

Figure 3

VI. Enantioselective Acetate Aldol Reaction Using Chiral Auxiliaries

It often happens that one type of chiral auxiliary works well in propionate aldol reactions but gives poor results in acetate aldol reactions. Scheme 7 shows four examples in which acetate aldol products are formed in high yield and high enantiomeric excess.

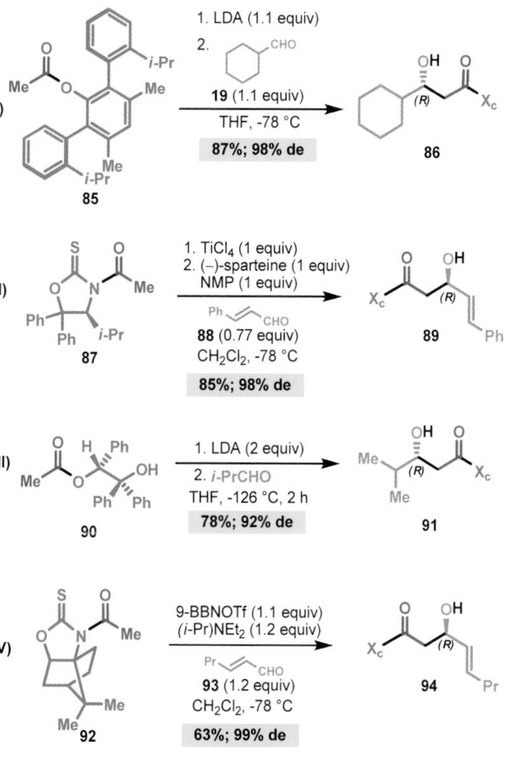

Reference key for equations: (I)[31]; (II)[32]; (III)[33]; (IV)[34]

Scheme 7

VII. Enantio- and Diastereoselective Aldol Reaction Using Chiral Ligands

Effective enantio- and diastereocontrol of aldol reactions can be achieved by the use of chiral reagents such as **84**, **96**, **98** and **101** (Scheme 8). Examples IV and V illustrate that either *syn* or *anti* aldol products may be prepared via an *E* or *Z* ester enolate, depending on the specific ester derivative and base catalyst employed.[35]

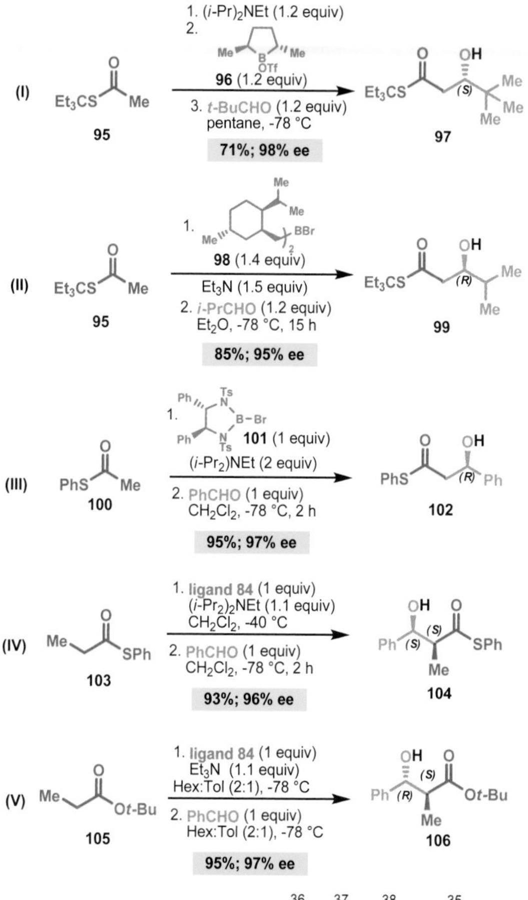

Reference key for equations: (I)[36]; (II)[37]; (III)[38]; (IV-V)[35]

Scheme 8

VIII. Enantio- and Diastereoselective Propionate and Higher Aldol Reaction Using Chiral Controllers

The use of chiral derivatives of carboxylic acids to generate chiral enolates for aldol coupling reactions provides a valuable method for the control of stereochemistry. A number of chiral controller groups have emerged as the most useful for several reasons, including: (1) the controller group is readily available and also recoverable for reuse; (2) a high degree of diastereoselectivity can be achieved and (3) the product is often crystalline and readily purified by recrystallization. Some examples are shown in Scheme 9.

STEREOCONTROLLED ALDOL REACTION

Enantio- and Diastereoselective Propionate and Higher Aldol Reaction Using Chiral Controllers (Continued)

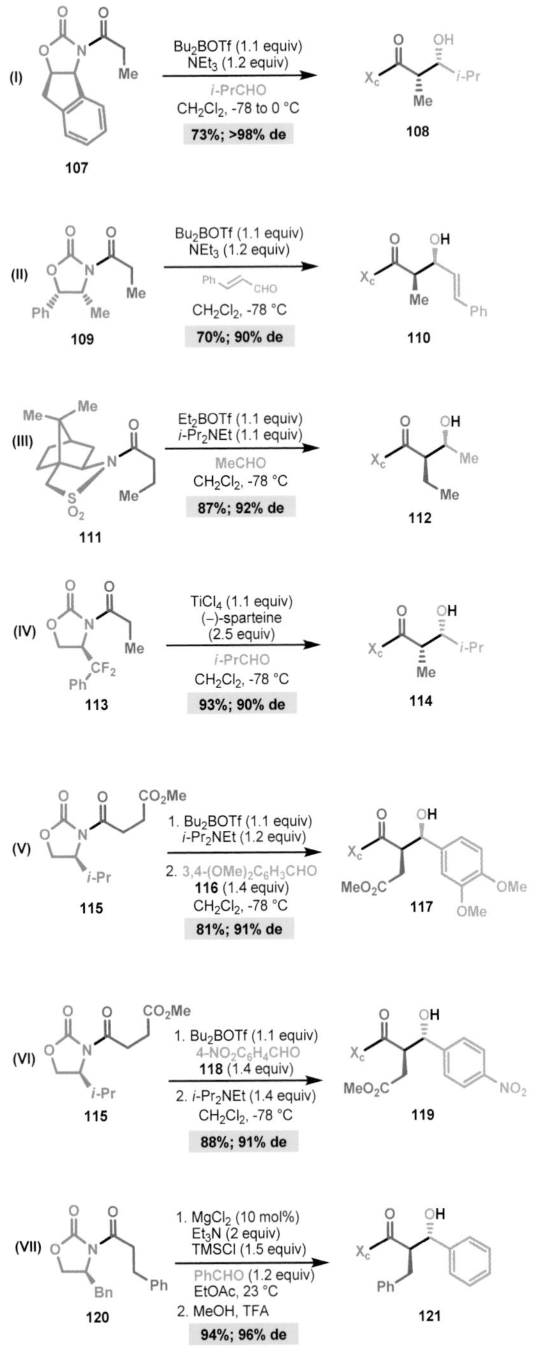

(I)

107

Bu$_2$BOTf (1.1 equiv)
NEt$_3$ (1.2 equiv)

i-PrCHO
CH$_2$Cl$_2$, -78 to 0 °C

73%; >98% de

108

(II)

109

Bu$_2$BOTf (1.1 equiv)
NEt$_3$ (1.2 equiv)

Ph⁀CHO
CH$_2$Cl$_2$, -78 °C

70%; 90% de

110

(III)

111

Et$_2$BOTf (1.1 equiv)
i-Pr$_2$NEt (1.1 equiv)

MeCHO
CH$_2$Cl$_2$, -78 °C

87%; 92% de

112

(IV)

113

TiCl$_4$ (1.1 equiv)
(−)-sparteine
(2.5 equiv)

i-PrCHO
CH$_2$Cl$_2$, -78 °C

93%; 90% de

114

(V)

115

1. Bu$_2$BOTf (1.1 equiv)
   i-Pr$_2$NEt (1.2 equiv)
2. 3,4-(OMe)$_2$C$_6$H$_3$CHO
   116 (1.4 equiv)
   CH$_2$Cl$_2$, -78 °C

81%; 91% de

117

(VI)

115

1. Bu$_2$BOTf (1.1 equiv)
   4-NO$_2$C$_6$H$_4$CHO
   118 (1.4 equiv)
2. *i*-Pr$_2$NEt (1.4 equiv)
   CH$_2$Cl$_2$, -78 °C

88%; 91% de

119

(VII)

120

1. MgCl$_2$ (10 mol%)
   Et$_3$N (2 equiv)
   TMSCl (1.5 equiv)
   PhCHO (1.2 equiv)
   EtOAc, 23 °C
2. MeOH, TFA

94%; 96% de

121

Reference key for equations: (I)[39]; (II)[40]; (III)[41]
(IV)[42]; (V-VI)[43]; (VII)[44]

Scheme 9

IX. Aldol Reaction as Key Steps in the Total Synthesis of Natural Products

There are many examples of stereocontrolled synthesis of natural products[1-2] that depend on the stereoselective aldol process a key step. Shown in Scheme 10 are two typical applications to the target structures pictured within the boxes below the key aldol reactions.

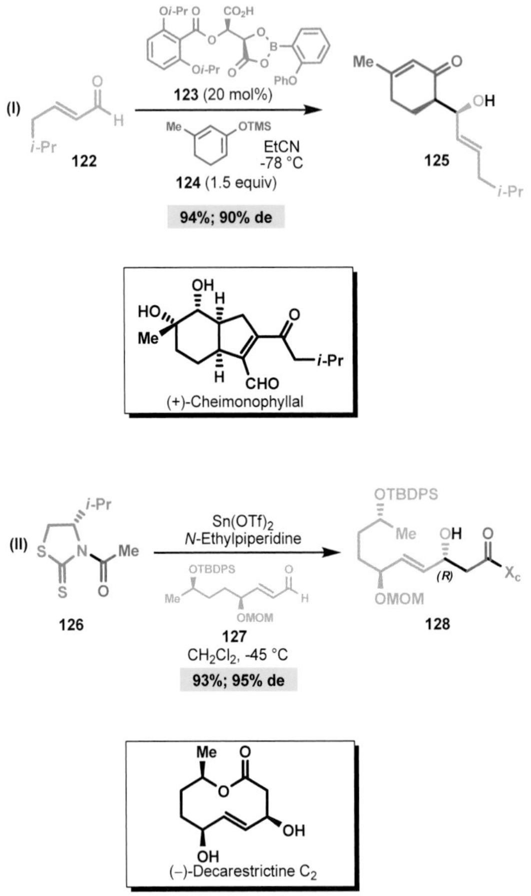

(I)

122

123 (20 mol%)

124 (1.5 equiv)
EtCN
-78 °C

94%; 90% de

125

(+)-Cheimonophyllal

(II)

126

Sn(OTf)$_2$
N-Ethylpiperidine

127
CH$_2$Cl$_2$, -45 °C

93%; 95% de

128

(−)-Decarestrictine C$_2$

Reference key for equations: (I)[45]; (II)[46]

Scheme 10

Enantio-and diastereoselective aldol reactions have been used so extensively in recent years, especially for the synthesis of polypropionate-type natural products, that a vast amount of information has been accumulated with regard to controller groups and metal enolate used (Li, Na, K, Zn, Ti(IV), Sn(II) or Sn(IV), BR$_2$ or AlR$_2$), reaction conditions, or in the case of Mukaiyama aldol reactions, the Lewis acids employed. This information can be found in various published reviews,[1-3] as well as literature cited therein. The results of an interesting comparative study are shown in Scheme 11 in which modular access to all possible *syn-* and *anti*-aldol products was obtained by careful choice of reaction conditions.[47]

STEREOCONTROLLED ALDOL REACTION

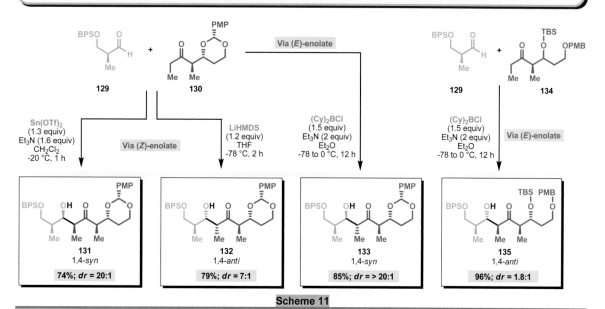

Scheme 11

Often the aldol reaction has been used several times in the same synthesis. For instance, in the synthesis of macbecin I, the three bonds, indicated as **a**, **b** and **c** (red marks), were formed sequentially using an oxazolidinone controller and boron enolate as shown in Schemes 12 and 13.[48,40]

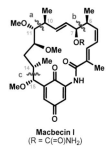

Macbecin I
(R = C(=O)NH$_2$)

Construction of bond a:

n-Bu$_2$BOTf
Et$_3$N

70%; dr = 20:1

136   137   138

Construction of bond b:

n-Bu$_2$BOTf
Et$_3$N

77%; dr = 20:1

136   139   140

Scheme 12

Construction of bond c:

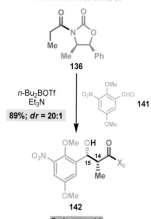

n-Bu$_2$BOTf
Et$_3$N

89%; dr = 20:1

136   141

142

Scheme 13

References

1. (a) Cowden, C. J. & Paterson, I. Asymmetric aldol reactions using boron enolates. *Org. React.* **1997**, 51, 1-200; (b) *Modern Aldol Reactions, Volume 1: Enolates, Organocatalysis, Biocatalysis and Natural Product Synthesis* (ed. Mahrwald, R.) (Wiley-VCH, **2004**).

2. *Modern Aldol Reactions, Volume 2: Metal catalysis* (ed. Mahrwald, R.) (Wiley-VCH, **2004**).

3. For reviews on the Mukaiyama aldol reaction, see: (a) Carreira, E. M. Mukaiyama aldol reaction in *Compr. Asymmetric Catal. I-III* 997-1065 (**1999**); (b) Kalesse, M. & Hassfeld, J. Asymmetric vinylogous Mukaiyama aldol reaction. *Asymmetric Synth.* **2007**, 105-109; (c) Carreira, E. M. & Kvaerno, L. Aldol Reactions in *Classics in Stereoselective Synthesis* 103-151 (Wiley, **2008**).

4. For selected recent reviews on catalytic enantioselective aldol reactions, see: (a) Carreira, E. M., Fettes, A. & Marti, C. Catalytic enantioselective aldol addition reactions. *Organic Reactions* **2006**, 67, 1-216; (b) Mlynarski, J. & Paradowska, J. Catalytic asymmetric aldol reactions in aqueous media. *Chem. Soc. Rev.* **2008**, 37, 1502-1511; (c) Pan, S. C. & List, B. New concepts for organocatalysis. *Ernst Schering Foundation Symposium Proceedings* **2008**, 1-43; (d)

Guillena, G., Najera, C. & Ramon, D. J. Enantioselective direct aldol reaction: the blossoming of modern organocatalysis. *Tetrahedron Asymmetry* **2007**, 18, 2249-2293; (e) Pellissier, H. Asymmetric organocatalysis. *Tetrahedron* **2007**, 63, 9267-9331; (f) Mukherjee, S., Yang, J. W., Hoffmann, S. & List, B. Asymmetric Enamine Catalysis. *Chem. Rev.* **2007**, 107, 5471-5569.

5. Eder, U., Wiechert, R. & Sauer, G. Optically active 1,5-indandione and 1,6-naphthalenedione derivatives. DE 2013757 (**1971**, Patent filed on March 20, 1970).

6. Hajos, Z. G. & Parrish, D. R. Asymmetric synthesis of optically active polycyclic organic compounds. DE 2102623 (**1971**, Patent filed on January 20, 1971).

7. Hoang, L., Bahmanyar, S., Houk, K. N. & List, B. Kinetic and Stereochemical Evidence for the Involvement of Only One Proline Molecule in the Transition States of Proline-Catalyzed Intra- and Intermolecular Aldol Reactions. *J. Am. Chem. Soc.* **2003**, 125, 16-17.

8. (a) Hajos, Z. G. & Parrish, D. R. Synthesis and conversion of 2-methyl-2-(3-oxobutyl)-1,3-cyclopentanedione to the isomeric racemic ketols of the [3.2.1]bicyclooctane and of the perhydroindane series. *J. Org. Chem.* **1974**, 39, 1612-1615; (b) Hajos, Z. G. & Parrish, D. R. Asymmetric synthesis of bicyclic intermediates of natural product chemistry. *J. Org. Chem.* **1974**, 39, 1615-1621; (c) Kürti, L. & Czakó, B. Hajos-Parrish Reaction in *Strategic Applications of Named Reactions in Organic Synthesis* 192-193 (Academic Press/Elsevier, San Diego, **2005**).

9. List, B., Lerner, R. A. & Barbas, C. F., III. Proline-Catalyzed Direct Asymmetric Aldol Reactions. *J. Am. Chem. Soc.* **2000**, 122, 2395-2396.

10. Nakadai, M., Saito, S. & Yamamoto, H. Diversity-based strategy for discovery of environmentally benign organocatalyst: diamine-protonic acid catalysts for asymmetric direct aldol reaction. *Tetrahedron* **2002**, 58, 8167-8177.

11. Trost, B. M. & Ito, H. A direct catalytic enantioselective aldol reaction via a novel catalyst design. *J. Am. Chem. Soc.* **2000**, 122, 12003-12004.

12. Suzuki, T., Yamagiwa, N., Matsuo, Y., Sakamoto, S., Yamaguchi, K., Shibasaki, M. & Noyori, R. Catalytic asymmetric aldol reaction of ketones and aldehydes using chiral calcium alkoxides. *Tetrahedron Lett.* **2001**, 42, 4669-4671.

13. Trost, B. M., Ito, H. & Silcoff, E. R. Asymmetric aldol reaction via a dinuclear zinc catalyst: α-Hydroxy ketones as donors. *J. Am. Chem. Soc.* **2001**, 123, 3367-3368.

14. Pihko, P. M. & Erkkila, A. Enantioselective synthesis of prelactone B using a proline-catalyzed crossed-aldol reaction. *Tetrahedron Lett.* **2003**, 44, 7607-7609.

15. Pidathala, C., Hoang, L., Vignola, N. & List, B. Direct catalytic asymmetric enolexo aldolizations. *Angew. Chem., Int. Ed.* **2003**, 42, 2785-2788.

16. Enders, D., Palecek, J. & Grondal, C. A direct organocatalytic entry to sphingoids: asymmetric synthesis of D-arabino- and L-ribo-phytosphingosine. *Chem. Commun. (Cambridge, U. K.)* **2006**, 655-657.

17. Itagaki, N., Kimura, M., Sugahara, T. & Iwabuchi, Y. Organocatalytic Entry to Chiral Bicyclo[3.n.1]alkanones via Direct Asymmetric Intramolecular Aldolization. *Org. Lett.* **2005**, 7, 4185-4188.

18. Kotrusz, P., Kmentova, I., Gotov, B., Toma, S. & Solcaniova, E. Proline-catalyzed asymmetric aldol reaction in the room temperature ionic liquid [bmim]PF6. *Chem. Commun. (Cambridge, U. K.)* **2002**, 2510-2511.

19. He, L., Tang, Z., Cun, L.-F., Mi, A.-Q., Jiang, Y.-Z. & Gong, L.-Z. L-Proline amide-catalyzed direct asymmetric aldol reaction of aldehydes with chloroacetone. *Tetrahedron* **2005**, 62, 346-351.

20. Itsuno, S. & Komura, K. Highly stereoselective synthesis of chiral aldol polymers using repeated asymmetric Mukaiyama aldol reaction. *Tetrahedron* **2002**, 58, 8237-8246.

21. Evans, D. A., Kozlowski, M. C., Burgey, C. S. & MacMillan, D. W. C. C2-Symmetric Copper(II) Complexes as Chiral Lewis Acids. Catalytic Enantioselective Aldol Additions of Enol Silanes to Pyruvate Esters. *J. Am. Chem. Soc.* **1997**, 119, 7893-7894.

22. Kobayashi, S. & Furuta, T. Use of heterocycles as chiral ligands and auxiliaries in asymmetric syntheses of sphingosine, sphingofungins B and F. *Tetrahedron* **1998**, 54, 10275-10294.

23. Singer, R. A. & Carreira, E. M. An in situ procedure for catalytic, enantioselective acetate aldol addition. Application to the synthesis of (R)-(-)-epinephrine. *Tetrahedron Lett.* **1997**, 38, 927-930.

24. Yamashita, Y., Ishitani, H., Shimizu, H. & Kobayashi, S. Highly anti-Selective Asymmetric Aldol Reactions Using Chiral Zirconium Catalysts. Improvement of Activities, Structure of the Novel Zirconium Complexes, and Effect of a Small Amount of Water for the Preparation of the Catalysts. *J. Am. Chem. Soc.* **2002**, 124, 3292-3302.

25. Furuta, K., Maruyama, T. & Yamamoto, H. Catalytic asymmetric aldol reactions. Use of a chiral (acyloxy)borane complex as a versatile Lewis-acid catalyst. *J. Am. Chem. Soc.* **1991**, 113, 1041-1042.

26. Evans, D. A., Burgey, C. S., Kozlowski, M. C. & Tregay, S. W. C2-Symmetric Copper(II) Complexes as Chiral Lewis Acids. Scope and Mechanism of the Catalytic Enantioselective Aldol Additions of

Enolsilanes to Pyruvate Esters. *J. Am. Chem. Soc.* **1999**, 121, 686-699.

27. Kobayashi, S., Hayashi, T., Iwamoto, S., Furuta, T. & Matsumura, M. Asymmetric synthesis of antifungal agents sphingofungins using catalytic asymmetric aldol reactions. *Synlett* **1996**, 672-674.

28. Mikami, K. & Matsukawa, S. Enantioselective and diastereoselective catalysis of the Mukaiyama aldol reaction: ene mechanism in titanium-catalyzed aldol reactions of silyl enol ethers. *J. Am. Chem. Soc.* **1993**, 115, 7039-7040.

29. Denmark, S. E., Wynn, T. & Beutner, G. L. Lewis Base Activation of Lewis Acids. Addition of Silyl Ketene Acetals to Aldehydes. *J. Am. Chem. Soc.* **2002**, 124, 13405-13407.

30. For reviews on asymmetric aldol reactions utilizing chiral auxiliaries, see: (a) Cowden, C. J. & Paterson, I. Asymmetric aldol reactions using boron enolates. *Org. React.* **1997**, 51, 1-200; (b) Palomo, C., Oiarbide, M. & Garcia, J. M. The aldol addition reaction: an old transformation at constant rebirth. *Chem.--Eur. J.* **2002**, 8, 36-44; (c) Velazquez, F. & Olivo, H. F. The application of chiral oxazolidinethiones and thiazolidinethiones in asymmetric synthesis. *Curr. Org. Chem.* **2002**, 6, 303-340; (d) Geary, L. M. & Hultin, P. G. The state of the art in asymmetric induction: the aldol reaction as a case study. *Tetrahedron Asymmetry* **2009**, 20, 131-173; (e) Li, J. & Menche, D. Direct methods for stereoselective polypropionate synthesis: a survey. *Synthesis* **2009**, 2293-2315.

31. Saito, S., Hatanaka, K., Kano, T. & Yamamoto, H. Diastereoselective aldol reaction with an acetate enolate: 2,6-bis(2-isopropylphenyl)-3,5-dimethylphenol as an extremely effective chiral auxiliary. *Angew. Chem., Int. Ed.* **1998**, 37, 3378-3381.

32. Guz, N. R. & Phillips, A. J. Practical and Highly Selective Oxazolidinethione-Based Asymmetric Acetate Aldol Reactions with Aliphatic Aldehydes. *Org. Lett.* **2002**, 4, 2253-2256.

33. Braun, M. & Graf, S. Stereoselective aldol reaction of doubly deprotonated (R)-(+)-2-hydroxy-1,2,2-triphenylethyl acetate (HYTRA): (R)-3-hydroxy-4-methylpentanoic acid (pentanoic acid, 3-hydroxy-4-methyl-, (R)-). *Org. Synth.* **1995**, 72, 38-47.

34. Yan, T.-H., Hung, A.-W., Lee, H.-C. & Chang, C.-S. An Unusual Enantioselective Aldol-Type Reaction of an Acetate Boryl Enolate Derived from a Chiral Thioimide. *J. Org. Chem.* **1994**, 59, 8187-8191.

35. Corey, E. J. & Kim, S. S. Versatile chiral reagent for the highly enantioselective synthesis of either anti or syn ester aldols. *J. Am. Chem. Soc.* **1990**, 112, 4976-4977.

36. Masamune, S., Sato, T., Kim, B. & Wollmann, T. A. Organoboron compounds in organic synthesis. 4. Asymmetric aldol reactions. *J. Am. Chem. Soc.* **1986**, 108, 8279-8281.

37. Gennari, C., Moresca, D., Vieth, S. & Vulpetti, A. Computer-aided design of chiral boron enolates: a new, highly enantioselective aldol reaction of thioacetates and thiopropionates. *Angew. Chem.* **1993**, 105, 1717-1719

38. Corey, E. J., Imwinkelried, R., Pikul, S. & Xiang, Y. B. Practical enantioselective Diels-Alder and aldol reactions using a new chiral controller system. *J. Am. Chem. Soc.* **1989**, 111, 5493-5495.

39. Ghosh, A. K., Duong, T. T. & McKee, S. P. Highly enantioselective aldol reaction: development of a new chiral auxiliary from cis-1-amino-2-hydroxyindan. *J. Chem. Soc., Chem. Commun.* **1992**, 1673-1674.

40. Evans, D. A., Miller, S. J. & Ennis, M. D. Asymmetric synthesis of the benzoquinoid ansamycin antitumor antibiotics: total synthesis of (+)-macbecin. *J. Org. Chem.* **1993**, 58, 471-485.

41. Oppolzer, W., Blagg, J., Rodriguez, I. & Walther, E. Bornane sultam-directed asymmetric synthesis of crystalline, enantiomerically pure syn aldols. *J. Am. Chem. Soc.* **1990**, 112, 2767-2772.

42. Fustero, S., Piera, J., Sanz-Cervera, J. F., Bello, P. & Mateu, N. Synthesis of a new fluorinated oxazolidinone and its reactivity as a chiral auxiliary in aldol reactions. *J. Fluorine Chem.* **2007**, 128, 647-653.

43. Hajra, S., Giri, A. K., Karmakar, A. & Khatua, S. Asymmetric aldol reactions under normal and inverse addition modes of the reagents. *Chem. Commun. (Cambridge, U. K.)* **2007**, 2408-2410.

44. Evans, D. A., Tedrow, J. S., Shaw, J. T. & Downey, C. W. Diastereoselective Magnesium Halide-Catalyzed anti-Aldol Reactions of Chiral N-Acyloxazolidinones. *J. Am. Chem. Soc.* **2002**, 124, 392-393.

45. Takao, K., Tsujita, T., Hara, M. & Tadano, K. Asymmetric Total Syntheses of (+)-Cheimonophyllon E and (+)-Cheimonophyllal. *J. Org. Chem.* **2002**, 67, 6690-6698.

46. Arai, M., Morita, N., Aoyagi, S. & Kibayashi, C. Total synthesis of (-)-(3R,6S,9R)-decarestrictine C2. *Tetrahedron Lett.* **2000**, 41, 1199-1203.

47. (a) Arikan, F., Li, J. & Menche, D. Diastereodivergent Aldol Reactions of β-Alkoxy Ethyl Ketones: Modular Access to (1,4)-*syn* and -*anti* Polypropionates. *Org. Lett.* **2008**, 10, 3521-3524; (b) Li, J. & Menche, D. Direct methods for stereoselective polypropionate synthesis: a survey. *Synthesis* **2009**, 2293-2315.

48. Evans, D. A., Miller, S. J., Ennis, M. D. & Ornstein, P. L. Asymmetric synthesis of macbecin I. *J. Org. Chem.* **1992**, 57, 1067-1069.

ADDITION OF sp³ C-NUCLEOPHILES TO C=O

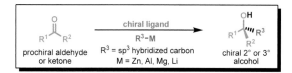

prochiral aldehyde or ketone

chiral 2° or 3° alcohol

R^3–M
R^3 = sp³ hybridized carbon
M = Zn, Al, Mg, Li

Introduction

Early studies showed that the traditional addition of an organolithium or organomagnesium reagent to an aldehyde or a prochiral ketone to form a racemic alcohol, which occurs rapidly and without appreciable acceleration by chiral ligands such as 1,2-amino alcohols or 1,2-diamines, was not practical for enantioselective synthesis. Subsequent advances for enantioselective carbonyl addition with RLi or RMgX entailed the use of one or more equivalents of chiral ligand.[1] In contrast, the generally slow addition of dialkylzinc reagents to carbonyl is strongly accelerated by a catalytic amount of chiral 1,2-amino alcohols (e.g. leucinol, **2**) and has become a valuable enantioselective reaction, as first demonstrated by Oguni in 1984 with examples such as **1→3**[2] (Scheme 1). The modest enantioselectivities observed by Oguni were greatly improved by Noyori who introduced the use of the camphor-derived chiral amino alcohol DAIB (**5**) in 1986 with reactions such as **4→6**[3].

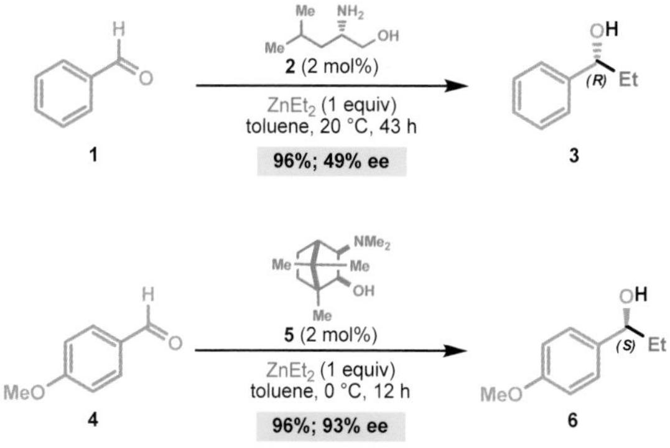

Scheme 1

1 → 3: 96%; 49% ee
2 (2 mol%), ZnEt₂ (1 equiv), toluene, 20 °C, 43 h

4 → 6: 96%; 93% ee
5 (2 mol%), ZnEt₂ (1 equiv), toluene, 0 °C, 12 h

The Oguni-Noyori process has been much used and periodically reviewed.[4] The simple model defined by **8** (Scheme 2) can account for the sense of the absolute stereochemistry for observed reactions catalyzed by **5** (see page 165 for pathway).[5]

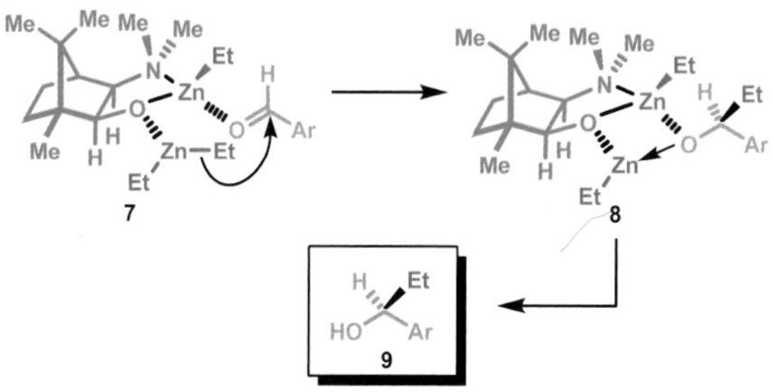

Scheme 2

I. Addition of Simple Dialkylzinc Reagents to Aldehydes

The most commonly used catalysts/ligands for enantioselective 1,2-addition reactions are shown in Figure 1.

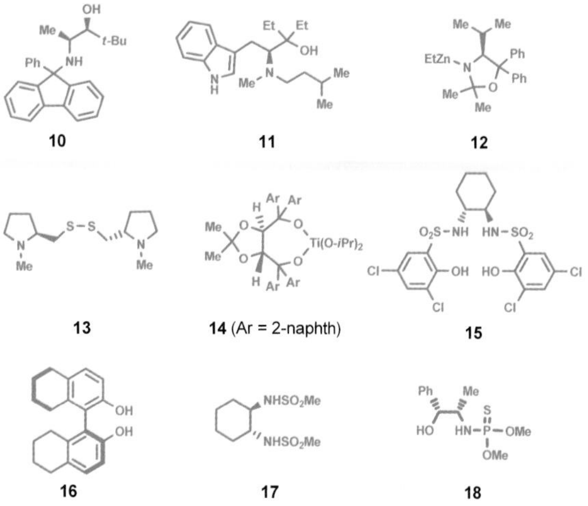

Figure 1

10 11 12

13 14 (Ar = 2-naphth) 15

16 17 18

The eight examples shown in Schemes 3 and 4 suggest that a large variety of chiral ligands (e.g., **10** to **18**) can catalyze the enantioselective 1,2-addition of dialkylzinc reagents to aldehydes.

(I) **19** → **20**
ligand 10 (3 mol%), ZnEt₂ (1 equiv), toluene, 18 °C, 18 h
98%; 98% ee

(II) **21** → **22**
ligand 11 (10 mol%), ZnEt₂ (1 equiv), toluene, 23 °C, 96 h
84%; 95% ee

(III) **23** → **24**
ligand 12 (10 mol%), ZnEt₂ (2 equiv), toluene, 0 °C, 4 h
75%; 97% ee

(IV) **25** → **26**
ligand 13 (2.5 mol%), ZnEt₂ (1 equiv), toluene, 0 °C, 48 h
88%; 90% ee

(V) **27** → **28**
catalyst 14 (20 mol%), Ti(O-iPr)₄ (1.2 equiv), ZnEt₂ (1.8 equiv), toluene, -27 °C, 30 h
77%; 99% ee

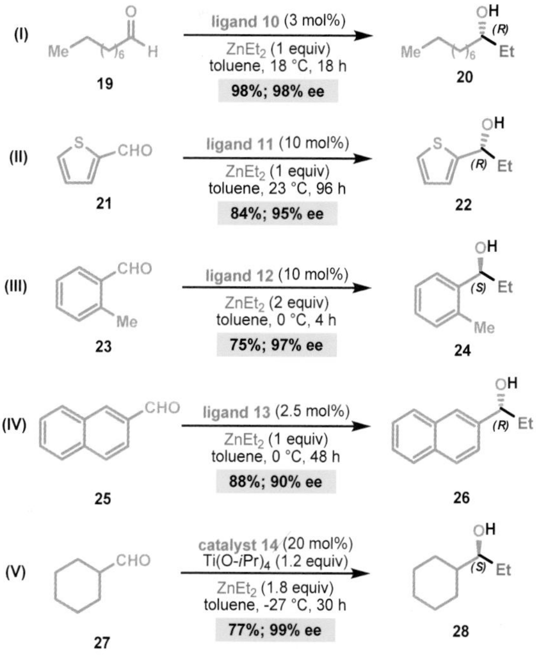

Reference key for equations: (I)[6]; (II)[7]; (III)[8]; (IV)[9]; (V)[10]

Scheme 3

ADDITION OF sp^3 C-NUCLEOPHILES TO C=O

Addition of Simple Dialkylzinc Reagents to Aldehydes (Continued)

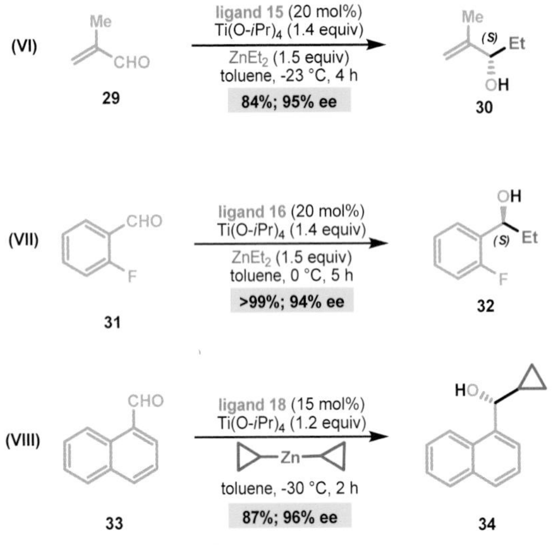

(VI)
Me
CHO
29
ligand 15 (20 mol%)
Ti(O-*i*Pr)$_4$ (1.4 equiv)
ZnEt$_2$ (1.5 equiv)
toluene, -23 °C, 4 h
84%; 95% ee
Me
(S)
Et
OH
30
(VII)
CHO
F
31
ligand 16 (20 mol%)
Ti(O-*i*Pr)$_4$ (1.4 equiv)
ZnEt$_2$ (1.5 equiv)
toluene, 0 °C, 5 h
>99%; 94% ee
OH
(S)
Et
F
32
(VIII)
CHO
33
ligand 18 (15 mol%)
Ti(O-*i*Pr)$_4$ (1.2 equiv)
Zn
toluene, -30 °C, 2 h
87%; 96% ee
HO
34

Reference key for equations: (VI)[11]; (VII)[12]; (VIII)[13]

Scheme 4

II. Addition of Functionalized Dialkylzinc Reagents to Aldehydes

Unfunctionalized dialkylzinc reagents can be obtained by transmetallation of organolithium or organomagnesium reagents with zinc salts. Functionalized dialkylzinc reagents can be prepared by reaction of primary alkyl iodides with diethylzinc in the presence of a catalytic amount of a Cu(I)-salt. Transmetallation of triorganoboranes (R_3B) using diethylzinc with removal of Et_3B has also been reported.[4a] This latter method has been utilized for the preparation of functionalized R_2Zn reagents **36** and **39** (Scheme 5) for catalytic enantioselective 1,2-addition to aldehydes **35** and **38**, respectively, in the presence of chiral ligand **17**.[14]

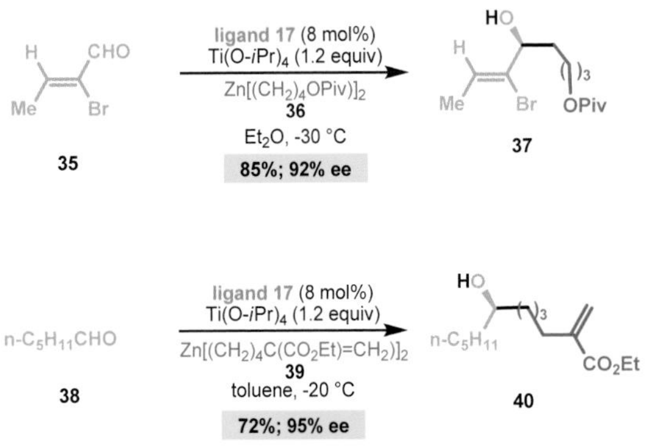

H
CHO
Me
Br
35
ligand 17 (8 mol%)
Ti(O-*i*Pr)$_4$ (1.2 equiv)
Zn[(CH$_2$)$_4$OPiv)]$_2$
36
Et$_2$O, -30 °C
85%; 92% ee
HO
H
Me
Br
OPiv
37
n-C$_5$H$_{11}$CHO
38
ligand 17 (8 mol%)
Ti(O-*i*Pr)$_4$ (1.2 equiv)
Zn[(CH$_2$)$_4$C(CO$_2$Et)=CH$_2$)]$_2$
39
toluene, -20 °C
72%; 95% ee
HO
n-C$_5$H$_{11}$
CO$_2$Et
40

Scheme 5

III. Addition of Ti or Al Reagents to Ketones

A large number of chiral catalysts promote the highly enantioselective 1,2-addition of R_2Zn to aldehydes.[4] However, this is not the case for ketones which give rise to 1,2-addition products in lower yield (because of C=O reduction or enolization) and with significantly lower enantioselectivity.[15] Highly enantioselective examples in this area are rare, because ketones are less reactive than aldehydes and the steric difference between the groups attached to the carbonyl group is generally smaller. The use of R_2Zn or RMgX and stoichiometric amounts of Ti(O-iPr)$_4$ or Al(Oi-Pr)$_3$ along with catalytic amounts of a chiral ligand (e.g., **49-51**) effects complete conversion of ketones to 1,2-addition products with excellent enantioselectivity (Scheme 6).[16]

The highly enantioselective and catalytic 1,2-addition of dimethylzinc to the α-ketoester **45** directly (without Al or Ti additives) can be achieved using the amino acid-based ligand **50**.[17] A similar transformation (**47**→**48**) was shown to be catalyzed by hydroxyproline derivative **51**.[18]

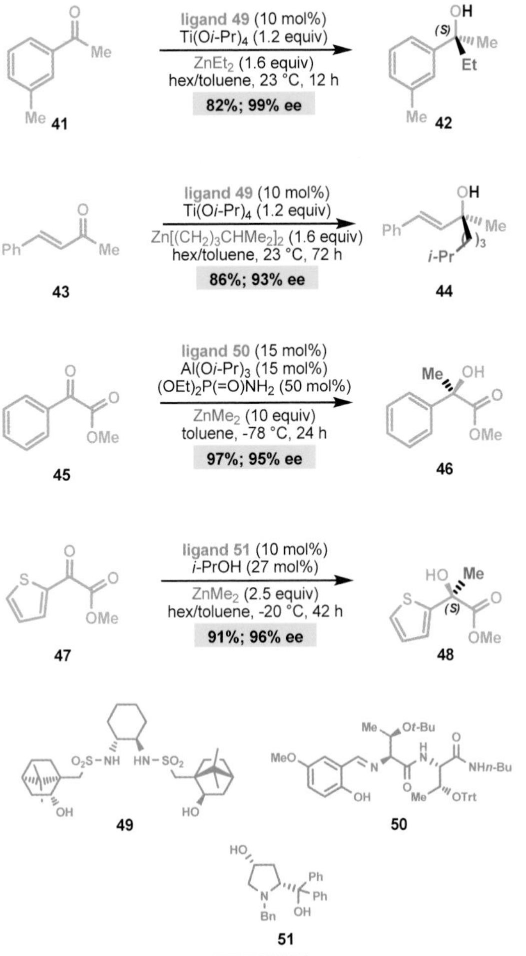

O
Me
Me
41
ligand 49 (10 mol%)
Ti(O*i*-Pr)$_4$ (1.2 equiv)
ZnEt$_2$ (1.6 equiv)
hex/toluene, 23 °C, 12 h
82%; 99% ee
OH
(S)
Me
Et
Me
42
O
Ph
Me
43
ligand 49 (10 mol%)
Ti(O*i*-Pr)$_4$ (1.2 equiv)
Zn[(CH$_2$)$_3$CHMe$_2$]$_2$ (1.6 equiv)
hex/toluene, 23 °C, 72 h
86%; 93% ee
OH
Ph
Me
i-Pr
44
O
O
OMe
45
ligand 50 (15 mol%)
Al(O*i*-Pr)$_3$ (15 mol%)
(OEt)$_2$P(=O)NH$_2$ (50 mol%)
ZnMe$_2$ (10 equiv)
toluene, -78 °C, 24 h
97%; 95% ee
Me
OH
O
OMe
46
S
O
O
OMe
47
ligand 51 (10 mol%)
i-PrOH (27 mol%)
ZnMe$_2$ (2.5 equiv)
hex/toluene, -20 °C, 42 h
91%; 96% ee
HO
Me
S
(S)
O
OMe
48
O$_2$S-NH
HN-SO$_2$
OH
HO
49
Me
O*t*-Bu
MeO
N
H
N
O
NH*n*-Bu
OH
O
Me
OTrt
50
HO
N
Bn
Ph
Ph
OH
51

Scheme 6

ADDITION OF sp³ C-NUCLEOPHILES TO C=O

IV. Addition of Chiral RMgX and RLi Complexes to Aldehydes and Ketones

As mentioned in the introduction, enantioselective addition of RMgX or RLi to aldehydes or ketones requires 1 equivalent or more of chiral ligand to generate a chiral complex and the use of temperatures between -100 to -116 °C.[19] Several recent examples of this process are shown in Scheme 7.[20]

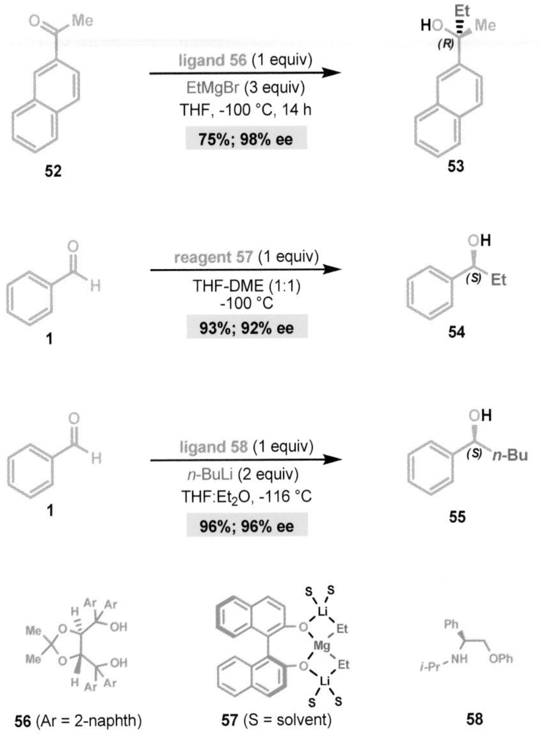

Scheme 7

References

1. Mukaiyama, T., Soai, K., Sato, T., Shimizu, H. & Suzuki, K. Enantioface-differentiating (asymmetric) addition of alkyllithium and dialkylmagnesium to aldehydes by using (2S,2'S)-2-hydroxymethyl-1-[(1-alkylpyrrolidin-2-yl)methyl]pyrrolidines as chiral ligands. J. Am. Chem. Soc. 1979, 101, 1455-1460.
2. Oguni, N. & Omi, T. Enantioselective addition of diethylzinc to benzaldehyde catalyzed by a small amount of chiral 2-amino-1-alcohols. Tetrahedron Lett. 1984, 25, 2823-2824.
3. Kitamura, M., Suga, S., Kawai, K. & Noyori, R. Catalytic asymmetric induction. Highly enantioselective addition of dialkylzincs to aldehydes. J. Am. Chem. Soc. 1986, 108, 6071-6072.
4. For leading reviews on the enantioselective additions of dialkylzincs to carbonyl compounds, see: (a) Knochel, P., Millot, N., Rodriguez, A. L. & Tucker, C. E. Preparation and applications of functionalized organozinc compounds. Org. React. 2001, 58, 417-731; (b) Pu, L. & Yu, H.-B. Catalytic Asymmetric Organozinc Additions to Carbonyl Compounds. Chem. Rev. 2001, 101, 757-824; (c) Soai, K. & Niwa, S. Enantioselective addition of organozinc reagents to aldehydes. Chem. Rev. 1992, 92, 833-856; (d) Hatano, M. & Ishihara, K. Catalytic enantioselective organozinc addition toward optically active tertiary alcohol synthesis. Chem. Rec. 2008, 8, 143-155.

5. Corey, E. J., Yuen, P. W., Hannon, F. J. & Wierda, D. A. Polyfunctional, structurally defined catalysts for the enantioselective addition of dialkylzinc reagents to aldehydes. J. Org. Chem. 1990, 55, 784-786.
6. Paleo, M. R., Cabeza, I. & Sardina, F. J. Enantioselective Addition of Diethylzinc to Aldehydes Catalyzed by N-(9-Phenylfluoren-9-yl) β-Amino Alcohols. J. Org. Chem. 2000, 65, 2108-2113.
7. (a) Dai, W.-M., Zhu, H. J. & Hao, X.-J. Chiral ligands derived from Abrine. 1. Synthesis of sec- and tert-β-amino alcohols and catalysis for enantioselective addition of diethylzinc toward aromatic aldehydes. Tetrahedron Asymmetry 1995, 6, 1857-1860; (b) Dai, W. M., Zhu, H. J. & Hao, X. J. Chiral ligands derived from abrine. Part 6: Importance of a bulky N-alkyl group in indole-containing chiral β-tertiary amino alcohols for controlling enantioselectivity in addition of diethylzinc toward aldehydes. Tetrahedron Asymmetry 2000, 11, 2315-2337.
8. Prasad, K. R. K. & Joshi, N. N. Chiral Zinc Amides as Catalysts for the Enantioselective Addition of Diethylzinc to Aldehydes. J. Org. Chem. 1997, 62, 3770-3771.
9. (a) Cran, G. A., Gibson, C. L. & Handa, S. Synthesis of chiral β-amino sulfides and β-amino thiols from α-amino acids. Tetrahedron Asymmetry 1995, 6, 1553-1556; (b) Gibson, C. L. Enantioselective addition of diethylzinc to aldehydes catalyzed by a β-amino disulfide derived from L-proline. Chem. Commun. 1996, 645-646.
10. Schmidt, B. & Seebach, D. Catalytic and stoichiometric enantioselective addition of methyllithium and magnesium and zinc compounds to aldehydes with the aid of a novel chiral spirotitanate. Angew. Chem. Int. Ed. Engl. 1991, 30, 99-101.
11. (a) Zhang, X. & Guo, C. Enantioselective addition of diethylzinc to aldehydes catalyzed by chiral titanate complexes with a tetradentate ligand. Tetrahedron Lett. 1995, 36, 4947-4950; (b) Qiu, J., Guo, C. & Zhang, X. Enantioselective Addition of Diethylzinc to Aldehydes Catalyzed by a Titanate Complex with a Chiral Tetradentate Ligand. J. Org. Chem. 1997, 62, 2665-2668.
12. Zhang, F.-Y. & Chan, A. S. C. Enantioselective addition of diethylzinc to aromatic aldehydes catalyzed by titanium-5,5',6,6',7,7',8,8'-octahydro-1,1'-bi-2-naphthol complex. Tetrahedron Asymmetry 1997, 8, 3651-3655.
13. Shibata, T., Tabira, H. & Soai, K. A catalytic and enantioselective cyclopropylation of aldehydes using dicyclopropylzinc. J. Chem. Soc., Perkin Trans. 1 1998, 177-178.
14. Langer, F., Schwink, L., Devasagayaraj, A., Chavant, P.-Y. & Knochel, P. Preparation of Functionalized Dialkylzincs via a Boron-Zinc Exchange. Reactivity and Catalytic Asymmetric Addition to Aldehydes. J. Org. Chem. 1996, 61, 8229-8243.
15. For a recent review on the asymmeric addition of organometallic reagents to ketones, see: Garcia, C. & Martin, V. S. Asymmetric addition to ketones: enantioselective formation of tertiary alcohols. Curr. Org. Chem. 2006, 10, 1849-1889.
16. Jeon, S.-J., Li, H., Garcia, C., LaRochelle, L. K. & Walsh, P. J. Catalytic asymmetric addition of alkylzinc and functionalized alkylzinc reagents to ketones. J. Org. Chem. 2005, 70, 448-455.
17. Wieland, L. C., Deng, H., Snapper, M. L. & Hoveyda, A. H. Al-Catalyzed Enantioselective Alkylation of α-Keto Esters by Dialkylzinc Reagents. Enhancement of Enantioselectivity and Reactivity by an Achiral Lewis Base Additive. J. Am. Chem. Soc. 2005, 127, 15453-15456.
18. Funabashi, K., Jachmann, M., Kanai, M. & Shibasaki, M. Multicenter strategy for the development of catalytic enantioselective nucleophilic alkylation of ketones: Me₂Zn addition to α-ketoesters. Angew. Chem., Int. Ed. 2003, 42, 5489-5492.
19. For a recent review, see: Luderer, M. R., Bailey, W. F., Luderer, M. R., Fair, J. D., Dancer, R. J. & Sommer, M. B. Asymmetric addition of achiral organomagnesium reagents or organolithiums to achiral aldehydes or ketones: a review. Tetrahedron Asymmetry 2009, 20, 981-998.
20. (a) Noyori, R., Suga, S., Kawai, K., Okada, S. & Kitamura, M. Enantioselective alkylation of carbonyl compounds. From stoichiometric to catalytic asymmetric induction. Pure Appl. Chem. 1988, 60, 1597-1606; (b) Weber, B. & Seebach, D. Highly enantioselective addition of primary alkyl Grignard reagents to carbocyclic and heterocyclic aryl ketones in the presence of magnesium TADDOLate. Tetrahedron 1994, 50, 6117-6128; (c) Roennholm, P., Soedergren, M. & Hilmersson, G. Improved and efficient synthesis of chiral N,P-ligands via cyclic sulfamidates for asymmetric addition of butyllithium to benzaldehyde. Org. Lett. 2007, 9, 3781-3783.

ADDITION OF sp² C-NUCLEOPHILES TO C=O

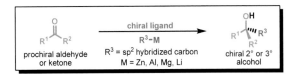

Introduction

Although the highly enantioselective and catalytic 1,2-addition of divinylzinc to aldehydes using the chiral ligand **2** was reported by W. Oppolzer et al.[1] in 1988, the reaction was not widely used because the preparation of divinylzinc was impractical (Scheme 1).[2]

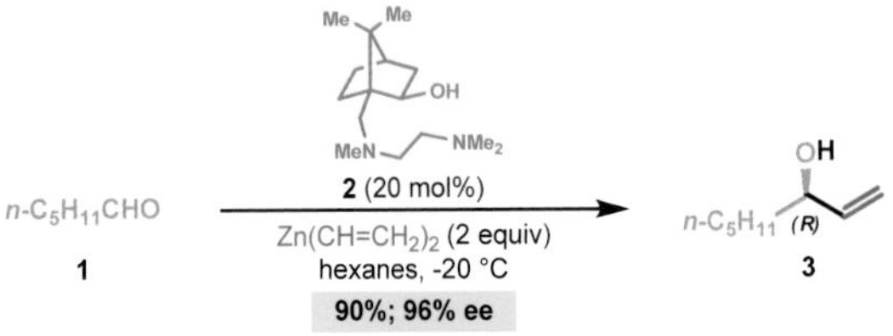

Scheme 1

I. Addition of Alkenylzinc Reagents to Aldehydes

There are currently two useful methods for the generation of alkenylzinc reagents: (a) reaction of ZnCl₂ with alkenyl Grignard reagents and (b) reaction of an alkyne with dialkylborane followed by boron-zinc exchange. Representative examples are shown in Scheme 2.[2]

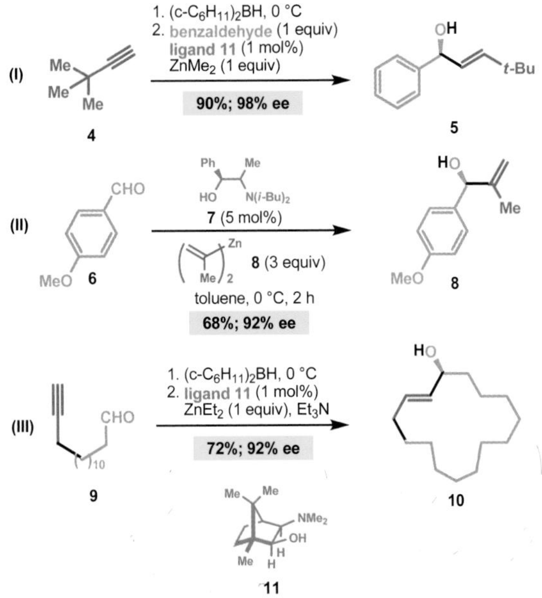

Reference key for equations: (I)[1b]; (II)[3]; (III)[4]

Scheme 2

II. Addition of Alkenylzinc Reagents to Ketones

The catalytic enantioselective vinylation of ketones emerged after the vinylation of aldehydes. The required alkenylzinc reagents were prepared via the hydrozirconation[5] of terminal alkynes followed by transmetallation to zinc. Examples of ketone vinylation are shown in Scheme 3.[6]

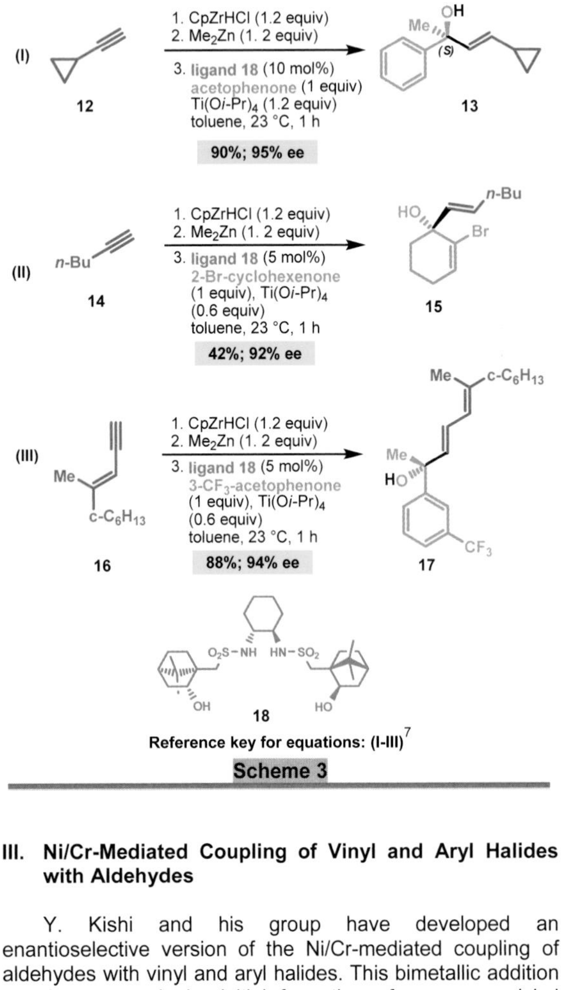

Reference key for equations: (I-III)[7]

Scheme 3

III. Ni/Cr-Mediated Coupling of Vinyl and Aryl Halides with Aldehydes

Y. Kishi and his group have developed an enantioselective version of the Ni/Cr-mediated coupling of aldehydes with vinyl and aryl halides. This bimetallic addition reaction proceeds by initial formation of an organonickel species which is converted in situ to an organochromium reagent, which then adds to RCHO after complexation with a chiral ligand.[8] The Kishi enantioselective vinylation/ arylation process involves the use of the phenanthroline complex **22** (Scheme 4) and the chiral ligand **21** for the synthesis of chiral secondary alcohols from aldehydes as shown in Scheme 4. The molecule shown below, which contains both the Ni (**A**) and Cr (**B**) binding sites can also be used to effect enantioselective addition of C(sp²) to RCHO.[9]

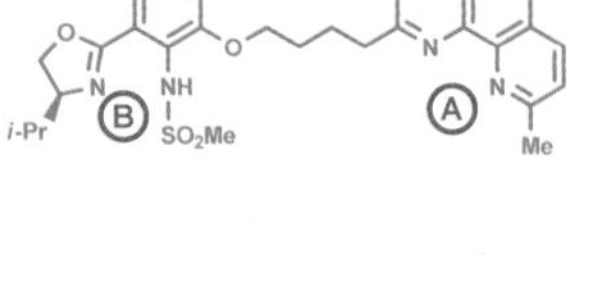

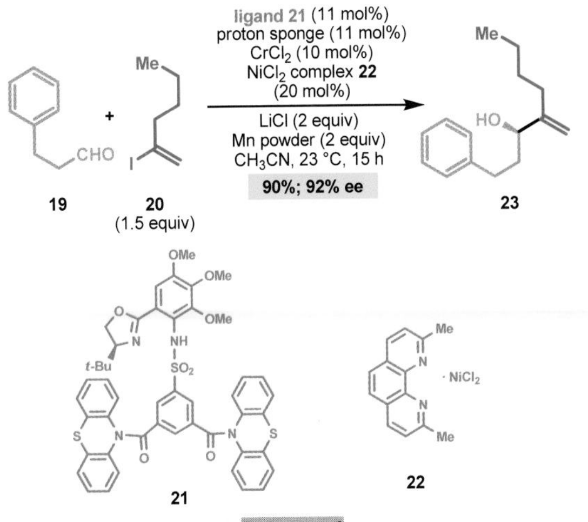

ADDITION OF sp² C-NUCLEOPHILES TO C=O

Scheme 4

IV. Addition of Arylorganometallic Reagents to Aldehydes and Ketones

Shown in Scheme 5 are some reported examples of enantioselective addition of aryl groups to C=O.

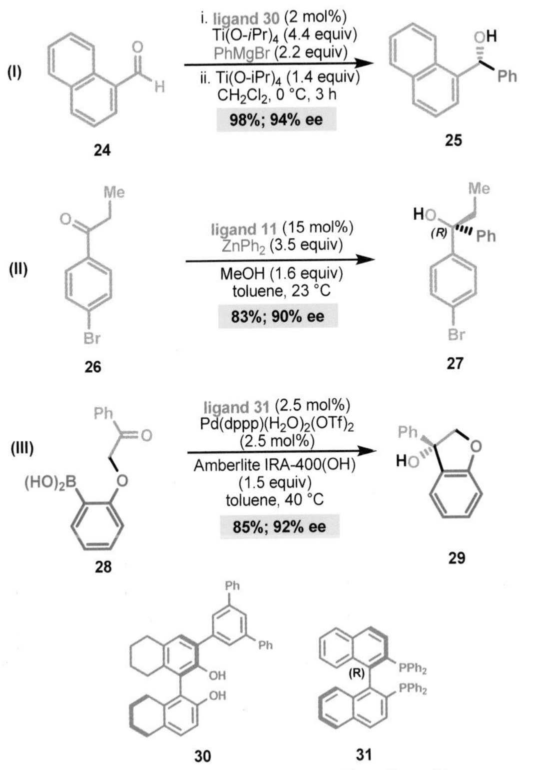

Reference key for equations: (I)[10]; (II)[11]; (III)[12]

Scheme 5

References

1. For seminal contributions, see: (a) Oppolzer, W. & Radinov, R. N. Enantioselective synthesis of *sec*-allyl alcohols by catalytic asymmetric addition of divinylzinc to aldehydes. *Tetrahedron Lett.* **1988**, 29, 5645-5648; (b) Oppolzer, W. & Radinov, R. N. Catalytic asymmetric synthesis of secondary (E)-allyl alcohols from acetylenes and aldehydes via (1-alkenyl)zinc intermediates. *Helv. Chim. Acta* **1992**, 75, 170-173.

2. For relevant reviews on enantioselective carbonyl alkenylations, see: (a) Knochel, P., Millot, N., Rodriguez, A. L. & Tucker, C. E. Preparation and applications of functionalized organozinc compounds. *Org. React.* **2001**, 58, 417-731; (b) Garcia, C. & Martin, V. S. Asymmetric addition to ketones: enantioselective formation of tertiary alcohols. *Curr. Org. Chem.* **2006**, 10, 1849-1889; (c) Luderer, M. R., Bailey, W. F., Luderer, M. R., Fair, J. D., Dancer, R. J. & Sommer, M. B. Asymmetric addition of achiral organomagnesium reagents or organolithiums to achiral aldehydes or ketones: a review. *Tetrahedron Asymmetry* **2009**, 20, 981-998.

3. Soai, K. & Takahashi, K. Asymmetric alkenylation of chiral and prochiral aldehydes catalyzed by chiral or achiral amino alcohols: catalytic diastereoselective synthesis of protected erythro-sphingosine and enantioselective synthesis of chiral diallyl alcohols. *J. Chem. Soc., Perkin Trans. 1* **1994**, 1257-1258.

4. Oppolzer, W. & Radinov, R. N. Synthesis of (R)-(-)-muscone by an asymmetrically catalyzed macrocyclization of an ω-alkynal. *J. Am. Chem. Soc.* **1993**, 115, 1593-1594.

5. For a review, see: Wipf, P. & Nunes, R. L. Selective carbon-carbon bond formations with alkenylzirconocenes. *Tetrahedron* **2004**, 60, 1269-1279.

6. For examples of other carbonyl vinylations, see: (a) Yuan, Y., Harrison-Marchand, A. & Maddaluno, J. Enantioselective hydroxyvinylation with lithioethenes. *Synlett* **2005**, 1555-1558; (b) Salvi, L., Jeon, S.-J., Fisher, E. L., Carroll, P. J. & Walsh, P. J. Catalytic Asymmetric Generation of (Z)-Disubstituted Allylic Alcohols. *J. Am. Chem. Soc.* **2007**, 129, 16119-16125; (c) Muramatsu, Y. & Harada, T. Catalytic asymmetric alkylation of aldehydes with Grignard reagents. *Angew. Chem., Int. Ed.* **2008**, 47, 1088-1090; (d) Biradar, D. B. & Gau, H.-M. Highly Enantioselective Vinyl Additions of Vinylaluminum to Ketones Catalyzed by a Titanium(IV) Catalyst of (S)-BINOL. *Org. Lett.* **2009**, 11, 499-502.

7. (a) Li, H. & Walsh, P. J. Catalytic asymmetric vinylation of ketones. *J. Am. Chem. Soc.* **2004**, 126, 6538-6539; (b) Li, H. & Walsh, P. J. Catalytic asymmetric vinylation and dienylation of ketones. *J. Am. Chem. Soc.* **2005**, 127, 8355-8361.

8. (a) For a review, see: Hargaden, G. C. & Guiry, P. J. The development of the asymmetric Nozaki-Hiyama-Kishi reaction. *Adv. Synth. Catal.* **2007**, 349, 2407-2424; (b) Namba, K., Cui, S., Wang, J. & Kishi, Y. A New Method for Translating the Asymmetric Ni/Cr-Mediated Coupling Reactions from Stoichiometric to Catalytic. *Org. Lett.* **2005**, 7, 5417-5419.

9. (a) Guo, H., Dong, C.-G., Kim, D.-S., Urabe, D., Wang, J., Kim, J. T., Liu, X., Sasaki, T. & Kishi, Y. A Toolbox Approach to the Search for Effective Ligands for Catalytic Asymmetric Cr-Mediated Coupling Reaction. *J. Am. Chem. Soc.* **2009**, 131, 15387-15393; (b) Liu, X., Henderson, J. A., Sasaki, T. & Kishi, Y. Dramatic Improvement on Catalyst Loadings and Molar Ratios of Coupling Partners for Ni/Cr-Mediated Coupling Reactions: Heterobimetallic Catalysts. *J. Am. Chem. Soc.* **2009**, 131, 16678-16680.

10. Muramatsu, Y. & Harada, T. Catalytic asymmetric Aryl transfer reactions to aldehydes with Grignard reagents as the Aryl source. *Chem.--Eur. J.* **2008**, 14, 10560-10563.

11. Dosa, P. I. & Fu, G. C. Catalytic Asymmetric Addition of ZnPh₂ to Ketones: Enantioselective Formation of Quaternary Stereocenters. *J. Am. Chem. Soc.* **1998**, 120, 445-446.

12. Liu, G. & Lu, X. Cationic Palladium Complex Catalyzed Highly Enantioselective Intramolecular Addition of Arylboronic Acids to Ketones. A Convenient Synthesis of Optically Active Cycloalkanols. *J. Am. Chem. Soc.* **2006**, 128, 16504-16505.

ADDITION OF sp C-NUCLEOPHILES TO C=O

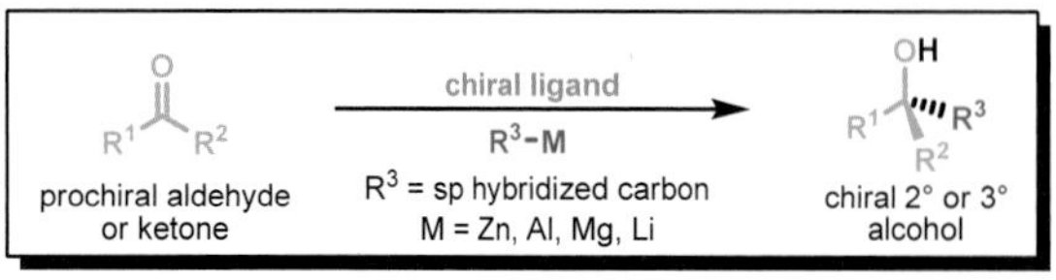

prochiral aldehyde or ketone
R^3 = sp hybridized carbon
M = Zn, Al, Mg, Li
chiral 2° or 3° alcohol

Introduction

The enantioselective addition of alkynylmetal reagents to aldehydes and ketones to form synthetically valuable propargylic alcohols generally starts with terminal alkynes which readily undergo metalation ($pK_a \sim 25$) with strong bases such as n-BuLi, EtMgBr, LHMDS, etc.[1] The acidity of the C(sp)-H bond in terminal alkynes can be significantly increased in the presence of Cu(I) or Ag(I) salts, so weaker bases such as tertiary amines can effect deprotonation. Dialkylzinc reagents such as $ZnEt_2$ deprotonate terminal alkynes in situ in the presence of an aldehyde or ketone and a chiral ligand. The first enantioselective addition of an alkyne to an aldehyde was described by Mukaiyama and co-workers using several equivalents of chiral ligand and lithium acetylide.[2] The first practical catalytic and highly enantioselective alkynylation of aldehydes, reported by Corey et al., employed an oxazaborolidine catalyst (2) and an alkynylborane (3) to convert benzaldehyde to the chiral alcohol (4).[3] This reaction affords the (R) propargylic alcohols as expected from the transition state model 5 (Figure 1).

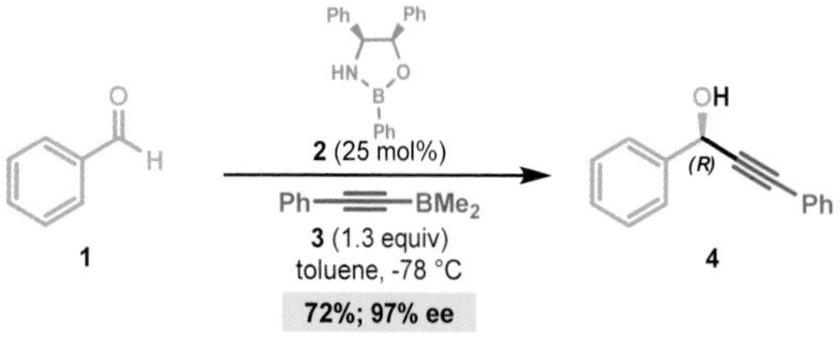

Scheme 1

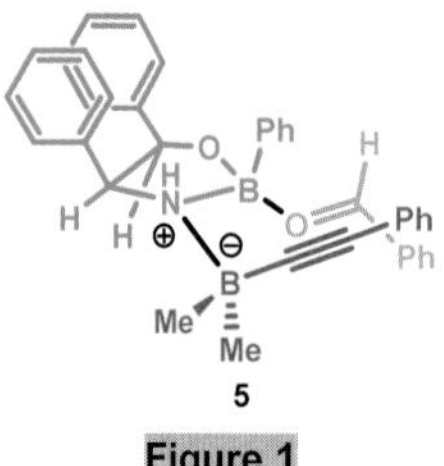

Figure 1

I. Addition of Alkynylzinc Reagents to Aldehydes

As is the case for alkyl and vinylzinc reagents, alkynylzinc reagents react only slowly with aldehydes and ketones in the absence of a Lewis base. Alkynylzinc reagents are ideal reaction partners because they tolerate many functional groups (e.g., esters, amides, nitro groups and nitriles) which react readily with stronger alkynyl nucleophiles.

Carreira et al showed that the chiral amino alcohol N-methylephedrine 20, when used in stoichiometric amounts in the presence of $Zn(OTf)_2$ and triethylamine, mediates the addition of alkynylzinc reagents in excellent yield and enantioselectivity (Entries I-III, Scheme 2). High yields were obtained for aliphatic and aromatic aldehydes, however, α,β-unsaturated aldehydes proved to be poor substrates for the transformation. The reactions can be conducted in air using reagent grade solvents that do not require pre-drying. The process could be made catalytic by raising the temperature to 60 °C (Entry IV, Scheme 2).[4] Aromatic aldehydes are not suitable substrates since they undergo the Canizzaro reaction under these reaction conditions.

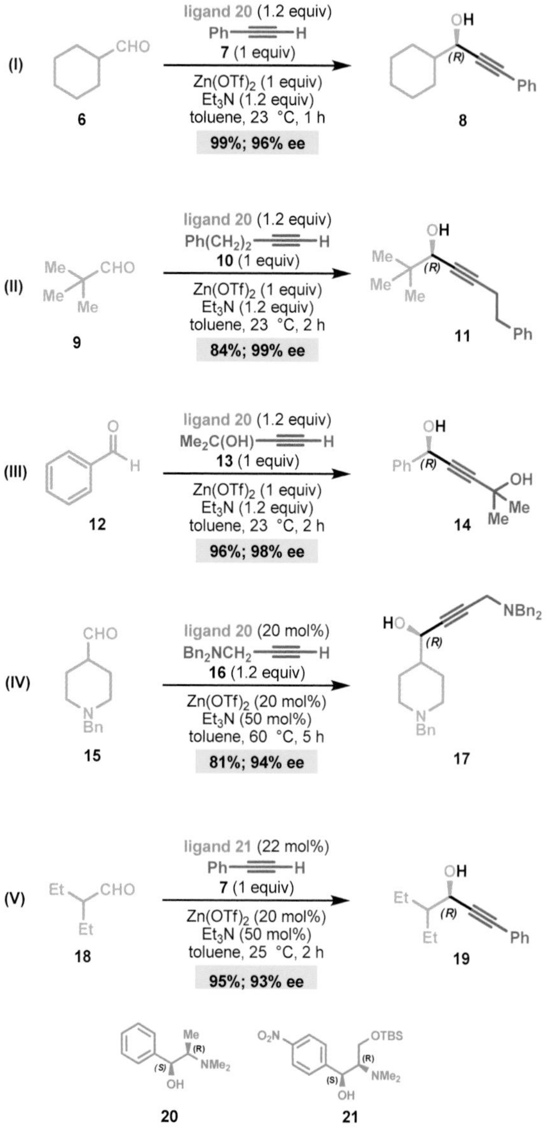

Reference key for equations: (I-III)[5]; (IV)[4]; (V)[6]

Scheme 2

ADDITION OF sp C-NUCLEOPHILES TO C=O

II. Addition of Alkynylzinc Reagents to Aldehydes Catalyzed by Titanium Complexes

The groups of Pu and Chan independently developed the alkynylation of aldehydes catalyzed by titanium-BINOL complexes. Two highly efficient examples are shown in Scheme 3. The simultaneous use of two ligands **27** and **28** for the conversion of **24** to **25** (Entry II) resulted in a significant improvement of the enantioselectivity.

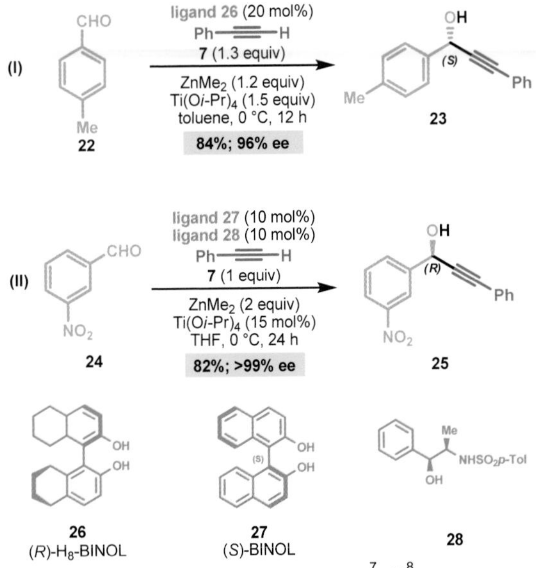

Scheme 3

III. Addition of Alkynylzinc Reagents to Ketones

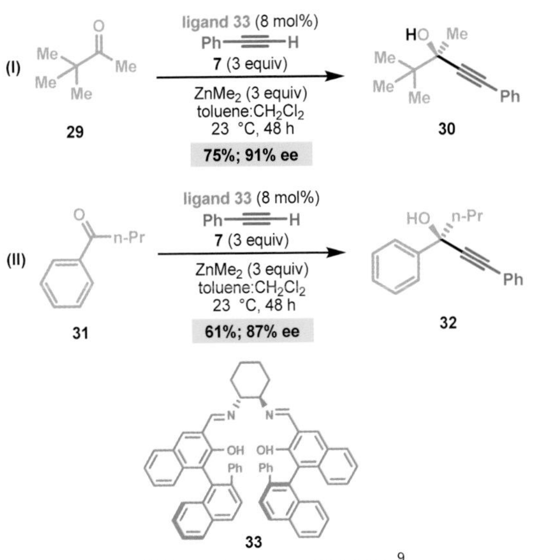

Scheme 4

Wang and coworkers described a highly efficient addition of a zinc derivative of phenylacetylene to aromatic ketones catalyzed by Schiff-base amino alcohols at very low catalyst loadings (Scheme 5).[10]

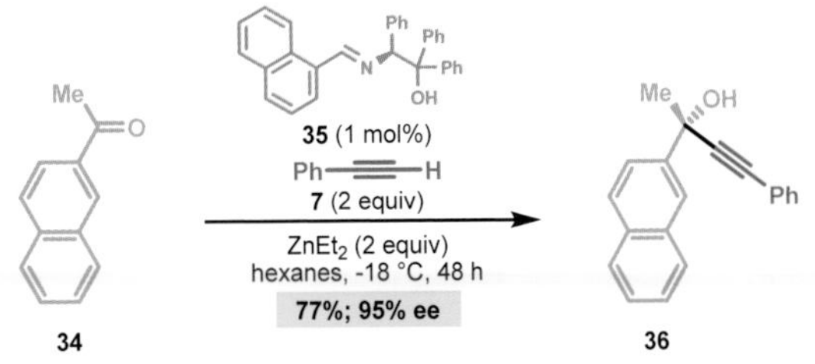

Scheme 5

References

1. For selected reviews, see: (a) Pu, L. Asymmetric alkynylzinc additions to aldehydes and ketones. *Tetrahedron* **2003**, 59, 9873-9886; (b) Cozzi, P. G., Hilgraf, R. & Zimmermann, N. Acetylenes in catalysis. Enantioselective additions to carbonyl groups and imines and applications beyond. *Eur. J. Org. Chem.* **2004**, 4095-4105; (c) Shibasaki, M. & Kanai, M. Asymmetric Synthesis of Tertiary Alcohols and α-Tertiary Amines via Cu-Catalyzed C-C Bond Formation to Ketones and Ketimines. *Chem. Rev.* **2008**, 108, 2853-2873; (d) Trost, B. M. & Weiss, A. H. The enantioselective addition of alkyne nucleophiles to carbonyl groups. *Adv. Synth. Catal.* **2009**, 351, 963-983.
2. (a) Mukaiyama, T., Suzuki, K., Soai, K. & Sato, T. Enantioselective addition of acetylene to aldehyde. Preparation of optically active alkynyl alcohols. *Chem. Lett.* **1979**, 447-448; (b) Mukaiyama, T. & Suzuki, K. Asymmetric addition of acetylide to aliphatic aldehydes. Preparation of optically active 5-octyl-2(5H)-furanone. *Chem. Lett.* **1980**, 255-256.
3. Corey, E. J. & Cimprich, K. A. Highly Enantioselective Alkynylation of Aldehydes Promoted by Chiral Oxazaborolidines. *J. Am. Chem. Soc.* **1994**, 116, 3151-3152.
4. Anand, N. K. & Carreira, E. M. A Simple, Mild, Catalytic, Enantioselective Addition of Terminal Acetylenes to Aldehydes. *J. Am. Chem. Soc.* **2001**, 123, 9687-9688.
5. (a) Frantz, D. E., Fassler, R., Tomooka, C. S. & Carreira, E. M. The discovery of novel reactivity in the development of C-C bond-forming reactions: in situ generation of zinc acetylides with Zn(II)/R₃N. *Acc. Chem. Res.* **2000**, 33, 373-381; (b) Frantz, D. E., Faessler, R. & Carreira, E. M. Facile enantioselective Synthesis of propargylic alcohols by direct addition of terminal alkynes to aldehydes. *J. Am. Chem. Soc.* **2000**, 122, 1806-1807; (c) Boyall, D., Lopez, F., Sasaki, H., Frantz, D. & Carreira, E. M. Enantioselective addition of 2-methyl-3-butyn-2-ol to aldehydes: preparation of 3-hydroxy-1-butynes. *Org. Lett.* **2000**, 2, 4233-4236.
6. (a) Jiang, B., Chen, Z. & Xiong, W. Highly enantioselective alkynylation of aldehydes catalyzed by a readily available chiral amino alcohol-based ligand. *Chem. Commun.* **2002**, 1524-1525; (b) Chen, Z., Xiong, W. & Jiang, B. Zn(OTf)₂: preparation and application in asymmetric alkynylation of aldehydes. *Chem. Commun.* **2002**, 2098-2099.
7. Lu, G., Li, X., Chan, W. L. & Chan, A. S. C. Titanium-catalyzed enantioselective alkynylation of aldehydes. *Chem. Commun.* **2002**, 172-173.
8. Li, X., Lu, G., Kwok, W. H. & Chan, A. S. C. Highly enantioselective alkynylzinc addition to aromatic aldehydes catalyzed by self-assembled titanium catalysts. *J. Am. Chem. Soc.* **2002**, 124, 12636-12637.
9. Saito, B. & Katsuki, T. Zn(salen)-catalyzed asymmetric alkynylation of ketones. *Synlett* **2004**, 1557-1560.
10. (a) Chen, C., Hong, L., Xu, Z.-Q., Liu, L. & Wang, R. Low Ligand Loading, Highly Enantioselective Addition of Phenylacetylene to Aromatic Ketones Catalyzed by Schiff-Base Amino Alcohols. *Org. Lett.* **2006**, 8, 2277-2280; (b) Wang, Q., Zhang, B., Hu, G., Chen, C., Zhao, Q. & Wang, R. Asymmetric addition of 1-ethynylcyclohexene to both aromatic and heteroaromatic ketones catalyzed by a chiral Schiff base-zinc complex. *Org. Biomol. Chem.* **2007**, 5, 1161-1163.

ADDITION OF ALLYLIC NUCLEOPHILES TO C=O

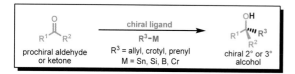

chiral ligand
R^3–M
prochiral aldehyde or ketone
R^3 = allyl, crotyl, prenyl
M = Sn, Si, B, Cr
chiral 2° or 3° alcohol

Introduction

The stereoselective addition of various allylmetal reagents (tin, silicon and boron) to ketones and aldehydes has been extensively studied since the late 1970s.[1] The first highly enantioselective catalytic 1,2-addition of allyl groups was reported by H. Yamamoto in 1991.[2] Allylsilanes react with aldehydes in the presence of the chiral (acyloxy)borane complex **7** to afford enantiomerically enriched homoallylic alcohols (e.g., **3**) in high yield (Equation I, Scheme 1). The first catalytic enantioselective allylation of the less reactive ketone class was developed by Tagliviani et al., utilizing tetrallyltin nucleophiles in the presence of BINOL-Ti(IV) catalysts (Equation II, Scheme 1).[3]

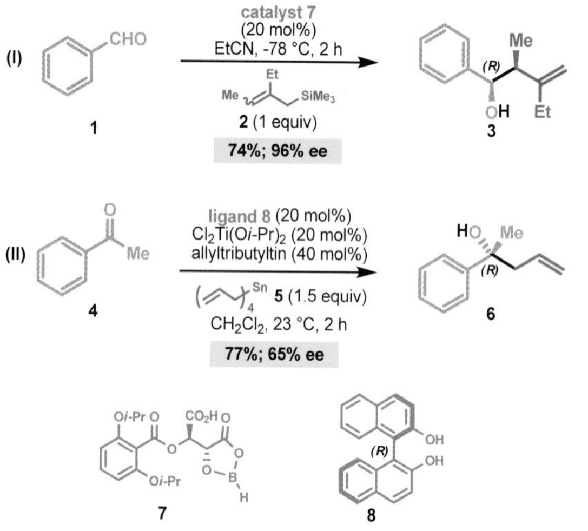

(I)
catalyst 7 (20 mol%)
EtCN, -78 °C, 2 h
2 (1 equiv)
74%; 96% ee
(II)
ligand 8 (20 mol%)
$Cl_2Ti(Oi\text{-}Pr)_2$ (20 mol%)
allyltributyltin (40 mol%)
5 (1.5 equiv)
CH_2Cl_2, 23 °C, 2 h
77%; 65% ee
1
3
4
6
7
8

Scheme 1

The diastereoselectivity of the enantioselective addition of different allylic organometallic reagents to carbonyl compounds, catalyzed by either chiral Lewis acids or bases, or chelating ligands, varies with the specific metal as follows:

(a) Si, Sn and B allylic reagents lead predominantly to *syn* diastereoselectivity, independent of the starting allylic geometry;

(b) Cr, Zn and In allylic reagents give rise predominantly to *anti* diastereoselectivity, independent of the starting allylic geometry;

(c) the *syn/anti* diastereoselectivity of allylic trichlorosilanes ($SiCl_3$) reflects the *E/Z* ratio of the starting allylic geometry.

Lewis acid
or
Lewis base
or
chelating
ligand
or
syn
anti

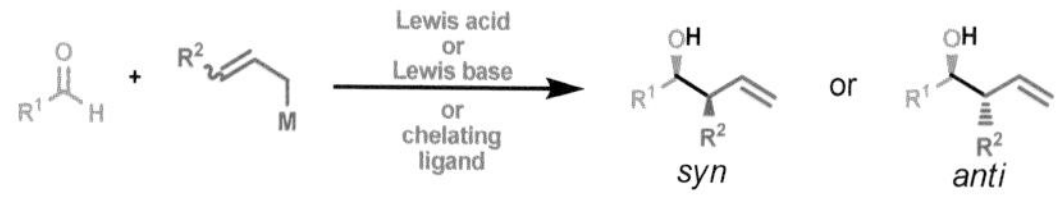

Scheme 2

I. Addition of Allylmetal (Si, Sn, Cr) Reagents to Aldehydes

A chiral Cr-salen reagent, derived from the salen ligand **23**, was used in the highly enantioselective coupling of cyclohexanecarboxaldehyde and allyl chloride (Equation IV, Scheme 3), but the yield was only moderate due to the competing pinacol coupling reaction. The coordination of trichlorosilanes with chiral phosphoramides enhances both the Lewis acidity of the silicon as well as the reactivity of the allylic group. This approach was used in the prenylation of **18** (Equation V).

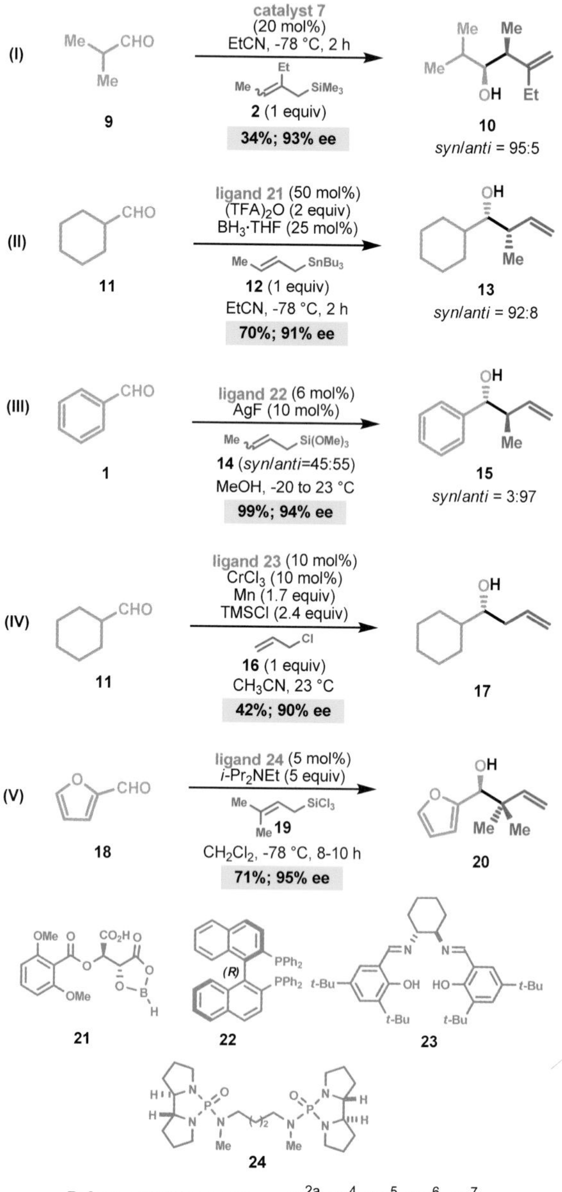

(I)
catalyst 7 (20 mol%)
EtCN, -78 °C, 2 h
2 (1 equiv)
34%; 93% ee
9
10
syn/anti = 95:5
(II)
ligand 21 (50 mol%)
$(TFA)_2O$ (2 equiv)
$BH_3\cdot THF$ (25 mol%)
12 (1 equiv)
EtCN, -78 °C, 2 h
70%; 91% ee
11
13
syn/anti = 92:8
(III)
ligand 22 (6 mol%)
AgF (10 mol%)
14 (syn/anti=45:55)
MeOH, -20 to 23 °C
99%; 94% ee
1
15
syn/anti = 3:97
(IV)
ligand 23 (10 mol%)
$CrCl_3$ (10 mol%)
Mn (1.7 equiv)
TMSCl (2.4 equiv)
16 (1 equiv)
CH_3CN, 23 °C
42%; 90% ee
11
17
(V)
ligand 24 (5 mol%)
$i\text{-}Pr_2NEt$ (5 equiv)
19
CH_2Cl_2, -78 °C, 8-10 h
71%; 95% ee
18
20
21
22
23
24

Reference key for equations: (I)[2a]; (II)[4]; (III)[5]; (IV)[6]; (V)[7]

Scheme 3

ADDITION OF ALLYLIC NUCLEOPHILES TO C=O

II. Addition of Allylmetal (Si, Sn, B) Reagents to Ketones

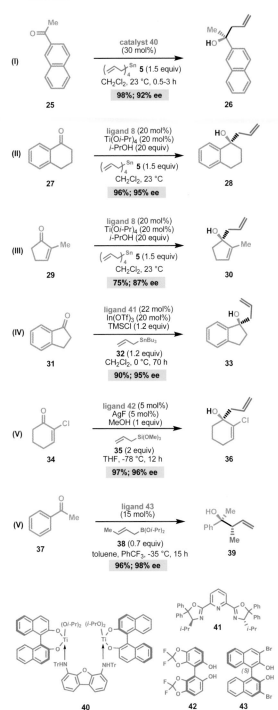

Reference key for equations: (I)[8]; (II-III)[9]; (IV)[10]; (V)[11]; (VI)[12]

Scheme 4

In addition to the catalytic enantioselective allylations demonstrated in Schemes 3 and 4, chiral allyl-, crotyl and prenylboron reagents (e.g. **44-46**) are also available for the highly enantioselective (>98% ee) allylation and crotylation of aldehydes. In the case of boronate **44**, the chiral ligand, diisopropyl tartrate, may be recovered and reused.[13]

References

1. For selected reviews on the enantioselective allylation of carbonyl compounds, see: (a) Yamamoto, Y. & Asao, N. Selective reactions using allylic metals. *Chem. Rev.* **1993**, 93, 2207-2293; (b) Denmark, S. E. & Fu, J. Catalytic enantioselective addition of allylic organometallic reagents to aldehydes and ketones. *Chem. Rev.* **2003**, 103, 2763-2793; (c) Garcia, C. & Martin, V. S. Asymmetric addition to ketones: enantioselective formation of tertiary alcohols. *Curr. Org. Chem.* **2006**, 10, 1849-1889; (d) Walsh, P. J., Li, H. & de Parrodi, C. A. A Green Chemistry Approach to Asymmetric Catalysis: Solvent-Free and Highly Concentrated Reactions. *Chem. Rev.* **2007**, 107, 2503-2545; (e) Hatano, M. & Ishihara, K. Recent progress in the catalytic synthesis of tertiary alcohols from ketones with organometallic reagents. *Synthesis* **2008**, 1647-1675; (f) Shibasaki, M. & Kanai, M. Asymmetric Synthesis of Tertiary Alcohols and α-Tertiary Amines via Cu-Catalyzed C-C Bond Formation to Ketones and Ketimines. *Chem. Rev.* **2008**, 108, 2853-2873.
2. (a) Furuta, K., Mouri, M. & Yamamoto, H. Chiral (acyloxy)borane catalyzed asymmetric allylation of aldehydes. *Synlett* **1991**, 561-562; (b) Ishihara, K., Mouri, M., Gao, Q., Maruyama, T., Furuta, K. & Yamamoto, H. Catalytic asymmetric allylation using a chiral (acyloxy)borane complex as a versatile Lewis acid catalyst. *J. Am. Chem. Soc.* **1993**, 115, 11490-11495.
3. Casolari, S., D'Addario, D. & Tagliavini, E. BINOL-Ti-Catalyzed Synthesis of Tertiary Homoallylic Alcohols: The First Catalytic Asymmetric Allylation of Ketones. *Org. Lett.* **1999**, 1, 1061-1063.
4. Marshall, J. A. & Palovich, M. R. Enantioselective and Diastereoselective Additions of Allylic Stannanes to Aldehydes Promoted by a Chiral (Acyloxy)borane Catalyst. *J. Org. Chem.* **1998**, 63, 4381-4384.
5. Yanagisawa, A., Kageyama, H., Nakatsuka, Y., Asakawa, K., Matsumoto, Y. & Yamamoto, H. Enantioselective addition of allylic trimethoxysilanes to aldehydes catalyzed by p-Tol-BINAP-AgF. *Angew. Chem., Int. Ed.* **1999**, 38, 3701-3703.
6. (a) Bandini, M., Cozzi, P. G., Melchiorre, P. & Umani-Ronchi, A. The first catalytic enantioselective Nozaki-Hiyama reaction. *Angew. Chem., Int. Ed.* **1999**, 38, 3357-3359; (b) Bandini, M., Cozzi, P. G. & Umani-Ronchi, A. Asymmetric synthesis with "privileged" ligands. *Pure Appl. Chem.* **2001**, 73, 325-329.
7. Nakajima, M., Saito, M., Shiro, M. & Hashimoto, S.-i. (S)-3,3'-Dimethyl-2,2'-biquinoline N,N'-Dioxide as an Efficient Catalyst for Enantioselective Addition of Allyltrichlorosilanes to Aldehydes. *J. Am. Chem. Soc.* **1998**, 120, 6419-6420.
8. Hanawa, H., Kii, S. & Maruoka, K. New chiral bis-titanium(IV) catalyst with dibenzofuran spacer for catalytic asymmetric allylation of aldehydes and aryl ketones. *Adv. Synth. Catal.* **2001**, 343, 57-60.
9. (a) Waltz, K. M., Gavenonis, J. & Walsh, P. J. A simple, reliable, catalytic asymmetric allylation of ketones. *Angew. Chem., Int. Ed.* **2002**, 41, 3697-3699; (b) Kim, J. G., Waltz, K. M., Garcia, I. F., Kwiatkowski, D. & Walsh, P. J. Catalytic asymmetric allylation of ketones and a tandem asymmetric allylation/diastereoselective epoxidation of cyclic enones. *J. Am. Chem. Soc.* **2004**, 126, 12580-12585.
10. Lu, J., Ji, S.-J., Teo, Y.-C. & Loh, T.-P. Highly Enantioselective Allylation of Aldehydes Catalyzed by Indium(III)-PYBOX Complex. *Org. Lett.* **2005**, 7, 159-161.
11. Wadamoto, M. & Yamamoto, H. Silver-Catalyzed Asymmetric Sakurai-Hosomi Allylation of Ketones. *J. Am. Chem. Soc.* **2005**, 127, 14556-14557.
12. Lou, S., Moquist, P. N. & Schaus, S. E. Asymmetric allylboration of ketones catalyzed by chiral diols. *J. Am. Chem. Soc.* **2006**, 128, 12660-12661.
13. For a recent comprehensive review, see: Lachance, H. & Hall, D. G. Allylboration of carbonyl compounds. *Organic Reactions* **2008**, 73, 1-573.

PROPARGYLATION AND ALLENYLATION OF C=O

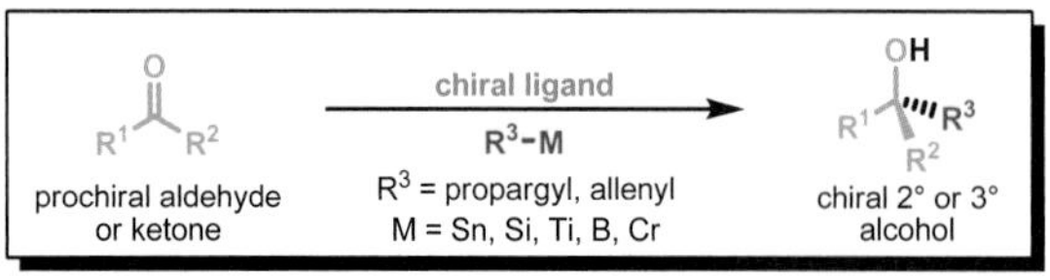

prochiral aldehyde or ketone

R^3 = propargyl, allenyl
M = Sn, Si, Ti, B, Cr

chiral 2° or 3° alcohol

Introduction

The enantioselective addition of propargylic or allenic metal reagents to aldehydes is mechanistically analogous to that of allylic metal reagents, but slower.[1] The first synthetically useful preparation of homopropargylic alcohols from aldehydes and allenyl tri-n-butylstannane **3**, a propragylic anion equivalent, was reported by Keck et al. in the early 1990s using a BINOL/Ti(IV) catalyst (Scheme 1).[2]

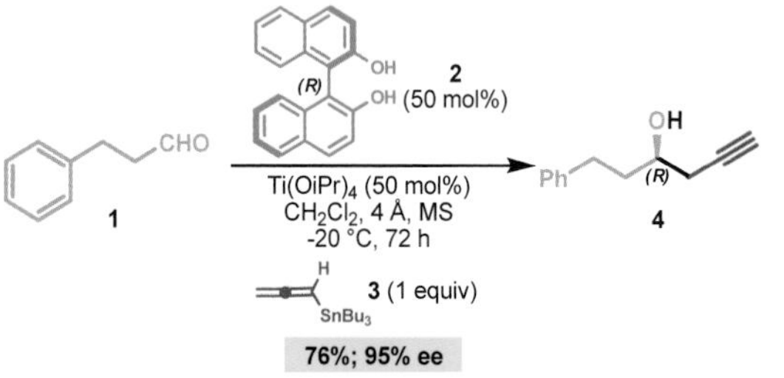

Scheme 1

I. Addition of Allenylmetal (Si, Sn, Cr) Reagents to Aldehydes

The rate of propargylation may be increased by the inclusion of stoichiometric quantities of additives such as boron ethers (e.g. B(OMe)$_3$ and i-PrSBEt$_2$) to break up the catalyst-product complex, as shown in Scheme 2 (Equation I).

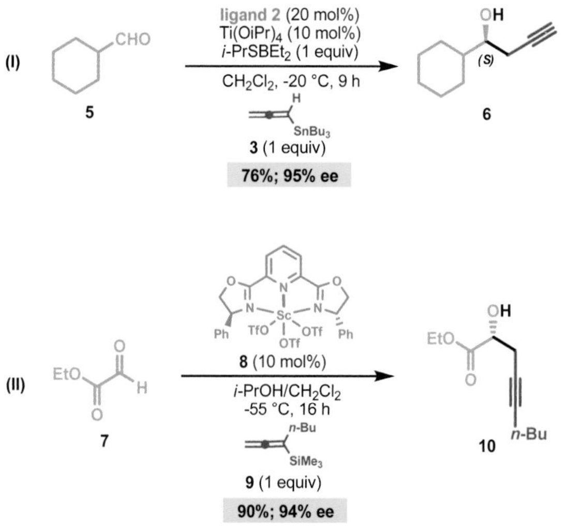

Reference key for equations: (I)[3]; (II)[4]

Scheme 2

An efficient organochromium mediated enantioselective homoallenylation is shown in Scheme 3 (Equation I). Enantiomerically pure allenylstannes react with achiral α-branched aldehydes in the presence of equimolar BF$_3$·Et$_2$O, to form predominantly syn adducts (Equation II, Scheme 3).

If the silicon center of an allenylsilane reagent is sterically congested, the normal addition pathway to aldehydes will no longer be dominant and functionalized dihydrofurans are produced. When the silyl group is $tert$-butyldiphenylsilyl (TBDPS), the formation of homoallylic alcohol is completely suppressed and the enantioselectivity is much improved (Equation III, Scheme 3).

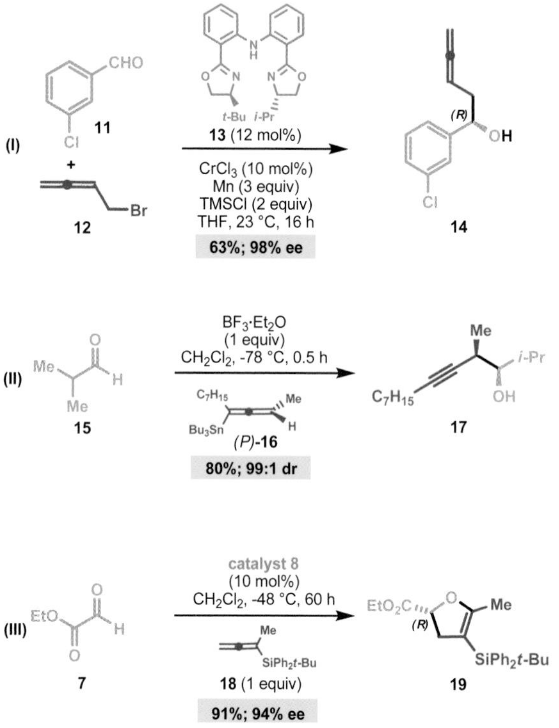

Reference key for equations: (I)[5]; (II)[6]; (III)[4]

Scheme 3

II. Addition of Propargylmetal (Si, Sn, Cr) Reagents to Aldehydes

The addition of propargylic metal reagents to carbonyl compounds leads to allenyl alcohols but can be complicated by low regiochemical selectivity if equilibration (e.g., **20**↔**21**, Scheme 4) occurs.[1,7]

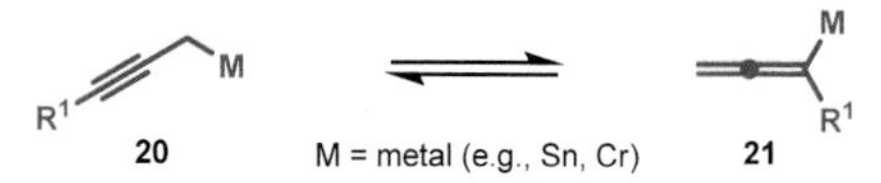

Scheme 4

PROPARGYLATION AND ALLENYLATION OF C=O

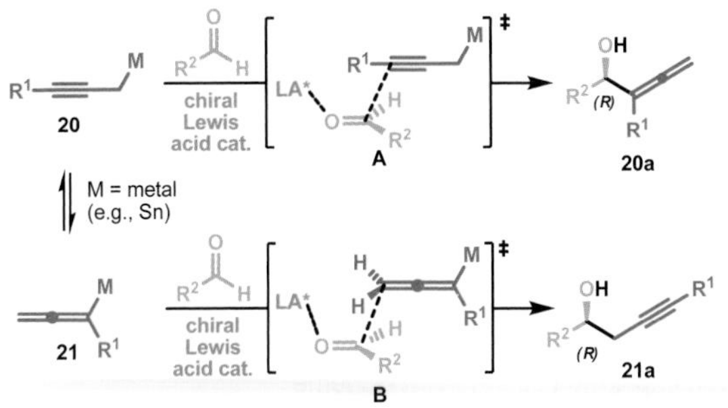

Scheme 5

Scheme 5 shows that regiochemistry can depend on the reaction conditions, if interconversion occurs between propargylmetal (**20**) and allenylmetal (**21**) reagents. An S_E2' attack leads to products **20a** or **21a** via transition states **A** or **B**, respectively.[7] The major reaction pathway depends on the stability of these transition states and not on the stability of the products. Representative examples are shown in Scheme 6.

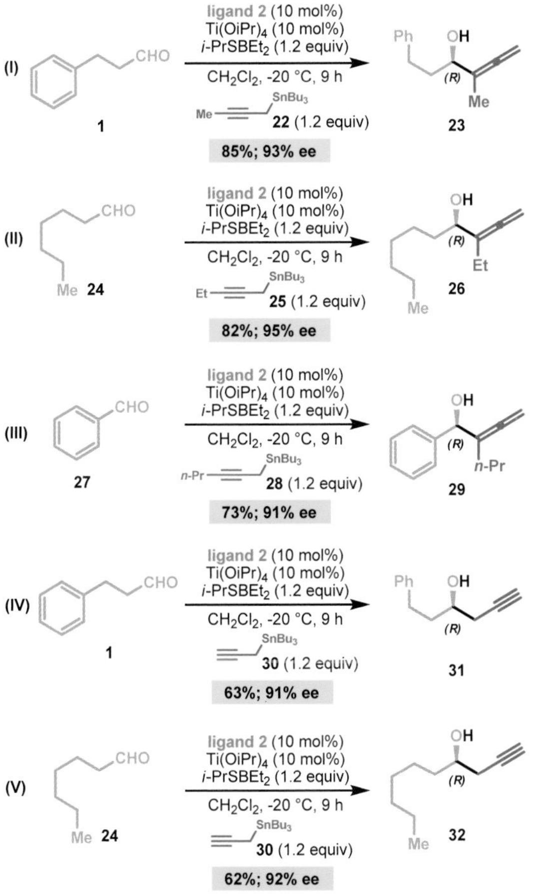

Reference key for equations: (I-V)[7]

Scheme 6

From a synthetic viewpoint, the catalytic and highly enantioselective synthesis of homopropragylic alcohols using low-valent metals, such as Cr(II), and propargylic halides appears preferable to the use of Lewis acid-mediated addition of allenylmetal compounds since pure allenylchromium reagents reagents are formed *in situ*.[8] A few examples are shown in Scheme 7.

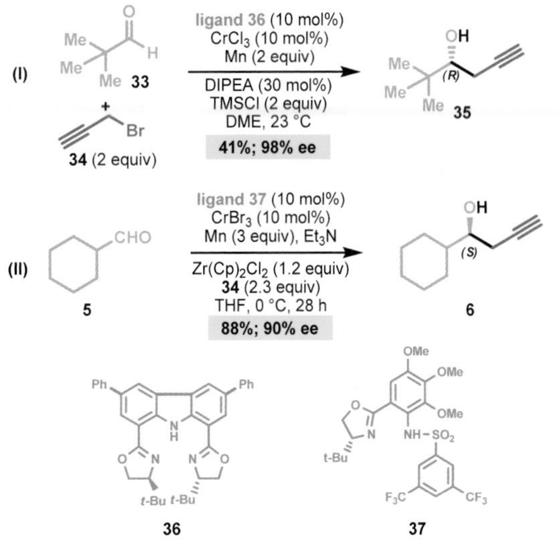

Scheme 7

References

1. For selected reviews on the enantioselective propargylation and allenylation of carbonyl compounds, see: (a) Curtis-Long, M. J. & Aye, Y. Vinyl-, Propargyl-, and Allenylsilicon Reagents in Asymmetric Synthesis: A Relatively Untapped Resource of Environmentally Benign Reagents. *Chem.--Eur. J.* **2009**, 15, 5402-5416; (b) Botuha, C., Chemla, F., Ferreira, F., Perez-Luna, A. & Roy, B. Allenylzinc reagents: new trends and synthetic applications. *New J. Chem.* **2007**, 31, 1552-1567; (c) Gung, B. W. Additions of allyl, allenyl, and propargylstannanes to aldehydes and imines. *Org. React.* **2004**, 64, 1-113; (d) Denmark, S. E. & Fu, J. Catalytic enantioselective addition of allylic organometallic reagents to aldehydes and ketones. *Chem. Rev.* **2003**, 103, 2763-2793.
2. Keck, G. E., Krishnamurthy, D. & Chen, X. Asymmetric synthesis of homopropargylic alcohols from aldehydes and allenyltri-n-butylstannane. *Tetrahedron Lett.* **1994**, 35, 8323-8324.
3. Yu, C.-M., Yoon, S.-K., Choi, H.-S. & Baek, K. Catalytic asymmetric prop-2-ynylation involving the use of the bifunctional synergetic reagent Et$_2$BSPr-i. *Chem. Commun.* **1997**, 763-764.
4. Evans, D. A., Sweeney, Z. K., Rovis, T. & Tedrow, J. S. Highly Enantioselective Syntheses of Homopropargylic Alcohols and Dihydrofurans Catalyzed by a Bis(oxazolinyl)pyridine-Scandium Triflate Complex. *J. Am. Chem. Soc.* **2001**, 123, 12095-12096.
5. Coeffard, V., Aylward, M. & Guiry, P. J. First Regio- and Enantioselective Chromium-Catalyzed Homoallenylation of Aldehydes. *Angew. Chem., Int. Ed.* **2009**, 48, 9152-9155.
6. Marshall, J. A. & Wang, X. J. Diastereoselective additions of allenylstannanes to aldehydes. *J. Org. Chem.* **1990**, 55, 6246-6248.
7. Yu, C.-M., Yoon, S.-K., Baek, K. & Lee, J.-Y. Catalytic asymmetric allenylation: regulation of the equilibrium between propargyl- and allenylstannanes during the catalytic process. *Angew. Chem., Int. Ed.* **1998**, 37, 2392-2395.
8. (a) Inoue, M. & Nakada, M. Studies on catalytic asymmetric Nozaki-Hiyama propargylation. *Org. Lett.* **2004**, 6, 2977-2980; (b) Liu, S., Kim, J. T., Dong, C.-G. & Kishi, Y. Catalytic Enantioselective Cr-Mediated Propargylation: Application to Halichondrin Synthesis. *Org. Lett.* **2009**, 11, 4520-4523.

ADDITION OF STABILIZED CARBANIONS TO C=O

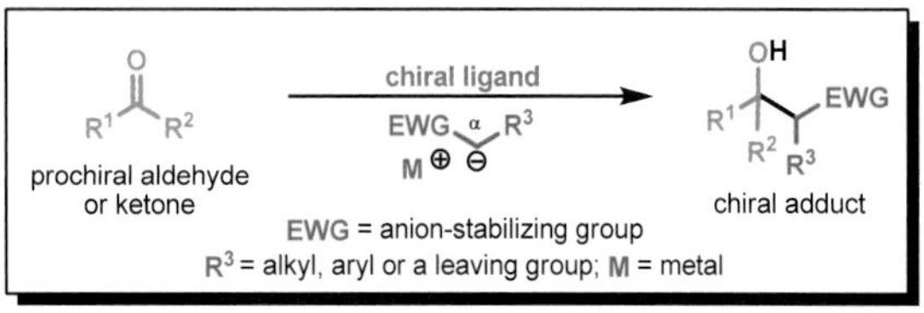

prochiral aldehyde or ketone

chiral adduct

EWG = anion-stabilizing group
R³ = alkyl, aryl or a leaving group; M = metal

Introduction

Nucleophilic addition of carbanions, stabilized by electron-withdrawing groups and also bearing a leaving group, to aldehydes and ketones can lead to epoxides rather than Wittig-type alkene formation, as outlined in Scheme 1.[1]

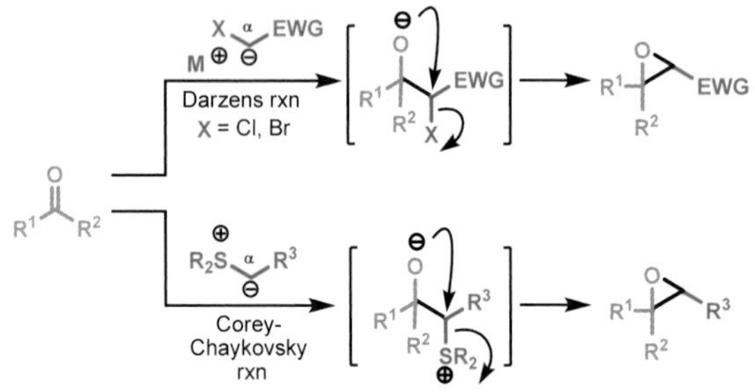

Scheme 1

I. Enantioselective Darzens Reaction

An enantioselective version of the Darzens reaction has been described using chiral quaternary ammonium salt phase transfer catalysts (PTC)[2] as shown in Scheme 2, Equation I. Enantioselectivity can also be achieved by the *in situ* formation of chiral sulfur ylides from diazoacetamides and a chiral binaphthylsulfide (Scheme 2, Equation II).[3] The enantiomerically pure product **8** was obtained by a single recrystallization.

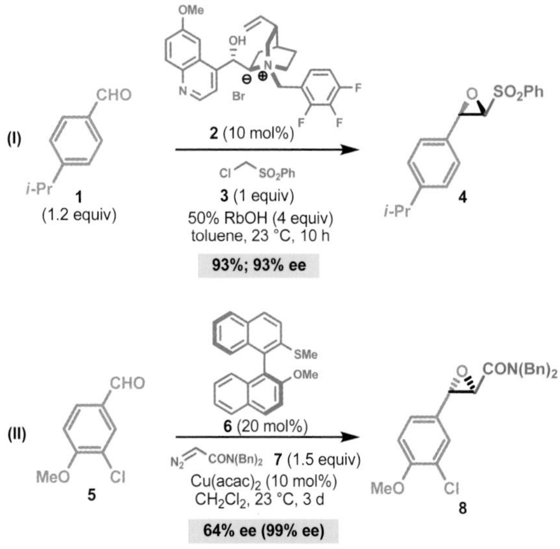

Scheme 2

A different and novel enantioselective Darzens-type reaction has been demonstrated in which a chiral Ti(IV)-BINOL complex is used as a Lewis acid to activate the carbonyl group of an aldehyde so as to promote the addition of an α-diazo carboxylic acid amide as shown in Scheme 3.[4]

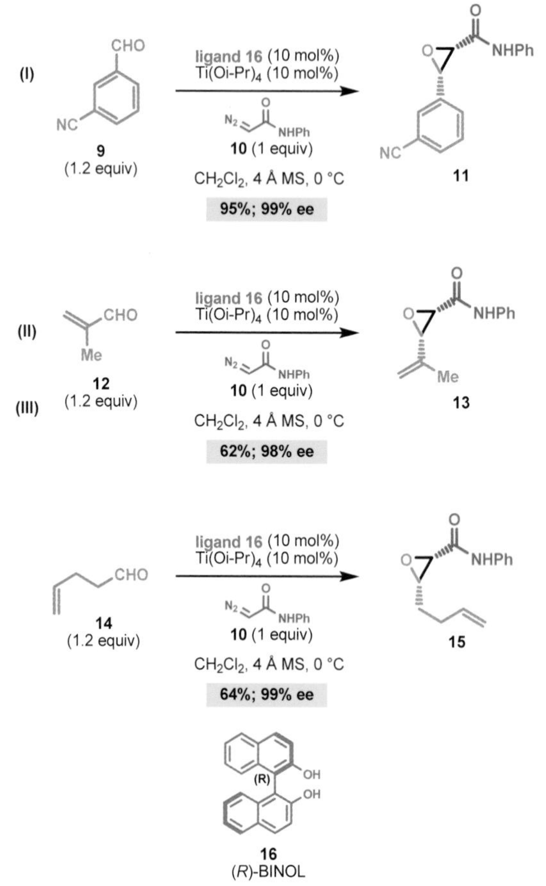

Reference key for equations: (I-III)[4]

Scheme 3

II. Enantioselective Corey-Chaykovsky Reaction

The preparation of epoxides by the reaction of sulfur ylides with aldehydes and ketones (Corey-Chaykovsky reaction) affords predominantly *trans* epoxides when substituted sulfur ylides are used.[5] This *trans*-selectivity contrasts with the *cis*-preference for the reactions shown in Scheme 3.

The sulfur ylide[5e] can be generated either by deprotonation of a preformed sulfonium salt or by reaction of a sulfide with a carbene quivalent. Enantioselective epoxide formation from carbonyl precursors has been demonstrated by two methods: (1) use of a chiral sulfide (e.g., **30-32**, Scheme 4) as precursor to a chiral S-ylide and (2) reaction of a prochiral aldehyde or ketone with an achiral S-ylide such as $Me_2S=CH_2$ and the Ti/lanthanide complex **33** as catalyst (Scheme 4).

ADDITION OF STABILIZED CARBANIONS TO C=O

Enantioselective (Continued)

Corey-Chaykovsky Epoxidation

(I)

17

30 (20 mol%)
n-Bu$_4$NI (1 equiv)
NaOH (2 equiv)
PhCH$_2$Br 18 (2 equiv)
catechol (0.5 mol%)
t-BuOH/H$_2$O (9:1), 23 °C

92%; 92% ee

19
dr = 83: 17

(II)

20

31 (10 mol%)
NaOH (2 equiv)
PhCH$_2$Br 18 (2 equiv)
catechol (0.5 mol%)
MeCN/H$_2$O (9:1), 23 °C

41%; 97% ee

21
dr = 91: 9

(III)

22

32 (5 mol%)
Rh$_2$(OAc)$_4$ (1 mol%)
BnEt$_3$NCl (10 mol%)
PhCH=NNTsNa
23 (1 equiv)
CH$_3$CN, 40 °C, 48 h

77%; 92% ee

24
dr = 91: 9

(IV)

25

33 (5 mol%)
La(OTf)$_3$ (5 mol%)
Ph$_3$As=O, 5 Å MS
$^-$CH$_2$S$^+$(=O)Me$_2$
26 (1.2 equiv)
THF, 23 °C, 12 h

97%; 92% ee

27

(V)

28

33 (5 mol%)
La(OTf)$_3$ (5 mol%)
Ph$_3$As=O, 5 Å MS
$^-$CH$_2$S$^+$(=O)Me$_2$
26 (1.2 equiv)
THF, 23 °C, 12 h

99%; 93% ee

29

30   31   32   33

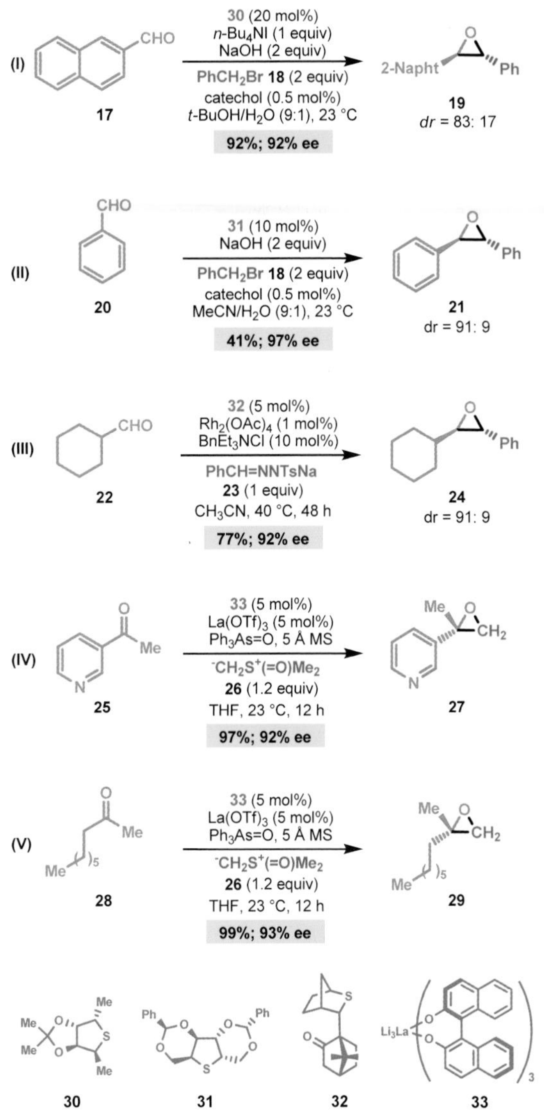

Reference key for equations: (I)[6]; (II)[7]; (III)[8]; (IV-V)[9]

Scheme 4

34

35 (4 mol%)
Rb$_2$CO$_3$ (4 mol%)
THF, 0 °C, 20 h

91%; 92% ee

36

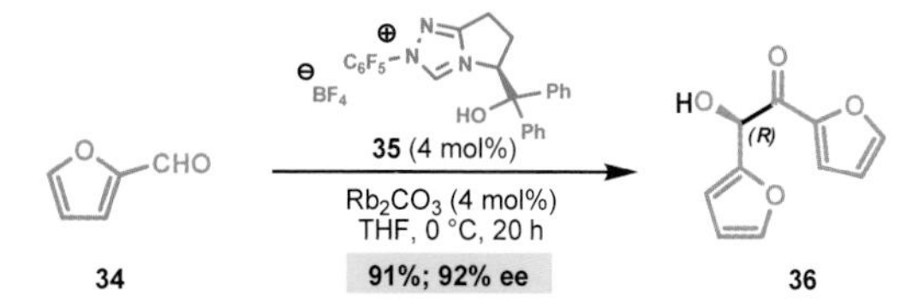

Scheme 5

References

1. (a) Li, A.-H., Dai, L.-X. & Aggarwal, V. K. Asymmetric Ylide Reactions: Epoxidation, Cyclopropanation, Aziridination, Olefination, and Rearrangement. *Chem. Rev.* **1997**, 97, 2341-2372; (b) Aggarwal, V. K., Badine, M. D. & Moorthie, V. A. Asymmetric Synthesis of Epoxides and Aziridines from Aldehydes and Imines in *Aziridines and Epoxides in Organic Synthesis* 1-35 (**2006**).

2. (a) Guillena, G. & Ramon, D. J. Enantioselective α-heterofunctionalization of carbonyl compounds: organocatalysis is the simplest approach. *Tetrahedron Asymmetry* **2006**, 17, 1465-1492; (b) Ku, J.-M., Yoo, M.-S., Park, H.-G., Jew, S.-S. & Jeong, B.-S. Asymmetric synthesis of α,β-epoxysulfones via phase-transfer catalytic Darzens reaction. *Tetrahedron* **2007**, 63, 8099-8103.

3. Imashiro, R., Yamanaka, T. & Seki, M. Catalytic asymmetric synthesis of glycidic amides via chiral sulfur ylides. *Tetrahedron Asymmetry* **1999**, 10, 2845-2851.

4. Liu, W.-J., Lv, B.-D. & Gong, L.-Z. An Asymmetric Catalytic Darzens Reaction between Diazoacetamides and Aldehydes Generates cis-Glycidic Amides with High Enantiomeric Purity. *Angew. Chem., Int. Ed.* **2009**, 48, 6503-6506.

5. (a) Aggarwal, V. K. & Winn, C. L. Catalytic, Asymmetric Sulfur Ylide-Mediated Epoxidation of Carbonyl Compounds: Scope, Selectivity, and Applications in Synthesis. *Acc. Chem. Res.* **2004**, 37, 611-620; (b) Kürti, L. & Czakó, B. Corey-Chaykovsky Epoxidation and Cyclopropanation in *Strategic Applications of Named Reactions in Organic Synthesis* 102-103 (Academic Press/Elsevier, San Diego, **2005**); (c) McGarrigle, E. M. & Aggarwal, V. K. Ylide-based reactions. *Enantioselective Organocatalysis* **2007**, 357-389; (d) McGarrigle, E. M., Myers, E. L., Illa, O., Shaw, M. A., Riches, S. L. & Aggarwal, V. K. Chalcogenides as Organocatalysts. *Chem. Rev. (Washington, DC, U. S.)* **2007**, 107, 5841-5883; (e) Briere, J.-F. & Metzner, P. Synthesis and use of chiral sulfur ylides. *Organosulfur Chemistry in Asymmetric Synthesis* **2008**, 179-208; (f) Aggarwal, V. K., Crimmin, M. & Riches, S. Synthesis of epoxides by carbonyl epoxidation. *Sci. Synth.* **2008**, 37, 321-406.

6. Davoust, M., Briere, J.-F., Jaffres, P.-A. & Metzner, P. Design of Sulfides with a Locked Conformation as Promoters of Catalytic and Asymmetric Sulfonium Ylide Epoxidation. *J. Org. Chem.* **2005**, 70, 4166-4169.

7. Winn, C. L., Bellenie, B. R. & Goodman, J. M. A highly enantioselective one-pot sulfur ylide epoxidation reaction. *Tetrahedron Lett.* **2002**, 43, 5427-5430.

8. Aggarwal, V. K., Bae, I., Lee, H.-Y., Richardson, J. & Williams, D. T. Sulfur-ylide-mediated synthesis of functionalized and trisubstituted epoxides with high enantioselectivity; application to the synthesis of CDP-840. *Angew. Chem., Int. Ed.* **2003**, 42, 3274-3278.

9. Sone, T., Yamaguchi, A., Matsunaga, S. & Shibasaki, M. Catalytic Asymmetric Synthesis of 2,2-Disubstituted Terminal Epoxides via Dimethyloxosulfonium Methylide Addition to Ketones. *J. Am. Chem. Soc.* **2008**, 130, 10078-10079.

10. (a) Zeitler, K. Extending mechanistic routes in heterazolium catalysis - promising concepts for versatile synthetic methods. *Angew. Chem., Int. Ed.* **2005**, 44, 7506-7510; (b) Marion, N., Diez-Gonzalez, S. & Nolan, S. P. N-heterocyclic carbenes as organocatalysts. *Angew. Chem., Int. Ed.* **2007**, 46, 2988-3000; (c) Enders, D., Niemeier, O. & Henseler, A. Organocatalysis by N-Heterocyclic Carbenes. *Chem. Rev. (Washington, DC, U. S.)* **2007**, 107, 5606-5655.

11. (a) O'Toole, S. E. & Connon, S. J. The enantioselective benzoin condensation promoted by chiral triazolium precatalysts: stereochemical control via hydrogen bonding. *Org. Biomol. Chem.* **2009**, 7, 3584-3593; (b) Baragwanath, L., Rose, C. A., Zeitler, K. & Connon, S. J. Highly Enantioselective Benzoin Condensation Reactions Involving a Bifunctional Protic Pentafluorophenyl-Substituted Triazolium Precatalyst. *J. Org. Chem.* **2009**, 74, 9214-9217.

III. Enantioselective Benzoin Condensation

The cyanide-ion catalyzed coupling (self-condensation) of two molecules of an aldehyde that lack α-hydrogens to afford an α-hydroxy ketone (benzoin condensation)[10] can be performed enantioselectively by utilizing the pentafluorophenyl-substituted triazolium precatalyst 35 as shown in Scheme 5.[11]

ENANTIOSELECTIVE CYANOSILYLATION OF C=O

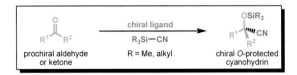

prochiral aldehyde or ketone	chiral ligand / R_3Si-CN / R = Me, alkyl	chiral O-protected cyanohydrin

Introduction

Cyanohydrins, which can be elaborated in multiple ways to other compounds (for examples, see Scheme 1), can now be accessed from aldehydes or methyl ketones by a variety of catalytic enantioselective processes involving HCN or TMSCN as reagents.[1] An impressive number of approaches have been described using a wide range of chiral catalysts. These include metal-containing and metal-free catalytic molecules, some of which are co-activators such as triphenylphosphine oxide (Ph_3P=O).

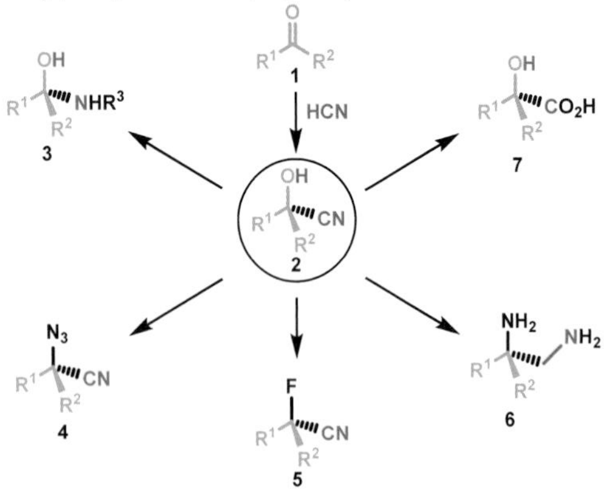

Scheme 1

Because cyanohydrin formation is readily reversible under even mildly basic conditions, it is essential to conduct enantioselective addition of HCN under irreversible conditions. In order to favor product-formation, various cyanating agents have been developed that trap the initially formed cyanohydrins alkoxide as an O-protected derivative.

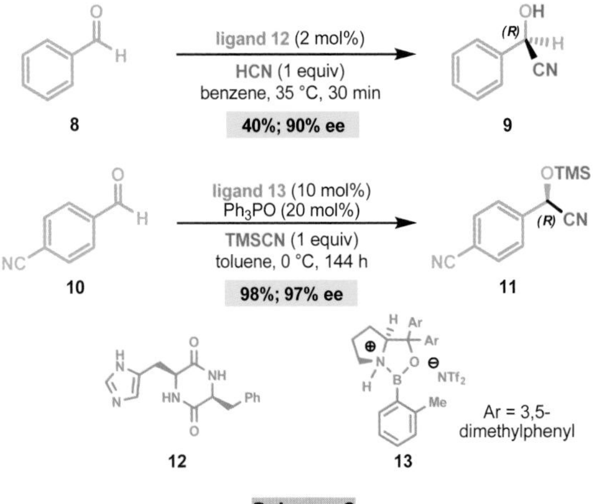

Scheme 2

The first non-enzyme catalyzed and highly enantioselective cyanohydrin formation between benzaldehyde and HCN (8→9, Scheme 2) was reported by S. Inoue et al. using a cyclic dipeptide (12) catalyst derived from phenylalanine and histidine.[2] A more recent and highly efficient example (10→11) by Corey and co-workers exploited chiral protonated oxazaborolidines as catalysts.[3]

A number of small organic molecules (e.g., O-arylated cinchona alkaloids 14, bifunctional thioureas 15 and bis-N-oxides 16, Figure 1) as well as Lewis acidic metal complexes (of ligands 17-18, Figure 2) have been developed as catalysts for enantioselective cyanosilylation. Some effective enantioselective cyanosilylation catalysts can activate both the carbonyl substrate and the cyanide source.

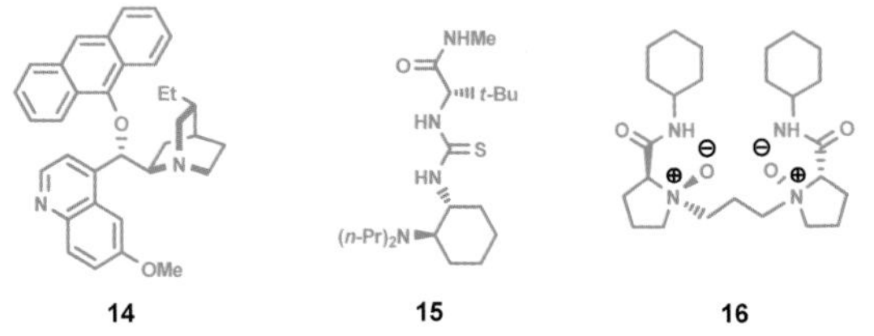

Figure 1

The most widely used metals for aldehyde substrates are titanium (Ti), vanadium (V) and aluminum (Al) in conjuction with ligands 17 or 18 (Figure 2). Lanthanide metal complexes are very effective for ketones.

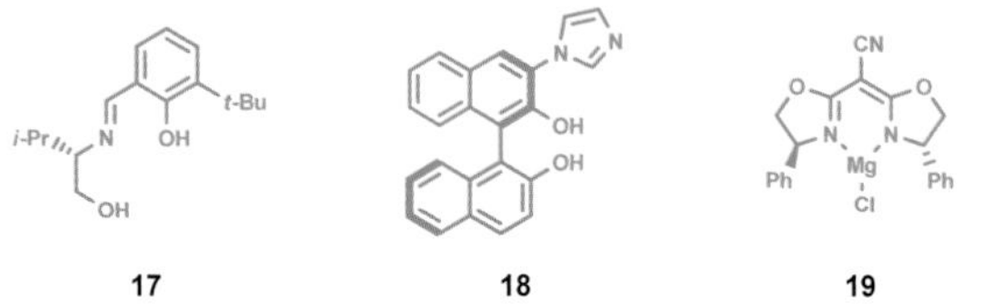

Figure 2

I. Enantioselective Addition of HCN or its Equivalents to Prochiral Aldehydes

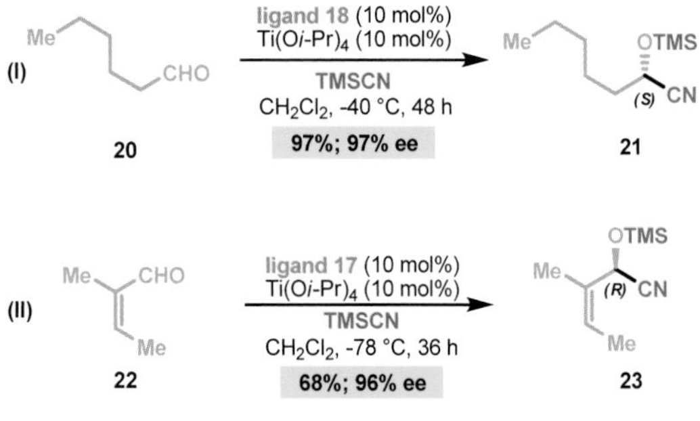

(I)

ligand 18 (10 mol%) / Ti(Oi-Pr)_4 (10 mol%) / TMSCN / CH_2Cl_2, -40 °C, 48 h

97%; 97% ee

(II)

ligand 17 (10 mol%) / Ti(Oi-Pr)_4 (10 mol%) / TMSCN / CH_2Cl_2, -78 °C, 36 h

68%; 96% ee

Reference key for equations: (I)[4]; (II)[5]

Scheme 3

ENANTIOSELECTIVE CYANOSILYLATION OF C=O

Enantioselective Addition of HCN or its Equivalents to Prochiral Aldehydes (Continued)

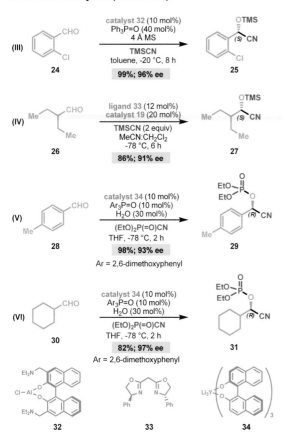

Scheme 4

II. Enantioselective Addition of HCN or its Equivalents to Prochiral Ketones

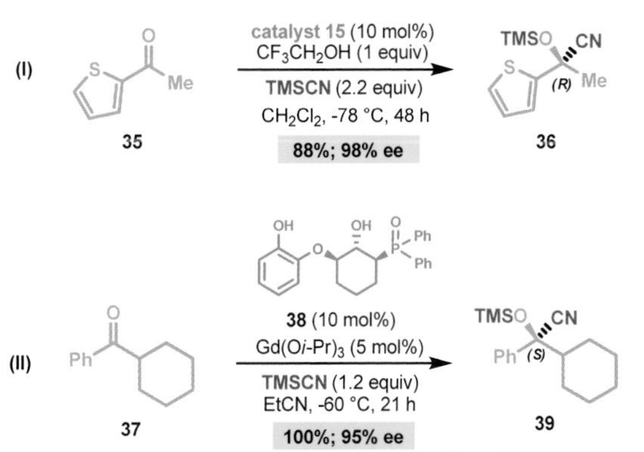

(See Scheme 5)

Scheme 5

Reference key for equations: (I)[9]; (II)[10]; (III)[11]

Enantioselective Addition of HCN or its Equivalents to Prochiral Ketones (Continued)

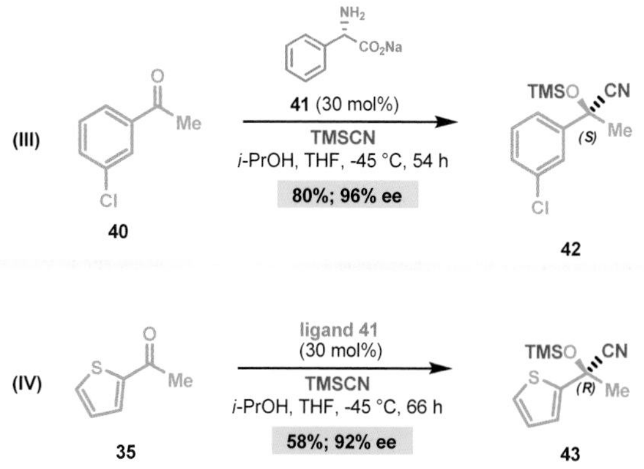

Scheme 6

Reference key for equations: (III-IV)[11]

References

1. For recent reviews on the enantioselective preparation and synthetic utility of cyanohydrins, see: (a) North, M., Usanov, D. L. & Young, C. Lewis Acid Catalyzed Asymmetric Cyanohydrin Synthesis. *Chem. Rev.* **2008**, 108, 5146-5226; (b) Khan, N.-u. H., Kureshy, R. I., Abdi, S. H. R., Agrawal, S. & Jasra, R. V. Metal-catalyzed asymmetric cyanation reactions. *Coord. Chem. Rev.* **2008**, 252, 593-623; (c) Brunel, J.-M. & Holmes, I. P. Chemically catalyzed asymmetric cyanohydrin syntheses. *Angew. Chem., Int. Ed.* **2004**, 43, 2752-2778; (d) Vachal, P. & Jacobsen, E. N. Cyanation of carbonyl and imino groups. *Compr. Asymmetric Catal., Suppl.* **2004**, 1, 117-130; (e) North, M. Synthesis and applications of non-racemic cyanohydrins. *Tetrahedron Asymmetry* **2003**, 14, 147-176.
2. Oku, J. & Inoue, S. Asymmetric cyanohydrin synthesis catalyzed by a synthetic cyclic dipeptide. *J. Chem. Soc., Chem. Commun.* **1981**, 229-230.
3. (a) Ryu, D. H. & Corey, E. J. Highly Enantioselective Cyanosilylation of Aldehydes Catalyzed by a Chiral Oxazaborolidinium Ion. *J. Am. Chem. Soc.* **2004**, 126, 8106-8107; (b) Ryu, D. H. & Corey, E. J. Enantioselective Cyanosilylation of Ketones Catalyzed by a Chiral Oxazaborolidinium Ion. *J. Am. Chem. Soc.* **2005**, 127, 5384-5387.
4. Yang, F., Wei, S., Chen, C.-A., Xi, P., Yang, L., Lan, J., Gau, H.-M. & You, J. A new strategy for designing non-C2-symmetric monometallic bifunctional catalysts and their application in enantioselective cyanation of aldehydes. *Chem.--Eur. J.* **2008**, 14, 2223-2231.
5. Hayashi, M., Miyamoto, Y., Inoue, T. & Oguni, N. Enantioselective trimethylsilylcyanation of some aldehydes catalyzed by chiral Schiff base-titanium alkoxide complexes. *J. Org. Chem.* **1993**, 58, 1515-1522.
6. Anthony, J. E., Eaton, D. L. & Parkin, S. R. A Road Map to Stable, Soluble, Easily Crystallized Pentacene Derivatives. *Org. Lett.* **2002**, 4, 15-18.
7. Corey, E. J. & Wang, Z. Enantioselective conversion of aldehydes to cyanohydrins by a catalytic system with separate chiral binding sites for aldehyde and cyanide components. *Tetrahedron Lett.* **1993**, 34, 4001-4004.
8. Abiko, Y., Yamagiwa, N., Sugita, M., Tian, J., Matsunaga, S. & Shibasaki, M. Catalytic asymmetric cyano-phosphorylation of aldehydes promoted by heterobimetallic YLi3tris(binaphthoxide) (YLB) complex. *Synlett* **2004**, 2434-2436.
9. Fuerst, D. E. & Jacobsen, E. N. Thiourea-Catalyzed Enantioselective Cyanosilylation of Ketones. *J. Am. Chem. Soc.* **2005**, 127, 8964-8965.
10. Masumoto, S., Suzuki, M., Kanai, M. & Shibasaki, M. Catalytic asymmetric synthesis of (S)-oxybutynin and a versatile intermediate for antimuscarinic agents. *Tetrahedron* **2004**, 60, 10497-10504.
11. Liu, X., Qin, B., Zhou, X., He, B. & Feng, X. Catalytic Asymmetric Cyanosilylation of Ketones by a Chiral Amino Acid Salt. *J. Am. Chem. Soc.* **2005**, 127, 12224-12225.

ENANTIOSELECTIVE CARBONYL ENE REACTION

prochiral aldehyde + substituted olefin → chiral Lewis acid catalyst / ene reaction → chiral homo-allylic alcohol

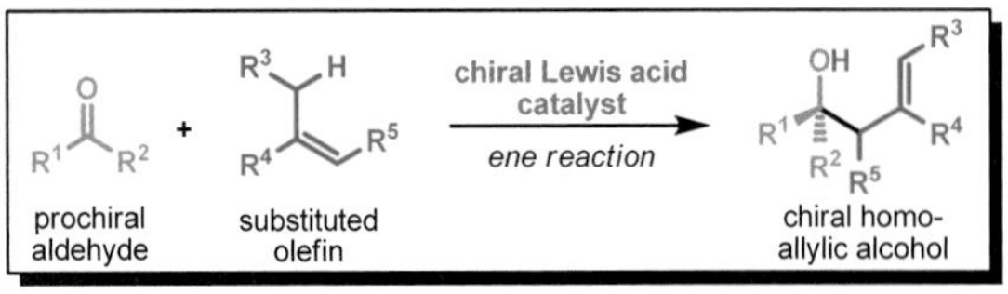

Introduction

The ene coupling of an aldehyde or ketone with an olefinic substrate can occur either by heating or catalysis by a Lewis acid and affords a homoallylic alcohol product as shown above.[1] Although thermal ene reactions usually proceed through a cyclic pre-transition state structure (Figure 1), Lewis acid-catalyzed ene reactions are mechanistically more complex and may in some cases occur via a stepwise pathway involving a carbocationic intermediate.[2]

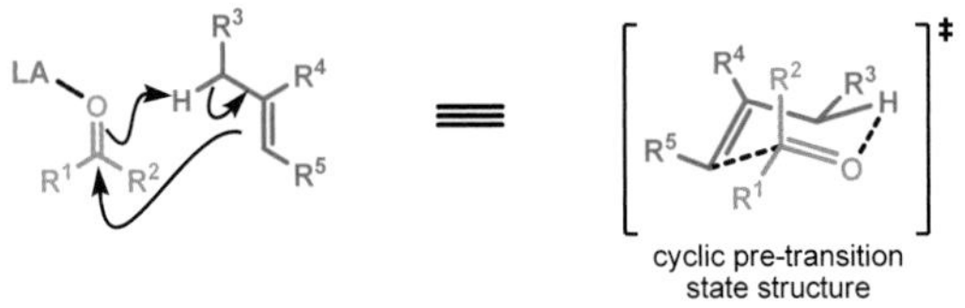

cyclic pre-transition state structure

Figure 1

The reactivity of the carbonyl component (enophile) depends on the electron deficiency of the carbonyl group, the most reactive being aldehydes such as CCl_3CHO as well as glyoxylates and pyruvates. Electron-rich 1,1-disubstituted and trisubstituted alkenes are the most reactive ene reaction partners, while mono-, 1,2- and tetrasubstituted alkenes are significantly less reactive. The first catalytic and highly enantioselective carbonyl ene reaction (Scheme 1) was reported by H. Yamamoto who utilized an chiral organoaluminum catalyst **2** for the coupling of pentafluorobenzaldehyde (**1**) and 2-(phenylthio)propene (**3**).[3]

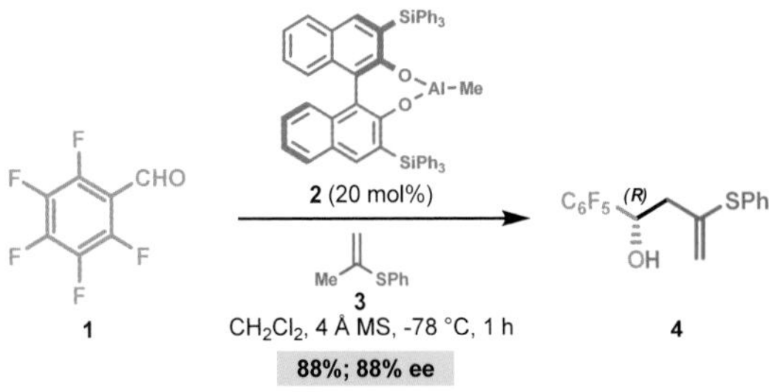

Scheme 1

Chiral Lewis acids derived from titanium (Ti), cobalt (Co), copper (Cu) and scandium (Sc) can also catalyze enantioselective carbonyl ene reactions. It is important to note that the long-known thermal carbonyl ene reactions require relatively high temperatures (>100 °C) and are not enantioselective.[1g]

I. Intermolecular Carbonyl Ene Reactions Between Activated Aldehydes and Substituted Alkenes

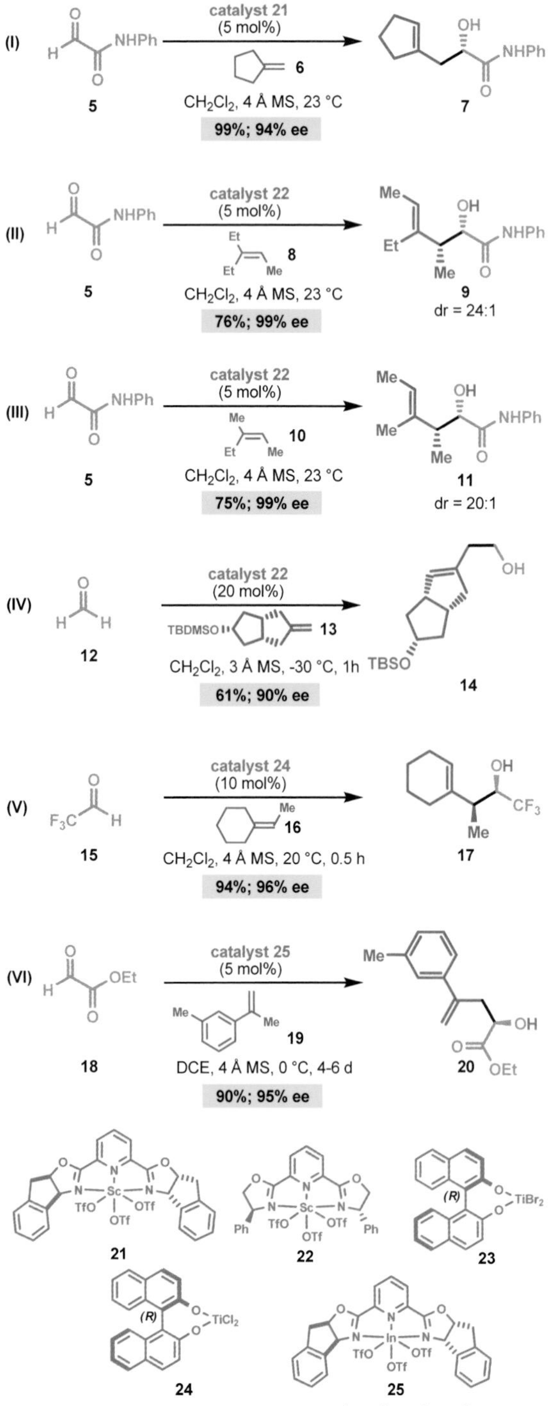

(I) **5** + **6** → catalyst 21 (5 mol%), CH_2Cl_2, 4 Å MS, 23 °C → **7**
99%; 94% ee

(II) **5** + **8** → catalyst 22 (5 mol%), CH_2Cl_2, 4 Å MS, 23 °C → **9**
76%; 99% ee
dr = 24:1

(III) **5** + **10** → catalyst 22 (5 mol%), CH_2Cl_2, 4 Å MS, 23 °C → **11**
75%; 99% ee
dr = 20:1

(IV) **12** + **13** → catalyst 22 (20 mol%), CH_2Cl_2, 3 Å MS, -30 °C, 1h → **14**
61%; 90% ee

(V) **15** + **16** → catalyst 24 (10 mol%), CH_2Cl_2, 4 Å MS, 20 °C, 0.5 h → **17**
94%; 96% ee

(VI) **18** + **19** → catalyst 25 (5 mol%), DCE, 4 Å MS, 0 °C, 4-6 d → **20**
90%; 95% ee

21   **22**   **23**

24   **25**

Reference key for equations: (I-III)[4]; (IV)[5]; (V)[6]; (VI)[7]

Scheme 2

ENANTIOSELECTIVE CARBONYL ENE REACTION

II. Intermolecular Carbonyl Ene Reactions Between Unactivated Aldehydes and Activated Alkenes

The carbonyl ene reaction usually does not occur with unactivated aldehydes. However, if highly electron rich alkenes (e.g. enol ethers) are used as reaction partners, this transformation can be accomplished enantioselectively under catalytic conditions. A number of examples are shown in Scheme 3.

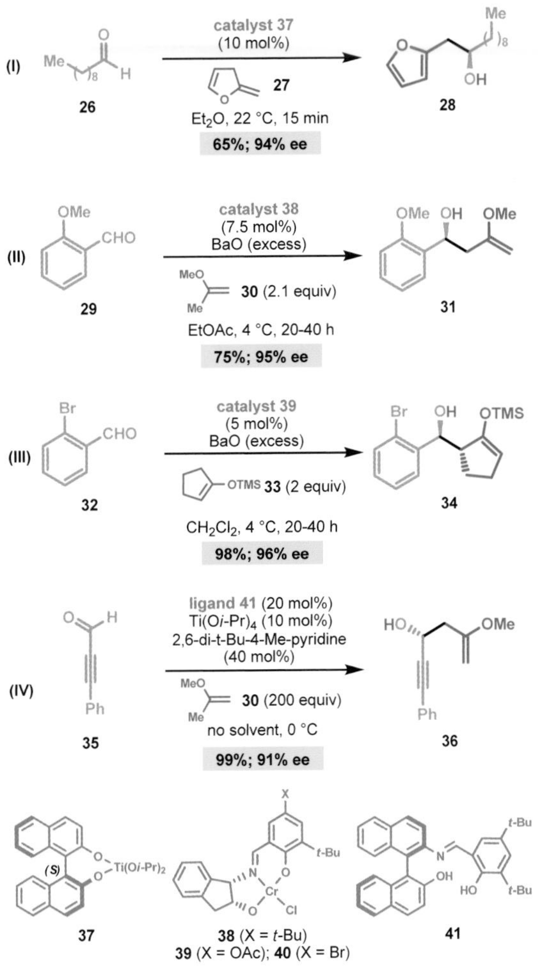

Scheme 3

Reference key for equations: (I)[8]; (II)[9]; (III)[10]; (IV)[11]

Reference key for equations: (I-II)[12]

Scheme 4

III. Intramolecular Carbonyl Ene Reactions

The intramolecular carbonyl ene reaction can take place more rapidly and under milder reaction conditions than its intermolecular counterpart when, as with 5- or 6-membered ring formation, it is entropically favored. For this reason the intramolecular coupling of unactivated olefins and aldehydes is possible and as a result stereochemically complex and synthetically useful carbocyclic structures are formed.[1g]

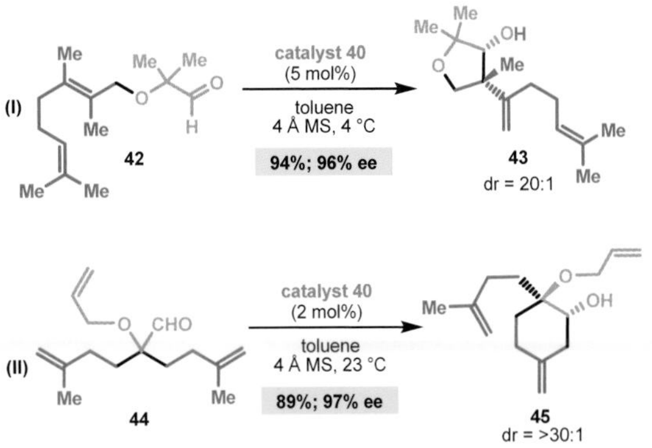

References

1. (a) Mikami, K., Terada, M., Narisawa, S. & Nakai, T. Asymmetric catalysis for carbonyl-ene reaction. *Synlett* **1992**, 255-265; (b) Mikami, K. & Shimizu, M. Asymmetric ene reactions in organic synthesis. *Chem. Rev.* **1992**, 92, 1021-1050; (c) Mikami, K. Asymmetric catalysis of carbonyl-ene reactions and related carbon-carbon bond forming reactions. *Pure Appl. Chem.* **1996**, 68, 639-644; (d) Mikami, K. & Terada, M. Ene-type reactions. *Compr. Asymmetric Catal. I-III* **1999**, 3, 1143-1174; (e) Dias, L. C. Chiral Lewis acid catalyzed ene reactions. *Curr. Org. Chem.* **2000**, 4, 305-342; (f) Johnson, J. S. & Evans, D. A. Chiral Bis(oxazoline) Copper(II) Complexes: Versatile Catalysts for Enantioselective Cycloaddition, Aldol, Michael, and Carbonyl Ene Reactions. *Acc. Chem. Res.* **2000**, 33, 325-335; (g) Clarke, M. L. & France, M. B. The carbonyl ene reaction. *Tetrahedron* **2008**, 64, 9003-9031.
2. Kürti, L. & Czakó, B. Prins Reaction in *Strategic Applications of Named Reactions in Organic Synthesis* 364-365 (Academic Press/Elsevier, San Diego, **2005**).
3. Maruoka, K., Hoshino, Y., Shirasaka, T. & Yamamoto, H. Asymmetric ene reaction catalyzed by chiral organoaluminum reagent. *Tetrahedron Lett.* **1988**, 29, 3967-3970.
4. Evans, D. A. & Wu, J. Enantioselective Syn-Selective Scandium-Catalyzed Ene Reactions. *J. Am. Chem. Soc.* **2005**, 127, 8006-8007.
5. Mikami, K. & Yoshida, A. Asymmetric addition reaction to non-prochiral carbonyl compound: asymmetric catalytic formaldehyde-ene reaction for isocarbacyclin synthesis. *Synlett* **1995**, 29-31.
6. Mikami, K., Yajima, T., Takasaki, T., Matsukawa, S., Terada, M., Uchimaru, T. & Maruta, M. Asymmetric catalysis of carbonyl-ene and aldol reactions with fluoral by chiral binaphthol-derived titanium complex. *Tetrahedron* **1996**, 52, 85-98.
7. Zhao, J.-F., Tsui, H.-Y., Wu, P.-J., Lu, J. & Loh, T.-P. Highly Enantioselective Carbonyl-ene Reactions Catalyzed by In(III)-PyBox Complex. *J. Am. Chem. Soc.* **2008**, 130, 16492-16493.
8. Miles, W. H., Dethoff, E. A., Tuson, H. H. & Ulas, G. Kishner's Reduction of 2-Furylhydrazone Gives 2-Methylene-2,3-dihydrofuran, a Highly Reactive Ene in the Ene Reaction. *J. Org. Chem.* **2005**, 70, 2862-2865.
9. Ruck, R. T. & Jacobsen, E. N. Asymmetric Catalysis of Hetero-Ene Reactions with Tridentate Schiff Base Chromium(III) Complexes. *J. Am. Chem. Soc.* **2002**, 124, 2882-2883.
10. Ruck, R. T. & Jacobsen, E. N. Asymmetric hetero-ene reactions of trimethylsilyl enol ethers catalyzed by tridentate Schiff base chromium(III) complexes. *Angew. Chem., Int. Ed.* **2003**, 42, 4771-4774.
11. Carreira, E. M., Lee, W. & Singer, R. A. Catalytic, Enantioselective Acetone Aldol Additions with 2-Methoxypropene. *J. Am. Chem. Soc.* **1995**, 117, 3649-3650.
12. Grachan, M. L., Tuidge, M. T. & Jacobsen, E. N. Enantioselective catalytic carbonyl-ene cyclization reactions. *Angew. Chem., Int. Ed.* **2008**, 47, 1469-1472.

ENANTIOSELECTIVE MANNICH REACTION

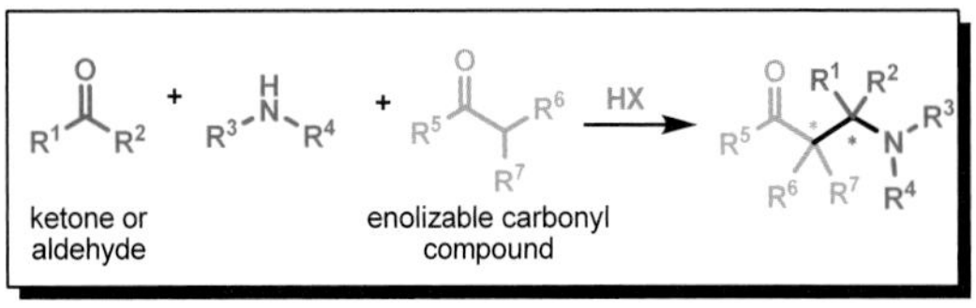

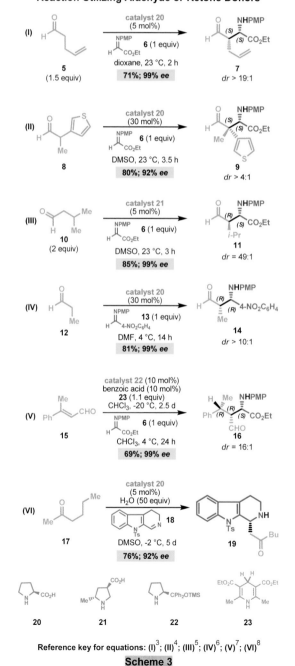

Introduction

The classical Mannich reaction is a nitrogen analog of the aldol process (see pages 64-70) in which C-C bond-formation occurs by the attack of a carbon nucleophile (e.g., enol or enolate equivalent) on a reactive C=N subunit that usually is a protonated or charged iminium group.[1] A simple example of the proton-accelerated form of this process is the conversion of acetophenone to phenyl β-dimethylaminoethyl ketone following the equation (Scheme 1):

Scheme 1

The Mannich reaction is a very powerful tool for simultaneous C-C bond formation and introduction of N. It has been used successfully in many syntheses of alkaloids and other nitrogen-containing organic substances, and also as a major biosynthetic process of nature. Highly efficient variants of this reaction are known with simple enols, silyl enol ethers, silyl ketene acetals, metal enolates and electron-rich aromatic compounds. In addition, masked forms of the imines or iminum salts (e.g., aminals and hemi-aminals) can be used to generate the reactive electrophilic species *in situ*.

Iminium salts are generally more reactive than imines. However, the use of Lewis acids can substantially increase the reactivity of imines. An early example of an enantioselective catalytic Mannich reaction was reported by Kobayashi et al. (Scheme 2).[2]

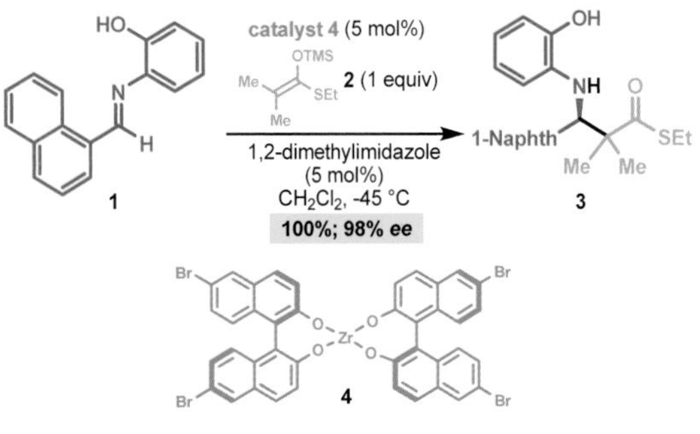

Scheme 2

Highly enantioselective Mannich reactions of aldehydes and methyl ketones have been demonstrated using chiral secondary amines (e.g., proline) which can generate chiral enamines *in situ*. These enamines can undergo Mannich coupling to reactive imines to form products with high enantio- and diastereoselectivity in favorable cases (See Scheme 3).

There are many examples in which a chiral protic acid activates the imine component to form a chiral anion-iminium tight ion pair which then attacks the enol, enolate or enolate equivalent (See Schemes 4 and 5 on page 89).

I. Catalytic Enantioselective Direct Mannich Reaction Utilizing Aldehyde or Ketone Donors

Reference key for equations: (I)[3]; (II)[4]; (III)[5]; (IV)[6]; (V)[7]; (VI)[8]

Scheme 3

ENANTIOSELECTIVE MANNICH REACTION

II. Catalytic Enantioselective Indirect Mannich Reaction Utilizing a Preformed Enolate Equivalent

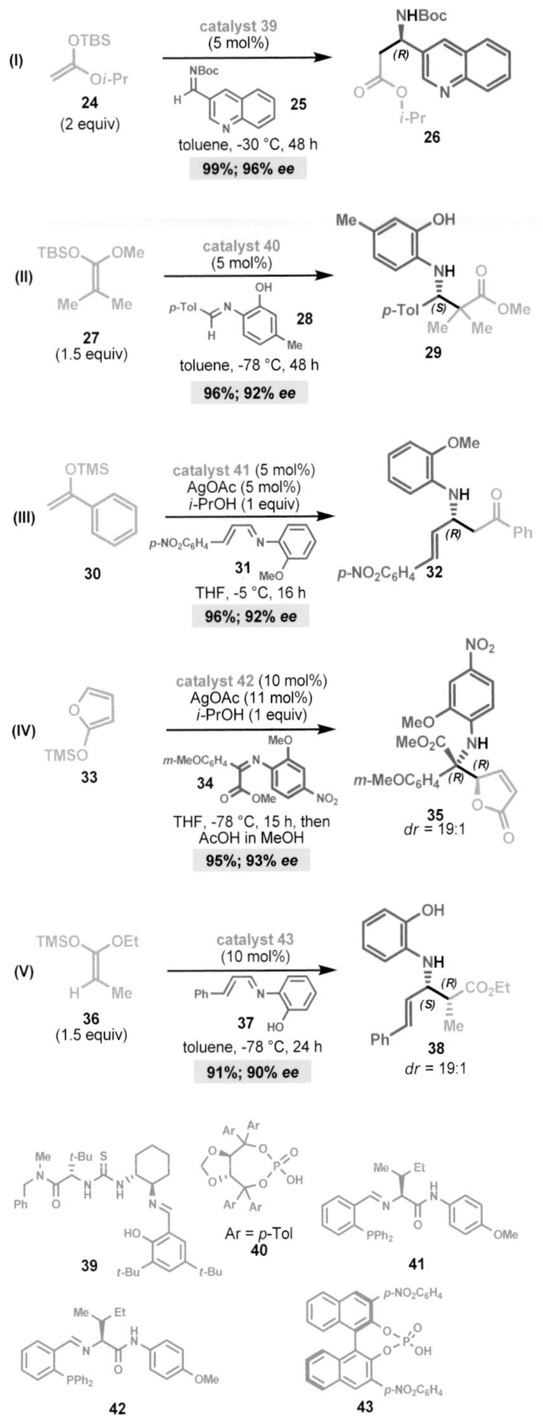

(I)
24 (2 equiv) — catalyst 39 (5 mol%); toluene, -30 °C, 48 h → **26**
99%; 96% ee

(II)
27 (1.5 equiv) — catalyst 40 (5 mol%); toluene, -78 °C, 48 h → **29**
96%; 92% ee

(III)
30, **31** — catalyst 41 (5 mol%), AgOAc (5 mol%), i-PrOH (1 equiv); THF, -5 °C, 16 h → **32**
96%; 92% ee

(IV)
33, **34** — catalyst 42 (10 mol%), AgOAc (11 mol%), i-PrOH (1 equiv); THF, -78 °C, 15 h, then AcOH in MeOH → **35** dr = 19:1
95%; 93% ee

(V)
36 (1.5 equiv), **37** — catalyst 43 (10 mol%); toluene, -78 °C, 24 h → **38** dr = 19:1
91%; 90% ee

Catalysts: **39**, **40** (Ar = p-Tol), **41**, **42**, **43**

Reference key for equations: (I)[9]; (II)[10]; (III)[11]; (IV)[12]; (V)[13]

Scheme 4

III. Catalytic Enantioselective Mannich Reaction With a Range of Substrates

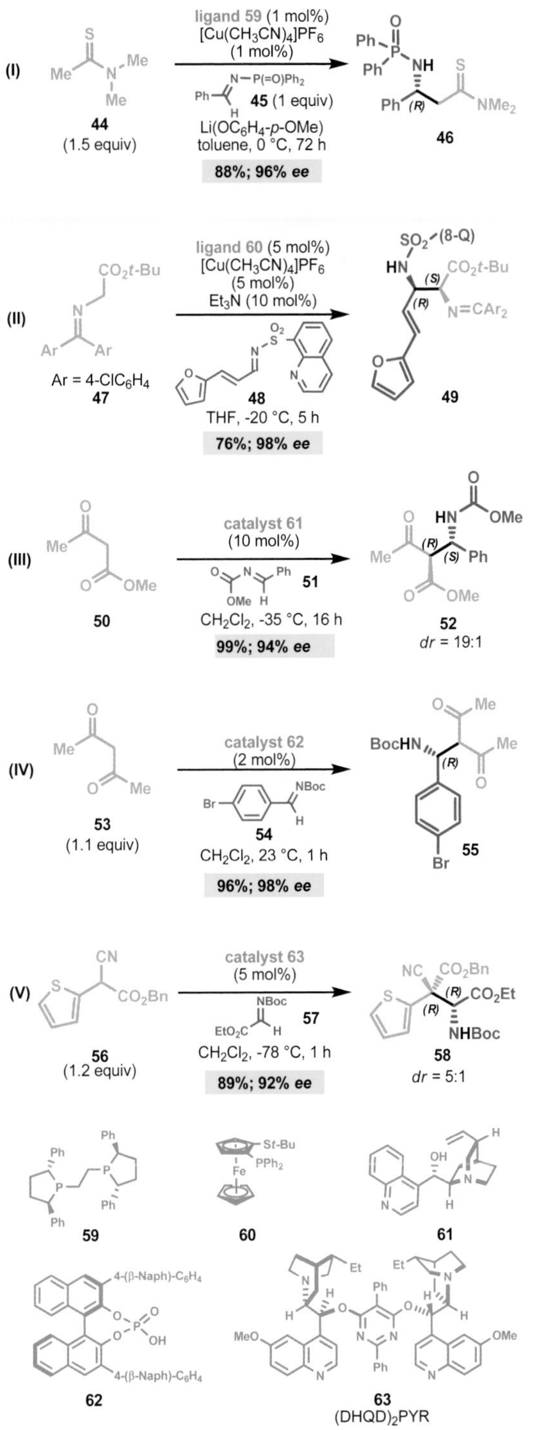

(I)
44 (1.5 equiv), **45** (1 equiv) — ligand 59 (1 mol%), [Cu(CH$_3$CN)$_4$]PF$_6$ (1 mol%), Li(OC$_6$H$_4$-p-OMe); toluene, 0 °C, 72 h → **46**
88%; 96% ee

(II)
47 (Ar = 4-ClC$_6$H$_4$), **48** — ligand 60 (5 mol%), [Cu(CH$_3$CN)$_4$]PF$_6$ (5 mol%), Et$_3$N (10 mol%); THF, -20 °C, 5 h → **49**
76%; 98% ee

(III)
50, **51** — catalyst 61 (10 mol%); CH$_2$Cl$_2$, -35 °C, 16 h → **52** dr = 19:1
99%; 94% ee

(IV)
53 (1.1 equiv), **54** — catalyst 62 (2 mol%); CH$_2$Cl$_2$, 23 °C, 1 h → **55**
96%; 98% ee

(V)
56 (1.2 equiv), **57** — catalyst 63 (5 mol%); CH$_2$Cl$_2$, -78 °C, 1 h → **58** dr = 5:1
89%; 92% ee

Ligands/catalysts: **59**, **60**, **61**, **62**, **63** (DHQD)$_2$PYR

Reference key for equations: (I)[14]; (II)[15]; (III)[16]; (IV)[17]; (V)[18]

Scheme 5

89

ENANTIOSELECTIVE MANNICH REACTION

IV. Catalytic Enantioselective Aza-Henry (Nitro-Mannich) Reaction[19]

The analog of the nitro aldol coupling reaction (Henry reaction) with an imine acceptor rather than a carbonyl group has also been converted to an enantioselective version by the use of a chiral protic or Lewis acidic catalyst to activate the imine component,[19] as illustrated by examples I-III in Scheme 6. Examples IV and V may involve reaction of the aldimine with a complex of nitronate with the catalyst 82 or 83.

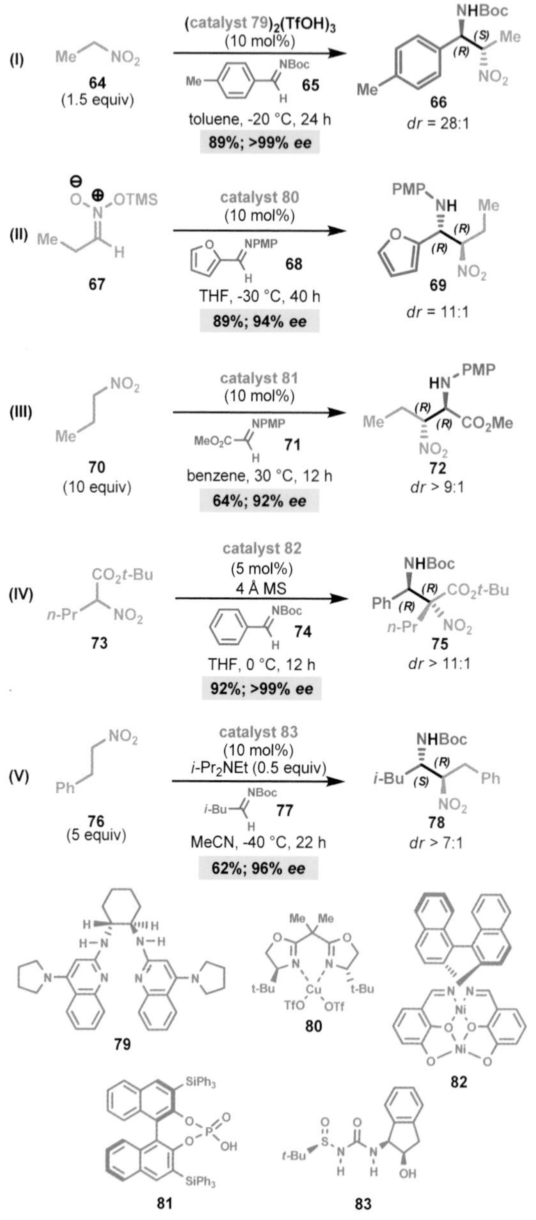

V. Application of Stereoselective Mannich Reactions to Complex Molecule Synthesis

The enantioselective total synthesis of (−)-actinophyllic acid by Overman et al.[25] employed a highly efficient tandem aza-Cope rearrangement / intramolecular Mannich reaction (see Example I, Scheme 6). The secondary amine hydrochloride 84, when exposed to paraformaldehyde formed the iminium salt 85 that underwent [3,3]-rearrangement to afford 86, which in turn gave the cyclized product 87 by an intramolecular Mannich reaction. Example II in Scheme 7 illustrates the use of the aza-Henry reaction during the total synthesis of (−)-nakadomarin A.[26]

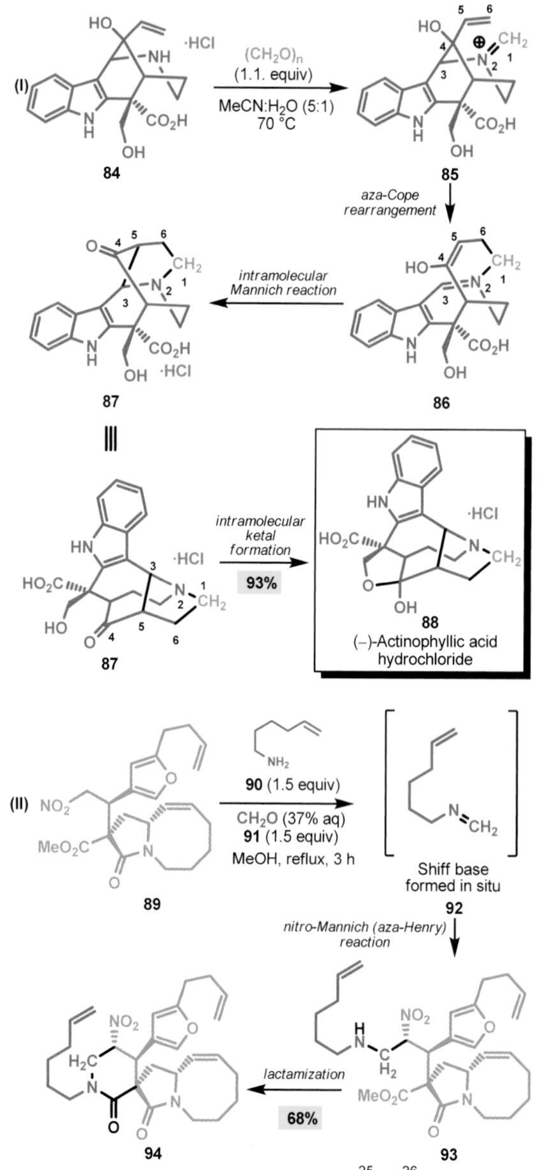

Reference key for equations: (I)[20]; (II)[21]; (III)[22]; (IV)[23]; (V)[24]

Scheme 6

Reference key for equations: (I)[25]; (II)[26]

Scheme 7

ENANTIOSELECTIVE MANNICH REACTION

Applications of Stereoselective Mannich Reactions in Complex Molecule Synthesis (Continued)

The alkaloid porantherine **97** possesses an interesting tetracyclic framework (Scheme 8). A short and efficient total synthesis of this natural product utilized sequential intramolecular Mannich reactions. The initial Mannich reaction formed **94** (Scheme 8), which after hydrolysis and protonation cyclized to **96** (via unstable enamine **95**). The orientation of the acetyl group can change under the reaction conditions by enolization.

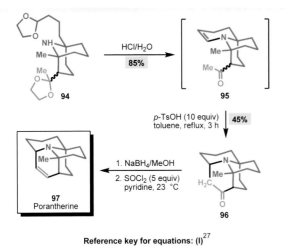

Reference key for equations: (I)[27]

Scheme 8

References

1. (a) Kobayashi, S. & Ishitani, H. Catalytic Enantioselective Addition to Imines. *Chem. Rev.* **1999**, 99, 1069-1094; (b) Marques, M. M. B. Catalytic enantioselective cross-Mannich reaction of aldehydes. *Angew. Chem., Int. Ed.* **2006**, 45, 348-352; (c) Tanaka, F. & Barbas, C. F., III. Aldol and Mannich-type reactions in *Enantioselective Organocatalysis* (ed. Dalko, P. I.), 19-55 (Wiley-VCH, **2007**); (d) Friestad, G. K. & Mathies, A. K. Recent developments in asymmetric catalytic addition to C=N bonds. *Tetrahedron* **2007**, 63, 2541-2569; (e) Mukherjee, S., Yang, J. W., Hoffmann, S. & List, B. Asymmetric Enamine Catalysis. *Chem. Rev.* **2007**, 107, 5471-5569; (f) Ting, A. & Schaus, S. E. Organocatalytic asymmetric Mannich reactions. New methodology, catalyst design, and synthetic applications. *Eur. J. Org. Chem.* **2007**, 5797-5815; (g) Verkade, J. M. M., vanHemert, L. J. C., Quaedflieg, P. J. L. M. & Rutjes, F. P. J. T. Organocatalysed asymmetric Mannich reactions. *Chem. Soc. Rev.* **2008**, 37, 29-41; (h) Cordova, A. & Rios, R. Direct catalytic asymmetric Mannich reactions and surroundings. *Amino Group Chemistry* **2008**, 185-205; (i) Ipaktschi, J. & Saidi, M. R. Synthesis of alkyl- and cycloalkylamines by Mannich function. *Sci. Synth.* **2009**, 40a, 435-478; (j) Arrayas, R. G. & Carretero, J. C. Catalytic asymmetric direct Mannich reaction: A powerful tool for the synthesis of α,β-diamino acids. *Chem. Soc. Rev.* **2009**, 38, 1940-1948; (k) Kazmaier, U. Direct Michael, Aldol, and Mannich Additions Catalyzed by Alkaline Earth Metals. *Angew. Chem., Int. Ed.* **2009**, 48, 5790-5792.
2. Ishitani, H., Ueno, M. & Kobayashi, S. Catalytic Enantioselective Mannich-Type Reactions Using a Novel Chiral Zirconium Catalyst. *J. Am. Chem. Soc.* **1997**, 119, 7153-7154.
3. Cordova, A., Watanabe, S.-i., Tanaka, F., Notz, W. & Barbas, C. F., III. A Highly Enantioselective Route to Either Enantiomer of Both α- and β-Amino Acid Derivatives. *J. Am. Chem. Soc.* **2002**, 124, 1866-1867.
4. Chowdari, N. S., Suri, J. T. & Barbas, C. F., III. Asymmetric Synthesis of Quaternary α- and β-Amino Acids and β-Lactams via Proline-Catalyzed Mannich Reactions with Branched Aldehyde Donors. *Org. Lett.* **2004**, 6, 2507-2510.
5. Mitsumori, S., Zhang, H., Cheong, P. H.-Y., Houk, K. N., Tanaka, F. & Barbas, C. F., III. Direct Asymmetric anti-Mannich-Type Reactions Catalyzed by a Designed Amino Acid. *J. Am. Chem. Soc.* **2006**, 128, 1040-1041.
6. Notz, W., Tanaka, F., Watanabe, S., Chowdari, N. S., Turner, J. M., Thayumanavan, R. & Barbas, C. F. The Direct Organocatalytic Asymmetric Mannich Reaction: Unmodified Aldehydes as Nucleophiles. *J. Org. Chem.* **2003**, 68, 9624-9634.
7. Zhao, G.-L. & Cordova, A. Direct organocatalytic asymmetric reductive Mannich-type reactions. *Tetrahedron Lett.* **2006**, 47, 7417-7421.
8. Itoh, T., Yokoya, M., Miyauchi, K., Nagata, K. & Ohsawa, A. Proline-Catalyzed Asymmetric Addition Reaction of 9-Tosyl-3,4-dihydro-β-carboline with Ketones. *Org. Lett.* **2003**, 5, 4301-4304.
9. Wenzel, A. G. & Jacobsen, E. N. Asymmetric catalytic mannich reactions catalyzed by urea derivatives: enantioselective synthesis of β-aryl-β-amino acids. *J. Am. Chem. Soc.* **2002**, 124, 12964-12965.
10. Akiyama, T., Saitoh, Y., Morita, H. & Fuchibe, K. Enantioselective Mannich-type reaction catalyzed by a chiral Bronsted acid derived from TADDOL. *Adv. Synth. Catal.* **2005**, 347, 1523-1526.
11. Josephsohn, N. S., Snapper, M. L. & Hoveyda, A. H. Ag-Catalyzed Asymmetric Mannich Reactions of Enol Ethers with Aryl, Alkyl, Alkenyl, and Alkynyl Imines. *J. Am. Chem. Soc.* **2004**, 126, 3734-3735.
12. Wieland, L. C., Vieira, E. M., Snapper, M. L. & Hoveyda, A. H. Ag-Catalyzed Diastereo- and Enantioselective Vinylogous Mannich Reactions of α-Ketoimine Esters. Development of a Method and Investigation of its Mechanism. *J. Am. Chem. Soc.* **2009**, 131, 570-576.
13. Akiyama, T., Itoh, J., Yokota, K. & Fuchibe, K. Enantioselective Mannich-type reaction catalyzed by a chiral Bronsted acid. *Angew. Chem., Int. Ed.* **2004**, 43, 1566-1568.
14. Suzuki, Y., Yazaki, R., Kumagai, N. & Shibasaki, M. Direct Catalytic Asymmetric Mannich-Type Reaction of Thioamides. *Angew. Chem., Int. Ed.* **2009**, 48, 5026-5029.
15. Hernandez-Toribio, J., Gomez Arrayas, R. & Carretero, J. C. Substrate-Controlled Diastereoselectivity Switch in Catalytic Asymmetric Direct Mannich Reaction of Glycine Derivatives with Imines: From anti- to syn-α,β-Diamino Acids. *Chem.--Eur. J.* **2010**, 16, 1153-1157.
16. Lou, S., Taoka, B. M., Ting, A. & Schaus, S. E. Asymmetric Mannich Reactions of β-Keto Esters with Acyl Imines Catalyzed by Cinchona Alkaloids. *J. Am. Chem. Soc.* **2005**, 127, 11256-11257.
17. Uraguchi, D. & Terada, M. Chiral Bronsted Acid-Catalyzed Direct Mannich Reactions via Electrophilic Activation. *J. Am. Chem. Soc.* **2004**, 126, 5356-5357.
18. Poulsen, T. B., Alemparte, C., Saaby, S., Bella, M. & Jorgensen, K. A. Direct organocatalytic and highly enantio- and diastereoselective Mannich reactions of α-substituted α-cyanoacetates. *Angew. Chem., Int. Ed.* **2005**, 44, 2896-2899.
19. Marques-Lopez, E., Merino, P., Tejero, T. & Herrera, R. P. Catalytic Enantioselective Aza-Henry Reactions. *Eur. J. Org. Chem.* **2009**, 2401-2420.
20. Davis, T. A., Wilt, J. C. & Johnston, J. N. Bifunctional Asymmetric Catalysis: Amplification of Bronsted Basicity Can Orthogonally Increase the Reactivity of a Chiral Bronsted Acid. *J. Am. Chem. Soc.* **2010**, 132, 2880-2882.
21. Anderson, J. C., Howell, G. P., Lawrence, R. M. & Wilson, C. S. An Asymmetric Nitro-Mannich Reaction Applicable to Alkyl, Aryl, and Heterocyclic Imines. *J. Org. Chem.* **2005**, 70, 5665-5670.
22. Chen, Z., Morimoto, H., Matsunaga, S. & Shibasaki, M. A Bench-Stable Homodinuclear Ni$_2$-Schiff Base Complex for Catalytic Asymmetric Synthesis of α-Tetrasubstituted anti-α,β-Diamino Acid Surrogates. *J. Am. Chem. Soc.* **2008**, 130, 2170-2171.
23. Rueping, M. & Antonchick, A. P. Bronsted-acid-catalyzed activation of nitroalkanes: a direct enantioselective aza-Henry reaction. *Org. Lett.* **2008**, 10, 1731-1734.
24. Robak, M. T., Trincado, M. & Ellman, J. A. Enantioselective Aza-Henry Reaction with an N-Sulfinyl Urea Organocatalyst. *J. Am. Chem. Soc.* **2007**, 129, 15110-15111.
25. Martin, C. L., Overman, L. E. & Rohde, J. M. Total Synthesis of (±)- and (-)-Actinophyllic Acid. *J. Am. Chem. Soc.* **2010**, 132, 4894-4906.
26. Jakubec, P., Cockfield, D. M. & Dixon, D. J. Total Synthesis of (-)-Nakadomarin A. *J. Am. Chem. Soc.* **2009**, 131, 16632-16633.
27. Corey, E. J. & Balanson, R. D. Total synthesis of (±)-porantherine. *J. Am. Chem. Soc.* **1974**, 96, 6516-6517.

ENANTIOSELECTIVE STRECKER REACTION

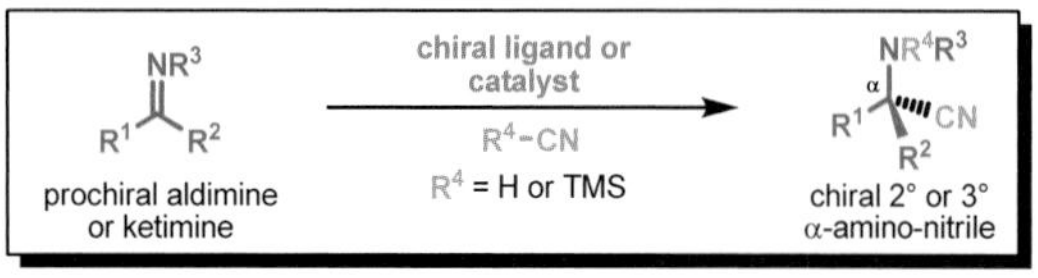

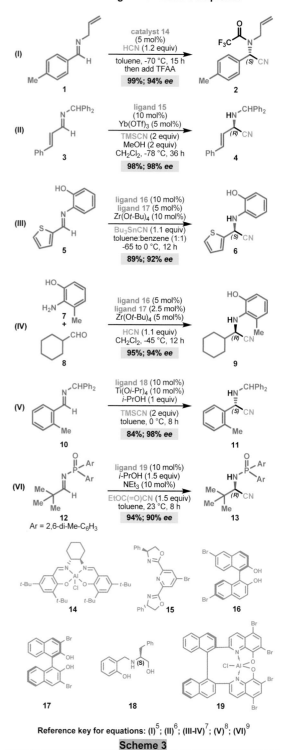

Introduction

The first laboratory synthesis of an α-amino acid (alanine by A. Strecker, ca. 1850) involved the reaction of acetaldehyde with aqueous ammonia and hydrogen cyanide and subsequent hydrolysis of the resulting α-amino-nitrile with aqueous acid.[1] The Strecker reaction has since been widely used for the preparation of chiral α-amino acids, either by (1) starting with imines that bear a chiral controller group on the nitrogen atom[2] or by (2) employing chiral catalysts that allow face-selective addition of cyanide to a prochiral imine.[3]

The asymmetric Strecker reaction may also be carried out as a three-component process in which the aldehyde, the 1° or 2° amine and HCN are combined in the presence of a chiral catalyst. An HCN equivalent such as TMSCN or an alkali metal cyanide may also be used in place of the gaseous (and toxic) HCN. α-Amino-nitriles may be converted to the corresponding α-amino acids under a variety of hydrolytic conditions (Scheme 1).[4]

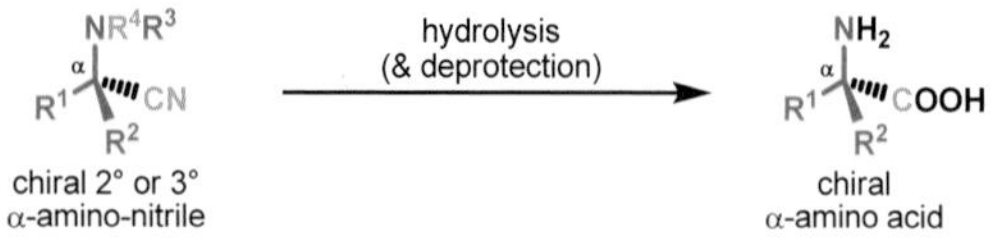

Scheme 1

The efficiency of the catalyst varies with the substrate and the steric properties of the substituent on the imine nitrogen atom (R^3) are also critical. The nitrogen substituent should also allow facile removal from the Strecker product.

The pathway of the Strecker reaction, via intermediates **A** and **B**, is summarized in Scheme 2.[3c].

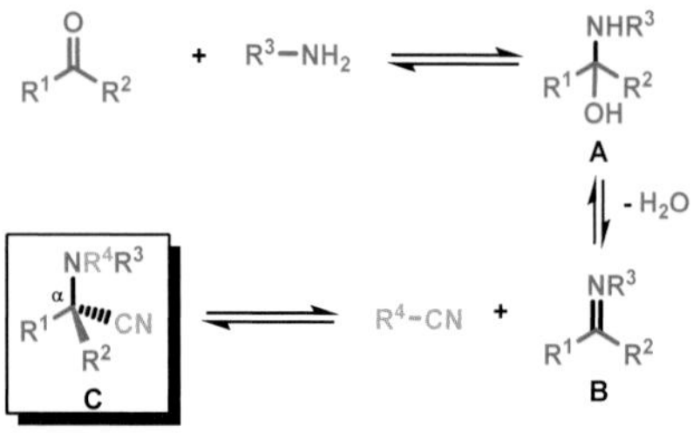

Scheme 2

There are two types of catalysts for enantioselective Strecker reactions, chiral metal complexes and proton-donating chiral organic molecules. The examples that follow in Schemes 3-8 are organized according to the substrate (i.e., aldimine or ketimine) and the type of catalyst used.

I. Catalytic Enantioselective Strecker Reaction of Aldimines Utilizing Chiral Metal Complexes

Reference key for equations: (I)[5]; (II)[6]; (III-IV)[7]; (V)[8]; (VI)[9]

Scheme 3

ENANTIOSELECTIVE STRECKER REACTION

Catalytic Enantioselective Strecker Reaction of Aldimines Utilizing Chiral Metal Complexes (Continued)

Recently, G. Li et al. have utilized achiral *N*-phosphonyl imines as substrates for the Strecker reaction and Et$_2$AlCN as the cyanide source (Scheme 4).[10] The use of the *N*-phosphonyl group is advantageous because: (a) the α-amino-nitrile product can be isolated in high purity by simply washing the crude product with hexanes; (b) *N*-deprotection can be effected under mildly acidic conditions (MeOH/HCl) and (c) the *N,N*-dialkyl diamine can be recovered quantitatively and reused.

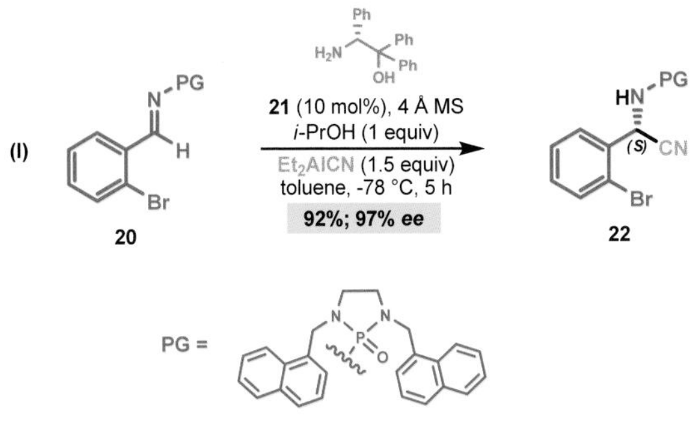

Scheme 4

II. Catalytic Enantioselective Strecker Reaction of Ketimines Utilizing Chiral Metal Complexes

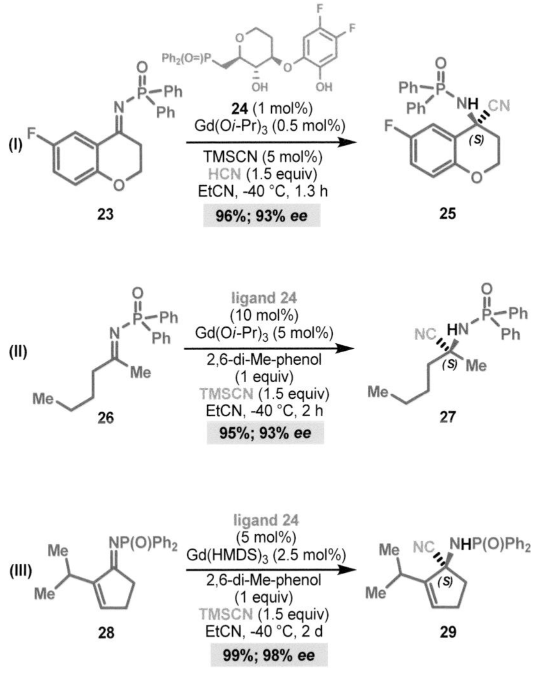

Reference key for equations: (I)[11]; (II)[12]; (III)[13]

Scheme 5

III. Catalytic Enantioselective Reissert Reaction Utilizing Chiral Metal Complexes

The Reissert reaction[14] involves the concomitant acylation/cyanation of *N*-heteroaromatic compounds. The reaction is usually conducted in the simultaneous presence of the acylating agent (e.g., acyl halide) and a cyanating agent (e.g., NaCN, TMSCN). This transformation has been rendered enantioselective and, without exception, all the examples utilize chiral metal complexes as catalysts.

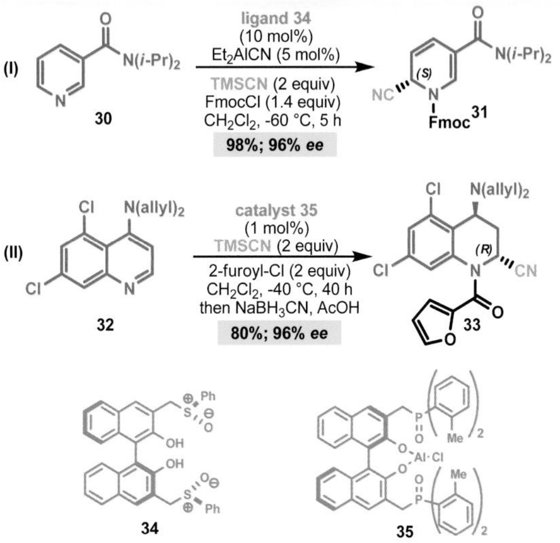

Reference key for equations: (I)[15]; (II)[16]

Scheme 6

IV. Catalytic Enantioselective Strecker Reaction of Aldimines Utilizing Metal-Free Molecular Catalysts

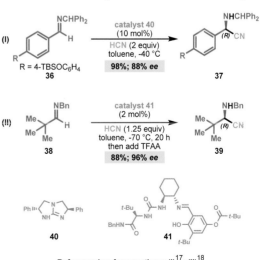

Reference key for equations: (I)[17]; (II)[18]

Scheme 7

ENANTIOSELECTIVE STRECKER REACTION

V. Catalytic Enantioselective Strecker Reaction of Ketimines Utilizing Organocatalysts

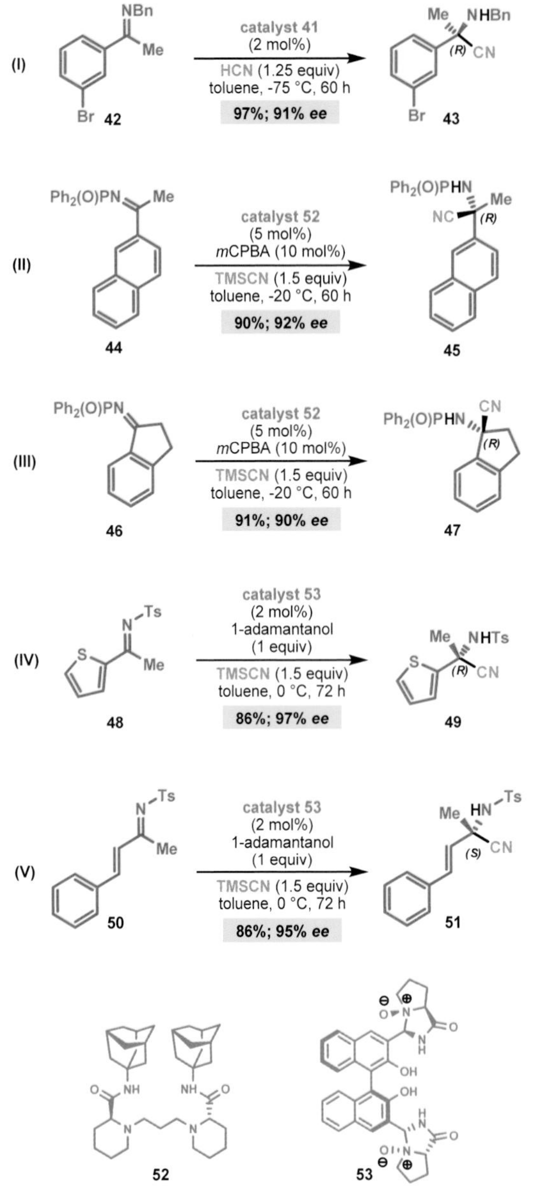

Reference key for equations: (I)[18]; (II-III)[19]; (IV-V)[20]

Scheme 8

References

1. (a) Strecker, A. The artificial synthesis of lactic acid and a new homologue of glycine. *Liebigs Ann. Chem.* **1850**, 75, 27-45; (b) Strecker, A. The preparation of a new material by the reaction of acetaldehyde-ammonia imine and hydrogen cyanide. *Liebigs Ann. Chem.* **1854**, 91, 349-351.
2. (a) Duthaler, R. O. Recent developments in the stereoselective synthesis of α-amino acids. *Tetrahedron* **1994**, 50, 1539-1650; (b)

Ohfune, Y. & Shinada, T. Enantio- and diastereoselective construction of α,α-disubstituted α-amino acids for the synthesis of biologically active compounds. *Eur. J. Org. Chem.* **2005**, 5127-5143; (c) Galatsis, P. Strecker amino acid synthesis. *Name Reactions for Functional Group Transformations* **2007**, 477-499.
3. (a) Yet, L. Recent developments in catalytic asymmetric Strecker-type reactions. *Angew. Chem., Int. Ed.* **2001**, 40, 875-877; (b) Groeger, H. Catalytic enantioselective strecker reactions and analogous syntheses. *Chem. Rev.* **2003**, 103, 2795-2827; (c) Shibasaki, M., Kanai, M. & Mita, T. The catalytic asymmetric Strecker reaction. *Org. React.* **2008**, 70, 1-119; (d) Connon, S. J. The catalytic asymmetric Strecker reaction: ketimines continue to join the fold. *Angew. Chem., Int. Ed.* **2008**, 47, 1176-1178; (e) Merino, P., Marques-Lopez, E., Tejero, T. & Herrera, R. P. Organocatalyzed Strecker reactions. *Tetrahedron* **2009**, 65, 1219-1234.
4. (a) Najera, C. & Sansano, J. M. Catalytic Asymmetric Synthesis of α-Amino Acids. *Chem. Rev.* **2007**, 107, 4584-4671; (b) Ohfune, Y., Sakaguchi, K. & Shinada, T. Asymmetric synthesis of α-substituted α-amino acids: Strecker and Claisen approaches. *ACS Symp. Ser.* **2009**, 1009, 57-71; (c) Ager, D. J. Synthesis of unnatural/nonproteinogenic α-amino acids. *Amino Acids, Peptides and Proteins in Organic Chemistry* **2009**, 1, 495-526.
5. Sigman, M. S. & Jacobsen, E. N. Enantioselective Addition of Hydrogen Cyanide to Imines Catalyzed by a Chiral (Salen)Al(III) Complex. *J. Am. Chem. Soc.* **1998**, 120, 5315-5316.
6. Karimi, B. & Maleki, A. Catalytic asymmetric Strecker hydrocyanation of imines using Yb(OTf)$_3$-pybox catalysts. *Chem. Commun. (Cambridge, U. K.)* **2009**, 5180-5182.
7. Ishitani, H., Komiyama, S., Hasegawa, Y. & Kobayashi, S. Catalytic Asymmetric Strecker Synthesis. Preparation of Enantiomerically Pure α-Amino Acid Derivatives from Aldimines and Tributyltin Cyanide or Achiral Aldehydes, Amines, and Hydrogen Cyanide Using a Chiral Zirconium Catalyst. *J. Am. Chem. Soc.* **2000**, 122, 762-766.
8. Banphavichit, V., Mansawat, W., Bhanthumnavin, W. & Vilaivan, T. A highly enantioselective Strecker reaction catalyzed by titanium-N-salicyl-β-amino alcohol complexes. *Tetrahedron* **2004**, 60, 10559-10568.
9. (a) Abell, J. P. & Yamamoto, H. Dual-Activation Asymmetric Strecker Reaction of Aldimines and Ketimines Catalyzed by a Tethered Bis(8-quinolinolato) Aluminum Complex. *J. Am. Chem. Soc.* **2009**, 131, 15118-15119; (b) Abell, J. P. & Yamamoto, H. Development and applications of tethered bis(8-quinolinolato) metal complexes (TBO x M). *Chem. Soc. Rev.* **2010**, 39, 61-69.
10. Kaur, P., Pindi, S., Wever, W., Rajale, T. & Li, G. Asymmetric Catalytic Strecker Reaction of N-Phosponyl Imines with Et$_2$AlCN Using Amino Alcohols and BINOLs as Catalysts. *Chem. Commun.* **2010**, 46, 4330-4332.
11. Kato, N., Suzuki, M., Kanai, M. & Shibasaki, M. Catalytic enantioselective Strecker reaction of ketimines using catalytic amount of TMSCN and stoichiometric amount of HCN. *Tetrahedron Lett.* **2004**, 45, 3153-3155.
12. Kato, N., Suzuki, M., Kanai, M. & Shibasaki, M. General and practical catalytic enantioselective Strecker reaction of keto-imines: significant improvement through catalyst tuning by protic additives. *Tetrahedron Lett.* **2004**, 45, 3147-3151.
13. Fukuda, N., Sasaki, K., Sastry, T. V. R. S., Kanai, M. & Shibasaki, M. Catalytic Asymmetric Total Synthesis of (+)-Lactacystin. *J. Org. Chem.* **2006**, 71, 1220-1225.
14. Reissert, A. Introduction of the Benzoyl Group to Tertiary Cyclic Bases. *Ber. Dtsch. Chem. Ges.* **1905**, 38, 1603-1614.
15. Ichikawa, E., Suzuki, M., Yabu, K., Albert, M., Kanai, M. & Shibasaki, M. New Entries in Lewis Acid-Lewis Base Bifunctional Asymmetric Catalyst: Catalytic Enantioselective Reissert Reaction of Pyridine Derivatives. *J. Am. Chem. Soc.* **2004**, 126, 11808-11809.
16. Takamura, M., Funabashi, K., Kanai, M. & Shibasaki, M. Catalytic enantioselective Reissert-type reaction: development and application to the synthesis of a potent NMDA receptor antagonist (-)-L-689,560 using a solid-supported catalyst. *J. Am. Chem. Soc.* **2001**, 123, 6801-6808.
17. Corey, E. J. & Grogan, M. J. Enantioselective Synthesis of α-Amino Nitriles from N-Benzhydryl Imines and HCN with a Chiral Bicyclic Guanidine as Catalyst. *Org. Lett.* **1999**, 1, 157-160.
18. Sigman, M. S., Vachal, P. & Jacobsen, E. N. A general catalyst for the asymmetric Strecker reaction. *Angew. Chem., Int. Ed.* **2000**, 39, 1279-1281.
19. Huang, J., Liu, X., Wen, Y., Qin, B. & Feng, X. Enantioselective Strecker Reaction of Phosphinoyl Ketoimines Catalyzed by in Situ Prepared Chiral N,N'-Dioxides. *J. Org. Chem.* **2007**, 72, 204-208.
20. Hou, Z., Wang, J., Liu, x. & Feng, X. Highly enantioselective Strecker reaction of ketoimines catalyzed by an organocatalyst from (S)-BINOL and L-prolinamide. *Chem.--Eur. J.* **2008**, 14, 4484-4486.

ADDITION OF sp³ C-NUCLEOPHILES TO C=N

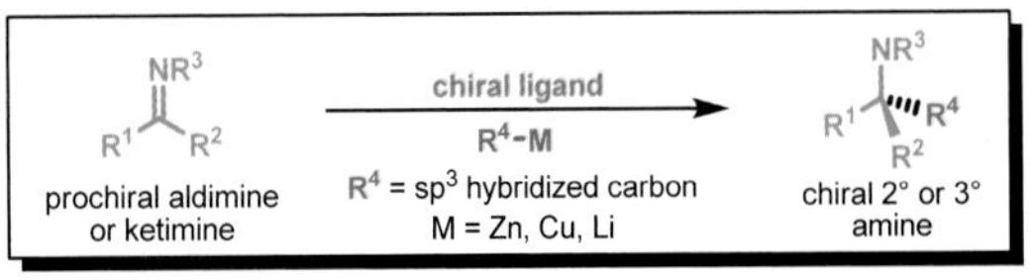

prochiral aldimine or ketimine

chiral ligand
R⁴-M
R⁴ = sp³ hybridized carbon
M = Zn, Cu, Li

chiral 2° or 3° amine

Introduction

Chiral amines with nitrogen attached either to a secondary or tertiary stereocenter are important targets for synthesis because there are so many of great importance as natural products or medicines. Considerable effort has been directed at their synthesis from imines by reduction[1] (see pages 29-34) or by addition of carbon nucleophiles.[2] The enantioselective Mannich reaction (see pages 88-91) represents a case of the latter which involves stabilized C-nucleophiles. Another class has been developed with non-stabilized organometallic reagents which are made chiral by coordination with chiral ligands. Imines that bear a chiral auxiliary on the imine nitrogen atom are also used as substrates for the preparation of chiral amines.[3] The various types of imine substrates which have been studied are summarized in Figure 1. These imines display differing degrees of reactivity towards C-nucleophiles. This section focuses on the enantioselective addition of unstabilized sp³-hybridized C- and allylic nucleophiles to aldimines and ketimines (for the arylation and alkynylation of imines see pages 98-100). Charged iminium salts and nitrones react readily even with weakly nucleophilic organometallic reagents, while weakly electrophilic imines require the use of strongly nucleophilic organometallic reagents (e.g., organomagnesium, organolithium).

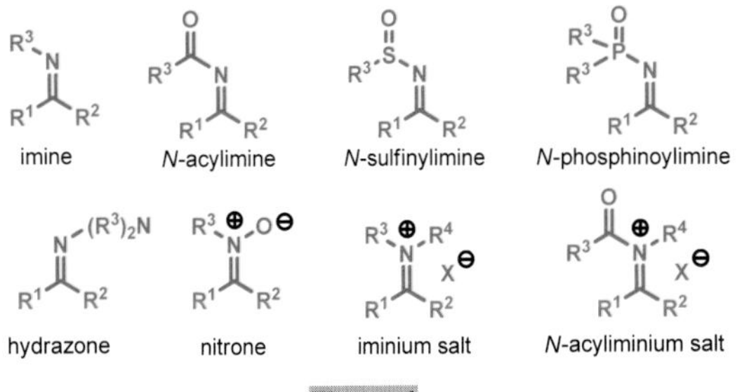

imine · N-acylimine · N-sulfinylimine · N-phosphinoylimine

hydrazone · nitrone · iminium salt · N-acyliminium salt

Figure 1

The high reactivity of organolithium reagents necessitates complete coordination to a chiral ligand in order to provide good enantioselectivity, as shown by early studies by K. Tomioka (see Scheme 1).[4] This work paved the way for the development of improved methods that utilize chiral ligands in catalytic quantities in the presence of chiral alkyl Cu(II) or Zr(IV) reagents, as shown in Scheme 2.

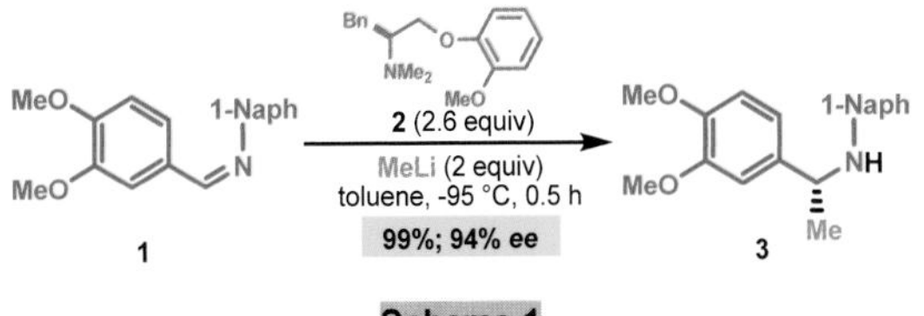

For catalytic enantioselective additions to C=N, the chiral Cu, Zr or Ti reagents are commonly generated from R₂Zn or R₃Al species rather than the more reactive RLi or Grignard precursors (see Scheme 2). The enantioselective allylation of imine derivatives is very readily achieved with a variety of allylic reagents because of the availability of a six-membered cyclic transition state for allyl transfer to the carbon of the C=N (see Schemes 4 and 5).

I. Catalytic Enantioselective Addition of Organozinc and Organoaluminum Reagents to Ketimines and Aldimines

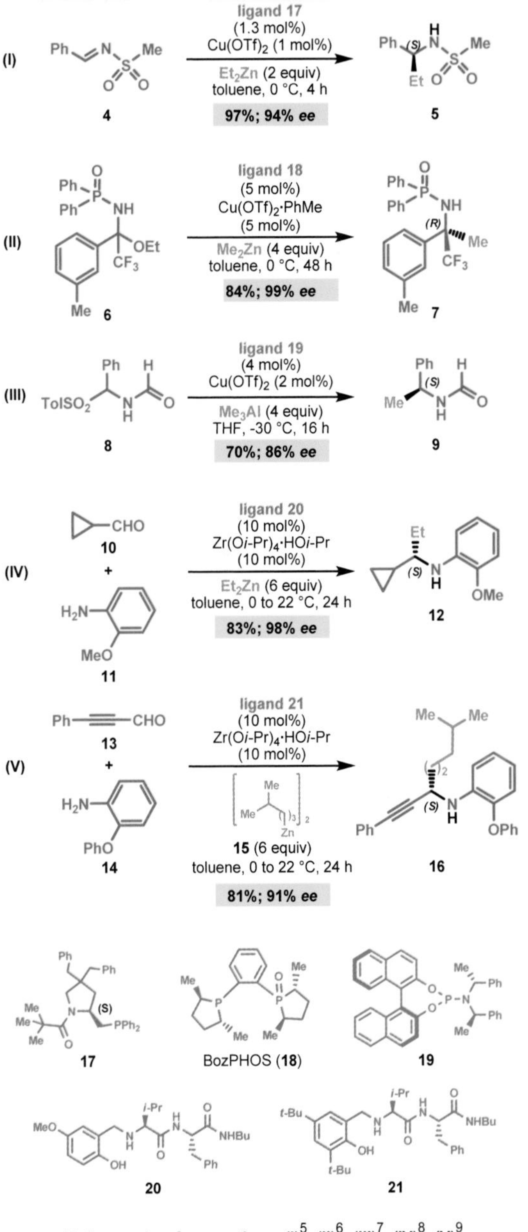

(I)
ligand 17 (1.3 mol%)
Cu(OTf)₂ (1 mol%)
Et₂Zn (2 equiv)
toluene, 0 °C, 4 h
97%; 94% ee
4 → 5

(II)
ligand 18 (5 mol%)
Cu(OTf)₂·PhMe (5 mol%)
Me₂Zn (4 equiv)
toluene, 0 °C, 48 h
84%; 99% ee
6 → 7

(III)
ligand 19 (4 mol%)
Cu(OTf)₂ (2 mol%)
Me₃Al (4 equiv)
THF, -30 °C, 16 h
70%; 86% ee
8 → 9

(IV)
10 + 11
ligand 20 (10 mol%)
Zr(Oi-Pr)₄·HOi-Pr (10 mol%)
Et₂Zn (6 equiv)
toluene, 0 to 22 °C, 24 h
83%; 98% ee
12

(V)
13 + 14
ligand 21 (10 mol%)
Zr(Oi-Pr)₄·HOi-Pr (10 mol%)
15 (6 equiv)
toluene, 0 to 22 °C, 24 h
81%; 91% ee
16

17 · BozPHOS (18) · 19

20 · 21

Reference key for equations: (I)[5]; (II)[6]; (III)[7]; (IV)[8]; (V)[9]

Scheme 2

Scheme 1

1
2 (2.6 equiv)
MeLi (2 equiv)
toluene, -95 °C, 0.5 h
99%; 94% ee
3

Scheme 1

ADDITION OF sp³ C-NUCLEOPHILES TO C=N

II. Stoichiometric Ligand-Mediated Enantioselective Addition of Organolithium Reagents to Aldimines

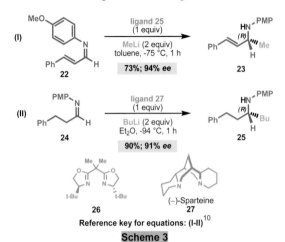

Reference key for equations: (I-II)[10]

Scheme 3

III. Catalytic Enantioselective Addition of Allyl Nucleophiles to Imino Compounds

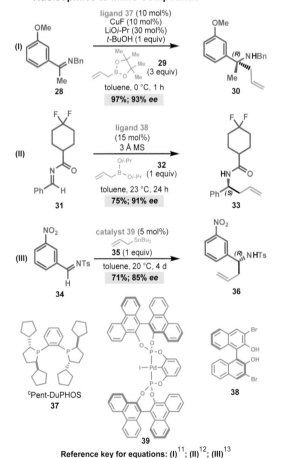

Reference key for equations: (I)[11]; (II)[12]; (III)[13]

Scheme 4

Catalytic Enantioselective Addition of Allyl Nucleophiles to Imino Compounds (Continued)

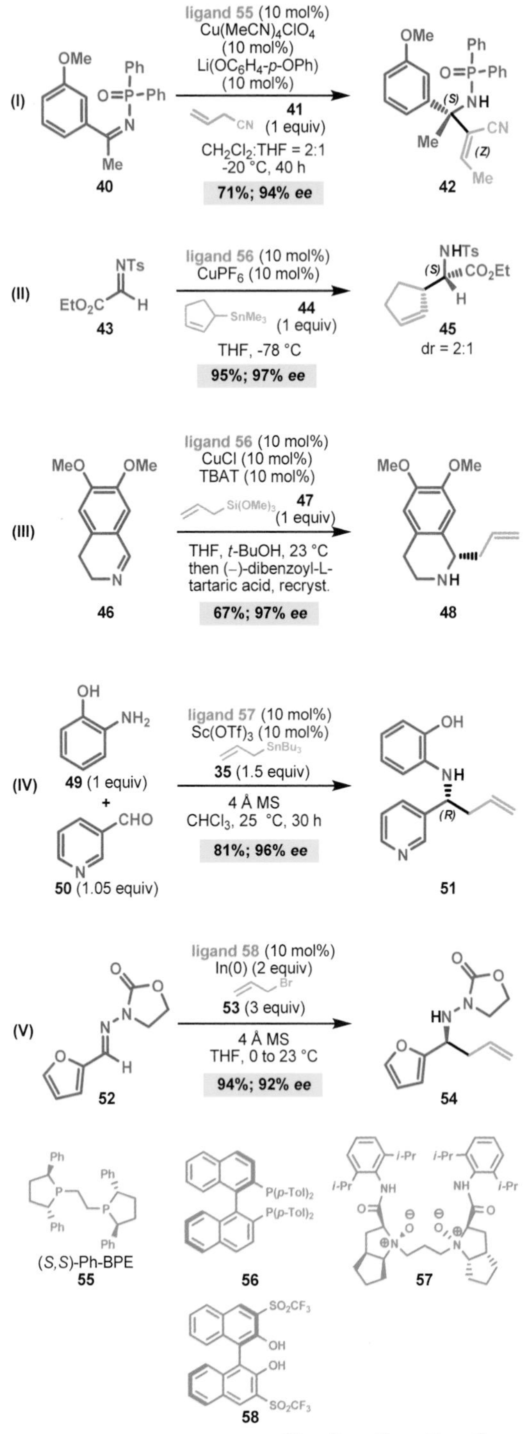

Reference key for equations: (I)[14]; (II)[15]; (III)[16]; (IV)[17]; (V)[18]

Scheme 5

ADDITION OF sp³ C-NUCLEOPHILES TO C=N

Enantioselective Addition of Allyl Nucleophiles to Imino Compounds (Continued)

Enantioselective allylation of C=N has also been developed utilizing chiral Pd(II) complexes as reactive intermediates (Scheme 6).

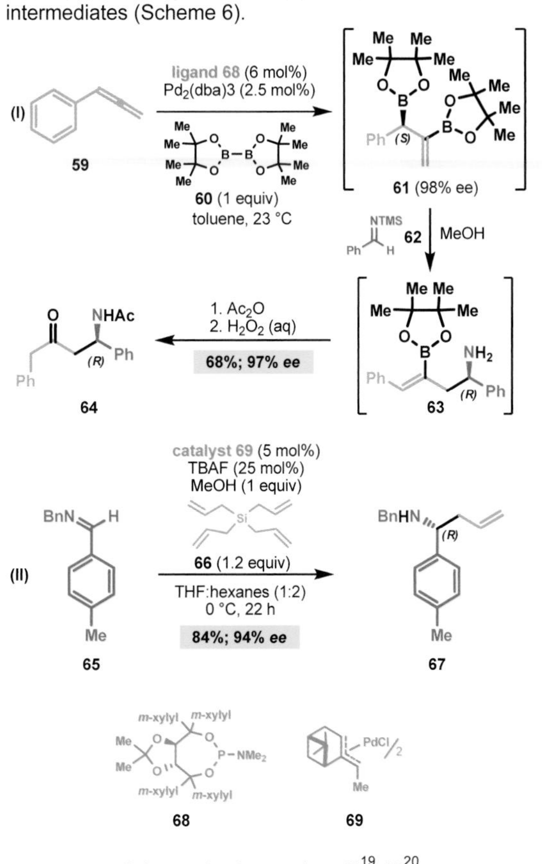

Reference key for equations: (I)[19]; (II)[20]

Scheme 6

References

1. Nugent, T. C. & El-Shazly, M. Chiral Amine Synthesis - Recent Developments and Trends for Enamide Reduction, Reductive Amination, and Imine Reduction. *Adv. Synth. Catal.* **2010**, 352, 753-819.

2. (a) Denmark, S. E. & Nicaise, O. J. C. Alkylation of Imino Groups in *Comprehensive Asymmetric Catalysis* (eds. Jacobsen, E. N., Pfaltz, A. & Yamamoto, H.), 923-964 (Springer Verlag, Heidelberg, **1999**); (b) Iguchi, M., Yamada, K.-i. & Tomioka, K. Enantioselective conjugate addition and 1,2-addition to C=N of organolithium reagents. *Top. Organomet. Chem.* **2003**, 5, 37-59; (c) Charette, A. B., Boezio, A. A., Cote, A., Moreau, E., Pytkowicz, J., Desrosiers, J.-N. & Legault, C. Asymmetric catalytic addition of diorganozinc reagents to imines: Scope and application. *Pure Appl. Chem.* **2005**, 77, 1259-1267; (d) Vilaivan, T., Bhanthumnavin, W. & Sritana-Anant, Y. Recent advances in catalytic asymmetric addition to imines and related C=N systems. *Curr. Org. Chem.* **2005**, 9, 1315-1392; (e) Ding, H. & Friestad, G. K. Asymmetric addition of allylic nucleophiles to imino compounds. *Synthesis* **2005**, 2815-2829; (f) Braese, S., Baumann, T., Dahmen, S. & Vogt, H. Enantioselective catalytic syntheses of α-branched chiral amines. *Chem. Commun. (Cambridge, U. K.)* **2007**, 1881-1890; (g) Ferraris, D. Catalytic, asymmetric alkylation of imines. *Tetrahedron* **2007**, 63, 9581-9597; (h) Dickstein, J. S. & Kozlowski,

M. C. Organometal additions to α-imino esters: N-alkylation via umpolung. *Chem. Soc. Rev.* **2008**, 37, 1166-1173; (i) Yamada, K.-i. & Tomioka, K. Copper-Catalyzed Asymmetric Alkylation of Imines with Dialkylzinc and Related Reactions. *Chem. Rev. (Washington, DC, U. S.)* **2008**, 108, 2874-2886; (j) Friestad, G. K. Synthesis of alkyl- and cycloalkylamines by addition of carbanions to azomethines. *Sci. Synth.* **2009**, 40a, 305-342; (k) Blay, G., Monleon, A. & Pedro, J. R. Recent developments in asymmetric alkynylation of imines. *Curr. Org. Chem.* **2009**, 13, 1498-1539.

3. (a) Alvaro, G. & Savoia, D. Addition of organometallic reagents to imines bearing stereogenic N-substituents. Stereochemical models explaining the 1,3-asymmetric induction. *Synlett* **2002**, 651-673; (b) Ellman, J. A., Owens, T. D. & Tang, T. P. N-tert-Butanesulfinyl Imines: Versatile Intermediates for the Asymmetric Synthesis of Amines. *Acc. Chem. Res.* **2002**, 35, 984-995; (c) Zhou, P., Chen, B.-C. & Davis, F. A. Recent advances in asymmetric reactions using sulfinimines (N-sulfinyl imines). *Tetrahedron* **2004**, 60, 8003-8030; (d) Friestad, G. K. Chiral N-acylhydrazones: Versatile imino acceptors for asymmetric amine synthesis. *Eur. J. Org. Chem.* **2005**, 3157-3172.

4. Taniyama, D., Hasegawa, M. & Tomioka, K. A facile and efficient asymmetric synthesis of (+)-salsolidine. *Tetrahedron Lett.* **2000**, 41, 5533-5536.

5. Fujihara, H., Nagai, K. & Tomioka, K. Copper-Amidophosphine Catalyst in Asymmetric Addition of Organozinc to Imines. *J. Am. Chem. Soc.* **2000**, 122, 12055-12056.

6. Lauzon, C. & Charette, A. B. Catalytic Asymmetric Synthesis of α,α,α-Trifluoromethyl Amines by the Copper-Catalyzed Nucleophilic Addition of Diorganozinc Reagents to Imines. *Org. Lett.* **2006**, 8, 2743-2745.

7. Pizzuti, M. G., Minnaard, A. J. & Feringa, B. L. Catalytic Enantioselective Addition of Organometallic Reagents to N-Formyllimines Using Monodentate Phosphoramidite Ligands. *J. Org. Chem.* **2008**, 73, 940-947.

8. Porter, J. R., Traverse, J. F., Hoveyda, A. H. & Snapper, M. L. Three-Component Catalytic Asymmetric Synthesis of Aliphatic Amines. *J. Am. Chem. Soc.* **2001**, 123, 10409-10410.

9. Akullian, L. C., Snapper, M. L. & Hoveyda, A. H. Three-component enantioselective synthesis of propargylamines through Zr-catalyzed additions of alkylzinc reagents to alkynylimines. *Angew. Chem., Int. Ed.* **2003**, 42, 4244-4247.

10. Denmark, S. E., Nakajima, N., Stiff, C. M., Nicaise, O. J. C. & Kranz, M. Studies on the bisoxazoline- and (-)-sparteine-mediated enantioselective addition of organolithium reagents to imines. *Adv. Synth. Catal.* **2008**, 350, 1023-1045.

11. Kanai, M., Wada, R., Shibuguchi, T. & Shibasaki, M. Cu(I)-catalyzed asymmetric allylation of ketones and ketimines. *Pure Appl. Chem.* **2008**, 80, 1055-1062.

12. Lou, S., Moquist, P. N. & Schaus, S. E. Asymmetric Allylboration of Acyl Imines Catalyzed by Chiral Diols. *J. Am. Chem. Soc.* **2007**, 129, 15398-15404.

13. Aydin, J., Kumar, K. S., Sayah, M. J., Wallner, O. A. & Szabo, K. J. Synthesis and catalytic application of chiral 1,1'-bi-2-naphthol- and biphenanthrol-based pincer complexes: selective allylation of sulfonimines with allyl stannane and allyl trifluoroborate. *J. Org. Chem.* **2007**, 72, 4689-4697.

14. Yazaki, R., Nitabaru, T., Kumagai, N. & Shibasaki, M. Direct Catalytic Asymmetric Addition of Allylic Cyanides to Ketoimines. *J. Am. Chem. Soc.* **2008**, 130, 14477-14479.

15. Fang, X., Johannsen, M., Yao, S., Gathergood, N., Hazell, R. G. & Jorgensen, K. A. Catalytic Approach for the Formation of Optically Active Allyl α-Amino Acids by Addition of Allylic Metal Compounds to α-Imino Esters. *J. Org. Chem.* **1999**, 64, 4844-4849.

16. Itoh, T., Miyazaki, M., Fukuoka, H., Nagata, K. & Ohsawa, A. Formal Total Synthesis of (-)-Emetine Using Catalytic Asymmetric Allylation of Cyclic Imines as a Key Step. *Org. Lett.* **2006**, 8, 1295-1297.

17. Li, X., Liu, X., Fu, Y., Wang, L., Zhou, L. & Feng, X. Direct allylation of aldimines catalyzed by C₂-symmetric N,N-dioxide-Sc(III)-complexes: highly enantioselective synthesis of homoallylic amines. *Chem.--Eur. J.* **2008**, 14, 4796-4798.

18. Kargbo, R., Takahashi, Y., Bhor, S., Cook, G. R., Lloyd-Jones, G. C. & Shepperson, I. R. Readily Accessible, Modular, and Tuneable BINOL 3,3'-Perfluoroalkylsulfones: Highly Efficient Catalysts for Enantioselective In-Mediated Imine Allylation. *J. Am. Chem. Soc.* **2007**, 129, 3846-3847.

19. Sieber, J. D. & Morken, J. P. Sequential Pd-Catalyzed Asymmetric Allene Diboration/α-Aminoallylation. *J. Am. Chem. Soc.* **2006**, 128, 74-75.

20. Fernandes, R. A. & Yamamoto, Y. The first catalytic asymmetric allylation of imines with the tetraallylsilane-TBAF-MeOH system, using the chiral bis-π-allylpalladium complex. *J. Org. Chem.* **2004**, 69, 735-738.

ADDITION OF sp and sp² C-NUCLEOPHILES TO C=N

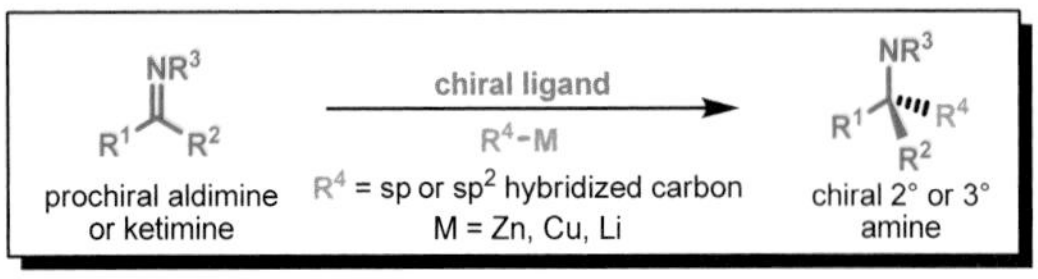

Introduction

The frequent occurrence of chiral amines as natural products and pharmaceutically active compounds has stimulated much research on their preparation using the catalytic enantioselective reduction of imines as has been discussed in detail on pages 29-34. This section outlines methods for the catalytic enantioselective synthesis of propargylic-, homoallylic- and arylamines by addition of sp- and sp²-hybridized C-nucleophiles to imines. Several recent reviews provide useful information on imine alkynylation[1], alkenylation[2] and arylation[3]. The schemes which follow illustrate a variety of such transformations.

I. Catalytic Enantioselective Direct Addition of Alkynes to Aldimines

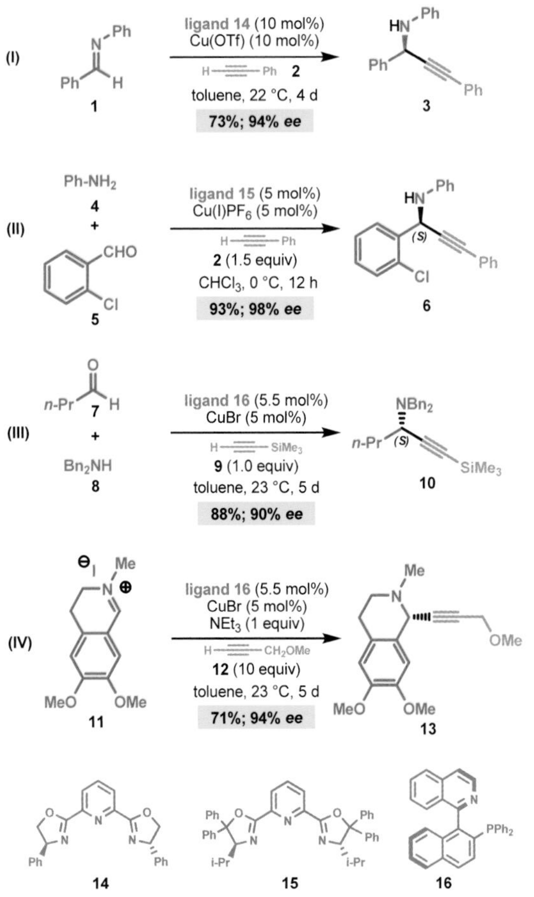

Reference key for equations: (I)[4]; (II)[5]; (III)[6]; (IV)[7]

Scheme 1

II. Enantioselective Addition of Alkynylmetal (Zn, Li) Reagents to Imine-Type Substrates

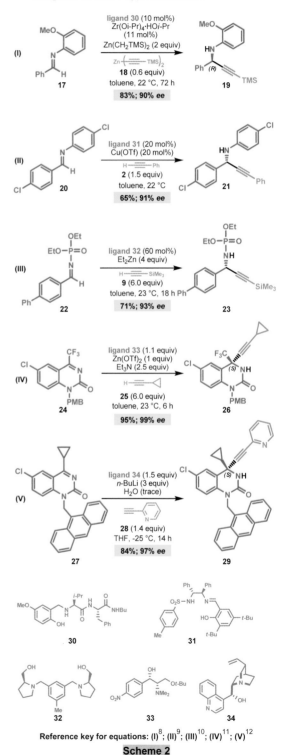

Reference key for equations: (I)[8]; (II)[9]; (III)[10]; (IV)[11]; (V)[12]

Scheme 2

ADDITION OF sp and sp^2 C-NUCLEOPHILES TO C=N

III. Catalytic Enantioselective Alkenylation (Vinylation) of Imines

The highly enantioselective transfer of alkenyl groups from pre-formed organometallic reagents to the carbon atom of imines (C=N), to form allylic amines, has not yet been realized in contrast to the analogous addition to carbonyl (C=O) compounds (see pages 74-75) mainly because of the diminished reactivity of the C=N bond towards nucleophiles. The transition metal-catalyzed coupling of imines with disubstituted alkynes under conditions that allow concomitant hydrometallation of the C-C triple bond provides a workable alternative for the case of (Z)-allylic amines.[2] Scheme 3 illustrates some examples of this approach.

The catalytic enantioselective version of the Lewis acid-catalyzed addition of vinylboronic acids to imines[13] has not been developed.

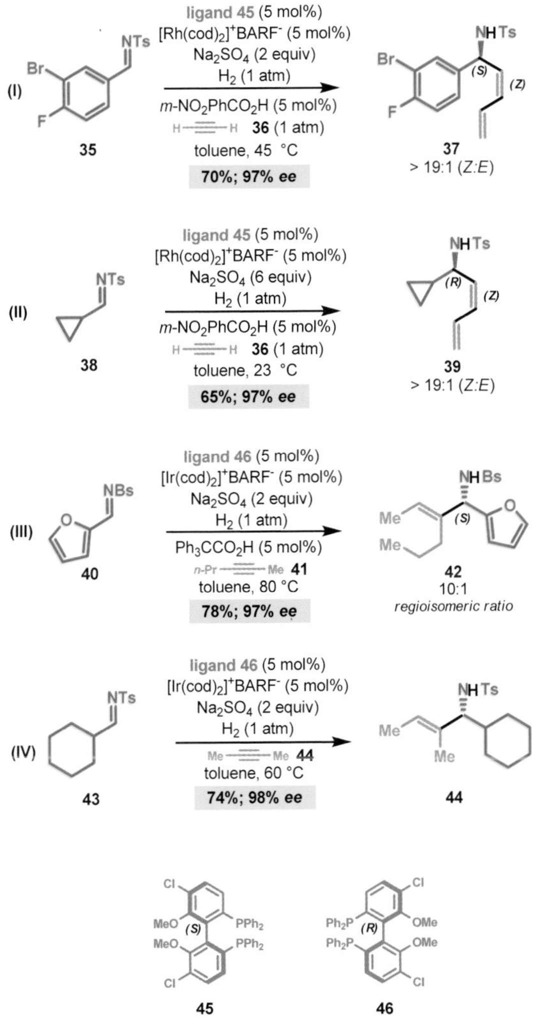

IV. Catalytic Enantioselective Arylation of Imines Utilizing Arylorganometallic Reagents (Sn, Ti, B)

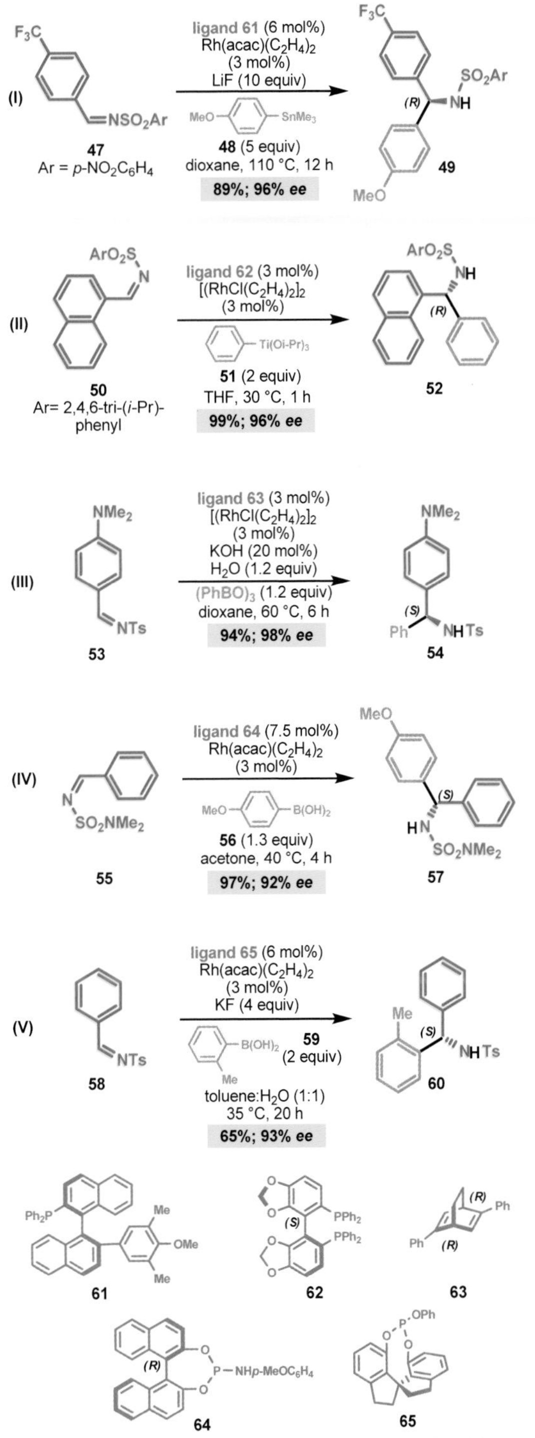

Scheme 3

Reference key for equations: (I)[14]; (II)[15]

Scheme 4

Reference key for equations: (I)[16]; (II)[17]; (III)[18]; (IV)[19]; (V)[20]

ADDITION OF sp and sp² C-NUCLEOPHILES TO C=N

V. Acid-Catalyzed Enantioselective Arylation of Imines by Electron-Rich Aromatic Compounds[3b-d]

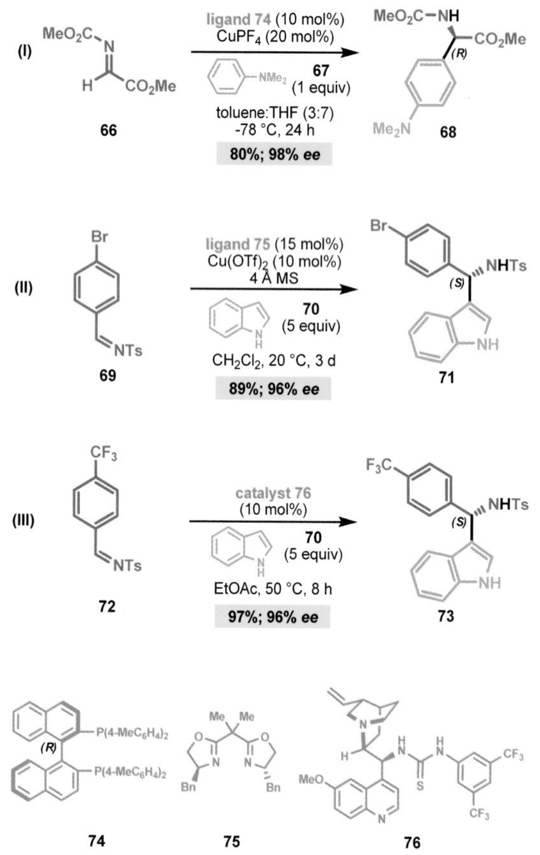

Reference key for equations: (I)[21]; (II)[22]; (III)[23]

Scheme 5

References

1. (a) Cozzi, P. G., Hilgraf, R. & Zimmermann, N. Acetylenes in catalysis. Enantioselective additions to carbonyl groups and imines and applications beyond. *Eur. J. Org. Chem.* **2004**, 4095-4105; (b) Zani, L. & Bolm, C. Direct addition of alkynes to imines and related C:N electrophiles: A convenient access to propargylamines. *Chem. Commun. (Cambridge, U. K.)* **2006**, 4263-4275; (c) Friestad, G. K. & Mathies, A. K. Recent developments in asymmetric catalytic addition to C=N bonds. *Tetrahedron* **2007**, 63, 2541-2569; (d) Blay, G., Monleon, A. & Pedro, J. R. Recent developments in asymmetric alkynylation of imines. *Curr. Org. Chem.* **2009**, 13, 1498-1539.
2. (a) Iida, H. & Krische, M. J. Catalytic reductive coupling of alkenes and alkynes to carbonyl compounds and imines mediated by hydrogen. *Top. Curr. Chem.* **2007**, 279, 77-104; (b) Skucas, E., Ngai, M.-Y., Komanduri, V. & Krische, M. J. Enantiomerically Enriched Allylic Alcohols and Allylic Amines via C-C Bond-Forming Hydrogenation: Asymmetric Carbonyl and Imine Vinylation. *Acc. Chem. Res.* **2007**, 40, 1394-1401.
3. (a) Batey, R. A. Nucleophilic addition reactions of aryl and alkenylboronic acids and their derivatives to imines and iminium ions. *Boronic Acids* **2005**, 279-304; (b) Poulsen, T. B. & Jorgensen, K. A. Catalytic Asymmetric Friedel-Crafts Alkylation Reactions-Copper Showed the Way. *Chem. Rev. (Washington, DC, U. S.)* **2008**, 108, 2903-2915; (c) Blay, G., Pedro, J. R. & Vila, C. Catalytic asymmetric Friedel-Crafts alkylations in total synthesis. *Catalytic Asymmetric*

Friedel-Crafts Alkylations **2009**, 223-270; (d) You, S.-L., Cai, Q. & Zeng, M. Chiral Bronsted acid catalyzed Friedel-Crafts alkylation reactions. *Chem. Soc. Rev.* **2009**, 38, 2190-2201.
4. Wei, C. & Li, C.-J. Enantioselective Direct-Addition of Terminal Alkynes to Imines Catalyzed by Copper(I)pybox Complex in Water and in Toluene. *J. Am. Chem. Soc.* **2002**, 124, 5638-5639.
5. Bisai, A. & Singh, V. K. Enantioselective One-Pot Three-Component Synthesis of Propargylamines. *Org. Lett.* **2006**, 8, 2405-2408.
6. Gommermann, N. & Knochel, P. Practical highly enantioselective synthesis of terminal propargylamines. An expeditious synthesis of (S)-(+)-coniine. *Chem. Commun. (Cambridge, U. K.)* **2004**, 2324-2325.
7. Taylor, A. M. & Schreiber, S. L. Enantioselective Addition of Terminal Alkynes to Isolated Isoquinoline Iminiums. *Org. Lett.* **2006**, 8, 143-146.
8. Akullian, L. C., Snapper, M. L. & Hoveyda, A. H. Three-component enantioselective synthesis of propargylamines through Zr-catalyzed additions of alkylzinc reagents to alkynylimines. *Angew. Chem., Int. Ed.* **2003**, 42, 4244-4247.
9. Liu, B., Liu, J., Jia, X., Huang, L., Li, X. & Chan, A. S. C. The synthesis of chiral N-tosylated aminoimine ligands and their application in enantioselective addition of phenylacetylene to imines. *Tetrahedron Asymmetry* **2007**, 18, 1124-1128.
10. Zhu, S., Yan, W., Mao, B., Jiang, X. & Wang, R. Enantioselective nucleophilic addition of trimethylsilylacetylene to N-phosphinoylimines promoted by C2-symmetric proline-derived β-amino alcohol. *J. Org. Chem.* **2009**, 74, 6980-6985.
11. Jiang, B. & Si, Y.-G. Highly enantioselective construction of a chiral tertiary carbon center by alkynylation of a cyclic N-acyl ketimine: an efficient preparation of HIV therapeutics. *Angew. Chem., Int. Ed.* **2003**, 43, 216-218.
12. Huffman, M. A., Yasuda, N., DeCamp, A. E. & Grabowski, E. J. J. Lithium Alkoxides of Cinchona Alkaloids as Chiral Controllers for Enantioselective Acetylide Addition to Cyclic N-Acyl Ketimines. *J. Org. Chem.* **1995**, 60, 1590-1594.
13. (a) McReynolds, M. D. & Hanson, P. R. The three-component boronic acid Mannich reaction: structural diversity and stereoselectivity. *Chemtracts* **2001**, 14, 796-801; (b) Kürti, L. & Czakó, B. Petasis Boronic Acid-Mannich Reaction in *Strategic Applications of Named Reactions in Organic Synthesis* 340-341 (Academic Press/Elsevier, San Diego, **2005**).
14. Skucas, E., Kong, J. R. & Krische, M. J. Enantioselective Reductive Coupling of Acetylene to N-Arylsulfonyl Imines via Rhodium Catalyzed C-C Bond-Forming Hydrogenation: (Z)-Dienyl Allylic Amines. *J. Am. Chem. Soc.* **2007**, 129, 7242-7243.
15. Ngai, M.-Y., Barchuk, A. & Krische, M. J. Enantioselective Iridium-Catalyzed Imine Vinylation: Optically Enriched Allylic Amines via Alkyne-Imine Reductive Coupling Mediated by Hydrogen. *J. Am. Chem. Soc.* **2007**, 129, 12644-12645.
16. Hayashi, T. & Ishigedani, M. Rhodium-Catalyzed Asymmetric Arylation of Imines with Organostannanes. Asymmetric Synthesis of Diarylmethylamines. *J. Am. Chem. Soc.* **2000**, 122, 976-977.
17. Hayashi, T., Kawai, M. & Tokunaga, N. Asymmetric synthesis of diarylmethyl amines by rhodium-catalyzed asymmetric addition of aryl titanium reagents to imines. *Angew. Chem., Int. Ed.* **2004**, 43, 6125-6128.
18. Tokunaga, N., Otomaru, Y., Okamoto, K., Ueyama, K., Shintani, R. & Hayashi, T. C2-Symmetric bicyclo[2.2.2]octadienes as chiral ligands: Their high performance in rhodium-catalyzed asymmetric arylation of N-tosylarylimines. *J. Am. Chem. Soc.* **2004**, 126, 13584-13585.
19. Jagt, R. B. C., Toullec, P. Y., Geerdink, D., de Vries, J. G., Feringa, B. L. & Minnaard, A. J. A ligand-library approach to the highly efficient rhodium/phosphoramidite-catalyzed asymmetric arylation of N,N-dimethylsulfamoyl-protected aldimines. *Angew. Chem., Int. Ed.* **2006**, 45, 2789-2791.
20. Duan, H.-F., Jia, Y.-X., Wang, L.-X. & Zhou, Q.-L. Enantioselective Rh-Catalyzed Arylation of N-Tosylarylimines with Arylboronic Acids. *Org. Lett.* **2006**, 8, 2567-2569.
21. Saaby, S., Bayon, P., Aburel, P. S. & Jorgensen, K. A. Optically Active Aromatic and Heteroaromatic α-Amino Acids by a One-Pot Catalytic Enantioselective Addition of Aromatic and Heteroaromatic C-H Bonds to α-Imino Esters. *J. Org. Chem.* **2002**, 67, 4352-4361.
22. Jia, Y.-X., Xie, J.-H., Duan, H.-F., Wang, L.-X. & Zhou, Q.-L. Asymmetric Friedel-Crafts Addition of Indoles to N-Sulfonyl Aldimines: A Simple Approach to Optically Active 3-Indolyl-methanamine Derivatives. *Org. Lett.* **2006**, 8, 1621-1624.
23. Wang, Y.-Q., Song, J., Hong, R., Li, H. & Deng, L. Asymmetric Friedel-Crafts Reaction of Indoles with Imines by an Organic Catalyst. *J. Am. Chem. Soc.* **2006**, 128, 8156-8157.

METAL-CATALYZED CONJUGATE ADDITION

EWG = COR, CHO, CO_2R, CONHR, COSR, CN, SO_2R, NO_2

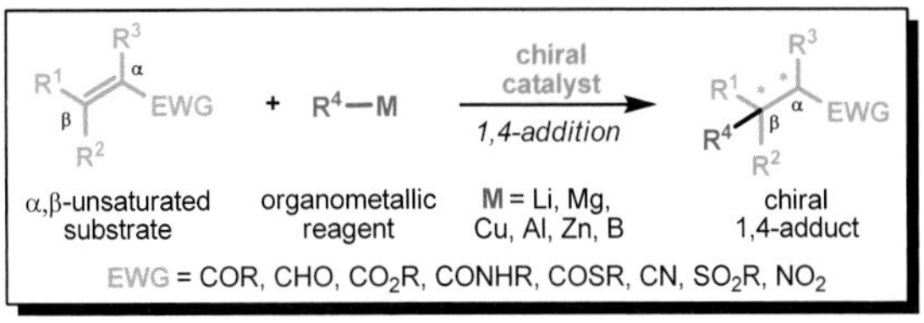

Introduction

Metal-catalyzed conjugate addition of carbon to electron-withdrawing group (EWG)-substituted carbon π-bonds, especially with copper as metal, has been extensively developed for enantioselective C-C bond formation.[1] Although highly enantioselective conjugate addition of R_2CuLi/chiral ligand complexes to α,β-enones was demonstrated more than two decades ago (Scheme 1, Equation 1),[2] the metal cuprates in current use are derived from dialkylzinc, Grignard or trialkylaluminum reagents which suffer less from the adverse effects of background reactions or alkoxide impurities. A useful catalytic process with Et_2Zn was described by Feringa using the BINOL-derived phosphoramidite ligand 4 (Scheme 1, Equation 2).[3]

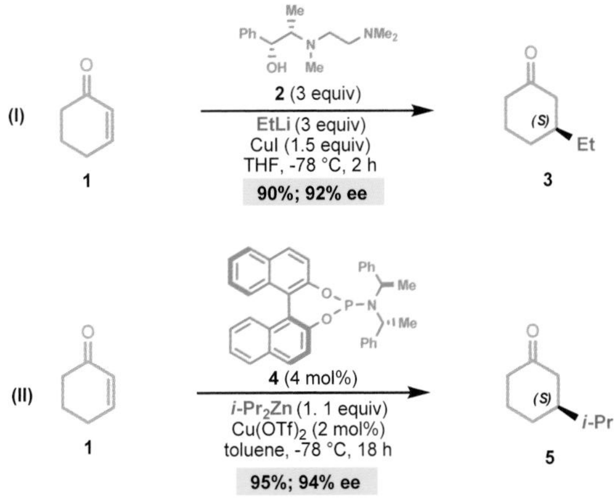

Scheme 1

A very large number and assortment of chiral metal ligands have been employed. Although none of these is universally applicable, a selection of the most practical ligands is shown in the examples which follow.

I. Catalytic Enantioselective Intermolecular Conjugate Addition of Chiral Copper Reagents

The crucial role of Cu(I) in the catalysis of the conjugate addition of cuprate reagents to electron-poor C-C double bonds has been clarified experimentally for R_2CuLi reagents. The available evidence indicates that copper acts as a metal nucleophile and attaches to C=C to form a dπ*-complex, e.g., 6 in Scheme 2.[4] The experiment depicted in Scheme 2 led to the isolation of a reactive intermediate (6) which gave the enone (1) upon treatment with H_2O but in contrast was fully

converted to the silyl enol ether 7 upon treatment with TMSCl.

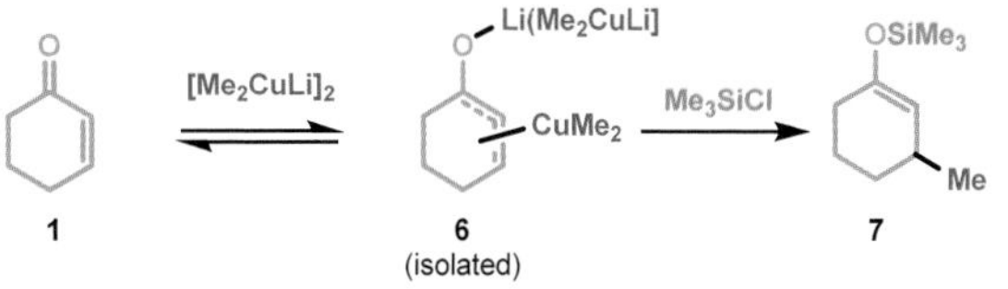

Scheme 2

Unfortunately, the pre-transition state assemblies for the enantioselective conjugate addition reaction of various chiral Cu(I) reagents which follow are unclear.[1f] It is possible that the product-determining step is reductive elimination from a β-Cu(III)adduct rather than C=C/Cu-π*-complexation.

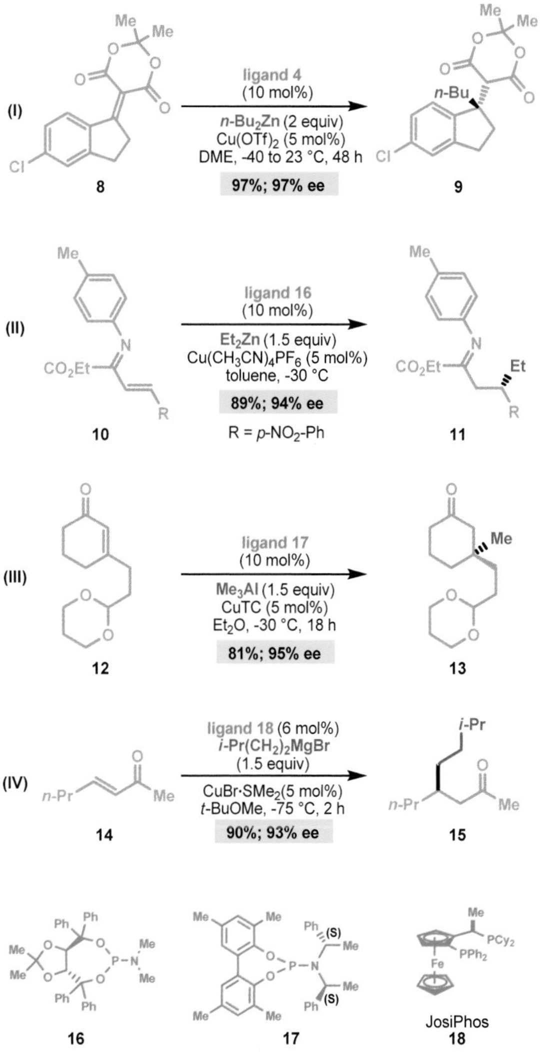

Reference key for equations: (I)[5]; (II)[6]; (III)[7]; (IV)[8]

Scheme 3

METAL-CATALYZED CONJUGATE ADDITION

Highly Diastereoselective Conjugate Addition of Achiral Copper Reagents

High levels of diastereoselectivity have been achieved in Cu-promoted conjugate additions either by the use of a chiral controller or with certain types of substrates.[9] The synthetic reactions shown in Scheme 4 exemplify both approaches.

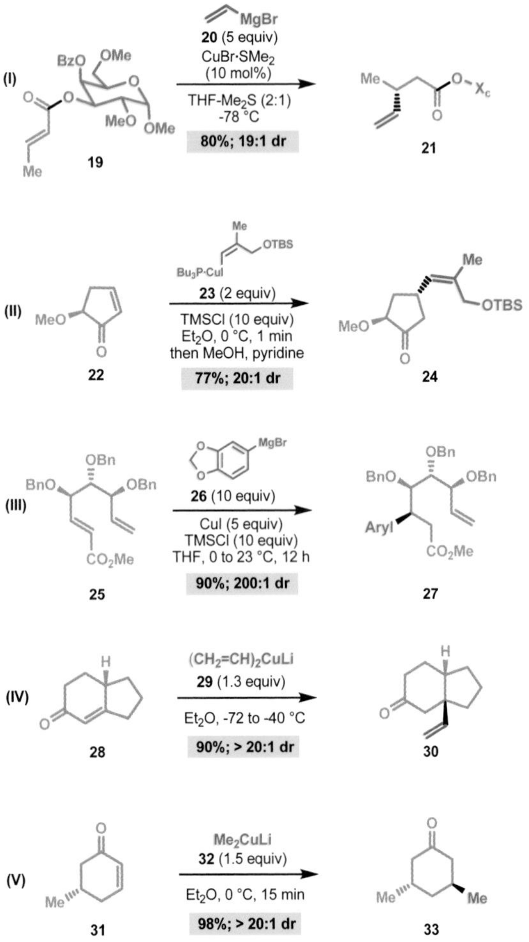

Reference key for equations: (I)[10]; (II)[11]; (III)[12]; (IV)[13]; (V)[14]

Scheme 4

Application of Catalytic Enantioselective Conjugate Addition of Chiral Cuprate Reagents in Natural Product Synthesis

A tandem 1,4-addition/enolate-trapping reaction was used as the key step during the synthesis of prostaglandin PGE$_1$ methyl ester **39** by Feringa et al. (see Sheme 5).[15] The initial adduct **37** was subjected to diastereoselective hydride reduction using Zn(BH$_4$)$_2$, and the major diastereomer **38** was converted to **39** in four steps.

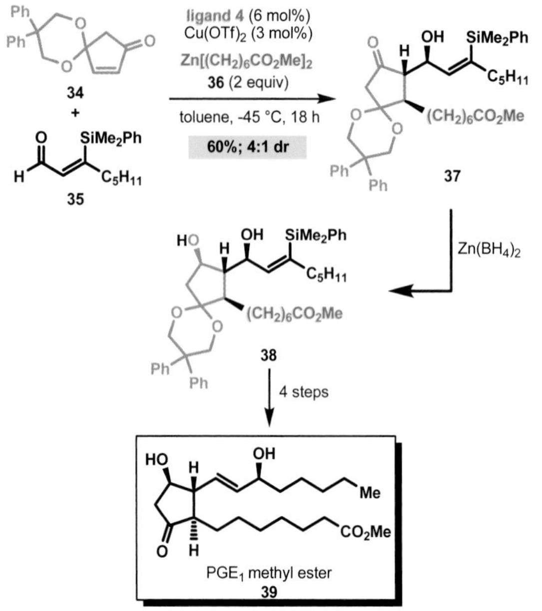

Scheme 5

II. Enantioselective Intermolecular Conjugate Addition of Chiral Nickel, Boron and Zinc Reagents

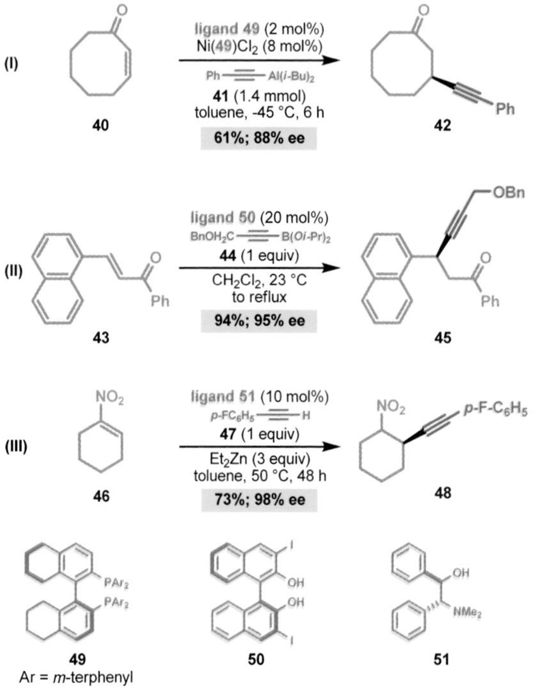

Reference key for equations: (I)[16]; (II)[17]; (III)[18]

Scheme 7

METAL-CATALYZED CONJUGATE ADDITION

III. Catalytic Enantioselective Intermolecular Conjugate Addition of Chiral Rhodium Reagents

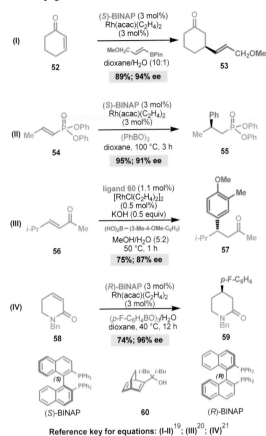

Reference key for equations: (I-II)[19]; (III)[20]; (IV)[21]

Scheme 3

References

1. (a) Lopez, F., Minnaard, A. J. & Feringa, B. L. Catalytic enantioselective conjugate addition with Grignard reagents. *Acc. Chem. Res.* **2007**, 40, 179-188; (b) Fujimori, S., Knopfel, T. F., Zarotti, P., Ichikawa, T., Boyall, D. & Carreira, E. M. Stereoselective conjugate addition reactions using in situ metalated terminal alkynes and the development of novel chiral P,N-ligands. *Bull. Chem. Soc. Jpn.* **2007**, 80, 1635-1657; (c) Alexakis, A., Vuagnoux-d'Augustin, M., Martin, D., Kehrli, S., Palais, L., Henon, H. & Hawner, C. Formation of all-carbon quaternary centers by copper-catalyzed asymmetric conjugate addition. *Chimia* **2008**, 62, 461-464; (d) Harutyunyan, S. R., den Hartog, T., Geurts, K., Minnaard, A. J. & Feringa, B. L. Catalytic Asymmetric Conjugate Addition and Allylic Alkylation with Grignard reagents. *Chem. Rev.* **2008**, 108, 2824-2852; (e) Gutnov, A. Palladium-catalyzed asymmetric conjugate addition of aryl-metal species. *Eur. J. Org. Chem.* **2008**, 4547-4554; (f) Jerphagnon, T., Pizzuti, M. G., Minnaard, A. J. & Feringa, B. L. Recent advances in enantioselective copper-catalyzed 1,4-addition. *Chem. Soc. Rev.* **2009**, 38, 1039-1075; (g) Shibasaki, M., Kanai, M., Matsunaga, S. & Kumagai, N. Recent Progress in Asymmetric Bifunctional Catalysis Using Multimetallic Systems. *Acc. Chem. Res.* **2009**, 42, 1117-1127.
2. Corey, E. J., Naef, R. & Hannon, F. J. Enantioselective conjugate addition of rationally designed chiral cuprate reagents to 2-cycloalkenones. *J. Am. Chem. Soc.* **1986**, 108, 7114-7116.
3. Feringa, B. L., Pineschi, M., Arnold, L. A., Imbos, R. & De Vries, A. H. M. Highly enantioselective catalytic conjugate addition and tandem conjugate addition - aldol reactions of organozinc reagents. *Angew. Chem., Int. Ed. Engl.* **1997**, 36, 2620-2623.
4. (a) Krauss, S. R. & Smith, S. G. Kinetics and mechanism of the conjugate addition of lithium dimethylcuprate to α,β-unsaturated

ketones. *J. Am. Chem. Soc.* **1981**, 103, 141-148; (b) Corey, E. J. & Boaz, N. W. d-Orbital stereoelectronic control of the stereochemistry of SN2' displacements by organocuprate reagents. *Tetrahedron Lett.* **1984**, 25, 3063-3066; (c) Hallnemo, G., Olsson, T. & Ullenius, C. An NMR investigation of the reaction between cinnamic acid esters and lithium dimethylcuprate. Initial formation of an olefin-copper π-complex. *J. Organomet. Chem.* **1985**, 282, 133-144; (d) Corey, E. J., Hannon, F. J. & Boaz, N. W. Coordinately induced 1,4-diastereoselection in the reaction of acyclic α,β-enones with organocopper reagents. A new type of organocopper reaction. *Tetrahedron* **1989**, 45, 545-555; (e) Corey, E. J. & Hannon, F. J. A possible transition state assembly for highly diastereoselective conjugate addition reactions of lithium dimethylcuprate with α,β-enones. *Tetrahedron Lett.* **1990**, 31, 1393-1396.
5. Fillion, E. & Wilsily, A. Asymmetric Synthesis of All-Carbon Benzylic Quaternary Stereocenters via Cu-Catalyzed Conjugate Addition of Dialkylzinc Reagents to 5-(1-Arylalkylidene) Meldrum's Acids. *J. Am. Chem. Soc.* **2006**, 128, 2774-2775.
6. Palacios, F. & Vicario, J. Copper-catalyzed asymmetric conjugate addition of diethylzinc to α,β-unsaturated imines derived from α-amino acids. Enantioselective synthesis of γ-substituted α-dehydroamino esters. *Org. Lett.* **2006**, 8, 5405-5408.
7. d'Augustin, M., Palais, L. & Alexakis, A. Enantioselective copper-catalyzed conjugate addition to trisubstituted cyclohexenones: Construction of stereogenic quaternary centers. *Angew. Chem., Int. Ed.* **2005**, 44, 1376-1378.
8. Lopez, F., Harutyunyan, S. R., Minnaard, A. J. & Feringa, B. L. Copper-Catalyzed Enantioselective Conjugate Addition of Grignard Reagents to Acyclic Enones. *J. Am. Chem. Soc.* **2004**, 126, 12784-12785.
9. (a) Lipshutz, B. H. & Sengupta, S. Organocopper reagents: substitution, conjugate addition, carbo/metallocupration, and other reactions. *Org. React.* **1992**, 41, 135-631; (b) Rossiter, B. E. & Swingle, N. M. Asymmetric conjugate addition. *Chem. Rev.* **1992**, 92, 771-806; (c) Breit, B. & Schmidt, Y. Directed Reactions of Organocopper Reagents. *Chem. Rev.* **2008**, 108, 2928-2951.
10. Totani, K., Nagatsuka, T., Yamaguchi, S., Takao, K.-i., Ohba, S. & Tadano, K.-i. Highly Diastereoselective 1,4-Addition of an Organocuprate to Methyl α-D-Gluco-, α-D-Manno-, or α-D-Galactopyranosides Tethering an α,β-Unsaturated Ester. Novel Asymmetric Access to β-C-Substituted Butanoic Acids. *J. Org. Chem.* **2001**, 66, 5965-5975.
11. Smith, A. B., III, Rano, T. A., Chida, N., Sulikowski, G. A. & Wood, J. L. Total synthesis of the cytotoxic macrocycle (+)-hitachimycin. *J. Am. Chem. Soc.* **1992**, 114, 8008-8022.
12. Nadein, O. N. & Kornienko, A. An Approach to Pancratistatins via Ring-Closing Metathesis: Efficient Synthesis of Novel 1-Aryl-1-deoxyconduritols F. *Org. Lett.* **2004**, 6, 831-834.
13. Corey, E. J. & Carney, R. L. New methods for carbocyclic synthesis applicable to the gibberellic acids. Stereoselective introduction of the angular vinyl grouping and pinacolic cyclization of keto aldehydes. *J. Am. Chem. Soc.* **1971**, 93, 7318-7319.
14. House, H. O. & Fischer, W. F., Jr. Chemistry of carbanions. XVII. Addition of methyl organometallic reagents to cyclohexenone derivatives. *J. Org. Chem.* **1968**, 33, 949-956.
15. Arnold, L. A., Naasz, R., Minnaard, A. J. & Feringa, B. L. Catalytic Enantioselective Synthesis of (-)-Prostaglandin E1 Methyl Ester Based on a Tandem 1,4-Addition-Aldol Reaction. *J. Org. Chem.* **2002**, 67, 7244-7254.
16. Larionov, O. V. & Corey, E. J. Ni(II)-Catalyzed Enantioselective Conjugate Addition of Acetylenes to α,β-Enones. *Org. Lett.* **2010**, 12, 300-302.
17. Wu, T. R. & Chong, J. M. Ligand-catalyzed asymmetric alkynylboration of enones: a new paradigm for asymmetric synthesis using organoboranes. *J. Am. Chem. Soc.* **2005**, 127, 3244-3245.
18. Yamashita, M., Yamada, K. & Tomioka, K. Chiral Amino Alcohol-Mediated Asymmetric Conjugate Addition of Arylalkynes to Nitroolefins. *Org. Lett.* **2005**, 7, 2369-2371.
19. (a) Takaya, Y., Ogasawara, M. & Hayashi, T. Rhodium-catalyzed asymmetric 1,4-addition of 2-alkenyl-1,3,2-benzodioxaboroles to α,β-unsaturated ketones. *Tetrahedron Lett.* **1998**, 39, 8479-8482; (b) Hayashi, T. & Yamasaki, K. Rhodium-Catalyzed Asymmetric 1,4-Addition and Its Related Asymmetric Reactions. *Chem. Rev.* **2003**, 103, 2829-2844.
20. Brown, M. K. & Corey, E. J. Catalytic enantioselective formation of chiral-bridged dienes which are themselves ligands for enantioselective catalysis. *Org. Lett.* **2009**, 12, 172-175.
21. Senda, T., Ogasawara, M. & Hayashi, T. Rhodium-Catalyzed Asymmetric 1,4-Addition of Organoboron Reagents to 5,6-Dihydro-2(1H)-pyridinones. Asymmetric Synthesis of 4-Aryl-2-piperidinones. *J. Org. Chem.* **2001**, 66, 6852-6856.

CONJUGATE ADDITION OF STABILIZED C-ANIONS

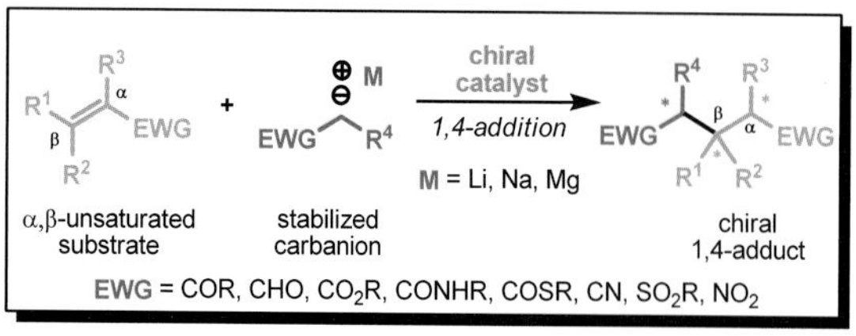

α,β-unsaturated substrate + stabilized carbanion → chiral 1,4-adduct

EWG = COR, CHO, CO$_2$R, CONHR, COSR, CN, SO$_2$R, NO$_2$

Introduction

The conjugate addition of stabilized carbanions to C=C bonds bearing one or more electron-withdrawing (EWG) groups (Michael reaction) is widely useful for synthesis, especially in newer enantioselective forms which may be catalyzed by chiral metal complexes or by metal-free neutral organic molecules ("organocatalysts").[1] The first example of asymmetric induction in a Michael reaction was reported by Wynberg et al. who used quinine (2) as a catalytic base for the 1,4-addition of an indanone 3 to methyl vinyl ketone (1), as shown in Scheme 1.[2]

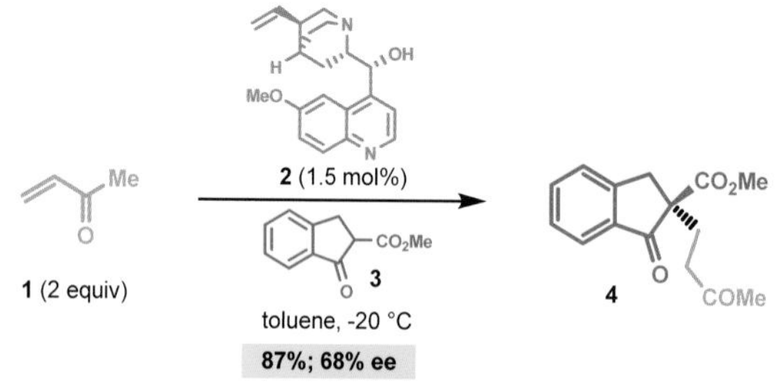

toluene, -20 °C
87%; 68% ee

Scheme 1

A selection of efficient catalytic enantioselective Michael reactions, organized according to the structure of the Michael acceptor, are shown in the Schemes that follow.

I. Catalytic Enantioselective Conjugate Addition to Enones and Enals

(I)

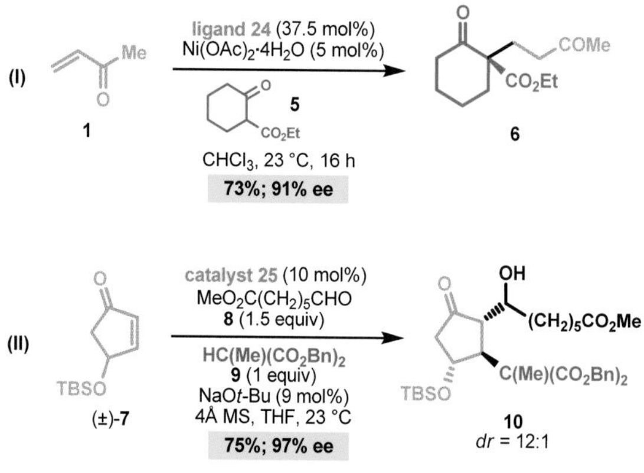

ligand 24 (37.5 mol%)
Ni(OAc)$_2$·4H$_2$O (5 mol%)

CHCl$_3$, 23 °C, 16 h
73%; 91% ee

(II)

catalyst 25 (10 mol%)
MeO$_2$C(CH$_2$)$_5$CHO
8 (1.5 equiv)

HC(Me)(CO$_2$Bn)$_2$
9 (1 equiv)
NaOt-Bu (9 mol%)
4Å MS, THF, 23 °C
75%; 97% ee

dr = 12:1

Reference key for equations: (I)[3]; (II)[4]

Scheme 2

Catalytic Enantioselective Conjugate Addition to Enones and Enals (Continued)

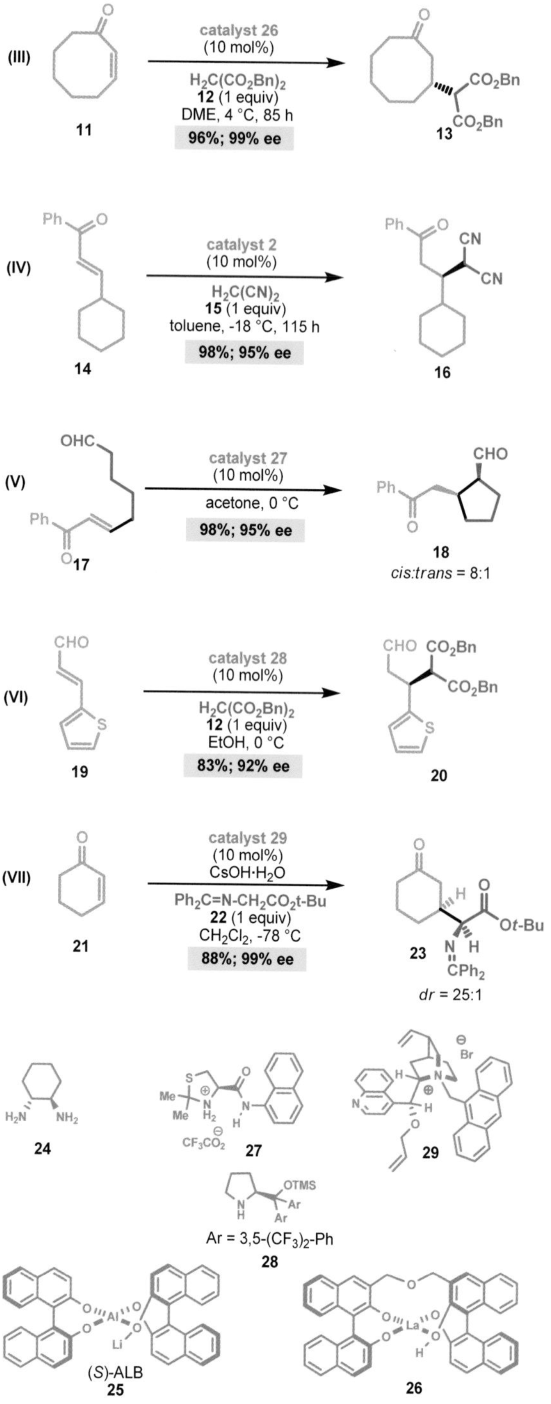

(III)

catalyst 26 (10 mol%)

H$_2$C(CO$_2$Bn)$_2$
12 (1 equiv)
DME, 4 °C, 85 h
96%; 99% ee

(IV)

catalyst 2 (10 mol%)

H$_2$C(CN)$_2$
15 (1 equiv)
toluene, -18 °C, 115 h
98%; 95% ee

(V)

catalyst 27 (10 mol%)

acetone, 0 °C
98%; 95% ee

cis:trans = 8:1

(VI)

catalyst 28 (10 mol%)

H$_2$C(CO$_2$Bn)$_2$
12 (1 equiv)
EtOH, 0 °C
83%; 92% ee

(VII)

catalyst 29 (10 mol%)
CsOH·H$_2$O

Ph$_2$C=N-CH$_2$CO$_2$t-Bu
22 (1 equiv)
CH$_2$Cl$_2$, -78 °C
88%; 99% ee

dr = 25:1

Ar = 3,5-(CF$_3$)$_2$-Ph
28

(S)-ALB
25

26

Reference key for equations: (III)[5]; (IV)[6]; (V)[7]; (VI)[8]; (VII)[9]

Scheme 3

CONJUGATE ADDITION OF STABILIZED C-ANIONS

II. Catalytic Enantioselective Conjugate Addition to Unsaturated Carboxylic Acid Derivatives

A variety of carboxylic acid derivatives have been shown to participate in catalytic enantioselective conjugate addition reactions. Some highly enantioselective examples are shown in Scheme 4.

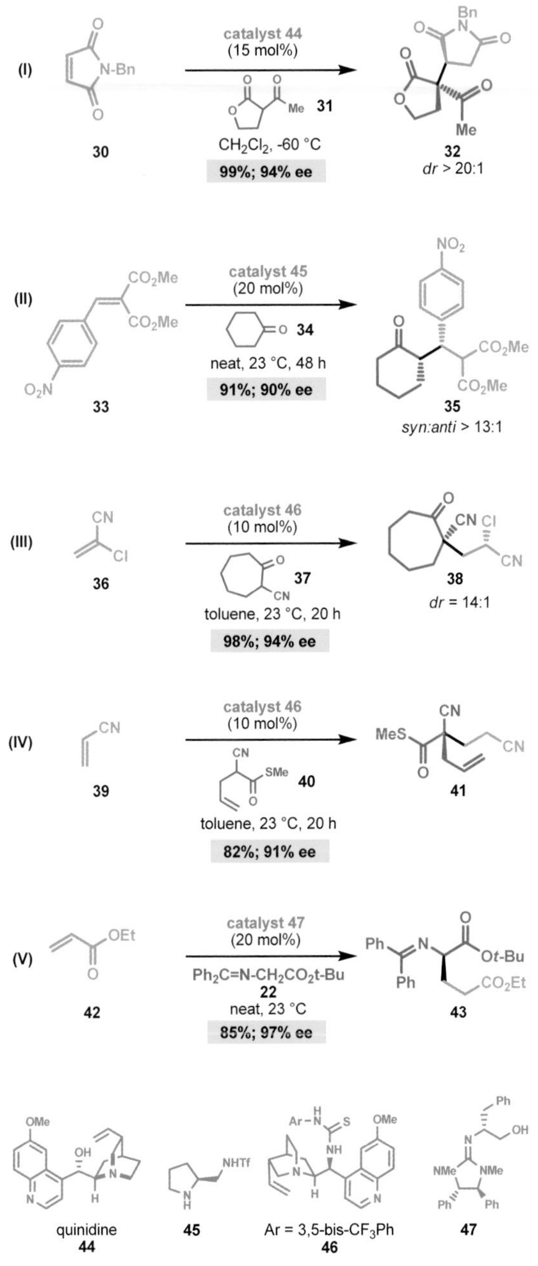

III. Catalytic Enantioselective Conjugate Addition to Various Electron-Deficient Alkenes

Nitroalkenes and vinylsulfones have been widely used as acceptors in conjugate addition reactions. The resulting nitro adducts are highly versatile and can be elaborated to a rich array of structures. Example III illustrates a three component reaction (54 + 55 + 56) which is highly chemo-, diastereo- and enantioselective, using the proline-derived catalyst 28. Three C-C single bonds, one ring and four contiguous stereogenic centers are generated in a single step.

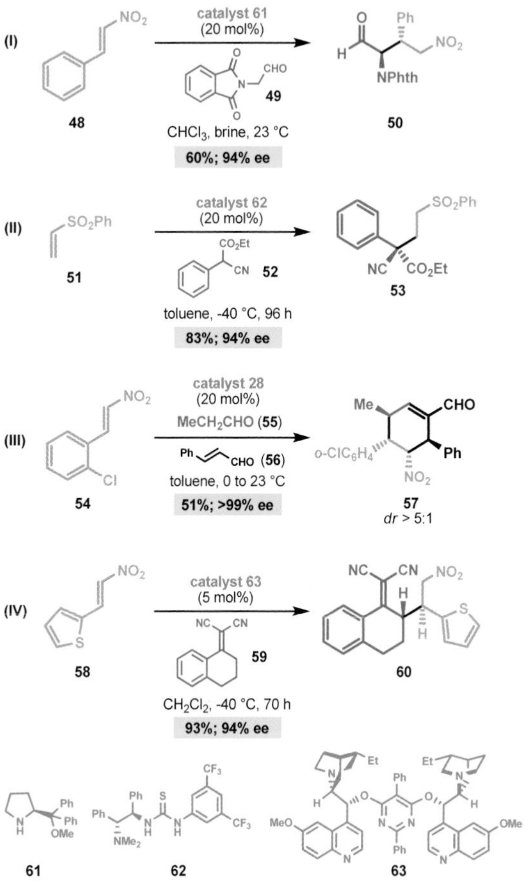

Reference key for equations: (I)[10]; (II)[11]; (III-IV)[12]; (V)[13]

Scheme 4

Reference key for equations: (I)[14]; (II)[15]; (III)[16]; (IV)[17]

Scheme 5

IV. Enantioselective Conjugate Addition with Neutral Nucleophiles

Silyl enol ethers as well as electron-rich aromatic compounds may serve as non-ionic nucleophiles in catalytic conjugate addition reactions (e.g., Mukaiyama-Michael conjugate addition) as demonstrated by several examples in Scheme 6.

CONJUGATE ADDITION OF STABILIZED C-ANIONS

Enantioselective Conjugate Addition with Neutral Nucleophiles (Continued)

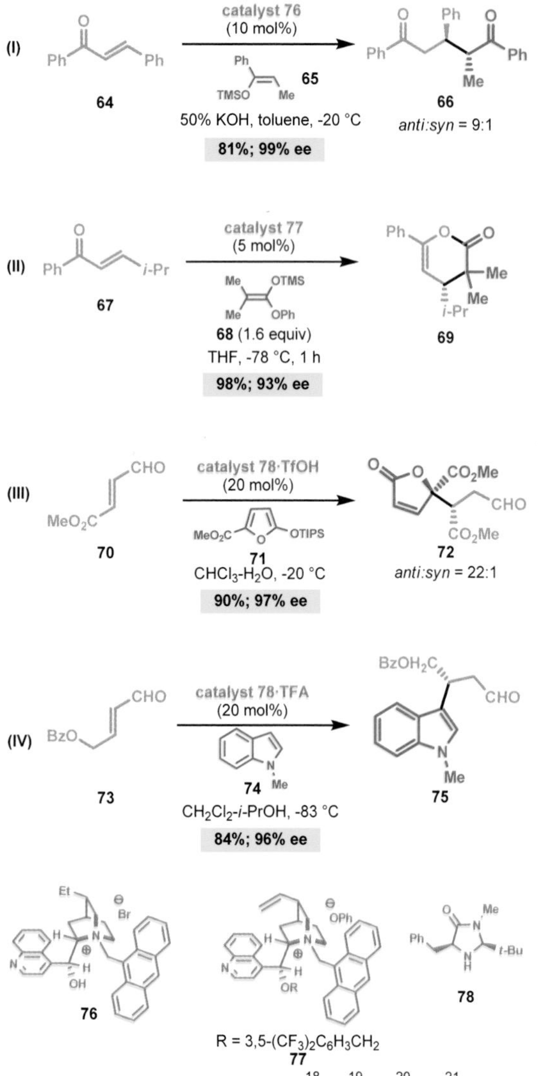

Scheme 6

Reference key for equations: (I)[18]; (II)[19]; (III)[20]; (IV)[21]

References

1. (a) Kanai, M. & Shibasaki, M. Asymmetric Michael reactions. *Catal. Asymmetric Synth. (2nd Ed.)* **2000**, 569-592; (b) Gil, M. V., Roman, E. & Serrano, J. A. Nitro compounds in asymmetric Michael reactions. *Trends Org. Chem.* **2001**, 9, 17-28; (c) Krause, N. & Hoffmann-Roder, A. Recent advances in catalytic enantioselective Michael additions. *Synthesis* **2001**, 171-196; (d) Jha, S. C. & Joshi, N. N. Catalytic, enantioselective Michael addition reactions. *ARKIVOC (Gainesville, FL, U. S.)* **2002**, 167-196; (e) Berner, O. M., Tedeschi, L. & Enders, D. Asymmetric Michael additions to nitroalkenes. *Eur. J. Org. Chem.* **2002**, 1877-1894; (f) Guo, H.-C. & Ma, J.-A. Catalytic asymmetric tandem transformations triggered by conjugate additions. *Angew. Chem., Int. Ed.* **2006**, 45, 354-366; (g) Almasi, D., Alonso, D. A. & Najera, C. Organocatalytic asymmetric conjugate additions. *Tetrahedron Asymmetry* **2007**, 18, 299-365; (h) Tsogoeva, S. B. Recent advances in asymmetric organocatalytic 1,4-conjugate additions. *Eur. J. Org. Chem.* **2007**, 1701-1716.

2. Wynberg, H. & Helder, R. Asymmetric induction in the alkaloid-catalyzed Michael reaction. *Tetrahedron Lett.* **1975**, 4057-4060.

3. Christoffers, J., Rossler, U. & Werner, T. Construction of quaternary stereocenters by Nickel-catalysis of asymmetric Michael reactions. *Eur. J. Org. Chem.* **2000**, 701-705.

4. Yamada, K.-i., Arai, T., Sasai, H. & Shibasaki, M. A Catalytic Asymmetric Synthesis of 11-Deoxy-PGF$_{1\alpha}$ Using ALB, a Heterobimetallic Multifunctional Asymmetric Complex. *J. Org. Chem.* **1998**, 63, 3666-3672.

5. Kim, Y. S., Matsunaga, S., Das, J., Sekine, A., Ohshima, T. & Shibasaki, M. Stable, storable, and reusable asymmetric catalyst: a novel La-linked-BINOL complex for the catalytic asymmetric Michael reaction. *J. Am. Chem. Soc.* **2000**, 122, 6506-6507.

6. Russo, A., Perfetto, A. & Lattanzi, A. Back to Natural Cinchona Alkaloids: Highly Enantioselective Michael Addition of Malononitrile to Enones. *Adv. Synth. Catal.* **2009**, 351, 3067-3071.

7. Hayashi, Y., Gotoh, H., Tamura, T., Yamaguchi, H., Masui, R. & Shoji, M. Cysteine-Derived Organocatalyst in a Highly Enantioselective Intramolecular Michael Reaction. *J. Am. Chem. Soc.* **2005**, 127, 16028-16029.

8. Brandau, S., Landa, A., Franzen, J., Marigo, M. & Jorgensen, K. A. Organocatalytic conjugate addition of malonates to α,β-unsaturated aldehydes: asymmetric formal synthesis of (-)-paroxetine, chiral lactams, and lactones. *Angew. Chem., Int. Ed.* **2006**, 45, 4305-4309.

9. Corey, E. J., Noe, M. C. & Xu, F. Highly enantioselective synthesis of cyclic and functionalized α-amino acids by means of a chiral phase transfer catalyst. *Tetrahedron Lett.* **1998**, 39, 5347-5350.

10. Bartoli, G., Bosco, M., Carlone, A., Cavalli, A., Locatelli, M., Mazzanti, A., Ricci, P., Sambri, L. & Melchiorre, P. Organocatalytic asymmetric conjugate addition of 1,3-dicarbonyl compounds to maleimides. *Angew. Chem., Int. Ed.* **2006**, 45, 4966-4970.

11. Cao, C.-L., Sun, X.-L., Zhou, J.-L. & Tang, Y. Enantioselectively organocatalytic Michael addition of ketones to alkylidene malonates. *J. Org. Chem.* **2007**, 72, 4073-4076.

12. Wang, B., Wu, F., Wang, Y., Liu, X. & Deng, L. Control of Diastereoselectivity in Tandem Asymmetric Reactions Generating Nonadjacent Stereocenters with Bifunctional Catalysis by Cinchona Alkaloids. *J. Am. Chem. Soc.* **2007**, 129, 768-769.

13. Ishikawa, T., Araki, Y., Kumamoto, T., Isobe, T., Seki, H. & Fukuda, K. Modified guanidines as chiral superbases: application to asymmetric Michael reaction of glycine imine with acrylate or its related compounds. *Chem. Commun. (Cambridge)* **2001**, 245-246.

14. Albertshofer, K., Thayumanavan, R., Utsumi, N., Tanaka, F. & Barbas, C. F. Amine-catalyzed Michael reactions of an aminoaldehyde derivative to nitroolefins. *Tetrahedron Lett.* **2007**, 48, 693-696.

15. Liu, T.-Y., Long, J., Li, B.-J., Jiang, L., Li, R., Wu, Y., Ding, L.-S. & Chen, Y.-C. Enantioselective construction of quaternary carbon centre catalyzed by bifunctional organocatalyst. *Org. Biomol. Chem.* **2006**, 4, 2097-2099.

16. Enders, D., Huettl, M. R. M., Grondal, C. & Raabe, G. Control of four stereocenters in a triple cascade organocatalytic reaction. *Nature (London, U. K.)* **2006**, 441, 861-863.

17. Xue, D., Chen, Y.-C., Wang, Q.-W., Cun, L.-F., Zhu, J. & Deng, J.-G. Asymmetric Direct Vinylogous Michael Reaction of Activated Alkenes to Nitroolefins Catalyzed by Modified Cinchona Alkaloids. *Org. Lett.* **2005**, 7, 5293-5296.

18. Zhang, F.-Y. & Corey, E. J. Enantio- and diastereoselective Michael reactions of silyl enol ethers and chalcones by catalysis using a chiral quaternary ammonium salt. *Org. Lett.* **2001**, 3, 639-641.

19. Tozawa, T., Nagao, H., Yamane, Y. & Mukaiyama, T. Enantioselective synthesis of 3,4-dihydropyran-2-ones by domino Michael addition and lactonization with new asymmetric organocatalysts: Cinchona-alkaloid-derived chiral quaternary ammonium phenoxides. *Chemistry--An Asian Journal* **2007**, 2, 123-134.

20. Brown, S. P., Goodwin, N. C. & MacMillan, D. W. C. The First Enantioselective Organocatalytic Mukaiyama-Michael Reaction: A Direct Method for the Synthesis of Enantioenriched γ-Butenolide Architecture. *J. Am. Chem. Soc.* **2003**, 125, 1192-1194.

21. Austin, J. F. & MacMillan, D. W. C. Enantioselective Organocatalytic Indole Alkylations. Design of a New and Highly Effective Chiral Amine for Iminium Catalysis. *J. Am. Chem. Soc.* **2002**, 124, 1172-1173.

COUPLING RXNS OF π-ALLYL-METAL COMPLEXES

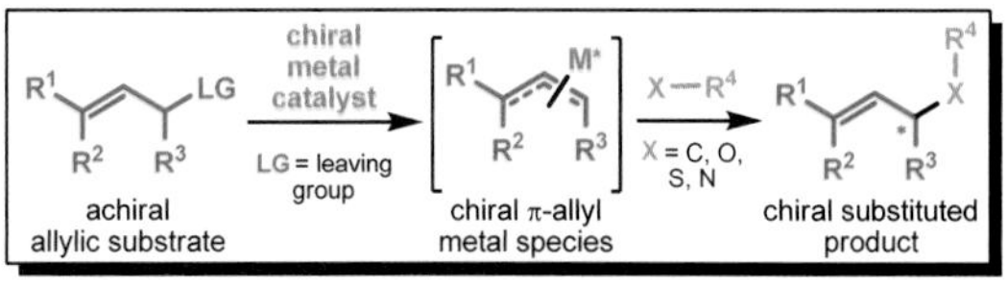

Introduction

Allylic carbon-carbon and carbon-heteroatom (C-O, C-N and C-S) bonds can be formed enantioselectively under catalysis via a suitable chiral Pd(0)-phosphine complex.[1] These reactions are especially useful in synthesis because they often proceed with high yield and enantioselectivity under mild conditions at 0 to 25 °C. One of the components of the coupling reaction is an allylic substrate which can react with the Pd(0) catalyst to form a π-allylmetal complex and the other is a compound containing an easily deprotonated C-H, N-H, O-H or S-H subunit. Because the conditions required for these Pd-catalyzed coupling processes are so mild, interference by functional groups is minimal. Catalysts of Rh(I) and Ir(I) have also been applied to C-O and C-N bond formation.

I. Catalytic Enantioselective Allylation: Formation of Carbon-Carbon (C-C) Bonds

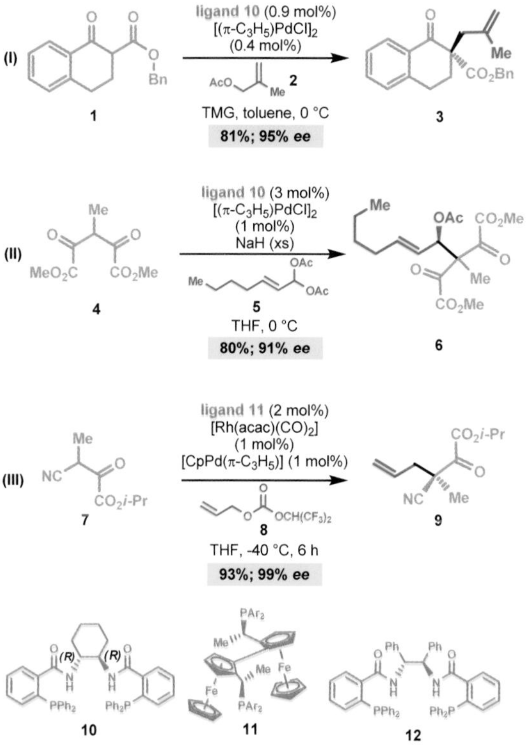

Reference key for equations: (I)[1b]; (II)[2]; (III)[3]

Scheme 1

II. Catalytic Enantioselective Allylation: Formation of Carbon-Heteroatom (C-X) Bonds

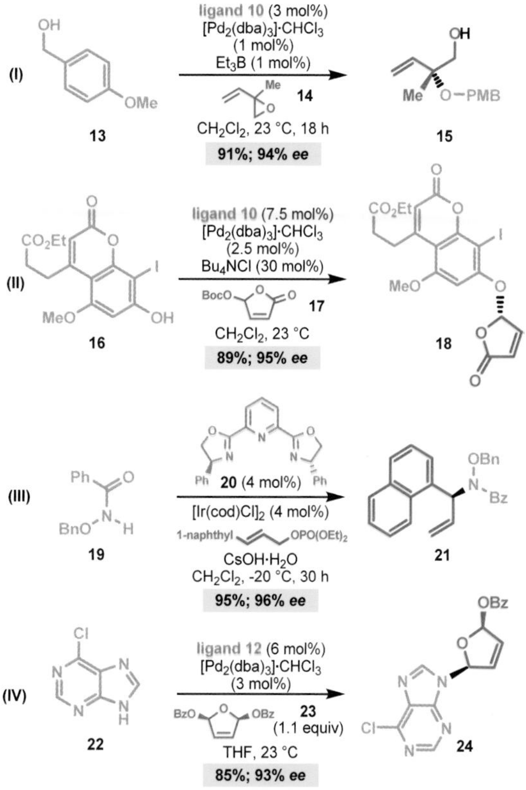

Reference key for equations: (I)[4]; (II)[5]; (III)[6]; (IV)[7]

Scheme 2

References

1. (a) Trost, B. M. & Van Vranken, D. L. Asymmetric Transition Metal-Catalyzed Allylic Alkylations. *Chem. Rev.* **1996**, 96, 395-422; (b) Trost, B. M. & Crawley, M. L. Asymmetric Transition-Metal-Catalyzed Allylic Alkylations: Applications in Total Synthesis. *Chem. Rev.* **2003**, 103, 2921-2943; (c) Trost, B. M. Asymmetric allylic alkylation, an enabling methodology. *J. Org. Chem.* **2004**, 69, 5813-5837; (d) Lu, Z. & Ma, S. Metal-catalyzed enantioselective allylation in asymmetric synthesis. *Angew. Chem., Int. Ed.* **2008**, 47, 258-297.
2. Trost, B. M. & Lee, C. B. Geminal Dicarboxylates as Carbonyl Surrogates for Asymmetric Synthesis. Part I. Asymmetric Addition of Malonate Nucleophiles. *J. Am. Chem. Soc.* **2001**, 123, 3671-3686.
3. Sawamura, M., Sudoh, M. & Ito, Y. An Enantioselective Two-Component Catalyst System: Rh-Pd-Catalyzed Allylic Alkylation of Activated Nitriles. *J. Am. Chem. Soc.* **1996**, 118, 3309-3310.
4. Trost, B. M. & Toste, F. D. A Catalytic Enantioselective Approach to Chromans and Chromanols. A Total Synthesis of (-)-Calanolides A and B and the Vitamin E Nucleus. *J. Am. Chem. Soc.* **1998**, 120, 9074-9075.
5. Trost, B. M. & Toste, F. D. Palladium-Catalyzed Kinetic and Dynamic Kinetic Asymmetric Transformation of 5-Acyloxy-2-(5H)-furanone. Enantioselective Synthesis of (-)-Aflatoxin B Lactone. *J. Am. Chem. Soc.* **1999**, 121, 3543-3544.
6. Miyabe, H., Matsumura, A., Moriyama, K. & Takemoto, Y. Utility of the Iridium Complex of the Pybox Ligand in Regio- and Enantioselective Allylic Substitution. *Org. Lett.* **2004**, 6, 4631-4634.
7. Trost, B. M. & Shi, Z. From Furan to Nucleosides. *J. Am. Chem. Soc.* **1996**, 118, 3037-3038.

METAL-CATALYZED C-H FUNCTIONALIZATION

Introduction

Chiral dinuclear Rh(II)-carboxylate complexes can catalyze the enantioselective reaction of α-diazocarbonyl compounds with suitable substrates to form C-H insertion products, even with unactivated hydrocarbons such as cyclohexane.[1] The resulting compounds may in principle also be obtained by the alkylation of chiral enolates. Benzylic and allylic C-H groups are especially susceptible to insertion and may react selectively. Allylic C-H insertion into silyl enol ethers leads to Michael-type adducts.

Selective C-H insertion may also occur at a carbon bearing an ether oxygen or amino nitrogen. These insertion reactions often take place with high enantioselectivity, and also with excellent diastereoselectivity. Insertion of α-diazocarbonyl-derived carbenes into ethers or amines can lead stereoselectively to aldol or Mannich type products. A number of synthetic examples are shown in Schemes 1-6.

Catalytic and highly enantioselective intramolecular C-H carbon- and heterofunctionalizations using chiral Rh(II) complexes are discussed on pages 139-142. For catalytic enantioselective α-heterofunctionalization of carbonyl compounds, see pages 54-55.

I. Enantioselective Intermolecular C-H Insertion Using Rh(II)-Carbene Complexes and a Saturated Hydrocarbon

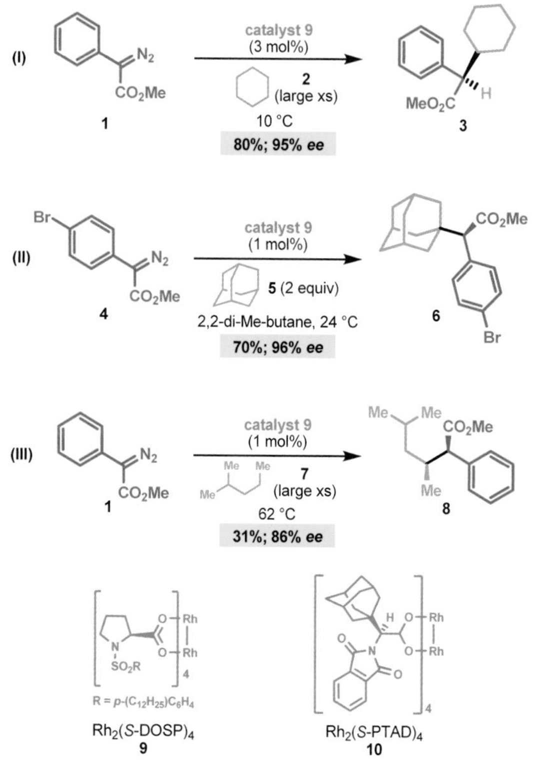

Reference key for equations: (I-III)[2]

Scheme 1

II. Enantioselective Intermolecular Allylic C-H Insertion Using Rh(II)-Carbene Complexes

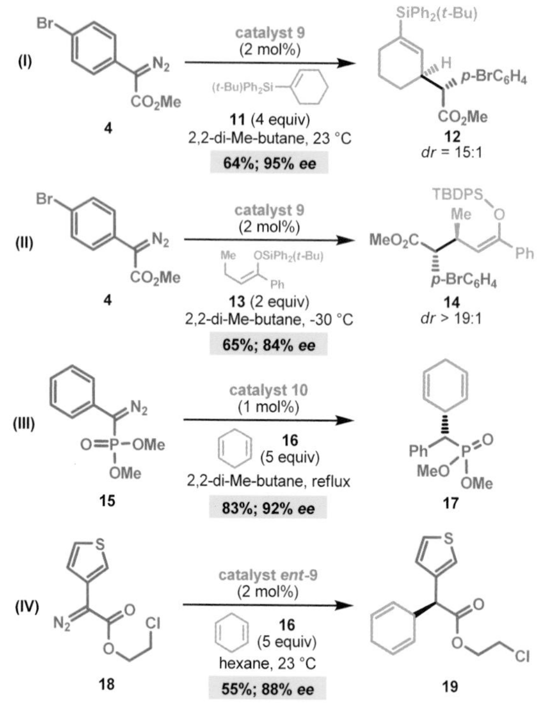

Reference key for equations: (I)[3]; (II)[4]; (III)[5]; (IV)[6]

Scheme 2

III. Catalytic Enantioselective Allylic C-H Insertion with Allylic Rearrangement

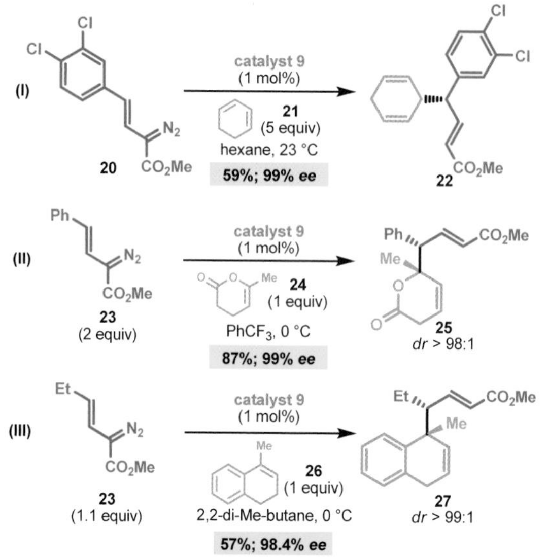

Reference key for equations: (I)[7]; (II)[8]; (III)[9]

Scheme 3

METAL-CATALYZED C-H FUNCTIONALIZATION

IV. Intermolecular Benzylic C-H Insertion by Rh(II)-Carbene Complexes

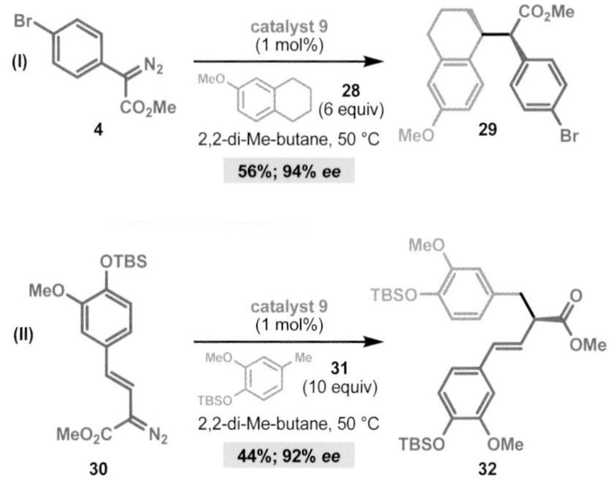

(I)

catalyst 9 (1 mol%)
28 (6 equiv)
2,2-di-Me-butane, 50 °C

4

29

56%; 94% ee

(II)

catalyst 9 (1 mol%)
31 (10 equiv)
2,2-di-Me-butane, 50 °C

30

32

44%; 92% ee

Reference key for equations: (I)[10]; (II)[11]

Scheme 4

V. Intermolecular C-H Insertion Alpha to a Heteroatom Using Rh(II)-Carbene Complexes

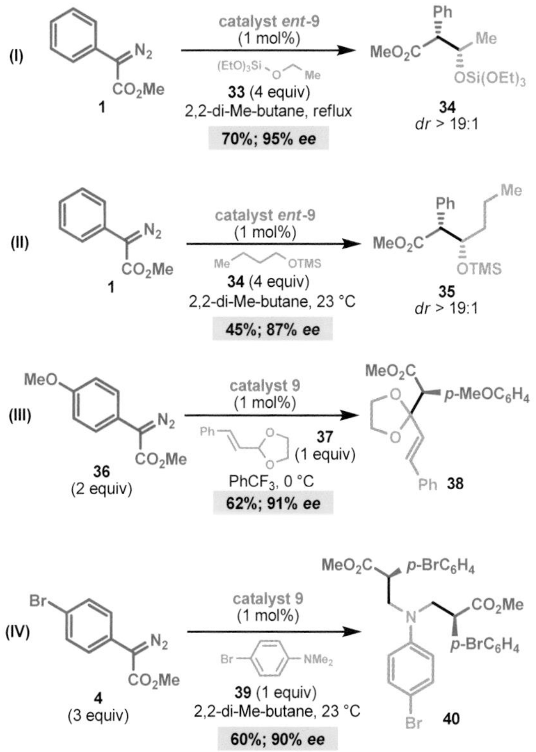

(I)

catalyst ent-9 (1 mol%)
33 (4 equiv)
2,2-di-Me-butane, reflux

1

34
dr > 19:1

70%; 95% ee

(II)

catalyst ent-9 (1 mol%)
34 (4 equiv)
2,2-di-Me-butane, 23 °C

1

35
dr > 19:1

45%; 87% ee

(III)

catalyst 9 (1 mol%)
37 (1 equiv)
PhCF₃, 0 °C

36 (2 equiv)

Ph 38

62%; 91% ee

(IV)

catalyst 9 (1 mol%)
39 (1 equiv)
2,2-di-Me-butane, 23 °C

4 (3 equiv)

40

60%; 90% ee

Reference key for equations: (I)[12]; (II)[13]; (III)[14]; (IV)[15]

Scheme 5

VI. Intermolecular C-H Insertion Using Mn-Salen-Complexes[16]

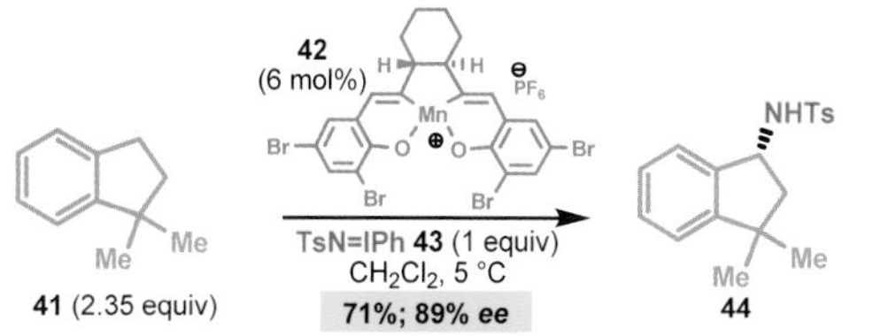

42 (6 mol%)

TsN=IPh 43 (1 equiv)
CH₂Cl₂, 5 °C

41 (2.35 equiv)

44

71%; 89% ee

Scheme 6

References

1. (a) Davies, H. M. L. Recent advances in catalytic enantioselective intermolecular C-H functionalization. *Angew. Chem., Int. Ed.* **2006**, 45, 6422-6425; (b) Davies, H. M. & Dai, X. Synthetic Reactions via C-H Bond Activation: Carbene and Nitrene C-H Insertion in *Comprehensive Organometallic Chemistry* (eds. Crabtree, R. & Mingos, D. M. P.), 167-212 (Elsevier, **2007**); (c) Davies, H. M. L. & Hansen, J. Asymmetric Synthesis Through C-H Activation in *Catalytic Asymmetric Synthesis, 3rd Ed.* (ed. Ojima, I.), 163-226 (John Wiley & Sons, Hoboken, NJ, **2010**); (d) Davies, H. M. L. & Dick, A. R. Functionalization of Carbon-Hydrogen Bonds Through Transition Metal Carbenoid Insertion in *C-H Activation (Topics of Current Chemistry)* (eds. Yu, J.-Q. & Shi, Z.), 303-345 (Springer-Verlag, Berlin, **2010**).

2. Davies, H. M. L., Hansen, T. & Churchill, M. R. Catalytic Asymmetric C-H Activation of Alkanes and Tetrahydrofuran. *J. Am. Chem. Soc.* **2000**, 122, 3063-3070.

3. Davies, H. M. L., Ren, P. & Jin, Q. Catalytic Asymmetric Allylic C-H Activation as a Surrogate of the Asymmetric Claisen Rearrangement. *Org. Lett.* **2001**, 3, 3587-3590.

4. Davies, H. M. L. & Ren, P. Catalytic Asymmetric C-H Activation of Silyl Enol Ethers as an Equivalent of an Asymmetric Michael Reaction. *J. Am. Chem. Soc.* **2001**, 123, 2070-2071.

5. Reddy, R. P., Lee, G. H. & Davies, H. M. L. Dirhodium Tetracarboxylate Derived from Adamantylglycine as a Chiral Catalyst for Carbenoid Reactions. *Org. Lett.* **2006**, 8, 3437-3440.

6. Davies, H. M. L., Walji, A. M. & Townsend, R. J. Catalytic asymmetric C-H activation by methyl thiophen-3-yldiazoacetate applied to the synthesis of (+)-cetiedil. *Tetrahedron Lett.* **2002**, 43, 4981-4983.

7. Davies, H. M. L., Stafford, D. G. & Hansen, T. Catalytic asymmetric synthesis of diarylacetates and 4,4-diarylbutanoates. A formal asymmetric synthesis of (+)-sertraline. *Org. Lett.* **1999**, 1, 233-236.

8. Davies, H. M. L. & Jin, Q. Catalytic asymmetric reactions for organic synthesis: the combined C-H activation/Cope rearrangement. *Proc. Natl. Acad. Sci. U. S. A.* **2004**, 101, 5472-5475.

9. Davies, H. M. L. & Jin, Q. Highly Diastereoselective and Enantioselective C-H Functionalization of 1,2-Dihydronaphthalenes: A Combined C-H Activation/Cope Rearrangement Followed by a Retro-Cope Rearrangement. *J. Am. Chem. Soc.* **2004**, 126, 10862-10863.

10. Davies, H. M. L., Jin, Q., Ren, P. & Kovalevsky, A. Y. Catalytic Asymmetric Benzylic C-H Activation by Means of Carbenoid-Induced C-H Insertions. *J. Org. Chem.* **2002**, 67, 4165-4169.

11. Davies, H. M. L. & Jin, Q. Intermolecular C-H activation at benzylic positions: synthesis of (+)-imperanene and (-)-α-conidendrin. *Tetrahedron Asymmetry* **2003**, 14, 941-949.

12. Davies, H. M. L. & Antoulinakis, E. G. Asymmetric Catalytic C-H Activation Applied to the Synthesis of Syn-Aldol Products. *Org. Lett.* **2000**, 2, 4153-4156.

13. Davies, H. M. L., Beckwith, R. E. J., Antoulinakis, E. G. & Jin, Q. New Strategic Reactions for Organic Synthesis: Catalytic Asymmetric C-H Activation α to Oxygen as a Surrogate to the Aldol Reaction. *J. Org. Chem.* **2003**, 68, 6126-6132.

14. Davies, H. M. L., Yang, J. & Nikolai, J. Asymmetric C-H insertion of Rh(II) stabilized carbenoids into acetals: A C-H activation protocol as a Claisen condensation equivalent. *J. Organomet. Chem.* **2005**, 690, 6111-6124.

15. Davies, H. M. L. & Jin, Q. Double C-H activation strategy for the asymmetric synthesis of C2-symmetric anilines. *Org. Lett.* **2004**, 6, 1769-1772.

16. (a) Kohmura, Y. & Katsuki, T. Mn(salen)-catalyzed enantioselective C-H amination. *Tetrahedron Lett.* **2001**, 42, 3339-3342; (b) Dauban, P. & Dodd, R. H. Catalytic C-H amination with nitrenes in *Amino Group Chemistry* 55-92 (Wiley, **2008**).

ENANTIOSELECTIVE CLAISEN REARRANGEMENT

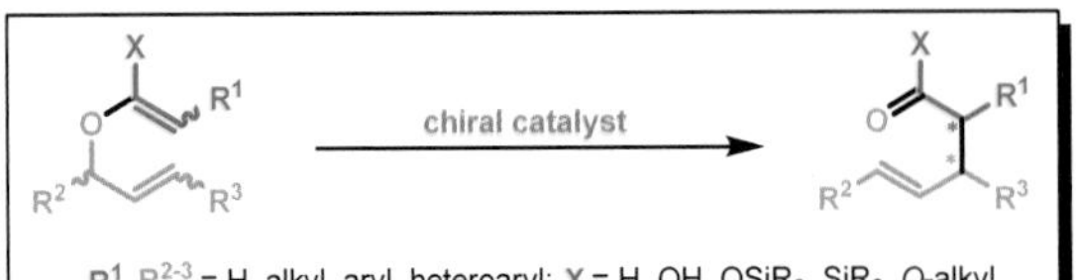

R[1], R[2-3] = H, alkyl, aryl, heteroaryl; X = H, OH, OSiR₃, SiR₃, O-alkyl

Introduction

The [3,3]-sigmatropic rearrangement of allyl vinyl ethers (Claisen rearrangement), discovered in 1912[1], is a powerful transformation for the stereocontrolled formation of C-C bonds that has been extensively used for the preparation of complex organic molecules and thoroughly reviewed.[2]

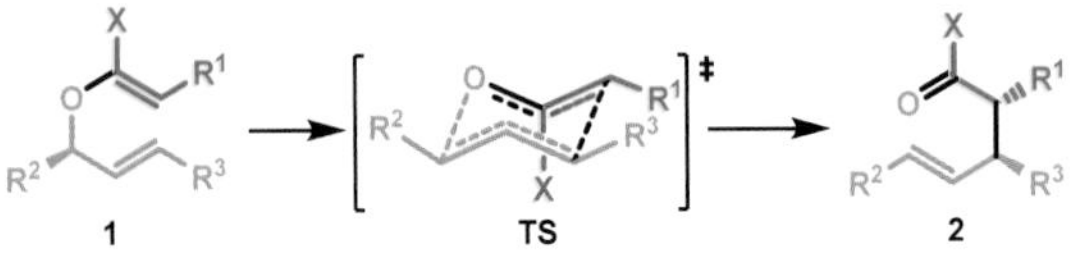

Scheme 1

The Claisen rearrangement is a symmetry-allowed pericyclic reaction that proceeds via a six-membered transition state (1→TS→2, Scheme 1).[3] Due to the general preference for a stereoelectronically favored chair-like transition state, the stereochemical information in the substrates is transferred to the products in a predictable manner. A non-concerted stepwise pathway is also possible (via e.g., ion pair or radical pair), depending on the substituents and reaction conditions. A number of variants (Figure 1) of this rearrangement have been developed, including the Carroll[4], Eschenmoser-Claisen[5], Johnson-Claisen[6], Ireland-Claisen[7], Bellus-ketene[8], Overman[9], anion-accelerated[10] and aza-Claisen[11] rearrangements.

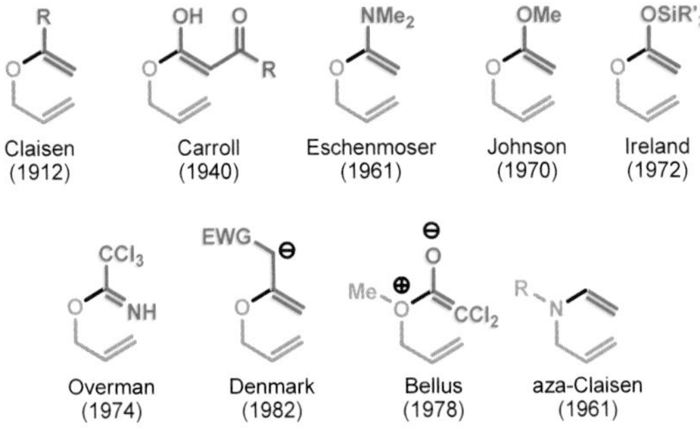

| Claisen (1912) | Carroll (1940) | Eschenmoser (1961) | Johnson (1970) | Ireland (1972) |

| Overman (1974) | Denmark (1982) | Bellus (1978) | aza-Claisen (1961) |

R = H, alkyl, aryl; R' = alkyl, aryl; R = alkyl, aryl

Figure 1

The Claisen rearrangement can proceed under thermal activation (usually at temperatures above 100 °C) or at lower temperatures in the presence of a catalyst[12] (e.g., a Lewis acid, protic acid or protic solvent).

Enantioselective variants have also been developed and widely applied in organic synthesis.[13,2a] High enantio-selectivity has been achieved in Claisen rearrangements by the use of chiral controllers or by employing stoichiomeric or catalytic amounts of a chiral reagent, often with recovery of

the chiral ligand for reuse. To date, the overwhelming majority of catalysts have been Lewis acidic metal complexes of Al, B, Mg and Cu. One drawback with these catalysts is the need for high loading due to strong product inhibition. Recently, thioureas and guanidinium ions have also been applied successfully as catalysts. The examples shown below illustrate the range of methodology now available.

Enantioselective Claisen Rearrangement Using Chiral Reagents or Chiral Catalysts

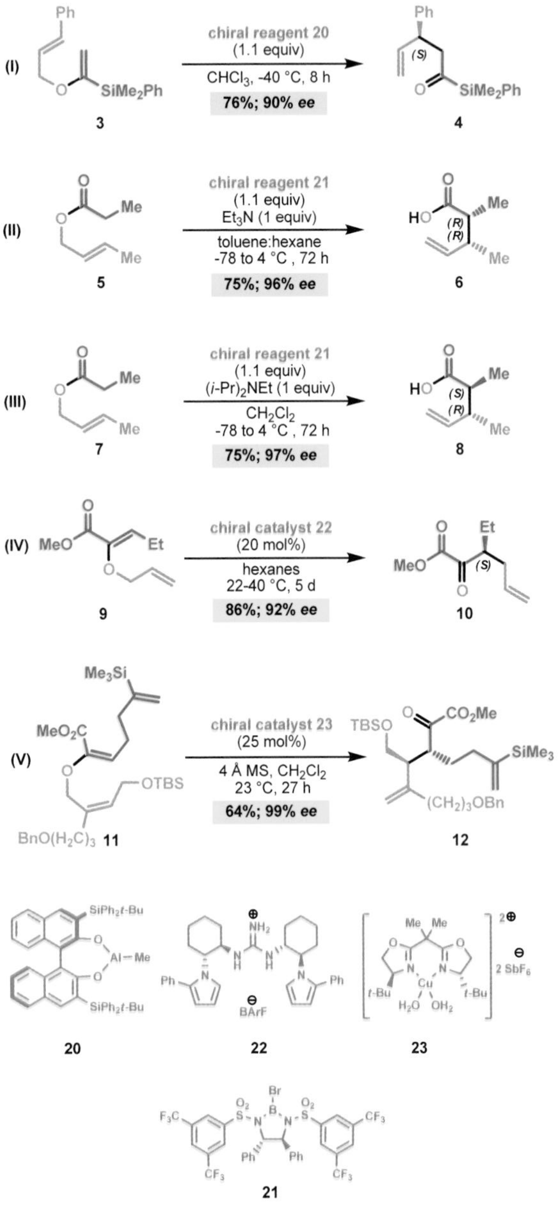

Reference key for equations: (I)[14]; (II-III)[15]; (IV)[16]; (V)[17]

Scheme 2

ENANTIOSELECTIVE CLAISEN REARRANGEMENT

Enantioselective Claisen Rearrangement Using Chiral Reagents or Chiral Catalysts (Continued)

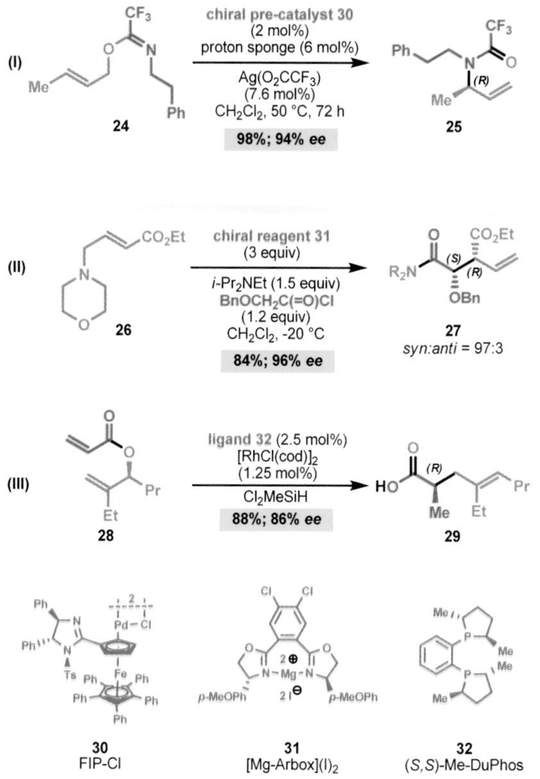

Reference key for equations: (I)[18]; (II)[19]; (III)[20]

Scheme 3

References

1. Claisen, L. Rearrangement of Phenol Allyl Ethers into C-Allylphenols. *Ber. Dtsch. Chem. Ges.* **1913**, 45, 3157-3166.
2. (a) Castro, A. M. M. Claisen Rearrangement over the Past Nine Decades. *Chem. Rev. (Washington, DC, U. S.)* **2004**, 104, 2939-3002; (b) *The Claisen Rearrangement: Methods and Applications* (eds. Hiersemann, M. & Nubbemeyer, U.) (Wiley-VCH, **2007**).
3. Rehbein, J. & Hiersemann, M. Mechanistic aspects of the aliphatic Claisen rearrangement in *The Claisen Rearrangement* 525-557 (Wiley-VCH, **2007**).
4. (a) Carroll, M. F. Addition of α,β-unsaturated alcohols to the active methylene group. I. The action of ethyl acetoacetate on linalo.ovrddot.ol and geraniol. *J. Chem. Soc.* **1940**, 704-706; (b) Carroll, M. F. Addition of β,γ-unsaturated alcohols to the active methylene group. II. The action of ethyl acetoacetate on cinnamyl alcohol and phenylvinylcarbinol. *J. Chem. Soc.* **1940**, 1266-1268; (c) Kürti, L. & Czakó, B. Carroll Rearrangement (Kimel-Cope Rearrangement) in *Strategic Applications of Named Reactions in Organic Synthesis* 76-77 (Academic Press/Elsevier, San Diego, **2005**); (d) Hatcher, M. A. & Posner, G. H. The Carroll rearrangement in *The Claisen Rearrangement* 397-430 (Wiley-VCH, **2007**); (e) Austeri, M., Buron, F., Constant, S., Lacour, J., Linder, D., Muller, J. & Tortoioli, S. Enantio- and regioselective CpRu-catalyzed Carroll rearrangement. *Pure Appl. Chem.* **2008**, 80, 967-977.
5. (a) Wick, A. E., Felix, D., Steen, K. & Eschenmoser, A. Claisen rearrangement of allyl and benzyl alcohols by N,N-dimethylacetamide acetals. *Helv. Chim. Acta* **1964**, 47, 2425-2429; (b) Kürti, L. & Czakó, B. Eschenmoser-Claisen Rearrangement in *Strategic Applications of*

Named Reactions in Organic Synthesis 156-157 (Academic Press/Elsevier, San Diego, **2005**); (c) Gradl, S. N. & Trauner, D. The Meerwein-Eschenmoser-Claisen rearrangement in *The Claisen Rearrangement* 367-396 (Wiley-VCH, **2007**).
6. (a) Johnson, W. S., Werthemann, L., Bartlett, W. R., Brocksom, T. J., Li, T.-T., Faulkner, D. J. & Petersen, M. R. Simple stereoselective version of the Claisen rearrangement leading to *trans*-trisubstituted olefinic bonds. Synthesis of squalene. *J. Am. Chem. Soc.* **1970**, 92, 741-743; (b) Kürti, L. & Czakó, B. Johnson-Claisen Rearrangement in *Strategic Applications of Named Reactions in Organic Synthesis* 226-227 (Academic Press/Elsevier, San Diego, **2005**); (c) Langlois, Y. Claisen-Johnson orthoester rearrangement in *The Claisen Rearrangement* 301-366 (Wiley-VCH, **2007**).
7. (a) Ireland, R. E. & Mueller, R. H. Claisen rearrangement of allyl esters. *J. Am. Chem. Soc.* **1972**, 94, 5897-5898; (b) Chai, Y., Hong, S.-p., Lindsay, H. A., McFarland, C. & McIntosh, M. C. New aspects of the Ireland and related Claisen rearrangements. *Tetrahedron* **2002**, 58, 2905-2928; (c) Kürti, L. & Czakó, B. Ireland-Claisen Rearrangement in *Strategic Applications of Named Reactions in Organic Synthesis* 90-91 (Academic Press/Elsevier, San Diego, **2005**); (d) McFarland, C. M. & McIntosh, M. C. The Ireland-Claisen rearrangement (1972-2004) in *The Claisen Rearrangement* 117-210 (Wiley-VCH, **2007**).
8. (a) Malherbe, R. & Bellus, D. A new type of Claisen rearrangement involving 1,3-dipolar intermediates. *Helv. Chim. Acta* **1978**, 61, 3096-3099; (b) Gonda, J. The Bellus-Claisen rearrangement. *Angew. Chem., Int. Ed.* **2004**, 43, 3516-3524.
9. (a) Overman, L. E. Thermal and mercuric ion catalyzed [3,3]-sigmatropic rearrangement of allylic trichloroacetimidates. 1,3 Transposition of alcohol and amine functions. *J. Am. Chem. Soc.* **1974**, 96, 597-599; (b) Overman, L. E. & Carpenter, N. E. The allylic trihaloacetimidate rearrangement. *Organic Reactions (Hoboken, NJ, United States)* **2005**, 66, 1-107; (c) Kürti, L. & Czakó, B. Overman Rearrangement in *Strategic Applications of Named Reactions in Organic Synthesis* 322-323 (Academic Press/Elsevier, San Diego, **2005**).
10. Denmark, S. E. & Harmata, M. A. Carbanion-accelerated Claisen rearrangements. *J. Am. Chem. Soc.* **1982**, 104, 4972-4974.
11. (a) Hill, R. K. & Gilman, N. W. Nitrogen analog of the Claisen rearrangement. *Tetrahedron Lett.* **1967**, 1421-1423; (b) Kürti, L. & Czakó, B. Aza-Claisen Rearrangement in *Strategic Applications of Named Reactions in Organic Synthesis* 20-21 (Academic Press/Elsevier, San Diego, **2005**); (c) Majumdar, K. C., Bhattacharyya, T., Chattopadhyay, B. & Sinha, B. Recent advances in the aza-Claisen rearrangement. *Synthesis* **2009**, 2117-2142.
12. Majumdar, K. C., Alam, S. & Chattopadhyay, B. Catalysis of the Claisen rearrangement. *Tetrahedron* **2007**, 64, 597-643.
13. (a) Enders, D., Knopp, M. & Schiffers, R. Asymmetric [3.3]-sigmatropic rearrangements in organic synthesis. *Tetrahedron Asymmetry* **1996**, 7, 1847-1882; (b) Ito, H. & Taguchi, T. Asymmetric Claisen rearrangement. *Chem. Soc. Rev.* **1999**, 28, 43-50; (c) Fleming, M., Rigby, J. H., Yoon, T. P. & MacMillan, D. W. C. Enantioselective Claisen rearrangements: Development of a first generation asymmetric acyl-Claisen reaction. *Chemtracts* **2001**, 14, 620-624; (d) Hiersemann, M. & Abraham, L. Catalysis of the Claisen rearrangement of aliphatic allyl vinyl ethers. *Eur. J. Org. Chem.* **2002**, 1461-1471; (e) Nubbemeyer, U. Recent advances in asymmetric [3,3]-sigmatropic rearrangements. *Synthesis* **2003**, 961-1008.
14. Maruoka, K., Banno, H. & Yamamoto, H. Asymmetric Claisen rearrangement catalyzed by chiral organoaluminum reagent. *J. Am. Chem. Soc.* **1990**, 112, 7791-7793.
15. Corey, E. J. & Lee, D. H. Highly enantioselective and diastereoselective Ireland-Claisen rearrangement of achiral allylic esters. *J. Am. Chem. Soc.* **1991**, 113, 4026-4028.
16. Uyeda, C. & Jacobsen, E. N. Enantioselective Claisen Rearrangements with a Hydrogen-Bond Donor Catalyst. *J. Am. Chem. Soc.* **2008**, 130, 9228-9229.
17. Pollex, A. & Hiersemann, M. Catalytic Asymmetric Claisen Rearrangement in Natural Product Synthesis: Synthetic Studies toward (-)-Xeniolide F. *Org. Lett.* **2005**, 7, 5705-5708.
18. Xin, Z.-q., Fischer, D. F. & Peters, R. Catalytic asymmetric formation of secondary allylic amines by aza-Claisen rearrangement of trifluoroacetimidates. *Synlett* **2008**, 1495-1499.
19. Yoon, T. P. & MacMillan, D. W. Enantioselective Claisen rearrangements: development of a first generation asymmetric acyl-Claisen reaction. *J. Am. Chem. Soc.* **2001**, 123, 2911-2912.
20. Fuller, N. O. & Morken, J. P. Studies on the Synthesis of the Inostamycin Natural Products: A Reductive Aldol/Reductive Claisen Approach to the C10-C24 Ketone Fragment. *Org. Lett.* **2005**, 7, 4867-4869.

STEREOSELECTIVE COPE REARRANGEMENT

R^1; R^{3-4} = H, alkyl, aryl, heteroaryl; R^2 = OH, NH_2, NHR

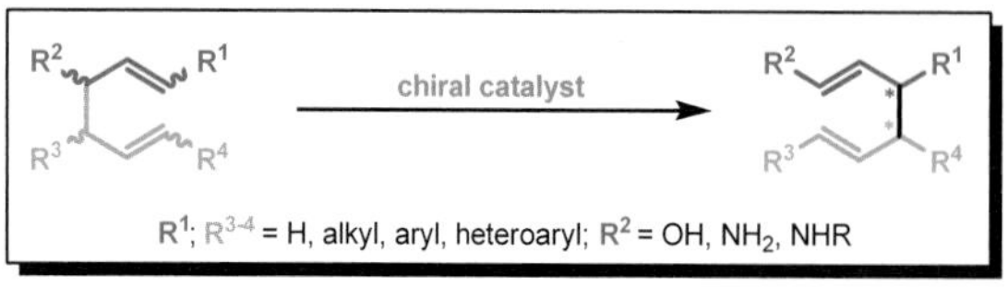

Introduction

The all-carbon [3,3]-sigmatropic rearrangement of 1,5-hexadienes (Cope rearrangement[1], Scheme 1) is mechanistically related to the Claisen rearrangement (see pages 110-111). It has been extensively exploited in the synthesis of complex molecules.[2] The Cope rearrangement is a symmetry-allowed pericyclic reaction that proceeds via that six-membered, chair-like transition state (1→TS→2, Scheme 1) in which the steric repulsion is minimal.[3]

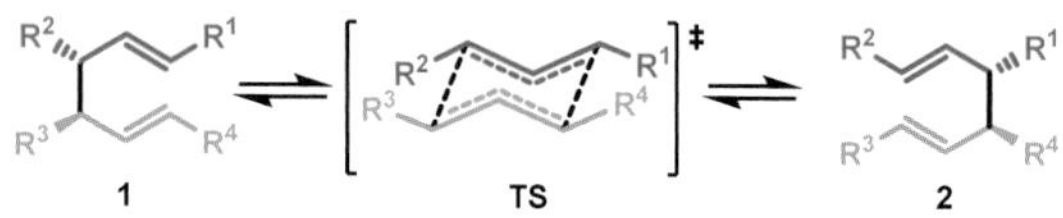

Scheme 1

Because the Cope rearrangement is reversible, synthetic utility requires a favorable equilibrium. In the variants of the Cope rearrangement shown in Figure 1, driving force arises from relief of ring-strain, cleavage of a weak N-N, or N-O bond or ion-stabilization.[4a,2c,2d,4b]

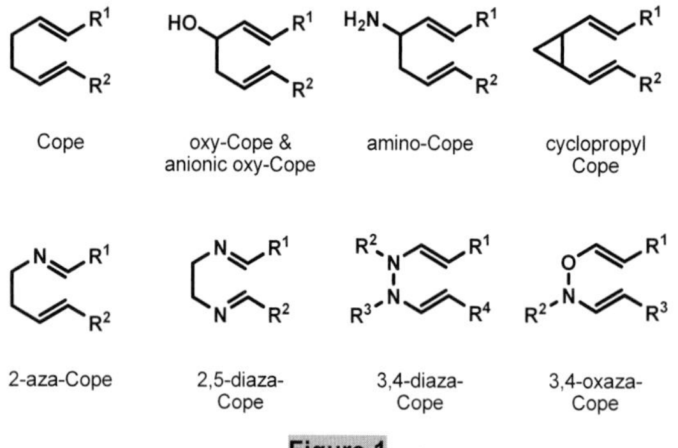

Cope   oxy-Cope & anionic oxy-Cope   amino-Cope   cyclopropyl Cope

2-aza-Cope   2,5-diaza-Cope   3,4-diaza-Cope   3,4-oxaza-Cope

Figure 1

There are many examples of stereoselective Cope rearrangement[5] in the total synthesis of natural products. Often the Cope rearrangement is part of a two or three-step sequence of reactions to ensure that it is irreversible. Most synthetic examples utilize starting materials with one or two chiral centers on the six-carbon path. Some examples involve a chiral controller group. Catalytic enantioselective Cope rearrangements are known only as part of a tandem process (see Example I, Scheme 2).[6]

Stereoselective Cope Rearrangement Utilizing Chiral Catalysts or Existing Chiral Centers

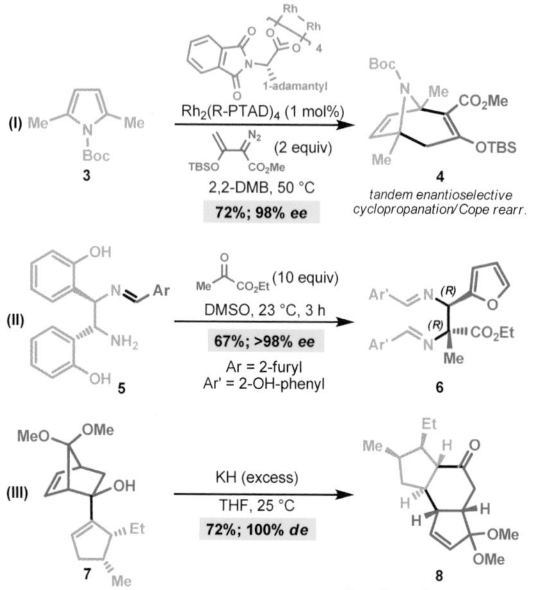

(I) 72%; 98% ee — tandem enantioselective cyclopropanation/Cope rearr.

(II) 67%; >98% ee — Ar = 2-furyl; Ar' = 2-OH-phenyl

(III) 72%; 100% de

Reference key for equations: (I)[6]; (II)[7]; (III)[8]

Scheme 2

References

1. (a) Cope, A. C. & Hardy, E. M. Introduction of substituted vinyl groups. V. A rearrangement involving the migration of an allyl group in a three-carbon system. J. Am. Chem. Soc. **1940**, 62, 441-444; (b) Cope, A. C., Hofmann, C. M. & Hardy, E. M. Rearrangement of allyl groups in three-carbon systems. II. J. Am. Chem. Soc. **1941**, 63, 1852-1857.

2. (a) Rhoads, S. J. & Raulins, N. R. Claisen and Cope rearrangements. Org. React. **1975**, 22, 1-252; (b) Lutz, R. P. Catalysis of the Cope and Claisen rearrangements. Chem. Rev. **1984**, 84, 205-247; (c) Hudlicky, T., Fan, R., Reed, J. W. & Gadamasetti, K. G. Divinylcyclopropane-cycloheptadiene rearrangement. Org. React. **1992**, 41, 1-133; (d) Wilson, S. R. Anion-assisted sigmatropic rearrangements. Org. React. **1993**, 43, 93-250; (e) Lukyanov, S. M. & Koblik, A. V. Rearrangements of dienes and polyenes. Chem. Dienes Polyenes **2000**, 2, 739-884; (f) Mullins, R. J. & McCracken, K. W. Cope and related rearrangements in Name Reactions for Homologations 88-135 (John Wiley & Sons, **2009**).

3. Staroverov, V. N. & Davidson, E. R. The Cope rearrangement in theoretical retrospect. THEOCHEM **2001**, 573, 81-89.

4. (a) Blechert, S. The hetero-Cope rearrangement in organic synthesis. Synthesis **1989**, 71-82; (b) Davies, H. M. L. Tandem cyclopropanation/Cope rearrangement: a general method for the construction of seven-membered rings. Tetrahedron **1993**, 49, 5203-5223.

5. (a) Enders, D., Knopp, M. & Schiffers, R. Asymmetric [3,3]-sigmatropic rearrangements in organic synthesis. Tetrahedron Asymmetry **1996**, 7, 1847-1882; (b) Nubbemeyer, U. Recent advances in asymmetric [3,3]-sigmatropic rearrangements. Synthesis **2003**, 961-1008.

6. Reddy, R. P. & Davies, H. M. L. Asymmetric Synthesis of Tropanes by Rhodium-Catalyzed [4 + 3] Cycloaddition. J. Am. Chem. Soc. **2007**, 129, 10312-10313.

7. Kim, H. & Chin, J. Stereospecific synthesis of α-substituted syn-α,β-diamino acids by the diaza-Cope rearrangement. Org. Lett. **2009**, 11, 5258-5260.

8. Paquette, L. A., Romine, J. L. & Lin, H. S. Diastereoselective π-facially controlled nucleophilic additions of chiral vinylorganometallics to chiral α,γ-unsaturated ketones. 2. A practical method for stereocontrolled elaboration of the decahydro-as-indacene subunit of ikarugamicin. Tetrahedron Lett. **1987**, 28, 31-34.

STEREOSELECTIVE [2,3]-REARRANGEMENT

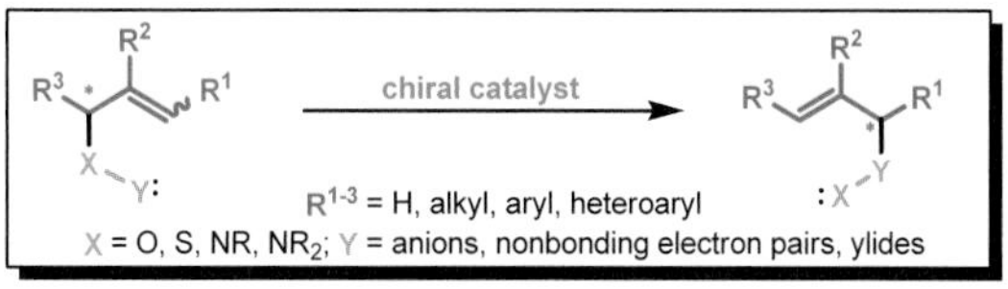

Introduction

[2,3]-Sigmatropic rearrangements (Scheme 1) are synthetically useful since they can lead to: (a) regiospecific carbon-carbon or carbon-heteroatom formation; (b) generation of specific olefin geometry; (c) generation of vicinal chiral centers with excellent diastereoselectivity and predictable transfer of chirality.[1] Mechanistically, [2,3]-rearrangements follow a concerted, 6-electron suprafacial sigmatropic pathway via a 5-membered transition state. The stereochemical outcome of these rearrangements can be rationalized by evaluating the steric repulsion ($A_{1,2}$ and $A_{1,3}$ allylic strains) between the substituents in the two alternative transition state geometries, **TS1** and **TS2** (Scheme 1) to estimate the lowest energy transition state.

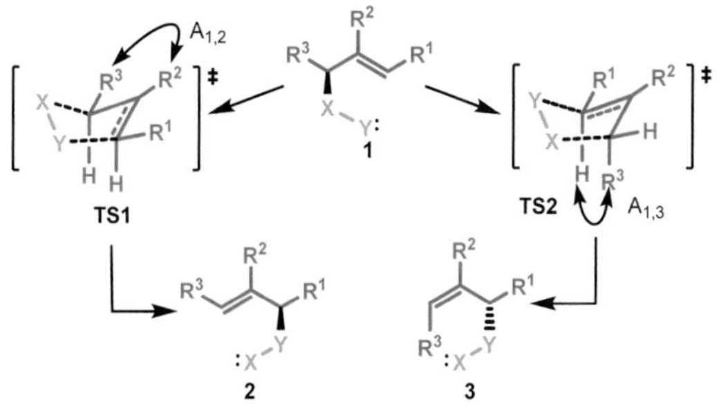

Scheme 1

Five useful classes of [2,3]-sigmatropic rearrangement (Wittig[2], aza-Wittig[3], Meisenheimer[4], Mislow-Evans[5] and Sommelet-Hauser[6]) are illustrated in Figure 1.

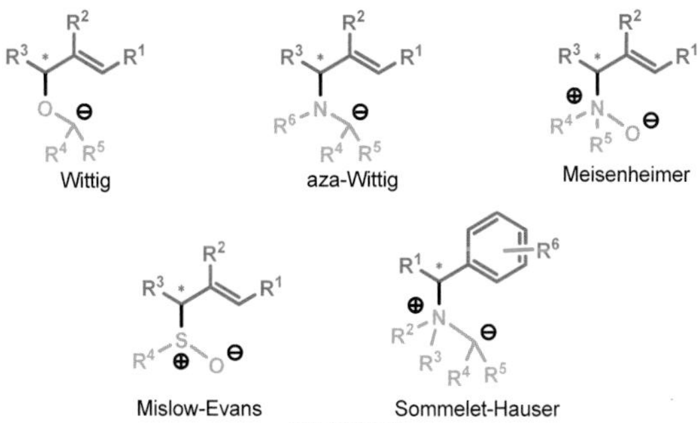

Figure 1

Many examples of stereoselective [2,3]-sigmatropic rearrangement are described in the reviews cited.[1]

Enantioselective variants of the [2,3]-Wittig, aza-Wittig as well as the Sommelet-Hauser rearrangement have been developed during the past decade.[1c,1e] A few highly efficient examples are shown in Scheme 2.

Stereoselective [2,3]-Rearrangements Utilizing Chiral Catalysts/Reagents and/or Existing Chiral Centers

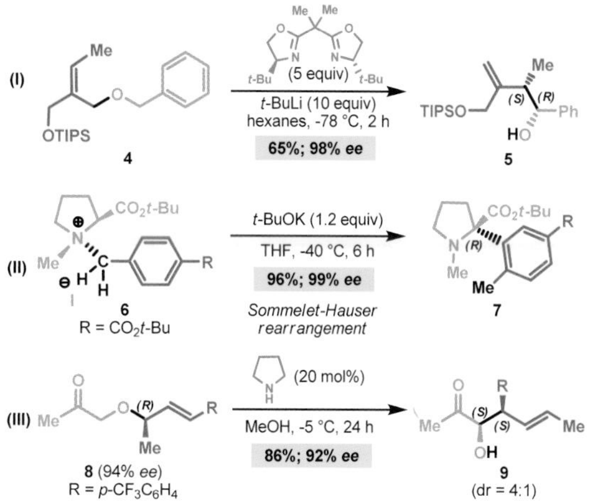

Reference key for equations: (I)[7]; (II)[8]; (III)[9]

Scheme 2

References

1. (a) Nakai, T. & Mikami, K. The [2,3]-Wittig rearrangement. *Org. React.* **1994**, 46, 105-209; (b) Vogel, C. The aza-Wittig rearrangement. *Synthesis* **1997**, 497-505; (c) Nakai, T. & Tomooka, K. Asymmetric [2,3]-Wittig rearrangement as a general tool for asymmetric synthesis. *Pure Appl. Chem.* **1997**, 69, 595-600; (d) McGowan, G. Applications of the [2,3]-Wittig rearrangement to total synthesis. *Aust. J. Chem.* **2002**, 55, 799; (e) Hodgson, D. M., Tomooka, K. & Gras, E. Enantioselective synthesis by lithiation adjacent to oxygen and subsequent rearrangement. *Top. Organomet. Chem.* **2003**, 5, 217-250; (f) Tomooka, K. Rearrangements of organolithium compounds. *Chem. Organolithium Compd.* **2004**, 2, 749-828; (g) MacNeil, S. Product subclass 19: sp³-hybridized α-lithio ethers and O-carbamates. *Sci. Synth.* **2006**, 8a, 637-660; (h) Sweeney, J. B. Sigmatropic rearrangements of 'onium' ylids. *Chem. Soc. Rev.* **2009**, 38, 1027-1038.
2. Whittig, G., Doser, H. & Lorenz, I. Isomerization of metallized fluorene ethers. *Liebigs Ann. Chem.* **1949**, 562, 192-205.
3. Durst, T., Van den Elzen, R. & LeBelle, M. J. Base-induced ring enlargements of 1-benzyl- and 1-allyl-2-azetidinones. *J. Am. Chem. Soc.* **1972**, 94, 9261-9263.
4. Meisenheimer, J. A peculiar rearrangement of methylallylaniline N-oxide. *Berichte der Deutschen Chemischen Gesellschaft [Abteilung] B: Abhandlungen* **1919**, 52B, 1667-1677.
5. (a) Bickart, P., Carson, F. W., Jacobus, J., Miller, E. G. & Mislow, K. Thermal racemization of allylic sulfoxides and interconversion of allylic sulfoxides and sulfenates. Mechanism and stereochemistry. *J. Am. Chem. Soc.* **1968**, 90, 4869-4876; (b) Evans, D. A., Andrews, G. C. & Sims, C. L. Reversible 1,3 transposition of sulfoxide and alcohol functions. Potential synthetic utility. *J. Am. Chem. Soc.* **1971**, 93, 4956-4957.
6. Kantor, S. W. & Hauser, C. R. Isomerizations of carbanions. II. Rearrangements of benzyltrimethylammonium ion and related quaternary ammonium ions by sodium amide involving migration into the ring. *J. Am. Chem. Soc.* **1951**, 73, 4122-4131.
7. Hirokawa, Y., Kitamura, M. & Maezaki, N. Highly enantioselective [2,3]-Wittig rearrangement of functionalized allyl benzyl ethers: a novel approach to lignan synthesis. *Tetrahedron Asymmetry* **2008**, 19, 1167-1170.
8. Tayama, E. & Kimura, H. Asymmetric Sommelet-Hauser rearrangement of N-benzylic ammonium salts. *Angew. Chem., Int. Ed.* **2007**, 46, 8869-8871.
9. McNally, A., Evans, B. & Gaunt, M. J. Organocatalytic sigmatropic reactions: development of a [2,3] Wittig rearrangement through secondary amine catalysis. *Angew. Chem., Int. Ed.* **2006**, 45, 2116-2119.

STEREOSELECTIVE ALL-CARBON ENE-REACTION

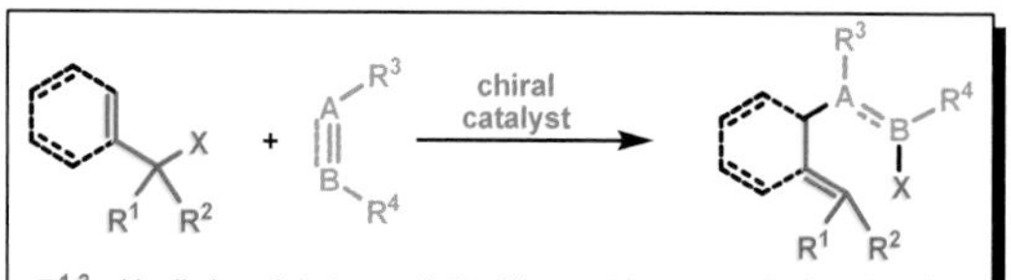

R^{1-2} = H, alkyl, aryl, heteroaryl; X = H or metal; A,B = substituted carbon or one the following combinations: C=S, C=N, N=O; R^{3-4} = usually EWG

Introduction

The ene reaction, first described by K. Alder in the 1940s[1], is formally the addition of alkenes to double or triple bonds with concomitant formation of a new C-C bond and the 1,5-transfer of an atom (X, typically H or a metal) in the allylic position of the alkene.[2] The fragment that contains the double or triple bond, called the enophile, usually bears an electron-withdrawing substituent (EWG). The ene reaction has a large number of variants (e.g., carbonyl-ene, imino-ene, metallo-ene)[3] in terms of the enophile used, however, in this section we will focus on the all-carbon version that utilizes alkenes or alkynes as enophiles. Mechanistically the ene reaction is related to the Diels-Alder reaction and it is a thermally-allowed, six-electron process that proceeds via a cyclic six-membered transition state (1→TS→2, Scheme 1). The Lewis-acid promoted ene reaction may be concerted or stepwise, depending on the nature of substituents on both of the reactants as well as on the reaction conditions.

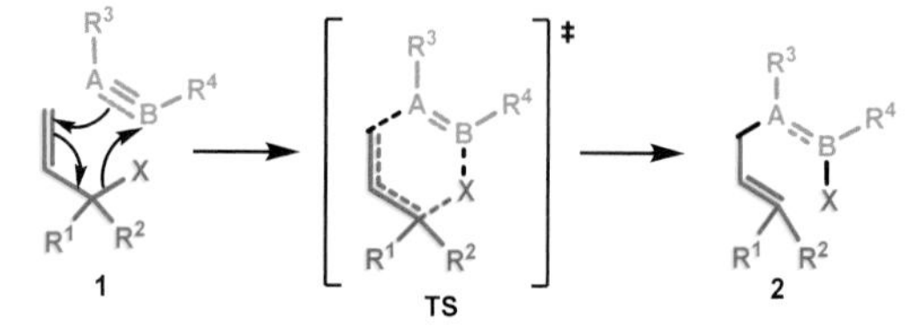

Scheme 1

Despite its similarity with the Diels-Alder reaction, the all-carbon ene reaction has been much less studied and used in organic synthesis as it requires somewhat harsher reaction conditions (>200 °C for non-activated or uncatalyzed systems). Intermolecular ene reactions tend to have higher activation energies than the corresponding Diels-Alder reactions, but the electronic nature of the reactants follow a similar trend: the enophile must be electron-deficient while the ene-component is usually electron-rich. However, intramolecular ene reactions occur more readily and they can be highly regio- and diastereoselective. Regioselectivity is determined by the following two factors: (a) the steric accessibility of the X group (usually an H atom) on the ene component, so primary and secondary hydrogens react significantly faster than tertiary ones and (b) the ability to adopt the reactive conformation.

Enantioselective versions of the the transition metal-promoted ene reaction, using a chiral catalyst or a chiral reagent, have been developed which follow a metallocycle pathway. Three highly efficient examples are shown in Scheme 2. Products 7 and 9 (examples II and III) were utilized in the total syntheses of (+)-pilocarpine and (-)-platensimycin, respectively.

Stereoselective All-Carbon Ene Reaction Using a Chiral Catalyst or a Chiral Reagent

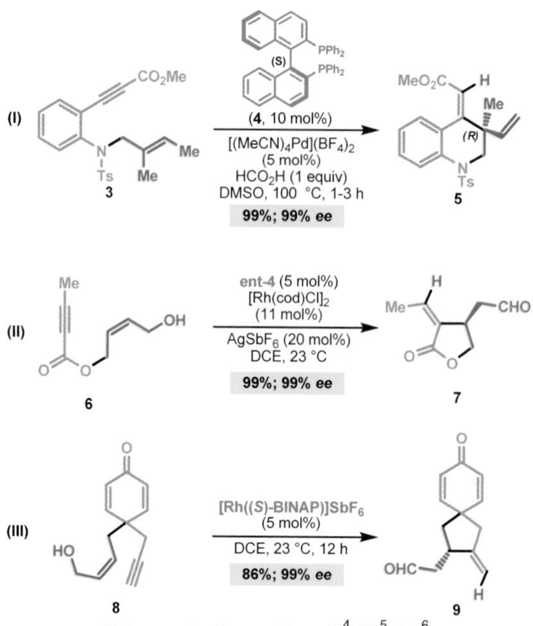

Reference key for equations: (I)[4]; (II)[5]; (III)[6]

Scheme 2

References

1. Alder, K., Pascher, F. & Schmitz, A. Substituting additions. I. Addition of maleic anhydride and azodicarboxylic esters to singly unsaturated hydrocarbons. Substitution processes in the allyl position. *Berichte der Deutschen Chemischen Gesellschaft [Abteilung] B: Abhandlungen* **1943**, 76B, 27-53.

2. (a) Snider, B. B. Ene Reactions with Alkenes as Enophiles in *Comp. Org. Synth.* (eds. Trost, B. M. & Fleming, I.), 1-27 (Pergamon Press, Oxford, **1991**); (b) Mikami, K. & Shimizu, M. Asymmetric ene reactions in organic synthesis. *Chem. Rev.* **1992**, 92, 1021-1050; (c) Mikami, K. & Terada, M. Ene-type reactions. *Compr. Asymmetric Catal. I-III* **1999**, 3, 1143-1174; (d) Dias, L. C. Chiral Lewis acid catalyzed ene reactions. *Curr. Org. Chem.* **2000**, 4, 305-342; (e) Michelet, V., Toullec, P. Y. & Genet, J.-P. Cycloisomerization of 1,n-enynes: challenging metal-catalyzed rearrangements and mechanistic insights. *Angew. Chem., Int. Ed.* **2008**, 47, 4268-4315; (f) Curran, T. T. Alder-ene reaction. *Name Reactions for Homologations* **2009**, 2-32; (g) Chiu, P. & Lam, S. K. Ene reactions. *Sci. Synth.* **2010**, 47b, 737-753.

3. (a) Oppolzer, W. Metallo-ene Reactions in *Comp. Org. Synth.* (eds. Trost, B. M. & Fleming, I.), 29-61 (Pergamon Press, Oxford, **1991**); (b) Borzilleri, R. M. & Weinreb, S. M. Imino ene reactions in organic synthesis. *Synthesis* **1995**, 347-360; (c) Adam, W. & Krebs, O. The Nitroso Ene Reaction: A Regioselective and Stereoselective Allylic Nitrogen Functionalization of Mechanistic Delight and Synthetic Potential. *Chem. Rev.* **2003**, 103, 4131-4146; (d) Clarke, M. L. & France, M. B. The carbonyl ene reaction. *Tetrahedron* **2008**, 64, 9003-9031.

4. Hatano, M. & Mikami, K. Highly Enantioselective Quinoline Synthesis via Ene-type Cyclization Catalyzed by a Cationic BINAP-Palladium(II) Complex. *J. Am. Chem. Soc.* **2003**, 125, 4704-4705.

5. Lei, A., He, M. & Zhang, X. Highly Enantioselective Syntheses of Functionalized α-Methylene-γ-butyrolactones via Rh(I)-catalyzed Intramolecular Alder Ene Reaction: Application to Formal Synthesis of (+)-Pilocarpine. *J. Am. Chem. Soc.* **2002**, 124, 8198-8199.

6. Nicolaou, K. C., Li, A., Ellery, S. P. & Edmonds, D. J. Rhodium-Catalyzed Asymmetric Enyne Cycloisomerization of Terminal Alkynes and Formal Total Synthesis of (-)-Platensimycin. *Angew. Chem., Int. Ed.* **2009**, 48, 6293-6295.

ENANTIOSELECTIVE STETTER REACTION

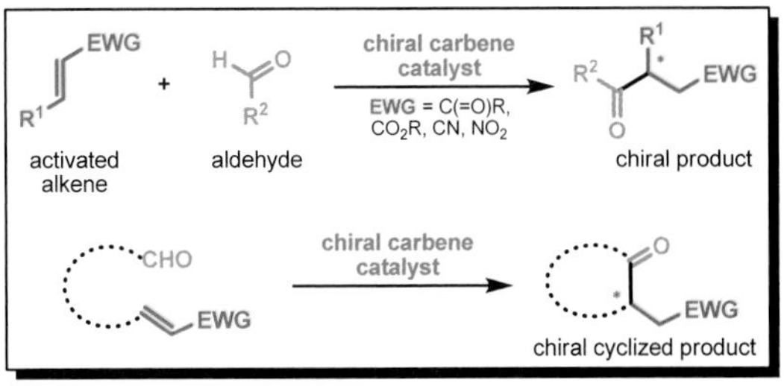

Introduction

N-substituted thiazoles or imidazoles when deprotonated at C(2) generate a dipolar species, e.g., **1** in Scheme 1, which can add to the CHO group of an aldehyde to initiate the pathway shown in Scheme 1.[1] In that sequence the CHO carbon is converted to a nucleophilic equivalent that can undergo 1,2-addition to C=O or 1,4-addition to C=C-C=O, as displayed in Scheme 1. That type of catalysis is directly analogous to the mode of action of the coenzyme thiamine diphosphate in biochemical reactions. Synthetic applications of such azolium dipolar ions to C-C bond formations, initially developed by Stetter,[1] have since been extended to enantioselective processes for inter- and intramolecular C-C bond formation. Four examples of enantioselective intermolecular C-C bond formation appear in Scheme 2.

Enantioselective cyclizations using catalysis by azolium dipolar ion activation can lead to the formation of 5-membered rings (Scheme 3) or 6-membered rings (Scheme 4) via conjugate addition pathways. Cyclization by way of nucleophilic 1,2-addition to C=O is illustrated in Scheme 5.

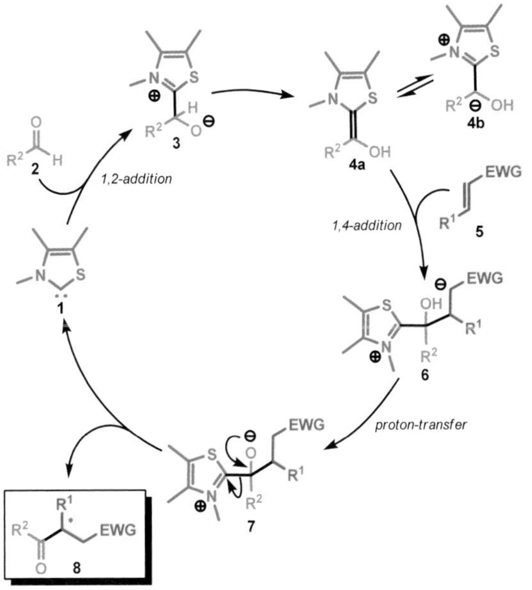

I. Catalytic Enantioselective Intermolecular Stetter Reaction

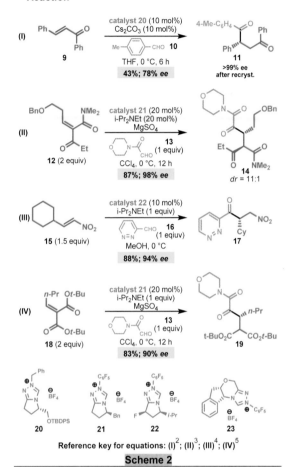

Reference key for equations: (I)[2]; (II)[3]; (III)[4]; (IV)[5]

Scheme 2

II. Catalytic Enantioselective Intramolecular Stetter Reaction: Formation of Five-Membered Rings

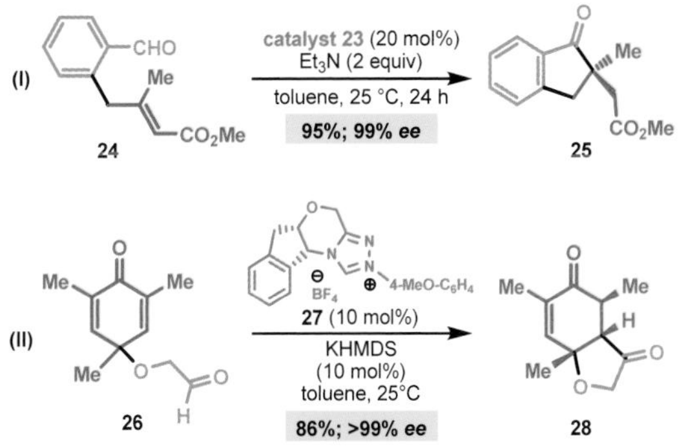

Reference key for equations: (I)[6]; (II)[7]

Scheme 3

Scheme 1

ENANTIOSELECTIVE STETTER REACTION

III. Catalytic Enantioselective Intramolecular Stetter Reaction: Formation of Six-Membered Rings

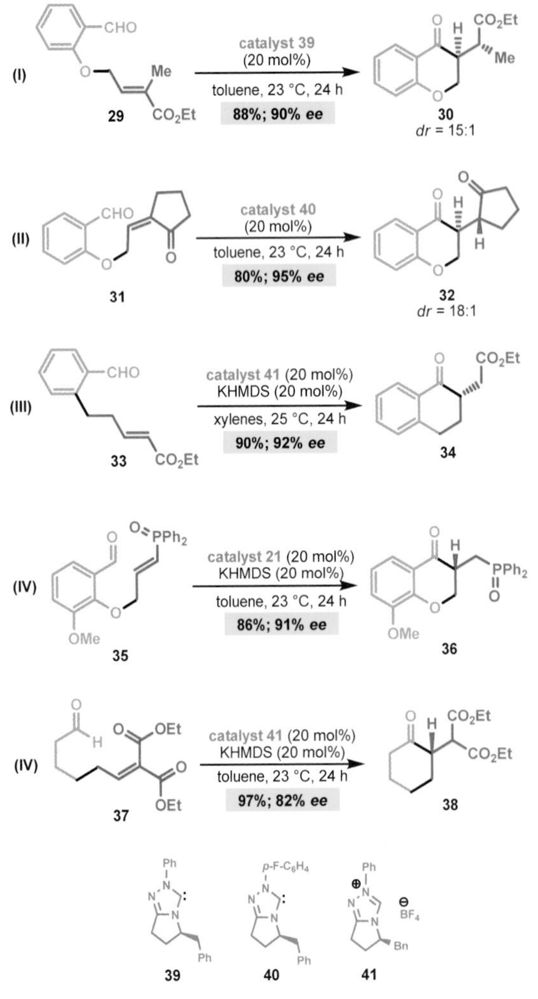

(I) 29 → 30
catalyst 39 (20 mol%), toluene, 23 °C, 24 h
88%; 90% ee
dr = 15:1

(II) 31 → 32
catalyst 40 (20 mol%), toluene, 23 °C, 24 h
80%; 95% ee
dr = 18:1

(III) 33 → 34
catalyst 41 (20 mol%), KHMDS (20 mol%), xylenes, 25 °C, 24 h
90%; 92% ee

(IV) 35 → 36
catalyst 21 (20 mol%), KHMDS (20 mol%), toluene, 23 °C, 24 h
86%; 91% ee

(IV) 37 → 38
catalyst 41 (20 mol%), KHMDS (20 mol%), toluene, 23 °C, 24 h
97%; 82% ee

39   40   41

Reference key for equations: (I-II)[8]; (III)[9]; (IV)[10]; (V)[1f]

Scheme 4

IV. Catalytic Enantioselective Intramolecular Cross-Benzoin Condensation

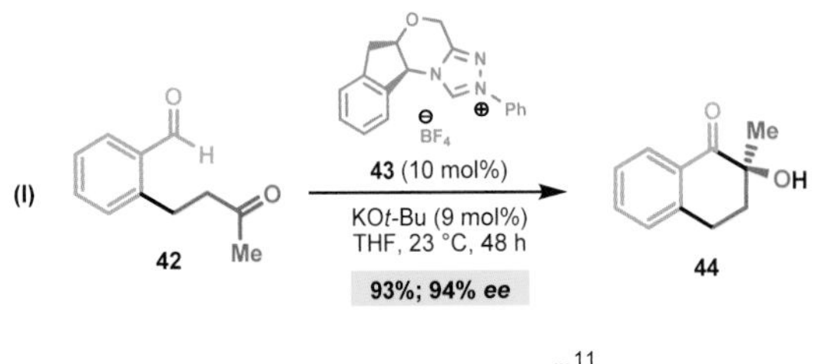

(I) 42 → 44
43 (10 mol%), KOt-Bu (9 mol%), THF, 23 °C, 48 h
93%; 94% ee

Reference key for equation: (I)[11]

Scheme 5

Catalytic Enantioselective Intramolecular Cross-Benzoin Condensation (Continued)

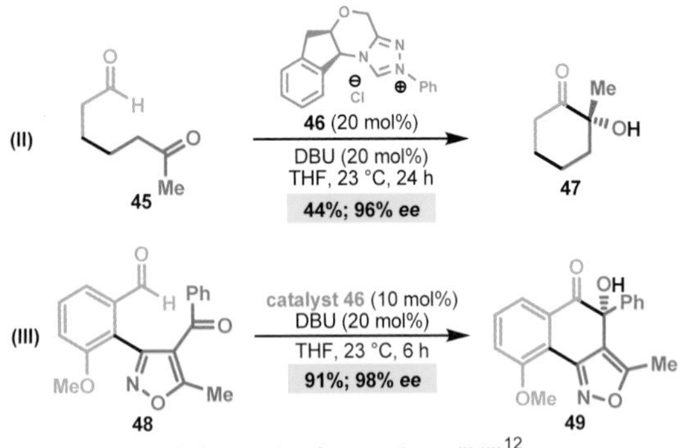

(II) 45 → 47
46 (20 mol%), DBU (20 mol%), THF, 23 °C, 24 h
44%; 96% ee

(III) 48 → 49
catalyst 46 (10 mol%), DBU (20 mol%), THF, 23 °C, 6 h
91%; 98% ee

Reference key for equations: (II-III)[12]

Scheme 5

References

1. (a) Christmann, M. New developments in the asymmetric Stetter reaction. *Angew. Chem., Int. Ed.* **2005**, 44, 2632-2634; (b) Dove, A. P., Pratt, R. C., Lohmeijet, B. G. G., Li, H., Hagberg, E. C., Waymouth, R. M. & Hedrick, J. L. N-Heterocyclic carbenes as Organic Catalysts in *N-Heterocyclic Carbenes in Synthesis* (ed. Nolan, S. P.), (Wiley-VCH, Weinheim, **2006**); (c) Webber, P. & Krische, M. J. The catalytic asymmetric intramolecular Stetter reaction. *Chemtracts* **2006**, 19, 262-269; (d) Enders, D., Niemeier, O. & Henseler, A. Organocatalysis by N-Heterocyclic Carbenes. *Chem. Rev. (Washington, DC, U. S.)* **2007**, 107, 5606-5655; (e) Zeitler, K. N-heterocyclic carbenes: organocatalysts displaying diverse modes of action. *Ernst Schering Foundation Symposium Proceedings* **2008**, 183-206; (f) Read de Alaniz, J. & Rovis, T. The catalytic asymmetric intramolecular Stetter reaction. *Synlett* **2009**, 1189-1207; (g) Moore, J. L. & Rovis, T. Carbene Catalysts in *Asymmetric Organocatalysis* (ed. List, B.), 77-144 (Springer-Verlag, Berlin, Heidelberg, **2010**).
2. Enders, D., Han, J. & Henseler, A. Asymmetric intermolecular Stetter reactions catalyzed by a novel triazolium derived N-heterocyclic carbene. *Chem. Commun. (Cambridge, U. K.)* **2008**, 3989-3991.
3. Liu, Q. & Rovis, T. Enantio- and Diastereoselective Intermolecular Stetter Reaction of Glyoxamide and Alkylidene Ketoamides. *Org. Lett.* **2009**, 11, 2856-2859.
4. DiRocco, D. A., Oberg, K. M., Dalton, D. M. & Rovis, T. Catalytic Asymmetric Intermolecular Stetter Reaction of Heterocyclic Aldehydes with Nitroalkenes: Backbone Fluorination Improves Selectivity. *J. Am. Chem. Soc.* **2009**, 131, 10872-10874.
5. Liu, Q., Perreault, S. & Rovis, T. Catalytic Asymmetric Intermolecular Stetter Reaction of Glyoxamides with Alkylidenemalonates. *J. Am. Chem. Soc.* **2008**, 130, 14066-14067.
6. Kerr, M. S. & Rovis, T. Enantioselective synthesis of quaternary stereocenters via a catalytic asymmetric Stetter reaction. *J. Am. Chem. Soc.* **2004**, 126, 8876-8877.
7. (a) Liu, Q. & Rovis, T. Asymmetric Synthesis of Hydrobenzofuranones via Desymmetrization of Cyclohexadienones Using the Intramolecular Stetter Reaction. *J. Am. Chem. Soc.* **2006**, 128, 2552-2553; (b) Liu, Q. & Rovis, T. Enantioselective Synthesis of Hydrobenzofuranones Using an Asymmetric Desymmetrizing Intramolecular Stetter Reaction of Cyclohexadienones. *Org. Process Res. Dev.* **2007**, 11, 598-604.
8. Read de Alaniz, J. & Rovis, T. A Highly Enantio- and Diastereoselective Catalytic Intramolecular Stetter Reaction. *J. Am. Chem. Soc.* **2005**, 127, 6284-6289.
9. Kerr, M. S., Read de Alaniz, J. & Rovis, T. A Highly Enantioselective Catalytic Intramolecular Stetter Reaction. *J. Am. Chem. Soc.* **2002**, 124, 10298-10299.
10. Cullen, S. C. & Rovis, T. Catalytic Asymmetric Stetter Reaction Onto Vinylphosphine Oxides and Vinylphosphonates. *Org. Lett.* **2008**, 10, 3141-3144.
11. Enders, D., Niemeier, O. & Balensiefer, T. Asymmetric intramolecular crossed-benzoin reactions by N-heterocyclic carbene catalysis. *Angew. Chem., Int. Ed.* **2006**, 45, 1463-1467.
12. Takikawa, H., Hachisu, Y., Bode, J. W. & Suzuki, K. Catalytic enantioselective crossed aldehyde-ketone benzoin cyclization. *Angew. Chem., Int. Ed.* **2006**, 45, 3492-3494.

BAYLIS-HILLMAN AND RELATED REACTIONS

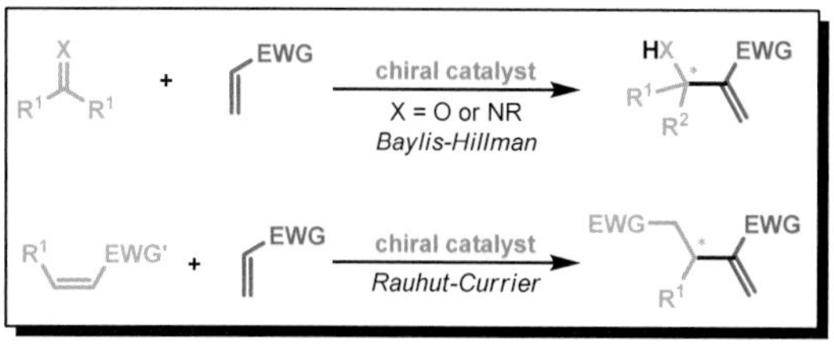

Introduction

α,β-Unsaturated carbonyl compounds generally are not subject to direct α-deprotonation and therefore do not undergo direct enolate addition to carbonyl or Michael acceptors. However, β-substituted enolates may be generated by conjugate addition of R_3P or R_3N, and these intermediates can behave as simple enolates in aldol or Michael addition. Subsequent β-elimination of R_3P or R_3N regenerates the α,β-double bond to form a Baylis-Hillman, or a conjugate Baylis Hillman, adduct.[1] Examples of enantioselective Baylis-Hillman coupling are shown in Schemes 1 and 2, and the conjugate Baylis-Hillman process is illustrated in Scheme 3.

I. Catalytic Enantioselective Intra- and Inter-molecular Baylis-Hillman Reactions

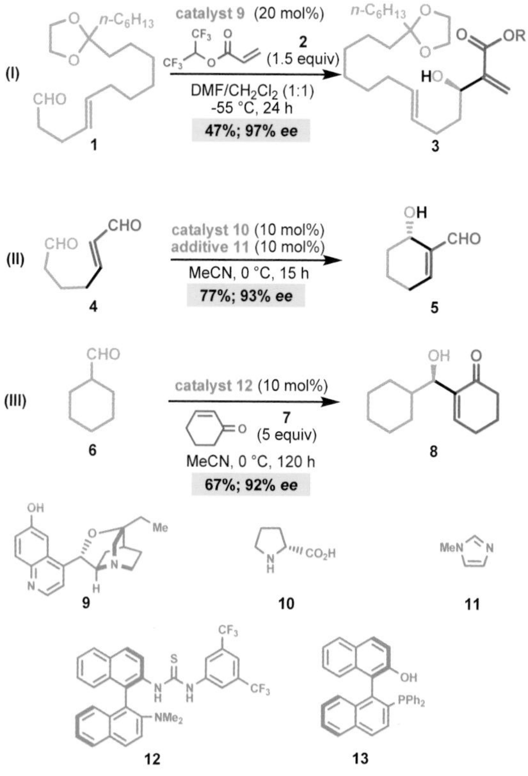

Reference key for equations: (I)[2]; (II)[3]; (III)[4]

Scheme 1

II. Catalytic Enantioselective Aza-Baylis-Hillman Reaction[1d]

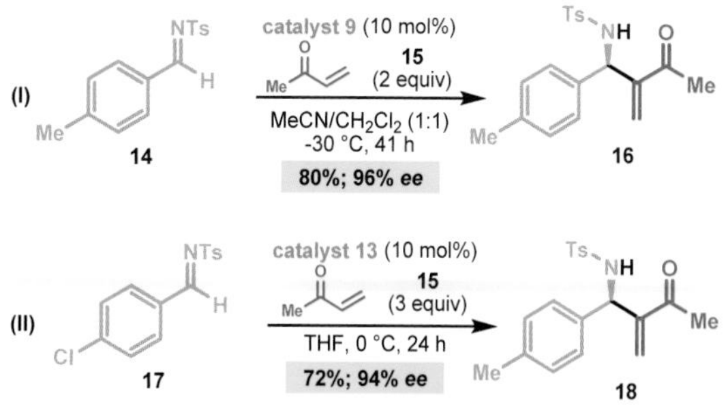

Reference key for equations: (I)[5]; (II)[6]

Scheme 2

III. Catalytic Enantioselective Conjugate Baylis-Hillman Reaction[7]

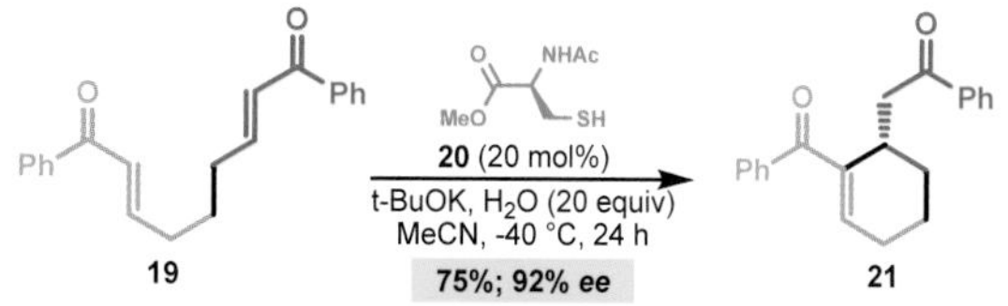

Scheme 3

References

1. (a) Langer, P. New strategies for the development of an asymmetric version of the Baylis - Hillman reaction. *Angew. Chem., Int. Ed.* **2000**, 39, 3049-3052; (b) Menozzi, C. & Dalko, P. I. Organocatalytic enantioselective Morita-Baylis-Hillman reactions in *Enantioselective Organocatalysis* (ed. Dalko, P. I.), 151-187 (Wiley-VCH, **2007**); (c) Masson, G., Housseman, C. & Zhu, J. The enantioselective Morita-Baylis-Hillman reaction and its aza counterpart. *Angew. Chem., Int. Ed.* **2007**, 46, 4614-4628; (d) Declerck, V., Martinez, J. & Lamaty, F. aza-Baylis-Hillman reaction. *Chem. Rev. (Washington, DC, U. S.)* **2009**, 109, 1-48; (e) Carrasco-Sanchez, V., Simirgiotis, M. J. & Santos, L. S. The Morita-Baylis-Hillman reaction: insights into asymmetry and reaction mechanisms by electrospray ionization mass spectrometry. *Molecules* **2009**, 14, 3989-4021.
2. Iwabuchi, Y., Furukawa, M., Esumi, T. & Hatakeyama, S. An enantio- and stereocontrolled synthesis of (-)-mycestericin E via cinchona alkaloid-catalyzed asymmetric Baylis-Hillman reaction. *Chem. Commun. (Cambridge, U. K.)* **2001**, 2030-2031.
3. Chen, S.-H., Hong, B.-C., Su, C.-F. & Sarshar, S. An unexpected inversion of enantioselectivity in the proline catalyzed intramolecular Baylis-Hillman reaction. *Tetrahedron Lett.* **2005**, 46, 8899-8903.
4. Wang, J., Li, H., Yu, X., Zu, L. & Wang, W. Chiral Binaphthyl-Derived Amine-Thiourea Organocatalyst-Promoted Asymmetric Morita-Baylis-Hillman Reaction. *Org. Lett.* **2005**, 7, 4293-4296.
5. Shi, M. & Xu, Y.-M. Catalytic, asymmetric Baylis-Hillman reaction of imines with methyl vinyl ketone and methyl acrylate. *Angew. Chem., Int. Ed.* **2002**, 41, 4507-4510.
6. Shi, M., Chen, L.-H. & Li, C.-Q. Chiral Phosphine Lewis Bases Catalyzed Asymmetric aza-Baylis-Hillman Reaction of N-Sulfonated Imines with Activated Olefins. *J. Am. Chem. Soc.* **2005**, 127, 3790-3800.
7. (a) Aroyan, C. E. & Miller, S. J. Enantioselective Rauhut-Currier Reactions Promoted by Protected Cysteine. *J. Am. Chem. Soc.* **2007**, 129, 256-257; (b) Aroyan, C. E., Dermenci, A. & Miller, S. J. The Rauhut-Currier reaction: a history and its synthetic application. *Tetrahedron* **2009**, 65, 4069-4084.

AROMATIC SUBSTITUTION REACTIONS

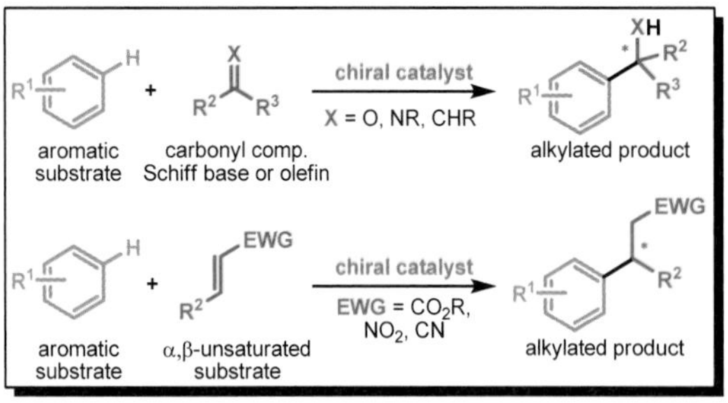

Introduction

The alkylation or acylation of an aromatic compound by an alkyl or acyl halide in the presence of a Lewis acid (C. Friedel and J.M. Crafts, 1877)[1] has been used for the synthesis of countless aromatic molecules.[2] The related Prins and Mannich reactions (with C=O and C=N electrophiles) have also been very useful.[3] During the period 1980-1990, enantioselective variants of the Prins reaction with aromatic substrates were reported using stoichiometric amounts of chiral aluminum complexes[4] or smaller amounts of a chiral zirconium catalyst 2 for the reaction between 1-naphthol and pyruvic esters (Equation I, Scheme 1).[5] More recently, catalysis of with a Cu(II)-BOX complex was described (Equation II of Scheme 1).[6]

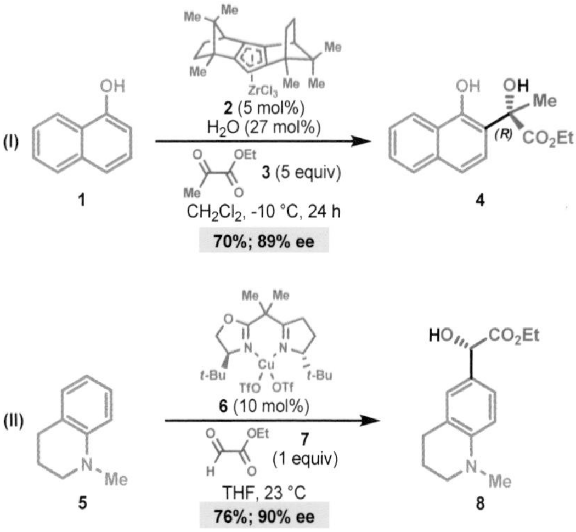

Reference key for equations: (I)[5]; (II)[6]

Scheme 1

Numerous other examples of enantioselective Prins reactions of aromatic and heteroaromatic substrates have been developed not only with chiral Lewis acids (Cu, Ti, Zr, Au and Pt complexes) and chiral Brønsted acids (e.g., BINOL-derived phosphoric acids, N-triflyl phosphoramides)[7] but also with a variety of organic molecules including thioureas, chiral primary or secondary amines and cinchona alkaloid derivatives.[8] The most favorable aromatic substrates are π-electron-rich systems such as indoles[9], pyrroles and phenols. These catalysts have been used with a range of electrophiles, including aldehydes,

ketones[10], imines, nitroolefins, α,β-unsaturated carbonyl compounds[11] and olefins.[12] The transition metal-catalyzed enantioselective alkylation of nucleophilic π-systems with allylic or benzylic electrophiles has also been reported.[13]

The utilization of chiral secondary amines[14] and chiral Brønsted acids[15] as catalysts for alkylation has increased the scope of this transformation and complemented existing methods that relied initially on chiral Lewis acid catalysts. Schemes 2-5 illustrate several efficient transformations that are grouped based on the type of electrophile used.

I. Catalytic Enantioselective Aromatic Substitution Reactions Utilizing Carbonyl Compounds As Electrophiles

Four examples of the catalytic enantioselective Prins reaction between a carbonyl compound and an electron-rich aromatic or heteroaromatic substrate to form an enantiomerically enriched chiral benzylic alcohol are displayed in Scheme 2.

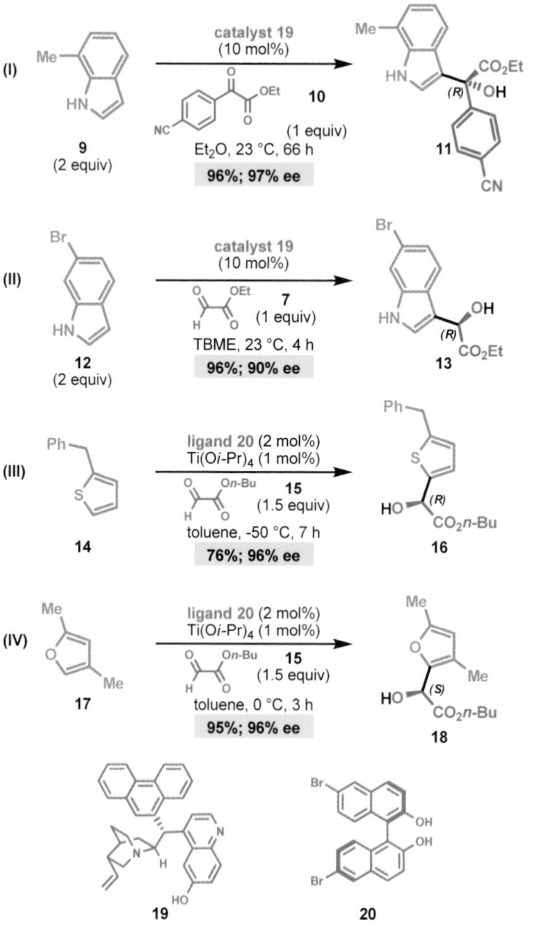

Reference key for equations: (I-II)[16]; (III)[17]; (IV)[18]

Scheme 2

AROMATIC SUBSTITUTION REACTIONS

II. Catalytic Enantioselective Aromatic Substitution Reactions Utilizing Imines as Electrophiles

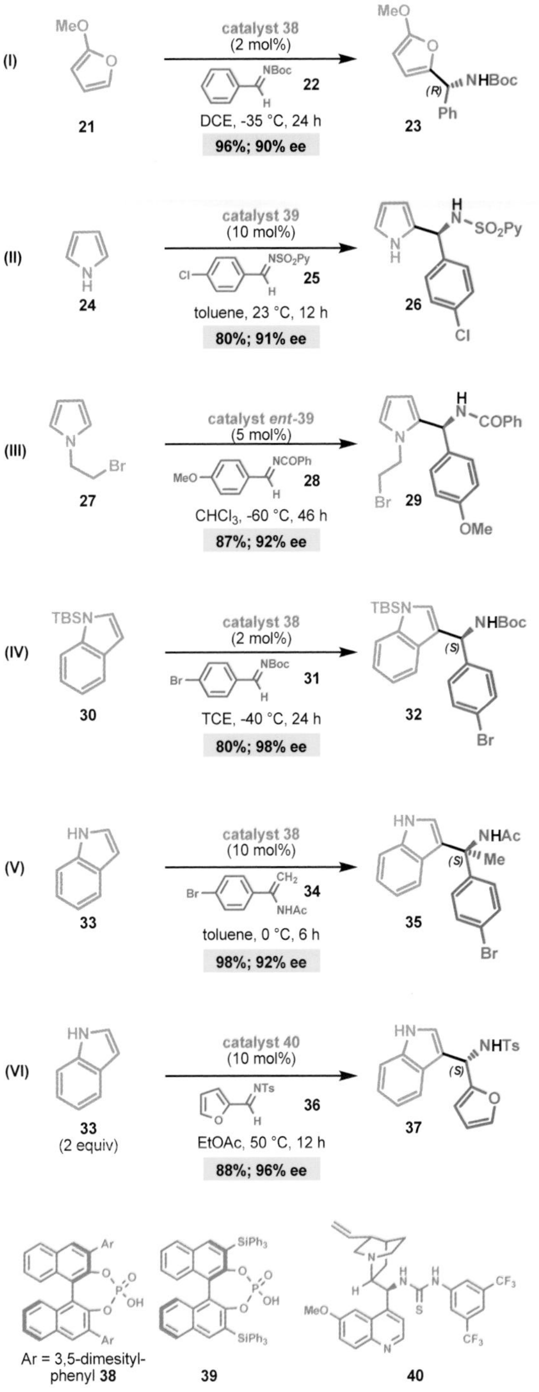

III. Catalytic Enantioselective Aromatic Substitution Reactions Utilizing Activated Alkenes as Electrophiles

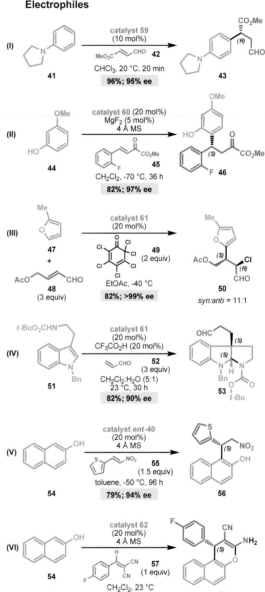

Reference key for equations: (I)[15]; (II)[19]; (III)[20]; (IV)[21]; (V)[22]; (VI)[23]

Scheme 3

Reference key for equations: (I)[24]; (II)[25]; (III)[26]; (IV)[27]; (V)[28]; (VI)[29]

Scheme 4

AROMATIC SUBSTITUTION REACTIONS

IV. Pd-Catalyzed Intramolecular Enantioselective Aromatic Substitution with an Allylic Substrate

Reference key for equation: (I)[30]

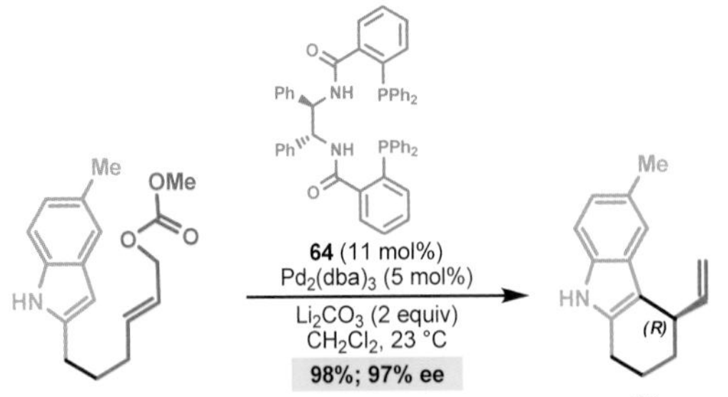

Scheme 5

References

1. (a) Friedel, C. & Crafts, J. M. A new general synthetical method for producing hydrocarbons. *J. Chem. Soc.* **1877**, 32, 725; (b) Crafts, J. M. & Ador, E. The reaction of phosgene with toluene in the presence of aluminum chloride. *Chem. Ber.* **1877**, 10, 2173-2176; (c) Crafts, J. M. & Ador, E. Effect of phthalic anhydride on naphthalin in the presence of aluminum trichloride. *Bull. Soc. Chim. France* **1880**, 531-532.

2. (a) Wan, Y., Ding, K., Dai, L., Ishii, A., Soloshonok, V. A., Mikami, K., Gathergood, N., Zhuang, W., Jorgensen, K. A., Jesen, K. B., Thorhauge, J. & Hazell, R. G. Enantioselective Friedel-Crafts reaction: from stoichiometric to catalytic. *Chemtracts* **2001**, 14, 610-615; (b) Bandini, M., Melloni, A. & Umani-Ronchi, A. New catalytic approaches in the stereoselective Friedel-Crafts alkylation reaction. *Angew. Chem., Int. Ed.* **2004**, 43, 550-556; (c) Bandini, M. General aspects and historical background of Friedel-Crafts alkylation reactions in *Catalytic Asymmetric Friedel-Crafts Alkylations* 1-16 (**2009**); (d) Blay, G., Pedro, J. R. & Vila, C. Catalytic asymmetric Friedel-Crafts alkylations in total synthesis in *Catalytic Asymmetric Friedel-Crafts Alkylations* 223-270 (**2009**); (e) Macquarrie, D. J. Industrial Friedel-Crafts chemistry in *Catalytic Asymmetric Friedel-Crafts Alkylations* 271-288 (**2009**); (f) Rueping, M. & Nachtsheim, B. J. A review of new developments in the Friedel-Crafts alkylation. From green chemistry to asymmetric catalysis. *Beilstein Journal of Organic Chemistry* **2010**, 6, No 6.

3. (a) Kürti, L. & Czakó, B. Mannich Reaction in *Strategic Applications of Named Reactions in Organic Synthesis* 274-275 (Academic Press/Elsevier, San Diego, **2005**); (b) Kürti, L. & Czakó, B. Prins Reaction in *Strategic Applications of Named Reactions in Organic Synthesis* 364-365 (Academic Press/Elsevier, San Diego, **2005**).

4. Bigi, F., Casiraghi, G., Casnati, G., Sartori, G., Gasparri Fava, G. & Ferrari Belicchi, M. Asymmetric electrophilic substitution on phenols. Enantioselective ortho-hydroxyalkylation mediated by chiral alkoxyaluminum chlorides. *J. Org. Chem.* **1985**, 50, 5018-5022.

5. (a) Erker, G. & van der Zeijden, A. A. H. Enantioselective Catalysis with a New Zirconium Trichloride Lewis Acid Containing a "Dibornacyclopentadienyl" Ligand. *Angew. Chem., Int. Ed.* **1990**, 29, 512-514; (b) Erker, G. & Van der Zeijden, A. A. H. Enantioselective catalysts having a new zirconium trichloride-Lewis acid with dibornaneannulated cyclopentadienyl ligand. *Angew. Chem.* **1990**, 102, 543-545.

6. Gathergood, N., Zhuang, W. & Jorgensen, K. A. Catalytic enantioselective Friedel-Crafts reactions of aromatic compounds with glyoxylate: a simple procedure for the synthesis of optically active aromatic mandelic acid esters. *J. Am. Chem. Soc.* **2000**, 122, 12517-12522.

7. (a) Poulsen, T. B. & Jorgensen, K. A. Catalytic Asymmetric Friedel-Crafts Alkylation Reactions-Copper Showed the Way. *Chem. Rev. (Washington, DC, U. S.)* **2008**, 108, 2903-2915; (b) You, S.-L., Cai, Q. & Zeng, M. Chiral Bronsted acid catalyzed Friedel-Crafts alkylation reactions. *Chem. Soc. Rev.* **2009**, 38, 2190-2201.

8. Terrasson, V., De Figueiredo, R. M. & Campagne, J.-M. Organocatalyzed Asymmetric Friedel-Crafts Reactions. *Eur. J. Org. Chem.* **2010**, 14, 2635-2655.

9. (a) Bandini, M., Melloni, A., Tommasi, S. & Umani-Ronchi, A. A journey across recent advances in catalytic and stereoselective alkylation of indoles. *Synlett* **2005**, 1199-1222; (b) Marques-Lopez, E., Diez-Martinez, A., Merino, P. & Herrera, R. P. The role of the indole in important organocatalytic enantioselective Friedel-Crafts alkylation reactions. *Curr. Org. Chem.* **2009**, 13, 1585-1609; (c) Bandini, M. & Eichholzer, A. Catalytic Functionalization of Indoles in a New Dimension. *Angew. Chem., Int. Ed.* **2009**, 48, 9608-9644.

10. Chen, J.-R. & Xiao, W.-J. Addition to carbonyl compounds: aldehydes/ketones in *Catalytic Asymmetric Friedel-Crafts Alkylations* 101-144 (**2009**).

11. Garcia, J. M., Oiarbide, M. & Palomo, C. Michael addition: chelating α,β-unsaturated compounds in *Catalytic Asymmetric Friedel-Crafts Alkylations* 17-99 (**2009**).

12. Widenhoefer, R. A. Unactivated alkenes in *Catalytic Asymmetric Friedel-Crafts Alkylations* 203-222 (**2009**).

13. (a) Bandini, M. & Umani-Ronchi, A. Nucleophilic allylic alkylation and hydroarylation of allenes in *Catalytic Asymmetric Friedel-Crafts Alkylations* 145-166 (**2009**); (b) Bandini, M. & Cozzi, P. G. Nucleophilic substitution on C-sp^3 carbon atoms in *Catalytic Asymmetric Friedel-Crafts Alkylations* 167-202 (**2009**).

14. Paras, N. A. & MacMillan, D. W. C. New Strategies in Organic Catalysis: The First Enantioselective Organocatalytic Friedel-Crafts Alkylation. *J. Am. Chem. Soc.* **2001**, 123, 4370-4371.

15. Uraguchi, D., Sorimachi, K. & Terada, M. Organocatalytic Asymmetric Aza-Friedel-Crafts Alkylation of Furan. *J. Am. Chem. Soc.* **2004**, 126, 11804-11805.

16. Li, H., Wang, Y.-Q. & Deng, L. Enantioselective Friedel-Crafts Reaction of Indoles with Carbonyl Compounds Catalyzed by Bifunctional Cinchona Alkaloids. *Org. Lett.* **2006**, 8, 4063-4065.

17. Majer, J., Kwiatkowski, P. & Jurczak, J. Highly Enantioselective Friedel-Crafts Reaction of Thiophenes with Glyoxylates: Formal Synthesis of Duloxetine. *Org. Lett.* **2009**, 11, 4636-4639.

18. Majer, J., Kwiatkowski, P. & Jurczak, J. Highly Enantioselective Synthesis of 2-Furanyl-hydroxyacetates from Furans via the Friedel-Crafts Reaction. *Org. Lett.* **2008**, 10, 2955-2958.

19. Nakamura, S., Sakurai, Y., Nakashima, H., Shibata, N. & Toru, T. Organocatalytic enantioselective aza-Friedel-Crafts alkylation of pyrroles with N-(heteroarenesulfonyl)imines. *Synlett* **2009**, 1639-1642.

20. Li, G., Rowland, G. B., Rowland, E. B. & Antilla, J. C. Organocatalytic enantioselective Friedel-Crafts reaction of pyrrole derivatives with imines. *Org. Lett.* **2007**, 9, 4065-4068.

21. Terada, M., Yokoyama, S., Sorimachi, K. & Uraguchi, D. Chiral phosphoric acid-catalyzed enantioselective aza-Friedel-Crafts reaction of indoles. *Adv. Synth. Catal.* **2007**, 349, 1863-1867.

22. Jia, Y.-X., Zhong, J., Zhu, S.-F., Zhang, C.-M. & Zhou, Q.-L. Chiral bronsted acid catalyzed enantioselective Friedel-Crafts reaction of indoles and α-aryl enamides: construction of quaternary carbon atoms. *Angew. Chem., Int. Ed.* **2007**, 46, 5565-5567.

23. Wang, Y.-Q., Song, J., Hong, R., Li, H. & Deng, L. Asymmetric Friedel-Crafts Reaction of Indoles with Imines by an Organic Catalyst. *J. Am. Chem. Soc.* **2006**, 128, 8156-8157.

24. Paras, N. A. & MacMillan, D. W. C. The Enantioselective Organocatalytic 1,4-Addition of Electron-Rich Benzenes to α,β-Unsaturated Aldehydes. *J. Am. Chem. Soc.* **2002**, 124, 7894-7895.

25. Lv, J., Li, X., Zhong, L., Luo, S. & Cheng, J.-P. Asymmetric Binary-Acid Catalysis with Chiral Phosphoric Acid and MgF$_2$: Catalytic Enantioselective Friedel-Crafts Reactions of β,γ-Unsaturated α-Ketoesters. *Org. Lett.* **2010**, 12, 1096-1099.

26. Huang, Y., Walji, A. M., Larsen, C. H. & MacMillan, D. W. C. Enantioselective Organo-Cascade Catalysis. *J. Am. Chem. Soc.* **2005**, 127, 15051-15053.

27. Austin, J. F., Kim, S.-G., Sinz, C. J., Xiao, W.-J. & MacMillan, D. W. C. Enantioselective organocatalytic construction of pyrroloindolines by a cascade addition-cyclization strategy: synthesis of (-)-flustramine B. *Proc. Natl. Acad. Sci. U. S. A.* **2004**, 101, 5482-5487.

28. Liu, T.-Y., Cui, H.-L., Chai, Q., Long, J., Li, B.-J., Wu, Y., Ding, L.-S. & Chen, Y.-C. Organocatalytic asymmetric Friedel-Crafts alkylation/cascade reactions of naphthols and nitroolefins. *Chem. Commun. (Cambridge, U. K.)* **2007**, 2228-2230.

29. Wang, X.-S., Yang, G.-S. & Zhao, G. Enantioselective synthesis of naphthopyran derivatives catalyzed by bifunctional thiourea-tertiary amines. *Tetrahedron Asymmetry* **2008**, 19, 709-714.

30. Bandini, M., Melloni, A., Piccinelli, F., Sinisi, R., Tommasi, S. & Umani-Ronchi, A. Highly Enantioselective Synthesis of Tetrahydro-β-Carbolines and Tetrahydro-γ-Carbolines Via Pd-Catalyzed Intramolecular Allylic Alkylation. *J. Am. Chem. Soc.* **2006**, 128, 1424-1425.

ENANTIOSELECTIVE DIELS-ALDER REACTION

substituted 1,4-diene | dienophile | chiral catalyst | endo-Diels-Alder adduct

EWG = electron-withdrawing group: CO_2R, CN, NO_2, CHO, COR

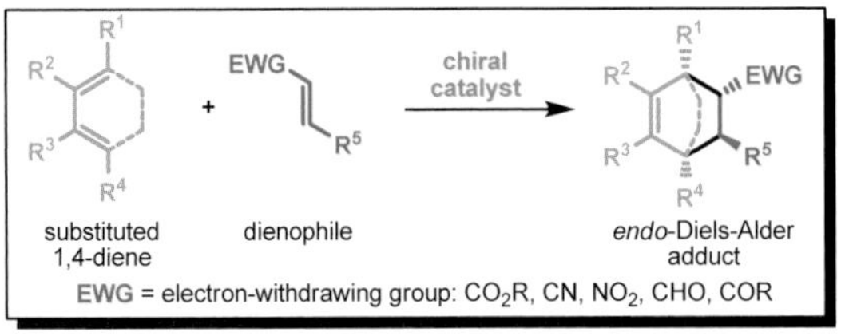

Introduction

The Diels-Alder reaction is arguably the most powerful and versatile construction in the arsenal of synthetic chemistry, and its position in the reaction hierarchy has been further advanced in recent years by the development of enantioselective variants, as documented in several reviews.[1] Diels-Alder reactions may be inter -or intramolecular and may form either carbocycles or heterocycles.[2] In addition, there are numerous subtypes depending on the type of diene or dienophile involved. Furthermore, the reaction may be thermal, pressure-accelerated or catalyzed by Lewis or protic acids, hydrogen bond donors or even chiral amines. The formation of chiral Diels-Alder products can be directed either by the application of a chiral controller group or the use of a chiral catalyst.

I. Stereoselective Diels-Alder Reactions Using a Chiral Controller Group

The first successful methods for obtaining chiral Diels-Alder products without the need for resolution of enantiomers employed chiral controller groups. The following example in Scheme 1 represents the earliest such case to achieve very high stereoselectivity, and an interesting application to the synthesis of prostaglandins without resolution.[3] The site of the coordination of $AlCl_3$ is indicated by the black arrow.

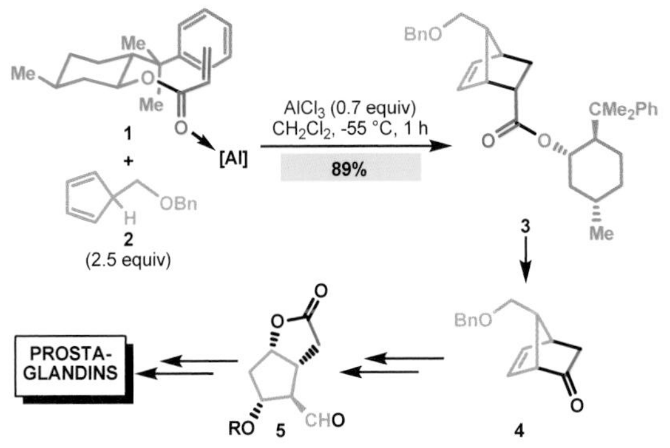

Scheme 1

A number of other chiral controllers have been found to be very effective in directing the face-selectivity of 1,2-diene addition to acrylate or fumarate type dienophiles. Some examples are shown in Scheme 2. The likely mode of Lewis acid coordination is indicated for each example (X_c = chiral controller).

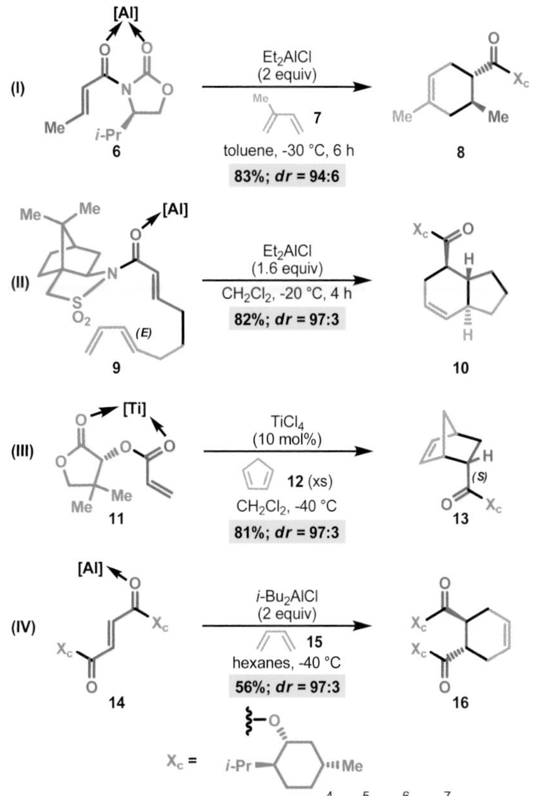

Reference key for equations: (I)[4]; (II)[5]; (III)[6]; (IV)[7]

Scheme 2

II. Enantioselective Diels-Alder Reactions Catalyzed by a Chiral Lewis Acid

An early example of a useful catalytic enantioselective Diels-Alder reaction was reported by K. Narasaka using N-butenoyl-1,3-oxazolidinone as the dienophile and a chiral diol-$TiCl_2$ complex (Scheme 3).[8] Replacement of the Ph groups in 18 by the more π-basic 3,5-dimethylphenyl group improves enantioselectivity.[1b]

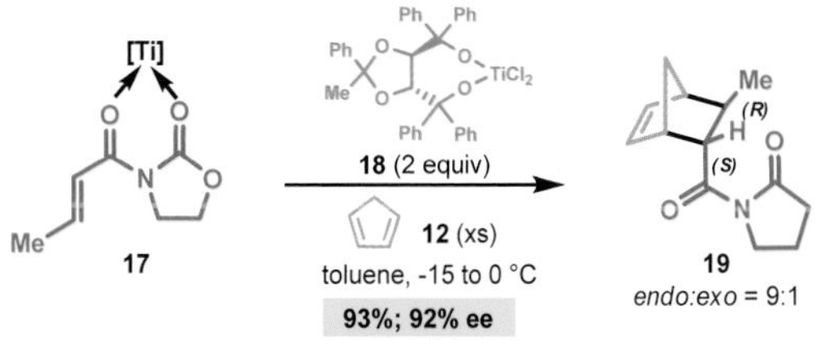

Scheme 3

The parent dienophile, N-acryloyl-1,3-oxazolidinone, has also been used in Diels-Alder reactions with a number of other chiral catalysts, especially those in the bisoxazoline (BOX) series. Highly enantioselective reactions have been observed for BOX complexes with FeI_3,[9] MgI_2[10] and $Cu(OTf)_2$[11] and N-acryloyl-1,3-oxazolidinone, which binds as a bidentate ligand in each case (Scheme 4).

ENANTIOSELECTIVE DIELS-ALDER REACTION

Enantioselective Diels-Alder Reactions Catalyzed by a Chiral Lewis Acid (Continued)

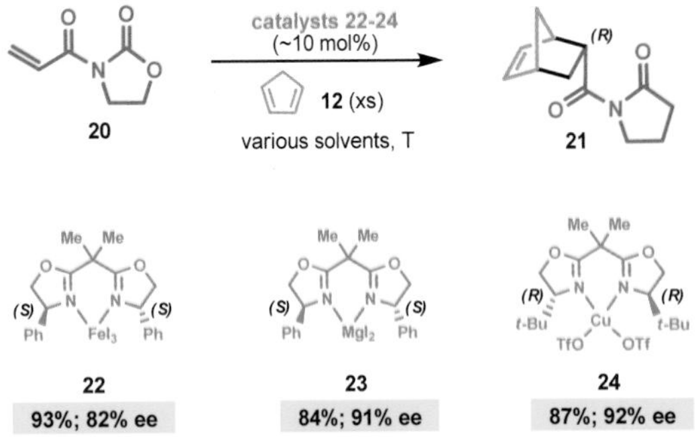

Scheme 4

The absolute stereocourse of the FeI_3 complex with the (S,S)-BOX ligand **22** corresponds to coordination of the ligand to an axial and an equatorial site of octahedral $L_2FeI_2^+$, whereas with MgI_2 the coordination about Mg^{2+} is tetrahedral. The same enantiomer of the Diels-Alder adduct (**21**) is formed when a (R,R)-bisoxazoline (**24**) is used as catalyst, indicating reaction via square-planar geometry for the catalyst.

The chiral aluminum-*bis*-sulfamide catalyst **25** is extremely effective in controlling enantioselective Diels-Alder reactions of **20** (Scheme 5). In this case, however, the reaction pathway appears to involve mono-coordination of **20** to Al and reaction via the pre-transition-state assembly **27** on the basis of NMR NOE studies.[12,1b]

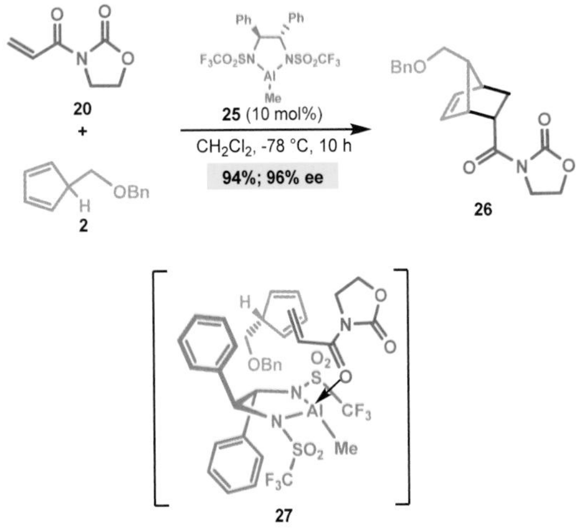

Scheme 5

An early experiment by K. Koga using a catalyst prepared from $AlCl_3$ and (–)-menthol revealed that the catalyzed reaction of cyclopentadiene with 2-methylacrolein produced the Diels-Alder adduct in ca. 70% ee.[13] This reaction was markedly improved by H. Yamamoto using a catalyst made from BH_3 and a mono ester of (R,R)-tartaric acid (presumed to be a cyclic boron-complex, **29**) as shown in Scheme 6.[14]

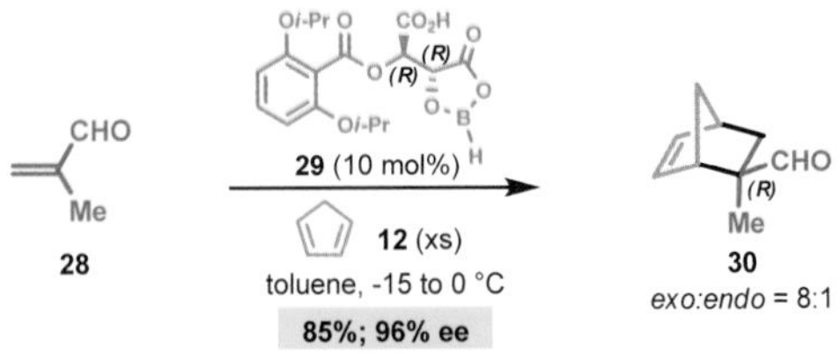

Scheme 6

The reaction of 2-bromoacrolein (**33**) with cyclopentadiene (**12**) proceeds with striking (200:1) enantioselectivity to afford the Diels-Alder adduct when promoted by the oxazolidine **32** as shown in Scheme 7.[1b] There is considerable evidence that this reaction proceeds via the highly organized pre-transition state assembly **35**.

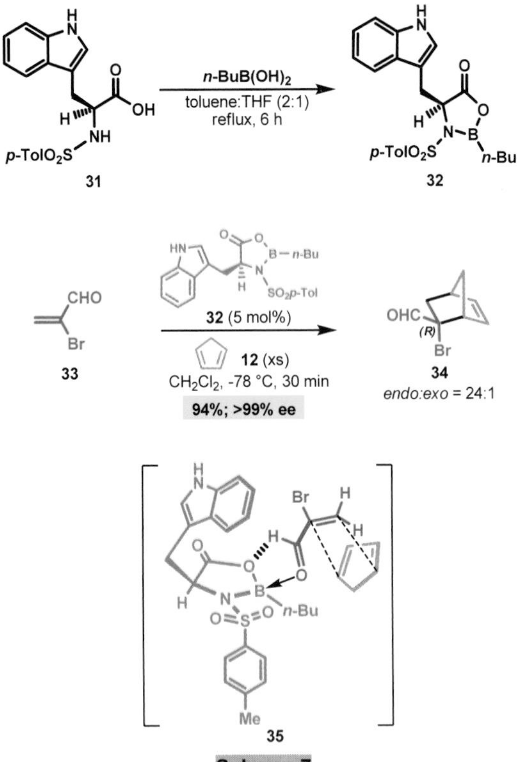

Scheme 7

2-Substituted acroleins are very amenable to enantioselective Diels-Alder reactions with various chiral Lewis acids. In contrast, β-substitution (e.g., with Me or Cl) greatly lowers the reactivity in this series, and acrolein itself generally leads to low enantioselectivities. The favorable selectivity found for 2-substituted acroleins has been ascribed to formyl C-H hydrogen bonding to one of the ligands on the chiral catalyst, the occurrence of which is suggested in part from X-ray crystallographic studies of simple complexes, for example **36**.

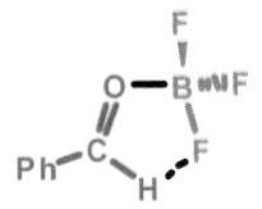

ENANTIOSELECTIVE DIELS-ALDER REACTION

III. Oxazaborolidinium Cations as Catalysts for Enantioselective Diels-Alder Reactions

The complexation of a chiral electron-donating ligand with a Lewis acid generally reduces the Lewis acidity at the metal. Thus, such complexes are much less activating than the free Lewis acids and have a greatly decreased catalytic range in terms of useful substrates. A solution to this problem was first demonstrated with the cationic boron Lewis acid **39**[15] which was a much more powerful chiral catalyst than those previously known. The reaction shown in Scheme 8 is complete within less than 15 minutes at -94 °C. The proposed transition state assembly **40** is also depicted in Scheme 8.

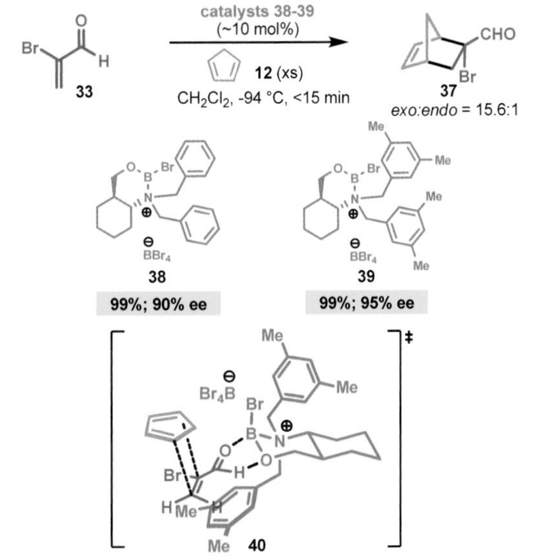

Scheme 8

The chiral cation **41**, which can be made simply by treatment of the corresponding oxazaborolidine with TfOH *in situ* is also a powerful catalyst for enantioselective Diels-Alder reactions, as illustrated along with the proposed pre-transition state assembly **43** in Scheme 9.[16,1d] Weaker protic acids than TfOH (e.g., MeSO$_3$H) are not strong enough to generate cation **41** cleanly, which emphasizes the electrophilicity of **41**.

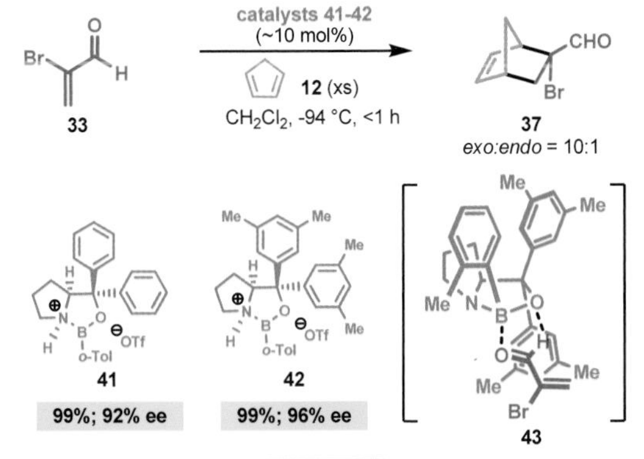

Scheme 9

The range of substrates for enantioselective Diels-Alder reactions with the oxazaborolidium cation – either as triflate (TfO⁻) or triflimide (Tf$_2$N⁻) salts – is broad and includes many types of α,β-unsaturated carbonyl compounds. A second type of highly organized pre-transition state assembly applies for α,β-unsaturated ketones and esters bearing an α-CH subunit as illustrated in Scheme 10 for trifluoroethyl acrylate.[1d]

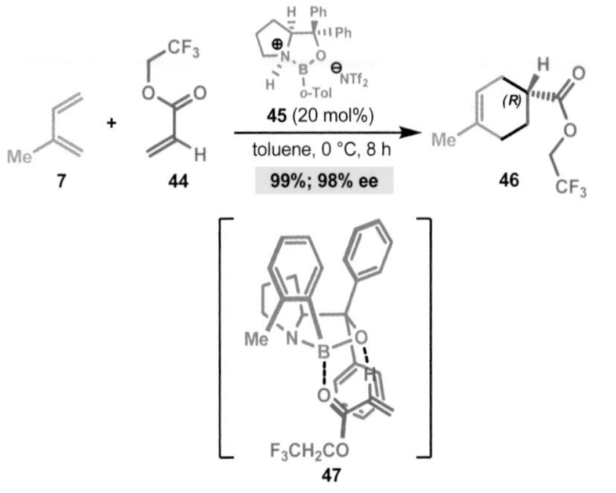

Scheme 10

A selection of some useful types of enantioselective Diels-Alder reactions that are catalyzed by a chiral oxazaborolidinium cation is shown in Schemes 11 and 12.[1d]

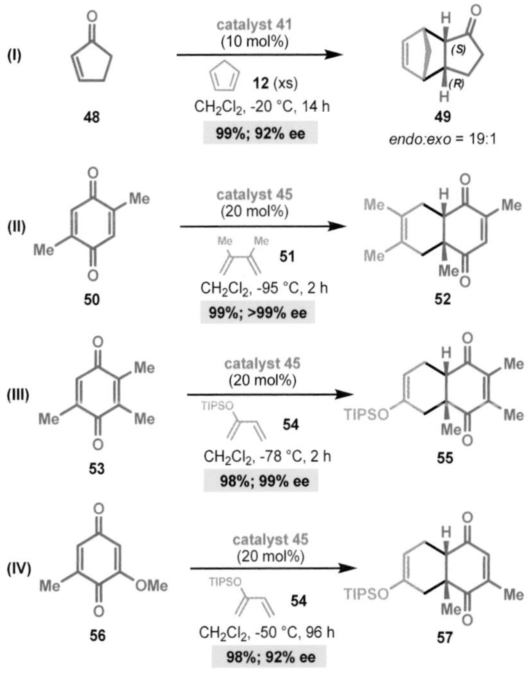

Reference key for equations: (I)[17]; (II)[18]; (III-IV)[19]

Scheme 11

123

ENANTIOSELECTIVE DIELS-ALDER REACTION

Oxazaborolidinium Cations as Catalysts for Enantioselective Diels-Alder Reactions (Continued)

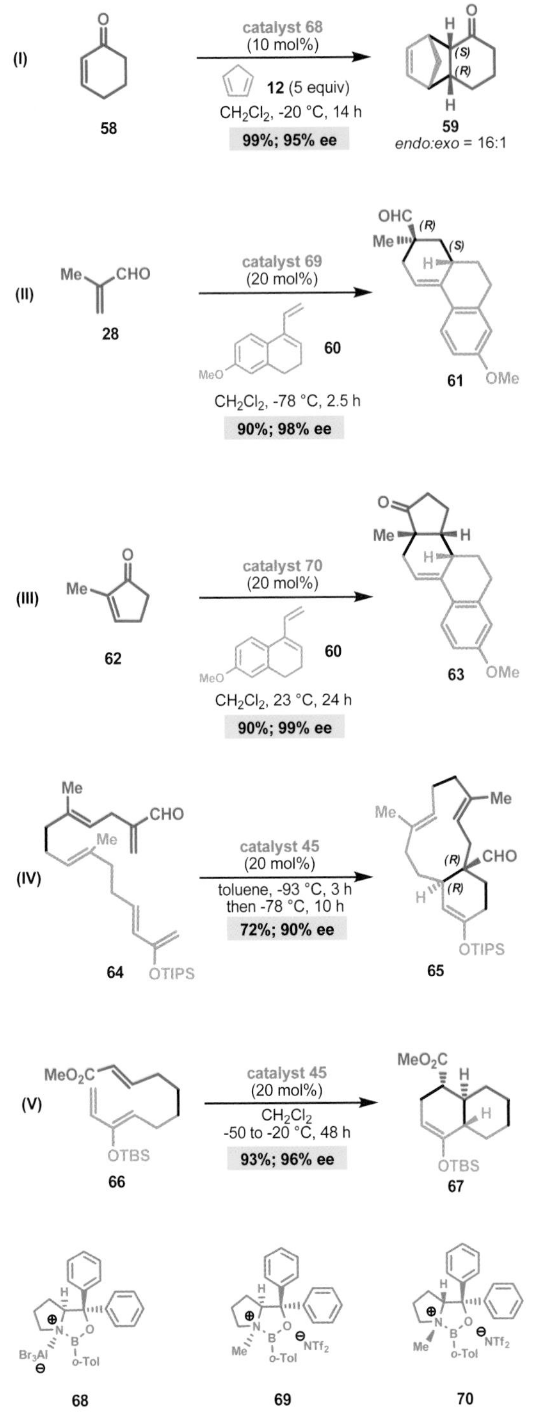

(I) 58 → catalyst 68 (10 mol%) / 12 (5 equiv) / CH$_2$Cl$_2$, -20 °C, 14 h → 59
99%; 95% ee *endo:exo* = 16:1

(II) Me–CHO 28 → catalyst 69 (20 mol%) / CH$_2$Cl$_2$, -78 °C, 2.5 h → 61 (from 60)
90%; 98% ee

(III) Me 62 → catalyst 70 (20 mol%) / CH$_2$Cl$_2$, 23 °C, 24 h → 63 (from 60)
90%; 99% ee

(IV) 64 OTIPS → catalyst 45 (20 mol%) / toluene, -93 °C, 3 h then -78 °C, 10 h → 65 OTIPS
72%; 90% ee

(V) MeO$_2$C 66 OTBS → catalyst 45 (20 mol%) / CH$_2$Cl$_2$ -50 to -20 °C, 48 h → 67 OTBS
93%; 96% ee

68 (Br$_3$Al, o-Tol)   69 (Me, o-Tol, NTf$_2$)   70 (Me, o-Tol, NTf$_2$)

Reference key for equations: (I)[20]; (II-III)[21]; (IV)[22]; (V)[23]

Scheme 12

IV. Enantioselective Diels-Alder Reactions of α,β-Unsaturated Carbonyl Compounds Catalyzed by a Chiral Amine

Appropriately structured chiral secondary amines react with α,β-unsaturated aldehydes to generate the corresponding iminium ions (**C**, Scheme 13) which are sufficiently reactive to combine with a variety of reactive to moderately reactive dienes (**D**) at ambient temperature to form Diels-Alder products (**F**).[1f]

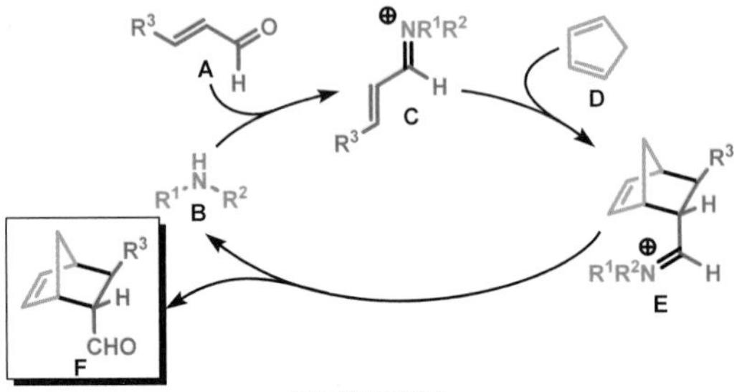

Scheme 13

(I) Ph 71 → 72 (5 mol%) / 12 (xs) / MeOH-H$_2$O, 23 °C, 21 h → 73
99%; 93% ee *exo:endo* = 1.3:1

(II) 74 → catalyst 72 (20 mol%) / 75 / MeOH-H$_2$O, 23 °C → 76
99%; 93% ee *endo:exo* = 14:1

(III) 77 → catalyst 72 (20 mol%) / CH$_3$CN, 5 °C → 78
76%; 94% ee *exo:endo* > 20:1

(IV) 79 → catalyst 72 (20 mol%) / p-TsOH (20 mol%) / CHCl$_3$, 25 °C → 80
65%; 98% ee *endo:exo* = 99:1

(V) 81 → 82 (15 mol%) / 83 / Et$_2$O, -50 °C, 24 h then NaBH$_4$, MeOH → 84
87%; 96% ee

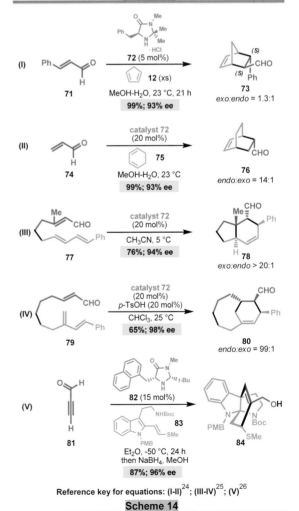

Reference key for equations: (I-II)[24]; (III-IV)[25]; (V)[26]

Scheme 14

ENANTIOSELECTIVE DIELS-ALDER REACTION

Enantioselective Diels-Alder Reactions of α,β-Unsaturated Carbonyl Compounds Catalyzed by a Chiral Amine (Continued)

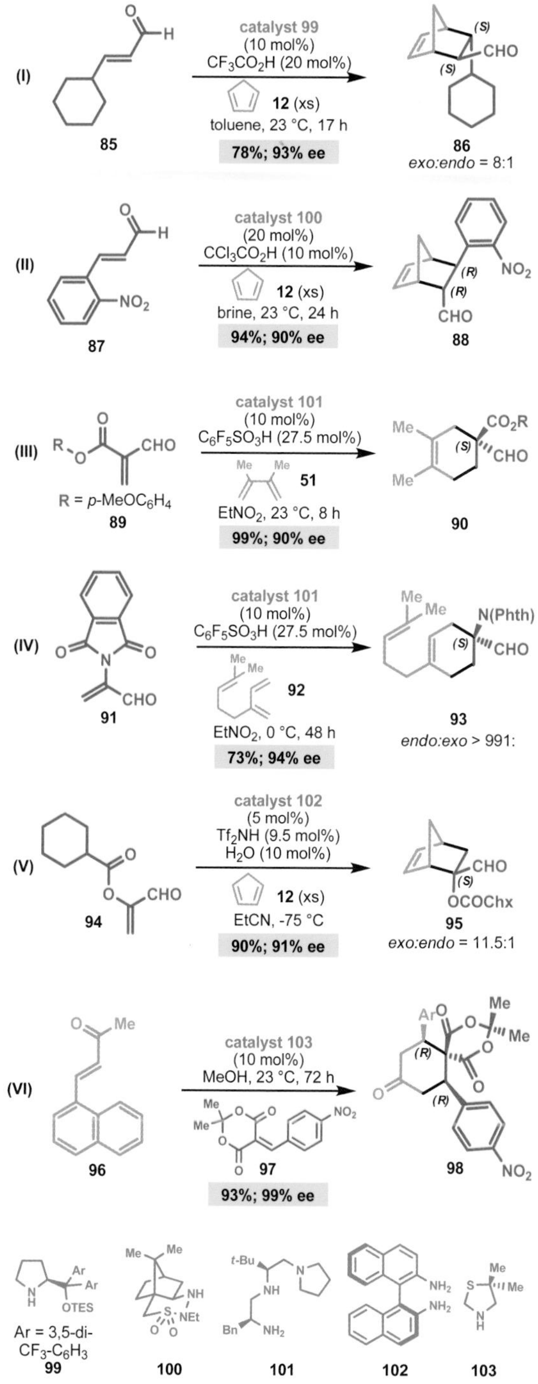

(I)
85 → catalyst 99 (10 mol%), CF₃CO₂H (20 mol%); 12 (xs); toluene, 23 °C, 17 h → **78%; 93% ee** → 86 (S)(S) CHO *exo:endo = 8:1*

(II)
87 → catalyst 100 (20 mol%), CCl₃CO₂H (10 mol%); 12 (xs); brine, 23 °C, 24 h → **94%; 90% ee** → 88 (R)(R) NO₂ CHO

(III)
89 R = p-MeOC₆H₄ → catalyst 101 (10 mol%), C₆F₅SO₃H (27.5 mol%); 51; EtNO₂, 23 °C, 8 h → **99%; 90% ee** → 90 (S) CO₂R CHO

(IV)
91 → catalyst 101 (10 mol%), C₆F₅SO₃H (27.5 mol%); 92; EtNO₂, 0 °C, 48 h → **73%; 94% ee** → 93 (S) N(Phth) CHO *endo:exo > 991:*

(V)
94 → catalyst 102 (5 mol%), Tf₂NH (9.5 mol%), H₂O (10 mol%); 12 (xs); EtCN, -75 °C → **90%; 91% ee** → 95 (S) CHO OCOChx *exo:endo = 11.5:1*

(VI)
96 → catalyst 103 (10 mol%), MeOH, 23 °C, 72 h → **93%; 99% ee** → 97, 98 (R)(R) NO₂

99 Ar = 3,5-di-CF₃-C₆H₃   100   101   102   103

Reference key for equation: (I)[27]; (II)[28]; (III)[29]; (IV)[30]; (V)[31]; (VI)[32]

Scheme 15

The transformation **96→98** (Equation VI, Scheme 15) proceeds via an enamine that is formed by the reaction between catalyst **103** and ketone **96**. Upon [4+2]-cycloaddition, the catalyst is regenerated by hydrolysis.

V. Enantioselective Diels-Alder Reactions of Michael Acceptors Catalyzed by a Chiral Tertiary Amine

Certain chiral tertiary amines can activate a moderately acidic diene by deprotonation (site of deprotonation is indicated by →H) to enhance its nucleophilicity by forming a chiral ion-pair. Such nucleophiles undergo reaction with strong Michael acceptors to form [4+2]-cycloaddition products – probably via a two-step pathway that is initiated by Michael addition.

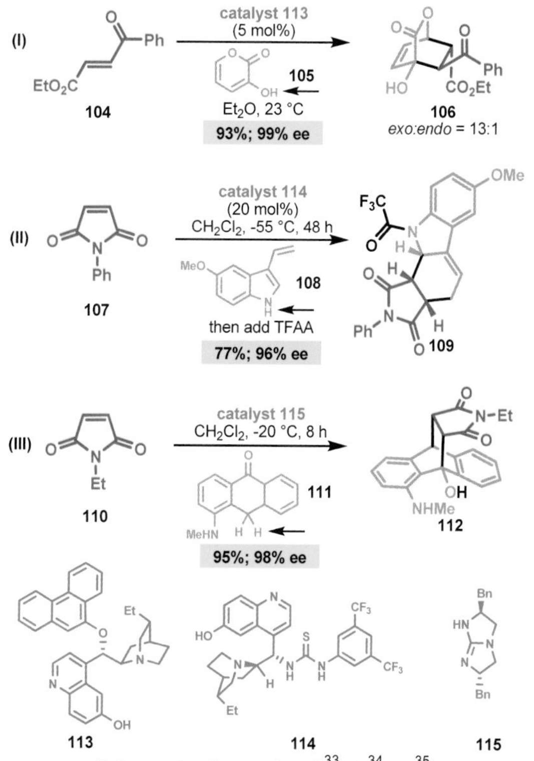

(I)
104 → catalyst 113 (5 mol%); 105; Et₂O, 23 °C → **93%; 99% ee** → 106 *exo:endo = 13:1*

(II)
107 → catalyst 114 (20 mol%), CH₂Cl₂, -55 °C, 48 h; 108; then add TFAA → **77%; 96% ee** → 109

(III)
110 → catalyst 115, CH₂Cl₂, -20 °C, 8 h; 111 → **95%; 98% ee** → 112

113   114   115

Reference key for equation: (I)[33]; (II)[34]; (III)[35]

Scheme 16

VI. Enantioselective Formation of Heterorings by [4+2]-Cycloaddition Reactions: The Hetero-Diels-Alder Reaction[2a-f,2i,2j]

The majority of known [4+2]-cycloadditions involve the addition of an electron-rich diene (e.g., silyloxy-1,3-diene) to the C=O group of an aldehyde or reactive ketone, a typical example being the Lewis acid-catalyzed addition of **116** and **118** to form **119** (Scheme 17).[36] A possible pre-transition state assembly (**120**) is also shown. Such reactions in some cases proceed via two-step pathways through an intermediate (**121**) that corresponds to a Mukaiyama aldol product.

ENANTIOSELECTIVE DIELS-ALDER REACTION

Enantioselective Formation of Heterorings by [4+2]-Cycloaddition Reactions: The Hetero-Diels-Alder Reaction (Continued)

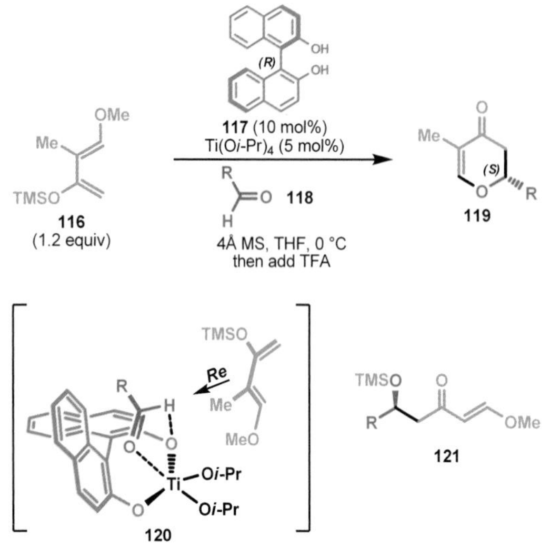

Scheme 17

There are fewer instances of hetero-dienophiles other than C=O in catalytic enantioselective Diels-Alder processes. However, such reactions involving chiral controller groups have been known for some time. Four examples of addition to N=O and C=N are shown in Scheme 18 (X_c = chiral controller).

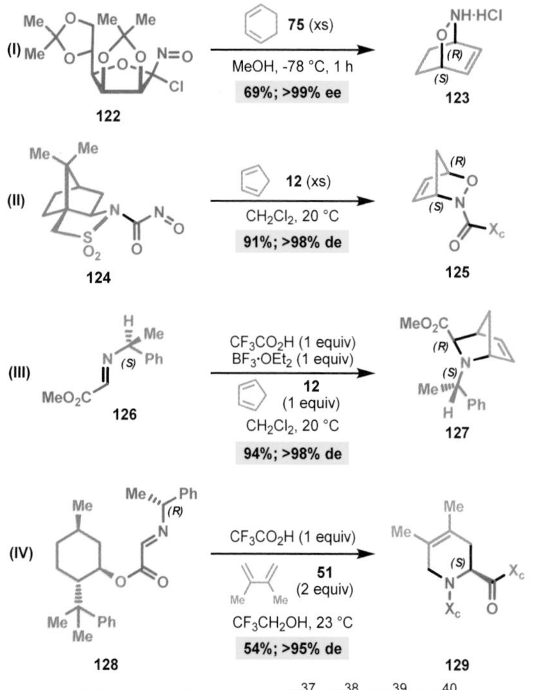

Reference key for equation: (I)[37]; (II)[38]; (III)[39]; (IV)[40]

Scheme 18

Some examples of catalytic enantioselective [4+2]-cycloaddition reactions, involving achiral hetero-dienes or imines, are shown in Scheme 19.

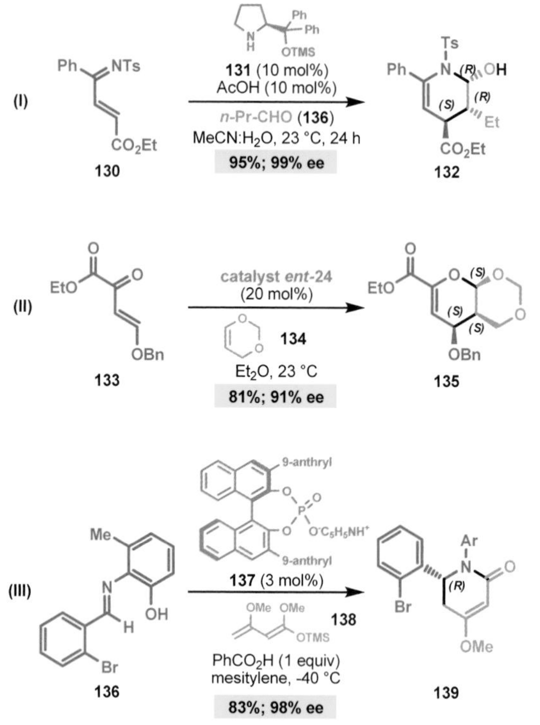

Reference key for equation: (I)[41]; (II)[42]; (III)[43]

Scheme 19

References

1. (a) Hayashi, Y. Catalytic asymmetric Diels-Alder reactions. *Cycloaddit. React. Org. Synth.* **2002**, 5-55; (b) Corey, E. J. Catalytic enantioselective Diels-Alder reactions: Methods, mechanistic fundamentals, pathways, and applications. *Angew. Chem., Int. Ed.* **2002**, 41, 1650-1667; (c) Takao, K., Munakata, R. & Tadano, K. Recent Advances in Natural Product Synthesis by Using Intramolecular Diels-Alder Reactions. *Chem. Rev. (Washington, DC, U. S.)* **2005**, 105, 4779-4807; (d) Corey, E. J. Enantioselective catalysis based on cationic oxazaborolidines. *Angew. Chem., Int. Ed.* **2009**, 48, 2100-2117; (e) Nicolaou, K. C. & Chen, J. S. The art of total synthesis through cascade reactions. *Chem. Soc. Rev.* **2009**, 38, 2993-3009; (f) Merino, P., Marques-Lopez, E., Tejero, T. & Herrera, R. P. Enantioselective organocatalytic Diels-Alder reactions. *Synthesis* **2010**, 1-26.
2. (a) Waldmann, H. Asymmetric hetero Diels-Alder reactions. *Synthesis* **1994**, 535-551; (b) Tietze, L. F. & Kettschau, G. Hetero Diels-Alder reactions in organic chemistry. *Top. Curr. Chem.* **1997**, 189, 1-120; (c) Kobayashi, S. Catalytic enantioselective aza Diels-Alder reactions. *Cycloaddit. React. Org. Synth.* **2002**, 187-209; (d) Jorgensen, K. A. Hetero-Diels-Alder reactions of ketones - a challenge for chemists. *Eur. J. Org. Chem.* **2004**, 2093-2102; (e) Heintzelman, G. R., Meigh, I. R., Mahajan, Y. R. & Weinreb, S. M. Diels-Alder reactions of imino dienophiles. *Org. React.* **2005**, 65, 141-599; (f) Gouverneur, V. & Reiter, M. Biocatalytic approaches to hetero-Diels-Alder adducts of carbonyl compounds. *Chem.--Eur. J.* **2005**, 11, 5806-5815; (g) Rowland, G. B., Rowland, E. B., Zhang, Q. & Antilla, J. C. Stereoselective Aza-Diels-Alder reactions. *Curr. Org. Chem.* **2006**, 10, 981-1005; (h) Paull, D. H., Wolfer, J., Grebinski, J. W., Weatherwax, A. & Lectka, T. Catalytic, asymmetric inverse electron

demand hetero Diels-Alder reactions of o-benzoquinone derivatives and ketene enolates. *Chimia* **2007**, 61, 240-246; (i) Lin, L., Liu, X. & Feng, X. Asymmetric hetero-Diels-Alder reactions of Danishefsky's and Brassard's dienes with aldehydes. *Synlett* **2007**, 2147-2157; (j) Pellissier, H. Asymmetric hetero-Diels-Alder reactions of carbonyl compounds. *Tetrahedron* **2009**, 65, 2839-2877.

3. Corey, E. J. & Ensley, H. E. Preparation of an optically active prostaglandin intermediate via asymmetric induction. *J. Am. Chem. Soc.* **1975**, 97, 6908-6909.

4. (a) Evans, D. A., Chapman, K. T. & Bisaha, J. Diastereofacial selectivity in intramolecular Diels-Alder reactions of chiral triene-N-acyloxazolidones. *Tetrahedron Lett.* **1984**, 25, 4071-4074; (b) Evans, D. A., Chapman, K. T. & Bisaha, J. New asymmetric Diels-Alder cycloaddition reactions. Chiral α,β-unsaturated carboximides as practical chiral acrylate and crotonate dienophile synthons. *J. Am. Chem. Soc.* **1984**, 106, 4261-4263.

5. (a) Oppolzer, W. & Dupuis, D. Asymmetric intramolecular Diels-Alder reactions of N-acylcamphorsultam trienes. *Tetrahedron Lett.* **1985**, 26, 5437-5440; (b) Oppolzer, W. Camphor derivatives as chiral auxiliaries in asymmetric synthesis. *Tetrahedron* **1987**, 43, 1969-2004.

6. Poll, T., Sobczak, A., Hartmann, H. & Helmchen, G. Diastereoface-discriminative metal coordination in asymmetric synthesis: D-pantolactone as practical chiral auxiliary for Lewis acid catalyzed Diels-Alder reactions. *Tetrahedron Lett.* **1985**, 26, 3095-3098.

7. Furuta, K., Iwanaga, K. & Yamamoto, H. Asymmetric Diels-Alder reaction. Cooperative blocking effect in organic synthesis. *Tetrahedron Lett.* **1986**, 27, 4507-4510.

8. (a) Narasaka, K., Inoue, M. & Okada, N. Asymmetric Diels-Alder reaction promoted by a chiral titanium reagent. *Chem. Lett.* **1986**, 1109-1112; (b) Narasaka, K., Inoue, M. & Yamada, T. Asymmetric Diels-Alder reaction with a catalytic amount of a chiral titanium reagent. *Chem. Lett.* **1986**, 1967-1968; (c) Narasaka, K., Inoue, M., Yamada, T., Sugimori, J. & Iwasawa, N. Asymmetric Diels-Alder reaction by the use of a chiral titanium catalyst with molecular sieves 4A. Remarkable solvent effect on the enantioselectivity. *Chem. Lett.* **1987**, 2409-2412; (d) Corey, E. J. & Matsumura, Y. Evidence for the importance of π-π-attractive interactions in enantioselective Diels-Alder reaction chiral catalysts of the type (RO)2TiCl2. *Tetrahedron Lett.* **1991**, 32, 6289-6292.

9. Corey, E. J., Imai, N. & Zhang, H. Y. Designed catalyst for enantioselective Diels-Alder addition from a C2-symmetric chiral bis(oxazoline)-iron(III) complex. *J. Am. Chem. Soc.* **1991**, 113, 728-729.

10. Corey, E. J. & Ishihara, K. Highly enantioselective catalytic Diels-Alder addition promoted by a chiral bis(oxazoline)-magnesium complex. *Tetrahedron Lett.* **1992**, 33, 6807-6810.

11. Evans, D. A., Lectka, T. & Miller, S. J. Bis(imine)-copper(II) complexes as chiral Lewis acid catalysts for the Biels-Alder reaction. *Tetrahedron Lett.* **1993**, 34, 7027-7030.

12. (a) Corey, E. J., Imwinkelried, R., Pikul, S. & Xiang, Y. B. Practical enantioselective Diels-Alder and aldol reactions using a new chiral controller system. *J. Am. Chem. Soc.* **1989**, 111, 5493-5495; (b) Corey, E. J., Sarshar, S. & Bordner, J. X-ray crystallographic and NMR studies on the origins of high enantioselectivity in Diels-Alder reactions catalyzed by a chiral diazaaluminolidine. *J. Am. Chem. Soc.* **1992**, 114, 7938-7939.

13. Hashimoto, S., Komeshima, N. & Koga, K. Asymmetric Diels-Alder reaction catalyzed by chiral alkoxyaluminum dichloride. *J. Chem. Soc., Chem. Commun.* **1979**, 437-438.

14. (a) Furuta, K., Shimizu, S., Miwa, Y. & Yamamoto, H. Chiral (acyloxy)borane (CAB): a powerful and practical catalyst for asymmetric Diels-Alder reactions. *J. Org. Chem.* **1989**, 54, 1481-1483; (b) Ishihara, K., Gao, Q. & Yamamoto, H. Mechanistic studies of a CAB-catalyzed asymmetric Diels-Alder reaction. *J. Am. Chem. Soc.* **1993**, 115, 10412-10413.

15. Hayashi, Y., Rohde, J. J. & Corey, E. J. A Novel Chiral Super-Lewis Acidic Catalyst for Enantioselective Synthesis. *J. Am. Chem. Soc.* **1996**, 118, 5502-5503.

16. Corey, E. J., Shibata, T. & Lee, T. W. Asymmetric Diels-Alder Reactions Catalyzed by a Triflic Acid Activated Chiral Oxazaborolidine. *J. Am. Chem. Soc.* **2002**, 124, 3808-3809.

17. Ryu, D. H., Lee, T. W. & Corey, E. J. Broad-Spectrum Enantioselective Diels-Alder Catalysis by Chiral, Cationic Oxazaborolidines. *J. Am. Chem. Soc.* **2002**, 124, 9992-9993.

18. Ryu, D. H. & Corey, E. J. Triflimide Activation of a Chiral Oxazaborolidine Leads to a More General Catalytic System for Enantioselective Diels-Alder Addition. *J. Am. Chem. Soc.* **2003**, 125, 6388-6390.

19. Ryu, D. H., Zhou, G. & Corey, E. J. Enantioselective and Structure-Selective Diels-Alder Reactions of Unsymmetrical Quinones Catalyzed by a Chiral Oxazaborolidinium Cation. Predictive Selection Rules. *J. Am. Chem. Soc.* **2004**, 126, 4800-4802.

20. Liu, D., Canales, E. & Corey, E. J. Chiral Oxazaborolidine-Aluminum Bromide Complexes Are Unusually Powerful and Effective Catalysts for Enantioselective Diels-Alder Reactions. *J. Am. Chem. Soc.* **2007**, 129, 1498-1499.

21. (a) Hu, Q.-Y., Rege, P. D. & Corey, E. J. Simple, Catalytic Enantioselective Syntheses of Estrone and Desogestrel. *J. Am.* *Chem. Soc.* **2004**, 126, 5984-5986; (b) Yeung, Y.-Y., Chein, R.-J. & Corey, E. J. Conversion of Torgov's Synthesis of Estrone into a Highly Enantioselective and Efficient Process. *J. Am. Chem. Soc.* **2007**, 129, 10346-10347; (c) Canales, E. & Corey, E. J. Highly Enantioselective [4 + 2] Cycloaddition Reactions Catalyzed by a Chiral N-Methyl-oxazaborolidinium Cation. *Org. Lett.* **2008**, 10, 3271-3273.

22. Snyder, S. A. & Corey, E. J. Concise Total Syntheses of Palominol, Dolabellatrienone, β-Araneosene, and Isoedunol via an Enantioselective Diels-Alder Macrobicyclization. *J. Am. Chem. Soc.* **2006**, 128, 740-742.

23. Zhou, G., Hu, Q.-Y. & Corey, E. J. Useful Enantioselective Bicyclization Reactions Using an N-Protonated Chiral Oxazaborolidine as Catalyst. *Org. Lett.* **2003**, 5, 3979-3982.

24. Ahrendt, K. A., Borths, C. J. & MacMillan, D. W. C. New Strategies for Organic Catalysis: The First Highly Enantioselective Organocatalytic Diels-Alder Reaction. *J. Am. Chem. Soc.* **2000**, 122, 4243-4244.

25. Wilson, R. M., Jen, W. S. & MacMillan, D. W. C. Enantioselective Organocatalytic Intramolecular Diels-Alder Reactions. The Asymmetric Synthesis of Solanapyrone D. *J. Am. Chem. Soc.* **2005**, 127, 11616-11617.

26. Jones, S. B., Simmons, B. & MacMillan, D. W. C. Nine-Step Enantioselective Total Synthesis of (+)-Minfiensine. *J. Am. Chem. Soc.* **2009**, 131, 13606-13607.

27. Gotoh, H. & Hayashi, Y. Diarylprolinol Silyl Ether as Catalyst of an exo-Selective, Enantioselective Diels-Alder Reaction. *Org. Lett.* **2007**, 9, 2859-2862.

28. He, H., Pei, B.-J., Chou, H.-H., Tian, T., Chan, W.-H. & Lee, A. W. M. Camphor Sulfonyl Hydrazines (CaSH) as Organocatalysts in Enantioselective Diels-Alder Reactions. *Org. Lett.* **2008**, 10, 2421-2424.

29. Ishihara, K. & Nakano, K. Design of an Organocatalyst for the Enantioselective Diels-Alder Reaction with α-Acyloxyacroleins. *J. Am. Chem. Soc.* **2005**, 127, 10504-10505.

30. Ishihara, K., Nakano, K. & Akakura, M. Organocatalytic Enantioselective Diels-Alder Reaction of Dienes with α-(N,N-Diacylamino)acroleins. *Org. Lett.* **2008**, 10, 2893-2896.

31. Sakakura, A., Suzuki, K., Nakano, K. & Ishihara, K. Chiral 1,1'-Binaphthyl-2,2'-diammonium Salt Catalysts for the Enantioselective Diels-Alder Reaction with α-Acyloxyacroleins. *Org. Lett.* **2006**, 8, 2229-2232.

32. Ramachary, D. B., Chowdari, N. S. & Barbas, C. F., III. Organocatalytic asymmetric domino Knoevenagel/Diels-Alder reactions: a bioorganic approach to the diastereospecific and enantioselective construction of highly substituted spiro[5,5]undecane-1,5,9-triones. *Angew. Chem., Int. Ed.* **2003**, 42, 4233-4237.

33. Wang, Y., Li, H., Wang, Y.-Q., Liu, Y., Foxman, B. M. & Deng, L. Asymmetric Diels-Alder Reactions of 2-Pyrones with a Bifunctional Organic Catalyst. *J. Am. Chem. Soc.* **2007**, 129, 6364-6365.

34. Gioia, C., Hauville, A., Bernardi, L., Fini, F. & Ricci, A. Organocatalytic asymmetric Diels-Alder reactions of 3-vinylindoles. *Angew. Chem., Int. Ed.* **2008**, 47, 9236-9239.

35. Shen, J., Nguyen, T. T., Goh, Y.-P., Ye, W., Fu, X., Xu, J. & Tan, C.-H. Chiral Bicyclic Guanidine-Catalyzed Enantioselective Reactions of Anthrones. *J. Am. Chem. Soc.* **2006**, 128, 13692-13693.

36. Keck, G. E., Li, X.-Y. & Krishnamurthy, D. Catalytic Enantioselective Synthesis of Dihydropyrones via Formal Hetero Diels-Alder Reactions of "Danishefsky's Diene" with Aldehydes. *J. Org. Chem.* **1995**, 60, 5998-5999.

37. Felber, H., Kresze, G., Braun, H. & Vasella, A. Asymmetric Diels-Alder reactions with α-chloronitroso compounds - II. The use of a carbohydrate derived α-chloro-α-nitroso ether. *Tetrahedron Lett.* **1984**, 25, 5381-5382.

38. Gouverneur, V., Dive, G. & Ghosez, L. Asymmetric Diels-Alder reactions of a nitroso compound derived from D-bornane-10,2-sultam. *Tetrahedron Asymmetry* **1991**, 2, 1173-1176.

39. Stella, L., Abraham, H., Feneau-Dupont, J., Tinant, B. & Declercq, J. P. Asymmetric aza-Diels-Alder reaction using the chiral 1-phenylethylimine of methyl glyoxylate. *Tetrahedron Lett.* **1990**, 31, 2603-2606.

40. Bailey, P. D., Londesbrough, D. J., Hancox, T. C., Heffernan, J. D. & Holmes, A. B. Highly enantioselective synthesis of pipecolic acid derivatives via an asymmetric aza-Diels-Alder reaction. *J. Chem. Soc., Chem. Commun.* **1994**, 2543-2544.

41. Han, B., Li, J.-L., Ma, C., Zhang, S.-J. & Chen, Y.-C. Organocatalytic asymmetric inverse-electron-demand aza-Diels-Alder reaction of N-sulfonyl-1-aza-1,3-butadienes and aldehydes. *Angew. Chem., Int. Ed.* **2008**, 47, 9971-9974.

42. Audrain, H., Thorhauge, J., Hazell, R. G. & Jorgensen, K. A. A Novel Catalytic and Highly Enantioselective Approach for the Synthesis of Optically Active Carbohydrate Derivatives. *J. Org. Chem.* **2000**, 65, 4487-4497.

43. Itoh, J., Fuchibe, K. & Akiyama, T. Chiral Bronsted acid catalyzed enantioselective aza-Diels-Alder reaction of Brassard's diene with imines. *Angew. Chem., Int. Ed.* **2006**, 45, 4796-4798.

ENANTIOSELECTIVE-[3+2]-CYCLOADDITION

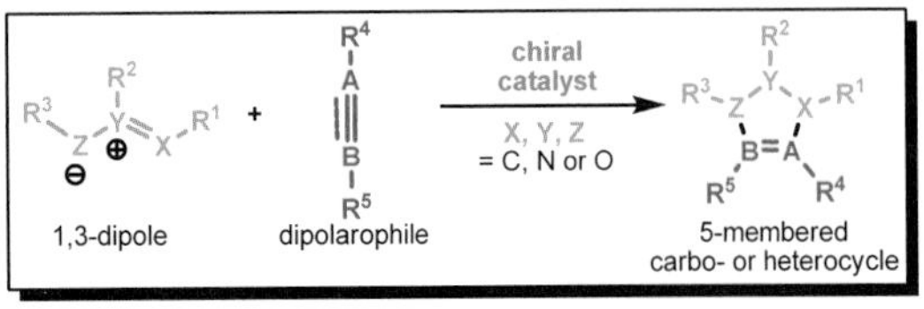

| 1,3-dipole | dipolarophile | 5-membered carbo- or heterocycle |

Introduction

There are many structural variants of [3+2]-cycloaddition involving as the three-atom component a dipolar triad (XYZ) or 1,3-diradical equivalent and a diad dipolarophile (A=B or A≡B). The transformation is a thermally allowed [4πs + 2πs] cycloaddition that leads to five-membered carbo- or heterocycles.[1] A selection of common 1,3-dipoles is shown in Figure 1. The dipolarophile may contain any double or triple bond, including C≡C, C=C, C≡N, C=N, C=O and C=S.

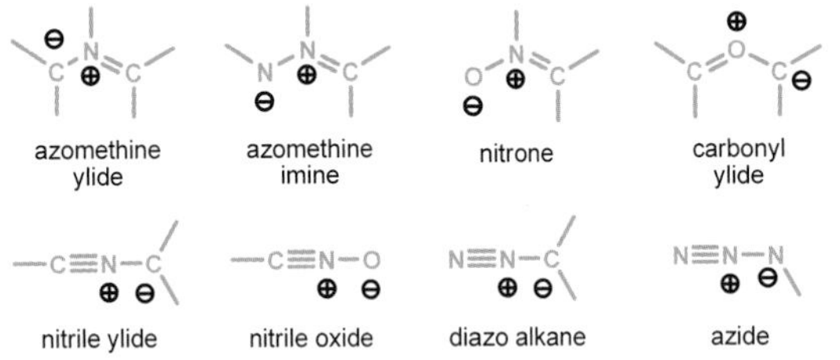

| azomethine ylide | azomethine imine | nitrone | carbonyl ylide |

| nitrile ylide | nitrile oxide | diazo alkane | azide |

Figure 1

The range of 1,3-dipolar cycloaddition reactions is very broad and includes both intramolecular and intermolecular cycloaddition with or without the formation of new stereocenters. The development of enantioselective variants of the [3+2]-cycloaddition has lagged the advances made in the enantioselective Diels-Alder reaction (see pages 121-127). However significant advances have been achieved during the past 15 years.[2]

The [3+2]-cycloaddition can be highly diastereoselective when new stereocenters are generated. The use of a removable chiral controller group may allow access to chiral [3+2]-adducts, as illustrated by the two examples in Scheme 1.

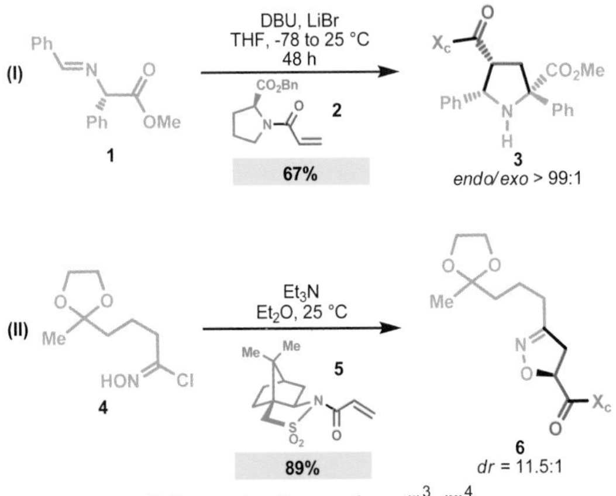

Reference key for equations: (I)[3]; (II)[4]

Scheme 1

Scheme 2 illustrates a number of intermolecular catalytic enantioselective 1,3-dipolar cycloadditions with a wide range of dipolarophiles. The catalysts may be chiral metal complexes or small organic molecules.

I. Catalytic Enantioselective Intermolecular [3+2]-Cycloaddition Reactions

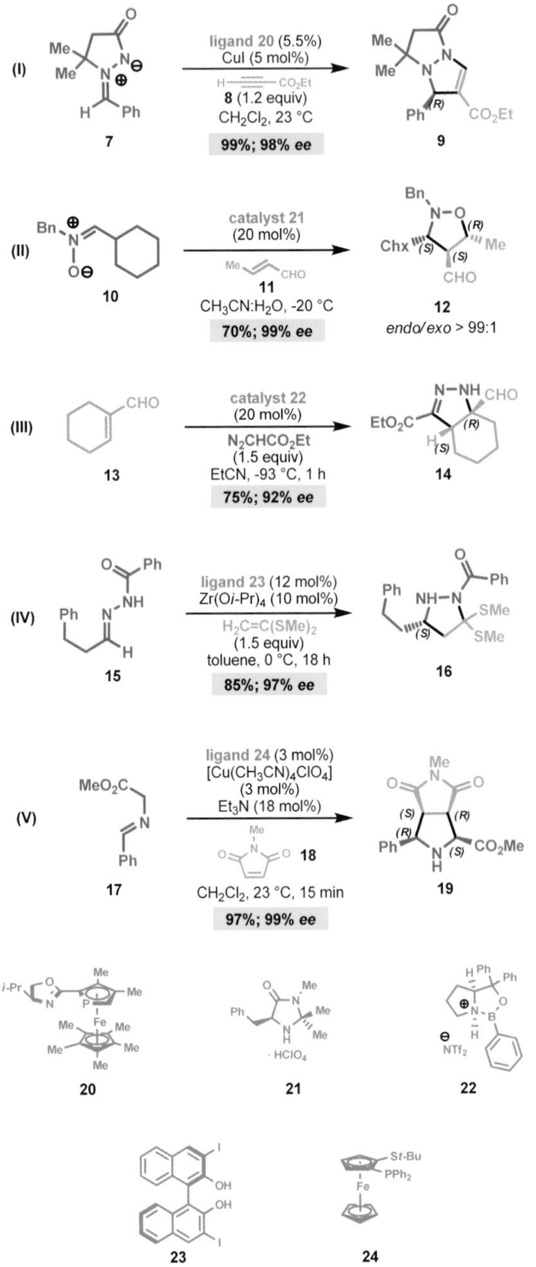

Reference key for equations: (I)[5]; (II)[6]; (III)[7]; (IV)[8]; (V)[9]

Scheme 2

ENANTIOSELECTIVE-[3+2]-CYCLOADDITION

II. Catalytic Enantioselective Inter- and Intramolecular [3+2]-Cycloaddition Reactions

The first example of a [3+2]-cycloaddition shown in Scheme 3 (below) involves a concomitant 1,2-rearrangement of the silyl subsituent. Examples II and III illustrate Rh-catalyzed intramolecular carbonyl-ylide formation and subsequent intramolecular [3+2]-cycloaddition. The last two examples (IV and V) occur via the addition of a trimethylene methane-Pd complex to electrophilic C=C and C=N, respectively.

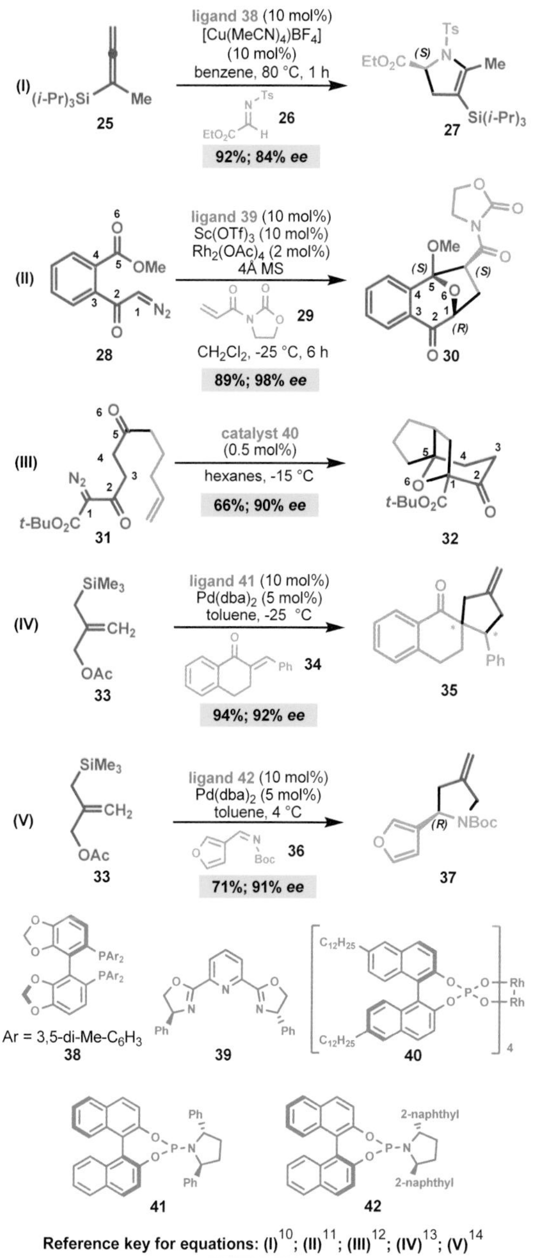

Reference key for equations: (I)[10]; (II)[11]; (III)[12]; (IV)[13]; (V)[14]

Scheme 3

References

1. (a) Huisgen, R. Adventures with heterocycles. *Chem. Pharm. Bull.* **2000**, 48, 757-765; (b) Molteni, G. Stereoselective cycloadditions of nitrilimines as a source of enantiopure heterocycles. *Heterocycles* **2005**, 65, 2513-2537; (c) Escolano, C., Duque, M. D. & Vazquez, S. Nitrile ylides: generation, properties and synthetic applications. *Curr. Org. Chem.* **2007**, 11, 741-772; (d) Nair, V. & Suja, T. D. Intramolecular 1,3-dipolar cycloaddition reactions in targeted syntheses. *Tetrahedron* **2007**, 63, 12247-12275; (e) Alvarez-Corral, M., Munoz-Dorado, M. & Rodriguez-Garcia, I. Silver-Mediated Synthesis of Heterocycles. *Chem. Rev.* **2008**, 108, 3174-3198; (f) Brandi, A., Cardona, F., Cicchi, S., Cordero, F. M. & Goti, A. Stereocontrolled Cyclic Nitrone Cycloaddition Strategy for the Synthesis of Pyrrolizidine and Indolizidine Alkaloids. *Chem.--Eur. J.* **2009**, 15, 7808-7821.

2. (a) Gothelf, K. V. & Jorgensen, K. A. Asymmetric 1,3-Dipolar Cycloaddition Reactions. *Chem. Rev.* **1998**, 98, 863-909; (b) Gothelf, K. V. & Jorgensen, K. A. Catalytic enantioselective 1,3-dipolar cycloaddition reactions of nitrones. *Chem. Commun. (Cambridge)* **2000**, 1449-1458; (c) Bonin, M., Chauveau, A. & Micouin, L. Asymmetric 1,3-dipolar cycloadditions of cyclic stabilized ylides derived from chiral 1,2-amino alcohols. *Synlett* **2006**, 2349-2363; (d) Pandey, G., Banerjee, P. & Gadre, S. R. Construction of Enantiopure Pyrrolidine Ring System via Asymmetric [3+2]-Cycloaddition of Azomethine Ylides. *Chem. Rev. (Washington, DC, U. S.)* **2006**, 106, 4484-4517; (e) Pellissier, H. Asymmetric 1,3-dipolar cycloadditions. *Tetrahedron* **2007**, 63, 3235-3285; (f) Stanley, L. M. & Sibi, M. P. Enantioselective Copper-Catalyzed 1,3-Dipolar Cycloadditions. *Chem. Rev.* **2008**, 108, 2887-2902; (g) Kissane, M. & Maguire, A. R. Asymmetric 1,3-dipolar cycloadditions of acryl amides. *Chem. Soc. Rev.* **2010**, 39, 845-883.

3. (a) Waldmann, H., Blaeser, E., Jansen, M. & Letschert, H. P. *Angew. Chem., Int. Ed.* **1994**, 33, 683-685; (b) Waldmann, H., Jansen, M. & Letschert, H. P. Asymmetric synthesis of highly substituted pyrrolidines by 1,3-dipolar cycloaddition of azomethine ylides to N-acrylylproline benzyl ester. *Angew. Chem.* **1994**, 106, 717-719.

4. Curran, D. P. & Heffner, T. A. On the scope of asymmetric nitrile oxide cycloadditions with Oppolzer's chiral sultam. Total syntheses of (+)-hepialone, (-)-(1R,3R,5S)-1,3-dimethyl-2,9-dioxabicyclo [3.3.1] nonane, and (-)-(1S)-7,7-dimethyl-6,8-dioxabicyclo [3.2.1] octane. *J. Org. Chem.* **1990**, 55, 4585-4595.

5. Shintani, R. & Fu, G. C. A New Copper-Catalyzed [3 + 2] Cycloaddition: Enantioselective Coupling of Terminal Alkynes with Azomethine Imines To Generate Five-Membered Nitrogen Heterocycles. *J. Am. Chem. Soc.* **2003**, 125, 10778-10779.

6. Jen, W. S., Wiener, J. J. M. & MacMillan, D. W. C. New Strategies for Organic Catalysis: The First Enantioselective Organocatalytic 1,3-Dipolar Cycloaddition. *J. Am. Chem. Soc.* **2000**, 122, 9874-9875.

7. Gao, L., Hwang, G.-S., Lee, M. Y. & Ryu, D. H. Catalytic enantioselective 1,3-dipolar cycloadditions of alkyl diazoacetates with α,β-disubstituted acroleins. *Chem. Commun. (Cambridge, U. K.)* **2009**, 5460-5462.

8. Yamashita, Y. & Kobayashi, S. Zirconium-Catalyzed Enantioselective [3+2] Cycloaddition of Hydrazones to Olefins Leading to Optically Active Pyrazolidine, Pyrazoline, and 1,3-Diamine Derivatives. *J. Am. Chem. Soc.* **2004**, 126, 11279-11282.

9. Daidouji, K., Fuchibe, K. & Akiyama, T. Cu(I)-Catalyzed Enantioselective [3 + 2] Cycloaddition Reaction of 1-Alkylallenylsilane with α-Imino Ester: Asymmetric Synthesis of Dehydroproline Derivatives. *Org. Lett.* **2005**, 7, 1051-1053.

10. Cabrera, S., Gomez Arrayas, R. & Carretero, J. C. Highly enantioselective copper(I)-fesulphos-catalyzed 1,3-dipolar cycloaddition of azomethine ylides. *J. Am. Chem. Soc.* **2005**, 127, 16394-16395.

11. Suga, H., Inoue, K., Inoue, S. & Kakehi, A. Highly Enantioselective 1,3-Dipolar Cycloaddition Reactions of 2-Benzopyrylium-4-olate Catalyzed by Chiral Lewis Acids. *J. Am. Chem. Soc.* **2002**, 124, 14836-14837.

12. Hodgson, D. M., Stupple, P. A. & Johnstone, C. Efficient RhII binaphthol phosphate catalysts for enantioselective intramolecular tandem carbonyl ylide formation-cycloaddition of α-diazo-β-keto esters. *Chem. Commun. (Cambridge)* **1999**, 2185-2186.

13. Trost, B. M., Stambuli, J. P., Silverman, S. M. & Schwoerer, U. Palladium-Catalyzed Asymmetric [3 + 2] Trimethylenemethane Cycloaddition Reactions. *J. Am. Chem. Soc.* **2006**, 128, 13328-13329.

14. Trost, B. M., Silverman, S. M. & Stambuli, J. P. Palladium-Catalyzed Asymmetric [3+2] Cycloaddition of Trimethylenemethane with Imines. *J. Am. Chem. Soc.* **2007**, 129, 12398-12399.

ENANTIOSELECTIVE ALL-C-[2+2]-CYCLOADDITION

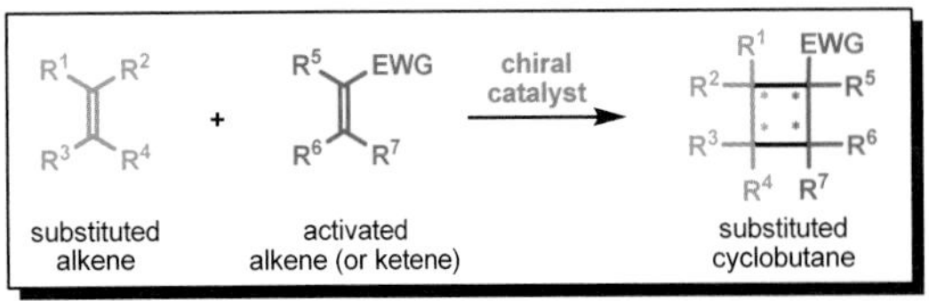

substituted alkene + activated alkene (or ketene) → substituted cyclobutane

I. Synthesis of Cyclobutanes via [2+2]-Cycloaddition Utilizing Chiral Controller Groups

The [2+2]-cycloaddition pathway to four-membered carbocyclic structures can be rendered enantioselective either by the application of a chiral controller group or by the use of a chiral Lewis acid catalyst.[1] The former approach (e.g., with phenylmenthyl, PM), which provided the earliest access to chiral cyclobutanes, is illustrated by the examples shown in Scheme 1.

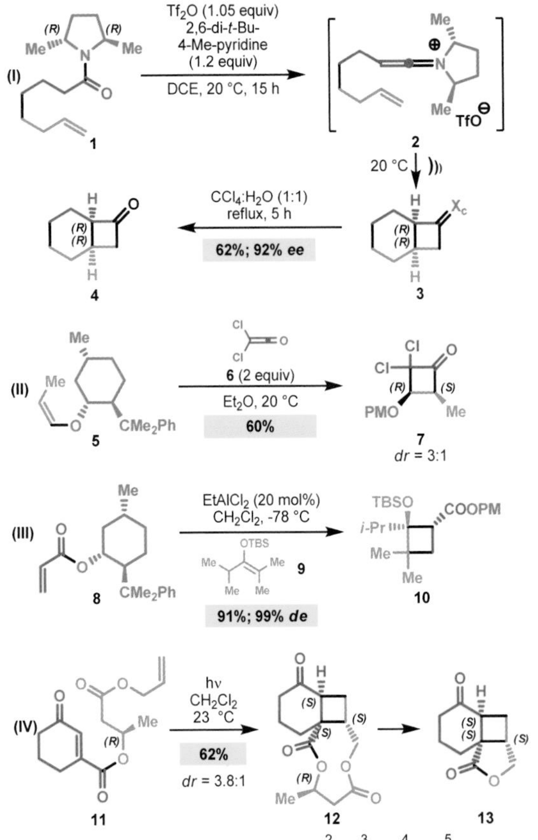

Reference key for equations: (I)[2]; (II)[3]; (III)[4]; (IV)[5]

Scheme 1

II. Catalytic Enantioselective [2+2]-Cycloaddition

The enantioselective formation of cyclobutanes by the use of chiral Lewis acids was first described by K. Narasaka.[6] Two typical reactions using his approach involving a chiral Ti(IV)-catalyst appear in Scheme 2 (Equations I and II). A chiral oxazaborolidine activated by

coordination with AlBr₃ is very effective for the catalytic enantioselective [2+2]-cycloaddition of enol ethers to trifluoroethyl acrylate, as illustrated by a typical example (Equation III) in Scheme 2.

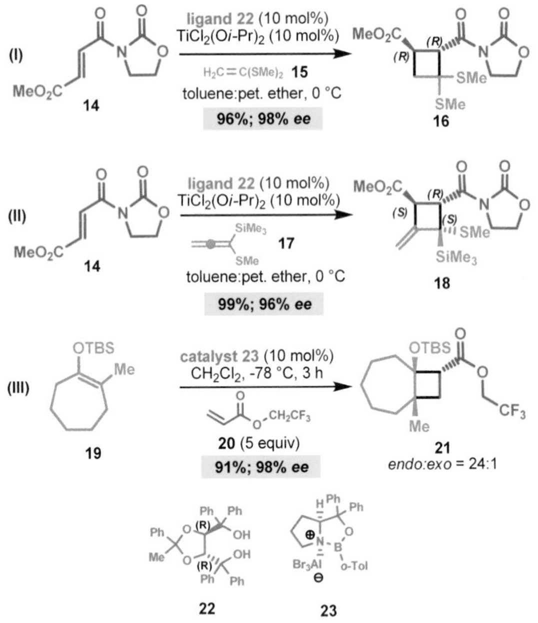

Reference key for equations: (I)[6]; (II)[7]; (III)[8]

Scheme 2

References

1. (a) Narasaka, K. & Hayashi, Y. Lewis acid catalyzed [2+2] cycloaddition reactions of vinyl sulfides and their analogs: catalytic asymmetric [2+2] cycloaddition reactions. *Adv. Cycloaddit.* **1997**, 4, 87-120; (b) Hayashi, Y. & Narasaka, K. [2+2] Cycloaddition reactions. *Compr. Asymmetric Catal. I-III* **1999**, 3, 1255-1269; (c) Lee-Ruff, E. & Mladenova, G. Enantiomerically Pure Cyclobutane Derivatives and Their Use in Organic Synthesis. *Chem. Rev.* **2003**, 103, 1449-1483; (d) Lee-Ruff, E. Synthesis of cyclobutanes. *Chemistry of Cyclobutanes* **2005**, 1, 281-355; (e) Carreira, E. M. & Kvaerno, L. [3+2]- and [2+2]-Cycloaddition Reactions in *Classics in Stereoselective Synthesis* 589-622 (Wiley-VCH, **2008**).
2. Chen, L. Y. & Ghosez, L. Study of chiral auxiliaries for the intramolecular [2+2] cycloaddition of a keteniminium salt to an olefinic double bond. A new asymmetric synthesis of cyclobutanones. *Tetrahedron Lett.* **1990**, 31, 4467-4470.
3. Greene, A. E. & Charbonnier, F. Asymmetric induction in the cycloaddition reaction of dichloroketene with chiral enol ethers. A versatile approach to optically active cyclopentenone derivatives. *Tetrahedron Lett.* **1985**, 26, 5525-5528.
4. Takasu, K., Nagao, S., Ueno, M. & Ihara, M. An auxiliary induced asymmetric synthesis of functionalized cyclobutanes by means of catalytic (2+2)-cycloaddition reaction. *Tetrahedron* **2004**, 60, 2071-2078.
5. Faure, S., Piva-Le Blanc, S., Piva, O. & Pete, J.-P. Hydroxy acids as efficient chiral spacers for asymmetric intramolecular [2+2] photocycloadditions. *Tetrahedron Lett.* **1997**, 38, 1045-1048.
6. Hayashi, Y. & Narasaka, K. Asymmetric [2 + 2] cycloaddition reaction catalyzed by a chiral titanium reagent. *Chem. Lett.* **1989**, 793-796.
7. Hayashi, Y., Niihata, S. & Narasaka, K. [2+2] Cycloaddition reaction between allenyl sulfides and electron deficient olefins promoted by Lewis acids. *Chem. Lett.* **1990**, 2091-2094.
8. Canales, E. & Corey, E. J. Highly Enantioselective [2+2]-Cycloaddition Reactions Catalyzed by a Chiral Aluminum Bromide Complex. *J. Am. Chem. Soc.* **2007**, 129, 12686-12687.

HETERO-[2+2]-CYCLOADDITION REACTIONS

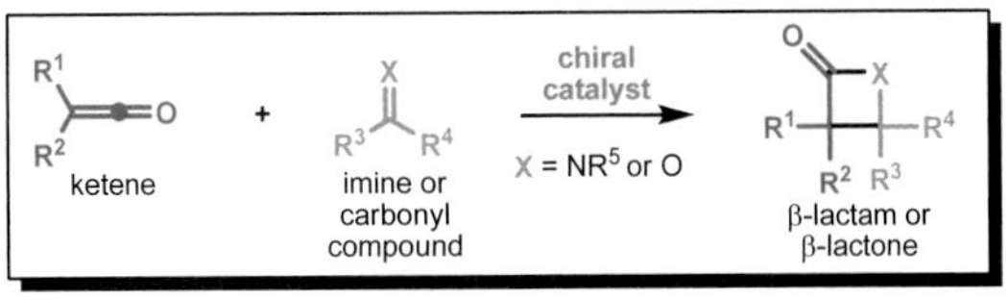

Introduction

The synthesis of β-lactams by the [2+2]-cycloaddition of a ketene to an imine (Staudinger reaction) is the principal route to this class of compounds, which includes important antibiotics of the penicillin, cephalosporin and monobactam families. After an earlier approach[1] to enantiomerically enriched β-lactams involving a removable chiral controller group (Scheme 1), a number of direct catalytic enantioselective syntheses were developed (see examples in Scheme 3).[2]

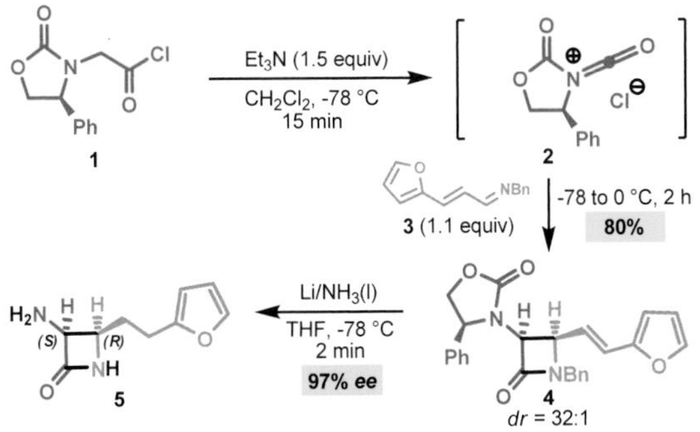

Scheme 1

I. Catalytic Enantioselective [2+2]-Cycloaddition of Ketenes with Imines

The uncatalyzed Staudinger reaction between a ketene (as electrophile) and an imine (as nucleophile) takes place rapidly at ambient temperature. Nonetheless, the reaction can be accelerated by the use of a tertiary amine which adds to the ketene (**6**) to form a dipolar enolate (**7**) that is a strong enough nucleophile to attack the imine (**8**) rapidly, as illustrated in Scheme 2. This catalytic process, a type of two-step [2+2]-cycloaddition, is exemplified by the six reactions shown in Scheme 3. The amine-catalyzed Staudinger reaction provides a practical route to β-lactams from imine derivatives made more electrophilic by an N-acyl or N-sulfonyl group.

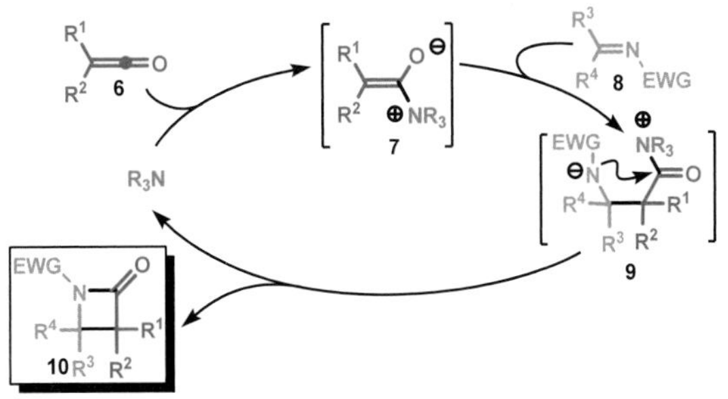

Scheme 2

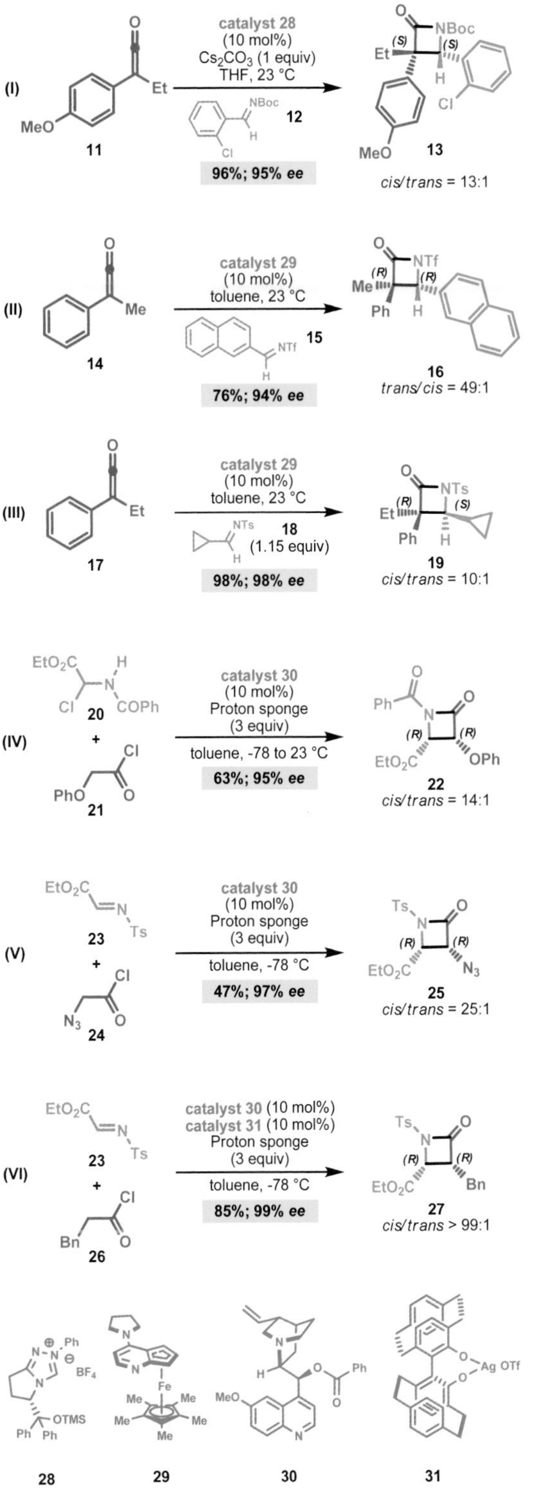

(I) 11 + 12 → 13; catalyst 28 (10 mol%), Cs$_2$CO$_3$ (1 equiv), THF, 23 °C; 96%; 95% ee; cis/trans = 13:1

(II) 14 + 15 → 16; catalyst 29 (10 mol%), toluene, 23 °C; 76%; 94% ee; trans/cis = 49:1

(III) 17 + 18 (1.15 equiv) → 19; catalyst 29 (10 mol%), toluene, 23 °C; 98%; 98% ee; cis/trans = 10:1

(IV) 20 + 21 → 22; catalyst 30 (10 mol%), Proton sponge (3 equiv), toluene, -78 to 23 °C; 63%; 95% ee; cis/trans = 14:1

(V) 23 + 24 → 25; catalyst 30 (10 mol%), Proton sponge (3 equiv), toluene, -78 °C; 47%; 97% ee; cis/trans = 25:1

(VI) 23 + 26 → 27; catalyst 30 (10 mol%), catalyst 31 (10 mol%), Proton sponge (3 equiv), toluene, -78 °C; 85%; 99% ee; cis/trans > 99:1

Reference key for equations: (I)[3]; (II)[4]; (III)[5]; (IV)[6]; (V)[7]; (VI)[8]

Scheme 3

HETERO-[2+2]-CYCLOADDITION REACTIONS

II. Catalytic Enantioselective [2+2]-Cycloaddition of Ketenes with Aldehydes and Ketones

The first enantioselective synthesis of a β-lactone was reported by H. Wynberg from ketene (**32**) and chloral (**33**) with quinidine (**33**) as a catalyst.[9] The most likely pathway involves the addition of the nucleophilic intermediate **32a** to chloral to give adduct **36** (Scheme 4).

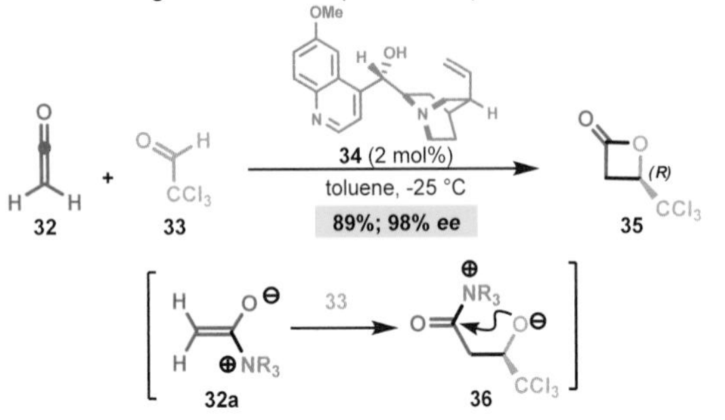

Scheme 4

A number of variants of this two-step [2+2]-cycloaddition to form an enantiomerically encriched β-lactone were subsequently developed, as illustrated by four examples in Schemes 5 and 6.

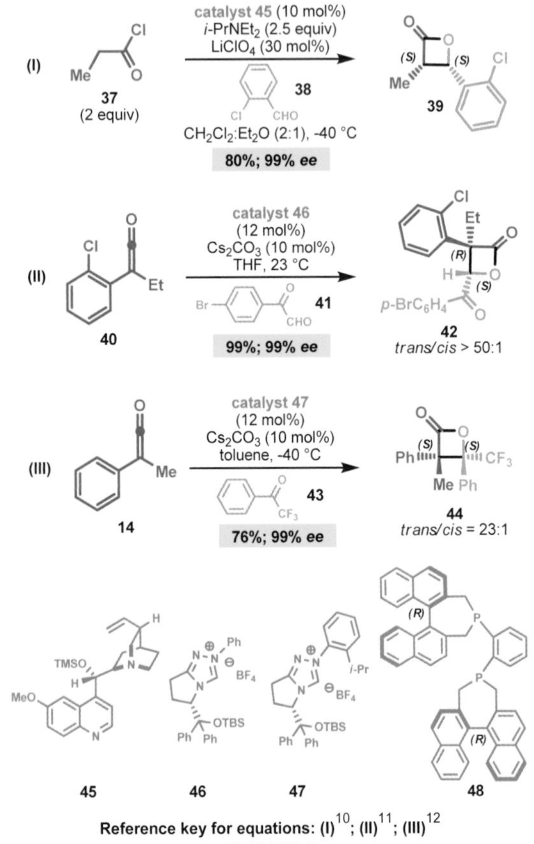

Reference key for equations: (I)[10]; (II)[11]; (III)[12]

Scheme 5

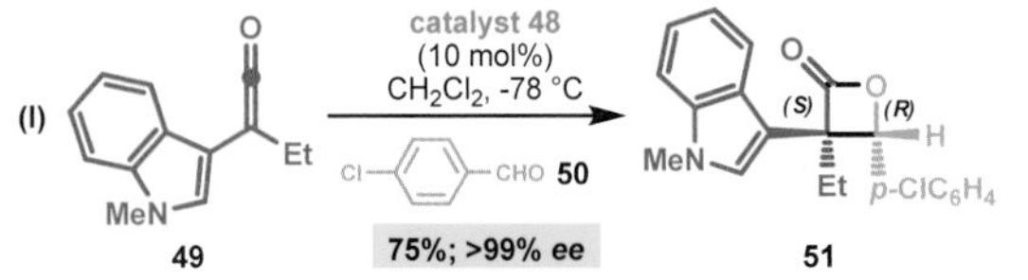

Reference key for equation: (I)[13]

Scheme 6

References

1. Evans, D. A. & Sjogren, E. B. The asymmetric synthesis of β-lactam antibiotics. I. Application of chiral oxazolidones in the Staudinger reaction. *Tetrahedron Lett.* **1985**, 26, 3783-3786.
2. (a) Hyatt, J. A. & Raynolds, P. W. Ketene cycloadditions. *Org. React.* **1994**, 45, 159-646; (b) Palomo, C., Aizpurua, J. M., Ganboa, I. & Oiarbide, M. Asymmetric synthesis of β-lactams through the Staudinger reaction and their use as building blocks of natural and non-natural products. *Curr. Med. Chem.* **2004**, 11, 1837-1872; (c) Ilyas, B., Ajmal, M., Imran, M. & Khan, S. A. Chemistry and biological activities of 2-azetidinones. *Oriental Journal of Chemistry* **2005**, 21, 511-524; (d) Tidwell, T. T. Product class 3: halogen-substituted ketenes. *Sci. Synth.* **2006**, 23, 101-168; (e) Gaunt, M. J. & Johansson, C. C. C. Recent Developments in the Use of Catalytic Asymmetric Ammonium Enolates in Chemical Synthesis. *Chem. Rev. (Washington, DC, U. S.)* **2007**, 107, 5596-5605; (f) Cardillo, G., Gentilucci, L. & Tolomelli, A. Asymmetric synthesis of three- and four-membered ring heterocycles. *Asymmetric Synthesis of Nitrogen Heterocycles* **2009**, 3-50; (g) Paull, D. H., Weatherwax, A. & Lectka, T. Catalytic, asymmetric reactions of ketenes and ketene enolates. *Tetrahedron* **2009**, 65, 6771-6803.
3. Zhang, Y.-R., He, L., Wu, X., Shao, P.-L. & Ye, S. Chiral N-Heterocyclic Carbene Catalyzed Staudinger Reaction of Ketenes with Imines: Highly Enantioselective Synthesis of N-Boc β-Lactams. *Org. Lett.* **2008**, 10, 277-280.
4. Lee, E. C., Hodous, B. L., Bergin, E., Shih, C. & Fu, G. C. Catalytic Asymmetric Staudinger Reactions to Form β-Lactams: An Unanticipated Dependence of Diastereoselectivity on the Choice of the Nitrogen Substituent. *J. Am. Chem. Soc.* **2005**, 127, 11586-11587.
5. Hodous, B. L. & Fu, G. C. Enantioselective Staudinger Synthesis of β-Lactams Catalyzed by a Planar-Chiral Nucleophile. *J. Am. Chem. Soc.* **2002**, 124, 1578-1579.
6. Hafez, A. M., Dudding, T., Wagerle, T. R., Shah, M. H., Taggi, A. E. & Lectka, T. A Multistage, One-Pot Procedure Mediated by a Single Catalyst: A New Approach to the Catalytic Asymmetric Synthesis of β-Amino Acids. *J. Org. Chem.* **2003**, 68, 5819-5825.
7. Taggi, A. E., Hafez, A. M., Wack, H., Young, B., Ferraris, D. & Lectka, T. The Development of the First Catalyzed Reaction of Ketenes and Imines: Catalytic, Asymmetric Synthesis of β-Lactams. *J. Am. Chem. Soc.* **2002**, 124, 6626-6635.
8. Wack, H., France, S., Hafez, A. M., Drury, W. J., III, Weatherwax, A. & Lectka, T. Development of a New Dimeric Cyclophane Ligand: Application to Enhanced Diastereo- and Enantioselectivity in the Catalytic Synthesis of β-Lactams. *J. Org. Chem.* **2004**, 69, 4531-4533.
9. (a) Wynberg, H. & Staring, E. G. J. Asymmetric synthesis of (S)- and (R)-malic acid from ketene and chloral. *J. Am. Chem. Soc.* **1982**, 104, 166-168; (b) Wynberg, H. & Staring, E. G. J. Catalytic asymmetric synthesis of chiral 4-substituted 2-oxetanones. *J. Org. Chem.* **1985**, 50, 1977-1979.
10. Zhu, C., Shen, X. & Nelson, S. G. Cinchona alkaloid-Lewis acid catalyst systems for enantioselective ketene-aldehyde cycloadditions. *J. Am. Chem. Soc.* **2004**, 126, 5352-5353.
11. He, L., Lv, H., Zhang, Y.-R. & Ye, S. Formal Cycloaddition of Disubstituted Ketenes with 2-Oxoaldehydes Catalyzed by Chiral N-Heterocyclic Carbenes. *J. Org. Chem.* **2008**, 73, 8101-8103.
12. Wang, X.-N., Shao, P.-L., Lv, H. & Ye, S. Enantioselective Synthesis of β-Trifluoromethyl-β-lactones via NHC-Catalyzed Ketene-Ketone Cycloaddition Reactions. *Org. Lett.* **2009**, 11, 4029-4031.
13. Mondal, M., Ibrahim, A. A., Wheeler, K. A. & Kerrigan, N. J. Phosphine-Catalyzed Asymmetric Synthesis of β-Lactones from Arylketoketenes and Aromatic Aldehydes. *Org. Lett.* **2010**, 12, 1664-1667.

ENANTIOSELECTIVE-[2+1]-CYCLOADDITION

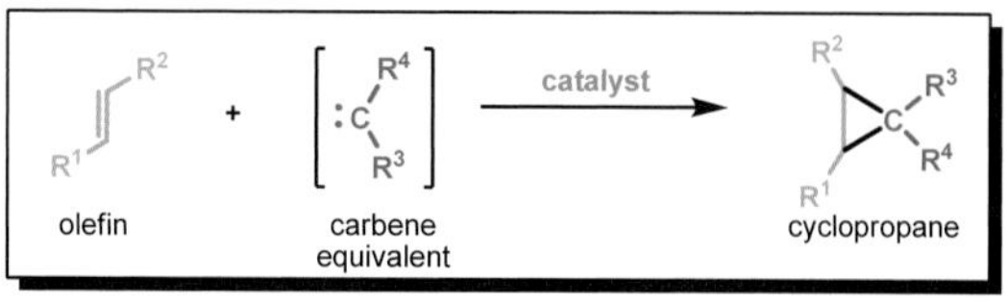

Introduction

Although the synthesis of cyclopropanes by the [2+1]-cycloaddition reaction of a carbene equivalent to a carbon-carbon double bond is conceptually a very simple construction,[1] it was not until the demonstration that dichlorocarbene (from CHCl₃ and a base) adds readily to olefins to form dichlorocyclopropanes (1954) that this approach found widespread use.[2] Even earlier, Buchner and his group had observed the formation of cyclopropylcarbonyl esters from ethyl diazoacetate and olefins in the presence of copper salts.[3] These methods and the Simmons-Smith reaction (1958) for adding ":CH₂" to C=C (using Zn and CH₂I₂) greatly increased the utility of the [2+1]-cycloaddition route to cyclopropanes.[4] In the ensuing years a number of highly diastereoselective inter- and intramolecular olefin cyclopropanation reactions have been demonstrated.[5] Subsequently, enantioselective [2+1]-cycloaddition reactions have been developed using complexes of the Simmons-Smith reagent (ICH₂CH₂ZnI) with various ligands and the reaction of α-diazocarbonyl compounds with olefins with chiral complexes of Cu (I) and Rh (II) as catalysts.[6]

I. Catalytic Enantioselective Simmons-Smith Cyclopropanation

Some examples of the catalytic enantioselective Simmons-Smith reaction are outlined in Scheme 1.

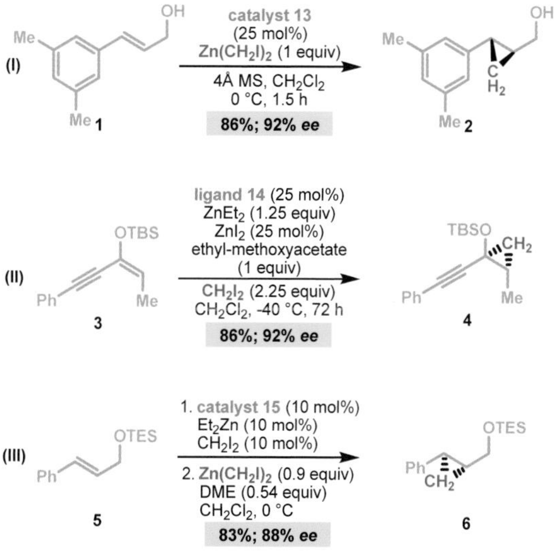

Reference key for equations: (I)[7]; (II)[8]; (III)[9]

Scheme 1

There are many examples of functional group-directed diastereoselective Simmons-Smith addition of ":CH₂" to olefins, e.g., the reactions shown in Scheme 2.

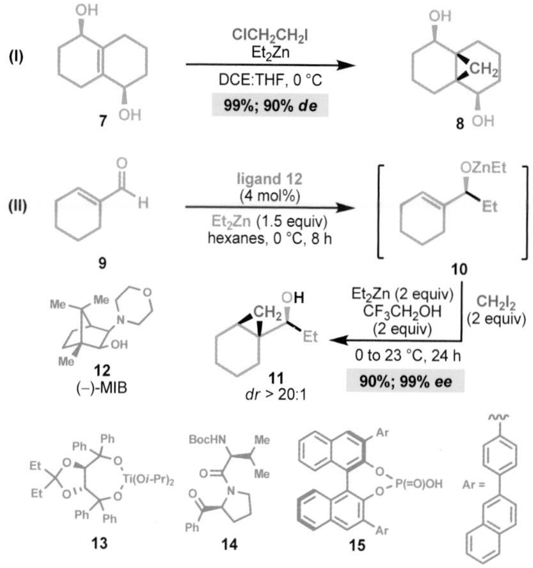

Reference key for equations: (I)[10]; (II)[11]

Scheme 2

II. Catalytic Enantioselective Cyclopropanation Utilizing Transition Metal Carbenoids Generated from Diazo Compounds

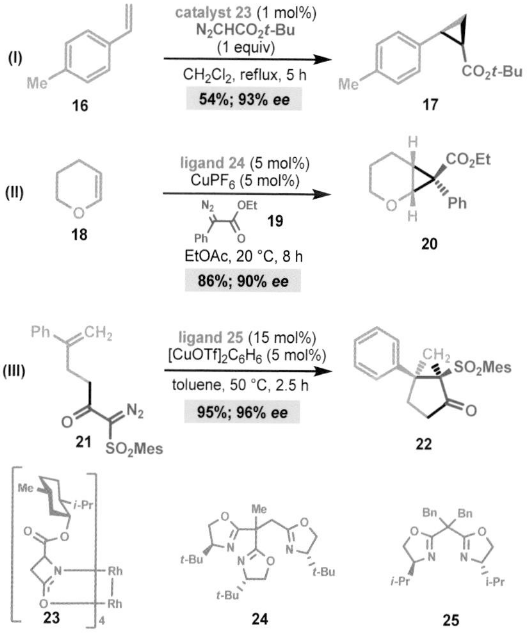

Reference key for equations: (I)[12]; (II)[13]; (III)[14]

Scheme 3

ENANTIOSELECTIVE-[2+1]-CYCLOADDITION

III. Catalytic Enantioselective Cyclopropanation Utilizing Ylides or α-Halo Carbanions

Catalytic enantioselective [2+1]-cycloaddition of electron-deficient alkenes (e.g., α,β-unsaturated carbonyl compounds, nitroolefins) and ylides or α-halo carbanions provides an enantioselective pathway to cyclopropanes as shown by the examples in Scheme 4. (For analogous [2+1]-cycloaddition to C=O, see pages 82-83).

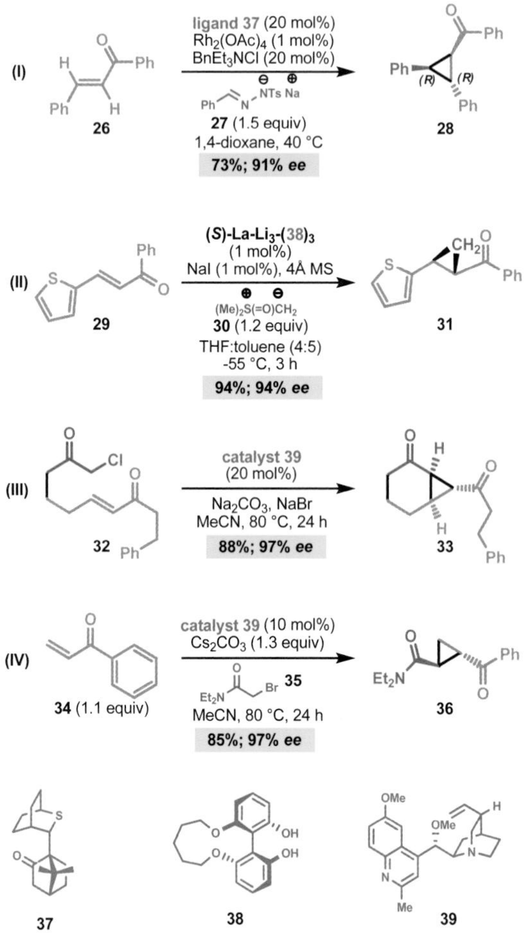

Reference key for equations: (I)[15]; (II)[16]; (III)[17]; (IV)[18]

Scheme 4

References

1. (a) Salaun, J. Cyclopropane derivatives and their diverse biological activities. *Top. Curr. Chem.* **2000**, 207, 1-67; (b) Donaldson, W. A. Synthesis of cyclopropane containing natural products. *Tetrahedron* **2001**, 57, 8589-8627; (c) de Meijere, A., Kozhushkov, S. I. & Schill, H. Three-Membered-Ring-Based Molecular Architectures. *Chem. Rev. (Washington, DC, U. S.)* **2006**, 106, 4926-4996.

2. Doering, W. v. E. & Hoffmann, A. K. The addition of dichlorocarbene to olefins. *J. Am. Chem. Soc.* **1954**, 76, 6162-6165.

3. Dave, V. & Warnhoff, E. W. Reactions of diazoacetic esters with alkenes, alkynes, heterocyclic and aromatic compounds. *Org. React.* **1970**, 18, 217-401.

4. (a) Simmons, H. E. & Smith, R. D. A new synthesis of cyclopropanes from olefins. *J. Am. Chem. Soc.* **1958**, 80, 5323-5324; (b) Simmons, H. E. & Smith, R. D. A new synthesis of cyclopropanes. *J. Am. Chem. Soc.* **1959**, 81, 4256-4264.

5. (a) Davies, H. M. L. & Antoulinakis, E. G. Intermolecular metal-catalyzed carbenoid cyclopropanations. *Org. React.* **2001**, 57, 1-326; (b) Denmark, S. E. & Beutner, G. Enantioselective [2+1] cycloaddition: cyclopropanation with zinc carbenoids. *Cycloaddit. React. Org. Synth.* **2002**, 85-150; (c) Pfaltz, A. Cyclopropanation. *Transition Metals for Organic Synthesis (2nd Edition)* **2004**, 1, 157-170; (d) Maas, G. Ruthenium-catalyzed carbenoid cyclopropanation reactions with diazo compounds. *Chem. Soc. Rev.* **2004**, 33, 183-190; (e) Li, F. & Reiser, O. Recent advances in catalytic intramolecular cyclopropanation reactions. *Chemtracts* **2006**, 19, 391-401.

6. (a) Singh, V. K., DattaGupta, A. & Sekar, G. Catalytic enantioselective cyclopropanation of olefins using carbenoid chemistry. *Synthesis* **1997**, 137-149; (b) Doyle, M. P. & Protopopova, M. N. New aspects of catalytic asymmetric cyclopropanation. *Tetrahedron* **1998**, 54, 7919-7946; (c) Muller, P., Allenbach, Y. F., Chappellet, S. & Ghanem, A. Asymmetric cyclopropanations and cycloadditions of dioxocarbenes. *Synthesis* **2006**, 1689-1696; (d) Nicolas, I., Le Maux, P. & Simonneaux, G. Asymmetric catalytic cyclopropanation reactions in water. *Coord. Chem. Rev.* **2008**, 252, 727-735; (e) Pellissier, H. Recent developments in asymmetric cyclopropanation. *Tetrahedron* **2008**, 64, 7041-7095; (f) Honma, M., Takeda, H., Takano, M. & Nakada, M. Development of catalytic asymmetric intramolecular cyclopropanation of α-diazo-β-keto sulfones and applications to natural product synthesis. *Synlett* **2009**, 1695-1712; (g) Goudreau, S. R. & Charette, A. B. Defying Ring Strain: New Approaches to Cyclopropanes. *Angew. Chem., Int. Ed.* **2010**, 49, 486-488.

7. Charette, A. B., Molinaro, C. & Brochu, C. Catalytic Asymmetric Cyclopropanation of Allylic Alcohols with Titanium-TADDOLate: Scope of the Cyclopropanation Reaction. *J. Am. Chem. Soc.* **2001**, 123, 12168-12175.

8. Du, H., Long, J. & Shi, Y. Catalytic Asymmetric Simmons-Smith Cyclopropanation of Silyl Enol Ethers. Efficient Synthesis of Optically Active Cyclopropanol Derivatives. *Org. Lett.* **2006**, 8, 2827-2829.

9. Lacasse, M.-C., Poulard, C. & Charette, A. B. Iodomethylzinc phosphates: powerful reagents for the cyclopropanation of alkenes. *J. Am. Chem. Soc.* **2005**, 127, 12440-12441.

10. El Sheikh, S., Kausch, N., Lex, J., Neudoerfl, J. M. & Schmalz, H.-G. Enantioselective synthesis of bicyclo[4.4.1]undecane-2,7-dione via samarium(II)-mediated fragmentation of a cyclopropane precursor. *Synlett* **2006**, 1527-1530.

11. Kim, H. Y., Lurain, A. E., Garcia-Garcia, P., Carroll, P. J. & Walsh, P. J. Highly Enantio- and Diastereoselective Tandem Generation of Cyclopropyl Alcohols with up to Four Contiguous Stereocenters. *J. Am. Chem. Soc.* **2005**, 127, 13138-13139.

12. Hu, W., Timmons, D. J. & Doyle, M. P. In Search of High Stereocontrol for the Construction of cis-Disubstituted Cyclopropane Compounds. Total Synthesis of a Cyclopropane-Configured Urea-PETT Analog That Is a HIV-1 Reverse Transcriptase Inhibitor. *Org. Lett.* **2002**, 4, 901-904.

13. Xu, Z.-H., Zhu, S.-N., Sun, X.-L., Tang, Y. & Dai, L.-X. Sidearm effects in the enantioselective cyclopropanation of alkenes with aryldiazoacetates catalyzed by trisoxazoline/Cu(I). *Chem. Commun.* **2007**, 1960-1962.

14. Sawada, T. & Nakada, M. Asymmetric catalysis of intramolecular cyclopropanation of 5-aryl-1-diazo-1-mesitylsulfonyl-5-hexen-2-ones. *Adv. Synth. Catal.* **2005**, 347, 1527-1532.

15. Aggarwal, V. K., Alonso, E., Fang, G., Ferrara, M., Hynd, G. & Porcelloni, M. Application of chiral sulfides to catalytic asymmetric aziridination and cyclopropanation with in situ generation of the diazo compound. *Angew. Chem., Int. Ed.* **2001**, 40, 1433-1436.

16. Kakei, H., Sone, T., Sohtome, Y., Matsunaga, S. & Shibasaki, M. Catalytic Asymmetric Cyclopropanation of Enones with Dimethyloxosulfonium Methylide Promoted by a La-Li3-(Biphenyldiolate)3 + NaI Complex. *J. Am. Chem. Soc.* **2007**, 129, 13410-13411.

17. Johansson, C. C. C., Bremeyer, N., Ley, S. V., Owen, D. R., Smith, S. C. & Gaunt, M. J. Enantioselective catalytic intramolecular cyclopropanation using modified cinchona alkaloid organocatalysts. *Angew. Chem., Int. Ed.* **2006**, 45, 6024-6028.

18. Papageorgiou, C. D., Cubillo de Dios, M. A., Ley, S. V. & Gaunt, M. J. Enantioselective organocatalytic cyclopropanation via ammonium ylides. *Angew. Chem., Int. Ed.* **2004**, 43, 4641-4644.

COMPLEX PSEUDO-CYCLOADDITION REACTIONS

Introduction: Catalytic Enantioselective Pseudo Cycloaddition

There are ring-forming reactions that have the appearance of cycloaddition processes, but actually involve a number of discrete steps and intermediates.[1]

I. Catalytic Enantioselective [4+3]-Cycloaddition

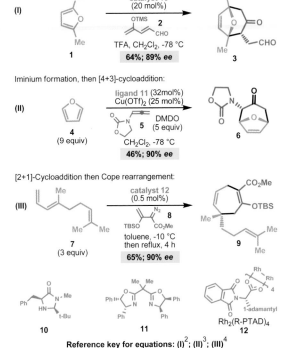

Reference key for equations: (I)[2]; (II)[3]; (III)[4]

Scheme 1

II. Catalytic Enantioselective [5+2]-Cycloaddition

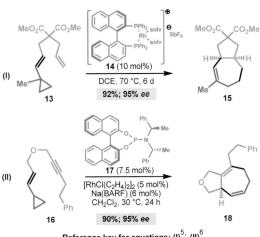

Reference key for equations: (I)[5]; (II)[6]

Scheme 2

III. Catalytic Enantioselective [2+2+2]-Cycloaddition

The reactions shown in Scheme 3 illustrate double cyclization via a 5,5-fused bicycle containing Rh followed by further addition to C≡C (Example I) or C=C (Example II).

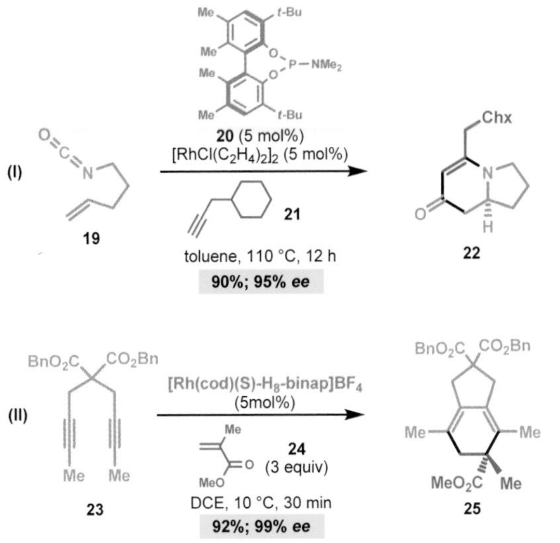

(I)

19 + 21

20 (5 mol%)
[RhCl(C₂H₄)₂]₂ (5 mol%)

toluene, 110 °C, 12 h
90%; 95% ee

22

(II)

23

[Rh(cod)(S)-H₈-binap]BF₄ (5mol%)

24 (3 equiv)

DCE, 10 °C, 30 min
92%; 99% ee

25

Reference key for equations: (I)[7]; (II)[8]

Scheme 3

References

1. (a) Evans, P. A. *Modern Rhodium-Catalyzed Organic Reactions* (Wiley-VCH, **2005**); (b) Harmata, M. Asymmetric catalytic [4+3] cycloaddition reactions. *Adv. Synth. Catal.* **2006**, 348, 2297-2306; (c) Zhang, J. & Xiao, Y. Cyclization of cyclopropane- or cyclopropene-containing compounds in *Handbook of Cyclization Reactions* (ed. Ma, S.-M.), 733-812 (Wiley-VCH, **2010**).

2. Harmata, M., Ghosh, S. K., Hong, X., Wacharasindhu, S. & Kirchhoefer, P. Asymmetric Organocatalysis of [4+3] Cycloaddition Reactions. *J. Am. Chem. Soc.* **2003**, 125, 2058-2059.

3. Huang, J. & Hsung, R. P. Chiral Lewis acid-catalyzed highly enantioselective [4 + 3] cycloaddition reactions of nitrogen-stabilized oxyallyl cations derived from allenamides. *J. Am. Chem. Soc.* **2005**, 127, 50-51.

4. Schwartz, B. D., Denton, J. R., Lian, Y., Davies, H. M. L. & Williams, C. M. Asymmetric [4 + 3] Cycloadditions between Vinylcarbenoids and Dienes: Application to the Total Synthesis of the Natural Product (-)-5-epi-Vibsanin E. *J. Am. Chem. Soc.* **2009**, 131, 8329-8332.

5. Wender, P. A., Haustedt, L. O., Lim, J., Love, J. A., Williams, T. J. & Yoon, J.-Y. Asymmetric Catalysis of the [5 + 2] Cycloaddition Reaction of Vinylcyclopropanes and π-Systems. *J. Am. Chem. Soc.* **2006**, 128, 6302-6303.

6. Shintani, R., Nakatsu, H., Takatsu, K. & Hayashi, T. Rhodium-Catalyzed Asymmetric [5+2] Cycloaddition of Alkyne-Vinylcyclopropanes. *Chem.--Eur. J.* **2009**, 15, 8692-8694.

7. Yu, R. T., Lee, E. E., Malik, G. & Rovis, T. Total synthesis of indolizidine alkaloid (-)-209D: overriding substrate bias in the asymmetric rhodium-catalyzed [2+2+2] cycloaddition. *Angew. Chem., Int. Ed.* **2009**, 48, 2379-2382.

8. Tsuchikama, K., Kuwata, Y. & Shibata, T. Highly Enantioselective Construction of a Chiral Spirocyclic Structure by the [2+2+2] Cycloaddition of Diynes and exo-Methylene Cyclic Compounds. *J. Am. Chem. Soc.* **2006**, 128, 13686-13687.

INTRAMOLECULAR MICHAEL ADDITION

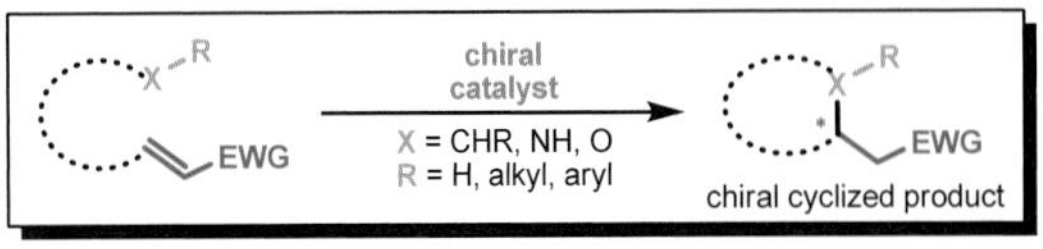

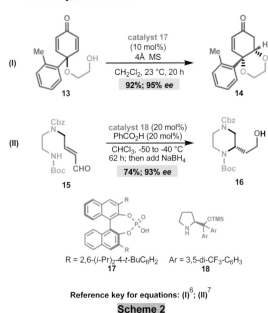

Introduction

A number of chiral pyrrolidine type bases are capable of catalyzing enantiocontrolled intramolecular Michael addition (conjugate addition) of an intermediate enamine to an α,β-enone acceptor.[1] Three examples (Equations I-III) of such reactions to generate a 5- or 6-membered carbocycle are shown in Scheme 1. Also shown in Scheme 1 is an example (Equation IV, **7→8**) of an intramolecular Michael cyclization combined with a vinylogous Stetter reaction (see pages 115-116). Two examples of heterocyclic ring-closure by enantioselective hetero-Michael pathways appear in Scheme 2.

I. Catalytic Enantioselective Intramolecular Conjugate Addition of C-Nucleophiles: Carbocycle Formation

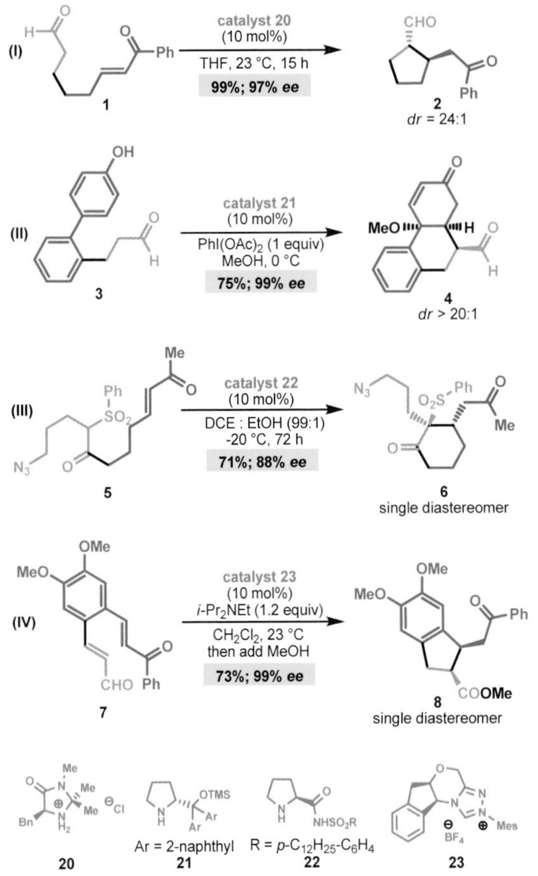

II. Catalytic Enantioselective Intramolecular Conjugate Addition of O- and N-Nucleophiles: Heterocycle Formation

Reference key for equations: (I)[6]; (II)[7]

Scheme 2

References

1. (a) Little, R. D., Masjedizadeh, M. R., Wallquist, O. & McLoughlin, J. I. The intramolecular Michael reaction. *Org. React.* **1995**, 47, 315-552; (b) Vicario, J. L., Badia, D., Carrillo, L., Etxebarria, J., Reyes, E. & Ruiz, N. The asymmetric aza-Michael reaction. A review. *Org. Prep. Proced. Int.* **2005**, 37, 513-538; (c) Vicario, J. L., Badia, D. & Carrillo, L. Organocatalytic enantioselective Michael and hetero-Michael reactions. *Synthesis* **2007**, 2065-2092; (d) Nising, C. F. & Braese, S. The oxa-Michael reaction: from recent developments to applications in natural product synthesis. *Chem. Soc. Rev.* **2008**, 37, 1218-1228; (e) Enders, D., Wang, C. & Liebich, J. X. Organocatalytic Asymmetric Aza-Michael Additions. *Chem.--Eur. J.* **2009**, 15, 11058-11076; (f) Lu, X. & Han, X. Intramolecular 1,2-addition and 1,4-addition reactions in *Handbook of Cyclization Reactions* (ed. Ma, S.), 169-226 (**2010**); (g) Brazier, J. B. & Tomkinson, N. C. O. Secondary and primary amine catalysts for iminium catalysis in *Topics in Current Chemistry (Asymmetric Organocatalysis)* (ed. List, B.), 281-347 (Springer-Verlag, **2010**); (h) Lu, X. & Han, X. Asymmetric Organocatalyzed Cyclization Reactions in *Handbook of Cyclization Reactions* (ed. Ma, S.), 1199-1241 (**2010**).
2. Hechavarria Fonseca, M. T. & List, B. Catalytic asymmetric intramolecular Michael reaction of aldehydes. *Angew. Chem., Int. Ed.* **2004**, 43, 3958-3960.
3. Vo, N. T., Pace, R. D. M., O'Hara, F. & Gaunt, M. J. An Enantioselective Organocatalytic Oxidative Dearomatization Strategy. *J. Am. Chem. Soc.* **2008**, 130, 404-405.
4. Yang, H. & Carter, R. G. Development of an Enantioselective Route Toward the Lycopodium Alkaloids: Total Synthesis of Lycopodine. *J. Org. Chem.* **2010**, 75, 4929-4938.
5. Phillips, E. M., Wadamoto, M., Chan, A. & Scheidt, K. A. A highly enantioselective intramolecular Michael reaction catalyzed by N-heterocyclic carbenes. *Angew. Chem., Int. Ed.* **2007**, 46, 3107-3110.
6. Gu, Q., Rong, Z.-Q., Zheng, C. & You, S.-L. Desymmetrization of Cyclohexadienones via Brønsted Acid-Catalyzed Enantioselective Oxo-Michael Reaction. *J. Am. Chem. Soc.* **2010**, 132, 4056-4057.
7. Fustero, S., Jimenez, D., Moscardo, J., Catalan, S. & del Pozo, C. Enantioselective Organocatalytic Intramolecular Aza-Michael Reaction: a Concise Synthesis of (+)-Sedamine, (+)-Allosedamine, and (+)-Coniine. *Org. Lett.* **2007**, 9, 5283-5286.

Reference key for equations: (I)[2]; (II)[3]; (III)[4]; (IV)[5]

Scheme 1

INTRAMOLECULAR 1,2-ADDITION TO C=O

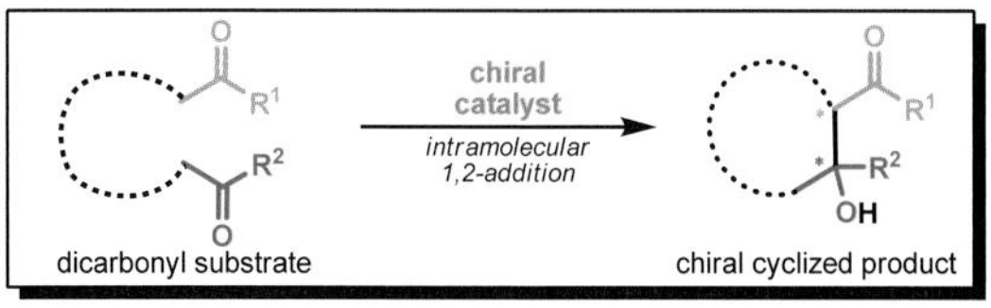

Introduction

The tactical combination of Michael and aldol reactions to form a 2-cyclohexenone subunit, generally called the Robinson annulation, has frequently been applied as a key step in the synthesis of carbocycles.[1] The use of catalysis by a chiral secondary amine, e.g., proline, for the intramolecular aldol step has opened up an enantioselective pathway that can be very useful (see pages 64-70 on stereocontrolled aldol reactions). Five examples of enantioselective intramolecular aldol processes are shown in Scheme 1. Three quite different enantioselective cyclizations via intramolecular addition of carbon to C=O under catalysis by chiral Rh or Pd complexes appear in Scheme 2.

I. Catalytic Enantioselective Robinson Annulation and Related Reactions

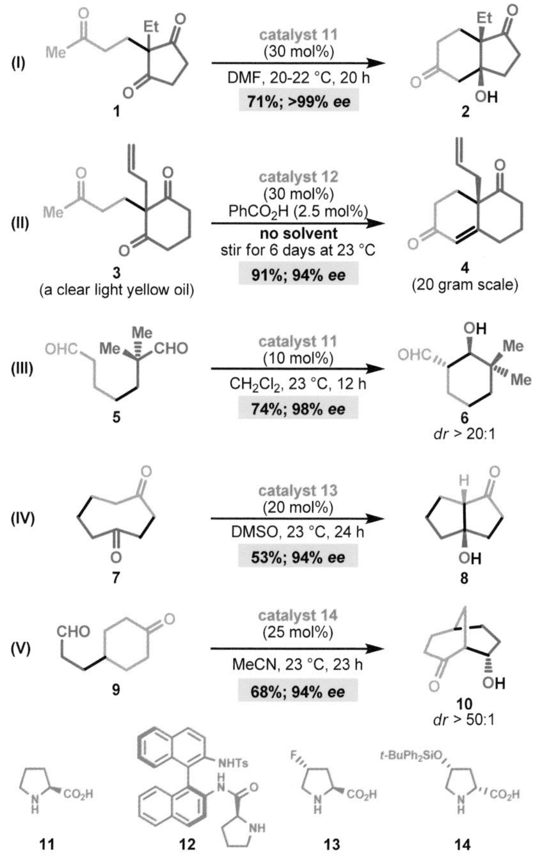

Reference key for equations: (I)[2]; (II)[3]; (III)[4]; (IV)[5]; (V)[6]

Scheme 1

II. Miscellaneous Catalytic Enantioselective Intramolecular 1,2-Addition Reactions

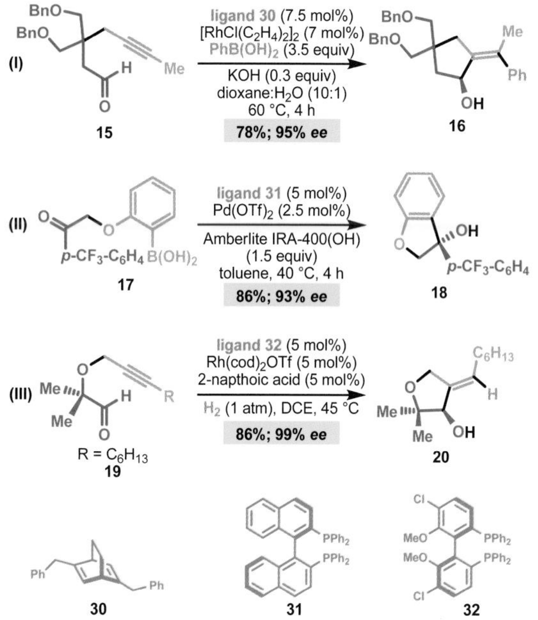

Reference key for equations: (I)[7]; (II)[8]; (III)[9]

Scheme 2

References

1. (a) Mukherjee, S., Yang, J. W., Hoffmann, S. & List, B. Asymmetric Enamine Catalysis. *Chem. Rev. (Washington, DC, U. S.)* **2007**, 107, 5471-5569; (b) *Asymmetric Organocatalysis (Topics in Current Chemistry)* (ed. List, B.) (Springer, **2010**); (c) Lu, X. & Han, X. Intramolecular 1,2-addition and 1,4-addition reactions in *Handbook of Cyclization Reactions* (ed. Ma, S.), 169-226 (**2010**); (d) Lu, X. & Han, X. Asymmetric Organocatalyzed Cyclization Reactions in *Handbook of Cyclization Reactions* (ed. Ma, S.), 1199-1241 (**2010**); (e) Trost, B. M. & Brindle, C. S. The direct catalytic asymmetric aldol reaction. *Chem. Soc. Rev.* **2010**, 39, 1600-1632.
2. Hajos, Z. G. & Parrish, D. R. Asymmetric synthesis of bicyclic intermediates of natural product chemistry. *J. Org. Chem.* **1974**, 39, 1615-1621.
3. Bradshaw, B., Etxebarria-Jardi, G. & Bonjoch, J. Total Synthesis of (-)-Anominine. *J. Am. Chem. Soc.* **2010**, 132, 5966-5967.
4. Pidathala, C., Hoang, L., Vignola, N. & List, B. Direct catalytic asymmetric enolexo aldolizations. *Angew Chem Int Ed Engl* **2003**, 42, 2785-2788.
5. Chandler, C. L. & List, B. Catalytic, Asymmetric Transannular Aldolizations: Total Synthesis of (+)-Hirsutene. *J. Am. Chem. Soc.* **2008**, 130, 6737-6739.
6. Itagaki, N., Kimura, M., Sugahara, T. & Iwabuchi, Y. Organocatalytic Entry to Chiral Bicyclo[3.n.1]alkanones via Direct Asymmetric Intramolecular Aldolization. *Org. Lett.* **2005**, 7, 4185-4188.
7. Shintani, R., Okamoto, K., Otomaru, Y., Ueyama, K. & Hayashi, T. Catalytic Asymmetric Arylative Cyclization of Alkynals: Phosphine-Free Rhodium/Diene Complexes as Efficient Catalysts. *J. Am. Chem. Soc.* **2005**, 127, 54-55.
8. Liu, G. & Lu, X. Cationic Palladium Complex Catalyzed Highly Enantioselective Intramolecular Addition of Arylboronic Acids to Ketones. A Convenient Synthesis of Optically Active Cycloalkanols. *J. Am. Chem. Soc.* **2006**, 128, 16504-16505.
9. Rhee, J. U. & Krische, M. J. Highly enantioselective reductive cyclization of acetylenic aldehydes via rhodium catalyzed asymmetric hydrogenation. *J. Am. Chem. Soc.* **2006**, 128, 10674-10675.

ENANTIOSELECTIVE PAUSON-KHAND REACTION

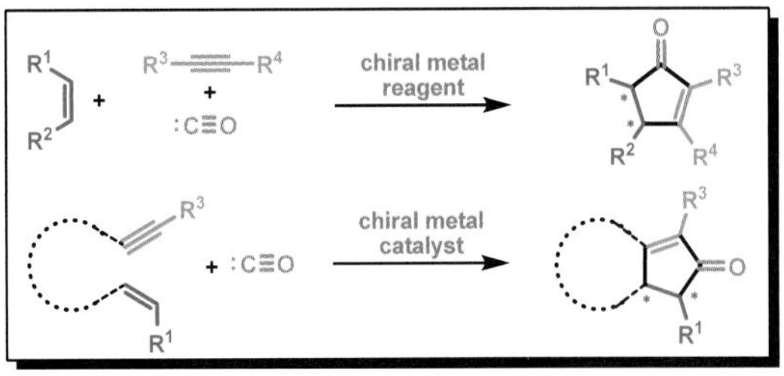

Introduction

Complexes of dicobalt octacarbonyl $[Co_2(CO)_8]$ with certain chiral diphosphines can catalyze the formation of bicyclic 2-cyclopentenones from enynes. This intramolecular version of the Pauson-Khand reaction,[1] which combines C=C, C≡C and C≡O subunits to form 2-cyclopentenones, constitutes an especially short and enantioselective entry into the series of 5,5-fused 2-cyclopentenones. The reaction may involve a six-membered cyclic acylcobalt intermediate, from which the cyclic enone is formed by reductive elimination.

Catalytic Enantioselective Pauson-Khand Reaction

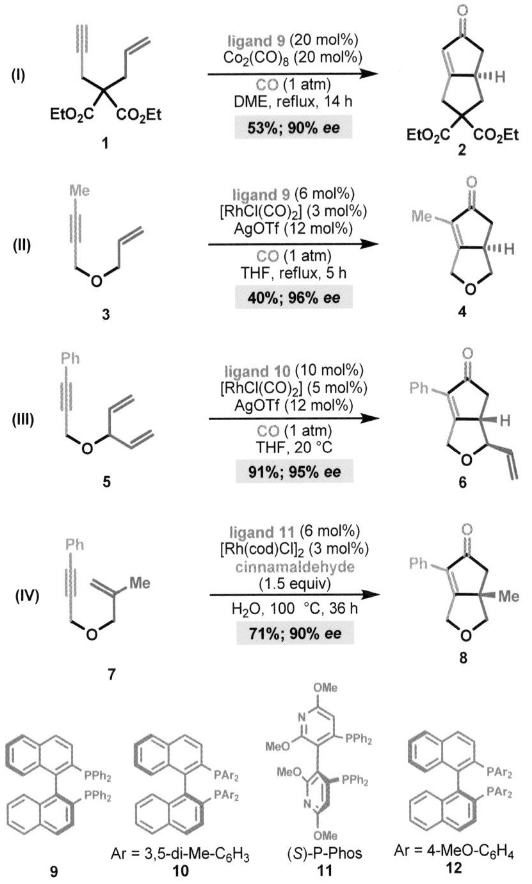

(I) ligand 9 (20 mol%) $Co_2(CO)_8$ (20 mol%) — CO (1 atm) DME, reflux, 14 h — **53%; 90% ee**

1 → 2

(II) ligand 9 (6 mol%) $[RhCl(CO)_2]$ (3 mol%) AgOTf (12 mol%) — CO (1 atm) THF, reflux, 5 h — **40%; 96% ee**

3 → 4

(III) ligand 10 (10 mol%) $[RhCl(CO)_2]$ (5 mol%) AgOTf (12 mol%) — CO (1 atm) THF, 20 °C — **91%; 95% ee**

5 → 6

(IV) ligand 11 (6 mol%) $[Rh(cod)Cl]_2$ (3 mol%) cinnamaldehyde (1.5 equiv) — H_2O, 100 °C, 36 h — **71%; 90% ee**

7 → 8

9

10 Ar = 3,5-di-Me-C_6H_3

11 (S)-P-Phos

12 Ar = 4-MeO-C_6H_4

Reference key for equations: (I)[2]; (II)[3]; (III)[4]; (IV)[5]

Scheme 1

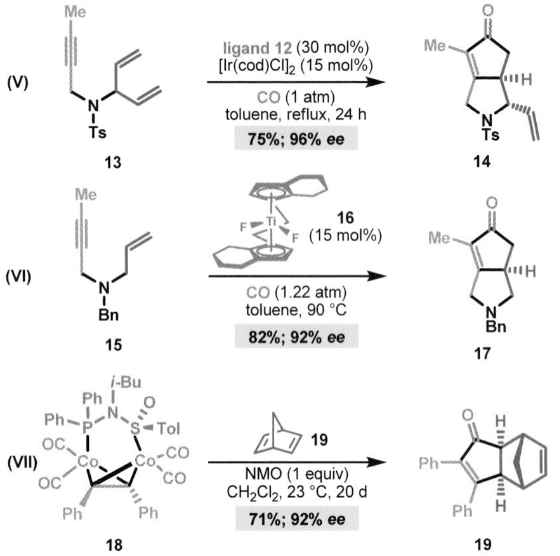

(V) ligand 12 (30 mol%) $[Ir(cod)Cl]_2$ (15 mol%) — CO (1 atm) toluene, reflux, 24 h — **75%; 96% ee**

13 → 14

(VI) 16 (15 mol%) — CO (1.22 atm) toluene, 90 °C — **82%; 92% ee**

15 → 17

(VII) 19 — NMO (1 equiv) CH_2Cl_2, 23 °C, 20 d — **71%; 92% ee**

18 → 19

Reference key for equations: (V)[6]; (VI)[7]; (VII)[8]

Scheme 1

References

1. (a) Pericas, M. A., Balsells, J., Castro, J., Marchueta, I., Moyano, A., Riera, A., Vazquez, J. & Verdaguer, X. Toward the understanding of the mechanism and enantioselectivity of the Pauson-Khand reaction. Theoretical and experimental studies. *Pure Appl. Chem.* **2002**, 74, 167-174; (b) Jeong, N. Rhodium(I)-catalyzed [2+2+1] and [4+1] Carbocyclization reactions in *Modern Rhodium-Catalyzed Organic Reactions* (ed. Evans, P. A.), 215-240 (Wiley-VCH, Weinheim, **2005**); (c) Strübing, D. & Beller, M. The Pauson-Khand reaction in *Catalytic Carbonylation Reactions (Topics of Organometallic Chemistry)* (ed. Beller, M.), 165-178 (Springer-verlag, **2006**); (d) Shibata, T. Recent advances in the catalytic Pauson-Khand-type reaction. *Adv. Synth. Catal.* **2006**, 348, 2328-2336; (e) Jeong, N. Pauson-Khand reaction in *Comprehensive Organometallic Chemistry* (eds. Crabtree, R. & Mingos, D. M. P.), 335-365 (Elsevier, **2007**); (f) Park, J. H., Chang, K.-M. & Chung, Y. K. Catalytic Pauson-Khand-type reactions and related carbonylative cycloaddition reactions. *Coord. Chem. Rev.* **2009**, 253, 2461-2480; (g) Lee, H.-W. & Kwong, F.-Y. A Decade of Advancements in Pauson-Khand-Type Reactions. *Eur. J. Org. Chem.* **2010**, 789-811.
2. Hiroi, K., Watanabe, T., Kawagishi, R. & Abe, I. Catalytic use of chiral phosphine ligands in asymmetric Pauson-Khand reactions. *Tetrahedron Asymmetry* **2000**, 11, 797-808.
3. Jeong, N., Sung, B. K. & Choi, Y. K. Rhodium(I)-Catalyzed Asymmetric Intramolecular Pauson-Khand-Type Reaction. *J. Am. Chem. Soc.* **2000**, 122, 6771-6772.
4. Kim, D. E., Kim, I. S., Ratovelomanana-Vidal, V., Genet, J.-P. & Jeong, N. Asymmetric Pauson-Khand-type Reaction Mediated by Rh(I) Catalyst at Ambient Temperature. *J. Org. Chem.* **2008**, 73, 7985-7989.
5. Kwong, F. Y., Li, Y. M., Lam, W. H., Qiu, L., Lee, H. W., Yeung, C. H., Chan, K. S. & Chan, A. S. C. Rhodium-catalyzed asymmetric aqueous Pauson-Khand-type reaction. *Chem.--Eur. J.* **2005**, 11, 3872-3880.
6. Jeong, N., Kim, D. H. & Choi, J. H. Desymmetrization of meso-dienyne by asymmetric Pauson-Khand type reaction catalysts. *Chem. Commun. (Cambridge, U. K.)* **2004**, 1134-1135.
7. Sturla, S. J. & Buchwald, S. L. Catalytic Asymmetric Cyclocarbonylation of Nitrogen-Containing Enynes. *J. Org. Chem.* **1999**, 64, 5547-5550.
8. Ji, Y., Riera, A. & Verdaguer, X. Asymmetric Intermolecular Pauson-Khand Reaction of Symmetrically Substituted Alkynes. *Org. Lett.* **2009**, 11, 4346-4349.

CYCLIZATION CATALYZED BY Rh-COMPLEXES

Introduction

Binuclear Rh(II) compounds with four bridging ligands (e.g., $Rh_2(OAc)_4$) are effective catalysts for the promotion of carbene-like reactions of diazo compounds, presumably via Rh-carbene reactive intermediates. The reactive Rh complexes can undergo intramolecular reactions with π-bonds to form [2+1]-cycloadducts or with reactive C-H linkages to form cyclic insertion products.[1] In addition, Rh(II)-complexes can catalyze intramolecular π-bond addition or C-H insertion of nitrogen starting from sulfonyl, phosphoryl or N-acyl iodosoimines. These reactions have been rendered enantioselective by the use of chiral Rh₂-complexes of the type Rh_2L_4, where L is a suitable chiral ligand – generally a chiral carboxylic acid or a lactam.

Scheme 1 which follows provides examples of enantioselective intramolecular addition to π-bonds and Scheme 2 illustrates intramolecular C-H insertion. Although the detailed mechanisms of these processes are not yet certain, there is evidence that it is not necessary that all four ligands be chiral and that de-bridging of one ligand may occur during reaction.[2]

I. Catalytic Enantioselective Inter- and Intramolecular Cyclization of Rh-Carbenoids via Insertion into C=C or C≡C

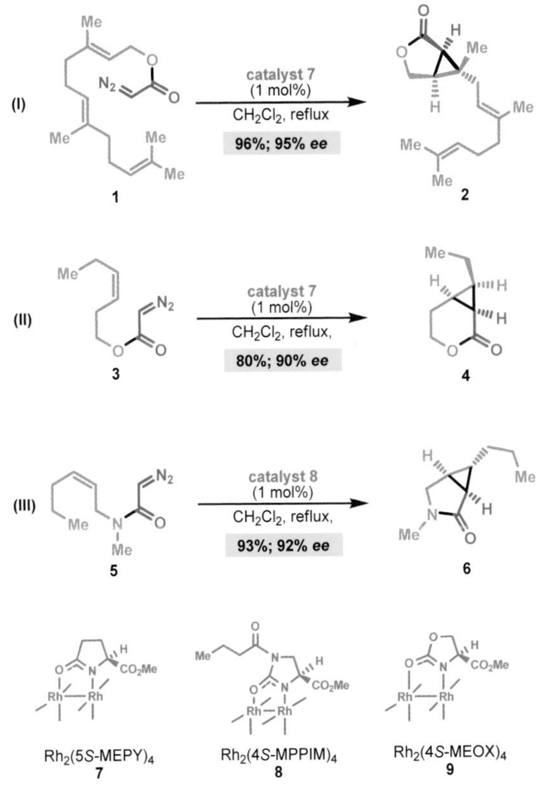

Rh₂(5S-MEPY)₄ **7**

Rh₂(4S-MPPIM)₄ **8**

Rh₂(4S-MEOX)₄ **9**

Reference key for equations: (I)[3]; (II)[4]; (III)[5]

Scheme 1

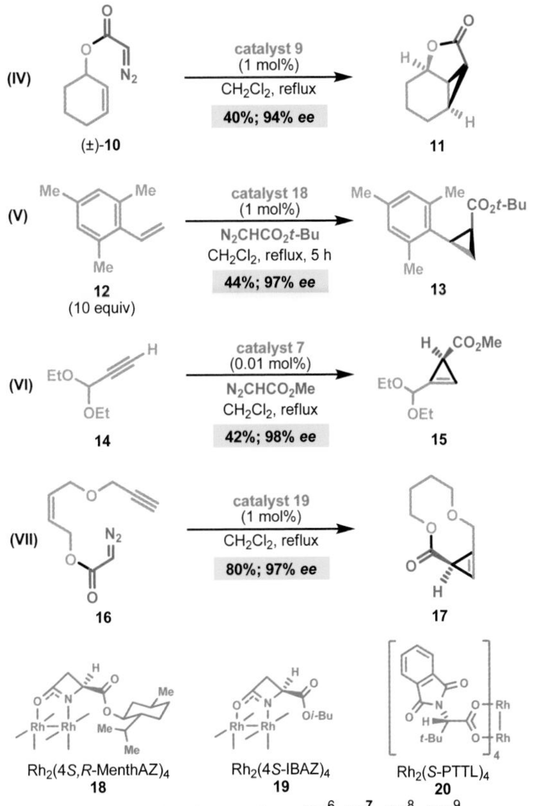

Rh₂(4S,R-MenthAZ)₄ **18**

Rh₂(4S-IBAZ)₄ **19**

Rh₂(S-PTTL)₄ **20**

Reference key for equations: (IV)[6]; (V)[7]; (IV)[8]; (VII)[9]

Scheme 1

II. Catalytic Enantioselective Cyclization of Rh-Carbenoids via C-H Insertion to form 4- and 5-Membered Rings

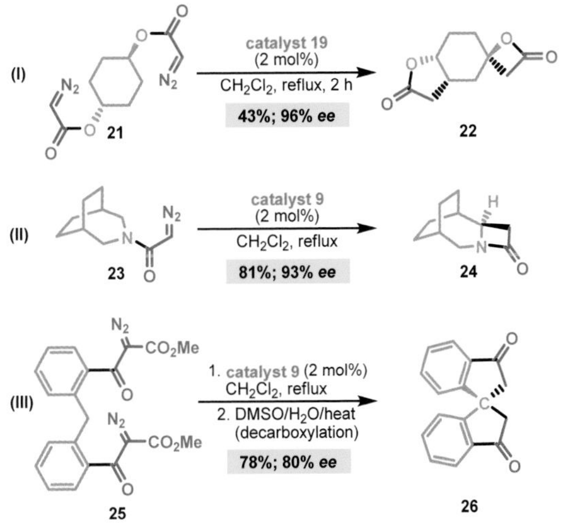

Reference key for equations: (I)[10]; (II)[11]; (III)[12]

Scheme 2

CYCLIZATION CATALYZED BY Rh-COMPLEXES

Catalytic Enantioselective Cyclization of Rh-Carbenoids via C-H Insertion to form 5- and 6-Membered Rings (Continued)

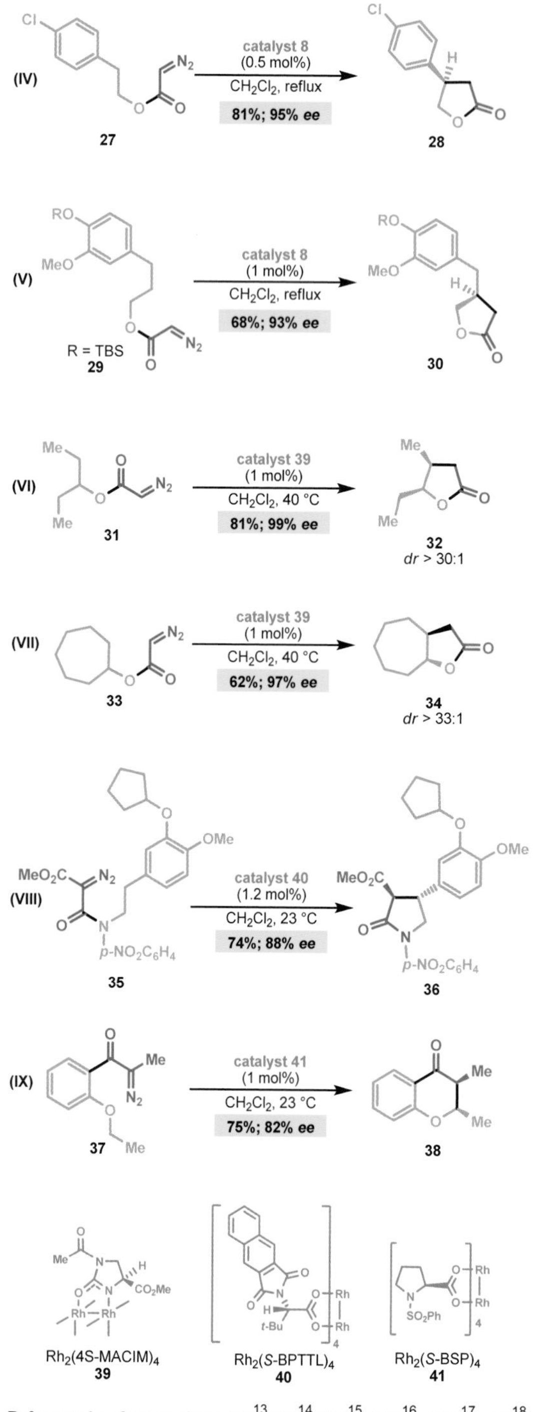

III. Catalytic Enantioselective Cyclization of Rh-Nitrenoids via C-H Insertion to form 5- and 6-Membered Rings

Intramolecular insertion of nitrogen into C-H bonds can be accomplished enantioselectively from N-sulfonylphenyliodosoimines using appropriate chiral Rh(II) complexes, as illustrated in Scheme 3.

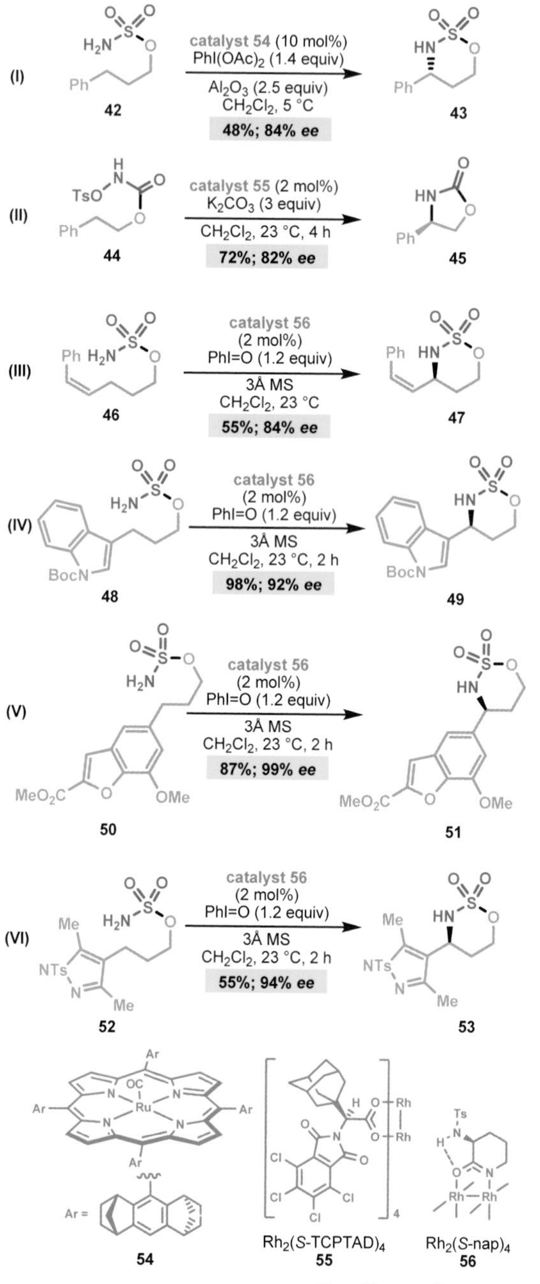

(IV)

27

catalyst 8
(0.5 mol%)
CH$_2$Cl$_2$, reflux
81%; 95% ee

28

(V)

R = TBS
29

catalyst 8
(1 mol%)
CH$_2$Cl$_2$, reflux
68%; 93% ee

30

(VI)

31

catalyst 39
(1 mol%)
CH$_2$Cl$_2$, 40 °C
81%; 99% ee

32
dr > 30:1

(VII)

33

catalyst 39
(1 mol%)
CH$_2$Cl$_2$, 40 °C
62%; 97% ee

34
dr > 33:1

(VIII)

35

catalyst 40
(1.2 mol%)
CH$_2$Cl$_2$, 23 °C
74%; 88% ee

36

(IX)

37

catalyst 41
(1 mol%)
CH$_2$Cl$_2$, 23 °C
75%; 82% ee

38

(I)

42

catalyst 54 (10 mol%)
PhI(OAc)$_2$ (1.4 equiv)
Al$_2$O$_3$ (2.5 equiv)
CH$_2$Cl$_2$, 5 °C
48%; 84% ee

43

(II)

44

catalyst 55 (2 mol%)
K$_2$CO$_3$ (3 equiv)
CH$_2$Cl$_2$, 23 °C, 4 h
72%; 82% ee

45

(III)

46

catalyst 56
(2 mol%)
PhI=O (1.2 equiv)
3Å MS
CH$_2$Cl$_2$, 23 °C
55%; 84% ee

47

(IV)

48

catalyst 56
(2 mol%)
PhI=O (1.2 equiv)
3Å MS
CH$_2$Cl$_2$, 23 °C, 2 h
98%; 92% ee

49

(V)

50

catalyst 56
(2 mol%)
PhI=O (1.2 equiv)
3Å MS
CH$_2$Cl$_2$, 23 °C, 2 h
87%; 99% ee

51

(VI)

52

catalyst 56
(2 mol%)
PhI=O (1.2 equiv)
3Å MS
CH$_2$Cl$_2$, 23 °C, 2 h
55%; 94% ee

53

Rh$_2$(4S-MACIM)$_4$
39

Rh$_2$(S-BPTTL)$_4$
40

Rh$_2$(S-BSP)$_4$
41

Rh$_2$(S-TCPTAD)$_4$
55

Rh$_2$(S-nap)$_4$
56

54

Reference key for equations: (IV)[13]; (V)[14]; (VI)[15]; (VII)[16]; (VIII)[17]; (IX)[18]

Scheme 2

Reference key for equations: (I)[19]; (II)[20]; (III-VI)[21]

Scheme 3

CYCLIZATION CATALYZED BY Rh-COMPLEXES

IV. Catalytic Enantioselective Cycloaddition of 1,3-Dipoles (Carbonyl Ylides) Generated via Chiral Rh-Carbenoids[22]

Chiral Rh(II) catalysts convert α-diazocarbonyl compounds which have another C=O separated from the diazo carbon by a 3- or 4-bond path to stabilized 5- or 6-membered cyclic 2-oxaallyl dipoles which remain attached to rhodium. These dipolar species can add intermolecularly to C=C or C=O to form chiral, bridged bicyclic structures with good enantioselectivity.

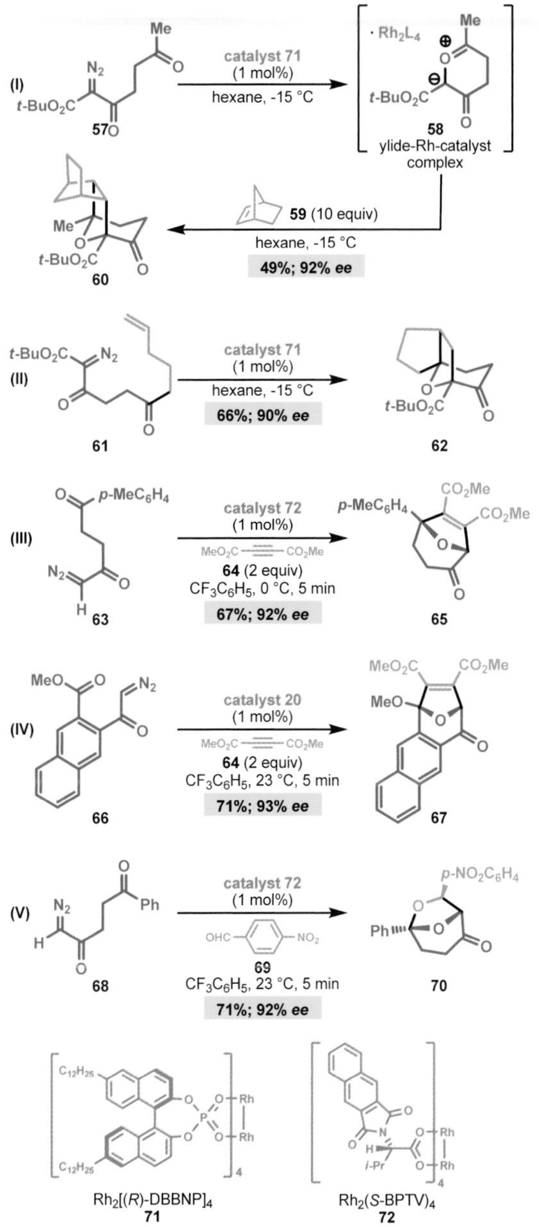

(I)
catalyst 71 (1 mol%)
hexane, -15 °C
57
58
ylide-Rh-catalyst complex
59 (10 equiv)
hexane, -15 °C
60
49%; 92% ee

(II)
catalyst 71 (1 mol%)
hexane, -15 °C
61
62
66%; 90% ee

(III)
catalyst 72 (1 mol%)
64 (2 equiv)
$CF_3C_6H_5$, 0 °C, 5 min
63
65
67%; 92% ee

(IV)
catalyst 20 (1 mol%)
64 (2 equiv)
$CF_3C_6H_5$, 23 °C, 5 min
66
67
71%; 93% ee

(V)
catalyst 72 (1 mol%)
69
$CF_3C_6H_5$, 23 °C, 5 min
68
70
71%; 92% ee

$Rh_2[(R)-DBBNP]_4$
71

$Rh_2(S-BPTV)_4$
72

Reference key for equations: (I)[23]; (II)[24]; (III)[25]; (IV)[26]; (V)[27]

Scheme 4

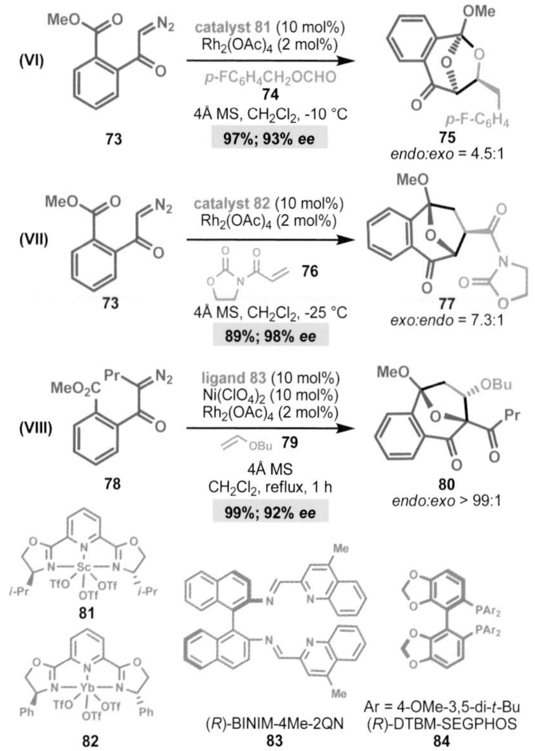

(VI)
catalyst 81 (10 mol%)
$Rh_2(OAc)_4$ (2 mol%)
p-$FC_6H_4CH_2OCHO$
74
4Å MS, CH_2Cl_2, -10 °C
73
97%; 93% ee
75
endo:exo = 4.5:1

(VII)
catalyst 82 (10 mol%)
$Rh_2(OAc)_4$ (2 mol%)
76
4Å MS, CH_2Cl_2, -25 °C
73
89%; 98% ee
77
exo:endo = 7.3:1

(VIII)
ligand 83 (10 mol%)
$Ni(ClO_4)_2$ (10 mol%)
$Rh_2(OAc)_4$ (2 mol%)
79
4Å MS
CH_2Cl_2, reflux, 1 h
78
99%; 92% ee
80
endo:exo > 99:1

81
82
(R)-BINIM-4Me-2QN
83
Ar = 4-OMe-3,5-di-t-Bu
(R)-DTBM-SEGPHOS
84

Reference key for equations: (VI)[28]; (VII)[29]; (VIII)[30]

Scheme 5

V. Catalytic Enantioselective Hydroacylation[31]

Chiral Rh(II)-bisphosphine complexes can catalyze intramolecular hydroacylation of unsaturated aldehydes to generate 5-membered cyclic ketones, or in special cases other ring sizes.

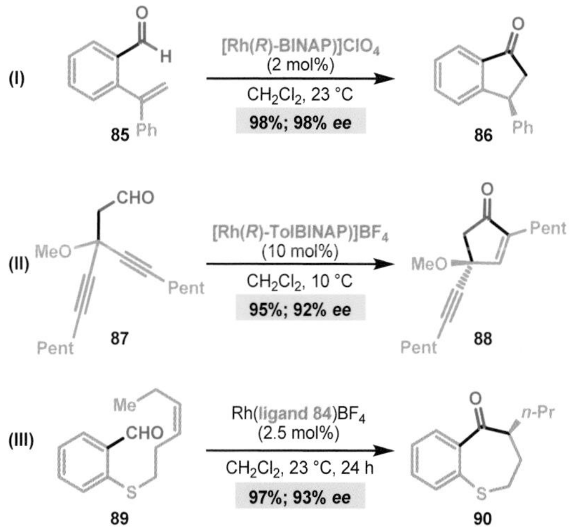

(I)
[Rh(R)-BINAP)]ClO_4
(2 mol%)
CH_2Cl_2, 23 °C
85
98%; 98% ee
86

(II)
[Rh(R)-TolBINAP)]BF_4
(10 mol%)
CH_2Cl_2, 10 °C
87
95%; 92% ee
88

(III)
Rh(ligand 84)BF_4
(2.5 mol%)
CH_2Cl_2, 23 °C, 24 h
89
97%; 93% ee
90

Reference key for equations: (I)[32]; (II)[33]; (III)[34]

Scheme 5

References

1. (a) Davies, H. M. L. & Hansen, J. Asymmetric Synthesis Through C-H Activation in *Catalytic Asymmetric Synthesis, 3rd Ed.* (ed. Ojima, I.), 163-226 (John Wiley & Sons, Hoboken, NJ, 2010); (b) Davies, H. M. L. & Dick, A. R. Functionalization of Carbon-Hydrogen Bonds Through Transition Metal Carbenoid Insertion in *C-H Activation (Topics of Current Chemistry)* (eds. Yu, J.-Q. & Shi, Z.), 303-345 (Springer-Verlag, Berlin, 2010); (c) Doyle, M. P., Duffy, R., Ratnikov, M. & Zhou, L. Catalytic Carbene Insertion into C-H Bonds. *Chem. Rev. (Washington, DC, U. S.)* 2010, 110, 704-724; (d) Capretto, D. A., Li, Z. & He, C. Cyclizations based on C-H activation in *Handbook of Cyclization Reactions* 991-1023 (2010).

2. (a) Lou, Y., Horikawa, M., Kloster, R. A., Hawryluk, N. A. & Corey, E. J. A New Chiral Rh(II) Catalyst for Enantioselective [2 + 1]-Cycloaddition. Mechanistic Implications and Applications. *J. Am. Chem. Soc.* 2004, 126, 8916-8918; (b) Lou, Y., Remarchuk, T. P. & Corey, E. J. Catalysis of Enantioselective [2+1]-Cycloaddition Reactions of Ethyl Diazoacetate and Terminal Acetylenes Using Mixed-Ligand Complexes of the Series Rh2(RCO2)n (L4-n). Stereochemical Heuristics for Ligand Exchange and Catalyst Synthesis. *J. Am. Chem. Soc.* 2005, 127, 14223-14230.

3. Rogers, D. H., Yi, E. C. & Poulter, C. D. Enantioselective Synthesis of (+)-Presqualene Diphosphate. *J. Org. Chem.* 1995, 60, 941-945.

4. Doyle, M. P., Austin, R. E., Bailey, A. S., Dwyer, M. P., Dyatkin, A. B., Kalinin, A. V., Kwan, M. M. Y., Liras, S., Oalmann, C. J. & et al. Enantioselective Intramolecular Cyclopropanations of Allylic and Homoallylic Diazoacetates and Diazoacetamides Using Chiral Dirhodium(II) Carboxamide Catalysts. *J. Am. Chem. Soc.* 1995, 117, 5763-5775.

5. Doyle, M. P. & Kalinin, A. V. Highly Enantioselective Intramolecular Cyclopropanation Reactions of N-Allylic-N-methyldiazoacetamides Catalyzed by Chiral Dirhodium(II) Carboxamidates. *J. Org. Chem.* 1996, 61, 2179-2184.

6. Doyle, M. P., Dyatkin, A. B., Kalinin, A. V., Ruppar, D. A., Martin, S. F., Spaller, M. R. & Liras, S. Highly selective enantiomer differentiation in intramolecular cyclopropanation reactions of racemic secondary allylic diazoacetates. *J. Am. Chem. Soc.* 1995, 117, 11021-11022.

7. Hu, W., Timmons, D. J. & Doyle, M. P. In Search of High Stereocontrol for the Construction of cis-Disubstituted Cyclopropane Compounds. Total Synthesis of a Cyclopropane-Configured Urea-PETT Analog That Is a HIV-1 Reverse Transcriptase Inhibitor. *Org. Lett.* 2002, 4, 901-904.

8. Doyle, M. P., Protopopova, M., Muller, P., Ene, D. & Shapiro, E. A. Effective Uses of Dirhodium(II) Tetrakis[methyl 2-oxopyrrolidine-5(R or S)-carboxylate] for Highly Enantioselective Intermolecular Cyclopropenation Reactions. *J. Am. Chem. Soc.* 1994, 116, 8492-8498.

9. Doyle, M. P., Ene, D. G., Peterson, C. S. & Lynch, V. Macrocyclic cyclopropenes by highly enantioselective intramolecular addition of metal carbenes to alkynes. *Angew. Chem., Int. Ed.* 1999, 38, 700-702.

10. Doyle, M. P., Wang, Y., Ghorbani, P. & Bappert, E. Amplification of Asymmetric Induction in Sequential Reactions of Bis-diazoacetates Catalyzed by Chiral Dirhodium(II) Carboxamidates. *Org. Lett.* 2005, 7, 5035-5038.

11. Doyle, M. P. & Kalinin, A. V. Highly enantioselective route to β-lactams via intramolecular C-H insertion reactions of diazoacetylazacycloalkanes catalyzed by chiral dirhodium(II) carboxamidates. *Synlett* 1995, 1075-1076.

12. Takahashi, T., Tsutsui, H., Tamura, M., Kitagaki, S., Nakajima, M. & Hashimoto, S. Catalytic asymmetric synthesis of 1,1'-spirobi[indan-3,3'-dione] via a double intramolecular C-H insertion process. *Chem. Commun. (Cambridge, U. K.)* 2001, 1604-1605.

13. Doyle, M. P. & Hu, W. Enantioselective carbon-hydrogen insertion is an effective and efficient methodology for the synthesis of (R)-(-)-baclofen. *Chirality* 2002, 14, 169-172.

14. Doyle, M. P., Hu, W. & Valenzuela, M. V. Total Synthesis of (S)-(+)-Imperanene. Effective Use of Regio- and Enantioselective Intramolecular Carbon-Hydrogen Insertion Reactions Catalyzed by Chiral Dirhodium(II) Carboxamidates. *J. Org. Chem.* 2002, 67, 2954-2959.

15. (a) Doyle, M. P., Dyatkin, A. B. & Tedrow, J. S. Synthesis of 2-deoxyxylolactone from glyceryl derivatives via highly enantioselective carbon-hydrogen insertion reactions. *Tetrahedron Lett.* 1994, 35, 3853-3856; (b) Doyle, M. P., Zhou, Q.-L., Dyatkin, A. B. & Ruppar, D. A. Enhancement of enantio-/diastereocontrol in catalytic intramolecular cyclopropanation and carbon-hydrogen insertion reactions of diazoacetates with Rh2(4S-MPPIM)4. *Tetrahedron Lett.* 1995, 36, 7579-7582; (c) Doyle, M. P., Tedrow, J. S., Dyatkin, A. B., Spaans, C. J. & Ene, D. G. Enantioselective Syntheses of 2-Deoxyxylono-1,4-lactone and 2-Deoxyribono-1,4-lactone from 1,3-Dioxan-5-yl Diazoacetates. *J. Org. Chem.* 1999, 64, 8907-8915.

16. Doyle, M. P., Dyatkin, A. B., Roos, G. H. P., Canas, F., Pierson, D. A., van Basten, A., Mueller, P. & Polleux, P. Diastereocontrol for

17. Anada, M., Mita, O., Watanabe, H., Kitagaki, S. & Hashimoto, S. Catalytic enantioselective synthesis of the phosphodiesterase type IV inhibitor (R)-(-)-rolipram via intramolecular C-H insertion process. *Synlett* 1999, 1775-1777.

Highly Enantioselective Carbon-Hydrogen Insertion Reactions of Cycloalkyl Diazoacetates. *J. Am. Chem. Soc.* 1994, 116, 4507-4508.

18. Ye, T., Fernandez Garcia, C. & McKervey, M. A. Chemoselectivity and stereoselectivity of cyclization of α-diazocarbonyls leading to oxygen and sulfur heterocycles catalyzed by chiral rhodium and copper catalysts. *J. Chem. Soc., Perkin Trans. 1* 1995, 1373-1379.

19. Liang, J.-L., Yuan, S.-X., Huang, J.-S., Yu, W.-Y. & Che, C.-M. Highly diastereo- and enantioselective intramolecular amidation of saturated C-H bonds catalyzed by ruthenium porphyrins. *Angew. Chem., Int. Ed.* 2002, 41, 3465-3468.

20. Davies, H. M. L., Dai, X. & Long, M. S. Combined C-H Activation/Cope Rearrangement as a Strategic Reaction in Organic Synthesis: Total Synthesis of (-)-Colombiasin A and (-)-Elisapterosin B. *J. Am. Chem. Soc.* 2006, 128, 2485-2490.

21. Zalatan, D. N. & Du Bois, J. A Chiral Rhodium Carboxamidate Catalyst for Enantioselective C-H Amination. *J. Am. Chem. Soc.* 2008, 130, 9220-9221.

22. (a) Savizky, R. M. & Austin, D. J. Rhodium(II)-catalyzed 1,3-dipolar cycloaddition reactions in *Modern Rhodium-Catalyzed Organic Reactions* 433-454 (2005); (b) Muthusamy, S. & Krishnamurthi, J. Heterocycles by cycloadditions of carbonyl ylides generated from diazo ketones in *Topics in Heterocyclic Chemistry* 147-192 (2008); (c) Hashimoto, T. & Maruoka, K. 1,3-Dipolar cycloaddition in *Handbook of Cyclization Reactions* 87-168 (2010).

23. Hodgson, D. M., Labande, A. H., Glen, R. & Redgrave, A. J. Catalytic enantioselective intermolecular cycloadditions of a 2-diazo-3,6-diketo ester-derived carbonyl ylide with alkyne and strained alkene dipolarophiles. *Tetrahedron Asymmetry* 2003, 14, 921-924.

24. Hodgson, D. M., Stupple, P. A. & Johnstone, C. Efficient RhII binaphthol phosphate catalysts for enantioselective intramolecular tandem carbonyl ylide formation-cycloaddition of α-diazo-β-keto esters. *Chem. Commun. (Cambridge)* 1999, 2185-2186.

25. Kitagaki, S., Anada, M., Kataoka, O., Matsuno, K., Umeda, C., Watanabe, N. & Hashimoto, S.-i. Enantiocontrol in Tandem Carbonyl Ylide Formation and Intermolecular 1,3-Dipolar Cycloaddition of α-Diazo Ketones Mediated by Chiral Dirhodium(II) Carboxylate Catalyst. *J. Am. Chem. Soc.* 1999, 121, 1417-1418.

26. Kitagaki, S., Yasugahira, M., Anada, M., Nakajima, M. & Hashimoto, S. Enantioselective intermolecular 1,3-dipolar cycloaddition via ester-derived carbonyl ylide formation catalyzed by chiral dirhodium(II) carboxylates. *Tetrahedron Lett.* 2000, 41, 5931-5935.

27. Tsutsui, H., Shimada, N., Abe, T., Anada, M., Nakajima, M., Nakamura, S., Nambu, H. & Hashimoto, S. Catalytic enantioselective tandem carbonyl ylide formation/1,3-dipolar cycloaddition reactions of α-diazo ketones with aromatic aldehydes using dirhodium(II) tetrakis[N-benzene-fused-phthaloyl-(S)-valinate]. *Adv. Synth. Catal.* 2007, 349, 521-526.

28. (a) Suga, H., Inoue, K., Inoue, S. & Kakehi, A. Highly Enantioselective 1,3-Dipolar Cycloaddition Reactions of 2-Benzopyrylium-4-olate Catalyzed by Chiral Lewis Acids. *J. Am. Chem. Soc.* 2002, 124, 14836-14837; (b) Suga, H., Inoue, K., Inoue, S., Kakehi, A. & Shiro, M. Chiral 2,6-Bis(oxazolinyl)pyridine-Rare Earth Metal Complexes as Catalysts for Highly Enantioselective 1,3-Dipolar Cycloaddition Reactions of 2-Benzopyrylium-4-olates. *J. Org. Chem.* 2005, 70, 47-56.

29. Suga, H., Suzuki, T., Inoue, K. & Kakehi, A. Asymmetric cycloaddition reactions between 2-benzopyrylium-4-olates and 3-(2-alkenoyl)-2-oxazolidinones in the presence of 2,6-bis(oxazolinyl)pyridine-lanthanoid complexes. *Tetrahedron* 2006, 62, 9218-9225.

30. Suga, H., Ishimoto, D., Higuchi, S., Ohtsuka, M., Arikawa, T., Tsuchida, T., Kakehi, A. & Baba, T. Dipole-LUMO/Dipolarophile-HOMO Controlled Asymmetric Cycloadditions of Carbonyl Ylides Catalyzed by Chiral Lewis Acids. *Org. Lett.* 2007, 9, 4359-4362.

31. (a) Tanaka, M., Sakai, K. & Suemune, H. Asymmetric rhodium-catalyzed intramolecular hydroacylation for five-membered ring ketone formation. *Curr. Org. Chem.* 2003, 7, 353-367; (b) Fu, G. C. Recent advances in rhodium(I)-catalyzed asymmetric olefin isomerization and hydroacylation reactions in *Modern Rhodium-Catalyzed Organic Reactions* 79-91 (2005); (c) Willis, M. C. Transition Metal Catalyzed Alkene and Alkyne Hydroacylation. *Chem. Rev. (Washington, DC, U. S.)* 2010, 110, 725-748.

32. Kundu, K., McCullagh, J. V. & Morehead, A. T., Jr. Hydroacylation of 2-Vinyl Benzaldehyde Systems: An Efficient Method for the Synthesis of Chiral 3-Substituted Indanones. *J. Am. Chem. Soc.* 2005, 127, 16042-16043.

33. Tanaka, K. & Fu, G. C. Enantioselective Synthesis of Cyclopentenones via Rhodium-Catalyzed Kinetic Resolution and Desymmetrization of 4-Alkynals. *J. Am. Chem. Soc.* 2002, 124, 10296-10297.

34. Coulter, M. M., Dornan, P. K. & Dong, V. M. Rh-Catalyzed Intramolecular Olefin Hydroacylation: Enantioselective Synthesis of Seven- and Eight-Membered Heterocycles. *J. Am. Chem. Soc.* 2009, 131, 6932-6933.

ENANTIOSELECTIVE HECK REACTION

I. Catalytic Enantioselective Intermolecular Heck Reaction

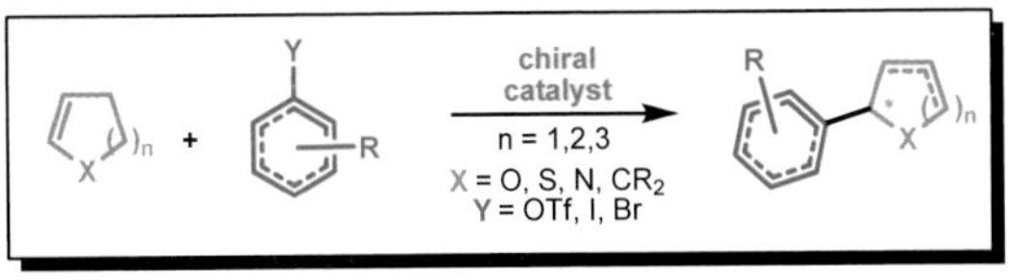

Introduction

The Pd(0)-catalyzed formation of C-C bonds by the sp^2-carbopalladation of C=C or C≡C (Heck reaction) is a very useful synthetic construction in either the inter- or intramolecular variant. The process can be rendered highly enantioselective by the use of various chiral phosphines as bidentate ligands on the Pd(0) catalyst.[1] The addition of the intermediate vinyl- or aryl-Pd-X species to C=C or C≡C occurs suprafacially (i.e., by a cis-addition pathway). The final stages of the Heck process involve an intramolecular cis-β-elimination of hydrogen to form product and to regenerate the Pd(0) catalyst (Scheme 1). The most useful enantioselective Heck reactions are those in which only one mode of cis-β-H elimination can occur from the carbopalladation product. The vast majority of enantioselective intermolecular Heck reactions involve the carbopalladation of endocyclic double bonds.

The mechanistic pathway of an enantioselective Heck reaction, as summarized in Scheme 1,[1f] involves an electrophilic Pd-species in the C=C addition step. The corresponding process with an uncharged Pd complex is both slower and considerably less enantioselective.

Six examples of highly enantioselective and efficient intermolecular Heck constructions are shown in Scheme 2. It should be mentioned that the products of reactions V and VI are formed by direct β-H elimination, whereas those of reactions I-IV are the result of an additional 1,2-hydrogen rearrangement.

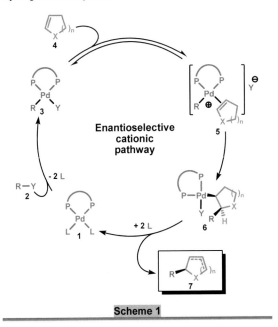

Enantioselective cationic pathway

Scheme 1

Scheme 2

(I) **9** + **10** (OTf) → **11**
ligand 21 (6 mol%)
Pd(OAc)$_2$ (3 mol%)
i-Pr$_2$NEt (2 equiv)
C$_6$H$_6$, 40 °C
71%; 93% ee

(II) **9** + **12** (CO$_2$Et, OTf) → **13** (EtO$_2$C)
ligand 21 (6 mol%)
Pd(OAc)$_2$ (3 mol%)
proton sponge (2 equiv)
C$_6$H$_6$, 40 °C
62%; 96% ee

(III) **14** (CO$_2$Me) + **12** (CO$_2$Et, OTf) → **15** (EtO$_2$C, CO$_2$Me)
ligand 21 (6 mol%)
Pd(OAc)$_2$ (3 mol%)
proton sponge (2 equiv)
C$_6$H$_6$, 60 °C
95%; 99% ee

(IV) **14** (CO$_2$Me) + **10** (OTf) → **16** (CO$_2$Me)
ligand 22 (12 mol%)
Pd$_2$(dba)$_3$ (6 mol%)
proton sponge (3 equiv)
DMF, 90 °C
84%; 93% ee

(V) **17** + **10** (OTf) → **18**
ligand 23 (6 mol%)
Pd$_2$(dba)$_3$ (3 mol%)
i-Pr$_2$NEt (3 equiv)
THF, 70 °C, 7 d
70%; 92% ee

(VI) **19** + **10** (OTf) → **20**
ligand 23 (6 mol%)
Pd$_2$(dba)$_3$ (3 mol%)
i-Pr$_2$NEt (3 equiv)
DMF, 70 °C, 5 d
96%; 91% ee

21 (PPh$_2$, PPh$_2$) **22** (PPh$_2$, PPh$_2$) **23** (PPh$_2$, t-Bu)

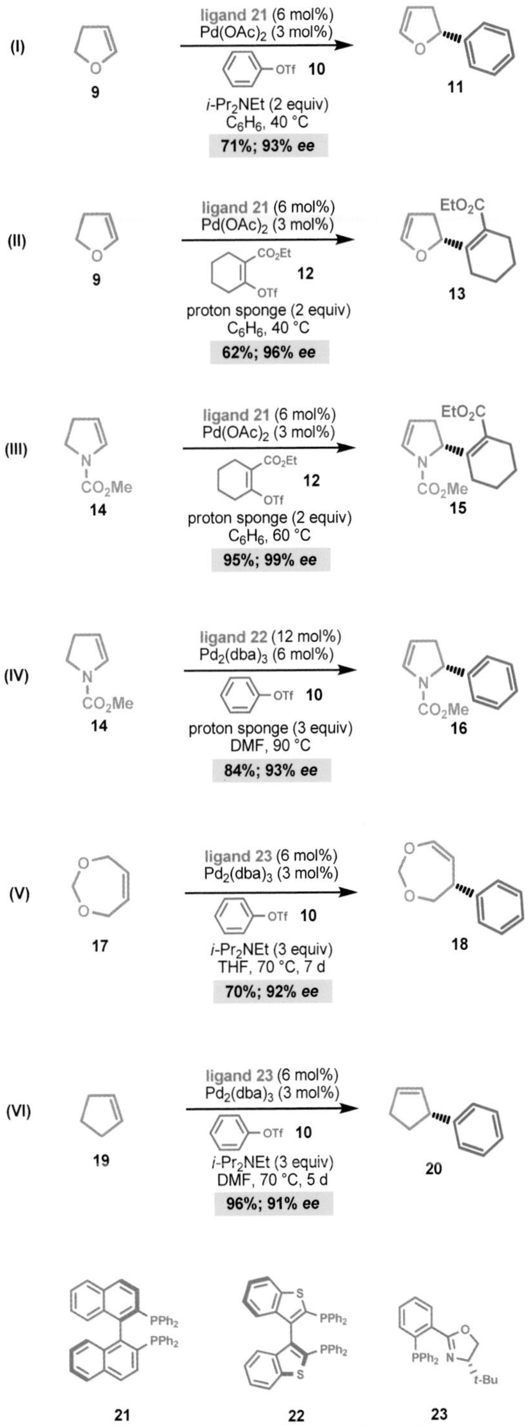

Reference key for equations: (I)[2]; (II)[3]; (III)[4]; (IV)[5]; (V, VI)[6]

Scheme 2

ENANTIOSELECTIVE HECK REACTION

II. Catalytic Enantioselective Intramolecular Heck Reaction

Five examples (I-V) of highly enantioselective intramolecular Heck cyclizations are shown in Scheme 3. Generally, the formation of six-membered rings tends to afford products with high enantiomeric excess. The last example (VI) in Scheme 3 and the example in Scheme 4 both fall in the category of de-symmetrization reactions.

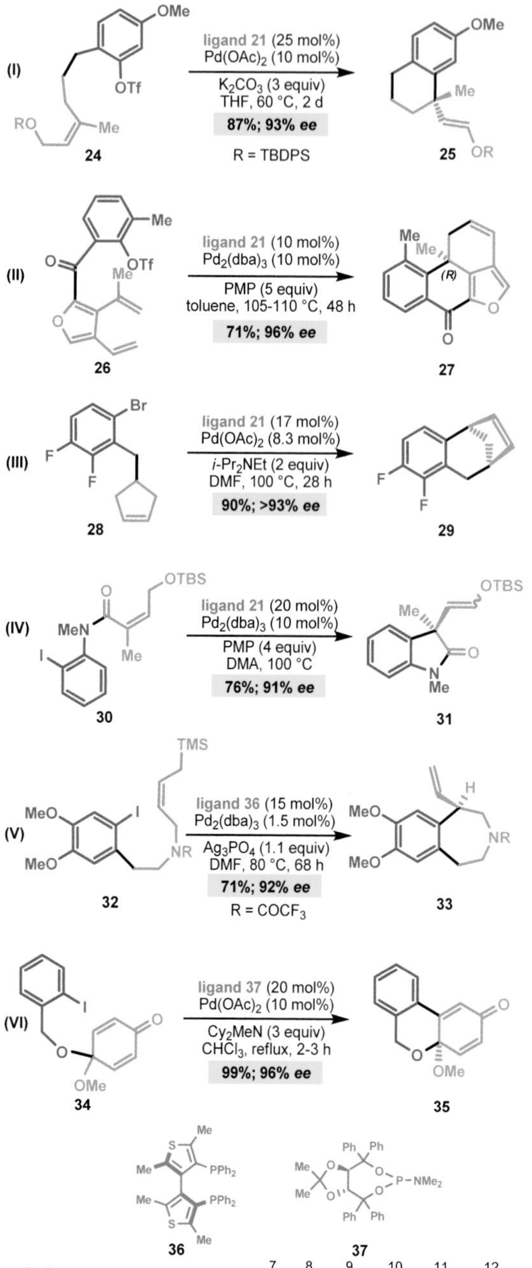

Reference key for equations: ; (I)[7]; (II)[8]; (III)[9]; (IV)[10]; (V)[11]; (VI)[12]

Scheme 3

Catalytic Enantioselective Intramolecular Heck Reaction (Continued)

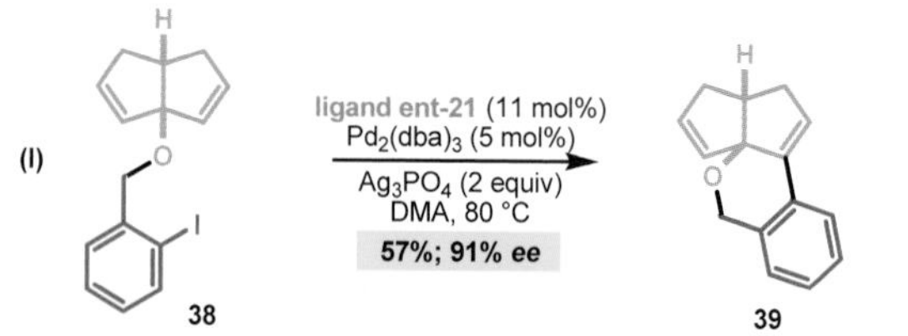

Reference key for equation: (I)[13]

Scheme 4

References

1. (a) Shibasaki, M., Boden, C. D. J. & Kojima, A. The asymmetric Heck reaction. *Tetrahedron* **1997**, 53, 7371-7395; (b) Donde, Y. & Overman, L. E. Asymmetric intramolecular Heck reactions in *Catal. Asymmetric Synth. (2nd Ed.)* 675-697 (**2000**); (c) Link, J. T. The intramolecular Heck reaction. *Org. React.* **2002**, 60, 157-534; (d) Shibasaki, M. & Miyazaki, F. Asymmetric Heck reactions in *Handb. Organopalladium Chem. Org. Synth.* 1283-1315 (**2002**); (e) Shibasaki, M., Vogl, E. M. & Ohshima, T. Asymmetric Heck reaction. *Adv. Synth. Catal.* **2004**, 346, 1533-1552; (f) Jutand, A. Mechanisms of the Mizoroki-Heck reaction in *Mizoroki-Heck Reaction* 1-50 (**2009**); (g) Coyne, A. G., Fitzpatrick, M. O. & Guiry, P. J. Ligand design for intermolecular asymmetric Mizoroki-Heck reactions in *Mizoroki-Heck Reaction* 405-431 (**2009**); (h) Shibasaki, M. & Ohshima, T. Desymmetrizing Heck reactions in *Mizoroki-Heck Reaction* 463-483 (**2009**); (i) Dounay, A. B. & Overman, L. E. The asymmetric intramolecular Mizoroki-Heck reaction in natural product total synthesis in *Mizoroki-Heck Reaction* 533-568 (**2009**).

2. Ozawa, F., Kubo, A. & Hayashi, T. Catalytic asymmetric arylation of 2,3-dihydrofuran with aryl triflates. *J. Am. Chem. Soc.* **1991**, 113, 1417-1419.

3. Ozawa, F., Kobatake, Y. & Hayashi, T. Palladium-catalyzed asymmetric alkenylation of cyclic olefins. *Tetrahedron Lett.* **1993**, 34, 2505-2508.

4. Ozawa, F. & Hayashi, T. Catalytic asymmetric arylation of N-substituted 2-pyrrolines with aryl triflates. *J. Organomet. Chem.* **1992**, 428, 267-277.

5. Tietze, L. F. & Thede, K. Highly regio- and enantioselective Heck reactions of N-substituted 2-pyrroline with the new chiral ligand BITIANP. *Synlett* **2000**, 1470-1472.

6. Loiseleur, O., Hayashi, M., Keenan, M., Schmees, N. & Pfaltz, A. Enantioselective Heck reactions using chiral P,N-ligands. *J. Organomet. Chem.* **1999**, 576, 16-22.

7. Takemoto, T., Sodeoka, M., Sasai, H. & Shibasaki, M. Catalytic asymmetric synthesis of benzylic quaternary carbon centers. An efficient synthesis of (-)-eptazocine. *J. Am. Chem. Soc.* **1993**, 115, 8477-8478.

8. Lau, S. Y. W. & Keay, B. A. Remote substituent effects on the enantiomeric excess of intramolecular asymmetric palladium-catalyzed polyene cyclizations. *Synlett* **1999**, 605-607.

9. Bashore, C. G., Vetelino, M. G., Wirtz, M. C., Brooks, P. R., Frost, H. N., McDermott, R. E., Whritenour, D. C., Ragan, J. A., Rutherford, J. L., Makowski, T. W., Brenek, S. J. & Coe, J. W. Enantioselective Synthesis of Nicotinic Receptor Probe 7,8-Difluoro-1,2,3,4,5,6-hexahydro-1,5-methano-3-benzazocine. *Org. Lett.* **2006**, 8, 5947-5950.

10. Overman, L. E. & Poon, D. J. Asymmetric Heck reactions via neutral intermediates: enhanced enantioselectivity with halide additives gives mechanistic insights. *Angew. Chem., Int. Ed. Engl.* **1997**, 36, 518-521.

11. (a) Tietze, L. F. & Schimpf, R. *Angew. Chem., Int. Ed.* **1994**, 33, 1089-1091; (b) Tietze, L. F. & Schimpf, R. Regio- and enantioselective silicon-terminated intramolecular Heck reactions. *Angew. Chem.* **1994**, 106, 1138-1139.

12. Imbos, R., Minnaard, A. J. & Feringa, B. L. A Highly Enantioselective Intramolecular Heck Reaction with a Monodentate Ligand. *J. Am. Chem. Soc.* **2002**, 124, 184-185.

13. Lautens, M. & Zunic, V. Sequential olefin metathesis - intramolecular asymmetric Heck reactions in the synthesis of polycycles. *Can. J. Chem.* **2004**, 82, 399-407.

CYCLIZATION VIA π-ALLYL-METAL COMPLEXES

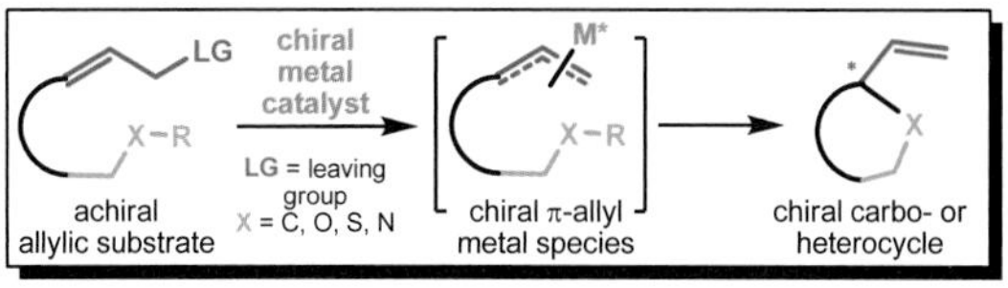

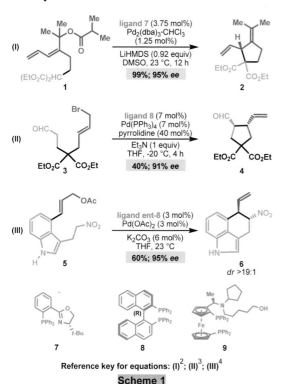

Introduction

The synthesis of medium-sized rings (most commonly 5- or 6-membered) via π-allyl-Pd intermediates by C-C, C-N or C-O bond formation is an especially powerful construction in the enantioselective version.[1] Such cyclizations generally are catalytic and employ a chiral bidentate ligand on Pd(0) as reagent. However, certain monodentate, but highly substituted, chiral phosphines can also be effective. The substrates for these ring-closures commonly contain an allylic ester subunit separated by an intervening chain from an acidic C-H, N-H or OH function. Bond formation occurs by nucleophilic attack by $C^{\ominus}$, $N^{\ominus}$ or $O^{\ominus}$ on the proximate carbon of the intermediate Pd-π-allyl complex. Instances of such ring-closures involving π-allyl complexes of Ni, Rh or Ir have been reported, but are relatively rare. Most of the reported examples involve the formation of a 5- or 6-membered ring. A variety of examples appear below to illustrate C-C bond formation (Schemes 1 and 2), C-N formation (Schemes 3 and 4) and C-O formation (Scheme 3).

I. Pd-Catalyzed Enantioselective Intramolecular Allylation Reaction to Form C-C Bonds

II. Pd- and Ir-Catalyzed Enantioselective Inter- and Intramolecular Allylation Reaction to Form C-C Bonds

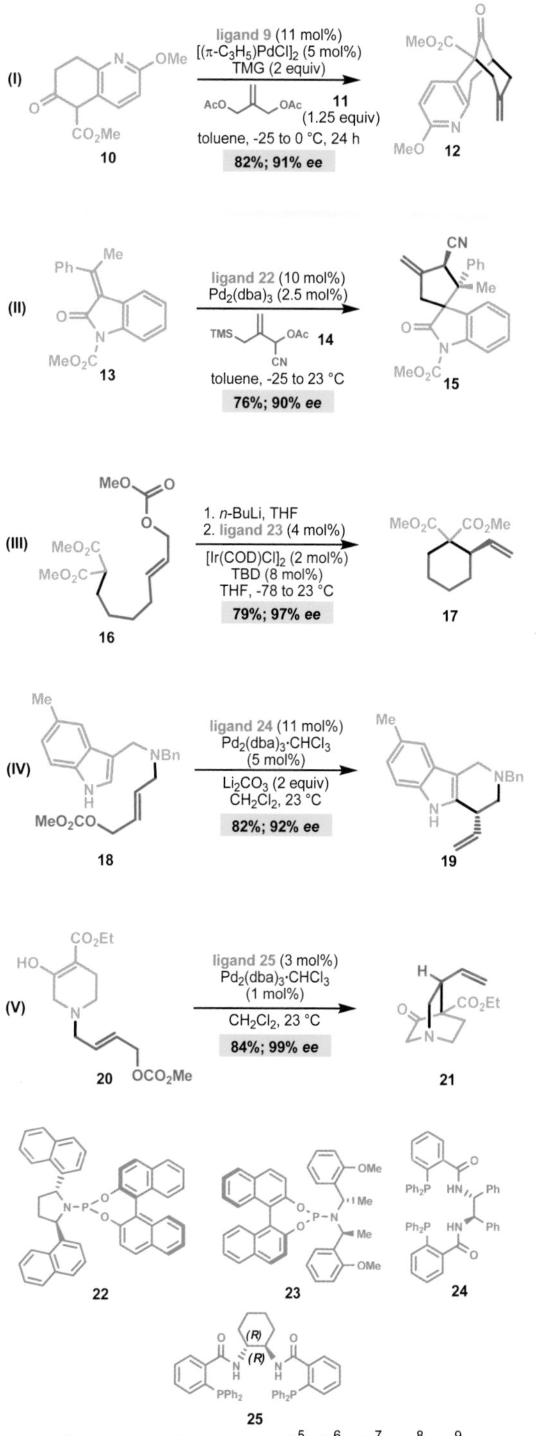

Reference key for equations: (I)[2]; (II)[3]; (III)[4]

Scheme 1

Reference key for equations: (I)[5]; (II)[6]; (III)[7]; (IV)[8]; (V)[9]

Scheme 2

CYCLIZATION VIA π-ALLYL-METAL COMPLEXES

III. Pd-Catalyzed Enantioselective Intramolecular Allylation Reaction to Form Carbon-Heteroatom (C-X) Bonds

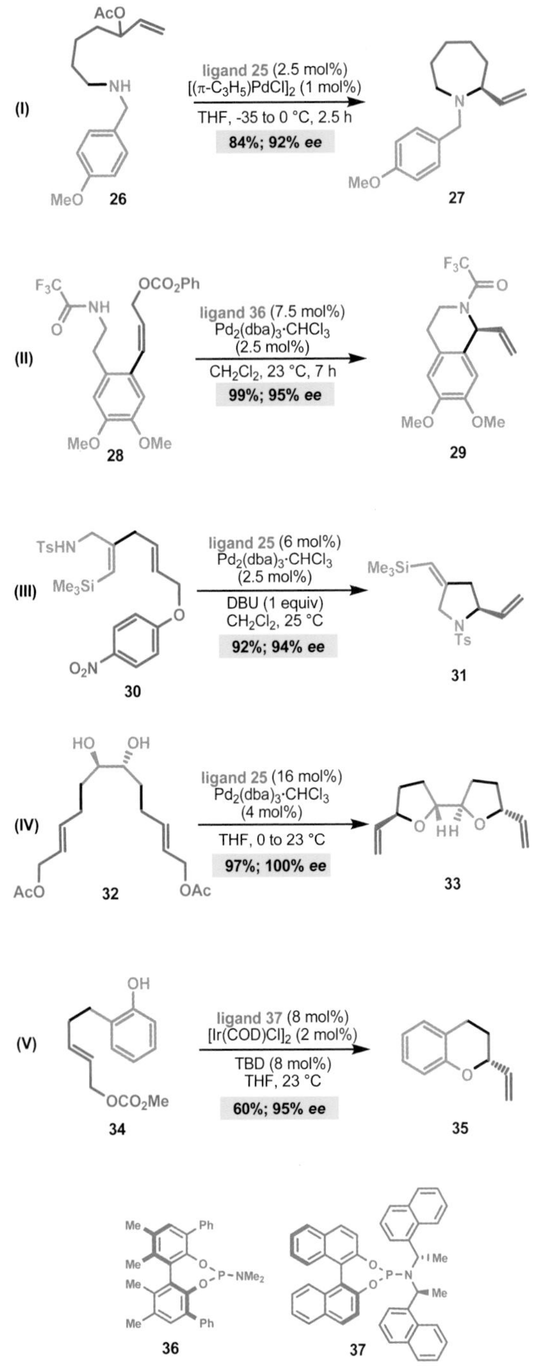

(I)

ligand 25 (2.5 mol%)
[(π-C₃H₅)PdCl]₂ (1 mol%)
THF, -35 to 0 °C, 2.5 h
84%; 92% ee

26 → 27

(II)

ligand 36 (7.5 mol%)
Pd₂(dba)₃·CHCl₃ (2.5 mol%)
CH₂Cl₂, 23 °C, 7 h
99%; 95% ee

28 → 29

(III)

ligand 25 (6 mol%)
Pd₂(dba)₃·CHCl₃ (2.5 mol%)
DBU (1 equiv)
CH₂Cl₂, 25 °C
92%; 94% ee

30 → 31

(IV)

ligand 25 (16 mol%)
Pd₂(dba)₃·CHCl₃ (4 mol%)
THF, 0 to 23 °C
97%; 100% ee

32 → 33

(V)

ligand 37 (8 mol%)
[Ir(COD)Cl]₂ (2 mol%)
TBD (8 mol%)
THF, 23 °C
60%; 95% ee

34 → 35

36   37

Reference key for equations: (I)[10]; (II)[11]; (III)[12]; (IV)[13]; (V)[14]

Scheme 3

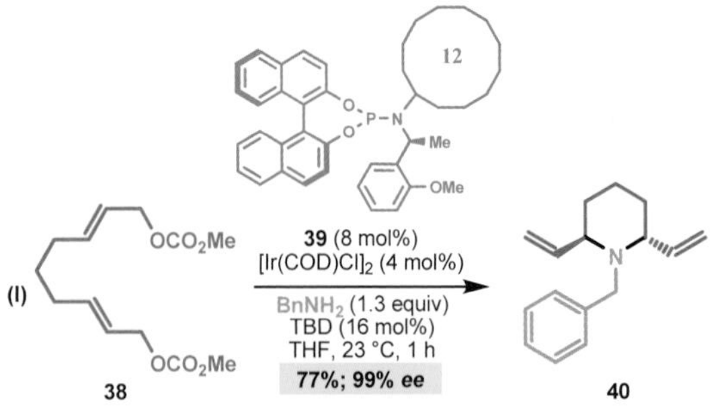

(I)

39 (8 mol%)
[Ir(COD)Cl]₂ (4 mol%)
BnNH₂ (1.3 equiv)
TBD (16 mol%)
THF, 23 °C, 1 h
77%; 99% ee

38 → 40

Reference key for equations: (I)[14]

Scheme 4

References

1. (a) Trost, B. M. & Crawley, M. L. Asymmetric Transition-Metal-Catalyzed Allylic Alkylations: Applications in Total Synthesis. *Chem. Rev. (Washington, DC, U. S.)* **2003**, 103, 2921-2943; (b) Trost, B. M. Asymmetric allylic alkylation, an enabling methodology. *J. Org. Chem.* **2004**, 69, 5813-5837; (c) Lu, Z. & Ma, S. Metal-catalyzed enantioselective allylation in asymmetric synthesis. *Angew. Chem., Int. Ed.* **2008**, 47, 258-297; (d) Lu, Z. & Ma, S. Transition metal-catalyzed intramolecular allylation reactions in *Handbook of Cyclization Reactions* 271-313 (**2010**).
2. Bian, J., Van Wingerden, M. & Ready, J. M. Enantioselective Total Synthesis of (+)- and (-)-Nigellamine A2. *J. Am. Chem. Soc.* **2006**, 128, 7428-7429.
3. Bihelovic, F., Matovic, R., Vulovic, B. & Saicic, R. N. Organocatalyzed Cyclizations of π-Allylpalladium Complexes: A New Method for the Construction of Five- and Six-Membered Rings. *Org. Lett.* **2007**, 9, 5063-5066.
4. Kardos, N. & Genet, J.-P. Synthesis of (-)-chanoclavine I. *Tetrahedron Asymmetry* **1994**, 5, 1525-1533.
5. He, X.-C., Wang, B., Yu, G. & Bai, D. Studies on the asymmetric synthesis of huperzine A. Part 2: Highly enantioselective palladium-catalyzed bicycloannulation of the β-keto-ester using new chiral ferrocenylphosphine ligands. *Tetrahedron Asymmetry* **2001**, 12, 3213-3216.
6. Trost, B. M., Cramer, N. & Silverman, S. M. Enantioselective construction of spirocyclic oxindolic cyclopentanes by palladium-catalyzed trimethylenemethane-[3+2]-cycloaddition. *J. Am. Chem. Soc.* **2007**, 129, 12396-12397.
7. Streiff, S., Welter, C., Schelwies, M., Lipowsky, G., Miller, N. & Helmchen, G. Carbocycles via enantioselective inter- and intramolecular iridium-catalyzed allylic alkylations. *Chem. Commun. (Cambridge, U. K.)* **2005**, 2957-2959.
8. Bandini, M., Melloni, A., Piccinelli, F., Sinisi, R., Tommasi, S. & Umani-Ronchi, A. Highly Enantioselective Synthesis of Tetrahydro-β-Carbolines and Tetrahydro-γ-Carbolines Via Pd-Catalyzed Intramolecular Allylic Alkylation. *J. Am. Chem. Soc.* **2006**, 128, 1424-1425.
9. Trost, B. M., Sacchi, K. L., Schroeder, G. M. & Asakawa, N. Intramolecular Palladium-Catalyzed Allylic Alkylation: Enantio- and Diastereoselective Synthesis of [2.2.2] Bicycles. *Org. Lett.* **2002**, 4, 3427-3430.
10. Trost, B. M., Krische, M. J., Radinov, R. & Zanoni, G. On Asymmetric Induction in Allylic Alkylation via Enantiotopic Facial Discrimination. *J. Am. Chem. Soc.* **1996**, 118, 6297-6298.
11. Ito, K., Akashi, S., Saito, B. & Katsuki, T. Asymmetric intramolecular allylic amination: Straightforward approach to chiral C1-substituted tetrahydroisoquinolines. *Synlett* **2003**, 1809-1812.
12. Trost, B. M., Machacek, M. R. & Faulk, B. D. Sequential Ru-Pd Catalysis: A Two-Catalyst One-Pot Protocol for the Synthesis of N- and O-Heterocycles. *J. Am. Chem. Soc.* **2006**, 128, 6745-6754.
13. Burke, S. D. & Jiang, L. Formal Synthesis of Uvaricin via Palladium-Mediated Double Cyclization. *Org. Lett.* **2001**, 3, 1953-1955.
14. Welter, C., Dahnz, A., Brunner, B., Streiff, S., Duebon, P. & Helmchen, G. Highly enantioselective syntheses of heterocycles via intramolecular Ir-catalyzed allylic amination and etherification. *Org. Lett.* **2005**, 7, 1239-1242.

ENANTIOSELECTIVE NAZAROV CYCLIZATION

dienone → 3-oxy-pentadienyl cation → cyclopentenone

chiral catalyst, 4π electro-cyclization

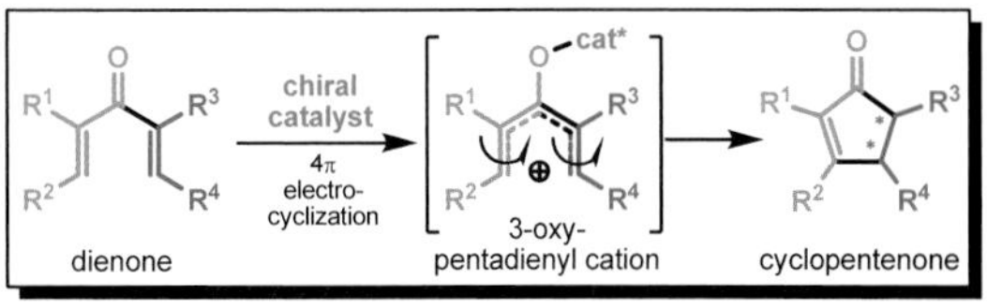

Introduction

Divinyl ketones undergo cyclization to 2-cyclopentenones (Nazarov reaction[1]) when activated by protonation or coordination to a Lewis acid. The process occurs via a 3-oxypentadienyl cation (see above) with conrotatory ring-closure. The reaction has been trans-formed into an enantioselective version by the application of certain chiral Lewis acids or chiral protic activation. A third mode of cyclization can occur via an intermediate 3-aminopentadienyl cation (e.g., 21→22). As will be apparent from the examples in Schemes 1 and 2 below, enantioselective cyclization has been demonstrated with a limited number of substrates. Control of absolute configuration has also been obtained in Nazarov cyclizations by the use of a chiral controller group.[1]

Catalytic Enantio- and Diastereoselective Nazarov Cyclization

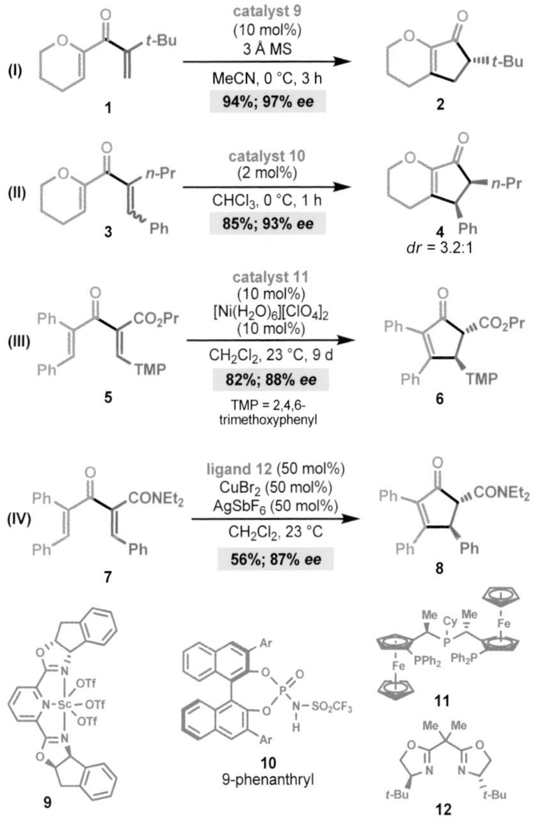

(I) catalyst 9 (10 mol%) 3 Å MS, MeCN, 0 °C, 3 h — **94%; 97% ee** — 1 → 2 (t-Bu)

(II) catalyst 10 (2 mol%) CHCl₃, 0 °C, 1 h — **85%; 93% ee** — 3 → 4 (Ph), dr = 3.2:1

(III) catalyst 11 (10 mol%) [Ni(H₂O)₆][ClO₄]₂ (10 mol%) CH₂Cl₂, 23 °C, 9 d — **82%; 88% ee** — 5 → 6, TMP = 2,4,6-trimethoxyphenyl

(IV) ligand 12 (50 mol%) CuBr₂ (50 mol%) AgSbF₆ (50 mol%) CH₂Cl₂, 23 °C — **56%; 87% ee** — 7 → 8

9
10 — 9-phenanthryl
11
12

Reference key for equations: (I)[2]; (II)[3]; (III)[4]; (IV)[5]

Scheme 1

Catalytic Enantio- and Diastereoselective Nazarov Cyclization (Continued)

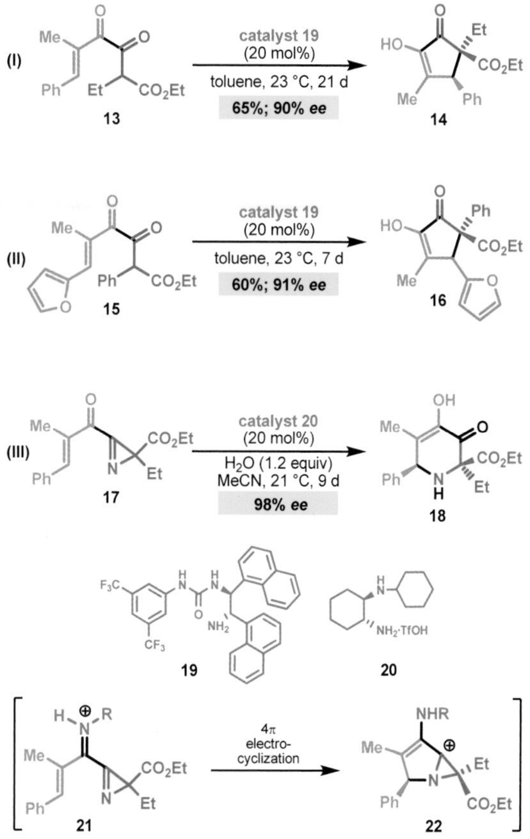

(I) catalyst 19 (20 mol%) toluene, 23 °C, 21 d — **65%; 90% ee** — 13 → 14

(II) catalyst 19 (20 mol%) toluene, 23 °C, 7 d — **60%; 91% ee** — 15 → 16

(III) catalyst 20 (20 mol%) H₂O (1.2 equiv) MeCN, 21 °C, 9 d — **98% ee** — 17 → 18

19
20

21 → 22, 4π electro-cyclization

Reference key for equations: (I-II)[6]; (III)[7]

Scheme 2

References

1. (a) Tius, M. A. Some new Nazarov chemistry. *Eur. J. Org. Chem.* **2005**, 2193-2206; (b) Pellissier, H. Recent developments in the Nazarov process. *Tetrahedron* **2005**, 61, 6479-6517; (c) Frontier, A. J. & Collison, C. The Nazarov cyclization in organic synthesis. Recent advances. *Tetrahedron* **2005**, 61, 7577-7606; (d) Nakanishi, W. & West, F. G. Advances in the Nazarov cyclization. *Curr. Opin. Drug Discovery Dev.* **2009**, 12, 732-751.
2. Liang, G. & Trauner, D. Enantioselective Nazarov reactions through catalytic asymmetric proton transfer. *J. Am. Chem. Soc.* **2004**, 126, 9544-9545.
3. Rueping, M., Ieawsuwan, W., Antonchick, A. P. & Nachtsheim, B. J. Chiral Broensted acids in the catalytic asymmetric Nazarov cyclization - the first enantioselective organocatalytic electrocyclic reaction. *Angew. Chem., Int. Ed.* **2007**, 46, 2097-2100.
4. Walz, I. & Togni, A. Ni(II)-catalyzed enantioselective Nazarov cyclizations. *Chem. Commun. (Cambridge, U. K.)* **2008**, 4315-4317.
5. Aggarwal, V. K. & Belfield, A. J. Catalytic Asymmetric Nazarov Reactions Promoted by Chiral Lewis Acid Complexes. *Org. Lett.* **2003**, 5, 5075-5078.
6. Basak, A. K., Shimada, N., Bow, W. F., Vicic, D. A. & Tius, M. A. An Organocatalytic Asymmetric Nazarov Cyclization. *J. Am. Chem. Soc.* **2010**, 132, 8266-8267.
7. Shimada, N., Ashburn, B. O., Basak, A. K., Bow, W. F., Vicic, D. A. & Tius, M. A. Organocatalytic asymmetric aza-Nazarov cyclization of an azirine. *Chem. Commun.* **2010**, 46, 3774-3775.

PICTET-SPENGLER AND RELATED CYCLIZATIONS

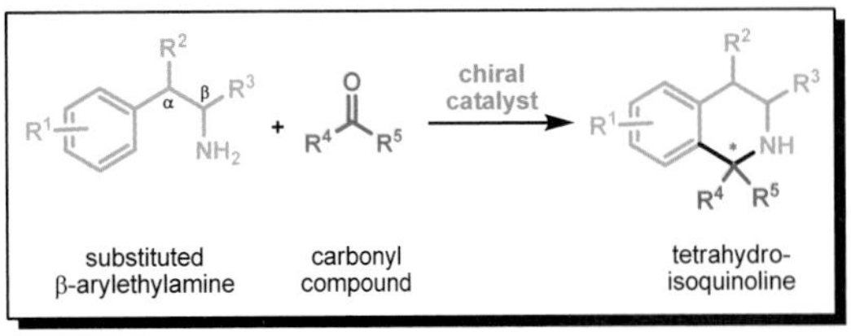

| substituted β-arylethylamine | carbonyl compound | tetrahydro-isoquinoline |

Introduction

The reaction of a β-arylethylamine with a carbonyl compound to form a substituted tetrahydroisoquinoline (Pictet-Spengler reaction)[1] and the related synthesis of tetrahydro-β-carbolines from substituted tryptamines (Figure 1) have been applied to the synthesis of many natural products and N-heterocyclic molecules.[2]

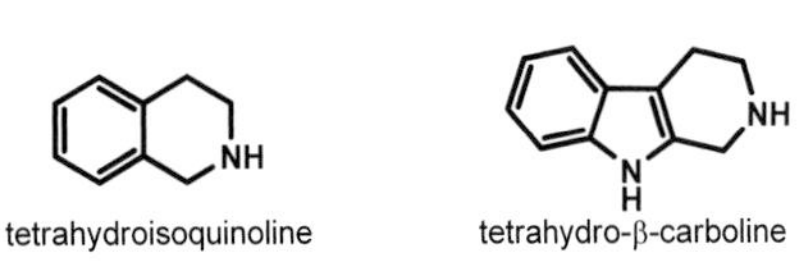

tetrahydroisoquinoline   tetrahydro-β-carboline

Figure 1

The Pictet-Spengler reaction is catalyzed by protic or Lewis acids and follows the pathway outlined in Scheme 1. An initially formed imine (**A**) undergoes activation by a Brønsted or Lewis acid, to form an iminium species (**B**) which cyclizes to product via intermediate **C**. The rate of the cyclization (**B**→**C**) depends on the nucleophilicity of the aromatic subunit.

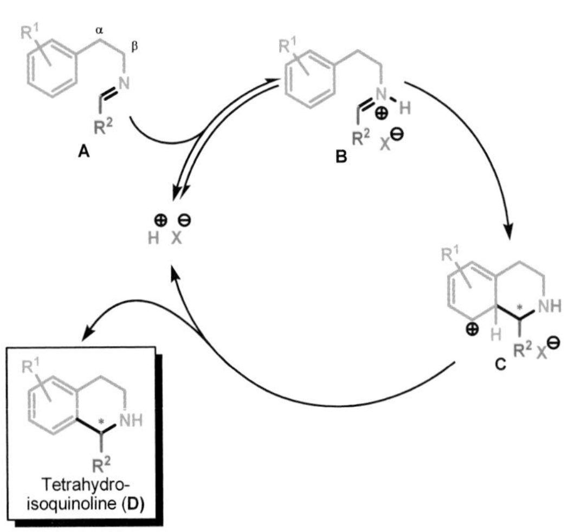

Tetrahydro-isoquinoline (**D**)

Scheme 1

Chiral Pictet-Spengler products are available either by the use of chiral controller groups,[2d] or by the application of chiral protic catalysts. The enantioselective version of the Pictet-Spengler reaction has been known since 2004, and subsequently reviewed.[3] Examples of the catalytic enantioselective Pictet-Spengler cyclizations appear in Schemes 2-4.

I. Enantioselective Pictet-Spengler Reaction With a Chiral Thiourea as Catalyst

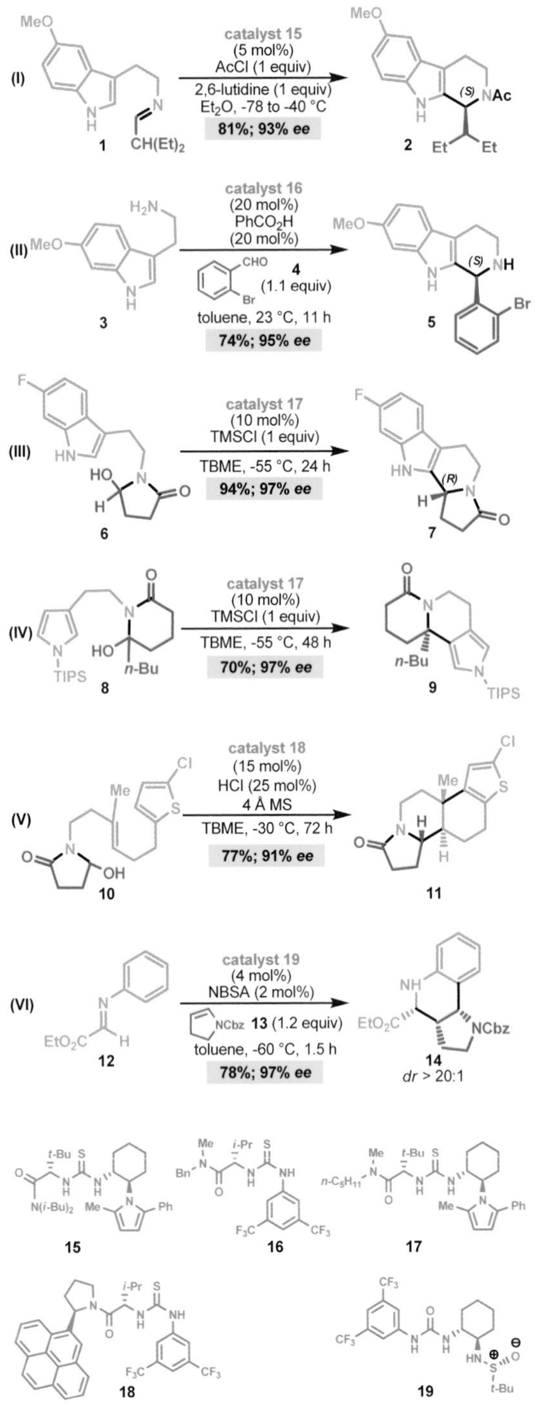

(I)
catalyst 15 (5 mol%)
AcCl (1 equiv)
2,6-lutidine (1 equiv)
Et₂O, -78 to -40 °C
81%; 93% ee

(II)
catalyst 16 (20 mol%)
PhCO₂H (20 mol%)
4 (1.1 equiv)
toluene, 23 °C, 11 h
74%; 95% ee

(III)
catalyst 17 (10 mol%)
TMSCl (1 equiv)
TBME, -55 °C, 24 h
94%; 97% ee

(IV)
catalyst 17 (10 mol%)
TMSCl (1 equiv)
TBME, -55 °C, 48 h
70%; 97% ee

(V)
catalyst 18 (15 mol%)
HCl (25 mol%)
4 Å MS
TBME, -30 °C, 72 h
77%; 91% ee

(VI)
catalyst 19 (4 mol%)
NBSA (2 mol%)
NCbz 13 (1.2 equiv)
toluene, -60 °C, 1.5 h
78%; 97% ee
dr > 20:1

15   16   17

18   19

Reference key for equations: (I)[4]; (II)[5]; (III)[6]; (IV)[7]; (V)[8]; (VI)[9]

Scheme 2

PICTET-SPENGLER AND RELATED CYCLIZATIONS

II. Enantioselective Pictet-Spengler Reaction Via a Chiral Ion Pair

The reaction of an imine or enamine intermediate in the presence of a chiral protic donor under conditions that favor contact ion pair formation can provide enantiocontrol of a Pictet-Spengler reaction.

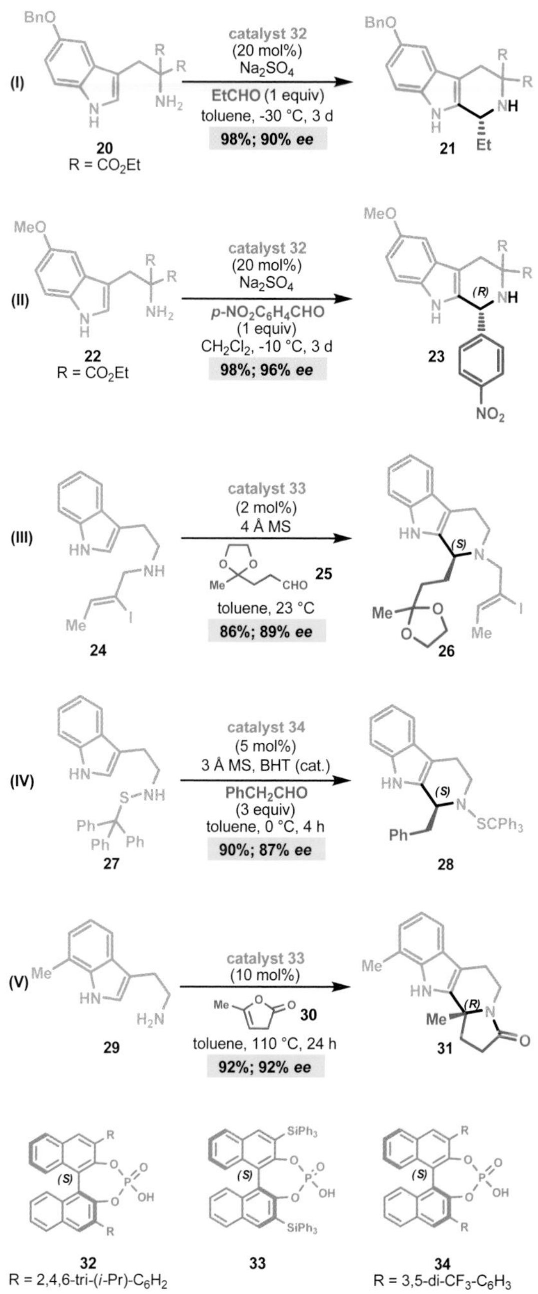

(I) **20** R = CO₂Et → catalyst 32 (20 mol%), Na₂SO₄, EtCHO (1 equiv), toluene, -30 °C, 3 d → **21** 98%; 90% ee

(II) **22** R = CO₂Et → catalyst 32 (20 mol%), Na₂SO₄, p-NO₂C₆H₄CHO (1 equiv), CH₂Cl₂, -10 °C, 3 d → **23** (R) 98%; 96% ee

(III) **24** → catalyst 33 (2 mol%), 4 Å MS, **25**, toluene, 23 °C → **26** (S) 86%; 89% ee

(IV) **27** → catalyst 34 (5 mol%), 3 Å MS, BHT (cat.), PhCH₂CHO (3 equiv), toluene, 0 °C, 4 h → **28** (S) 90%; 87% ee

(V) **29** → catalyst 33 (10 mol%), **30**, toluene, 110 °C, 24 h → **31** (R) 92%; 92% ee

32 R = 2,4,6-tri-(i-Pr)-C₆H₂   **33**   **34** R = 3,5-di-CF₃-C₆H₃

Reference key for equations: (I-II)[10]; (III)[11]; (IV)[12]; (V)[13]

Scheme 3

Enantioselective Pictet-Spengler Reaction Via a Chiral Ion Pair (Continued)

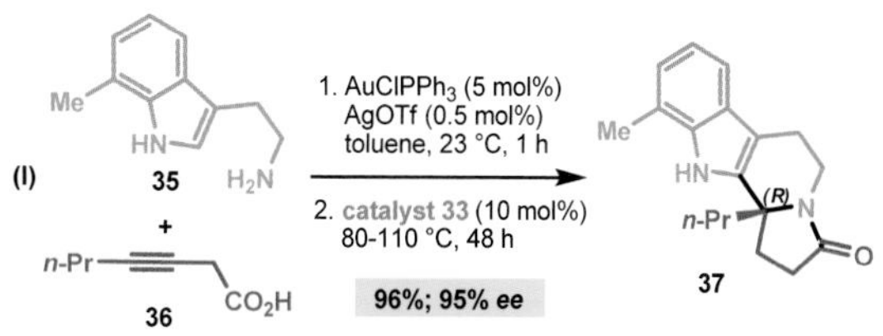

(I) **35** + **36** n-Pr ≡ CO₂H → 1. AuClPPh₃ (5 mol%), AgOTf (0.5 mol%), toluene, 23 °C, 1 h; 2. catalyst 33 (10 mol%), 80-110 °C, 48 h → **37** (R) 96%; 95% ee

Reference key for equation: (I)[13]

Scheme 4

References

1. Pictet, A. & Spengler, T. Formation of Isoquinoline Derivatives by the Action of Methylal on Phenylethylamine, Phenylalanine and Tyrosine. *Chem. Ber.* **1911**, 44, 2030-2036.
2. (a) Cox, E. D. & Cook, J. M. The Pictet-Spengler condensation: a new direction for an old reaction. *Chem. Rev.* **1995**, 95, 1797-1842; (b) Bentley, K. W. β-phenylethylamines and the isoquinoline alkaloids. *Nat. Prod. Rep.* **2004**, 21, 395-424; (c) Chrzanowska, M. & Rozwadowska, M. D. Asymmetric Synthesis of Isoquinoline Alkaloids. *Chem. Rev.* **2004**, 104, 3341-3370; (d) Kaufman, T. S. Synthesis of optically-active isoquinoline and indole alkaloids employing the Pictet-Spengler condensation with removable chiral auxiliaries bound to nitrogen. *New Methods for the Asymmetric Synthesis of Nitrogen Heterocycles* **2005**, 99-147; (e) Larghi, E. L. & Kaufman, T. S. The oxa-Pictet-Spengler cyclization: synthesis of isochromans and related pyran-type heterocycles. *Synthesis* **2006**, 187-220; (f) Youn, S. W. The Pictet-Spengler reaction: efficient carbon-carbon bond forming reaction in heterocyclic synthesis. *Org. Prep. Proced. Int.* **2006**, 38, 505-591; (g) Edwankar, C. R., Edwankar, R. V., Namjoshi, O. A., Rallapalli, S. K., Yang, J. & Cook, J. M. Recent progress in the total synthesis of indole alkaloids. *Curr. Opin. Drug Discovery Dev.* **2009**, 12, 752-771.
3. (a) Belyk, K. Highly enantioselective catalytic acyl Pictet-Spengler reactions. *Chemtracts* **2005**, 18, 57-64; (b) Bandini, M. & Eichholzer, A. Catalytic Functionalization of Indoles in a New Dimension. *Angew. Chem., Int. Ed.* **2009**, 48, 9608-9644.
4. Taylor, M. S. & Jacobsen, E. N. Highly enantioselective catalytic acyl-Pictet-Spengler reactions. *J. Am. Chem. Soc.* **2004**, 126, 10558-10559.
5. Klausen, R. S. & Jacobsen, E. N. Weak Bronsted Acid-Thiourea Co-catalysis: Enantioselective, Catalytic Protio-Pictet-Spengler Reactions. *Org. Lett.* **2009**, 11, 887-890.
6. Raheem, I. T., Thiara, P. S., Peterson, E. A. & Jacobsen, E. N. Enantioselective Pictet-Spengler-Type Cyclizations of Hydroxylactams: H-Bond Donor Catalysis by Anion Binding. *J. Am. Chem. Soc.* **2007**, 129, 13404-13405.
7. Raheem, I. T., Thiara, P. S. & Jacobsen, E. N. Regio- and enantioselective catalytic cyclization of pyrroles onto N-acyliminium ions. *Org. Lett.* **2008**, 10, 1577-1580.
8. Knowles, R. R., Lin, S. & Jacobsen, E. N. Enantioselective Thiourea-Catalyzed Cationic Polycyclizations. *J. Am. Chem. Soc.* **2010**, 132, 5030-5032.
9. Xu, H., Zuend, S. J., Woll, M. G., Tao, Y. & Jacobsen, E. N. Asymmetric Cooperative Catalysis of Strong Bronsted Acid-Promoted Reactions Using Chiral Ureas. *Science* **2010**, 327, 986-990.
10. Seayad, J., Seayad Abdul, M. & List, B. Catalytic asymmetric Pictet-Spengler reaction. *J. Am. Chem. Soc.* **2006**, 128, 1086-1087.
11. Wanner, M. J., Boots, R. N. A., Eradus, B., de Gelder, R., van Maarseveen, J. H. & Hiemstra, H. Organocatalytic Enantioselective Total Synthesis of (-)-Arboricine. *Org. Lett.* **2009**, 11, 2579-2581.
12. Wanner, M. J., van der Haas, R. N. S., de Cuba, K. R., van Maarseveen, J. H. & Hiemstra, H. Catalytic asymmetric Pictet-Spengler reactions via sulfenyliminium ions. *Angew. Chem., Int. Ed.* **2007**, 46, 7485-7487.
13. Muratore, M. E., Holloway, C. A., Pilling, A. W., Storer, R. I., Trevitt, G. & Dixon, D. J. Enantioselective Bronsted Acid-Catalyzed N-Acyliminium Cyclization Cascades. *J. Am. Chem. Soc.* **2009**, 131, 10796-10797.

ENANTIOSELECTIVE OLEFIN METATHESIS

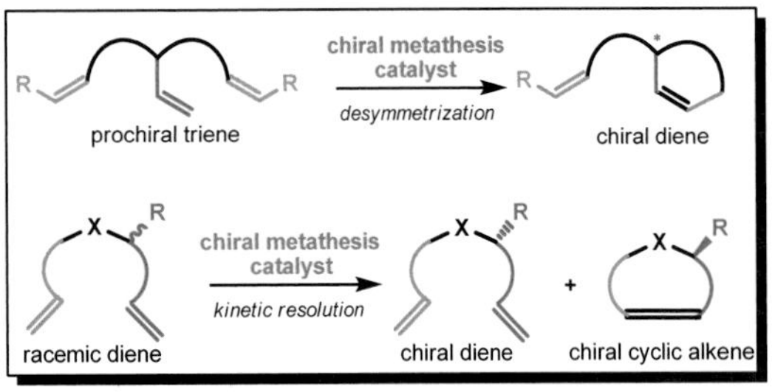

Introduction

Enantioselective olefin metathesis[1] has been achieved either by de-symmetrization of a prochiral triene or by kinetic resolution of a racemic diene using ring-closing metathesis (RCM, see above). The chiral metathesis catalysts, molybdenum (Mo)-based alkylidenes or ruthenium (Ru)-based carbenes, generally differ in reactivity and functional group tolerance. A selection of examples is shown in Schemes 1 and 2.

Symmetry-breaking RCM has been used for the total synthesis of several natural products.

RCM has also been conducted in tandem with ring-opening metathesis (ROM) and ring-opening cross metathesis (ROCM) for the syntheses of various complex carbocycles and heterocycles.

I. Catalytic Enantioselective Kinetic Resolution Using Ring Closing Metathesis (RCM) Reaction

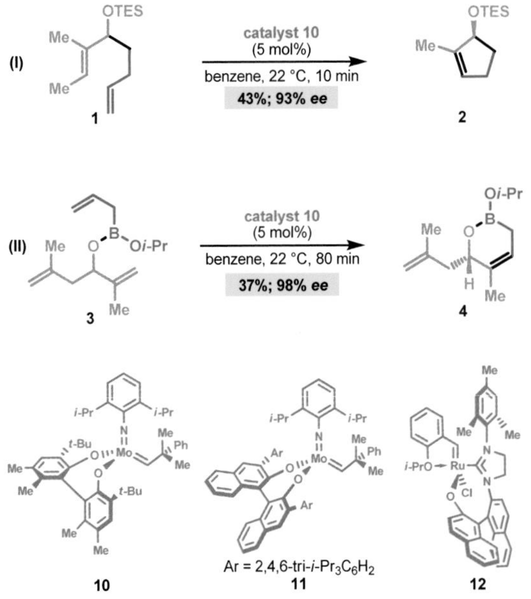

Reference key for equations: (I)[2]; (II)[3]

Scheme 1

II. Catalytic Enantioselective Desymmetrization Using RCM and ROM/RCM

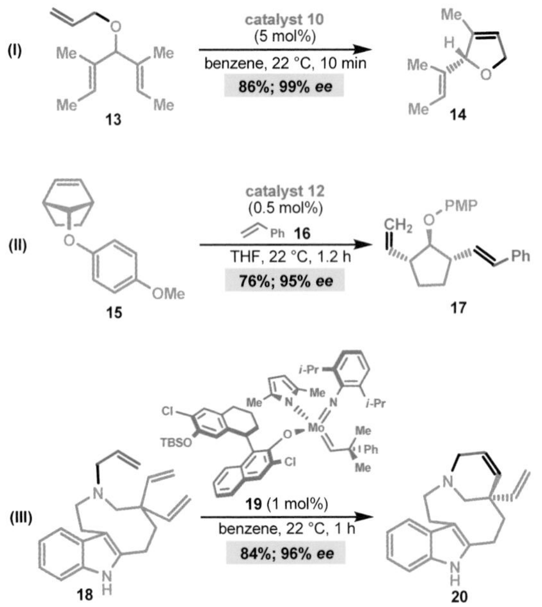

Reference key for equations: (I)[4]; (II)[5]; (III)[6]

Scheme 2

References

1. (a) Hoveyda, A. H. & Schrock, R. R. Catalytic asymmetric olefin metathesis. *Chem.--Eur. J.* **2001**, 7, 945-950; (b) Connon, S. J. & Blechert, S. Recent advances in alkene metathesis. *Top. Organomet. Chem.* **2004**, 11, 93-124; (c) Klare, H. F. T. & Oestreich, M. Asymmetric ring-closing metathesis with a twist. *Angew. Chem., Int. Ed.* **2009**, 48, 2085-2089; (d) Hoveyda, A. H., Malcolmson, S. J., Meek, S. J. & Zhugralin, A. R. Catalytic Enantioselective Olefin Metathesis Reactions in *Catalytic Asymmetric Synthesis, 3rd Ed.* (ed. Ojima, I.), (John Wiley & Sons, Hoboke, NJ, **2010**); (e) Hoveyda, A. H., Malcolmson, S. J., Meek, S. J. & Zhugralin, A. R. Catalytic Enantioselective Olefin Metathesis in Natural Product Synthesis. Chiral Metal-Based Complexes that Deliver High Enantioselectivity and More. *Angew. Chem., Int. Ed.* **2010**, 49, 34-44; (f) Mori, M. Ring-closing metathesis of dienes and enynes. *Handbook of Cyclization Reactions* **2010**, 1, 527-598.
2. Alexander, J. B., La, D. S., Cefalo, D. R., Hoveyda, A. H. & Schrock, R. R. Catalytic Enantioselective Ring-Closing Metathesis by a Chiral Biphen-Mo Complex. *J. Am. Chem. Soc.* **1998**, 120, 4041-4042.
3. Jernelius, J. A., Schrock, R. R. & Hoveyda, A. H. Enantioselective synthesis of cyclic allylboronates by Mo-catalyzed asymmetric ring-closing metathesis (ARCM). A one-pot protocol for net catalytic enantioselective cross metathesis. *Tetrahedron* **2004**, 60, 7345-7351.
4. Zhu, S. S., Cefalo, D. R., La, D. S., Jamieson, J. Y., Davis, W. M., Hoveyda, A. H. & Schrock, R. R. Chiral Mo-Binol Complexes: Activity, Synthesis, and Structure. Efficient Enantioselective Six-Membered Ring Synthesis through Catalytic Metathesis. *J. Am. Chem. Soc.* **1999**, 121, 8251-8259.
5. Van Veldhuizen, J. J., Garber, S. B., Kingsbury, J. S. & Hoveyda, A. H. A Recyclable Chiral Ru Catalyst for Enantioselective Olefin Metathesis. Efficient Catalytic Asymmetric Ring-Opening/Cross Metathesis in Air. *J. Am. Chem. Soc.* **2002**, 124, 4954-4955.
6. Malcolmson, S. J., Meek, S. J., Sattely, E. S., Schrock, R. R. & Hoveyda, A. H. Highly efficient molybdenum-based catalysts for enantioselective alkene metathesis. *Nature (London, U. K.)* **2008**, 456, 933-937.

SYMMETRY-BREAKING REACTIONS

Introduction

There are many ways by which chiral molecules can be generated from C_2-symmetric prochiral precursors.[1] Several such examples appear on the foregoing pages of this book [e.g., olefin metathesis (page 150), Stetter reaction (page 115), intramolecular 1,4-addition reactions (page 136)]. In some cases C_2-symmetry is broken simply by S_N2 displacement at sp^3-carbon of an epoxide or aziridine. In others, the symmetry-breaking step involves a change of sp^2-carbon to sp^3-carbon at C=O or C=C. A selection of interesting examples of synthesis of chiral molecules by de-symmetrizing processes is displayed in Schemes 1 and 2.

Reactions which break C2-symmetry in substrates with multiple prochiral centers are especially powerful because they generate multiple stereocenters (see Equation III, Scheme 1).

I. Catalytic Enantioselective De-symmetrization of Cyclic Systems by Ring-Opening

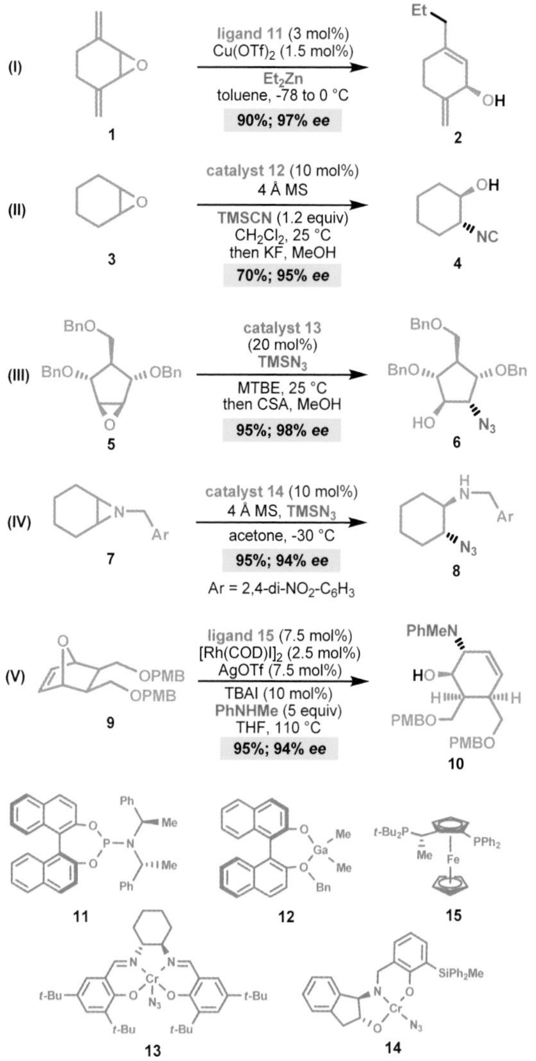

II. Miscellaneous Catalytic Enantioselective De-symmetrization Reactions

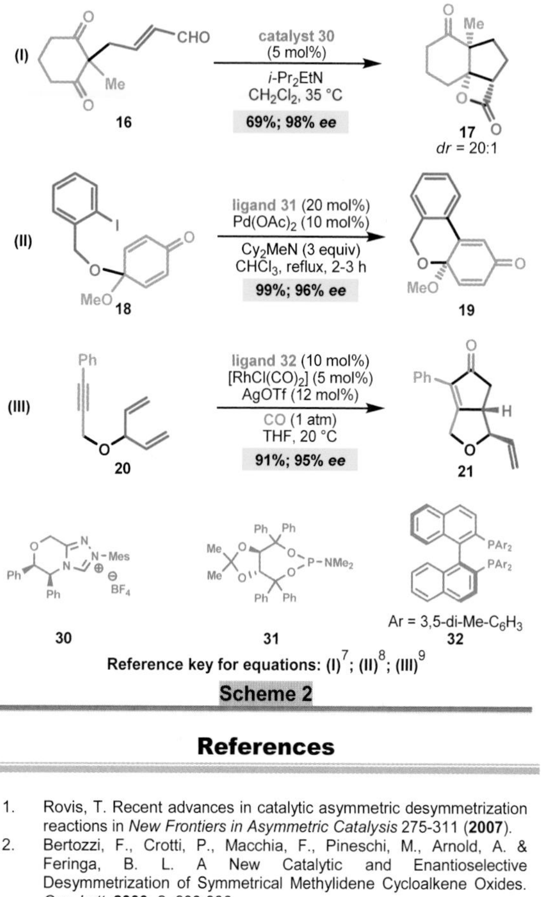

Reference key for equations: (I)[7]; (II)[8]; (III)[9]

Scheme 2

References

1. Rovis, T. Recent advances in catalytic asymmetric desymmetrization reactions in *New Frontiers in Asymmetric Catalysis* 275-311 (**2007**).
2. Bertozzi, F., Crotti, P., Macchia, F., Pineschi, M., Arnold, A. & Feringa, B. L. A New Catalytic and Enantioselective Desymmetrization of Symmetrical Methylidene Cycloalkene Oxides. *Org. Lett.* **2000**, 2, 933-936.
3. Zhu, C., Yuan, F., Gu, W. & Pan, Y. The first example of enantioselective isocyanosilylation of meso epoxides with TMSCN catalyzed by novel chiral organogallium and -indium complexes. *Chem. Commun. (Cambridge, U. K.)* **2003**, 692-693.
4. Kassab, D. J. & Ganem, B. An Enantioselective Synthesis of (-)-Allosamidin by Asymmetric Desymmetrization of a Highly Functionalized meso-Epoxide. *J. Org. Chem.* **1999**, 64, 1782-1783.
5. Li, Z., Fernandez, M. & Jacobsen, E. N. Enantioselective Ring Opening of Meso Aziridines Catalyzed by Tridentate Schiff Base Chromium(III) Complexes. *Org. Lett.* **1999**, 1, 1611-1613.
6. Lautens, M. & Fagnou, K. Effects of halide ligands and protic additives on enantioselectivity and reactivity in rhodium-catalyzed asymmetric ring-opening reactions. *J. Am. Chem. Soc.* **2001**, 123, 7170-7171.
7. Phillips, E. M., Roberts, J. M. & Scheidt, K. A. Catalytic Enantioselective Total Syntheses of Bakkenolides I, J, and S: Application of a Carbene-Catalyzed Desymmetrization. *Org. Lett.* **2010**, 12, 2830-2833.
8. Imbos, R., Minnaard, A. J. & Feringa, B. L. A Highly Enantioselective Intramolecular Heck Reaction with a Monodentate Ligand. *J. Am. Chem. Soc.* **2002**, 124, 184-185.
9. Jeong, N., Sung, B. K. & Choi, Y. K. Rhodium(I)-Catalyzed Asymmetric Intramolecular Pauson-Khand-Type Reaction. *J. Am. Chem. Soc.* **2000**, 122, 6771-6772.

Reference key for equations: (I)[2]; (II)[3]; (III)[4]; (IV)[5]; (V)[6]

Scheme 1

PART II.

PLANNING ENANTIOSELECTIVE SYNTHESES OF COMPLEX MOLECULES: LOGIC, STRATEGIES, STEREOCONTROL

PLANNING ENANTIOSELECTIVE SYNTHESES

Introduction

The naturally occurring molecules of molecular weight in the hundreds (centamolecules) generally are chiral and occur as only one of the two possible enantiomers, which can be named NE (natural enantiomer) or UE (unnatural). The production of only the NE form is a consequence of the synthetic control which results from the precise regulation of each reaction in the biosynthetic pathway by a chiral catalyst. Eons of evolution were required to assemble the catalytic machinery responsible for the exquisite control of biosynthetic processes by enzymes.

One acceptable chemical synthesis of a naturally occurring centamolecule (or other target substance), in contrast, must be devised and executed on a timescale of months or a few years in order to be practical. Until the mid-twentieth century the multi-step synthesis of complex natural products was generally beyond the reach of existing science. The present level of capability was reached only after countless developments and discoveries in multiple areas of chemistry. Modern synthetic chemistry would not exist were it not for these advances, including the following:

(a) Knowledge of detailed reaction mechanism and pathways;

(b) The development of new methods for determining structure, e.g., NMR and other forms of spectroscopy and single crystal X-ray diffraction analysis;

(c) Powerful new methods for analysis, separation and purification of reaction products;

(d) Methods and data for predicting the relationship between chemical structure and reactivity;

(e) The invention of countless new reagents, catalysts and reactions, especially those allowing selectivity in terms of molecular position, reaction site and stereochemistry;

(f) The development of a logic for simplifying and analyzing the structure of a target molecule to allow the rational derivation of possible synthetic pathways;

(g) The development of ways of controlling absolute stereochemistry so as to allow enantioselective synthesis.

It is the last, and most recent, of these advances in the past few decades that opened up the possibility of conducting the direct synthesis of a chiral complex molecule without the need for conventional resolution or physical separation of enantiomers. The availability of modern enantioselective reactions has added another layer of complexity to synthetic planning, which is the focus of this section of *Enantioselective Chemical Synthesis*. The strategies to deal with that complexity fall with the broad category of "stereochemical strategies" which can effectively guide a systematic search to derive simple and efficient synthetic pathways. The logical application of general strategies to the planning of multi-step chemical syntheses has been discussed in detail in *The Logic of Chemical Synthesis* (1989, John Wiley & Sons)[1] and reviewed more briefly in various places.[2]

Strategies for Synthetic Planning: A Brief Review

It is helpful to recapitulate some of the underlying ideas of logic in synthesis in order to set the stage for a survey of the specific strategies of enantioselective synthesis. Retrosynthetic analysis is a powerful technique for simplifying a complex target structure to generate a tree of simpler structures and lead eventually to compounds that are available or easily prepared. The search for valid retrosynthetic pathways is driven by the goal of reducing molecular complexity and guided by the concurrent application of more than one simplifying strategy. The most effective strategies will be those that match the structural elements responsible for complexity of the target molecule. Among the kinds of structural complexity that have to be considered are cyclic connectivity (topology), stereocenter content, molecular size, centers of reactivity, element and functional group content, chemical reactivity, structural instability and density of complicating elements. For each type of complexity there are appropriate strategies for reducing that complexity.

There are three different directional modes of analysis that can be applied to the planning of a synthesis. The oldest of these can be used when it is possible to discern from a target structure one or more starting materials that map on to a portion of the target and that can be connected to it by known synthetic reactions. Most of the multistep syntheses before 1950, which generally involved simpler target structures, were devised under guidance by identifiable starting material(s) in what is essentially a forward-looking approach, i.e., in the direction of experimental execution. The opposite mode of analysis is required when scrutiny of the target structure does not reveal a possible starting point. A classic illustration of that situation may be found in the structure of longifolene (Scheme 1), the structure that originally inspired retrosynthetic analysis in 1957.

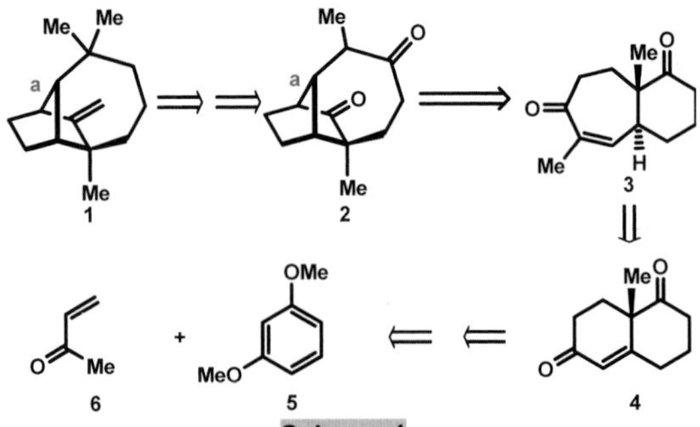

Scheme 1

The path of the retrosynthetic analysis shown in Scheme 1 was guided by the goals of disconnecting bond a in **1** and finding an appropriate process to enable C-C cleavage. That approach led to the generation of structure **2** in four steps from **1**. A Michael disconnection of **2** then generated the bicyclic diketone **3**. It was not until the target was simplified to **3** that the key intermediate **4** and the logical starting materials **5** and **6** were identified. The use of such retrosynthetic thinking is now universal in synthetic planning with complex target structures.

PLANNING ENANTIOSELECTIVE SYNTHESES

Retrosynthetic analysis can also be combined with ideas for possible available building blocks for constructing the target compound. In general, this becomes a workable approach when the structure of an available or known compound maps on to a subunit in the target structure. The success of such a matching process then allows a bidirectional analysis i.e., a starting structure-guided search. This kind of analysis – bidirectional, structure-guided search, represents a third directional mode for generating synthetic pathways. The industrial synthesis of antiflu agent oseltamivir exemplifies a synthesis planned by bidirectional search (Scheme 2).[3]

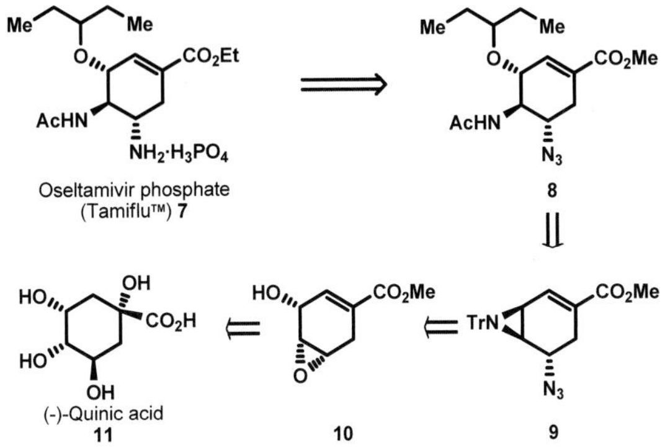

Oseltamivir phosphate
(Tamiflu™) 7

(-)-Quinic acid
11

Scheme 2

At this point it is useful to review the terminology that has been employed in retrosynthetic analysis. The retrosynthetic equivalent of a reaction is "*transform*". A transform operates on a target structure to generate a synthetic precursor, e.g.:

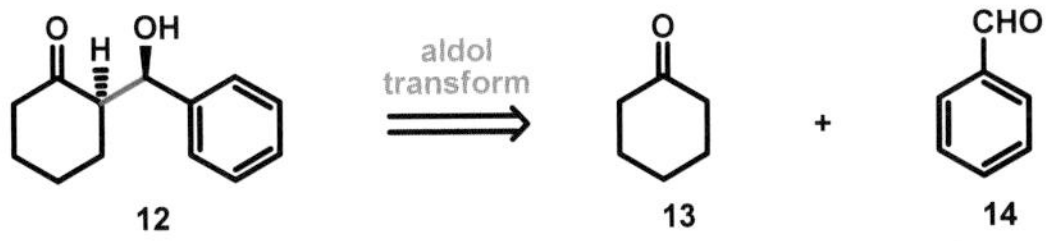

Scheme 3

The retron or substructural keying element for the aldol transform is the subunit: C(=O)-C-C-OH. The aldol transform is powerfully simplifying because it disconnects the target to simpler structures while also reducing stereochemical complexity. In retrosynthetic analysis it is important to place transforms in a hiercarchy with regard to the intrinsic power to simplify a structure. A disconnective transform may be so strongly simplifying that it can be fruitful to find a way to apply it even though several retrosynthetic steps (subgoals) may be necessary to establish the required retron and allow a valid disconnection. Thus, to return to the retrosynthetic analysis of longifolene (Scheme 1), the disconnection of bond **a** of longifolene using the *intramolecular Michael transform* cannot be accomplished directly, but is valid after retrosynthetic conversion to **2**, which has the complete retron (Scheme 3) for the Michael transform.

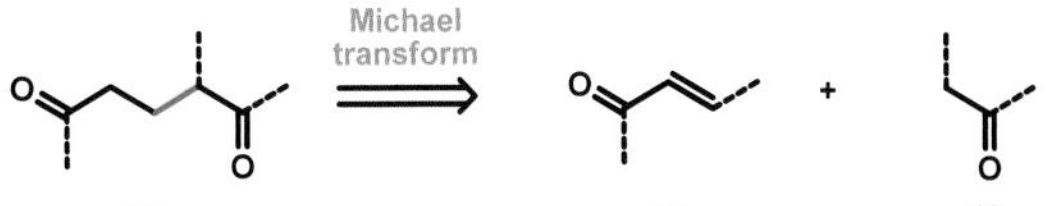

Scheme 4

This aspect of retrosynthetic analysis was presented earlier[2c] along with a discussion of strategies that are useful in the retrosynthetic design of a synthesis.

"Transforms vary in terms of their power to simplify a target structure. The most powerfully simplifying transforms, which reduce molecular complexity in the retrosynthetic direction, occupy a special position in the hierarchy of all transforms. Their application, even when the appropriate retron is absent, may justify the use of a number of non-simplifying transforms to generate that retron. In general, simplifying transforms function to modify structural elements which contribute to molecular complexity: molecular size, cyclic connectivity (topology), stereocenter content, element and functional group content, chemical reactivity, structural instability and density of complicating elements.

Molecular complexity is important to strategy selection. For each type of molecular complexity there is a collection of general strategies for dealing with that complexity. For instance, in the case of a complex polycyclic structure, strategies for the simplification of the molecular network, i.e. topological strategies, must play an important part in transform selection. However, the most efficient mode of retrosynthetic analysis lies not in the separate application of individual strategies, but in the concurrent application of as many different independent strategies as possible.

The major types of strategies which are of value in retrosynthetic analysis may be summarized briefly as follows (1-6):

(1) *Transform-based strategies* – long range search or look ahead to apply a powerfully simplifying transform (or a tactical combination of simplifying transforms) to a target (TGT) with certain appropriate keying features. The retron required for application of a powerful transform may not be present in a complex TGT and a number of antithetic steps (subgoals) may be needed to establish it.

(2) *Structure-goal strategies* – directed at the structure of a potential intermediate or potential starting material. Such a goal greatly narrows a retrosynthetic search and allows the application of bidirectional search techniques.

(3) *Topological strategies* – the identification of one or more individual bond disconnections or correlated bond-pair disconnections as strategic. Topological strategies may also lead to the recognition of a key substructure for disassembly or to the use of rearrangement transforms.

PLANNING ENANTIOSELECTIVE SYNTHESES

(4) *Stereochemical strategies* – general strategies which clear, i.e. remove, stereocenters and stereorelationships under stereocontrol. Such stereocontrol can arise from transform- mechanism control or substrate-structure control. In the case of the former the retron for a particular transform contains critical stereochemical information (absolute or relative) on one or more stereocenters. Stereochemical strategies may also dictate the retention of certain stereocenter(s) during retrosynthetic processing or the joining of atoms in three-dimensional proximity. A major function of stereochemical strategies is the achievement of an experimentally valid clearance of stereocenters, including clearance of molecular chirality.

(5) *Functional group-based strategies* – the retrosynthetic reduction of molecular complexity involving functional groups (FG's) takes various forms. Single FG's or pairs of FG's (and the interconnecting atom path) can key directly the disconnection of a TGT skeleton to form simpler molecules or signal the application of transforms which replace functional groups by hydrogen. Functional group interchange (FGI) is a commonly used tactic for generating the retrons of simplifying transforms from a TGT. FG's may key transforms which stereoselectively remove stereocenters, break topologically strategic bonds or join proximate atoms to form rings.

(6) *"Other" types of strategies* – the recognition of substructural units within a TGT which represent major obstacles to synthesis often provides major strategic input. Certain other strategies result from the requirements of a particular problem, for example a requirement that several related target structures be synthesized from a common intermediate. A TGT which resists retrosynthetic simplification may require the invention of new chemical methodology. The recognition of obstacles to synthesis provides a stimulus for the discovery of such novel processes. The application of a chain of hypotheses to guide the search for an effective line of retrosynthetic analysis is important.

Other strategies deal with optimization of a synthetic design *after* a set of pathways has been generated antithetically, specifically for the ordering of synthetic steps, the use of protection or activation steps, or the determination of alternate paths.

Systematic and rigorous retrosynthetic analysis is the *broad principle* of synthetic problem solving under which the individual strategies take their place. Another overarching idea is the use *concurrently* of as many independent strategies as possible to guide the search for retrosynthetic pathways.

The greater the number of strategies which are used in parallel to develop a line of analysis, the easier the analysis and the simpler the emerging synthetic plan is likely to be."

Stereochemical Strategies: Diastereoselectivity

A much more detailed discussion of the application of strategies to the design of chemical syntheses of complex target molecules has already been described by one of us in the book "The Logic of Chemical Synthesis"[1]. The focus of the discussion in this section is specifically on stereochemical strategies for use in synthetic design. Those strategies, in a broad sense, are required to deal with the very large problem of controlling stereochemistry during the execution of the synthesis. In synthetic practice, reactions which are not stereocontrolled are of limited value, whereas those which are stereocontrolled are crucial in modern synthesis. They also form the basis for the stereochemical strategies that guide the planning of a synthesis for stereochemically complex molecules. Reactions may be stereocontrolled for a number of reasons, including the following:

(a) Stereochemistry may be determined by reaction mechanism, for example *cis* (suprafacial) dihydroxylation of C=C by OsO_4, *trans* bromination of C=C by Br_2 or inversion of stereochemistry during S_N2 displacement at sp^3-carbon.

(b) Stereochemistry may be a consequence of steric effects of a nearby stereocenter which favor one pathway over another (diastereoselectivity), e.g.[4]

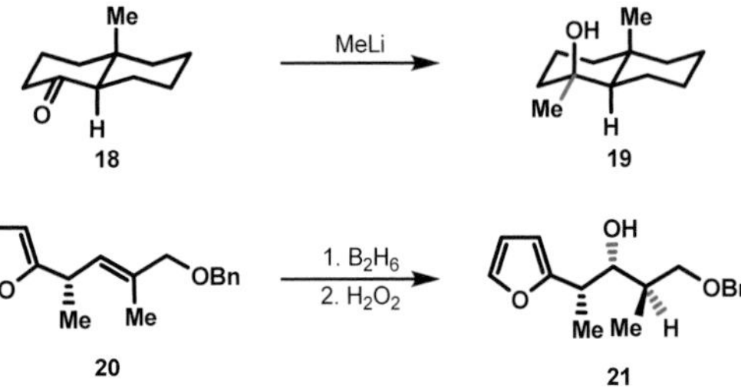

Scheme 5

(c) Diastereocontrol[5] may be influenced by a nearby functional group, for instance:[5]

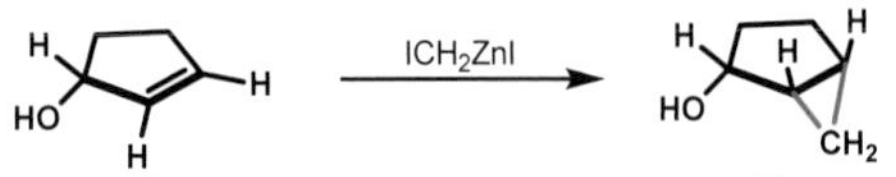

Scheme 6

(d) A remote chiral controller may lead to stereocontrol which is not otherwise observed as in the conjugate addition and alkylation sequence (**24**→**25**) shown in Scheme 6:[6] A second instance of this type of stereocontrol which also involves functional-group participation is the tartrate acetal displacement (**26**→**27**[7], Scheme 7):

PLANNING ENANTIOSELECTIVE SYNTHESES

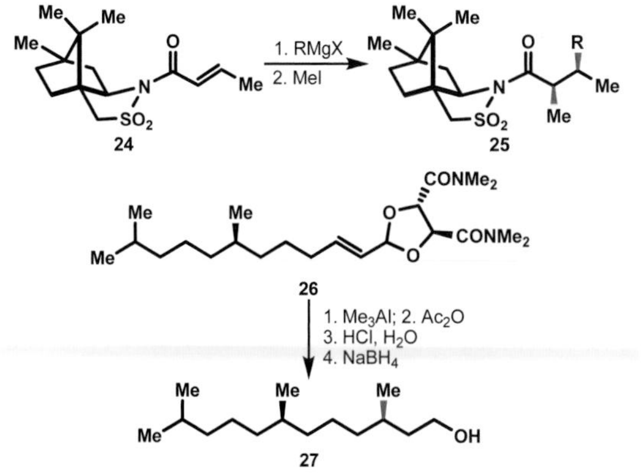

Scheme 7

1. RMgX 2. MeI (24 → 25)

26
1. Me$_3$Al; 2. Ac$_2$O
3. HCl, H$_2$O
4. NaBH$_4$
27

(e) A functional group can accept an external reagent to form an intermediate which allows the introduction of a new stereocenter with diastereocontrol, as illustrated by the three examples which are shown in Scheme 8.

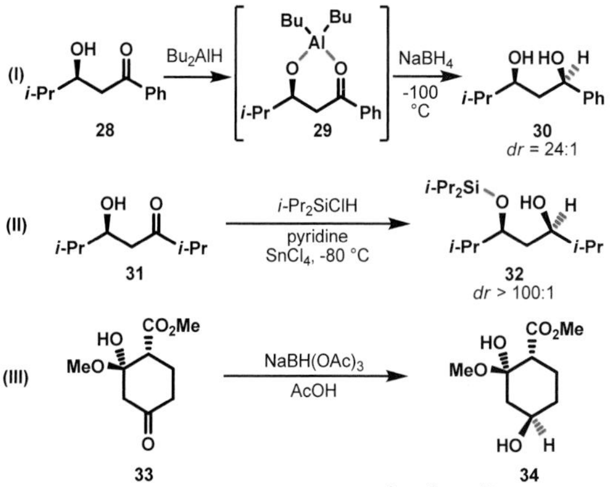

(I) i-Pr ... Bu$_2$AlH ... NaBH$_4$ -100 °C ... 30 dr = 24:1 (28 → 29 → 30)

(II) i-Pr ... i-Pr$_2$SiClH pyridine SnCl$_4$, -80 °C ... 32 dr > 100:1 (31 → 32)

(III) NaBH(OAc)$_3$ AcOH (33 → 34)

Reference key for equations: (I)[8]; (II)[9]; (III)[10]

Scheme 8

(f) Formation of an intermediate in which chelation occurs between a metal ion and two coordination sites in a substrate can result in diastereocontrol (chelation control) as shown by the transformations in Scheme 9.

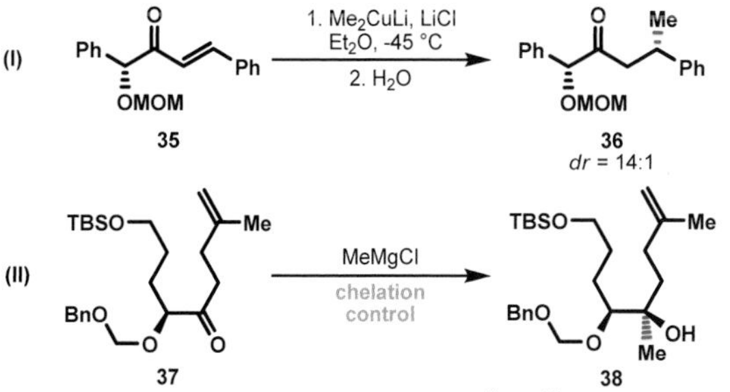

(I) Ph ... OMOM ... 1. Me$_2$CuLi, LiCl Et$_2$O, -45 °C 2. H$_2$O ... 36 dr = 14:1 (35 → 36)

(II) TBSO ... BnO ... MeMgCl chelation control ... 38 (37 → 38)

Reference key for equations: (I)[11]; (II)[12]

Scheme 9

(g) Stereoelectronic effects can give rise to string stereocontrol as shown by the examples in Scheme 10.

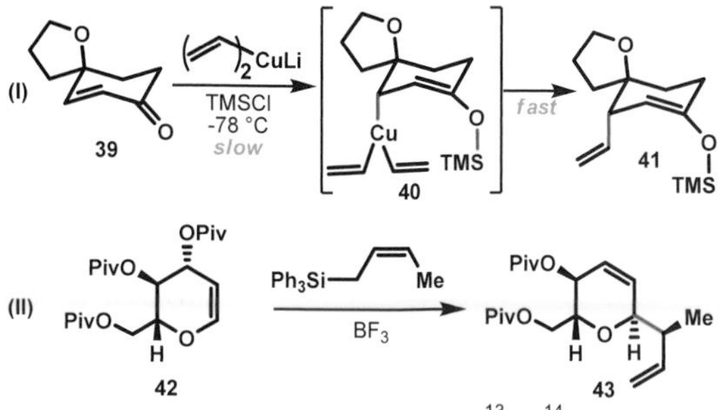

(I) 39 ... ($\quad$)$_2$CuLi TMSCl -78 °C slow ... 40 ... fast ... 41 TMS

(II) 42 ... Ph$_3$Si—Me BF$_3$ ... 43

Reference key for equations: (I)[13]; (II)[14]

Scheme 10

Control of Absolute Configuration

A different kind of stereochemical control can be expected by the use of an enantioselective chiral reagent or catalyst which provides absolute stereocontrol, the main subject of this book. It is a powerful tool in total synthesis because absolute stereocontrol can be used even to override the stereodirection of stereocenters which are present in a substrate or to allow stereocontrol in a transformation which would otherwise not be stereocontrolled.

The utility of stereocontrolled reactions in the construction of molecules with multiple stereocenters is mirrored in retrosynthetic analysis, because the corresponding stereocontrolled *transforms* are effective, and crucial, to the reduction of stereocomplexity in a target structure. In a retrosynthetic sense the removal of or "clearance" of stereocenters is the main goal of stereochemical strategies.

The most powerfully simplifying stereocontrolled transforms are those which concurrently reduce molecular size, topological or other elements of complexity or number of functional groups – especially centers of high chemical reactivity. In general, transforms that control stereochemistry in an absolute sense are more useful strategically than those that do not. A stereocontrolled transform becomes even more powerful when the reaction pathway is mechanistically well-defined because that allows a more reliable estimate of whether such a transform will be a valid synthetic step over a range of structures.

Clearing Stereocenters in Retrosynthetic Analysis

A retrosynthetic evaluation of the stereocenters which are *directly clearable* in a target, and which are not, is helpful to the selection of an effective retrosynthetic pathway. It is often possible to clear stereocenters within conformationally fixed rings by disconnecting an appendage attached to that center in what corresponds to substrate structure stereocontrol in the synthetic direction. Such opportunistic retrosynthetic simplification gains strategic value when it paves the way for further simplifying disconnections.

PLANNING ENANTIOSELECTIVE SYNTHESES

The recent advances in the field of enantioselective reactions and catalysts are reflected in a dramatic increase in the possibilities for clearing stereocenters in retrosynthetic analysis, especially with regard to acyclic stereocenters. In addition, the growing number of synthetically accessible chiral building blocks, both of natural origin and from enantioselective synthesis, has an ever growing impact on the planning of synthesis. If a stereocenter is not readily clearable retrosynthetically, there is a chance that it can be removed from the target by disconnections that lead to synthetically available chiral molecules.

Clearing Acyclic Stereocenters

Target molecules with numerous acyclic stereocenters arrayed along a chain present a different challenge in retrosynthetic analysis as compared to polycyclic structures with stereocenters embedded in the cyclic network. The skeletal disconnections which can be called strategic depend not only on whether they may be carried out with clearance of one or two stereocenters but also on whether they produce two or three fragments that are readily simplified further or are of comparable complexity. The degree to which such disconnections are "strategic" thus depends significantly on the nature of the fragments they produce. The analysis of the fragments of a disconnection, as well as the structural simplification inherent in that disconnection, demands a search through two or more levels of transform application. The concurrent evaluation of (1) the direct effect of transform application on the reduction of molecular size and stereochemical complexity and (2) the accessibility of the fragments which that application produces, is an effective procedure for synthetic planning.

Fragment-Guided Disconnections and Stereochemical Simplification

The implementation of this dual disconnection/ fragment analysis approach for targets with multiple acyclic stereocenters requires a detailed evaluation of the fragments which would result from various chain disconnections as well as the availability of suitable transforms for those disconnections. Such disconnections are strategically advantageous if (a) each of the fragments is less complex structurally than the target; (b) readily available in chiral form, or (c) readily disconnectable to structurally simple achiral precursors.

Bond disconnections that produce key fragments are strategically valuable even if their implementation requires prior application of one or more transforms to generate the required retron. For instance if a C-C single bond cannot be directly disconnected, but the corresponding C=C double bond can be disconnected by the Wittig transform, the extra step can actually simplify the synthetic pathway. Such an example is shown in Scheme 11. Structure **44** is an intermediate that can be generated retrosynthetically from the large and complex marine toxin halichondrin. It can be further disconnected to two synthetically accessible precursors by:

(1) introduction of a C=C double bond (a functional group addition transform that corresponds to C=C hydrogenation in the synthetic direction) and;

(2) application of the Wittig transform to effect the C=C double bond disconnection.[15]

(a) *The Wittig or related transforms (e.g., alkene and alkyne metathesis) that disconnect π-bonds*

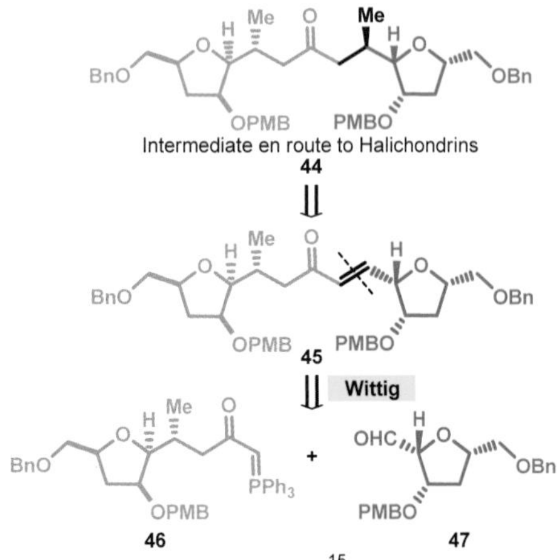

Intermediate en route to Halichondrins
44

45

Wittig

46 + **47**

Reference:[15]

Scheme 11

Fortunately there are now available many transforms that disconnect acyclic C-C single bonds to generate desirable simple precursors, not only for Csp^3-Csp^3 bonds, but also Csp^3-Csp^2, Csp^2-Csp^2 or Csp^2-Csp bonds, including those that involve catalysis by transition metals such as Pd and Cu. A typical example of a Csp^2-Csp^2 disconnection appears in Scheme 12.

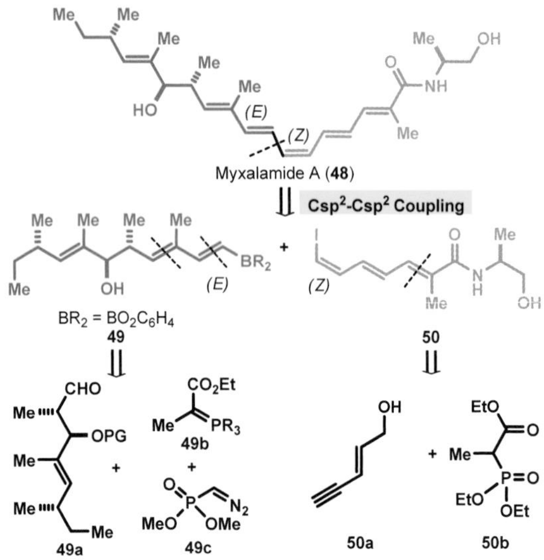

Myxalamide A (**48**)

Csp^2-Csp^2 Coupling

$BR_2 = BO_2C_6H_4$
49 + **50**

49a + **49b** + **49c**

50a + **50b**

Reference:[16]

Scheme 12

PLANNING ENANTIOSELECTIVE SYNTHESES

The analysis of fragments from retrosynthetic disconnection also provides strategic guidance for structures with both acyclic and cyclic stereocenters. Four such examples are shown in Schemes 13 and 14.

(b) *Various transforms that disconnect Csp^2-Csp, Csp^2-Csp^2, Csp^2-Csp^3 or Csp^3-Csp^3 linkages*

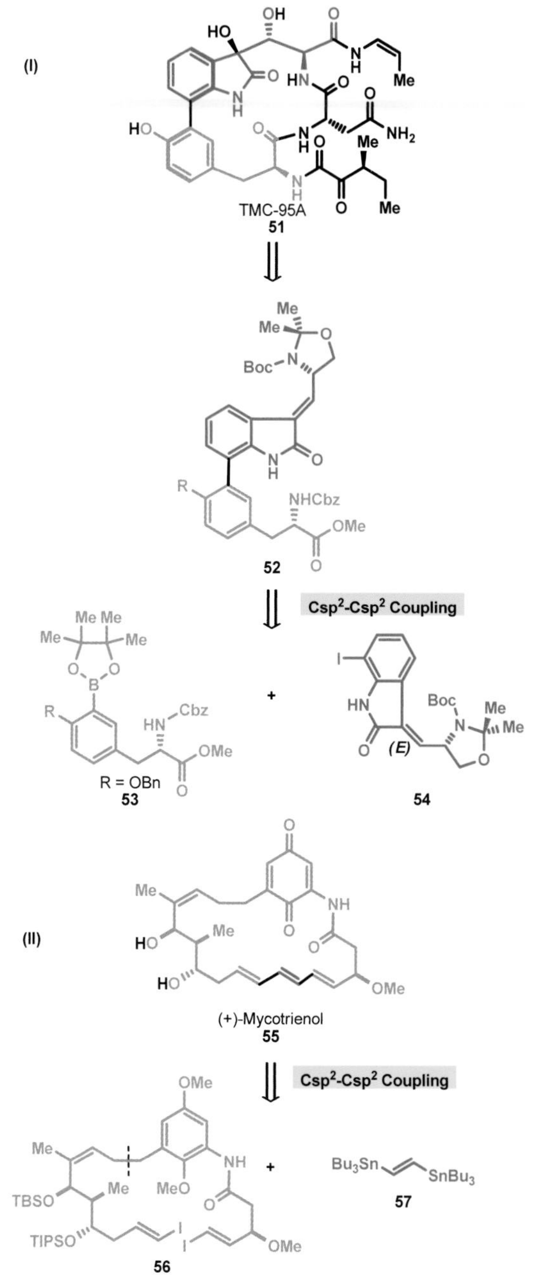

(I)

TMC-95A
51

Csp²-Csp² Coupling

52

R = OBn
53

+

54

(II)

(+)-Mycotrienol
55

Csp²-Csp² Coupling

56

+

Bu₃Sn——SnBu₃
57

Reference key for equations: (I)[17]; (II)[18]

Scheme 13

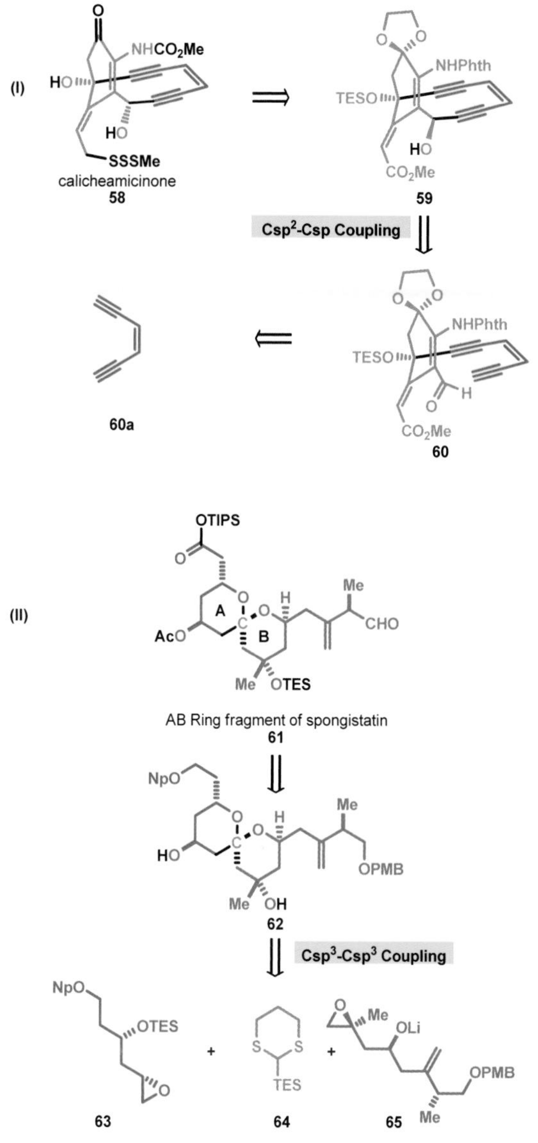

(I)

calicheamicinone
58

59

Csp²-Csp Coupling

60a

60

(II)

OTIPS

AB Ring fragment of spongistatin
61

62

Csp³-Csp³ Coupling

63

+

64

+

65

Reference key for equations: (I)[19]; (II)[20]

Scheme 14

Summary: Strategic Disconnection at Acyclic Bonds

The disconnection of stereochemically complex target structures at acyclic bonds can be guided by the concurrent application of multiple strategies. The first is the selection of the most simplifying stereocontrolled transforms applicable to the target with the goal of simultaneously clearing stereocenters and effecting bond disconnection. The second is to identify those disconnections that generate synthetically accessible key fragments. The third is to identify transforms that simplify stereo- and/or functional group complexity.

PLANNING ENANTIOSELECTIVE SYNTHESES

Transform-Based Strategies

As described in considerable detail in the "*Logic of Chemical Synthesis*", some transforms are so powerfully simplifying that their application to molecular simplification may be highly fruitful even though a number of retrosynthetic steps and intermediates (subgoals) may be required for generation of the full retron that keys the simplifying disconnection to the simplified precursor (goal). The prime example of such a process is the Diels-Alder transform. However, there are many others, especially for topological simplification.

It is also useful to think of transforms that correspond to enantiocontrolled reactions as having a special place in the hierarchy of transforms that reduce stereochemical complexity. The most powerfully simplifying of these are disconnective as well as being able to reduce the number of stereocenters by two or more. The schemes which follow illustrate transform-based retrosynthetic analysis for a number of powerfully simplifying transforms.

Application of the Enantioselective Diels-Alder Transform

The Diels-Alder reaction has served as a key step in the synthesis of a large number of complex molecules, and frequently used in the earliest stage to introduce not only a cyclic subunit, but also functionality and stereocenter(s). The availability of enantioselective methodology has markedly enhanced the synthetic power of the Diels-Alder process and further elevated it in the hierarchy of simplifying transforms.

Two examples of long-range retrosynthetic search to apply the enantioselective Diels-Alder transform to specific targets are shown in Schemes 15 and 16.

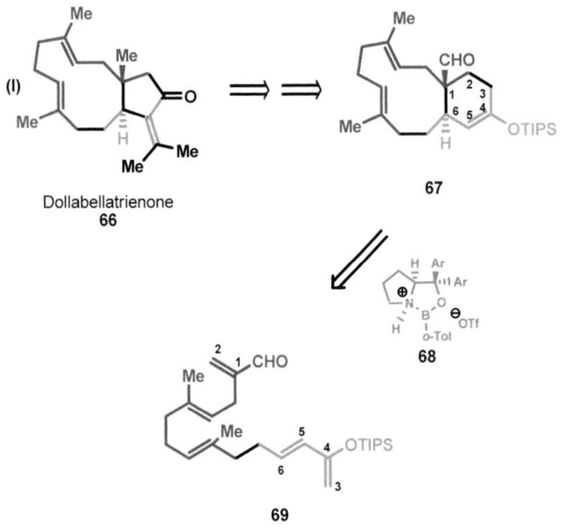

Dollabellatrienone
66

67

68

69

Reference:[21]

Scheme 15

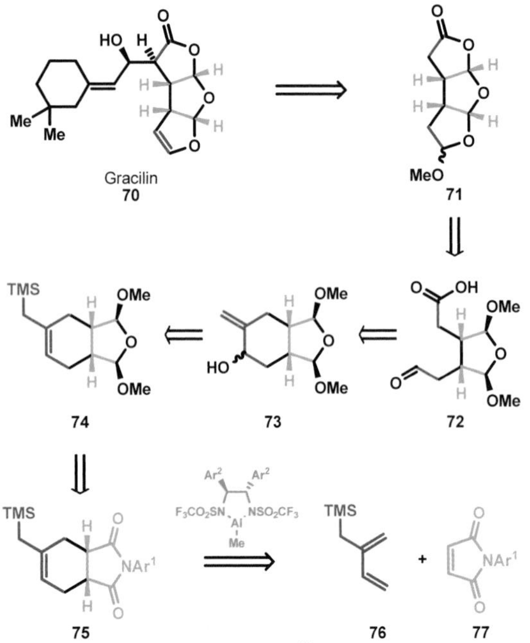

Gracilin
70

71

74   **73**   **72**

75   **76**   **77**

Reference:[22]

Scheme 16

Application of the Aldol Transform

The enantio- and diastereocontrolled aldol transform is powerfully simplifying in terms of clearing acyclic stereocenters and molecular disconnection to simpler fragments for further dissection. The extensive understanding of the mechanistic pathways involved in highly stereocontrolled aldol processes and the ability to anticipate the stereochemistry of an aldol product add further to its importance in retrosynthetic analysis and synthetic practice. Three examples of the application of the aldol transform to molecular simplification follow in Schemes 17 and 18.

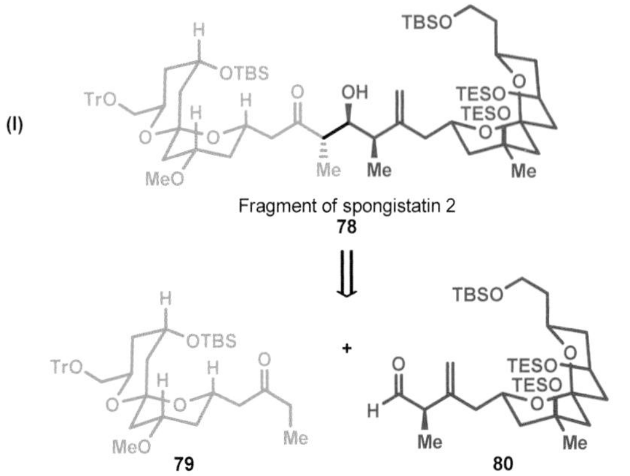

Fragment of spongistatin 2
78

79   **80**

Reference key for equation: (I)[23]

Scheme 17

PLANNING ENANTIOSELECTIVE SYNTHESES

Application of the Aldol Transform (Continued)

(II)

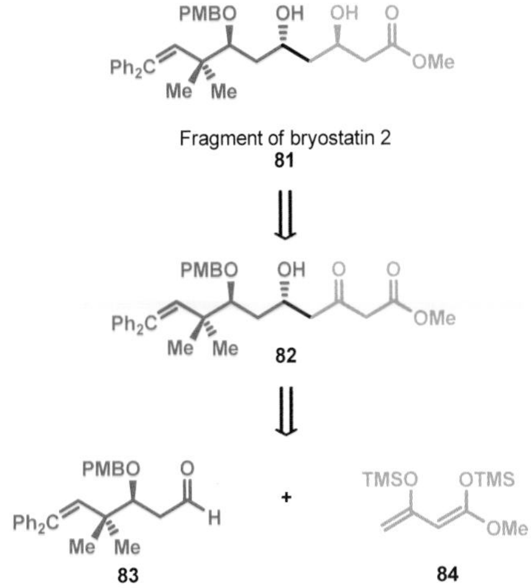

PMBO OH OH O

Ph₂C

Me Me

OMe

Fragment of bryostatin 2
81

⟹

PMBO OH O O

Ph₂C

Me Me 82

OMe

⟹

PMBO O

Ph₂C

Me Me
83

+

TMSO OTMS

OMe

84

(III)

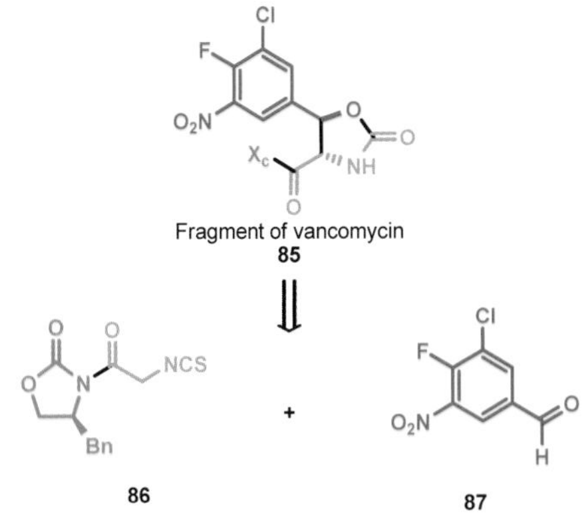

Cl

F

O₂N

Xc

O

O

NH

Fragment of vancomycin
85

⟹

O

O

N

Bn

NCS

86

+

F

Cl

O₂N

H

O

87

Reference key for equations: (II)[24]; (III)[25]

Scheme 18

Application of the Cation-π CyclizationTransform

The formation of polycyclic fused-ring structures by a cascade of cation-olefin addition steps can be of great value for suitable target molecules, because it can produce multiple rings and stereocenters in a single step. Two examples of the application of the corresponding transform are shown in Schemes 19 and 20.

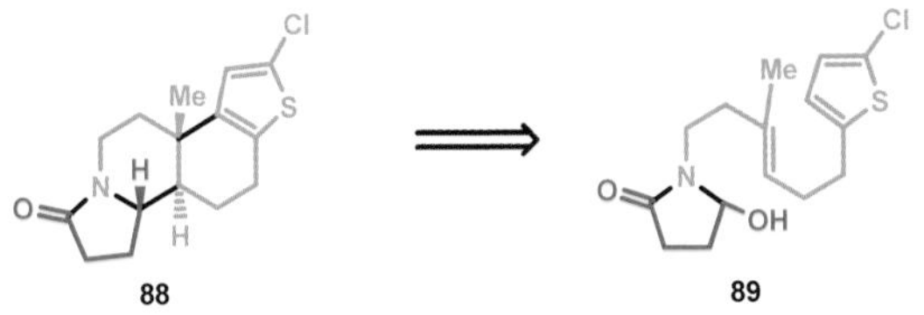

Cl

Me

S

O

N

H

H

88

⟹

Me

Cl

S

O

N

OH

89

Reference:[26]

Scheme 19

(I)

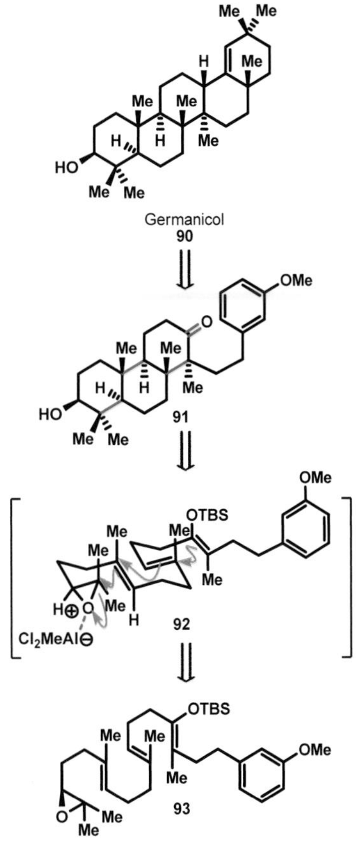

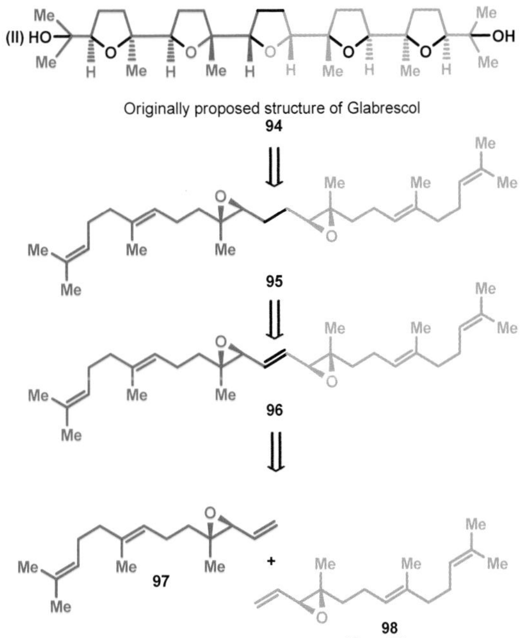

Me Me

H Me

Me Me

H H Me

HO

Me Me

Germanicol
90

⟹

OMe

Me Me O

H H Me

HO

Me Me

91

⟹

OTBS

Me Me

Me

H⊕O Me Me

H

Cl₂MeAl⊖

92

OMe

⟹

OTBS

Me Me

Me

O Me

Me

93

OMe

(II) HO

Me

Me

O

H

Me

O

H

Me

O

H

H

O

Me

H

O

Me

H

Me

Me

OH

Originally proposed structure of Glabrescol
94

⟹

Me

O

Me Me

Me Me Me

O

Me

95

⟹

Me

O

Me Me Me

O

Me Me

Me

96

⟹

Me

Me Me Me

O

97

+

Me Me Me

O

98

Reference key for equation: (I)[27]; (II)[28]

Scheme 20

PLANNING ENANTIOSELECTIVE SYNTHESES

Application of the Heck Cyclization Transform

The enantioselective, intramolecular version of the Heck reaction for Pd-catalyzed C-C bond formation is very useful for forming five- and six-membered rings. The corresponding transform can be used to simplify simultaneously stereo- and topological complexity in a target, as illustrated by the examples in Scheme 21.

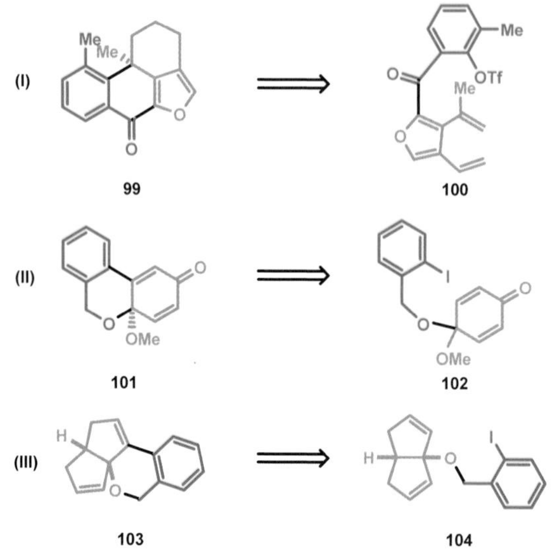

(I)

99   100

(II)

101   102

(III)

103   104

Reference key for equation: (I)[29]; (II)[30]; (III)[31]

Scheme 21

Application of the Pictet-Spengler Transform

The Pictet-Spengler reaction is a cation-π-cyclization process combined with C-N bond-formation (Schiff-base formation). In its enantioselective form it is often ideal for the synthesis of *N*-heterocyclic structures. The corresponding transform is disconnective to simpler fragments and reduces both stereo- and topological complexity, as shown in Scheme 22.

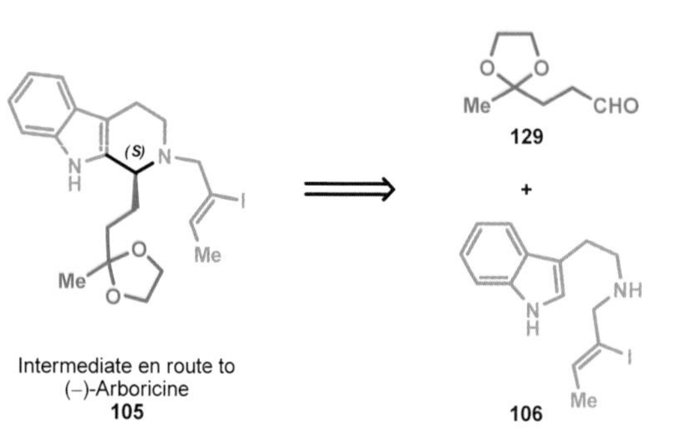

Intermediate en route to
(−)-Arboricine
105

129

106

Reference:[32]

Scheme 22

Application of Sigmatropic Rearrangement Transforms

Sigmatropic rearrangements can be used to generate vicinal stereocenters by C-C coupling reactions, and so the corresponding transforms can simplify both stereochemistry and molecular connectivity. They are especially powerful when their application sets the stage for further simplification by other transforms of stereochemistry and topology.

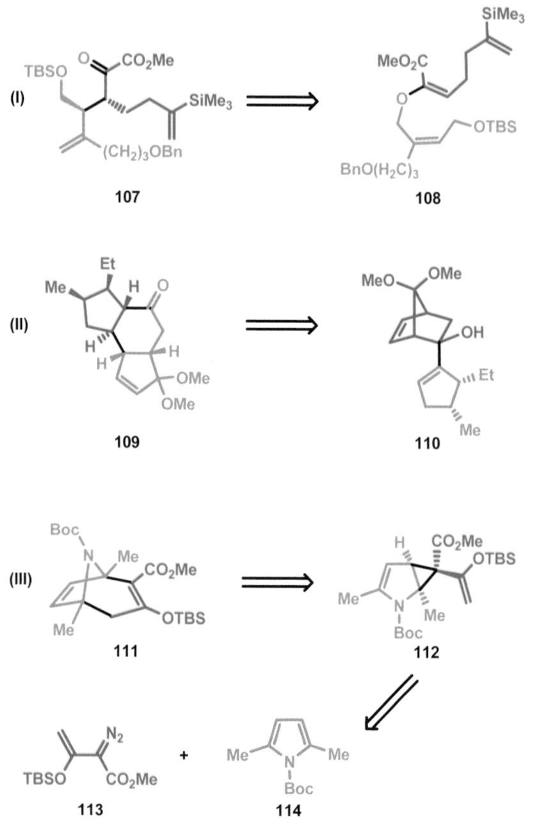

(I)

107   108

(II)

109   110

(III)

111   112

113   +   114

Reference key for equation: (I)[33]; (II)[34]; (III)[35]

Scheme 23

Application of C=C Oxidation Transforms

The enantioselective epoxidation and 1,2-dihydroxylation of C=C provide chiral products which are valuable because they can be elaborated in many ways to a wide range of structures. This versatility, together with their predictable enantioselectivity, place these C=C oxidation transforms at a level that justifies their selection for long-range retrosynthetic search.

Shown on the next page in Scheme 24 are three examples of the application of the epoxidation transform retrosynthetically. Scheme 25 outlines an application of the asymmetric 1,2-dihydroxylation transform.

PLANNING ENANTIOSELECTIVE SYNTHESES

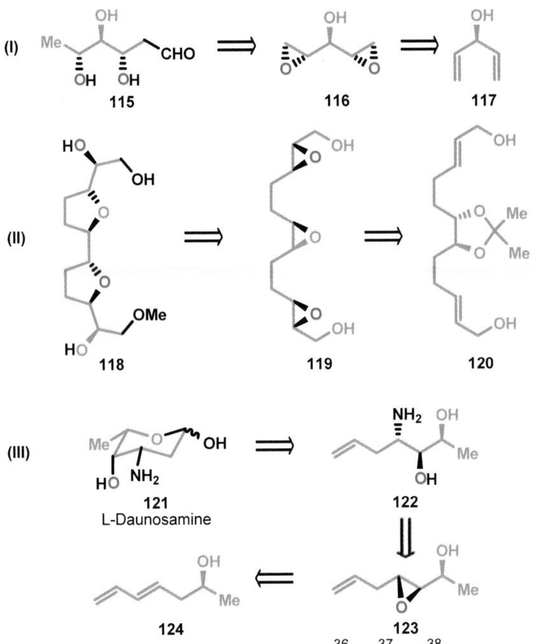

(I)

(II)

(III)

Reference key for equations: (I)[36]; (II)[37]; (III)[38]

Scheme 24

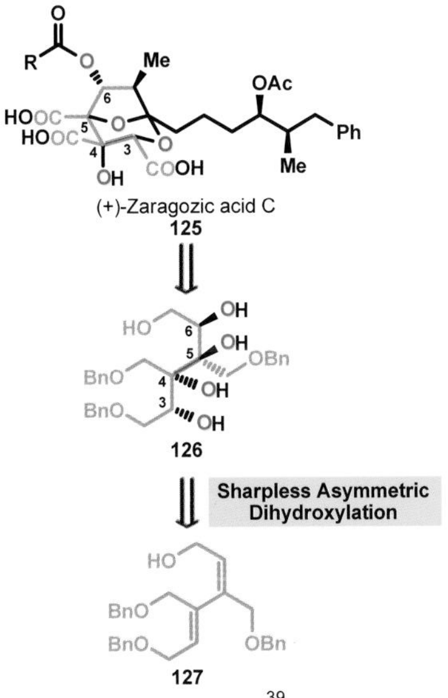

(+)-Zaragozic acid C
125

Sharpless Asymmetric Dihydroxylation

127

Reference:[39]

Scheme 25

Enantioselective Reaction Pathways

It was not long ago that the chemical synthesis of chiral molecules was approached in a very different way, because it has only been within the last 2-3 decades that enantiocontrolled reactions have been available.

Previously, there were basically just two ways of effecting the synthesis of a chiral substance. One of these involved the synthesis of a racemic intermediate, resolution to separate enantiomers (preferably at the earliest possible stage of the synthesis) and then completing the synthesis from the enantiomer with absolute configuration corresponding to the target. The other, more narrowly applicable method, required the availability of one or more chiral, naturally produced building blocks that could be elaborated to the target. The striking difference between these older, often elegant syntheses and the modern multistep processes that are now available with the advent of enantioselective transformations is now obvious. However, there is another and less appreciated aspect of the dramatic advances in enantiocontrolled synthetic chemistry which has to do with a greatly increased understanding of the way in which enantioselective chemical processes take place. This section examines that new knowledge first from a historical perspective and then in terms of future implications.

Our present ability to construct complex chemical structures by multistep sequences would not have been possible without a certain amount of serendipity, and also the revolutionary advances in our understanding of:

(a) structure-reactivity relationships;

(b) reaction mechanisms and pathways;

(c) electronic and steric effects on chemical reactions;

(d) fundamental bases of catalysis;

(e) energetic of molecular conformations and conformational dynamics and

(f) three-dimensional (stereochemical) aspects of molecular structure and transformations.

Pre-Transition State Assemblies

By the mid 1960s the theory of many of the basic stereocontrolled reactions (e.g., nucleophilic displacement at sp^3-carbon, addition to C=C or elimination to form C=C) was sufficiently understood to allow their application in complex situations with confidence. In addition, extensive information was available on the various reactive intermediates that intervene in chemical reactions, and it became possible even to generate and utilize such intermediates for synthetic purposes. The integration of the chemistry of the various elements, including transition metals, into synthetic practice allowed a further explosion in the variety of available reagents and reaction methodology. All of these developments helped to set the stage for the current era of chemical synthesis.

The existence of this book was compelled by the enormous range of the recent developments in enantioselective synthetic chemistry, and especially by the discovery or development of reactions that are very enantioselective (>95% ee). One fundamental reason why such processes exist is that they generally proceed via highly constrained transition states that are tightly organized in three-dimensions.

PLANNING ENANTIOSELECTIVE SYNTHESES

Pre-Transition State Assemblies (Continued)

That high degree of organization makes it much easier to understand highly enantioselective transformations in three-dimensional detail, at least with regard to the three-dimensional assembly of reactants in space just before entry into the transition state. It is very important to have this knowledge of "pre-transition state" assemblies because it improves a synthetic chemist's ability to predict reaction product(s) and evaluate the suitability of a reaction for a particular application, thus elevating the power and precision of modern synthetic chemistry. Highly enantioselective processes, both stoichiometric and catalytic, provide a unique arena to probe subtle steric, electrostatic and solvation effects that are very much more difficult to evaluate in other settings, even with modern computational methods.

The importance to synthetic science of understanding enantiocontrolled reactions very deeply cannot be overestimated. As the knowledge of detailed enantioselective pathways expands, the discovery of new enantioselective reactions is facilitated and the whole area becomes more rational and valuable. A selection of rationally derived pre-transition state assemblies for several useful enantiocontrolled processes is presented in the following sections.

Catalytic Enantioselective Hydrogenation of C=C

The heterogeneous transition metal-catalyzed addition of H_2 to π-bonds has been enormously useful for synthetic chemistry for more than a century. The reaction pathway is complex and not fully understood. Modern enantioselective processes involving soluble, chiral low-valent metal complexes as catalysts are even more useful because they are both practical and serviceable for the preparation of many chiral molecules that otherwise are not easily made. Paradoxically, the reaction pathways and the basis of enantioselectivity are easier to determine.

An excellent example is the enantioselective synthesis of α-amino acids by the Rh(I)- or Ru(II)-catalyzed hydrogenation of α-acetylaminoacrylic esters (see pages 6-13). A likely pre-transition state assembly for the Rh(I)-(S)-BINAP (129) catalyzed production of (S)-α-amino acid derivative 130 is shown in Scheme 26. The plane of the two phenyl groups that are colored magenta is parallel with the plane of the two naphthalene rings of the binaphthyl ligand. The plane of the other two phenyl rings attached to the phosphorous atoms (shown in green) is perpendicular to the plane of the phenyl rings that are shown in magenta. The two remaining coordination sites around the octahedral Rh-complex are tied up by the carbonyl group and the C=C π-bond of the substrate (128).[40]

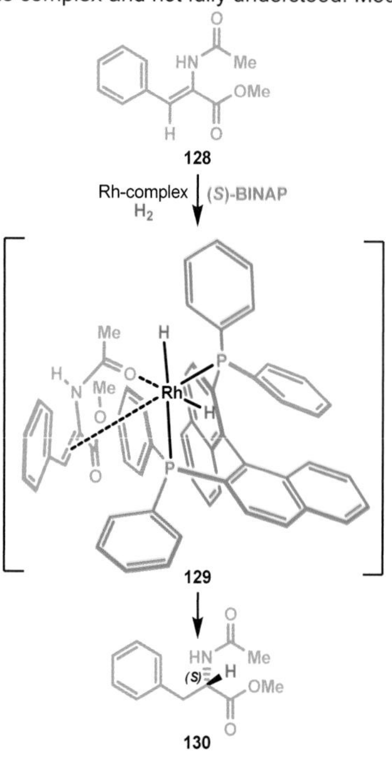

Enantioselective Reduction of C=O by Boranes Under Catalysis by a Chiral Oxazaborolidine

The enantioface-selective reduction of ketonic carbonyl groups by BH_3, catechol borane or other B-H reagents in the presence of chiral diphenylprolinol-derived catalysts of type 131 (Scheme 27) has been widely useful in synthesis. The basis for enantioselectivity and the reaction pathway are clear, and prediction of the absolute configuration of the products is reliable (see Scheme 27 and also pages 24-26). The process proceeds via the intermediates and the pre-transition state assembly 134 shown in Scheme 27.[41]

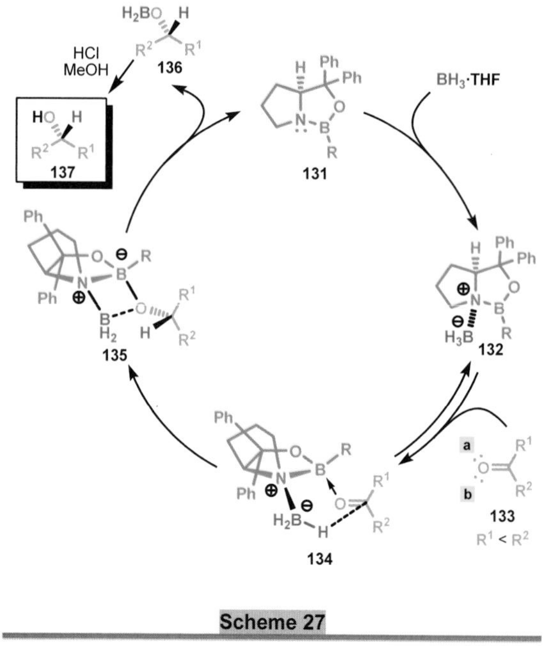

Scheme 26

Scheme 27

PLANNING ENANTIOSELECTIVE SYNTHESES

The Enantioselective Reduction of C=O by Chiral Ru-π-Aryl Complexes and *i*-PrOH

The highly selective and useful reduction of prochiral ketones using *i*-PrOH as the stoichiomeric reductant and Ru(II)-catalysts has been proposed to proceed via a six-membered cyclic pre-transition state assembly **139** by Noyori and co-workers.[42] The cyclic model corresponds to a low-energy transition state arrived at by B3LYP-level calculations (using CH_2O as substrate). Based on that result the pre-transition state assembly shown in Scheme 28 was assumed, since it leads to the observed predominating enantiomer, possibly because steric repulsion is minimal and some stabilization may result from an edge-face interaction of the π-aryl group on Ru and the aryl substituent attached to C=O.[42b]

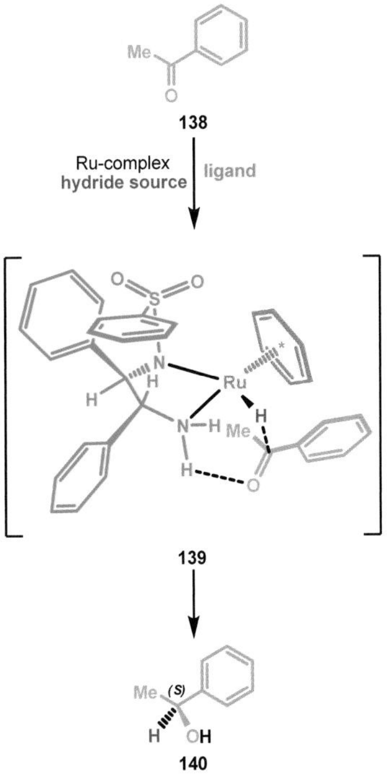

138

Ru-complex hydride source | ligand

139

140

Scheme 28

Catalytic Enantioselective Allylation of Aldehydes

The allylation of carbonyl groups by metallo-allyl reagents is a very facile process that generally occurs via a six-membered cyclic transition state. Consequently, it may be rendered enantioselective either by the use of chiral ligands on the metal reagent or by catalysis with a chiral Lewis acid. An early example of the latter was provided by Keck[43] et al. using (BINOL)$_2$Ti as catalyst, an allylstannane (R$_3$Sn-allyl) and an aldehyde. A likely pre-transition state assembly is shown in Scheme 29.

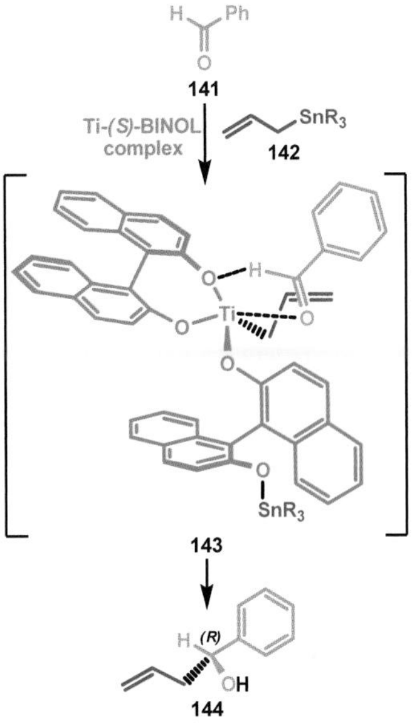

141

Ti-(*S*)-BINOL complex | **142**

143

144

Scheme 29

Enantioselective Addition of Dialkylzinc Reagents to Aldehydes

Dialkylzinc reagents react with ketones or aldehydes much more slowly than the corresponding lithium or magnesium species, and their reactions are subject to acceleration by the attachment of chiral donor ligands to the metal. Chiral amino alcohols have been found to be very effective catalysts for the enantioselective conversion of aldehydes to chiral secondary alcohols as shown in Scheme 30.[44]

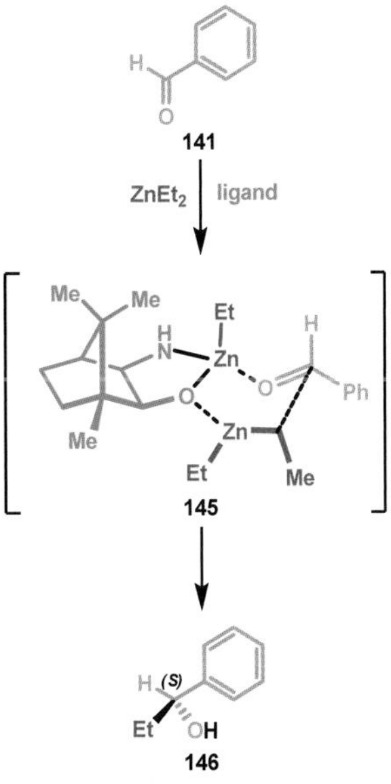

141

ZnEt$_2$ | ligand

145

146

Scheme 30

PLANNING ENANTIOSELECTIVE SYNTHESES

Enantioselective Intermolecular Diels-Alder Reaction Catalyzed by Chiral Oxazaborolidine Complexes

Chiral boron complexes have served as very effective Lewis acid catalysts in the Diels-Alder reaction and have been applied to the enantioselective synthesis of complex natural products. Mechanistic studies have led to an especially clear picture of the most likely pre-transition state assemblies for chiral oxazaborolidines, both neutral (Scheme 31) and cationic (Scheme 32).[45] The product in Scheme 31 is an intermediate for the synthesis of desogestrel, a third generation oral contraceptive (see pages 255-257).[46] The product in Scheme 32 was used for the synthesis of eunicenone (see pages 263-265).[47]

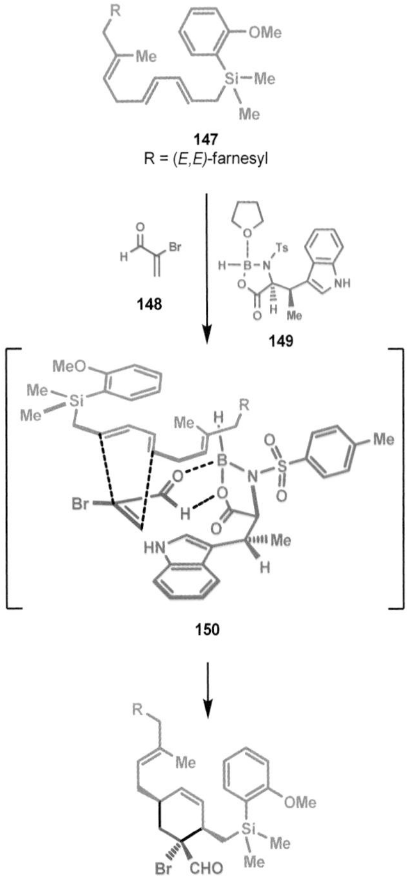

Scheme 31

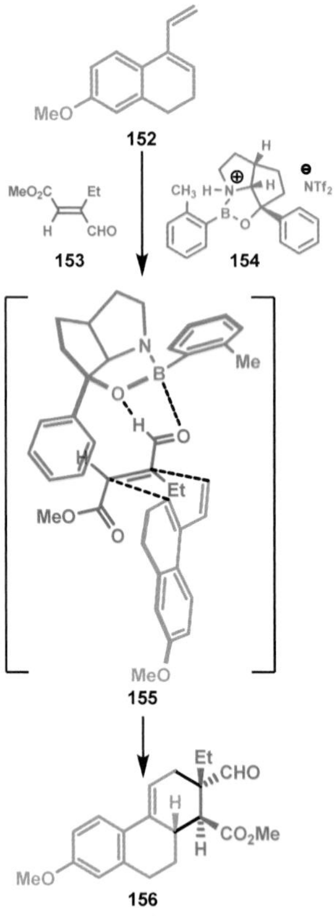

Another highly enantioselective Diels-Alder process for which the pre-transition state assembly is quite transparent involves the C_2-symmetric bisoxazoline **159** coordinated to a Lewis acidic metal and a bidentate dienophile (Scheme 33). The pre-transition state assembly **160** provides a clear explanation for the observed product (**161**) with Mg(II) as metal (from MgI_2) via tetrahedral coordination.[48]

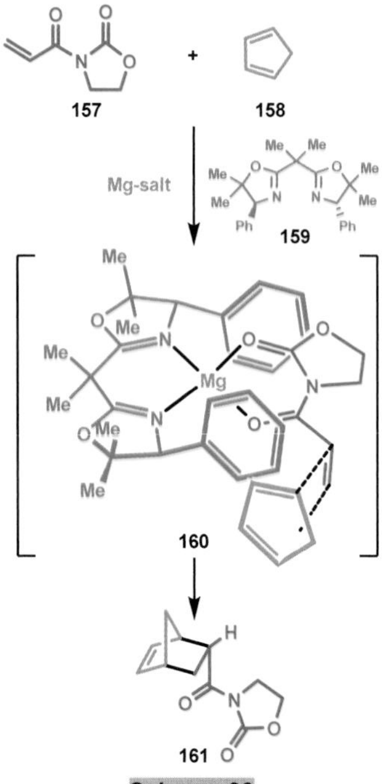

Scheme 32

Scheme 33

PLANNING ENANTIOSELECTIVE SYNTHESES

Enantioselective Jacobsen Epoxidation

The enantioselectivity of the Jacobsen epoxidation is generally highest with conjugated cyclic olefins as substrates. The epoxidation appears to be initiated by electrophilic attack by oxygen of a chiral Mn(V)-oxo salen complex on the more nucleophilic C=C terminus. The enantioselectivity is interpreted most easily in terms of a two-step process and the pre-transition state assembly shown in Scheme 34. Attack by oxygen on indene (**152**) leads to cationic character at the benzylic carbon and electrostatic stabilization by proximity of that center to one of the salen oxygens attached to Mn.[49] The pathway shown in Scheme 34 provides a simple explanation of the stereochemical course of Jacobsen epoxidations.

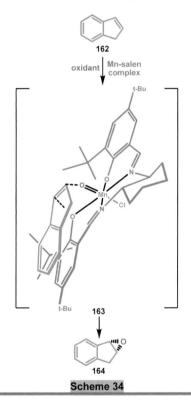

Scheme 34

Enantioselective Rh(I)-Catalyzed 1,4-Addition of Aryl Groups to Cyclic Enones

T. Hayashi found that N. Miyaura's Rh(I)-catalyzed conjugate addition of aryl- or alkenylboronic acids to α,β-enones can be converted to a highly enantioselective process by the use of the chiral BINAP ligand.[50] Hayashi's model of the pre-transition state assembly and the reaction pathway are summarized in Scheme 35. The transfer of an aryl or alkenyl group from Rh to the enone is considered to occur via a four-membered cyclic transition state. Chiral dienes, e.g., bicycle[2.2.2]-octadienes, can be used in place of chiral bisphosphines as ligands on the Rh(I).

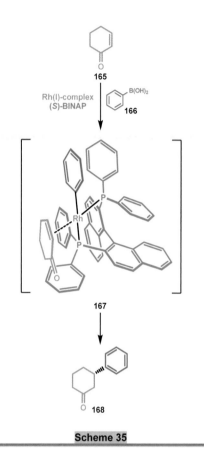

Scheme 35

Enantioselective Enolate Alkylation by Phase-Transfer Catalysis

Ion-pair-mediated reactions under phase-transfer conditions with a chiral quaternary ammonium salt can be used for enantioselective enolate alkylation. Early work by M.J. O'Donnell showed that enantioenriched α-amino acid esters could be prepared by the alkylation of a Schiff-base between benzophenone and a glycine ester, using as catalyst a cinchona alkaloid-derived quaternary ammonium salt.[51] The process was made highly efficient by the attachment of a rigid 9-anthracenylmethyl substituent on the bridgehead nitrogen and the preferred arrangement of the contact ion pair was clarified by X-ray crystallographic studies.[52] The pre-transition state model derived from this research is shown in Scheme 36. The enolate oxygen makes contact with the only tetrahedral face of the cationic nitrogen that is accessible (the remaining three are sterically buried) and the substrate nestles neatly in a binding pocket formed by the quinoline and quinuclidine rings. Only one face of the α-carbon of the enolate is accessible to the alkylating reagent.

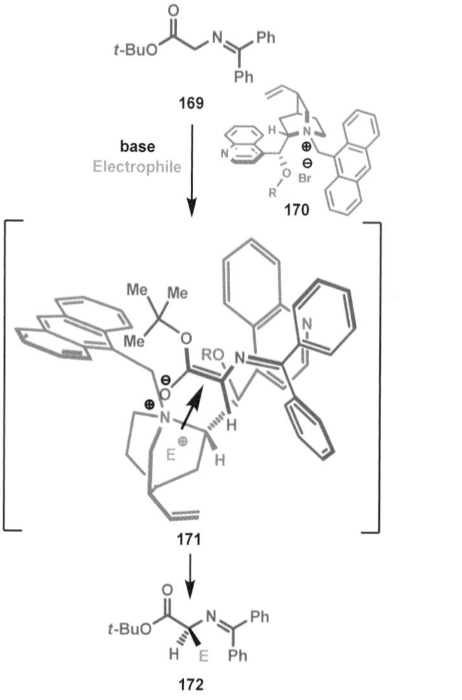

PLANNING ENANTIOSELECTIVE SYNTHESES

Scheme 36

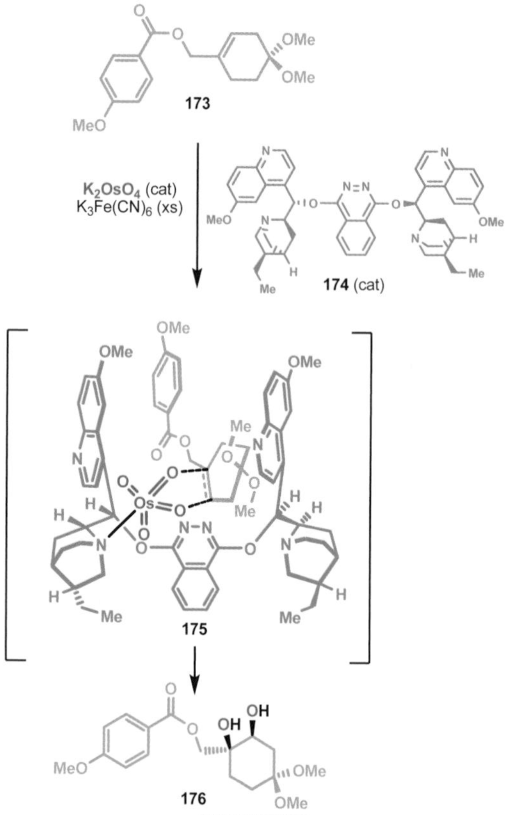

Scheme 37

Catalytic Enantioselective 1,2-Dihydroxylation of Olefins by OsO$_4$ Using Bis-Cinchona Alkaloids as Ligands

K.B. Sharpless and his co-workers developed various bis-cinchona alkaloids as ligands for the OsO$_4$-mediated 1,2-dihydroxylation of C=C.[53] Extensive studies of this process have revealed that the alkaloid-OsO$_4$ complex can adopt a U-shaped conformation which can provide a binding site for the olefinic substrate which positions the C=C subunit in proximity to one axial O and one equatorial O of OsO$_4$. The [3+2]-addition of OsO$_4$ to C=C occurs with the face-selectivity expected from the pre-transition state assembly shown in Scheme 37.[54]

Enantioselective C-C Bond Formation via the Heck Reaction

The enantioselective Heck reaction (see pages 143-144) is an important process for C-C bond formation both in the intermolecular and intramolecular subtypes. The catalyst, generally Pd(0) complexed with a chiral bidentate ligand, reacts with one of the coupling partners (Csp^2-X) to form L$_2$Pd(X)(Csp^2) and that intermediate coordinates to the olefinic partner with displacement of X. The key olefin-carbopalladation step occurs via a pre-transition state assembly such as **179** to form simultaneously C-C and C-Pd bonds. The coupling product is then formed with elimination of H and Pd, as shown in Scheme 38.[55]

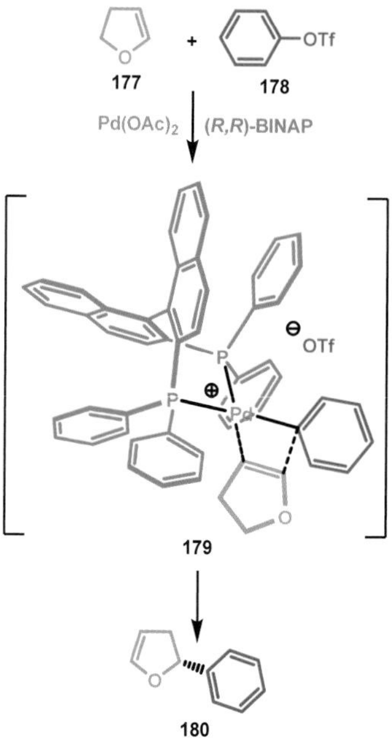

Scheme 38

PLANNING ENANTIOSELECTIVE SYNTHESES

Enantioselective Michael Addition Between Aldehydes and Nitroalkenes Catalyzed by Diphenylprolinol Silyl Ethers

Proline and various of its derivatives are remarkably effective chiral catalysts for reactions of aldehydes and other unhindered carbonyl compounds that can form reactive enamines or iminium ions from a pyrrolidine-type reagent. The trimethylsilyl ether of diphenylprolinol (**183**) has been extensively applied for such processes. The conjugate addition of nitromethane (**182**) to cinnamaldehyde (**181**) occurs with high enantioselectivity to form adduct **185** via an iminium cation-nitromethide ion-pair in a pathway that likely involves the pre-transition state assembly **184** (Scheme 39; *See* also pages 105-106).[56]

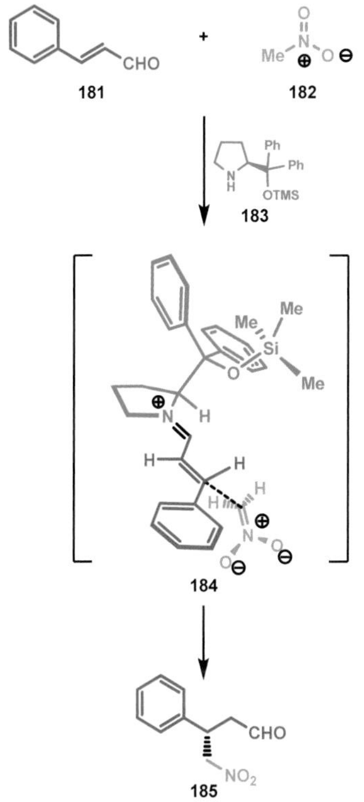

Scheme 39

Enantioselective Intermolecular Mannich Reaction Catalyzed by Diphenylprolinol Silyl Ethers

The reaction of the *N*-acylimine **186** with acetaldehyde (**187**), using **183** as catalyst, produces the Mannich-type adduct **188** with excellent enantioselectivity via the chiral enamine of **187** and **183** (Scheme 40; *See* also pages 89-91). The pre-transition state assembly for the formation of **189** represented by **187** shows the attack of the imine **186** at the lower face of the enamine and C=N of the imine which provides electrostatic stabilization in the transition state. The phenyl (Ph) and C(=O)Ph groups are positioned to minimize steric interactions.[57]

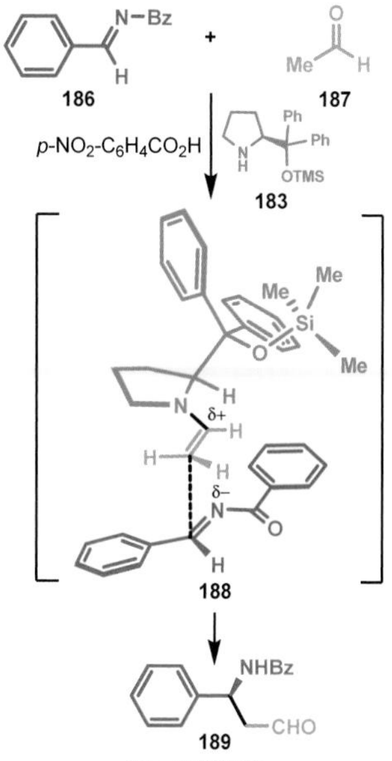

Scheme 40

Enantioselective Intramolecular Aldol Reaction Catalyzed by (*S*)-Proline

The enantioselective intramolecular aldol cyclization that is involved in the (*S*)-proline-catalyzed conversion of **190** to **193** (Scheme 41; *See* also pages 64-70) is rationalized very simply by the intermediacy of an enamine which cyclizes via the pre-transition state assembly **192**. The proximity of the proline COOH function to one of the diastereotopic ketonic carbonyl groups allows H-bond stabilization in the transition state, as shown in Scheme 41.[58]

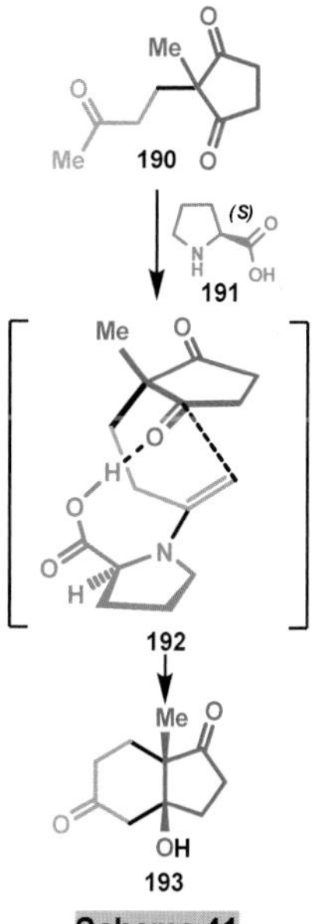

Scheme 41

PLANNING ENANTIOSELECTIVE SYNTHESES

Enantioselective Ireland-Claisen Rearrangement

Highly enantioselective Ireland-Claisen rearrangement of an achiral allyl ester can be effected via a chiral (cisoid) boron enolate as outlined in Scheme 42 (see also pages 110-111).[59] The absolute stereochemical course of the rearrangement can be explained by the pre-transition state assembly **196**, which results from: (1) the cisoid geometry of the boron enolate; (2) a chair-like six-membered ring structure; (3) maximum orbital overlap in the transition state and (4) minimal steric repulsion between the rearranging subunit **194** and the $PhSO_2$ subunit attached to nitrogen.

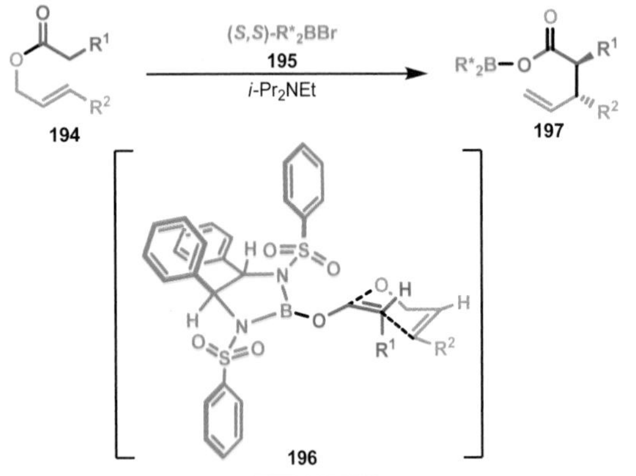

Scheme 42

Enantioselective Imine Reduction

The trichlorosilane/HCl-mediated enantioselective reduction of imines provides a practical route to many chiral amines.[60] The absolute stereochemical course of this useful process is most readily understood in terms of the pre-transition state assembly **199** (Scheme 43).

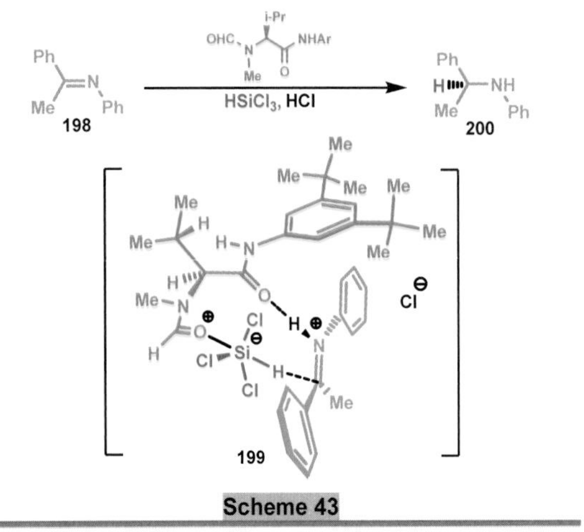

Scheme 43

The collection of pre-transition states for enantioselective chemical reactions which appears in the foregoing sections should not be construed as an indication that most highly enantioselective processes are understood in such detail. In fact, there are many important and highly useful transformations, several of which are illustrated in Part I of this book, that have not been rationalized. The examples shown above provide a preview of future progress and the tasks that lie ahead. That progress will likely accelerate in the decades to come as new discoveries and insights are gained, as our knowledge of reaction pathways deepens and as the ability to invent new reactions, reagents and catalysts grows. The future is both challenging and promising.

References

1. Corey, E. J. & Cheng, X. M. *The Logic of Chemical Synthesis* (John Wiley & Sons, Hoboken, NJ, **1989**).
2. (a) Corey, E. J. Computer-assisted analysis of complex synthetic problems. *Quart. Rev., Chem. Soc.* **1971**, 25, 455-482; (b) Corey, E. J. Robert Robinson lecture: retrosynthetic thinking-essentials and examples. *Chem. Soc. Rev.* **1988**, 17, 111-133; (c) Corey, E. J. The logic of chemical synthesis: multistep synthesis of complex carbogenic molecules. (Nobel lecture). *Angew. Chem., Int. Ed.* **1991**, 30, 455-465.
3. (a) Abrecht, S., Harrington, P., Iding, H., Karpf, M., Trussardi, R., Wirz, B. & Zutter, U. The synthetic development of the anti-influenza neuraminidase inhibitor oseltamivir phosphate (Tamiflu): A challenge for synthesis & process research. *Chimia* **2004**, 58, 621-629; (b) Abrecht, S., Federspiel, M. C., Estermann, H., Fischer, R., Karpf, M., Mair, H.-J., Oberhauser, T., Rimmler, G., Trussardi, R. & Zutter, U. The synthetic-technical development of oseltamivir phosphate, Tamiflu: a race against time. *Chimia* **2007**, 61, 93-99.
4. Schmid, G., Fukuyama, T., Akasaka, K. & Kishi, Y. Synthetic studies on polyether antibiotics. 4. Total synthesis of monensin. 1. Stereocontrolled synthesis of the left half of monensin. *J. Am. Chem. Soc.* **1979**, 101, 259-260.
5. Hoveyda, A. H., Evans, D. A. & Fu, G. C. Substrate-directable chemical reactions. *Chem. Rev.* **1993**, 93, 1307-1370.
6. Oppolzer, W., Poli, G., Kingma, A. J., Starkemann, C. & Bernardinelli, G. Asymmetric induction at C(β) and C(α) of N-enoylsultams by organomagnesium 1,4-addition/enolate trapping. *Helv. Chim. Acta* **1987**, 70, 2201-2214.
7. Fujiwara, J., Fukutani, Y., Hasegawa, M., Maruoka, K. & Yamamoto, H. Unprecedented regio- and stereochemical control in the addition of organoaluminum reagents to chiral α,β-unsaturated acetals. *J. Am. Chem. Soc.* **1984**, 106, 5004-5005.
8. (a) Narasaka, K. & Pai, H. C. Stereoselective synthesis of meso (or erythro) 1,3-diols from β-hydroxyketones. *Chem. Lett.* **1980**, 1415-1418; (b) Oishi, T. & Nakata, T. New aspects of stereoselective synthesis of 1,3-polyols. *Synthesis* **1990**, 635-645.
9. Anwar, S. & Davis, A. P. The application of difunctional organosilicon compounds to organic synthesis; 1,3-asymmetric induction in the reduction of β-hydroxy ketones. *J. Chem. Soc., Chem. Commun.* **1986**, 831-832.
10. Turnbull, M. D., Hatter, G. & Ledgerwood, D. E. Stereochemical control in the synthesis of the cyclohexyl portion of the milbemycin skeleton. *Tetrahedron Lett.* **1984**, 25, 5449-5452.
11. Corey, E. J., Hannon, F. J. & Boaz, N. W. Coordinatively induced 1,4-diastereoselection in the reaction of acyclic α,β-enones with organocopper reagents. A new type of organocopper reagent. *Tetrahedron* **1989**, 45, 545-555.
12. Still, W. C. & McDonald, J. H., III. Chelation-controlled nucleophilic additions. 1. A highly effective system for asymmetric induction in the reaction of organometallics with α-alkoxy ketones. *Tetrahedron Lett.* **1980**, 21, 1031-1034.
13. Corey, E. J. & Boaz, N. W. Evidence for a reversible d,π*-complexation, β-cupration sequence in the conjugate addition reaction of Gilman reagents with α,β-enones. *Tetrahedron Lett.* **1985**, 26, 6015-6018.
14. Danishefsky, S. J., Armistead, D. M., Wincott, F. E., Selnick, H. G. & Hungate, R. The total synthesis of avermectin A1a. *J. Am. Chem. Soc.* **1989**, 111, 2967-2980.
15. DiFranco, E., Ravikumar, V. T. & Salomon, R. G. Total synthesis of halichondrins: enantioselective construction of a homochiral tetracyclic KLMN-ring intermediate from D-mannitol. *Tetrahedron Lett.* **1993**, 34, 3247-3250.
16. Mapp, A. K. & Heathcock, C. H. Total Synthesis of Myxalamide A. *J. Org. Chem.* **1999**, 64, 23-27.

17. Lin, S. & Danishefsky, S. J. The total synthesis of proteasome inhibitors TMC-95A and TMC-95B: discovery of a new method to generate cis-propenyl amides. *Angew. Chem., Int. Ed.* **2002**, 41, 512-515.

18. Masse, C. E., Yang, M., Solomon, J. & Panek, J. S. Total Synthesis of (+)-Mycotrienol and (+)-Mycotrienin I: Application of Asymmetric Crotylsilane Bond Constructions. *J. Am. Chem. Soc.* **1998**, 120, 4123-4134.

19. Smith, A. L., Hwang, C. K., Pitsinos, E., Scarlato, G. R. & Nicolaou, K. C. Enantioselective total synthesis of (-)-calicheamicinone. *J. Am. Chem. Soc.* **1992**, 114, 3134-3136.

20. Smith, A. B., III, Doughty, V. A., Sfouggatakis, C., Bennett, C. S., Koyanagi, J. & Takeuchi, M. Spongistatin Synthetic Studies. An Efficient, Second-Generation Construction of an Advanced ABCD Intermediate. *Org. Lett.* **2002**, 4, 783-786.

21. Snyder, S. A. & Corey, E. J. Concise Total Syntheses of Palominol, Dolabellatrienone, β-Araneosene, and Isoedunol via an Enantioselective Diels-Alder Macrobicyclization. *J. Am. Chem. Soc.* **2006**, 128, 740-742.

22. Corey, E. J. & Letavic, M. A. Enantioselective Total Synthesis of Gracilins B and C Using Catalytic Asymmetric Diels-Alder Methodology. *J. Am. Chem. Soc.* **1995**, 117, 9616-9617.

23. (a) Evans, D. A., Trotter, B. W., Cote, B. & Coleman, P. J. Enantioselective synthesis of altohyrtin C (spongistatin 2): synthesis of the EF-bis(pyran) subunit. *Angew. Chem., Int. Ed. Engl.* **1997**, 36, 2741-2744; (b) Evans, D. A., Trotter, B. W., Cote, B., Coleman, P. J., Dias, L. C. & Tyler, A. N. Enantioselective synthesis of altohyrtin C (spongistatin 2): fragment assembly and revision of the spongistatin 2 stereochemical assignment. *Angew. Chem., Int. Ed. Engl.* **1997**, 36, 2744-2747.

24. Evans, D. A., Carter, P. H., Carreira, E. M., Prunet, J. A., Charette, A. B. & Lautens, M. Asymmetric synthesis of bryostatin 2. *Angew. Chem., Int. Ed.* **1998**, 37, 2354-2359.

25. Evans, D. A., Wood, M. R., Trotter, B. W., Richardson, T. I., Barrow, J. C. & Katz, J. L. Total syntheses of vancomycin and eremomycin aglycons. *Angew. Chem., Int. Ed.* **1998**, 37, 2700-2704.

26. Knowles, R. R., Lin, S. & Jacobsen, E. N. Enantioselective Thiourea-Catalyzed Cationic Polycyclizations. *J. Am. Chem. Soc.* **2010**, 132, 5030-5032.

27. Surendra, K. & Corey, E. J. Rapid and Enantioselective Synthetic Approaches to Germanicol and Other Pentacyclic Triterpenes. *J. Am. Chem. Soc.* **2008**, 130, 8865-8869.

28. Nicolaou, K. C., Bulger, P. G. & Sarlah, D. Metathesis reactions in total synthesis. *Angew. Chem., Int. Ed.* **2005**, 44, 4490-4527.

29. Lau, S. Y. W. & Keay, B. A. Remote substituent effects on the enantiomeric excess of intramolecular asymmetric palladium-catalyzed polyene cyclizations. *Synlett* **1999**, 605-607.

30. Imbos, R., Minnaard, A. J. & Feringa, B. L. A Highly Enantioselective Intramolecular Heck Reaction with a Monodentate Ligand. *J. Am. Chem. Soc.* **2002**, 124, 184-185.

31. Lautens, M. & Zunic, V. Sequential olefin metathesis - intramolecular asymmetric Heck reactions in the synthesis of polycycles. *Can. J. Chem.* **2004**, 82, 399-407.

32. Wanner, M. J., Boots, R. N. A., Eradus, B., de Gelder, R., van Maarseveen, J. H. & Hiemstra, H. Organocatalytic Enantioselective Total Synthesis of (-)-Arboricine. *Org. Lett.* **2009**, 11, 2579-2581.

33. Pollex, A. & Hiersemann, M. Catalytic Asymmetric Claisen Rearrangement in Natural Product Synthesis: Synthetic Studies toward (-)-Xeniolide F. *Org. Lett.* **2005**, 7, 5705-5708.

34. Paquette, L. A., Romine, J. L. & Lin, H. S. Diastereoselective π-facially controlled nucleophilic additions of chiral vinylorganometallics to chiral α,γ-unsaturated ketones. 2. A practical method for stereocontrolled elaboration of the decahydro-as-indacene subunit of ikarugamycin. *Tetrahedron Lett.* **1987**, 28, 31-34.

35. Reddy, R. P. & Davies, H. M. L. Asymmetric Synthesis of Tropanes by Rhodium-Catalyzed [4 + 3] Cycloaddition. *J. Am. Chem. Soc.* **2007**, 129, 10312-10313.

36. Babine, R. E. Asymmetric epoxidation of divinyl carbinol: a new approach to the synthesis of 2,6-dideoxyhexoses. *Tetrahedron Lett.* **1986**, 27, 5791-5794.

37. Hoye, T. R. & Suhadolnik, J. C. Symmetry-assisted synthesis of triepoxide stereoisomers of E,Z,E,-dodeca-2,6,10-trien-1,12-diol and their cascade reactions to 2,5-linked bistetrahydrofurans. *J. Am. Chem. Soc.* **1985**, 107, 5312-5313.

38. Ginesta, X., Pasto, M., Pericas, M. A. & Riera, A. New Stereodivergent Approach to 3-Amino-2,3,6-trideoxysugars. Enantioselective Synthesis of Daunosamine, Ristosamine, Acosamine, and Epi-daunosamine. *Org. Lett.* **2003**, 5, 3001-3004.

39. Armstrong, A., Barsanti, P. A., Jones, L. H. & Ahmed, G. Total Synthesis of (+)-Zaragozic Acid C. *J. Org. Chem.* **2000**, 65, 7020-7032.

40. (a) Tang, W. & Zhang, X. New Chiral Phosphorus Ligands for Enantioselective Hydrogenation. *Chem. Rev. (Washington, DC, U. S.)* **2003**, 103, 3029-3069; (b) Najera, C. & Sansano, J. M. Catalytic Asymmetric Synthesis of α-Amino Acids. *Chem. Rev. (Washington, DC, U. S.)* **2007**, 107, 4584-4671.

41. Corey, E. J. & Helal, C. J. Reduction of carbonyl compounds with chiral oxazaborolidine catalysts: A new paradigm for enantioselective catalysis and a powerful new synthetic method. *Angew. Chem., Int. Ed.* **1998**, 37, 1986-2012.

42. (a) Yamakawa, M., Ito, H. & Noyori, R. The Metal-Ligand Bifunctional Catalysis: A Theoretical Study on the Ruthenium(II)-Catalyzed Hydrogen Transfer between Alcohols and Carbonyl Compounds. *J. Am. Chem. Soc.* **2000**, 122, 1466-1478; (b) Noyori, R., Yamakawa, M. & Hashiguchi, S. Metal-Ligand Bifunctional Catalysis: A Nonclassical Mechanism for Asymmetric Hydrogen Transfer between Alcohols and Carbonyl Compounds. *J. Org. Chem.* **2001**, 66, 7931-7944.

43. Keck, G. E., Tarbet, K. H. & Geraci, L. S. Catalytic asymmetric allylation of aldehydes. *J. Am. Chem. Soc.* **1993**, 115, 8467-8468.

44. (a) Kitamura, M., Suga, S., Kawai, K. & Noyori, R. Catalytic asymmetric induction. Highly enantioselective addition of dialkylzincs to aldehydes. *J. Am. Chem. Soc.* **1986**, 108, 6071-6072; (b) Corey, E. J. & Hannon, F. J. Zinc complexes of chiral phenols as catalysts for enantioselective addition of organozinc reagents to aldehydes. *Tetrahedron Lett.* **1987**, 28, 5237-5240.

45. Corey, E. J. Catalytic enantioselective Diels-Alder reactions: Methods, mechanistic fundamentals, pathways, and applications. *Angew. Chem., Int. Ed.* **2002**, 41, 1650-1667.

46. (a) Corey, E. J. & Huang, A. X. A Short Enantioselective Total Synthesis of the Third-Generation Oral Contraceptive Desogestrel. *J. Am. Chem. Soc.* **1999**, 121, 710-714; (b) Hu, Q.-Y., Rege, P. D. & Corey, E. J. Simple, Catalytic Enantioselective Syntheses of Estrone and Desogestrel. *J. Am. Chem. Soc.* **2004**, 126, 5984-5986.

47. Lee, T. W. & Corey, E. J. Enantioselective Total Synthesis of Eunicenone A. *J. Am. Chem. Soc.* **2001**, 123, 1872-1877.

48. (a) Corey, E. J., Imai, N. & Zhang, H. Y. Designed catalyst for enantioselective Diels-Alder addition from a C2-symmetric chiral bis(oxazoline)-iron(III) complex. *J. Am. Chem. Soc.* **1991**, 113, 728-729; (b) Corey, E. J. & Ishihara, K. Highly enantioselective catalytic Diels-Alder addition promoted by a chiral bis(oxazoline)-magnesium complex. *Tetrahedron Lett.* **1992**, 33, 6807-6810.

49. Kürti, L., Blewett, M. & Corey, E. J. Origin of Enantioselectivity in the Jacobsen Epoxidation of Olefins. *Org. Lett.* **2009**, 11, 4592-4595.

50. Hayashi, T. Rhodium-catalyzed asymmetric 1,4-addition of organoboronic acids and their derivatives to electron deficient olefins. *Synlett* **2001**, 879-887.

51. O'Donnell, M. J., Bennett, W. D. & Wu, S. The stereoselective synthesis of α-amino acids by phase-transfer catalysis. *J. Am. Chem. Soc.* **1989**, 111, 2353-2355.

52. Corey, E. J., Xu, F. & Noe, M. C. A Rational Approach to Catalytic Enantioselective Enolate Alkylation Using a Structurally Rigidified and Defined Chiral Quaternary Ammonium Salt under Phase Transfer Conditions. *J. Am. Chem. Soc.* **1997**, 119, 12414-12415.

53. Sharpless, K. B., Amberg, W., Bennani, Y. L., Crispino, G. A., Hartung, J., Jeong, K. S., Kwong, H. L., Morikawa, K., Wang, Z. M. & et al. The osmium-catalyzed asymmetric dihydroxylation: a new ligand class and a process improvement. *J. Org. Chem.* **1992**, 57, 2768-2771.

54. Corey, E. J. & Noe, M. C. A Critical Analysis of the Mechanistic Basis of Enantioselectivity in the Bis-Cinchona Alkaloid Catalyzed Dihydroxylation of Olefins. *J. Am. Chem. Soc.* **1996**, 118, 11038-11053.

55. Ozawa, F., Kubo, A. & Hayashi, T. Catalytic asymmetric arylation of 2,3-dihydrofuran with aryl triflates. *J. Am. Chem. Soc.* **1991**, 113, 1417-1419.

56. Hayashi, Y., Gotoh, H., Hayashi, T. & Shoji, M. Diphenylprolinol silyl ethers as efficient organocatalysts for the asymmetric Michael reaction of aldehydes and nitroalkenes. *Angew. Chem., Int. Ed.* **2005**, 44, 4212-4215.

57. Hayashi, Y., Okano, T., Itoh, T., Urushima, T., Ishikawa, H. & Uchimaru, T. Direct organocatalytic Mannich reaction of acetaldehyde: an improved catalyst and mechanistic insight from a computational study. *Angew. Chem., Int. Ed.* **2008**, 47, 9053-9058.

58. (a) Hajos, Z. G. & Parrish, D. R. Synthesis and conversion of 2-methyl-2-(3-oxobutyl)-1,3-cyclopentanedione to the isomeric racemic ketols of the [3.2.1]bicyclooctane and of the perhydroindane series. *J. Org. Chem.* **1974**, 39, 1612-1615; (b) Hajos, Z. G. & Parrish, D. R. Asymmetric synthesis of bicyclic intermediates of natural product chemistry. *J. Org. Chem.* **1974**, 39, 1615-1621; (c) Hoang, L., Bahmanyar, S., Houk, K. N. & List, B. Kinetic and Stereochemical Evidence for the Involvement of Only One Proline Molecule in the Transition States of Proline-Catalyzed Intra- and Intermolecular Aldol Reactions. *J. Am. Chem. Soc.* **2003**, 125, 16-17.

59. (a) Corey, E. J. & Lee, D. H. Highly enantioselective and diastereoselective Ireland-Claisen rearrangement of achiral allylic esters. *J. Am. Chem. Soc.* **1991**, 113, 4026-4028; (b) Corey, E. J., Roberts, B. E. & Dixon, B. R. Enantioselective Total Synthesis of I²-Elemene and Fuscol Based on Enantiocontrolled Ireland-Claisen Rearrangement. *J. Am. Chem. Soc.* **1995**, 117, 193-196.

60. Malkov, A. V., Vrankova, K., Stoncius, S. & Kocovsky, P. Asymmetric Reduction of Imines with Trichlorosilane, Catalyzed by Sigamide, an Amino Acid-Derived Formamide: Scope and Limitations. *J. Org. Chem.* **2009**, 74, 5839-5849.

171

PART III.

ENANTIOSELECTIVE
MULTI-STEP SYNTHESIS: EXAMPLES

EXAMPLES OF ENANTIOSELECTIVE SYNTHESIS

Introduction

The third and last part of this book provides an overview of the practice of multistep enantioselective synthesis in the form of specific examples. Each of these molecular syntheses is described in a standardized format (see illustration below). A brief introduction to the target molecule is followed by a concise description of the thinking behind the retrosynthetic plan and comments on any key or non-standard aspects of the planning process. The reaction sequence used for the synthesis of the target molecule is then outlined in graphical form with key data on reaction conditions, yields and other details. Again, concise comments are made on unusual or non-obvious

transformations. A section on further developments, if any, closes the topic. All the examples in Part III are taken from research in the Corey laboratory at Harvard over the past two decades and the reasoning behind the plan is, therefore, not just surmise. Of course, there are many more synthetic sequences that have been developed in other laboratories all over the world that are at least as interesting and noteworthy. It would be both instructive and useful if in the future these are compiled in formula schemes of the type used in this collection.

Standard Arrangement in Part III

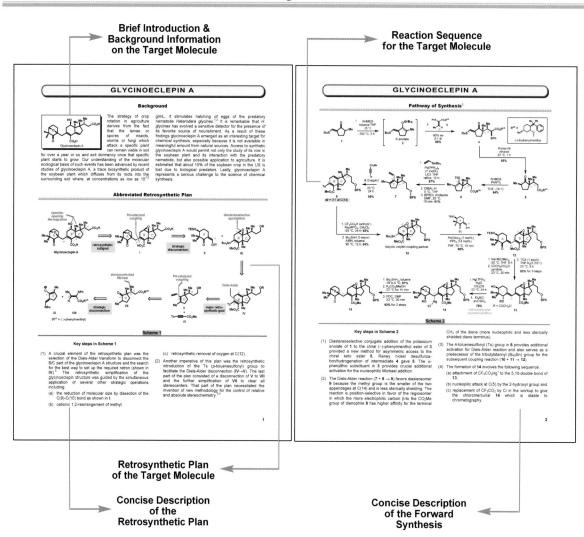

Brief Introduction & Background Information on the Target Molecule

Reaction Sequence for the Target Molecule

Retrosynthetic Plan of the Target Molecule

Concise Description of the Retrosynthetic Plan

Concise Description of the Forward Synthesis

EXAMPLES OF ENANTIOSELECTIVE SYNTHESIS
CONTENTS

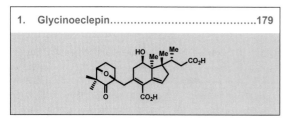

EXAMPLES OF ENANTIOSELECTIVE SYNTHESIS
CONTENTS

EXAMPLES OF ENANTIOSELECTIVE SYNTHESIS
CONTENTS

EXAMPLES OF ENANTIOSELECTIVE SYNTHESIS
CONTENTS

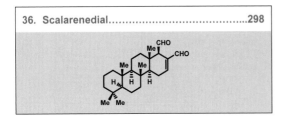

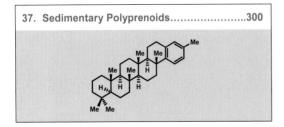

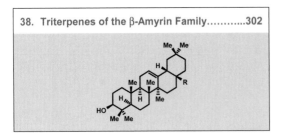

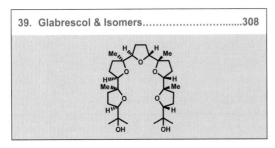

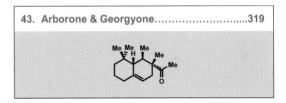

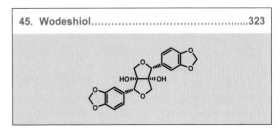

GLYCINOECLEPIN A

Background

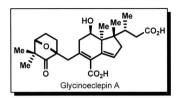

Glycinoeclepin A

The strategy of crop rotation in agriculture derives from the fact that the larvae or spores of insects, worms or fungi which attack a specific plant can remain viable in soil for over a year or so and exit dormancy once that specific plant starts to grow. Our understanding of the molecular ecological basis of such events has been advanced by recent studies of glycinoeclepin A, a trace biosynthetic product of the soybean plant which diffuses from its roots into the surrounding soil where, at concentrations as low as 10^{-12} g/mL, it stimulates hatching of eggs of the predatory nematode *Heterodera glycines*.[1] It is remarkable that *H. glycines* has evolved a sensitive detector for the presence of its favorite source of nourishment. As a result of these findings glycinoeclepin A emerged as an interesting target for chemical synthesis, especially because it is not available in meaningful amounts from natural sources. Access to synthetic glycinoeclepin A would permit not only the study of its role in the soybean plant and its interaction with the predatory nematode, but also possible application to agriculture. It is estimated that about 10% of the soybean crop in the US is lost due to biological predators. Lastly, glycinoeclepin A represents a serious challenge to the science of chemical synthesis.

Abbreviated Retrosynthetic Plan

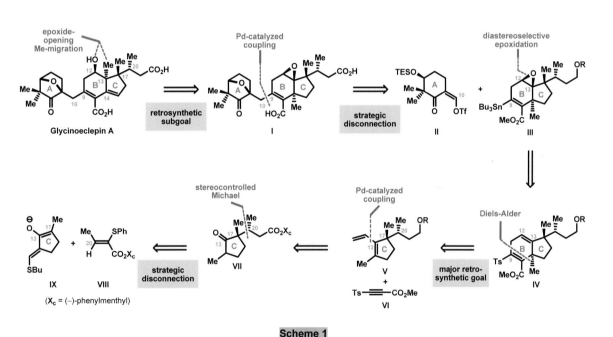

Scheme 1

Key steps in Scheme 1

(1) A crucial element of the retrosynthetic plan was the selection of the Diels-Alder transform to disconnect the B/C part of the glycinoeclepin A structure and the search for the best way to set up the required retron (shown in **IV**).[2] The retrosynthetic simplification of the glycinoeclepin structure was guided by the simultaneous application of several other strategic operations including:

  (a) the reduction of molecular size by dissection of the C(9)-C(10) bond as shown in **I**;

  (b) cationic 1,2-rearrangement of methyl;

  (c) retrosynthetic removal of oxygen at C(12).

(2) Another imperative of this plan was the retrosynthetic introduction of the Ts (*p*-toluenesulfonyl) group to facilitate the Diels-Alder disconnection (**IV→V**). The last part of the plan consisted of a disconnection of **V** to **VII** and the further simplification of **VII** to clear all stereocenters. That part of the plan necessitated the invention of new methodology for the control of relative and absolute stereochemistry.[2-3]

179

GLYCINOECLEPIN A

Pathway of Synthesis[2]

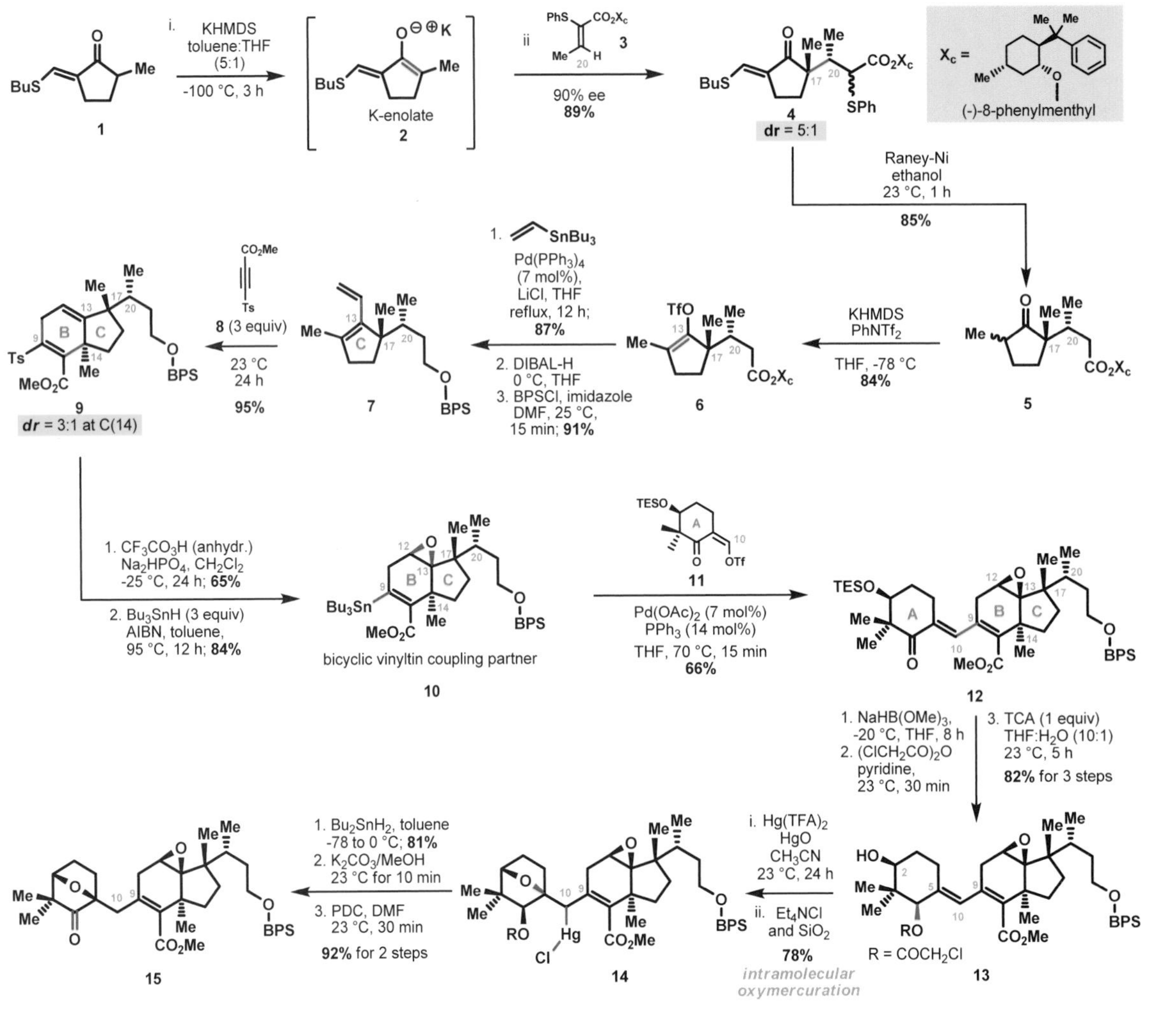

Scheme 2

Key steps in Scheme 2

(1) Diastereoselective conjugate addition of the potassium enolate of **1** to the chiral (−)-phenylmenthol ester of **3** provided a new method for asymmetric access to the chiral keto ester **5**. Raney nickel desulfurization/hydrogenation of intermediate **4** gave **5**. The α-phenylthio substituent in **3** provides crucial additional activation for the nucleophilic Michael addition.

(2) The Diels-Alder reaction (**7** + **8** → **9**) favors diastereomer **9** because the methyl group is the smaller of the two appendages at C(14) and is less sterically shielding. The reaction is position-selective in favor of the regioisomer in which the more electrophilic carbon β-to the CO_2Me group of dienophile **8** has higher affinity for the terminal CH_2 of the diene (more nucleophilic and less sterically shielded diene terminus).

(3) The 4-toluenesulfonyl (Ts) group in **8** provides additional activation for Diels-Alder reaction and also serves as a predecessor of the tributylstannyl (Bu₃Sn) group for the subsequent coupling reaction (**10** + **11** → **12**).

(4) The formation of **14** involves the following sequence:

(a) attachment of $CF_3CO_2Hg^+$ to the 5,10-double bond of **13**;

(b) nucleopilic attack at C(5) by the 2-hydroxyl group and

(c) replacement of CF_3CO_2 by Cl in the workup to give the chloromercurial **14**, which is stable to chromatography.

GLYCINOECLEPIN A

Pathway of Synthesis (continued)[2]

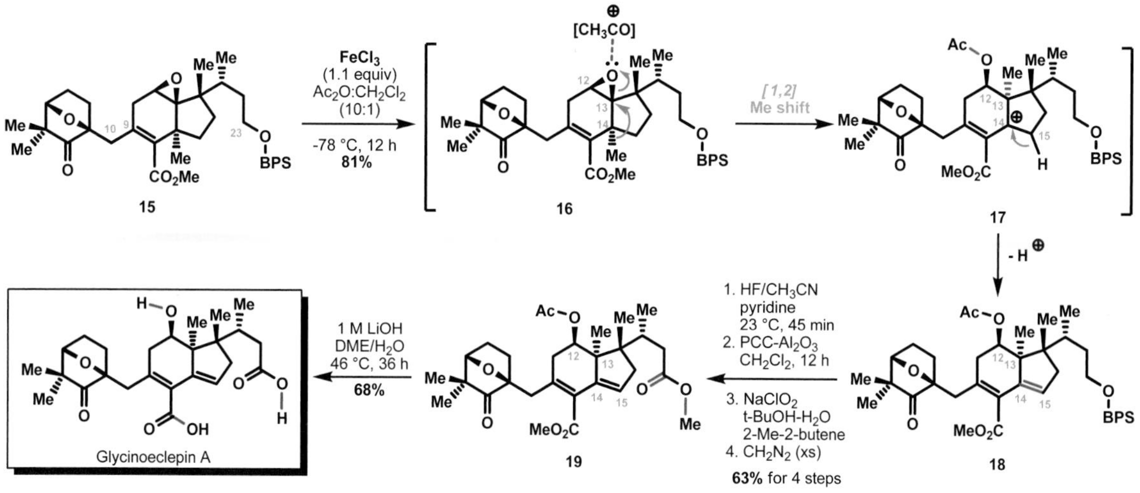

Scheme 3

Key steps in Scheme 3

(1) The rearrangement of the epoxide **16** to the acetate **19** was effected using the reagent Ac_2O-$FeCl_3$ which generates the equivalent of $[CH_3CO]^{+}$.[4] Attachment of CH_3CO^{+} to the C(12)-C(13) epoxide **16** triggers the generation of a C(13)-cation and induces migration of the methyl group from C(14) to C(13). Ordinary Lewis acids convert epoxide **16** mainly to the corresponding C(12) ketone.

(2) The rearrangement of **15** was carried out with a protected alcohol function at C(23) rather than a carboxylic ester to preclude interference by the latter in the rearrangement step **15 → 18**.

Further Developments

In subsequent work a similar, but simpler synthetic route was used to construct a benzenoid analogue of glycinoeclepin A (**20**). This analog was found to stimulate hatching of the eggs of the nematode *Heterodera glycines* at subnanomolar concentrations.[5] The pathway of synthesis of **20** is summarized by the following abbreviated sequence (Scheme 4).[5]

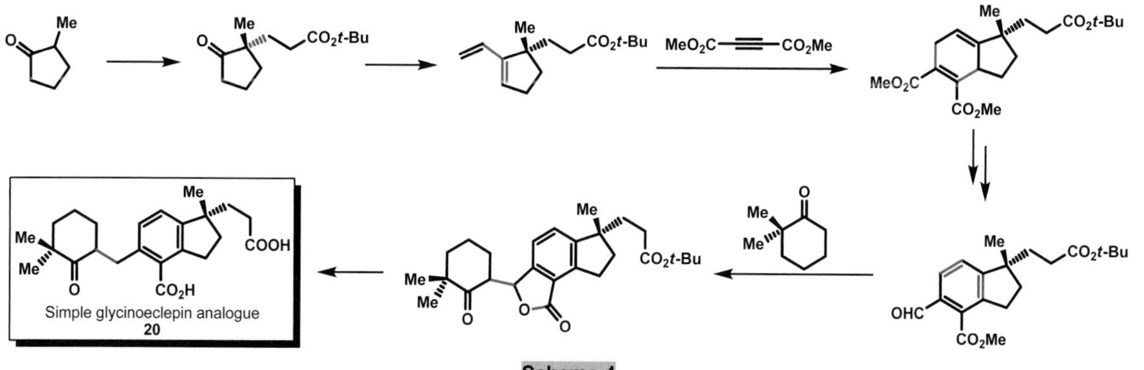

Scheme 4

GLYCINOECLEPIN A

Further Developments (continued)

The likely biosynthesis of glycinoeclepin A from a triterpenoid precursor suggested the possibility of a chemical synthesis from an abundantly available tetracyclic triterpenoid as starting material. Abietospiran (Scheme 4) was chosen as such a precursor because it is abundant in the bark of the fir tree *Abies alba* and easily obtained by simple extraction. Abietospiran was successfully transformed into 12-deoxyglycinoeclepin A by the condensed sequence shown in Scheme 5.[6] Among the most interesting steps in this synthetic route are the following:

(1) cationic, epoxide-initiated and acid-catalyzed ring-expansion of **22** to **23**, during which the oxide bridge across ring A is also established;

(2) cleavage of the B-ring, **24**→**25**;

(3) replacement of the formyl group in **25** by the 6-oxo function of 12-deoxyglycinoeclepin A.

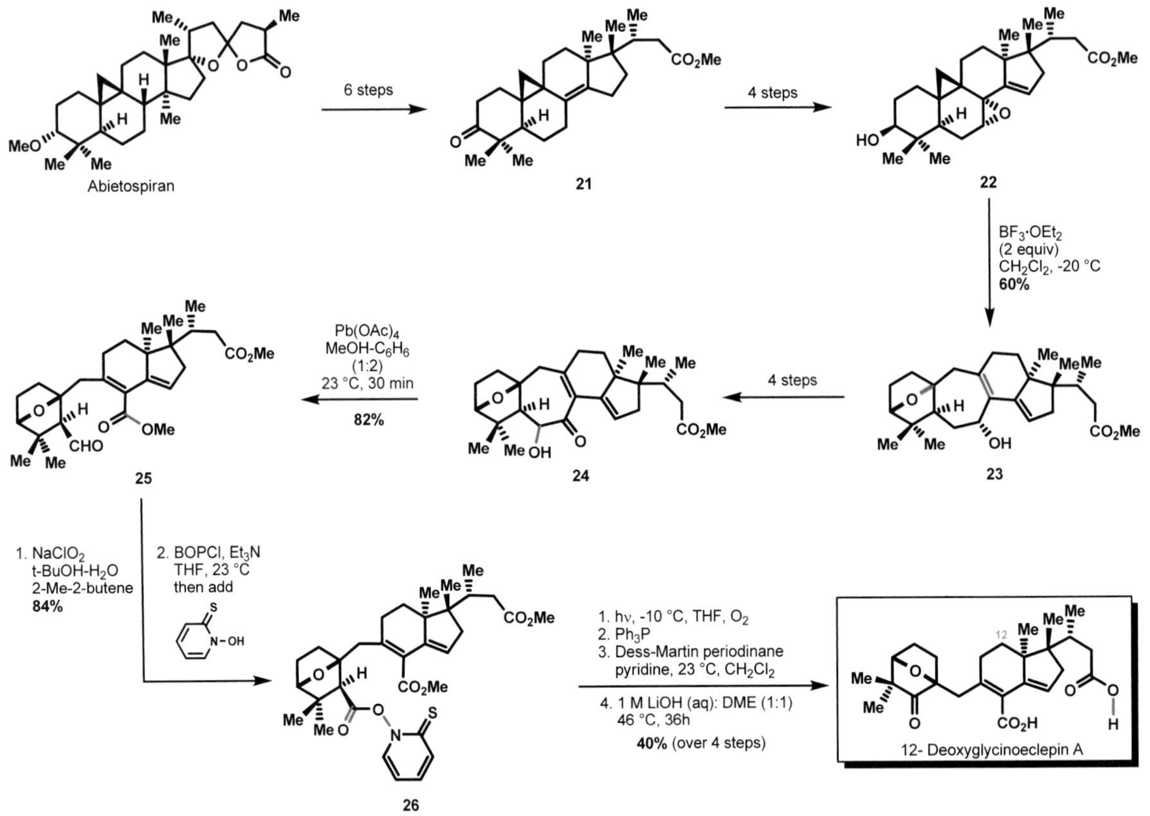

Scheme 5

References

1. (a) Masamune, T., Anetai, M., Takasugi, M. & Katsui, N. Isolation of a natural hatching stimulus, glycinoeclepin A, for the soybean cyst nematode. *Nature (London)* **1982**, 297, 495-496; (b) Fukuzawa, A., Furusaki, A., Ikura, M. & Masamune, T. Glycinoeclepin A, a natural hatching stimulus for the soybean cyst nematode. *J. Chem. Soc., Chem. Commun.* **1985**, 222-224.

2. Corey, E. J. & Houpis, I. N. Total synthesis of glycinoeclepin A. *J. Am. Chem. Soc.* **1990**, 112, 8997-8998.

3. Corey, E. J. & Houpis, I. N. The control of diastereoselectivity in the Michael reaction of ketonic enolates with crotonic acid derivatives. *Tetrahedron Lett.* **1993**, 34, 2421-2424.

4. Ganem, B. & Small, V. R., Jr. Ferric chloride in acetic anhydide. Mild and versatile reagent for the cleavage of ethers. *J. Org. Chem.* **1974**, 39, 3728-3730.

5. Giroux, S. & Corey, E. J. Enantioselective Synthesis of a Simple Benzenoid Analog of Glycinoeclepin A. *Org. Lett.* **2008**, 10, 5617-5619.

6. Corey, E. J. & Hong, B. Chemical Emulation of the Biosynthetic Route to Glycinoeclepin from a Cycloartenol Derivative. *J. Am. Chem. Soc.* **1994**, 116, 3149-3150.

ECTEINASCIDIN 743

Background

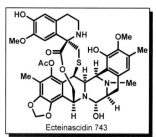

Ecteinascidin 743

Ecteinascidin 743 (Et-743) is a marine natural product from the sponge *Ecteinascidia turbinate* whose isolation and structural assignment are due to the pioneering studies of K.L. Rinehart and his group at the University of Illinois, Urbana. It is one of the most potent antitumor substances known, with subnanomolar antiproliferative activity against many tumor cell lines. The total synthesis of ecteinascidin 743[1] was an important step in enabling pre-clinical and clinical development studies to go forward. It has now been found to be effective against soft tissue sarcomas (for which there have been no therapeutic agents) as well as ovarian, breast and testicular tumors. Now approved for use in many countries, it is available under the generic name trabectedin. It is generally administered intravenously at 21-day intervals at a dose of ca. 1.5 mg. Trabectedin (Et-743) binds strongly to the minor groove of DNA, with multiple consequences arising from changes in gene transcription (ca. 250 down- and 90 upregulated genes). Among the downregulated genes are many cancer-activated genes including one associated with drug resistance and some involved in cell cycle progression and nucleotide-induced DNA repair.

The current commercial production method for trabectedin consists of a combination of microbial production of a molecule containing the pentacyclic framework (as in 8) and a slight modification of the last several steps of the synthesis outlined below.[2]

Abbreviated Retrosynthetic Plan

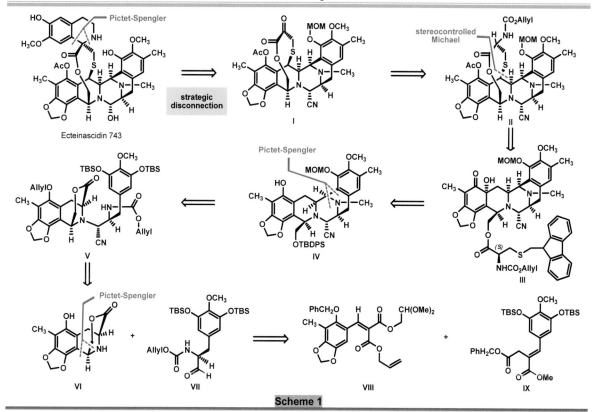

Scheme 1

(1) The initial disconnection to **I** and 2-hydroxy-3-methoxyphenylethylamine leads by a two-step retrosequence to the cysteinyl lactone **II**.

(2) The disconnection **II** → **III** corresponds to a Michael transform to give a quinone methide for which the hydroxydienone **III** is a precursor. In turn, **III** derives from the phenolic intermediate **IV**.

(3) Disconnection of **IV** leads first to **V** and from there to the fragments **VI** and **VII** which further simplify via the corresponding substituted phenylalanines to the achiral precursors **VIII** and **IX**.

(4) The above retrosynthetic sequence depends on three Pictet-Spengler disconnections: **Et-743 ⇒ I, IV ⇒ V** and **VI ⇒ VIII**. The application of the Pictet-Spengler transform for the disconnection of each of the three embedded isoquinoline subunits was a key transform-goal that guided and shaped the synthetic plan.

ECTEINASCIDIN 743

Pathway of Synthesis

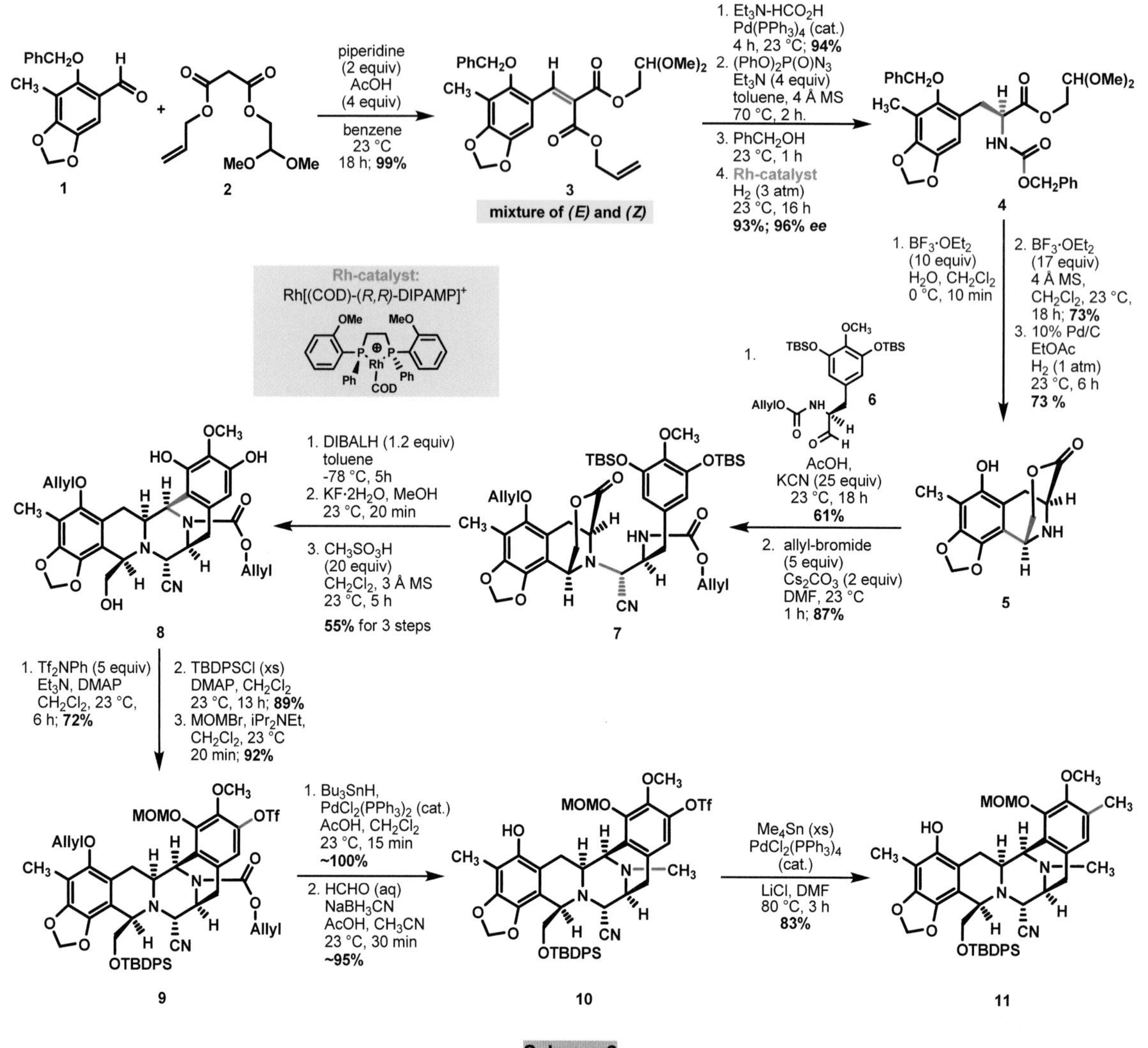

Scheme 2

Key steps in Scheme 2

(1) The last step of the enantioselective synthesis of the substituted phenylalanine derivative **4** utilized the Knowles hydrogenation of the corresponding (Z)-α,β-unsaturated ester with a cationic rhodium complex of (R,R)-1,2-bis-[phenyl,o-methoxyphenyl] phosphinoethane, (R,R)-DIPAMP⁺BF₄⁻.[3] The same kind of enantioselective hydrogenation was utilized for the synthesis of the chiral methyl ester corresponding to aldehyde **6** from an achiral α,β-unsaturated ester precursor.

(2) The conversion **4→7** involved the following elements:

(a) BF₃-promoted hydrolysis of the dimethyl acetal **4** to the corresponding aldehyde;

(b) intramolecular Pictet-Spengler cyclization and

(c) reductive removal of N-benzyloxycarbonyl protecting group.

(3) The transformation **7→8** consisted of the following sequence:

(a) reduction of lactone **7** to the corresponding lactol;

(b) cleavage of the two TBS protecting groups and

(c) diastereoselective Pictet-Spengler cyclization.

ECTEINASCIDIN 743

Pathway of Synthesis (continued)

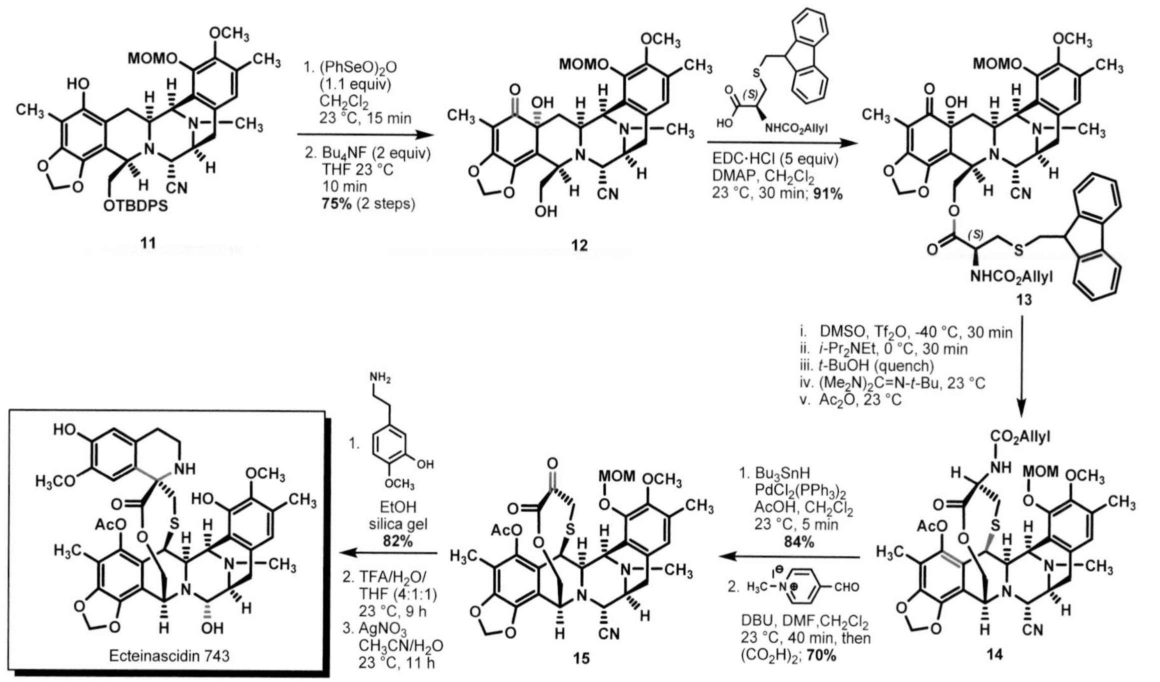

Scheme 3

Key steps in Scheme 3

(1) In the conversion of **13** to **14**:

   (a) step *i* converts the tertiary hydroxyl group of **13** to an *O*-dimethylsulfonium cation/triflate salt;

   (b) step *ii* effects elimination of dimethyl sulfoxide to form a bright yellow quinone methide **17** (stable at 0 °C);

   (c) step *iv* cleaves the *S*-fluorenylmethyl protecting group to form a thiolate ion which undergoes intramolecular Michael addition to the quinone methide subunit to form the transannular lactone/sulfide bridge.

(2) The formation of the α-keto lactone **15** from the corresponding α-amino-lactone utilized the base-catalyzed prototropic shift of a strongly electron-withdrawing imine intermediate.[4]

(3) The Pictet-Spengler reaction that leads to ecteinascidin 743 is highly diastereoselective.[5]

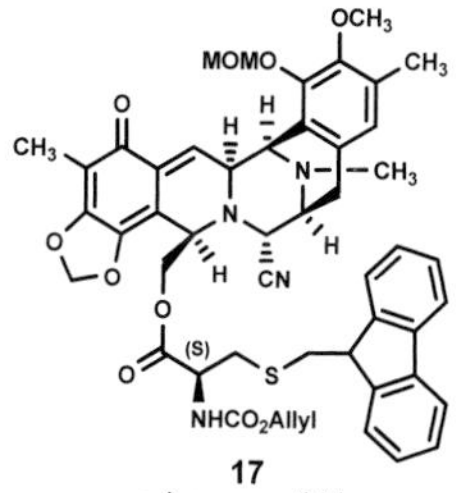

17
quinone methide

References

1. Corey, E. J., Gin, D. Y. & Kania, R. S. Enantioselective Total Synthesis of Ecteinascidin 743. *J. Am. Chem. Soc.* **1996**, 118, 9202-9203.
2. Cuevas, C., Perez, M., Martin, M. J., Chicharro, J. L., Fernandez-Rivas, C., Flores, M., Francesch, A., Gallego, P., Zarzuelo, M., de la Calle, F., Garcia, J., Polanco, C., Rodriguez, I. & Manzanares, I. Synthesis of Ecteinascidin ET-743 and Phthalascidin Pt-650 from Cyanosafracin B. *Org. Lett.* **2000**, 2, 2545-2548.
3. (a) Knowles, W. S., Sabacky, M. J., Vineyard, B. D. Catalytic asymmetric hydrogenation. *J. Chem. Soc., Chem. Commun.* **1972**, 10-11; (b) Knowles, W. S. Asymmetric hydrogenation. *Acc. Chem. Res.* **1983**, 16, 106-112.
4. Corey, E. J. & Achiwa, K. Oxidation of primary amines to ketones. *J. Am. Chem. Soc.* **1969**, 91, 1429-1432.
5. For subsequent synthetic work on Ecteinascidin 743 see: (a) Chen, J., Chen, X., Willot, M., Zhu, J. Asymmetric total syntheses of ecteinascidin 597 and ecteinascidin 583. *Angew. Chem., Int. Ed.* **2006**, 45, 8028-8032; (b) Zheng, S., Chan, C., Furuuchi, T., Wright, B. J. D., Zhou, B., Guo, J., Danishefsky, S. J. Stereospecific formal total synthesis of Ecteinascidin 743. *Angew. Chem., Int. Ed.* **2006**, 45, 1754-1759; (c) Chen, J., Chen, X., Bois-Choussy, M., Zhu, J. Total Synthesis of Ecteinascidin 743. *J. Am. Chem. Soc.* **2006**, 128, 87-89; (d) Endo, A., Yanagisawa, A., Abe, M., Tohma, S., Kan, T., Fukuyama, T. Total Synthesis of Ecteinascidin 743. *J. Am. Chem. Soc.* **2002**, 124, 6552-6554.

(+)-MIROESTROL

Background

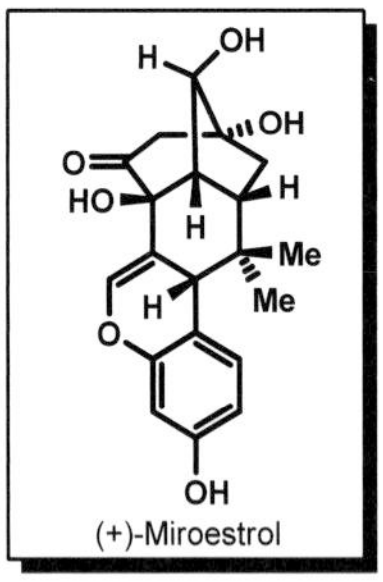

(+)-Miroestrol

The plant *Pueraria mirifica*, long used in the folk medicine of northern Thailand, was so prized for its rejuvenating properties that it was harvested almost to the point of extinction. It contains miroestrol, a very potent estrogen, whose structure was determined in 1960.[1] The synthesis outlined below represents the only one to date for this unusual structure, which had previously been studied in several laboratories.[2]

Abbreviated Retrosynthetic Plan

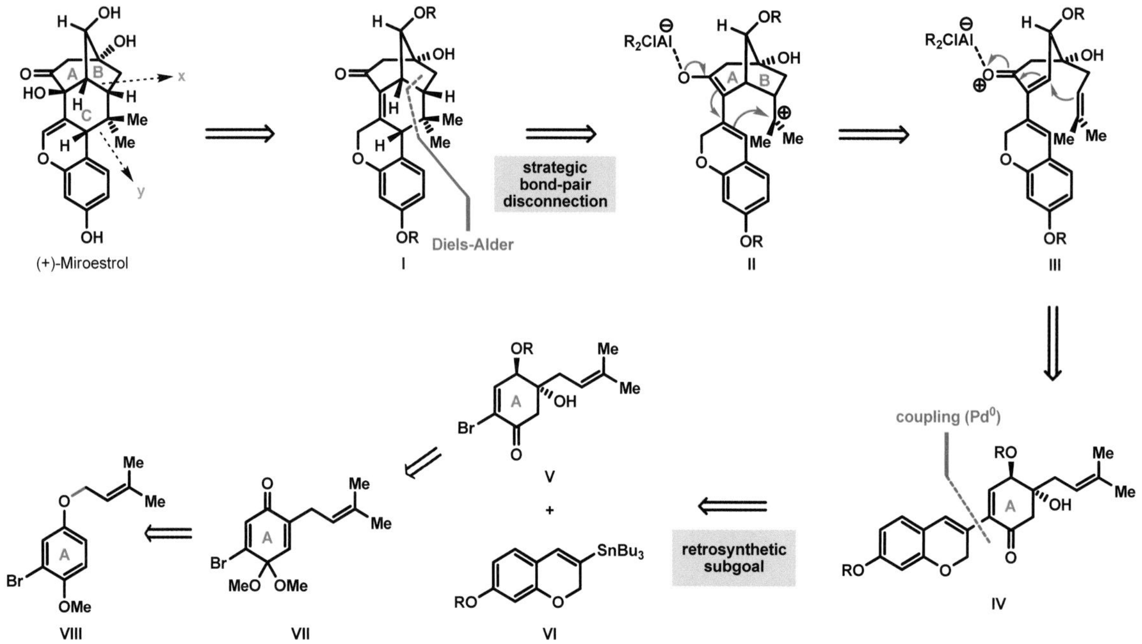

Scheme 1

Key Steps in Scheme 1

(1) The retrosynthetic plan was based on the idea that the strategic disconnection of bonds x and y could be effected after the generation of intermediate I which contains a full retron for the Diels-Alder transform. This simultaneous application of topological and transform-based strategies leads to the greatly simplified precursor IV.

(2) Because the thermal Diels-Alder cyclization of IV to A seemed improbable, a Lewis acid-catalyzed, cationic process was applied using mechanistic transform application.[3] Thus, I was disconnected via cationic metastable intermediates II and III to IV.

(3) Intermediate IV clearly simplifies to V and VI, and the former can be taken back to the achiral intermediate VII, provided that the stereocenters in V can be cleared (which at the outset was problematic).

(4) Dienone VII can be accessed from VIII in two steps:

  (a) cationic allylic rearrangement from O to C (ortho) and

  (b) phenol oxidation in methanol.

(+)-MIROESTROL

Pathway of Synthesis

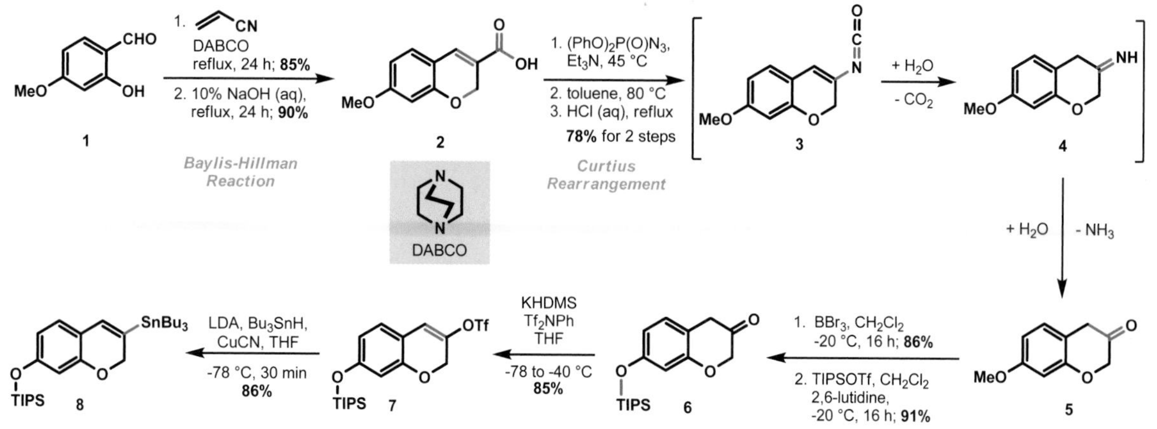

Scheme 2

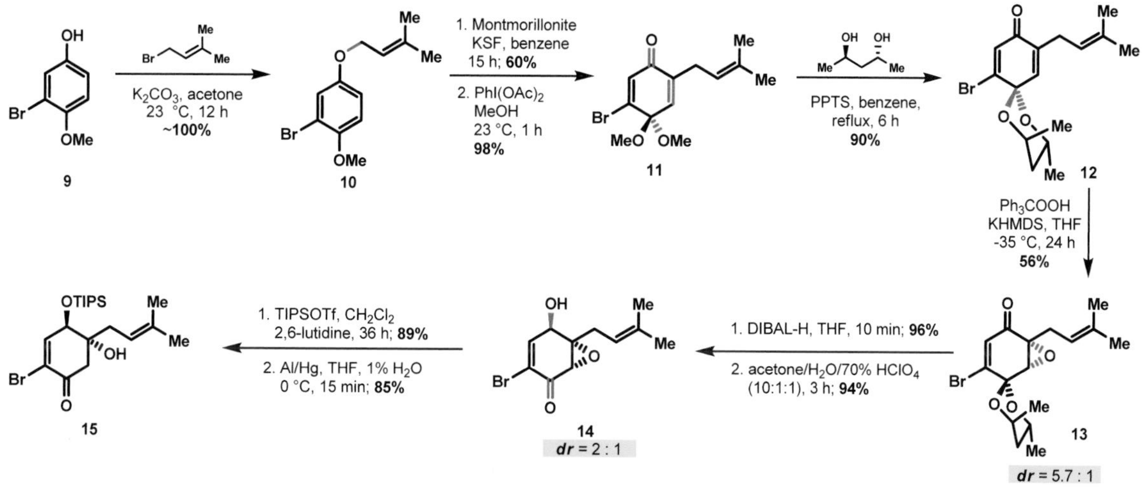

Scheme 3

Key steps in Scheme 2 and Scheme 3

(1) π-Face-selective epoxidation of **11** was effected via a carefully chosen chiral monoketal derivative (**12**) to form **13**. The bulky bromine substituent in **12** forms the chair conformer shown in Scheme 3. The epoxidation **12→13** occurs at the sterically more available π-face.

acyl azide

(2) The formation of **3** from **2** occurs by Curtius rearrangement of the intermediate acyl azide.

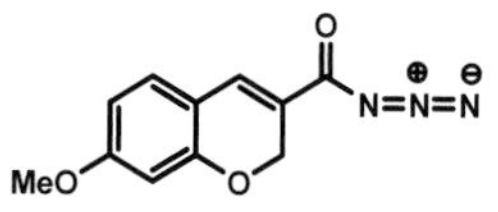

(3) Acidic hydrolysis of **3** affords ketone **5** via an intermediate ketimine **4**.

(+)-MIROESTROL

Pathway of Synthesis (continued)

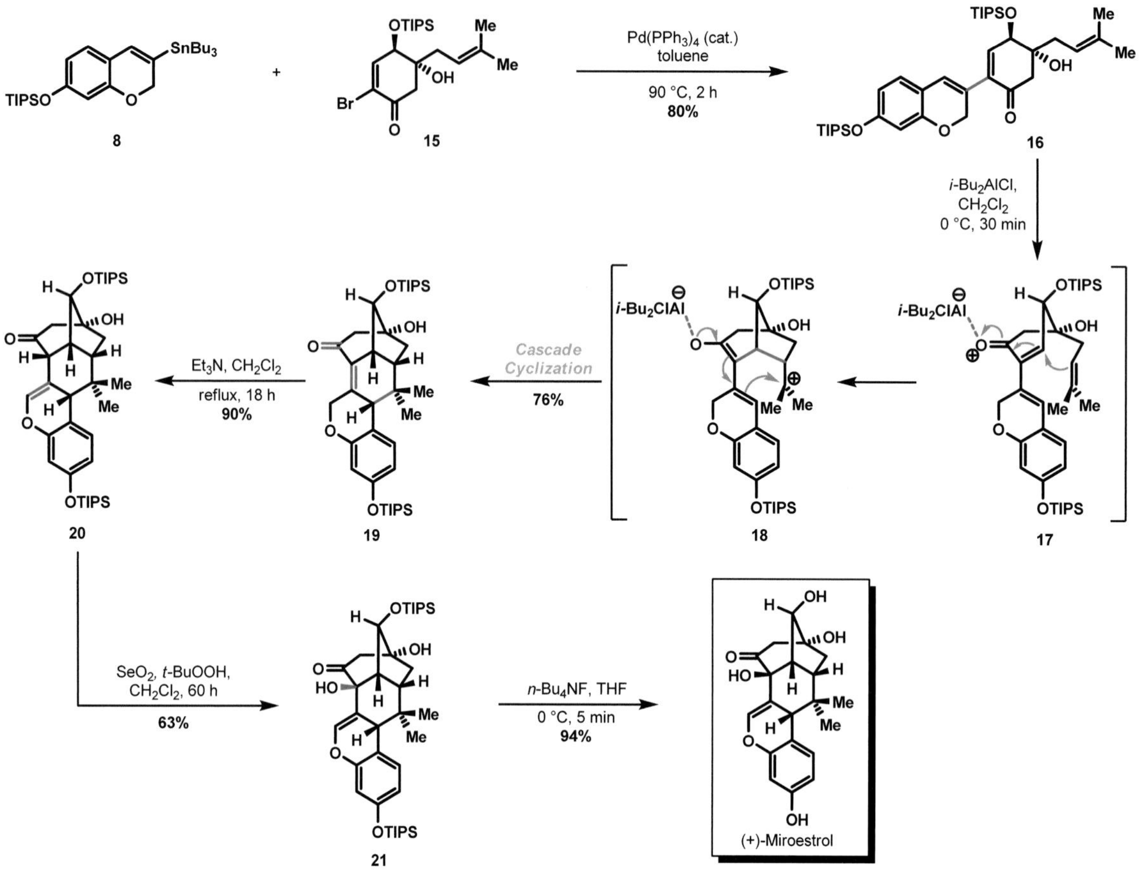

Scheme 4

Key steps in Scheme 4

(1) The transannular cyclization **16**→**19** proceeded readily using the soluble Lewis acid *i*-Bu₂AlCl in CH₂Cl₂, presumably via intermediates **17** and **18**. This process may also be regarded as a two-step cationic Diels-Alder cycloaddition.

(2) The synthesis of (+)-miroestrol was completed by allylic oxidation and silyl ether cleavage.

References

1. (a) Cain, J. C. Miroestrol-estrogen from the plant Pueraria mirifica. *Nature (London, U. K.)* **1960**, 188, 774-777; (b) Taylor, N. E., Hodgkin, D. C. & Rollett, J. S. The x-ray crystallographic determination of the structure of bromomirestrol. *J. Chem. Soc.* **1960**, 3685-3695.

2. Corey, E. J. & Wu, L. I. Enantioselective total synthesis of miroestrol. *J. Am. Chem. Soc.* **1993**, 115, 9327-9328.

3. Corey, E. J. & Cheng, X. M. *The Logic of Chemical Synthesis* (John Wiley & Sons, Hoboken, NJ, **1989**).

(−)-NEOTRIPTERIFORDIN

Background

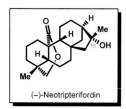

(−)-Neotripterifordin

The Chinese medicinal plant *Tripterygium wilfordii* Hook (celastraceae) has provided extracts with antitumor, anti-inflammatory, and immuno-suppressive activities and a number of bioactive compounds, including the antitumor diterpenoids triptolide and tripdiolide and the potent inhibitor of HIV replication, neotripterifordin (EC_{50} 25 nM).[1] The successful total synthesis of neotripterifordin[2] that is shown in this section allowed the revision of stereochemistry which had originally been misassigned with regard to the configuration at C(16).

Abbreviated Retrosynthetic Plan

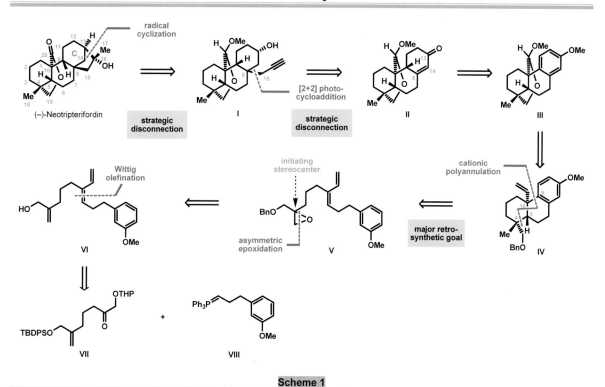

Scheme 1

Key Steps in Scheme 1

(1) The retrosynthetic analysis for neotripterifordin in Scheme 1 was guided by the simultaneous application of strategic bond disconnections that would enable eventual disconnection of the polycyclic framework by means of the cation-olefin annulation transform. In concert with these objectives, it was necessary to modify functional groups to prevent functional group interference at any stage.

(2) The goal of achieving an enantioselective synthesis dictated a design that insured valid clearance of all stereocenters, including that which determines the stereochemistry of the key cationic polyannulation process.

(3) The retrosynthetic step **neotripterifordin** ⇒ **I** disconnects the first of two strategic bonds to the two-carbon bridge across ring C by application of the radical

π-cyclization transform. The step **I** ⇒ **II** disconnects the second bond by a sequence that had to be determined experimentally.

(4) The step **II** ⇒ **III** replaced the 2-cyclohexenone subunit by an anisole equivalent (Birch reduction transform), and was selected because it provides a reactive π-electron-rich terminating aromatic group for the key cationic polyannulation step.

(5) Disconnection of **IV** was also enhanced by the retrosynthetic introduction of an angular vinyl group (expected to facilitate the cation-olefin cyclization reaction). Application of the polyannulation transform simplified **IV** to **V**, the latter being available from the achiral intermediate **VI** by enantioselective Sharpless-Katsuki epoxidation. A Wittig disconnection simplifies **VI** to the fragments **VII** and **VIII**.

(−)-NEOTRIPTERIFORDIN

Pathway of Synthesis

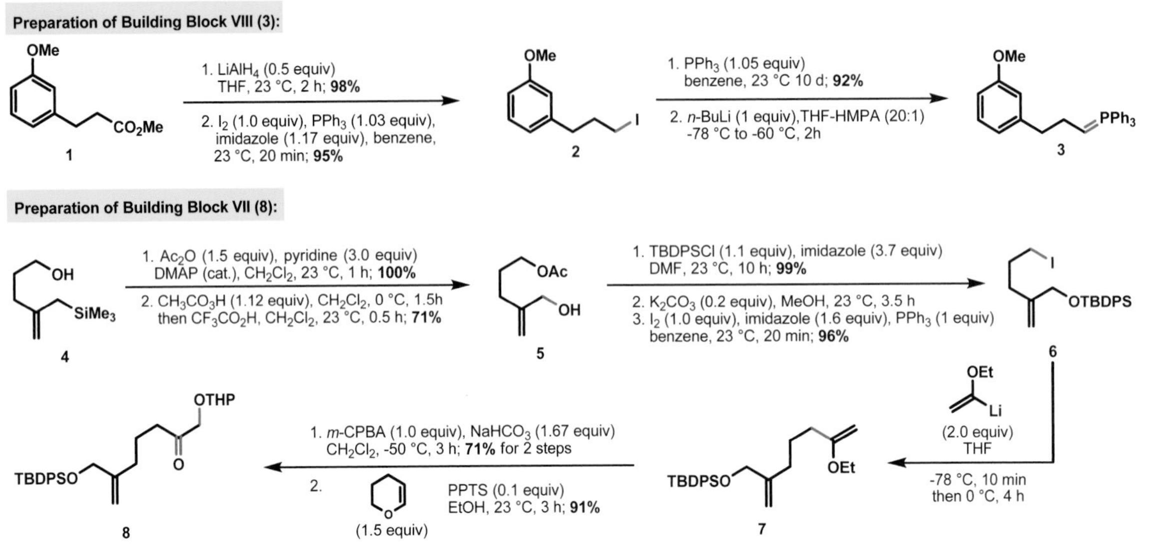

Preparation of Building Block VIII (3):

Preparation of Building Block VII (8):

Scheme 2

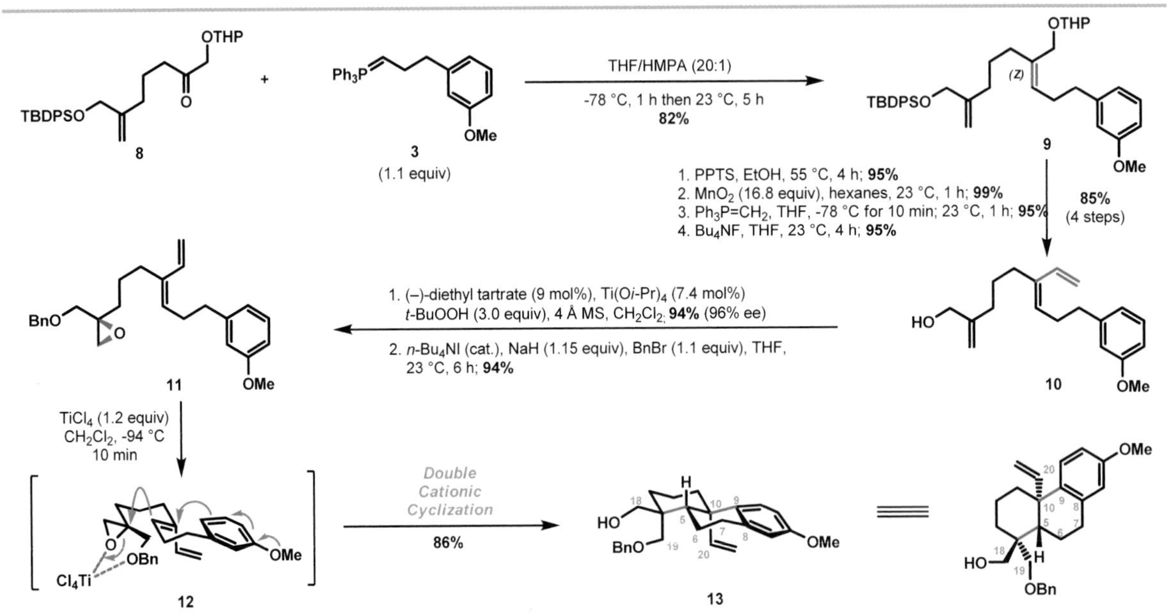

Scheme 3

Key steps in Scheme 3

(1) The Wittig coupling of **3** and **8** produced the Z isomer of **9** stereoselectively, as expected from prior reports.[3]

(2) The epoxy ether **11** was activated by bidentate coordination to TiCl$_4$ (see **12**) to provide the tricycle **13** efficiently and stereoselectively. This powerful construction represents the essential foundation of the synthesis and the main transform-based retrosynthetic objective.

(3) The synthesis shown above was adapted to provide not only (−)-neotripteriforden but also the diastereomer at the carbinol center, C(16). The latter structure had previously been incorrectly assigned to the natural product.

(–)-NEOTRIPTERIFORDIN

Pathway of Synthesis (continued)

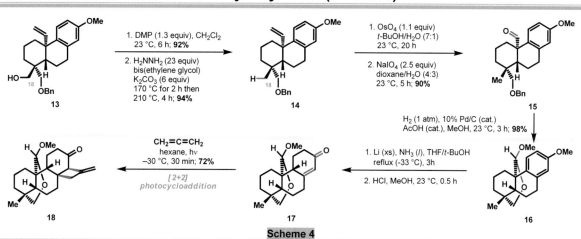

1. DMP (1.3 equiv), CH$_2$Cl$_2$
23 °C, 6 h; **92%**

2. H$_2$NNH$_2$ (23 equiv)
bis(ethylene glycol)
K$_2$CO$_3$ (6 equiv)
170 °C for 2 h then
210 °C, 4 h; **94%**

1. OsO$_4$ (1.1 equiv)
t-BuOH/H$_2$O (7:1)
23 °C, 20 h

2. NaIO$_4$ (2.5 equiv)
dioxane/H$_2$O (4:3)
23 °C, 5 h; **90%**

H$_2$ (1 atm), 10% Pd/C (cat.)
AcOH (cat.), MeOH, 23 °C, 3 h; **98%**

CH$_2$=C=CH$_2$
hexane, hν
–30 °C, 30 min; **72%**

*[2+2]
photocycloaddition*

1. Li (xs), NH$_3$ (l), THF/t-BuOH
reflux (-33 °C), 3h

2. HCl, MeOH, 23 °C, 0.5 h

13 **14** **15** **16** **17** **18**

Scheme 4

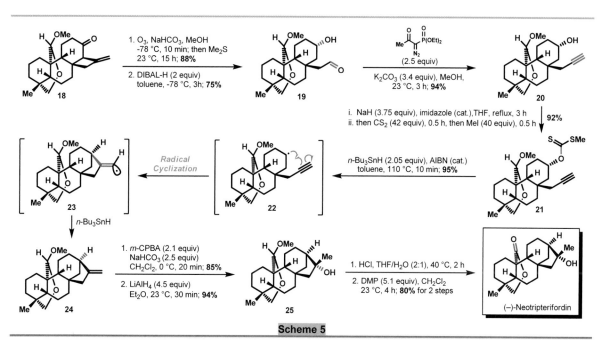

1. O$_3$, NaHCO$_3$, MeOH
-78 °C, 10 min; then Me$_2$S
23 °C, 15 h; **88%**

2. DIBAL-H (2 equiv)
toluene, -78 °C, 3h; **75%**

Me—(=O)—(=O)—P(OEt)$_2$
N$_2$
(2.5 equiv)

K$_2$CO$_3$ (3.4 equiv), MeOH,
23 °C, 3 h; **94%**

i. NaH (3.75 equiv), imidazole (cat.),THF, reflux, 3 h
ii. then CS$_2$ (42 equiv), 0.5 h, then MeI (40 equiv), 0.5 h

92%

*Radical
Cyclization*

n-Bu$_3$SnH (2.05 equiv), AIBN (cat.)
toluene, 110 °C, 10 min; **95%**

n-Bu$_3$SnH

1. m-CPBA (2.1 equiv)
NaHCO$_3$ (2.5 equiv)
CH$_2$Cl$_2$, 0 °C, 20 min; **85%**

2. LiAlH$_4$ (4.5 equiv)
Et$_2$O, 23 °C, 30 min; **94%**

1. HCl, THF/H$_2$O (2:1), 40 °C, 2 h

2. DMP (5.1 equiv), CH$_2$Cl$_2$
23 °C, 4 h; **80% for 2 steps**

(–)-Neotripteriforidin

18 **19** **20** **21** **22** **23** **24** **25**

Scheme 5

Key steps in Schemes 4 and 5

(1) Since standard conjugate addition reactions with enone **17** failed, allene photocycloaddition was employed (**17 → 18**).

(2) The angular propynyl group of **20** was generated from **18** by the sequence involving a retroaldol/selective reduction combination and a subsequent alkynylation with excess of [N$_2$C-P(=O)(OEt)$_2$]$^\ominus$.

References

1. (a) Chen, K., Shi, Q., Fujioka, T., Zhang, D., Hu, C., Jin, J., Kilkuskie, R. E. & Lee, K. H. Anti-aids agents, 4. Tripterifordin, a novel anti-HIV principle from Tripterygium wilfordii: isolation and structural elucidation. *J. Nat. Prod.* **1992**, 55, 88-92; (b) Chen, K., Shi, Q., Fujioka, T., Nakano, T., Hu, C.-Q., Jin, J.-Q., Kilkuskie, R. E. & Lee, K.-H. Anti-AIDS agents. XIX. Neotripterifordin, a novel anti-HIV principle from Tripterygium wilfordii: isolation and structural elucidation. *Bioorg. Med. Chem.* **1995**, 3, 1345-1348.

2. Corey, E. J. & Liu, K. Enantioselective Total Synthesis of the Potent Anti-HIV Agent Neotripterifordin. Reassignment of Stereochemistry at C(16). *J. Am. Chem. Soc.* **1997**, 119, 9929-9930.

3. Sreekumar, C., Darst, K. P. & Still, W. C. A direct synthesis of Z-trisubstituted allylic alcohols via the Wittig reaction. *J. Org. Chem.* **1980**, 45, 4260-4262.

AFLATOXIN B₂

Background

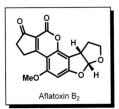

Aflatoxin B₂

The aflatoxins are a group of structurally similar compounds produced by many species of fungi in the *Aspergillus* family which can grow on stored crops such as cereals and nuts. They are both exceedingly toxic and carcinogenic and for this reason must be excluded from the food supply. The most potent, aflatoxins B₁ and B₂, are known to cause gene mutations. There have been many syntheses of racemic aflatoxins,[1] but the natural enantiomer has been synthesized by only two groups using very different routes.[2] By far the shortest of these is described in this section.[2b]

Abbreviated Retrosynthetic Plan

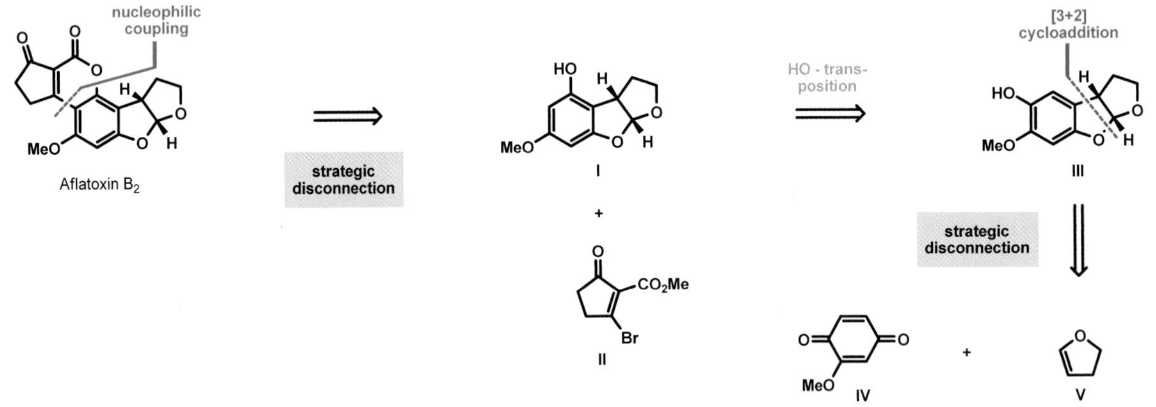

Scheme 1

Key Steps in Scheme 1

(1) The disconnection of aflatoxin B₂ to I and II, which had previously been established, reduces complexity and also suggests a further simplification to commercially available materials, IV and V.

(2) The final disconnection with clearance of all stereocenters depends on a new enantiocontrolled [3+2]-cycloaddition transform which, in turn, required the retrosynthetic translocation of the phenolic OH group as shown in I ⇒ III.

Pathway of Synthesis

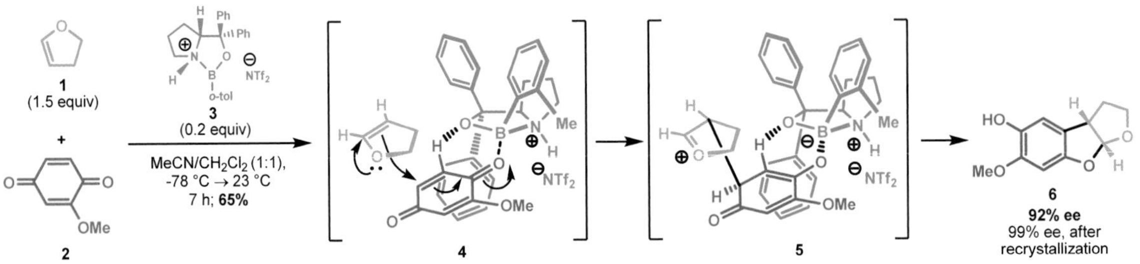

Scheme 2

Explanation of Scheme 2

The first, and key, step in the synthesis was the [3+2]-cycloaddition of **1** to **2** under the influence of the chiral oxazaborolidinium cation **3**.[2b] This novel process was proposed to proceed via the pre-transition-state assembly **4**, on the basis of analogy with many other known cycloaddition reactions that can be catalyzed by **3**. The intermediacy of **5** in the formation of the product **6** was demonstrated by trapping experiments.[2b,3]

AFLATOXIN B$_2$

Pathway of Synthesis (continued)

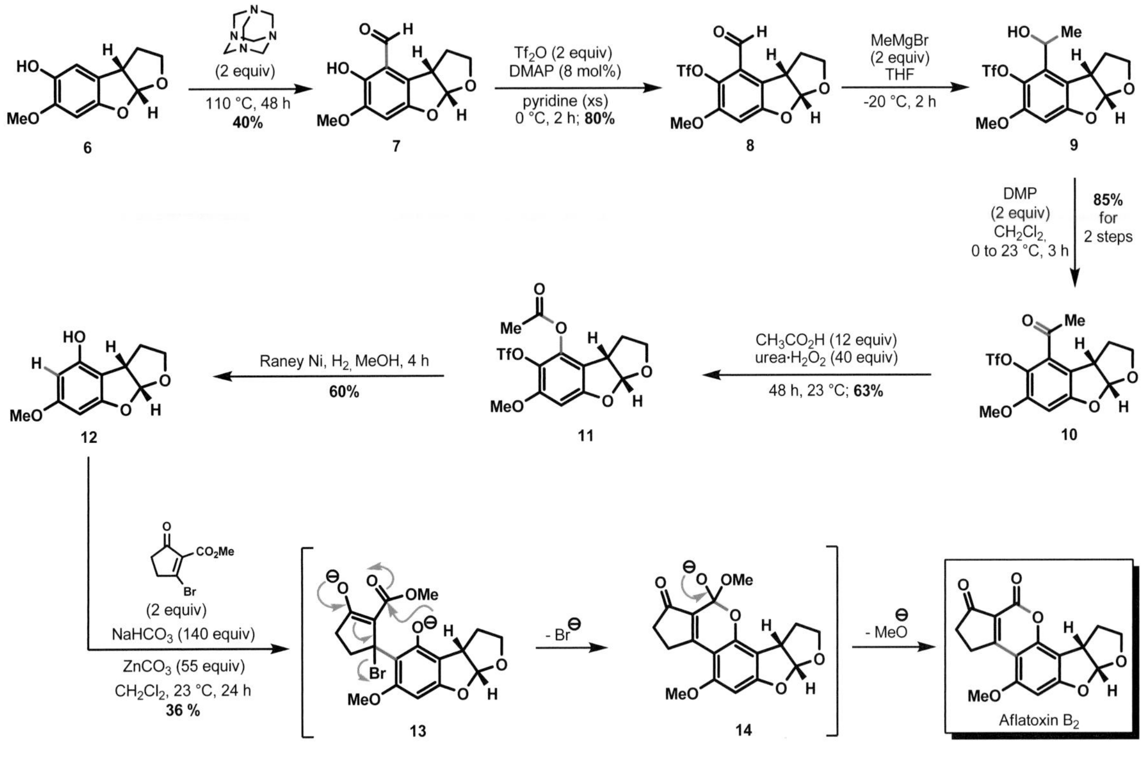

Scheme 3

Key steps in Scheme 3

(1) The vicinal transposition of the hydroxyl group on the aromatic ring was effected by the steps shown, leading overall from **6** to **12**. This sequence was important to bridge the structural gap between the [3+2]-cycloadduct **6** and the aflatoxin structure.

(2) The final step, in which two rings are added to **12**, involves a tactical combination of Michael addition-elimination (**12** →**13**) and lactonization (**13** → **14**) using mild base catalysis.

References

1. (a) Büchi, G., Foulkes, D. M., Kurono, M. & Mitchell, G. F. The total synthesis of racemic aflatoxin B$_1$. *J. Am. Chem. Soc.* **1966**, 88, 4534-4536; (b) Büchi, G., Foulkes, D. M., Kurono, M., Mitchell, G. F. & Schneider, R. S. The total synthesis of racemic aflatoxin B$_1$. *J. Am. Chem. Soc.* **1967**, 89, 6745-6753; (c) Castellino, A. J. & Rapoport, H. Syntheses of tetrahydrofuro[2,3-b]benzofurans: a synthesis of (±)-aflatoxin B2. *J. Org. Chem.* **1986**, 51, 1006-1011; (d) Weeratunga, G., Horne, S. & Rodrigo, R. A formal synthesis of aflatoxin B$_2$. *J. Chem. Soc., Chem. Commun.* **1988**, 721-722; (e) Horne, S., Weeratunga, G. & Rodrigo, R. The regiospecific p-deiodination of 2,4-diiodophenols; a new synthesis of aflatoxin B$_2$. *J. Chem. Soc., Chem. Commun.* **1990**, 39-41; (f) Koreeda, M., Dixon, L. A. & Hsi, J. D. An efficient formal synthesis of aflatoxin B$_2$ by use of the Kikuchi rearrangement reaction. *Synlett* **1993**, 555-556; (g) Pirrung, M. C. & Lee, Y. R. Formal total synthesis of (±)-

aflatoxin B$_2$ utilizing the rhodium carbenoid dipolar cycloaddition. *Tetrahedron Lett.* **1996**, 37, 2391-2394.

2. (a) Trost, B. M. & Toste, F. D. Palladium Catalyzed Kinetic and Dynamic Kinetic Asymmetric Transformations of γ-Acyloxybutenolides. Enantioselective Total Synthesis of (+)-Aflatoxin B$_1$ and B$_{2a}$. *J. Am. Chem. Soc.* **2003**, 125, 3090-3100; (b) Zhou, G. & Corey, E. J. Short, Enantioselective Total Synthesis of Aflatoxin B$_2$ Using an Asymmetric [3+2]-Cycloaddition Step. *J. Am. Chem. Soc.* **2005**, 127, 11958-11959.

3. For a review see: Corey E.J. Enantioselective Catalysis Based on Cationic Oxazaborolidines. *Angew. Chem. Intl. Ed*, **2009**, 48, 2100-2117.

(+)-PAEONIFLORIN

Background

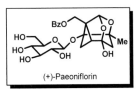

(+)-Paeoniflorin

Paeoniflorin, the β-glycoside of paeoniflorigenin, is a novel complex terpenoid from the roots of the Chinese paeony (*Paeonia lactiflora*), which is widely used in traditional Chinese medicine. There is a very extensive literature on the biological activities of paeoniflorin, with numerous studies even in the past few years.[1] It has been reported to have analgesic, anti-inflammatory, antihyperglycemic and antithrombotic effects and to enhance glucose uptake by muscle.

The structure of paeoniflorin can be disconnected to glucose and paeoniflorigenin, the latter being the critical target for synthesis. The intriguing highly oxygenated pinene constitution of paeoniflorigenin posed an unusual challenge for chemical synthesis for three decades.[2]

Abbreviated Retrosynthetic Plan

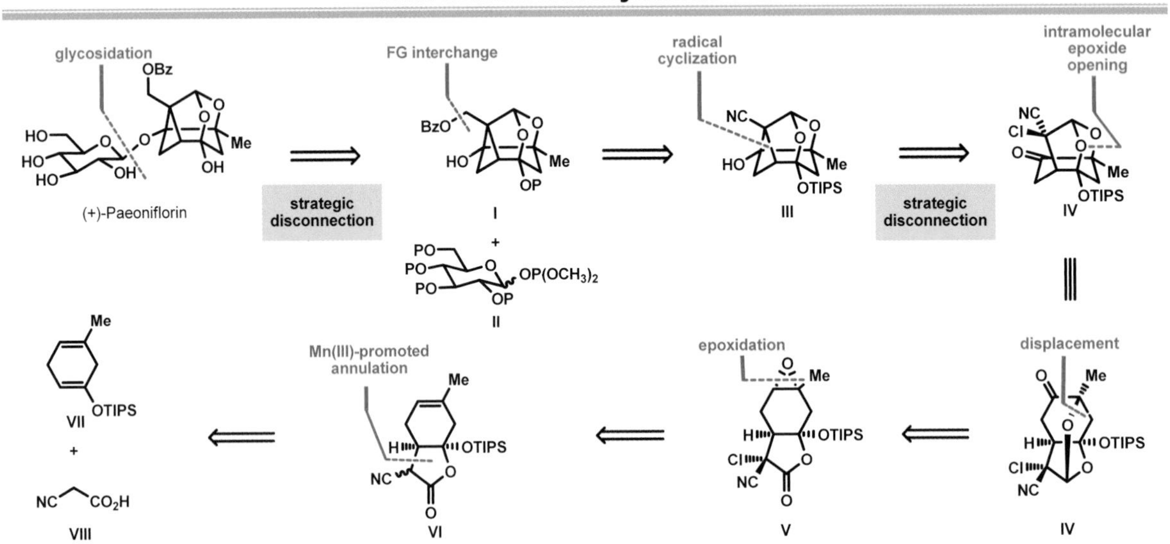

Scheme 1

Key steps in Scheme 1

(1) The high density of functionality and ring-bridging required that retrosynthetic simplification of these two elements of complexity be highly coordinated. Changes in functionality were essential to the disconnections of the cyclic network.

(2) The first disconnection **III** ⇒ **IV** required changes in the functionality of **I**, as did the second, **IV** ⇒ **V**.

(3) The disconnection, **VI** ⇒ **VII** + **VIII**, depended on recently developed methodology.[3]

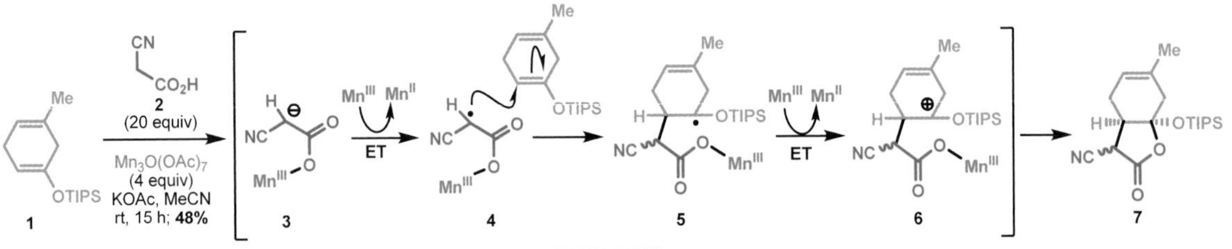

Scheme 2

Key steps in Scheme 2

The cycloaddition of cyanoacetic acid **2** to **1** by Mn(III) acetate probably occurs by the sequence:

(a) coordination of the carboxyl group to Mn(III);

(b) deprotonation of the α-methylene group and electron-transfer (ET) to Mn(III) to form first **3** and then radical **4**;

(c) addition of radical **4** to the double bond of **1** and

(d) a second electron-transfer to Mn(III) and lactonization via intermediate **6**.

(+)-PAEONIFLORIN

Pathway of Synthesis

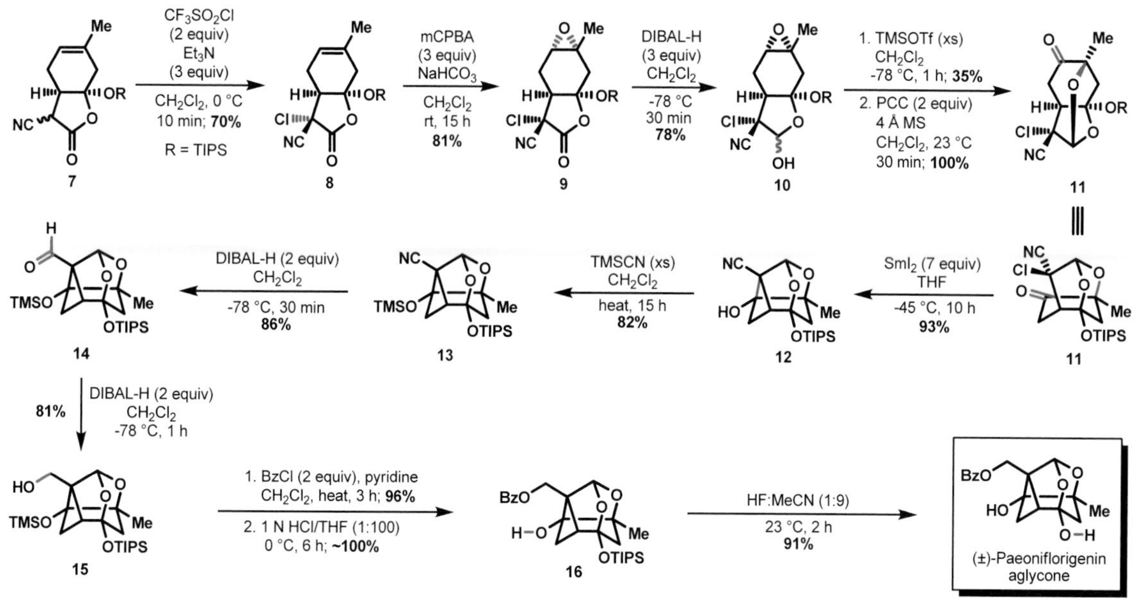

Scheme 3

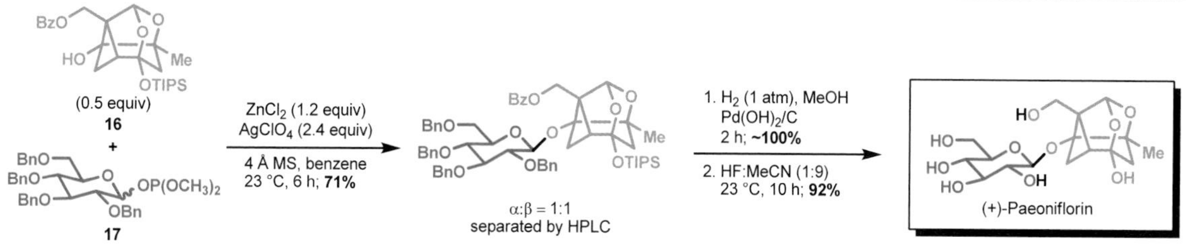

Scheme 4

Key steps in Schemes 3 and 4

The transformation **11→12** is considered to take place by radical-forming de-chlorination of **11** by SmI₂, addition to the nearby carbonyl group and subsequent one-electron reduction.

That process was selected for ring-closure rather than the standard alternative, aldol cyclization. The choice was made because of the possibility that the aldol process might be disfavored thermodynamically.

References

1. (a) Wu, H., Li, W., Wang, T., Shu, Y., Liu, P. Paeoniflorin suppress NF-kB activation through modulation of IkBalpha and enhances 5-fluorouracil-induced apoptosis in human gastric carcinoma cells. *Biomed. Pharmacother.* **2008**, 62, 659-666; (b) Radix Paeoniae in *WHO Monographs on Selected Medicinal Plants* **1999**, 1, pp 195-200 (World Health Organization, Geneva); (c) Tang, L.-M., Liu, I.-M., Cheng, J.-T. Stimulatory effect of paeoniflorin on adenosine release to increase the glucose uptake into white adipocytes of wistar rat. *Planta Med.* **2003**, 69, 332-336.

2. (a) Corey, E. J. & Wu, Y. J. Total synthesis of (±)-paeoniflorigenin and paeoniflorin. *J. Am. Chem. Soc.* **1993**, 115, 8871-8872; (b) For a subsequent synthesis of paeoniflorin see: Hatakeyama, S., Kawamura,

M., Takano, S. Total Synthesis of (-)-Paeoniflorin. *J. Am. Chem. Soc.* **1994**, 116, 4081-4082.

3. (a) Corey, E. J., Gross, A. W. Carbolactonization of olefins under mild conditions by cyanoacetic and malonic acids promoted by manganese(III) acetate. *Tetrahedron Lett.* **1985**, 26, 4291-4294; (b) Corey, E. J., Ghosh, A. K. Two-step synthesis of furans by manganese(III)-promoted annulation of enol ethers. *Chem. Lett.* **1987**, 223-226. For an excellent review see: Snider, B. B. Manganese(III)-Based Oxidative Free-Radical Cyclizations. *Chem. Rev.* **1996**, 96, 339-363.

(+)-GRACILIN B & (+)-GRACILIN C

Background

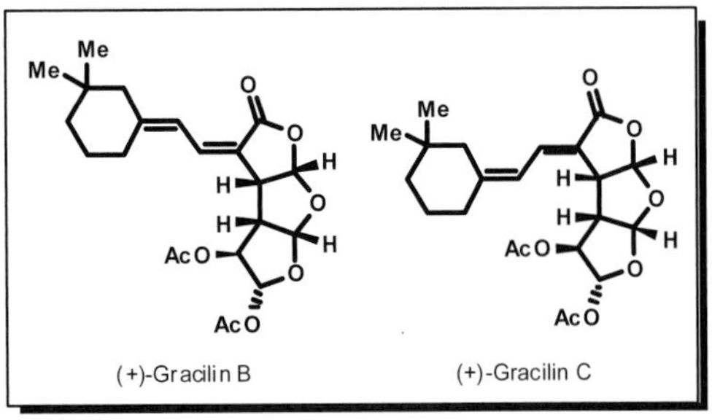

(+)-Gracilin B   (+)-Gracilin C

The gracilins are a group of structurally related bis-nor diterpenes in which both the B and C rings of a tricyclic 6,6,6-fused-ring precursor have been cleaved in biosynthesis. Gracilins B and C were originally isolated from the Mediterrenean sea sponge *Spongionella gracilis*.[1] These natural products are not only structurally interesting and novel, but also noteworthy for their antitumor activity. Gracilin B has been found to block cell adhesion, which is essential for tumor growth, by the broad-spectrum of integrin proteins.[2] The chemical synthesis of gracilins presents a challenge especially with regard to enantio- and diastereocontrol.[3]

Abbreviated Retrosynthetic Plan

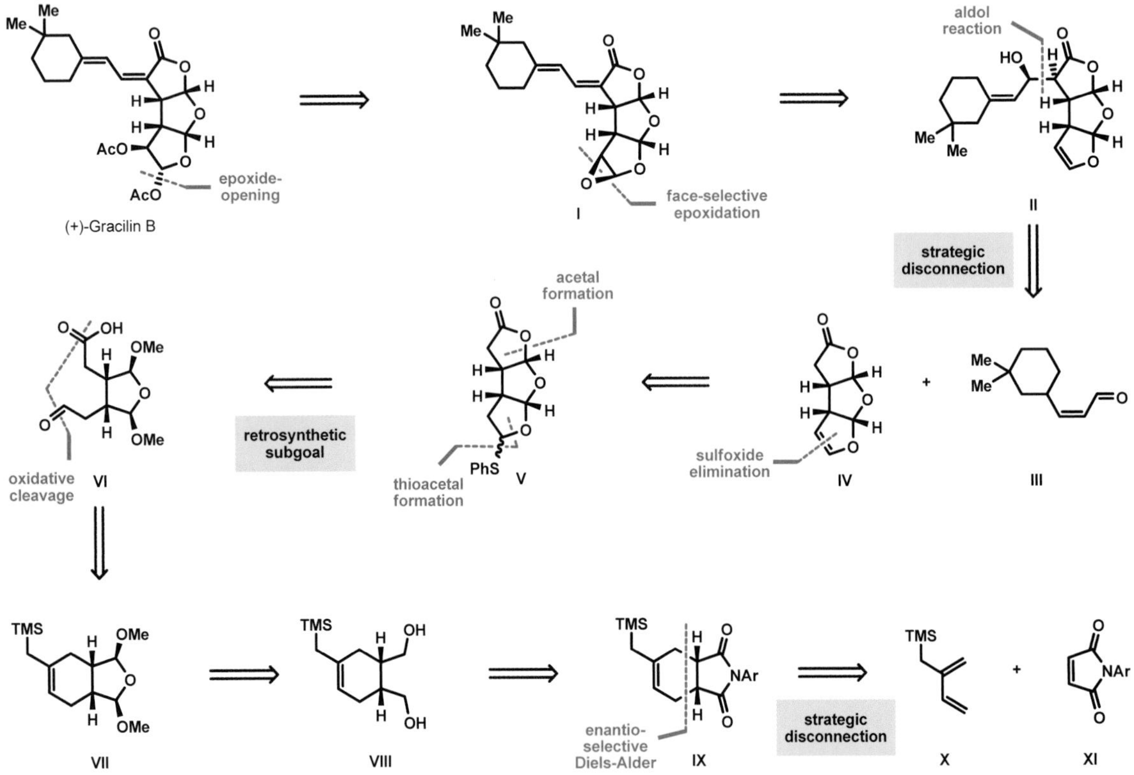

Scheme 1

Key Steps in Scheme 1

(1) The first retrosynthetic steps are required to clear reactive functionality and stereocenters from the terminal tetrahydrofuran ring, and lead to **II** (see Scheme 3). Aldol disconnection simplifies **II** to **III** and **IV**.

(2) The deeper strategy that guides this retrosynthetic analysis is the simplification of **IV** to an intermediate containing a valid Diels-Alder retron and functionality that is suitable for generating the gracilin tricyclic core

and appropriate functionality. That objective was accomplished by the disconnective retrosynthetic sequence: **IV** ⇒ **V** ⇒ **VI** and the further conversion to the key intermediate **IX** via **VII** and **VIII**.

(3) The stereocenters in **IX** can be cleared by a known catalytic enantioselective transform[4] leading to the achiral **X** and **XI** (see Scheme 2).[5]

(+)-GRACILIN B & (+)-GRACILIN C

Pathway of Synthesis

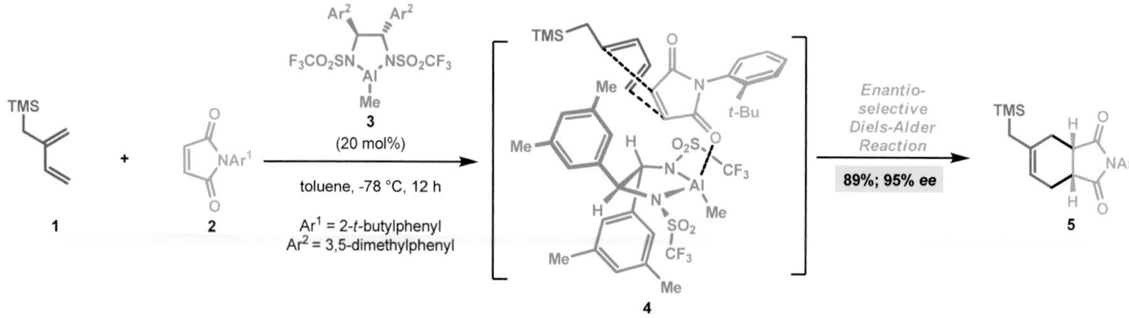

Scheme 2

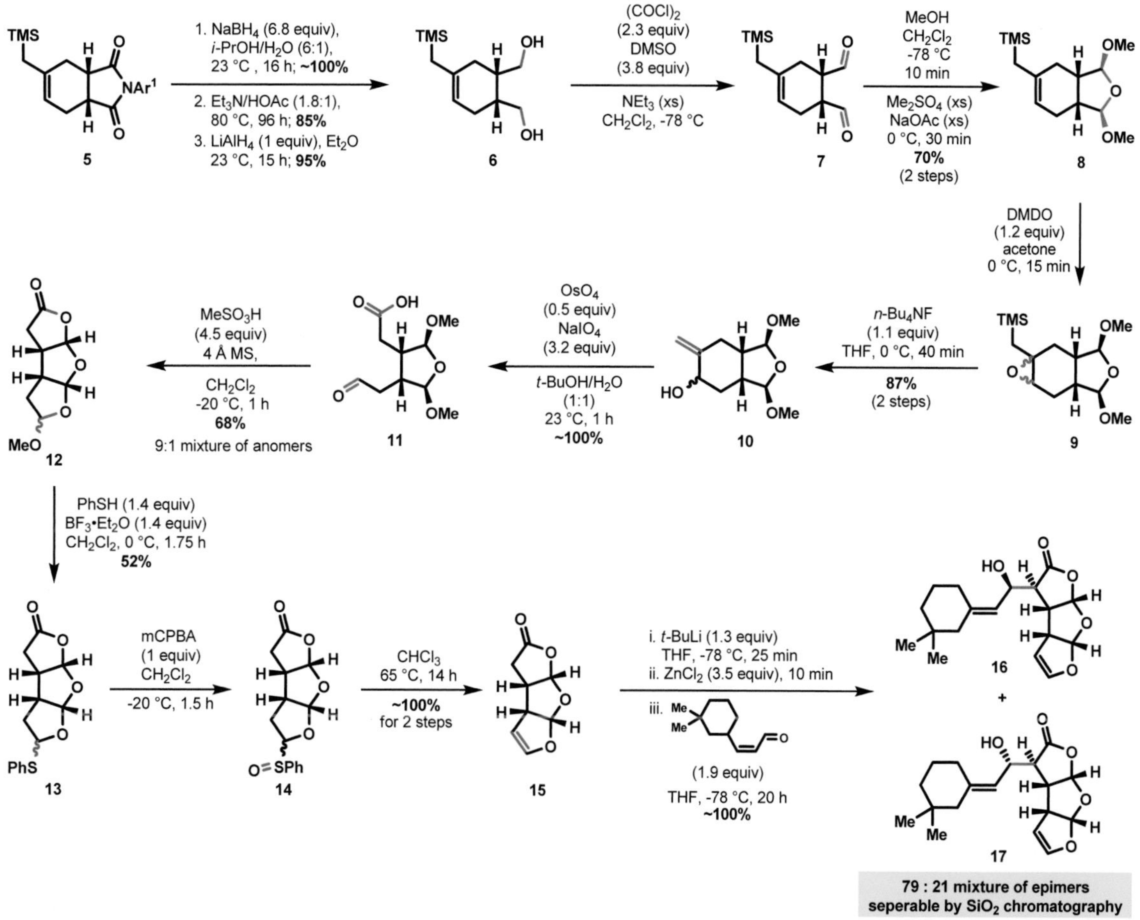

Scheme 3

(+)-GRACILIN B & (+)-GRACILIN C

Pathway of Synthesis (continued)

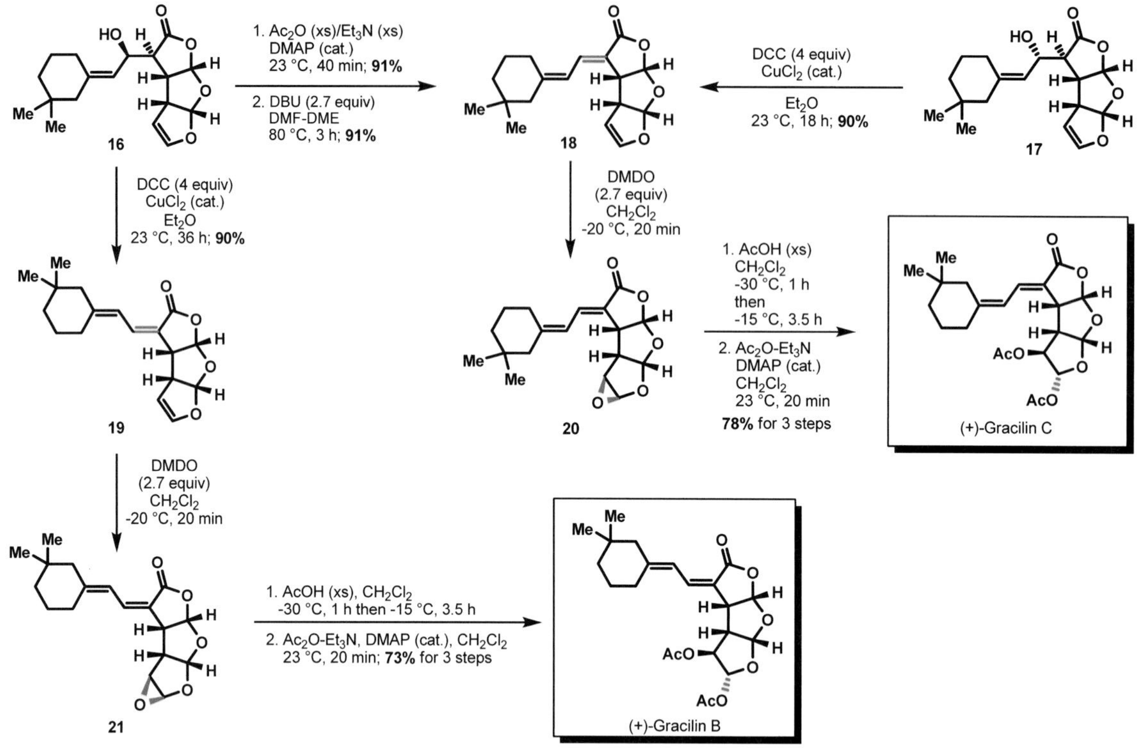

Scheme 4

Key steps in Scheme 4

The aldol product **16** could be converted stereoselectively into either **18** by antiperiplanar elimination or **19** by synperiplanar elimination. The latter process occurs by cycloelimination of an intermediate isourea derivative of **16** formed by Cu(II)-catalyzed addition of the hydroxyl group in **16** to dicyclohexylcarbodiimide (DCC).[6]

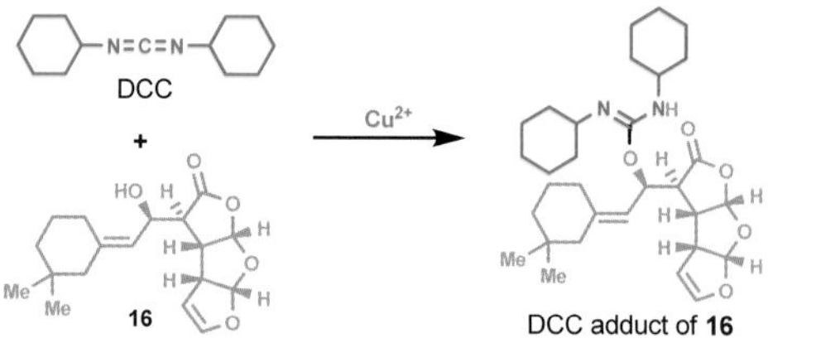

References

1. (a) Mayol, L., Piccialli, V., Sica, D. Minor bisnorditerpenes from the marine sponge Spongionella gracilis and revision of the Delta 6 configuration of gracilin B. *J. Nat. Prod.* **1986**, 49, 823-828; (b) Mayol, L., Piccialli, V., Sica, D. Gracilin A, an unique nor-diterpene metabolite from the marine sponge Spongionella gracilis. *Tetrahedron Lett.* **1985**, 26, 1357-1360; (c) Mayol, L., Piccialli, V., Sica, D. Application of 2D-NMR spectroscopy in the structural determination of gracilin B, a bis-nor-diterpene from the sponge Spongionella gracilis. *Tetrahedron Lett.* **1985**, 26, 1253-1256.

2. Rueda, A., Losada, A., Fernandez, R., Cabanas, C., Garcia-Fernandez, L. F., Reyes, F. & Cuevas, C. Gracilins G-I, cytotoxic bisnorditerpenes from Spongionella pulchella, and the anti-adhesive properties of gracilin B. *Letters in Drug Design & Discovery* **2006**, 3, 753-760.

3. Corey, E. J. & Letavic, M. A. Enantioselective Total Synthesis of Gracilins B and C Using Catalytic Asymmetric Diels-Alder Methodology. *J. Am. Chem. Soc.* **1995**, 117, 9616-9617.

4. (a) Corey, E. J., Sarshar, S., Lee, D.-H. First Example of a Highly Enantioselective Catalytic Diels-Alder Reaction of an Achiral C$_{2v}$-Symmetric Dienophile and an Achiral Diene. *J. Am. Chem. Soc.* **1994**,

116, 12089-12090; (b) Corey, E. J., Lee, D.-H., Sarshar, S. Convenient routes to symmetrical benzils and chiral 1,2-diaryl-1,2-diaminoethanes, useful controllers and probes for enantioselective synthesis. *Tetrahedron: Asymmetry* **1995**, 6, 3-6.

5. (a) Corey, E. J., Sarshar, S., Bordner, J. X-ray crystallographic and NMR studies on the origins of high enantioselectivity in Diels-Alder reactions catalyzed by a chiral diazaaluminolidine. *J. Am. Chem. Soc.* **1992**, 114, 7938-7939; (b) Corey, E. J., Imwinkelried, R., Pikul, S., Xiang, Y. B. Practical enantioselective Diels-Alder and aldol reactions using a new chiral controller system. *J. Am. Chem. Soc.* **1989**, 111, 5493-5495; (c) Corey, E. J., Imai, N., Pikul, S. Catalytic enantioselective synthesis of a key intermediate for the synthesis of prostanoids. *Tetrahedron Lett.* **1991**, 32, 7517-7520.

6. Corey, E. J., Andersen, N. H., Carlson, R. M., Paust, J., Vedejs, E., Vlattas, I. & Winter, R. E. K. Total synthesis of prostaglandins. Synthesis of the pure dl-E$_1$, -F$_{1\alpha}$, -F$_{1b}$, -A$_1$, and -B$_1$ hormones. *J. Am. Chem. Soc.* **1968**, 90, 3245-3247.

(+)-NICANDRENONE 1

Background

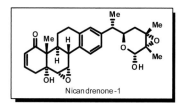

Nicandrenone-1

The nicandrenone (NIC) family of structurally complex, steroid-derived natural products includes the active principals of *Nicandra physaloides* (the Pyruvian "shoofly" plant) which give rise to its insect repellent and antifeedant properties.[1] The novel structures of the nicandrenones were elucidated independently by two groups in the 1970s.[2] The NIC family is structurally related to another even larger class of plant products, the withanolides.[3] The synthesis of nicandrenones requires not only the use of several modern synthetic methods for enantio- and diastereocontrol but also for C-C linkage and functional group manipulation.[4]

Abbreviated Retrosynthetic Plan

Scheme 1

Key Steps in Scheme 1

(1) The retrosynthetic simplification of the nicandrenone structure was mainly driven by two long-range goals:

  (a) disconnection of ring B using the Diels-Alder transform (**VI** ⇒ **VII**) and

  (b) disconnection of the appendage on ring D of **I** to simpler fragments **II** and **III**.

The enablement of these goals required the application of several less-simplifying transforms and the use of certain control elements.

(2) The introduction of the trialkylsilyl group (R₃Si) and the stereocenter to which it is attached in **I** served to convert the α,β-enone subunit of ring A to a saturated equivalent and also to provide a means of stereocontrol.

(3) The structural modifications made during the sequence **II** ⇒ **IV** ⇒ **V** ⇒ **VI** were required to generate the Diels-Alder retron in **VI**.

(+)-NICANDRENONE 1

Pathway of Synthesis

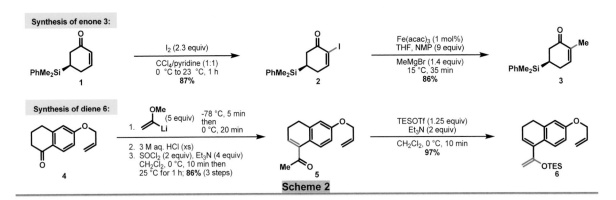

Synthesis of enone 3:

1 → I₂ (2.3 equiv), CCl₄/pyridine (1:1), 0 °C to 23 °C, 1 h, **87%** → 2 → Fe(acac)₃ (1 mol%), THF, NMP (9 equiv), MeMgBr (1.4 equiv), 15 °C, 35 min, **86%** → 3

Synthesis of diene 6:

4 → 1. OMe (5 equiv) Li, -78 °C, 5 min then 0 °C, 20 min; 2. 3 M aq. HCl (xs); 3. SOCl₂ (2 equiv), Et₃N (4 equiv), CH₂Cl₂, 0 °C, 10 min then 25 °C for 1 h; **86%** (3 steps) → 5 → TESOTf (1.25 equiv), Et₃N (2 equiv), CH₂Cl₂, 0 °C, 10 min, **97%** → 6

Scheme 2

3 + 6 → MeAlCl₂ (1.05 equiv), CH₂Cl₂, −78 °C, 2.5 h, **85%**, R = SiPhMe₂ → 7 *exo* TS → 8 ***exo-endo* selectivity > 15:1** | 9 disfavored *endo* TS

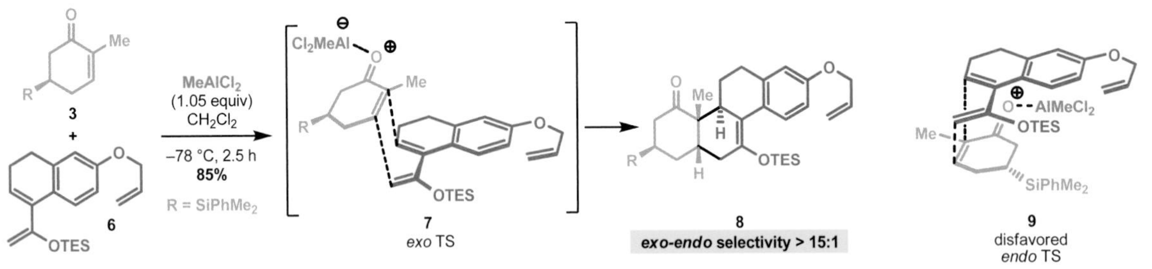

Scheme 3

Explanation of Schemes 2 and 3

The *exo*-Diels-Alder pathway 6→7→8→9 is favored over the normally preferred *endo* mode of reaction because of adverse steric repulsions in the latter that originate because the diene subunit is somewhat skewed rather than flat.

A detailed discussion of this interesting pathway has been presented.[5]

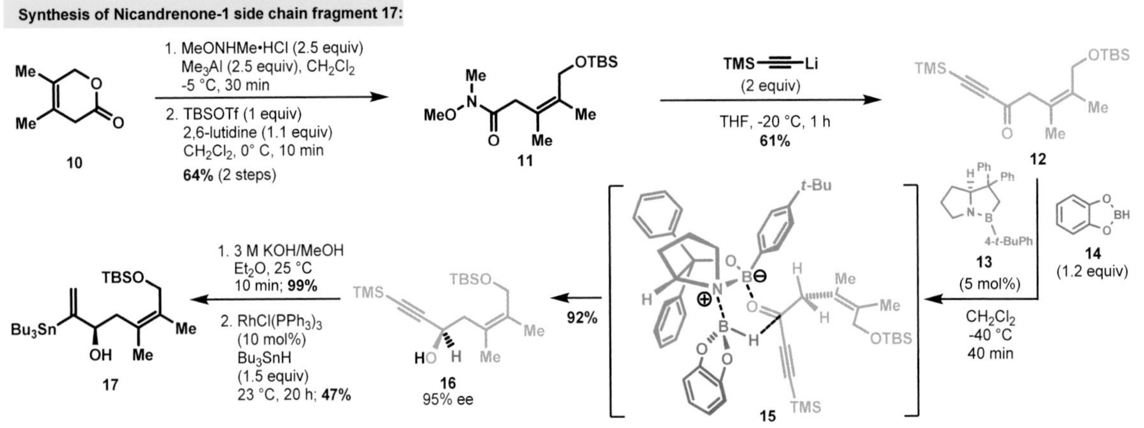

Synthesis of Nicandrenone-1 side chain fragment 17:

10 → 1. MeONHMe•HCl (2.5 equiv), Me₃Al (2.5 equiv), CH₂Cl₂, -5 °C, 30 min; 2. TBSOTf (1 equiv), 2,6-lutidine (1.1 equiv), CH₂Cl₂, 0° C, 10 min, **64%** (2 steps) → 11 → TMS≡Li (2 equiv), THF, -20 °C, 1 h, **61%** → 12

13 (5 mol%) + 14 (1.2 equiv) → CH₂Cl₂, -40 °C, 40 min → 15 → **92%** → 16 95% ee → 1. 3 M KOH/MeOH, Et₂O, 25 °C, 10 min; 99%; 2. RhCl(PPh₃)₃ (10 mol%), Bu₃SnH (1.5 equiv), 23 °C, 20 h; **47%** → 17

Scheme 4

(+)-NICANDRENONE 1

Pathway of Synthesis (continued)

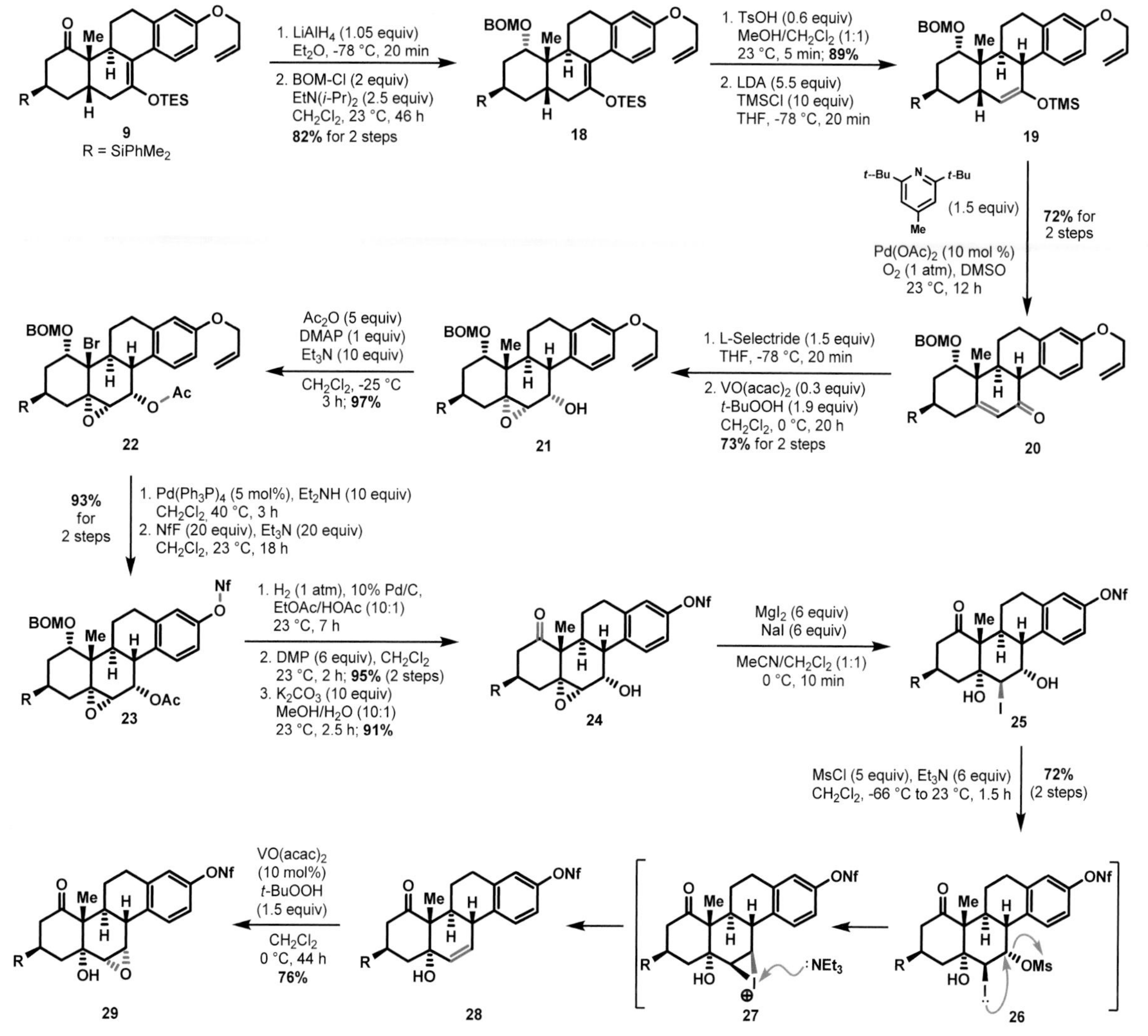

Scheme 5

Key steps in Scheme 5

(1) Reduction of **9** and protection of the resulting alcohol was required to remove interference in the subsequent steps **18→19** and **20→21**.

(2) The use of DMSO accelerated the Pd(II)-catalyzed oxidation of enol ether **19** to enone **20**.[6]

(3) The application of the Wharton reaction[7] to the 5,6-epoxide of **20** did not provide the required intermediate **28a**, and so a longer sequence had to be employed.

Attempted conversion of epoxyketone using the Wharton reaction:

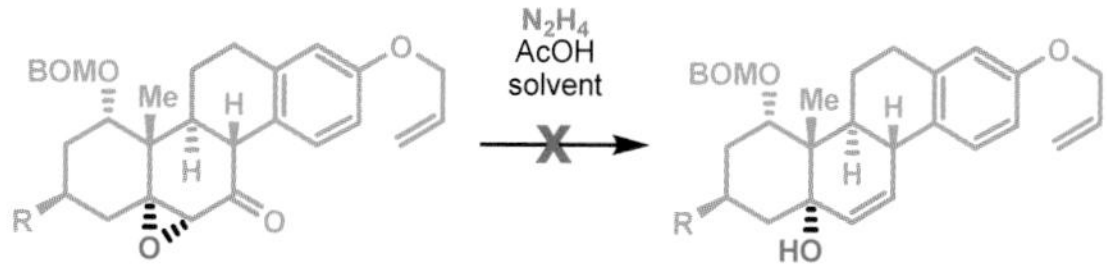

α,β-epoxyketone derived from **20**

28a
(not observed)

(+)-NICANDRENONE 1

Pathway of Synthesis (continued)

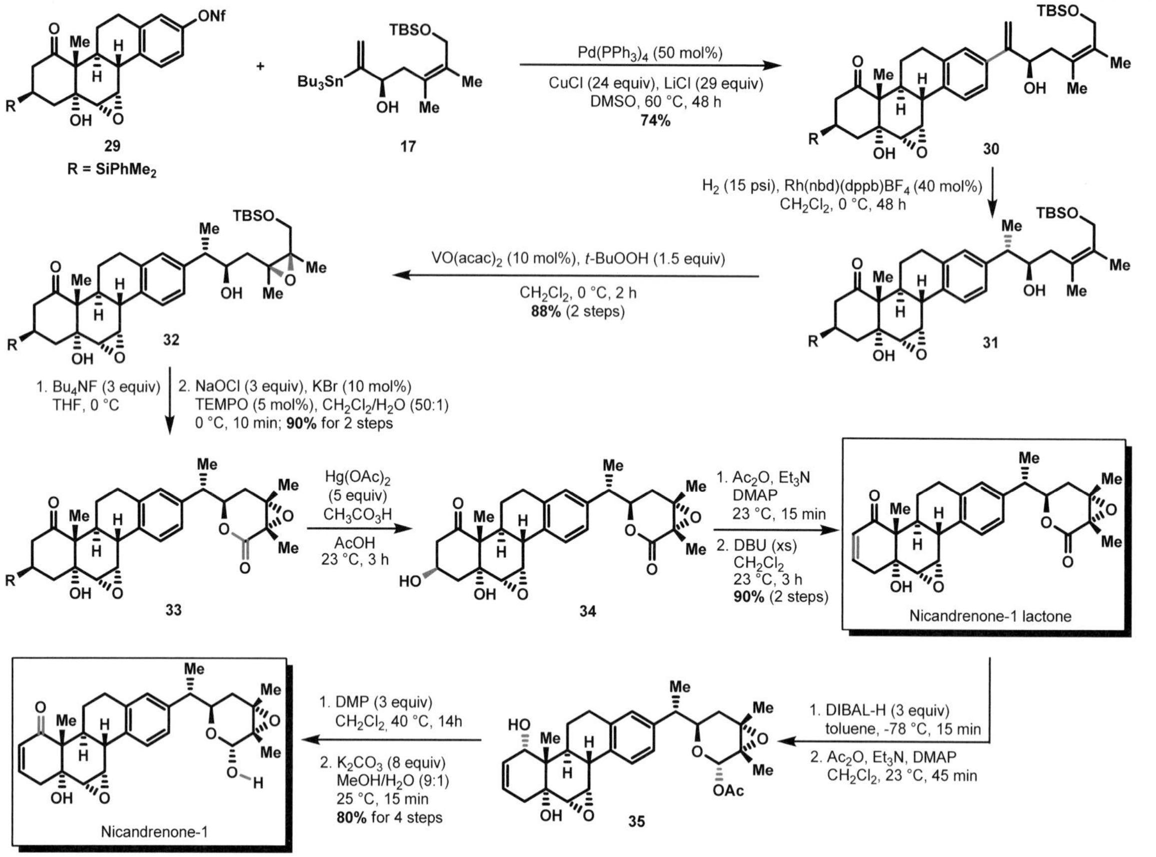

Reagents and conditions for the scheme:

29, R = SiPhMe$_2$ + 17 → Pd(PPh$_3$)$_4$ (50 mol%), CuCl (24 equiv), LiCl (29 equiv), DMSO, 60 °C, 48 h, **74%** → 30

30 → H$_2$ (15 psi), Rh(nbd)(dppb)BF$_4$ (40 mol%), CH$_2$Cl$_2$, 0 °C, 48 h → 31

31 → VO(acac)$_2$ (10 mol%), t-BuOOH (1.5 equiv), CH$_2$Cl$_2$, 0 °C, 2 h, **88%** (2 steps) → 32

32 → 1. Bu$_4$NF (3 equiv), THF, 0 °C 2. NaOCl (3 equiv), KBr (10 mol%), TEMPO (5 mol%), CH$_2$Cl$_2$/H$_2$O (50:1), 0 °C, 10 min; **90%** for 2 steps → 33

33 → Hg(OAc)$_2$ (5 equiv), CH$_3$CO$_3$H, AcOH, 23 °C, 3 h → 34

34 → 1. Ac$_2$O, Et$_3$N, DMAP, 23 °C, 15 min 2. DBU (xs), CH$_2$Cl$_2$, 23 °C, 3 h, **90%** (2 steps) → Nicandrenone-1 lactone

Nicandrenone-1 lactone → 1. DIBAL-H (3 equiv), toluene, −78 °C, 15 min 2. Ac$_2$O, Et$_3$N, DMAP, CH$_2$Cl$_2$, 23 °C, 45 min → 35

35 → 1. DMP (3 equiv), CH$_2$Cl$_2$, 40 °C, 14h 2. K$_2$CO$_3$ (8 equiv), MeOH/H$_2$O (9:1), 25 °C, 15 min, **80%** for 4 steps → Nicandrenone-1

Scheme 6

Key steps in Schemes 6

(1) The coupling of **17** and **29** under standard conditions was not successful and consequently a copper-accelerated modification was developed.[8]

(2) The diastereoselective hydrogenation **30**→**31** using a cationic Rh(I) catalyst is facilitated by the allylic hydroxyl group which also influences the face selectivity.

References

1. Nalbandov, O., Yamamoto, R. T. & Fraenkel, G. S. Nicandrenone, a new compound with insecticidal properties, isolated from Nicandra physalodes. *J. Agric. Food Chem.* **1964**, 12, 55-59.
2. (a) Bates, R. B., Eckert, D. J. Nicandrenone, an insecticidal plant steroid derivative with ring D aromatic. *J. Am. Chem. Soc.* **1972**, 94, 8258-8260; (b) Bates, R. B., Morehead, S. R. Structure of Nic-2, a major steroidal constituent of the insect repellent plant Nicandra physaloides. *J. Chem. Soc., Chem. Commun.* **1974**, 125-126; (c) Begley, M. J., Crombie, L., Ham, P. J., Whiting, D. A. Constitution of four novel methyl steroid relatives (ring-D aromatic) from the insect repellent plant Nicandra physaloides. X-ray analysis of Nic-10. *J. Chem. Soc., Chem. Commun.* **1972**, 1250-1251.
3. (a) Ray, A. B. Chemistry of Datura withanolides, a brief review. *J. Indian Chem. Soc.* **1998**, 75, 672-678; (b) Kirson, I., Glotter, E. Recent developments in naturally occurring ergostane-type steroids. A review. *J. Nat. Prod.* **1981**, 44, 633-647.
4. Stoltz, B. M., Kano, T. & Corey, E. J. Enantioselective Total Synthesis of Nicandrenones. *J. Am. Chem. Soc.* **2000**, 122, 9044-9045.
5. Ge, M., Stoltz, B. M. & Corey, E. J. Mechanistic Insights into the Factors Determining Exo-Endo Selectivity in the Lewis Acid-Catalyzed Diels-Alder Reaction of 1,3-Dienes with 2-Cycloalkenones. *Org. Lett.* **2000**, 2, 1927-1929.
6. (a) Kürti, L., Czakó, B. Saegusa Oxidation in *Strategic Applications of Named Reactions in Organic Synthesis* **2005**, pp 390-391 (Academic Press/Elsevier); (b) Larock, R. C., Hightower, T. R., Kraus, G. A., Hahn, P., Zheng, D. A simple effective new palladium-catalyzed conversion of enol silanes to enones and enals. *Tetrahedron Lett.* **1995**, 36, 2423-2426.
7. Kürti, L. & Czakó, B. Wharton Olefin Synthesis (Wharton Transposition) in *Strategic Applications of Named Reactions in Organic Synthesis* 482-483 (Academic Press/Elsevier, San Diego, **2005**).
8. Han, X., Stoltz, B. M. & Corey, E. J. Cuprous Chloride Accelerated Stille Reactions. A General and Effective Coupling System for Sterically Congested Substrates and for Enantioselective Synthesis. *J. Am. Chem. Soc.* **1999**, 121, 7600-7605.

(+)-OKARAMINE N

Background

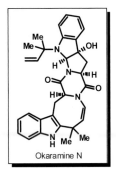

Okaramine N

The first synthesis of okaramine N, one of a family of polycyclic fungal natural products,[1] was demonstrated using tryptophan as a building block and Pd(II)-mediated ring-closure to form the 8-membered azocine subunit.[2]

Abbreviated Retrosynthetic Plan

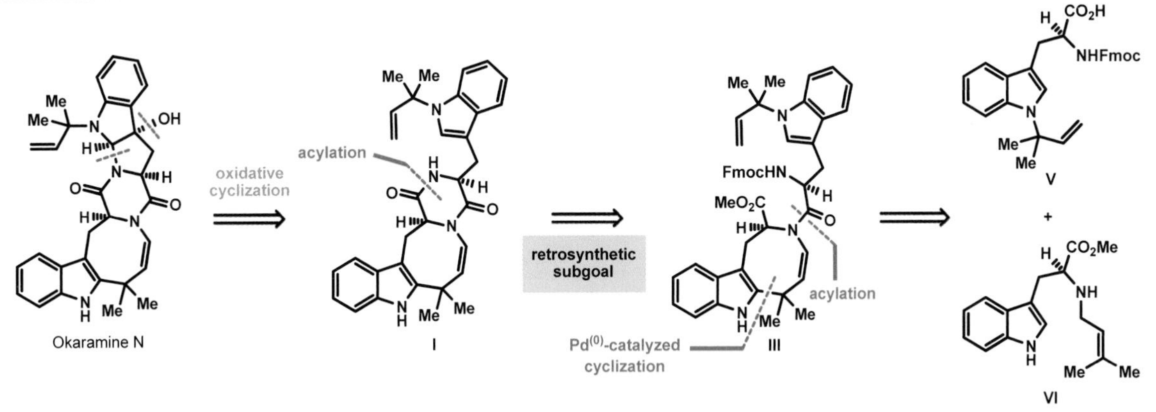

Okaramine N ⟹ (oxidative cyclization) I ⟹ (acylation, retrosynthetic subgoal) III (Pd(0)-catalyzed cyclization) ⟹ (acylation) V + VI

Scheme 1

Pathway of Synthesis

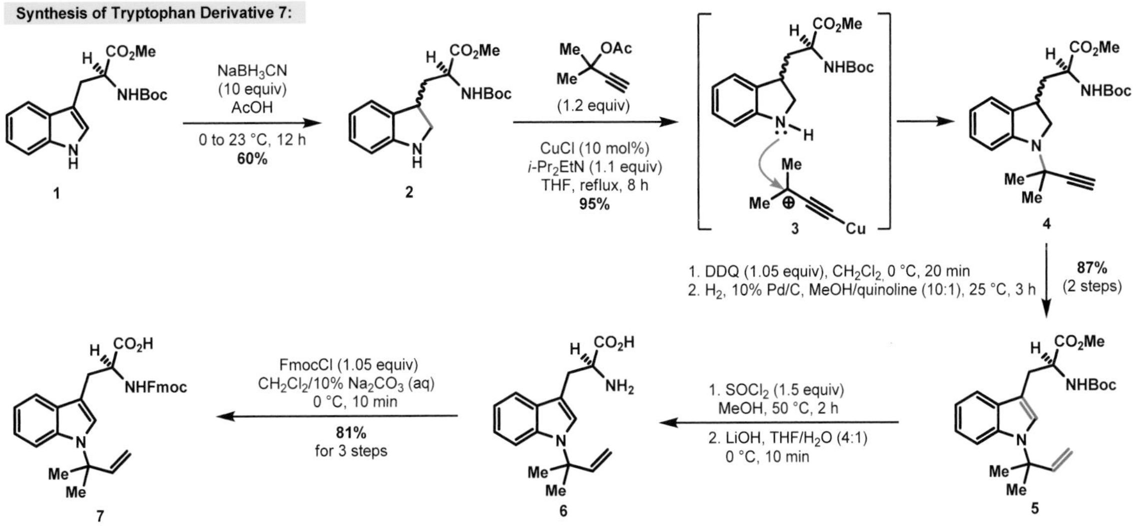

Synthesis of Tryptophan Derivative 7:

1 → NaBH₃CN (10 equiv) AcOH, 0 to 23 °C, 12 h, 60% → **2** → Me OAc / Me (1.2 equiv), CuCl (10 mol%), i-Pr₂EtN (1.1 equiv), THF, reflux, 8 h, 95% → [**3**] → **4**

4 → 1. DDQ (1.05 equiv), CH₂Cl₂, 0 °C, 20 min; 2. H₂, 10% Pd/C, MeOH/quinoline (10:1), 25 °C, 3 h, 87% (2 steps) → **5**

5 → 1. SOCl₂ (1.5 equiv), MeOH, 50 °C, 2 h; 2. LiOH, THF/H₂O (4:1), 0 °C, 10 min → **6** → FmocCl (1.05 equiv), CH₂Cl₂/10% Na₂CO₃ (aq), 0 °C, 10 min, 81% for 3 steps → **7**

Scheme 2

(+)-OKARAMINE N

Pathway of Synthesis (continued)

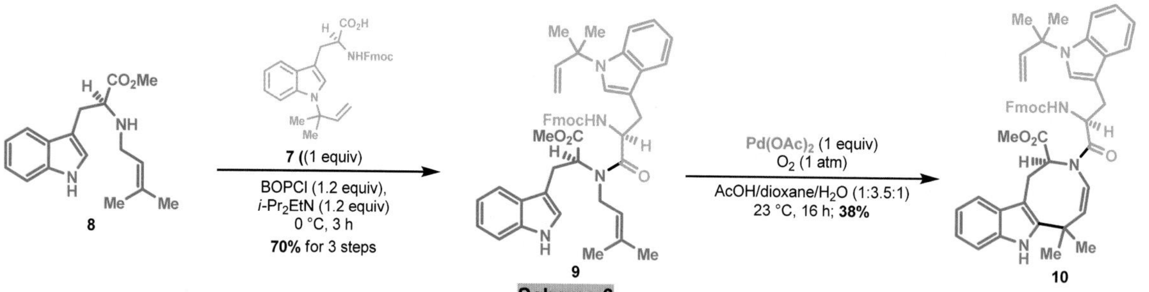

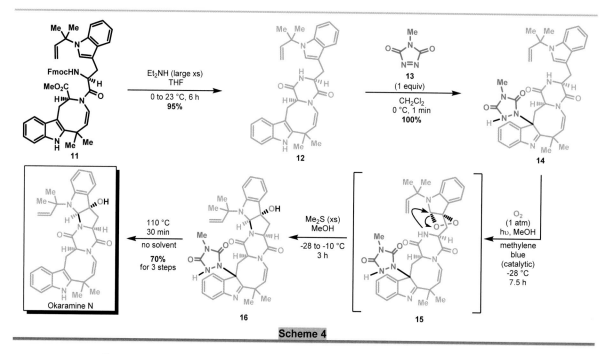

Scheme 3

Key steps in Schemes 2 and 3

(1) The tertiary prenyl group was attached to the nitrogen of **2** by use of the Cu-stabilized electrophilic reagent **3** (Scheme 2).[3]

(2) The Pd(II)-mediated ring-closure **9**→**10** was first utilized for the synthesis of austamide (see page 205).

Scheme 4

Key steps in Scheme 4

(1) Because of the susceptibility of the indole ring to photooxidation (attack by singlet oxygen), it had to be protected temporarily by addition of *N*-methyltriazolinedione **13** (ene-type reaction).

(2) The triazolinedione adduct **16** underwent a reverse ene process upon heating to form okaramine N.

References

1. Shiono, Y., Akiyama, K. & Hayashi, H. Okaramines N, O, P, Q and R, new okaramine congeners, from Penicillium simplicissimum ATCC 90288. *Biosci., Biotechnol., Biochem.* **2000**, 64, 103-110.
2. Baran, P. S., Guerrero, C. A. & Corey, E. J. Short, Enantioselective Total Synthesis of Okaramine N. *J. Am. Chem. Soc.* **2003**, 125, 5628-5629.
3. Hennion, G. F. & Hanzel, R. S. Alkylation of amines with tertiary, acetylenic chlorides. Preparation of sterically hindered amines. *J. Am. Chem. Soc.* **1960**, 82, 4908-4912.

(+)-AUSTAMIDE

Background

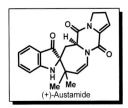

(+)-Austamide

Austamide is a fungal natural product with potent antibiotic, antihelminthic and insecticidal activity.[1] It is also a member of a structural class of natural products which consist of modified amino acid and prenyl subgroups. Kishi reported the first synthesis of racemic austamide in 29 steps.[2] A shorter route to the natural form of austamide is outlined in this section.[3]

Abbreviated Retrosynthetic Plan

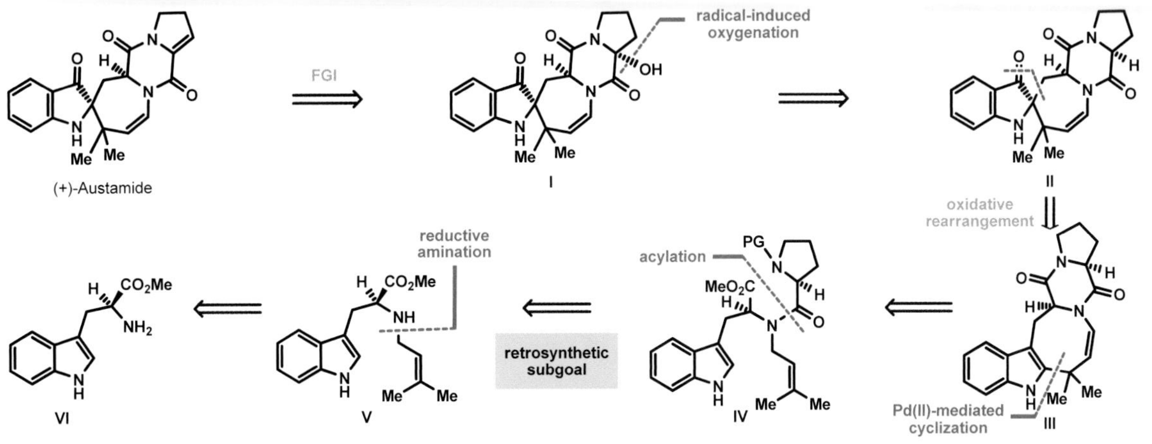

Scheme 1

Key steps in Scheme 1

(1) A key challenge in designing a concise synthesis of the natural enantiomer of austamide was to find a way to control absolute configuration through the use of the amino acids that serve as nature's building blocks as S-goals (structural subgoals) for retrosynthetic analysis. (S)-Proline and (S)-tryptophan were identified as such precursors.

(2) The stereocenter of (S)-proline which has been lost in austamide was introduced retrosynthetically by the sequence: austamide ⇒ **I** ⇒ **II**.

(3) The tryptophan subunit was generated retrosynthetically by an oxidative rearrangement transform.

(4) Successive disconnection of the 8-membered ring of **III** and an amide linkage gave **IV** which was further disconnected to **V** and **VI**.

Pathway of Synthesis

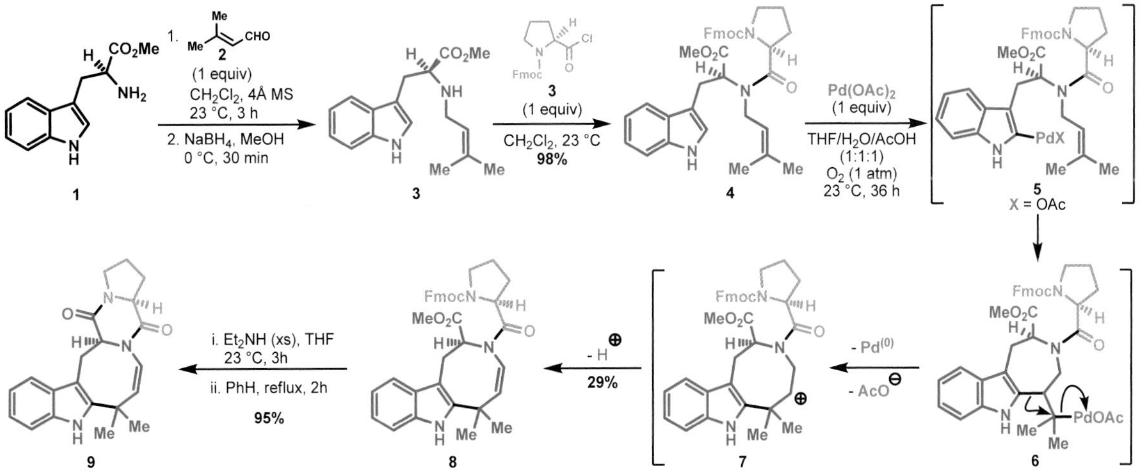

Scheme 2

205

(+)-AUSTAMIDE

Pathway of Synthesis (continued)

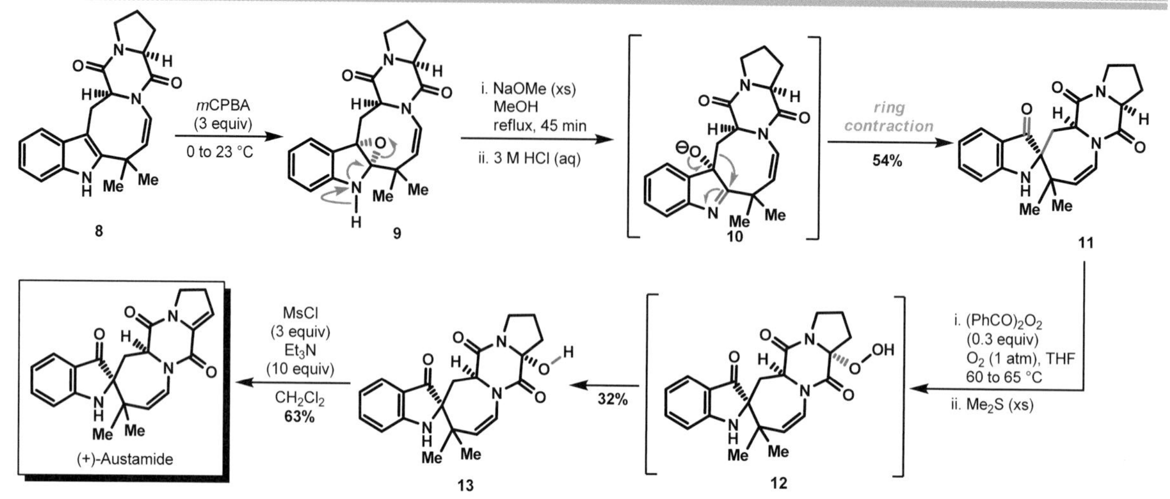

Scheme 3

Key steps in Schemes 2 and 3

(1) The 1-step conversion of **4** to **8** is thought to be the result of the sequence **4→5→6**, each step of which is precedented. The acetoxypalladation **4→5** is analogous to the known acetoxymercuration of electron-rich π-systems. The transformation **5→6** could occur either by direct Heck reaction or by Heck cyclization to form a 7-membered ring followed by carbocationic rearrangement affording 7-membered→8-membered ring-expansion.

(2) The cationic rearrangement of annulated epoxy indoles to spiro structures has long been known.[4]

Pathway to (+)-Deoxyaustamide

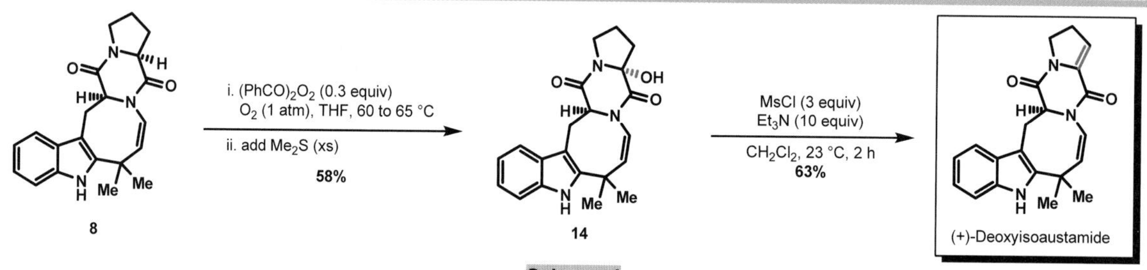

Scheme 4

Key steps in Scheme 4

An oxygenation-dehydration sequence was also applied to the synthesis of the related natural product (+)-deoxyaustamide. The process consisted of $PhCO_2 \cdot$ radical-initiated α-peroxidation of **8**, *in situ* reduction of the intermediate hydroperoxide to **14** and dehydration.

References

1. (a) Steyn, P. S., Vleggaar, R. 12,13-Dihydro-12-hydroxyaustamide, a new dioxopiperazine from Aspergillus ustus. *Phytochemistry* **1976**, 15, 355-356; (b) Steyn, P. S. Structure of five dioxopiperazines from Aspergillus ustus. *Tetrahedron* **1973**, 29, 107-120; (c) Steyn, P. S. Austamide, a new toxic metabolite from Aspergillus custus. *Tetrahedron Lett.* **1971**, 3331-3334.

2. Hutchison, A. J., Kishi, Y. Stereospecific total synthesis of dl-austamide. *J. Am. Chem. Soc.* **1979**, 101, 6786-6788.

3. Baran, P. S., Corey, E. J. A Short Synthetic Route to (+)-Austamide, (+)-Deoxyisoaustamide, and (+)-Hydratoaustamide from a Common Precursor by a Novel Palladium-Mediated Indole -> Dihydroindoloazocine Cyclization. *J. Am. Chem. Soc.* **2002**, 124, 7904-7905.

4. Witkop, B., Patrick, J. B. Gelsemine. II. The chemistry and rearrangements of spiroöxindoles. *J. Am. Chem. Soc.* **1953**, 75, 2572-2576.

(−)-ASPIDOPHYTINE

Background

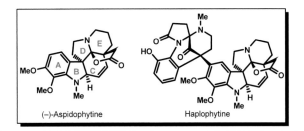

(−)-Aspidophytine

Haplophytine

Aspidophytine is a component of the plant *Haplophyton cimicidium* along with haplophytine, a dimeric indole alkaloid of which it constitutes a major subunit. The powder prepared from the dried leaves of this plant has been used in Mexico and Central America as an insecticide at least since the Aztec era.[1] The structures of these very potent agents were determined in the 1960s.[2] The first synthesis of aspidophytine was demonstrated in 1999.[3]

Abbreviated Retrosynthetic Plan

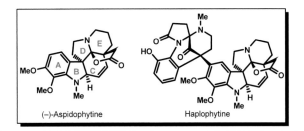

Scheme 1

Key steps in Scheme 1

(1) The retrosynthetic conversion of aspidophytine to **I**, **II** and **III** served the purpose of enabling a subsequent disconnection of the **C** and **D** rings (**III** ⇒ **IV** ⇒ **V**).

(2) The intermediate **V** can be disconnected directly to **VI** and **VII**, producing a profound molecular simplification.

(3) The initiating stereocenter for the synthesis which controls the formation of all the stereocenters in

aspidophytine, was cleared retrosynthetically by the enantiocontrolled pathway **VII** ⇒ **XI** ⇒ **XII** ⇒ **XIII**. The generation of this pathway was guided by two goals:

(a) generating the initiating stereocenter by enantioselective reduction of an achiral cyclopentanone (**XIII**) and

(b) using a [3,3]-sigmatropic rearrangement to transfer that chiral information to the quaternary stereocenter of **VII**.

(−)-ASPIDOPHYTINE

Pathway of Synthesis

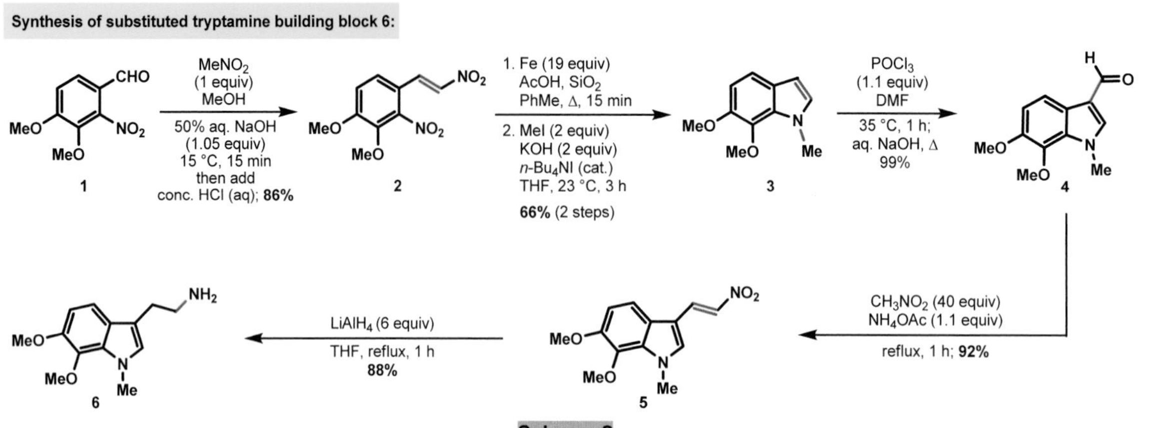

Synthesis of substituted tryptamine building block 6:

Scheme 2

Synthesis of dialdehyde building block 18:

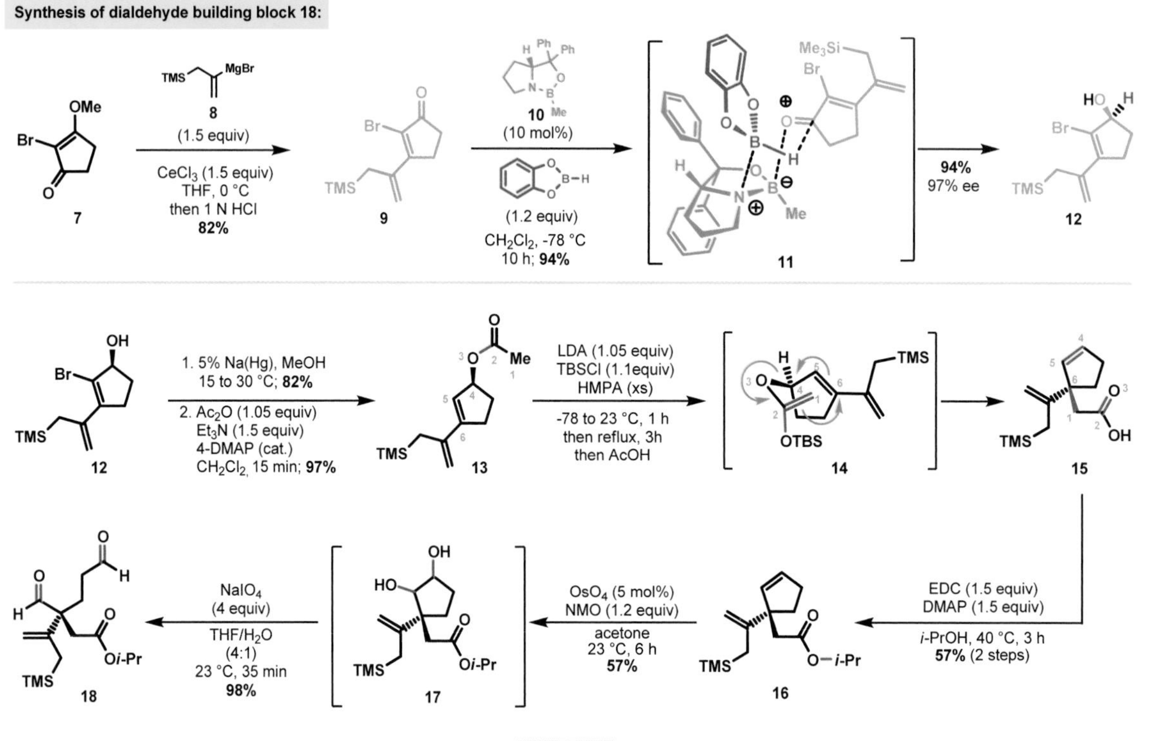

Scheme 3

Key steps in Schemes 2 and 3

(1) The α-bromine substituent in **9** serves to increase the enantioselectivity of the chiral oxazaborolidine-catalyzed reduction step, **9**→**12** (via **11**). This control element was then removed (**12**→**13**).

(2) The quaternary stereocenter was generated by a suprafacial [3,3]-sigmatropic rearrangement (Ireland-Claisen) **13**→**15**.[4]

(−)-ASPIDOPHYTINE

Pathway of Synthesis (continued)

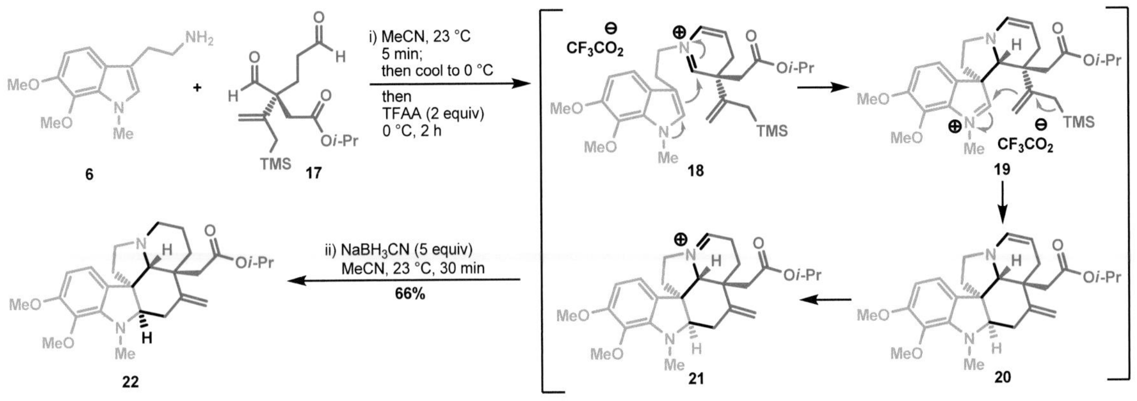

Scheme 4

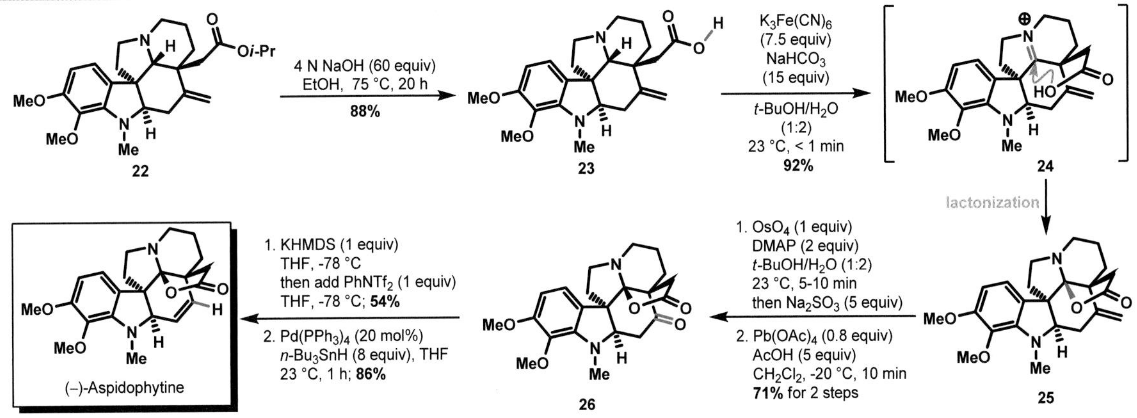

Scheme 5

Key steps in Schemes 4 and 5

(1) The allylic TMS group is required to facilitate the cyclization **19**→**20** (Scheme 4).[5]

(2) The exo-methylene group in **25** was very resistant to oxidation. 4-Dimethylaminopyridine greatly accelerated the reaction of **25** with OsO_4.[6]

References

1. (a) Crosby, D.G. in *Naturally Occuring Insecticides* (eds. Jacobson, M. & Crosby, D. G.),**1991**, pp 213 (Marcel Dekker, New York); (b) Sukh Dev; Koul, O. in *Insecticides of Natural Origin* **1997**, pp 250-251 (Harwood Academic Publishers, Amsterdam).
2. (a) Rae, I. D., et al. Haplophytine. *J. Am. Chem. Soc.* **1967**, 89, 3061-3062; (b) Yates, P., MacLachlan, F. N., Rae, I. D., Rosenberger, M., Szabo, A. G., Willis, C. R., Cava, M. P., Behforouz, M., Lakshmikantham, M. V., Zeiger, W. Haplophytine. Novel type of indole alkaloid. *J. Am. Chem. Soc.* **1973**, 95, 7842-7850.
3. He, F., Bo, Y., Altom, J. D. & Corey, E. J. Enantioselective Total Synthesis of Aspidophytine. *J. Am. Chem. Soc.* **1999**, 121, 6771-6772.
4. Kürti, L. & Czakó, B. Ireland-Claisen Rearrangement in *Strategic Applications of Named Reactions in Organic Synthesis* 90-91 (Academic Press/Elsevier, San Diego, **2005**).
5. A number of other syntheses of aspidophytine have been reported: (a) Sumi, S., Matsumoto, K., Tokuyama, H., Fukuyama, T. Enantioselective Total Synthesis of Aspidophytine. *Org. Lett.* **2003**, 5, 1891-1893; (b)

Mejia-Oneto, J. M., Padwa, A. Application of the Rh(II) Cyclization/Cycloaddition Cascade for the Total Synthesis of (±)-Aspidophytine. *Org. Lett.* **2006**, 8, 3275-3278; (c) Marino, J. P., Cao, G. Total synthesis of aspidophytine. *Tetrahedron Lett.* **2006**, 47, 7711-7713; (d) Mejia-Oneto, J. M., Padwa, A. Total synthesis of the alkaloid (±)-aspidophytine based on carbonyl ylide cycloaddition chemistry. *Helv. Chim. Acta* **2008**, 91, 285-302; (e) Nicolaou, K. C., Dalby, S. M., Majumder, U. A Concise Asymmetric Total Synthesis of Aspidophytine. *J. Am. Chem. Soc.* **2008**, 130, 14942-14943.
6. For subsequent studies of possible methods for the synthesis of halophytine from aspidophytine, see: (a) Rege, P. D., Tian, Y., Corey, E. J. Studies of New Indole Alkaloid Coupling Methods for the Synthesis of Haplophytine. *Org. Lett.* **2006**, 8, 3117-3120; (b) Corey, E. J., Tian, Y. Selective 4-arylation of pyridines by a nonmetalloorganic process. *Org. Lett.* **2005**, 7, 5535-5537.

DOLABELLANES (Part I)

Background

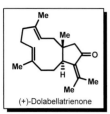

(+)-Dolabellatrienone

The dolabellanes are a large family of naturally occurring diterpenoids that share a common nucleus of fused five- and eleven-membered rings.[1] Although dolabellanes have been isolated from marine sources, they also occur in some terrestrial microorganisms. Dolabellatrienone is one of the more widely distributed members of this class.[2] The first enantioselective synthesis of dolabellatrienone or its enantiomer from a common achiral 15-membered macrocyclic lactone intermediate served also to revise the absolute configuration previously assigned to the dolabellane class.[3] The success of the synthesis depended on a recently developed enantioselective version of the Ireland-Claisen rearrangement.[4]

Abbreviated Retrosynthetic Plan

Scheme 1

Key steps in Scheme 1

(1) The retrosynthetic analysis was driven primarily by the stereochemical strategy of creating the two vicinal stereocenters of dolabellatrienone with enantiocontrol by Ireland-Claisen rearrangement of the achiral macrocyclic lactone **IV**.

(2) The retron for the rearrangement was put in place by disconnection of the five-membered ring of dolabellatrienone to the unsaturated acid **I** and subsequent appendage modification to generate the key intermediate **II**.

(3) The enantioselective version of the Ireland-Claisen rearrangement that depended on the intermediate vinyloxyborane **III** which derives from a chiral boron controller based on 1,2-diphenyl-1,2-diaminoethane.

DOLABELLANES (Part I)

Pathway of Synthesis

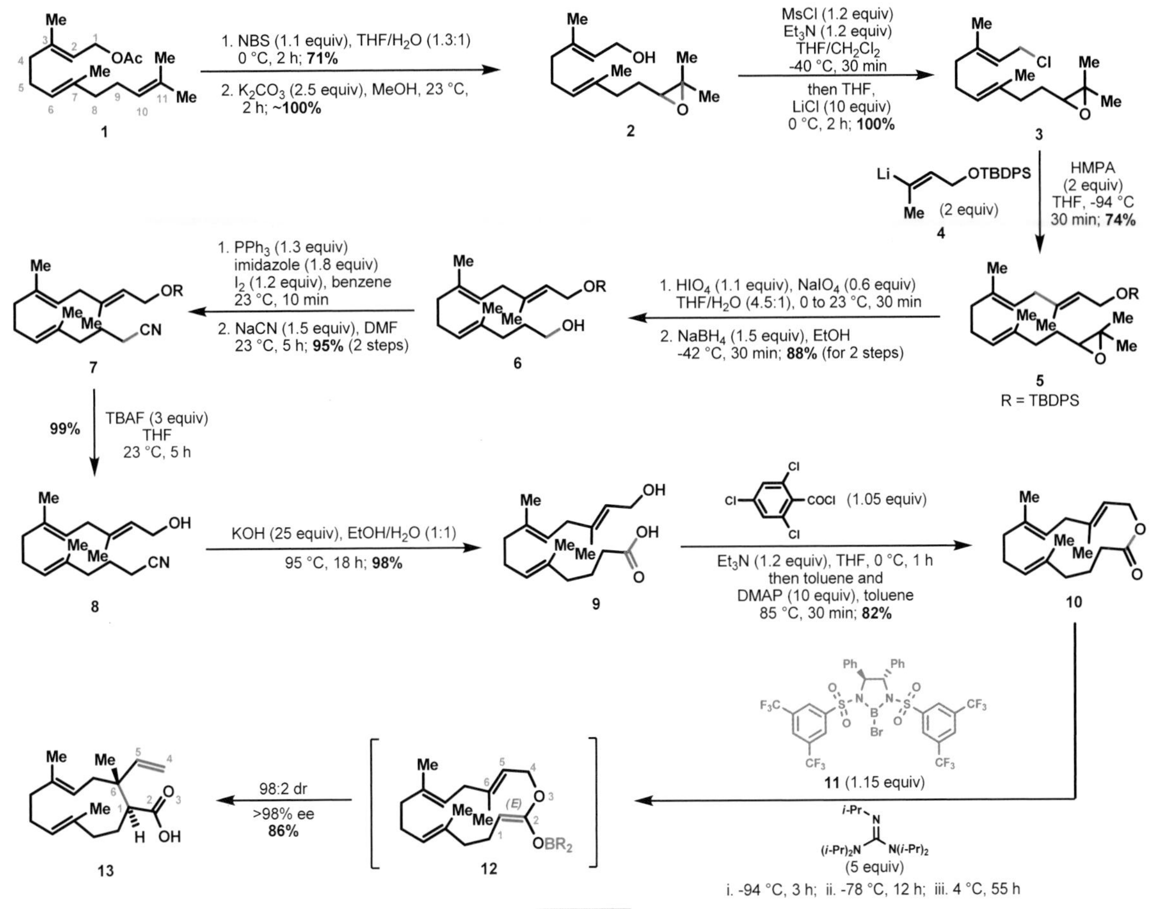

Scheme 2

Key steps in Scheme 2

(1) The conversion of **1** to **2** occurs via a bromohydrin intermediate, formed in good yield, although accompanied by a smaller amount of bromohydrin corresponding to addition to the central double bond of **1**. The selectivity is thought to be due to three factors:

(a) deactivation of the 2,3-double bond of **1** due to the electron-withdrawing inductive effect of the acetoxymethylene substituent;

(b) steric shielding of the central double bond [at C(6)-C(7)] by the C(1) to C(4) and C(9) to C(11) regions of **1** and;

(c) the favoring of folded conformers of **1** by a confining shell of solvent in the aqueous mixture.

(2) The key highly enantioselective Ireland-Claisen rearrangement is precedented by a number of examples that had previously been demonstrated with simpler substrates, according to the following equation:

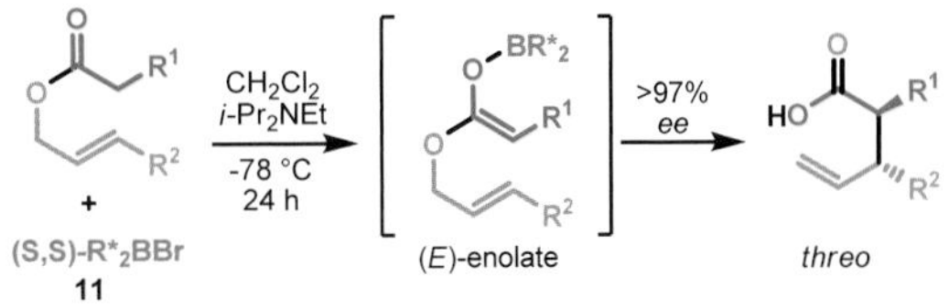

The absolute stereochemical course of this enantioselective rearrangement is thought to be the result of a strong preference for the pre-transition state assembly **A**:

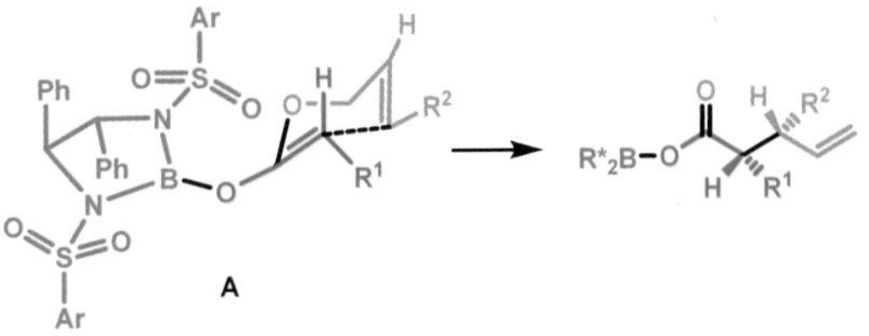

DOLABELLANES (Part I)

Pathway of Synthesis (continued)

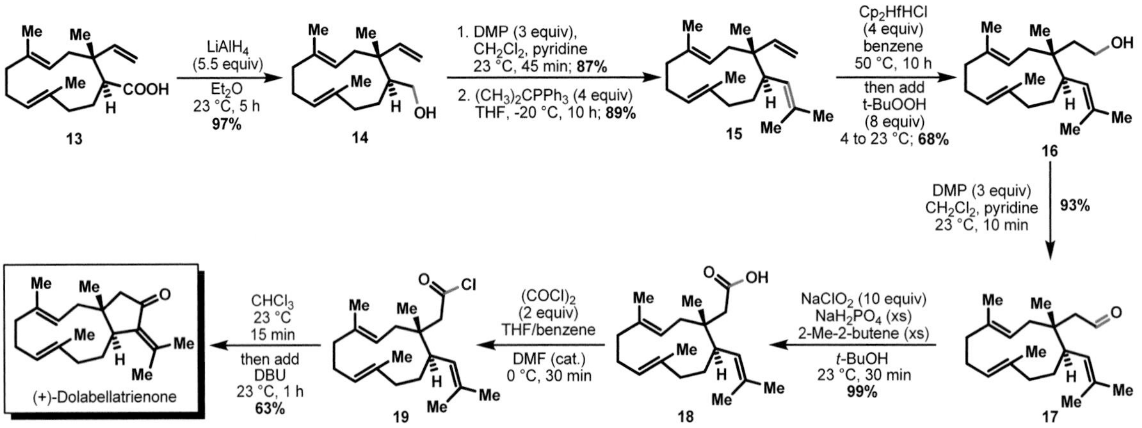

Scheme 3

Key steps in Scheme 3

The success of the cyclization of **19** to dolabellatrienone is due to the greater proximity of the electrophilic chlorocarbonyl group to the isopropylidene appendage as compared to the other two trisubstituted olefinic linkages.

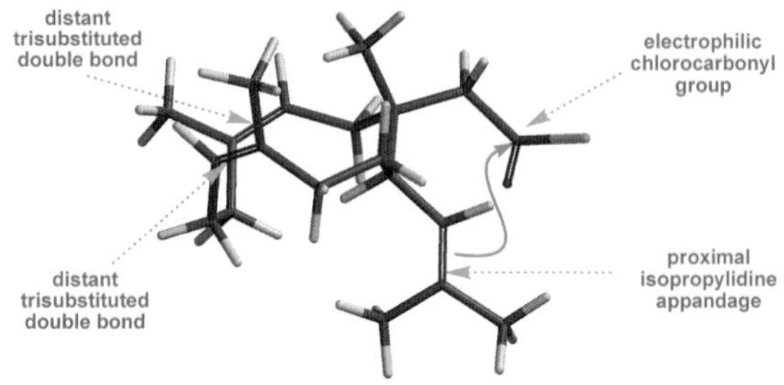

DOLABELLANES (Part II)

Background

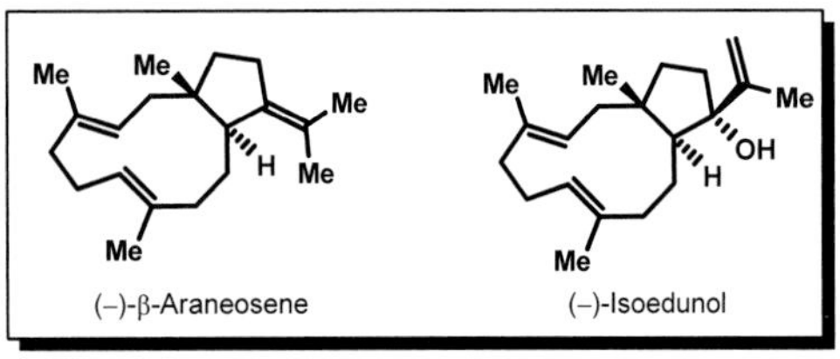

(−)-β-Araneosene   (−)-Isoedunol

β-Araneosene, isolated in the 1970s from the terrestrial mold *Sordaria Araneosa*, was one of the original members of the dolabellane family.[5] Its discovery was followed two decades later by the disclosure of the hydroxylated derivative isoedunol.[6] The enantioselective and stereoselective synthesis of these dolabellanes, which provided both a challenge and an opportunity for synthetic innovation, was reported in 2005.[7]

DOLABELLANES (Part II)

Abbreviated Retrosynthetic Plan

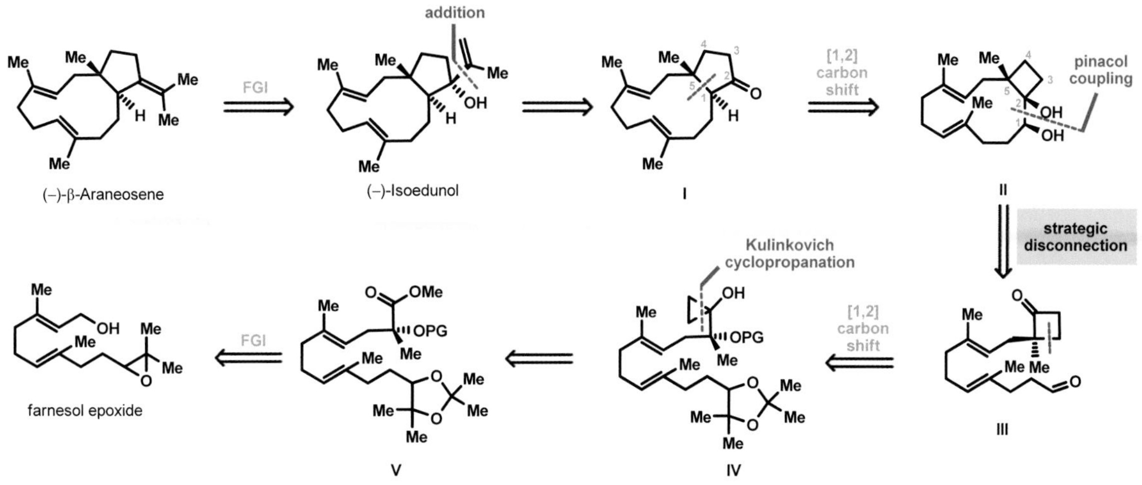

Scheme 4

Key steps in Scheme 4

(1) The overriding retrosynthetic strategy of the synthetic plan involved the use of the pinacol cyclization transform to disconnect the larger ring of the bicyclic dolabellane core (**II** ⇒ **III**). The retron required for the pinacol disconnection was generated by a prior application of the pinacol rearrangement transform (**I** ⇒ **II**). These two transforms when used in tandem exemplify the

application of "tactical combinations" as a strategy in retrosynthetic analysis.

(2) There was little precedent for establishing the quaternary stereocenter for **III** with control of absolute configuration. One option was the synthesis of **III** from an accessible chiral precursor **V**. The choice of **V** as a possible intermediate for the synthesis guided that retrosynthetic generation of **IV** by another pinacolic rearrangement.

Pathway of Synthesis

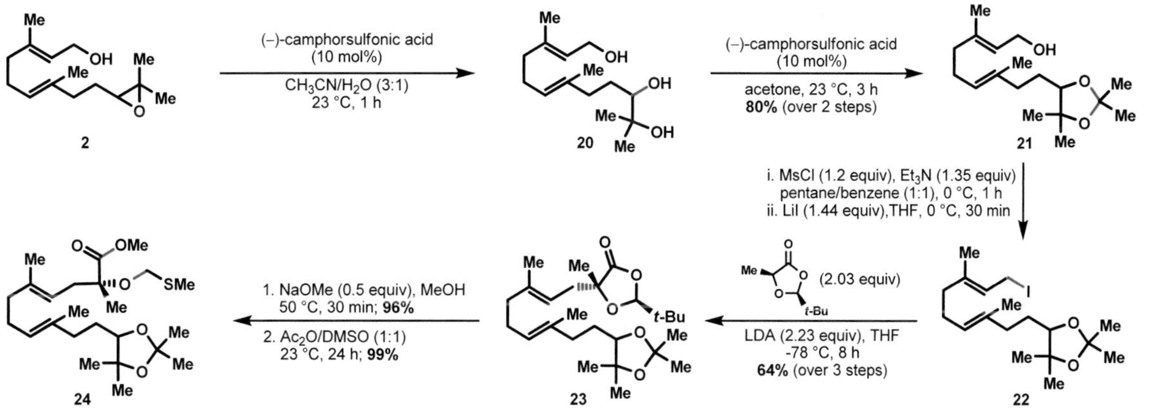

Scheme 5

Key steps in Scheme 5

The conversion of **22** to **23** by alkylation of the chiral dioxolanone derived from (S)-lactic acid and pivalaldehyde follows methodology pioneered by D. Seebach.[8]

The alkylation was completely diastereoselective because of steric shielding of one face of the enolate by the bulky t-butyl group.

DOLABELLANES (Part II)

Pathway of Synthesis (continued)

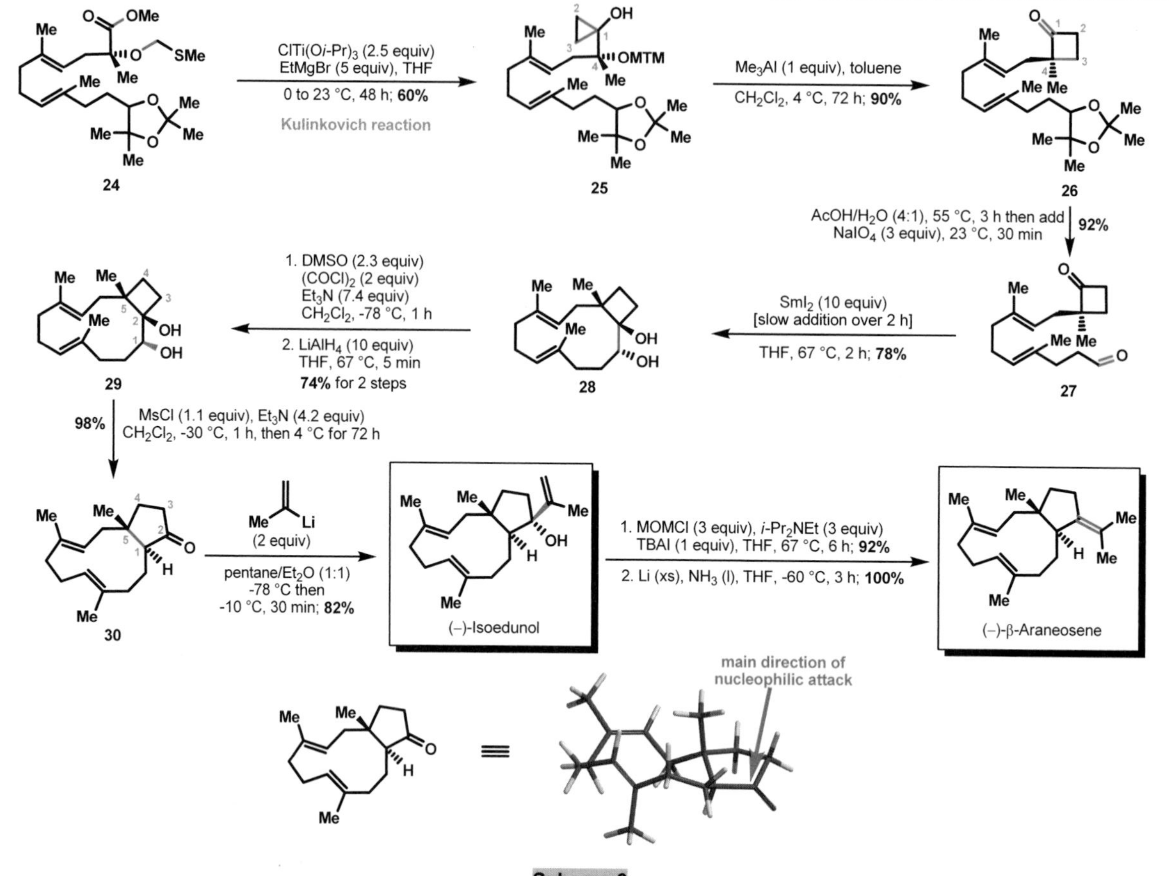

Scheme 6

Key steps in Scheme 6

(1) The methylthiomethyl group (MTM) was utilized for O-protection in **24** because it is easily introduced (even for hindered tertiary alcohols), stable under the conditions required for the conversion of **24** to **25** and removable in the pinacolic rearrangement **25→26**.

(2) The Kulinkovich reaction (**24→25**), the Ti(II)-mediated one-pot conversion of carboxylic esters to the corresponding 1-alkylcyclopropanols, has been reviewed.[9]

(3) Trimethylaluminum is very effective in promoting the pinacolic rearrangement **25** to **26** without loss of enantiomeric purity. Stronger Lewis acids lead to racemization, presumably because of the intervention of a discrete β-hydroxy-cyclopropylcarbinyl cation. It is thought that Me_3Al has a dual role in the rearrangement of **25** to **26**:

(a) deprotonation of the cyclopropanol group to form a Me_2Al-alkoxide and

(b) complexation of Al with the nearby MTM-oxygen to activate its departure in a concerted rearrangement (**A**).

(4) The pinacol cyclization **27→28** is a rare example of the application of this process to rings larger than five or six. For another rare example of macrocyclic ring-closure using pinacolic reduction, and a very different synthesis of (±)-dolabellatrienone, see **31→32**.[10] Also, during the synthesis of (±)-δ-araneosene, the five-membered ring was formed via McMurry coupling, see **33→34**.[11]

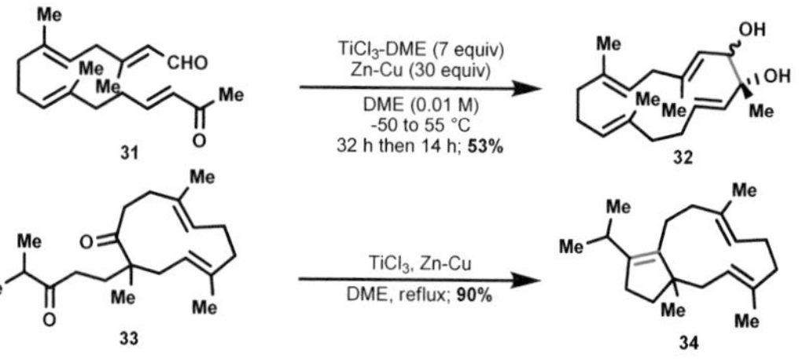

(5) The geometry of the *trans*-diol **28** was unfavorable for the required rearrangement to **30** and so it was epimerized to the *cis*-diol **29** which is stereoelectronically well disposed for formation of **30** by a concerted process.

(6) The addition of 2-lithiopropene to ketone **30** occurs with excellent diastereoselectivity for steric reasons.

DOLABELLANES (Part III)

Background

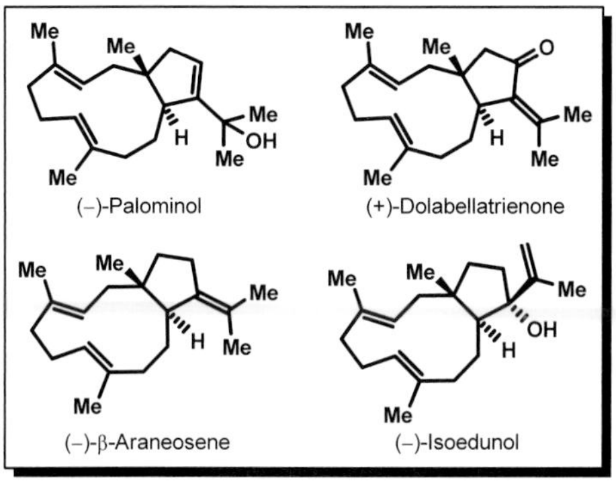

(−)-Palominol

(+)-Dolabellatrienone

(−)-β-Araneosene

(−)-Isoedunol

Another synthetic strategy provided an enantioselective route to the four dolabellanes shown at right from a common intermediate (**I**).[12] This approach depended on a recently developed chiral catalyst for a range of enantioselective cycloaddition reactions.[13]

Abbreviated Retrosynthetic Plan

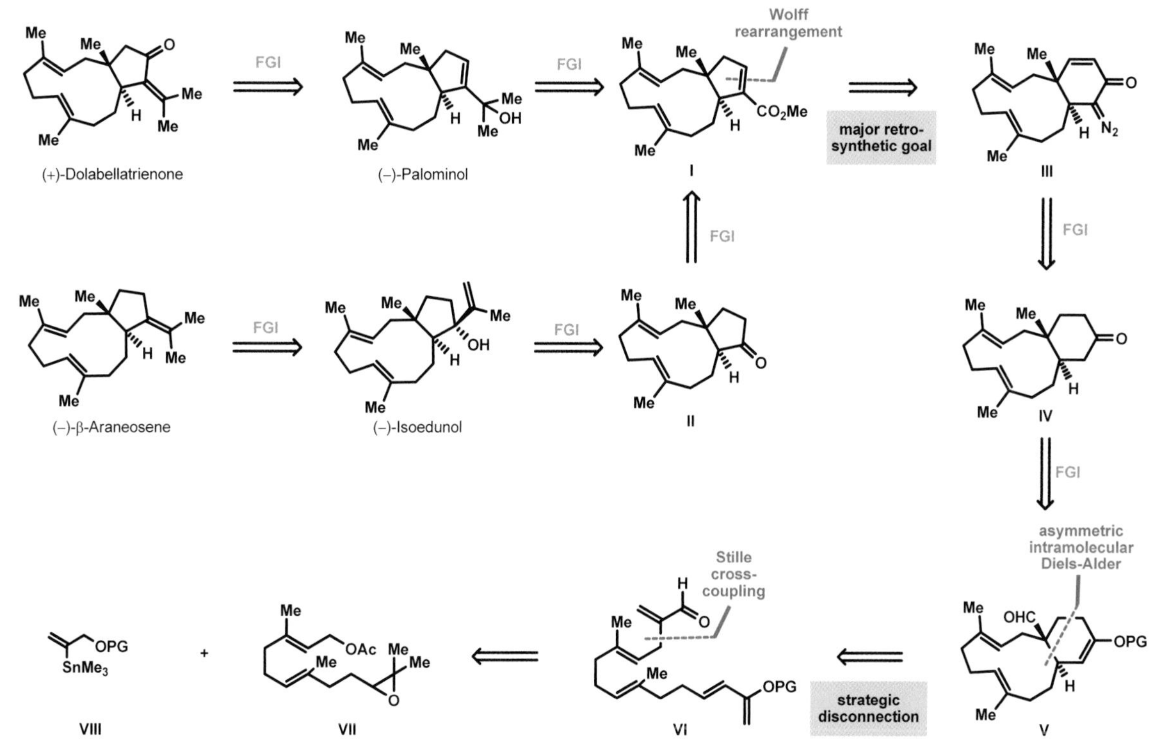

Scheme 7

Key steps in Scheme 7

(1) The dominant strategy which guided the retrosynthetic analysis behind this synthesis was transform-based and specifically aimed at the use of asymmetric [4+2]-cycloaddition to generate the bicyclic intermediate **V** from the acyclic precursor **VI**.

(2) The required acyclic precursor **VI** could be simplified to two readily obtainable precursors **VII** and **VIII**.

(3) The sequence for the synthetic connection between **V** and dolabellatrienone was developed by structure-**IV**-guided retrosynthetic analysis, an important feature of which was the application of the Wolff rearrangement[14] transform to connect the five-membered ring of the dolabellanes to the six-membered ring of the Diels-Alder retron. Functional group modification of dolabellatrienone gave **I** which contains the Wolff rearrangement retron. Application of the Wolff transform to **I** leads to **III** and **IV**.

DOLABELLANES (Part III)

Pathway of Synthesis

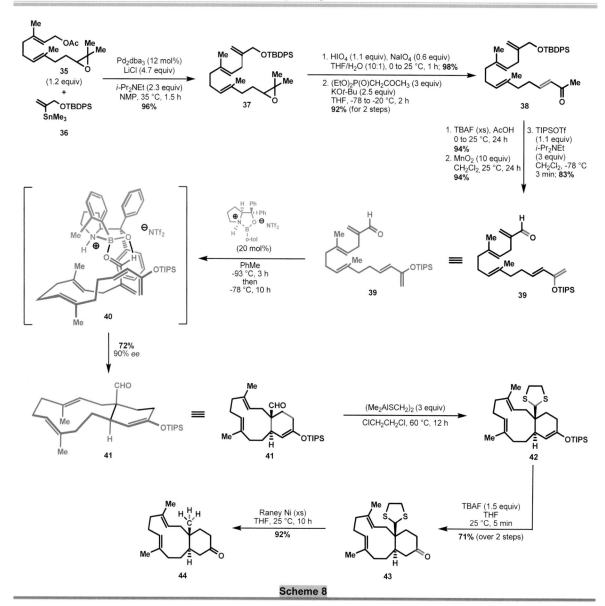

Scheme 8

Key steps in Scheme 8

(1) Excellent conditions were determined by experiment for the coupling of **35** and **36** to form **37** very selectively.

(2) Because the silyloxy diene **39** is a very sensitive compound, the exact conditions for its formation are critical.

(3) The oxazaborolidinium cation-catalyzed intramolecular Diels-Alder reaction to form **41** is remarkably effective and enantioselective. It is interesting that simple Lewis

acids such as BF$_3$·Et$_2$O, SnCl$_4$ or MeAlCl$_2$ did not afford even traces of racemic **41**.[12]

(4) The absolute configuration of **41** is that predicted by the mechanistic model and the pre-transition-state assembly **40**.[13]

(5) The conversion of **41** to **44** utilized novel methodology which made possible the selective conversion of the hindered formyl group in aldehyde **41** to the corresponding cyclic thioacetal without involvement of the enol silyl ether functionality.[12]

DOLABELLANES (Part III)

Pathway of Synthesis (continued)

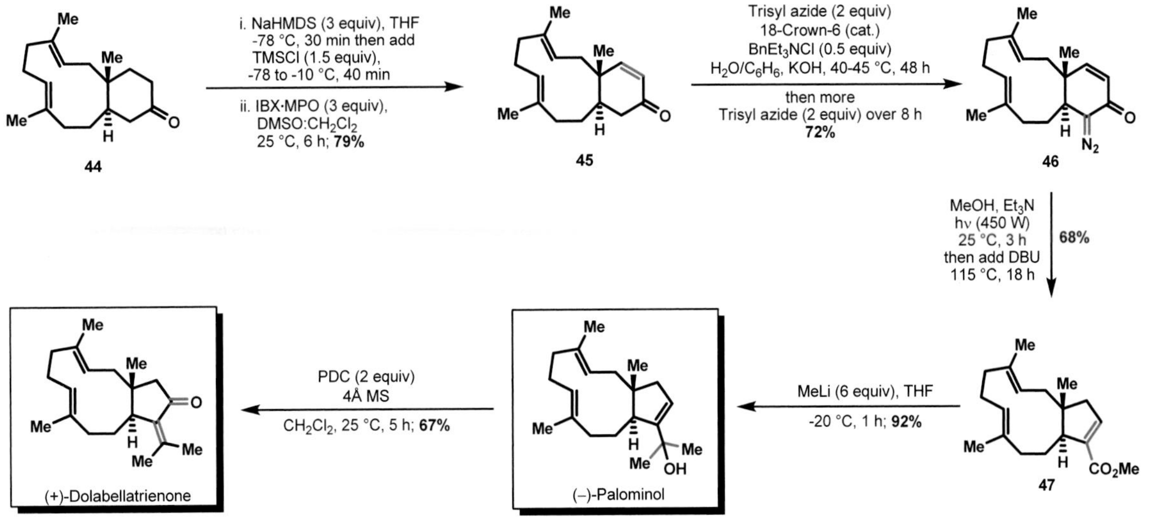

44

i. NaHMDS (3 equiv), THF
-78 °C, 30 min then add
TMSCl (1.5 equiv),
-78 to -10 °C, 40 min

ii. IBX·MPO (3 equiv),
DMSO:CH$_2$Cl$_2$
25 °C, 6 h; **79%**

45

Trisyl azide (2 equiv)
18-Crown-6 (cat.)
BnEt$_3$NCl (0.5 equiv)
H$_2$O/C$_6$H$_6$, KOH, 40-45 °C, 48 h
then more
Trisyl azide (2 equiv) over 8 h
72%

46

MeOH, Et$_3$N
hν (450 W)
25 °C, 3 h   **68%**
then add DBU
115 °C, 18 h

(+)-Dolabellatrienone

PDC (2 equiv)
4Å MS
CH$_2$Cl$_2$, 25 °C, 5 h; **67%**

(−)-Palominol

MeLi (6 equiv), THF
-20 °C, 1 h; **92%**

47

Scheme 9

47

1. L-Selectride (4 equiv)
THF, 0 °C, 1.5 h; **87%**

2. LDA (1.0 equiv), THF
0 °C, 30 min then bubble O$_2$ (1 atm)
0 °C, 10 min; **81%**

48

1. LiAlH$_4$ (3 equiv), THF
25 °C, 10 min

2. NaIO$_4$-silica gel (xs)
CH$_2$Cl$_2$, 25 °C, 15 min; **73%**

30

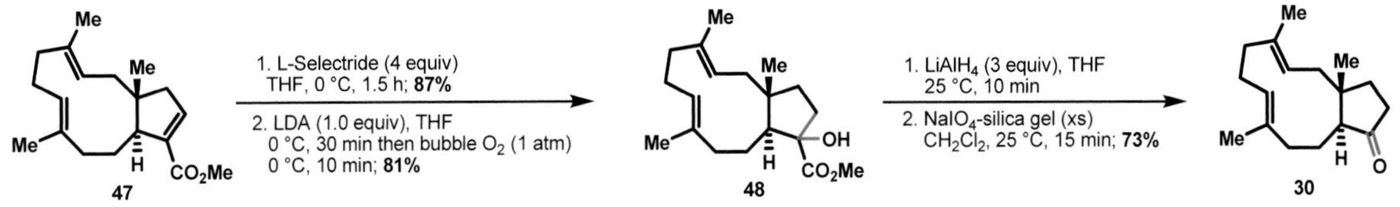

Scheme 10

References

1. For a review of dolabellane marine diterpenoids, see: Rodriguez, A. D. The natural products chemistry of West Indian gorgonian octocorals. *Tetrahedron* **1995**, 51, 4571-4618.
2. (a) Look, S. A., Fenical, W. New bicyclic diterpenoids from the Caribbean gorgonian octocoral Eunicea calyculata. *J. Org. Chem.* **1982**, 47, 4129-4134; (b) Shin, J., Fenical, W. Structures and reactivities of new dolabellane diterpenoids from the Caribbean gorgonian Eunicea laciniata. *J. Org. Chem.* **1991**, 56, 3392-3398.
3. Corey, E. J. & Kania, R. S. First Enantioselective Total Synthesis of a Naturally Occurring Dolabellane. Revision of Absolute Configuration. *J. Am. Chem. Soc.* **1996**, 118, 1229-1230.
4. (a) Corey, E. J., Lee, D. H. Highly enantioselective and diastereoselective Ireland-Claisen rearrangement of achiral allylic esters. *J. Am. Chem. Soc.* **1991**, 113, 4026-4028; (b) Corey, E. J., Kim, S. S. Versatile chiral reagent for the highly enantioselective synthesis of either anti or syn ester aldols. *J. Am. Chem. Soc.* **1990**, 112, 4976-4977; (c) Kürti, L., Czakó, B. Ireland-Claisen Rearrangement in *Strategic Applications of Named Reactions in Organic Synthesis* **2005**, pp 91-91 (Academic Press/Elsevier, San Diego).
5. Borschberg, H.-J. Ph.D. Thesis No. 5578 (Eidenössichen Technischen Hohchschule (ETH), Zürich, **1975**).
6. Rodriguez, A. D., Gonzalez, E. & Gonzalez, C. Additional dolabellane diterpenes from the Caribbean gorgonian octocoral Eunicea laciniata. *J. Nat. Prod.* **1995**, 58, 226-232.
7. Kingsbury, J. S. & Corey, E. J. Enantioselective Total Synthesis of Isoedunol and β-Araneosene Featuring Unconventional Strategy and Methodology. *J. Am. Chem. Soc.* **2005**, 127, 13813-13815.
8. Seebach, D., Naef, R. & Calderari, G. α-Alkylation of α-heterosubstituted carboxylic acids without racemization. EPC-syntheses of tertiary alcohols and thiols. *Tetrahedron* **1984**, 40, 1313-1324.
9. (a) Kulinkovich, O. G. The Chemistry of Cyclopropanols. *Chem. Rev.* **2003**, 103, 2597-2632; (b) Sato, F., Okamoto, S. The divalent titanium complex Ti(O-i-Pr)$_4$/2 i-PrMgX as an efficient and practical reagent for fine chemical synthesis. *Adv. Synth. Catal.* **2001**, 343, 759-784.
10. Corey, E. J. & Kania, R. S. Concise total synthesis of (±)-palominol and (±)-dolabellatrienone via a dianion-accelerated oxy-Cope rearrangement. *Tetrahedron Lett.* **1998**, 39, 741-744.
11. Hu, T. & Corey, E. J. Short Syntheses of (±)-δ-Araneosene and Humulene Utilizing a Combination of Four-Component Assembly and Palladium-Mediated Cyclization. *Org. Lett.* **2002**, 4, 2441-2443.
12. Snyder, S. A. & Corey, E. J. Concise Total Syntheses of Palominol, Dolabellatrienone, β-Araneosene, and Isoedunol via an Enantioselective Diels-Alder Macrobicyclization. *J. Am. Chem. Soc.* **2006**, 128, 740-742.
13. For a review see: Corey E.J. Enantioselective Catalysis Based on Cationic Oxazaborolidines. *Angew. Chem. Intl. Ed*, **2009**, 48, 2100-2117.
14. Kürti, L. & Czakó, B. Wolff Rearrangement in *Strategic Applications of Named Reactions in Organic Synthesis* 494-495 (Academic Press/Elsevier, San Diego, **2005**).

SALINOSPORAMIDES

Background

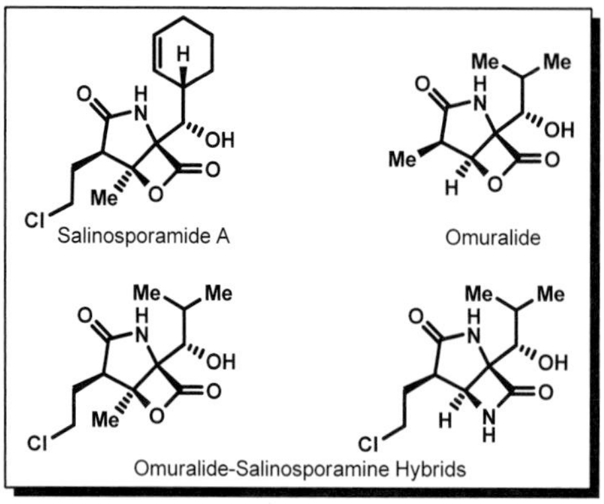

Salinosporamide A

Omuralide

Omuralide-Salinosporamine Hybrids

Subsequent to the development of synthetic lacacystin and omuralide (page 223) and the recognition that these compounds inhibit proteasome action, Fenical and co-workers discovered salinosporamide A, a marine microbial product obtained from sea bottom samples.[1] Biological studies of salinosporamide A showed it to be a more effective inhibitor of proteasome function and also to be cytotoxic to many tumor cell lines at nanomolar concentrations. The first synthesis of salinosporamide[2] A is discussed in this section.[3]

Abbreviated Retrosynthetic Plan

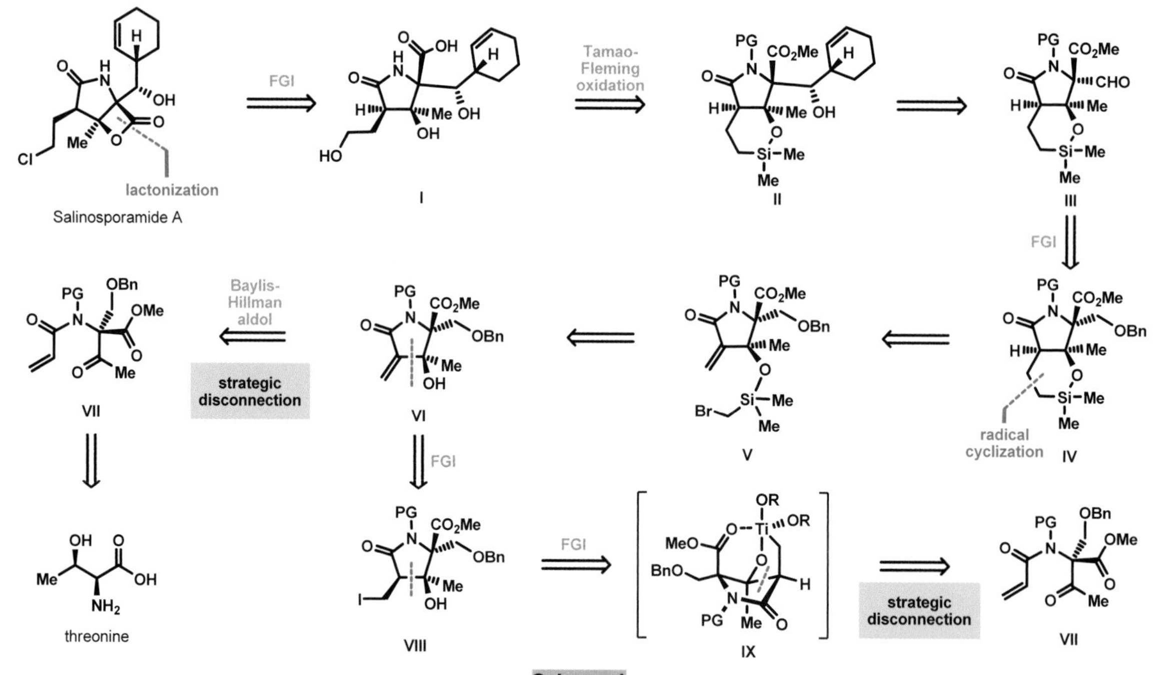

Scheme 1

Key steps in Scheme 1

(1) As was the case with the design of a synthetic pathway to lactacystin and omuralide (page 223), simplification of the synthetic problem was sought using a chiral starting material to establish absolute configuration and emplace the nitrogen-bearing quaternary stereocenter of the γ-lactam ring (γ-to the carbonyl, see VII).

(2) Bidirectional analysis using VII as a S-goal lead to two alternative paths for closure of the γ-lactam ring and formation of the other quaternary stereocenter (β-to the γ-lactam carbonyl).

(3) The strereocenter β-to the lactam carbonyl was used to control the orientation of the appendage on the C(α)-stereocenter.

(4) The retrosynthetic sequence from salinosporamide A to II sets up a doubly diastereoselective disconnection of the cyclohexenyl subunit to give the aldehyde III.

(5) The retrosynthetic pathway III ⇒ IV ⇒ V ⇒ VI was guided by the simplification of having VII as a S-goal and the need to clear the stereocenter at C(α).

SALINOSPORAMIDES

Pathway of Synthesis

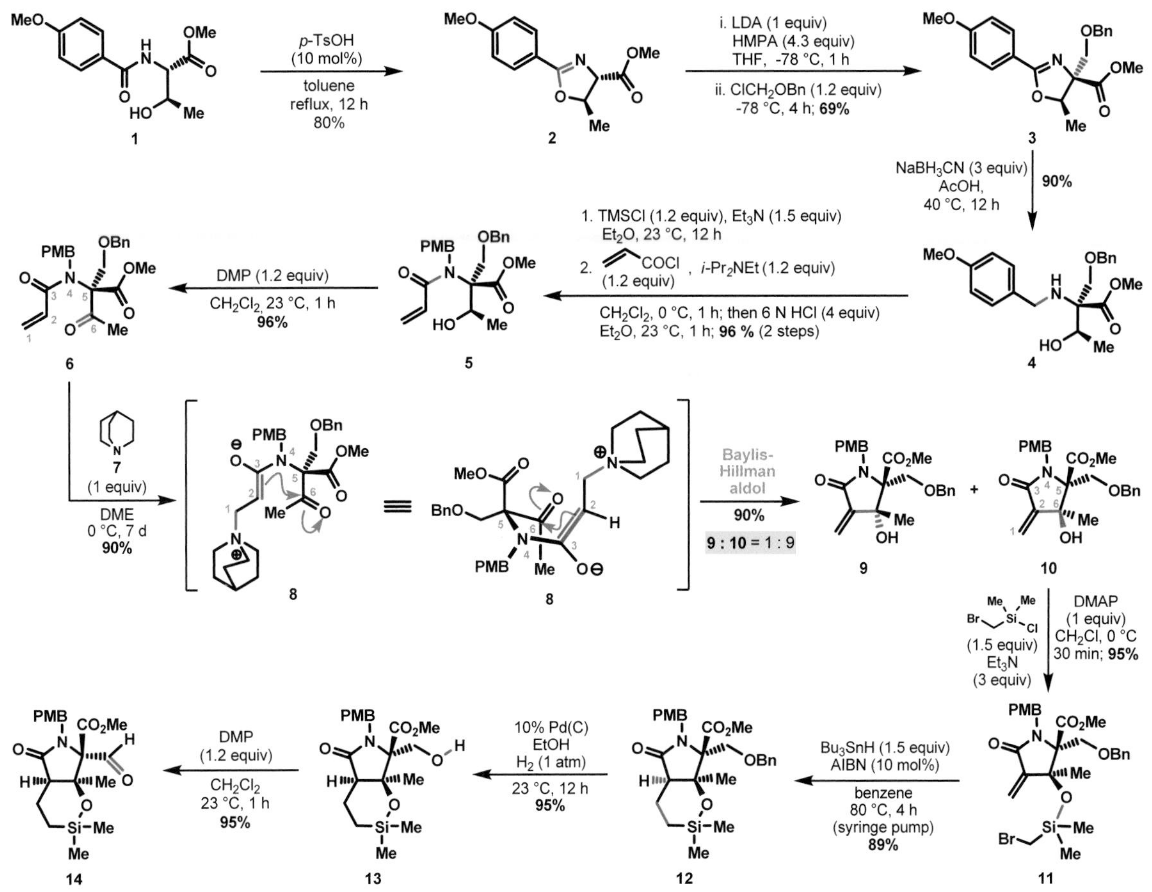

Scheme 2

Key steps in Schemes 2

(1) The alkylation **2→3** is diastereoselective because of steric screening by the adjacent methyl substituent of the intermediate enolate.

(2) The Baylis-Hillman cyclization **6→9** is also diastereoselective, possibly because of electrostatic effects in the transition state for the ring-closure.

(3) The stereocontrolled radical-mediated cyclization of the bromosilane **11** to **12** occurs very stereoselectively. The intermediate carbonyl-stabilized radical resulting after ring-closure abstracts a hydrogen atom from Bu₃SnH at the sterically less shielded face to form the *cis*-5,6-fused ring system of **12**.

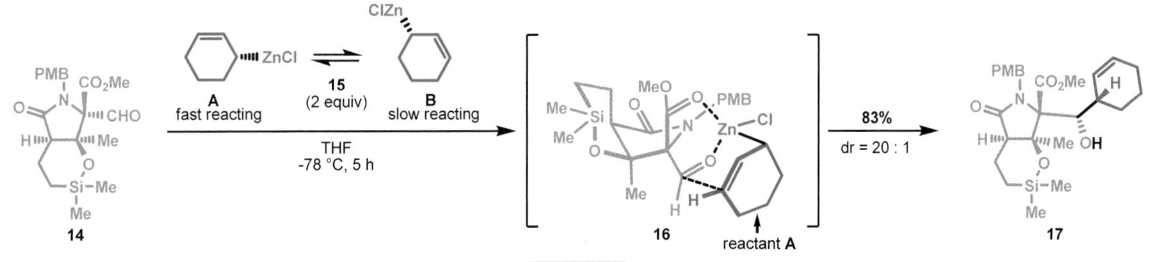

Scheme 3

SALINOSPORAMIDES

Key steps in Scheme 3

The doubly diastereoselective addition of the racemic zinc reagent **15** to the aldehyde **14** is thought to involve just one of the two enantiomers (**A**) of zinc reagent **15**. The two enantiomers are likely in rapid equilibrium on the time scale of the carbonyl addition reaction and it appears that one of these (**A**) reacts more rapidly than the other (**B**) because the pre-transition state assembly **16** lies on the lowest energy pathway to the carbonyl adduct. Diastereomer **17** predominates if the pre-transition state assembly has the chair-like geometry expressed in **16**.

Pathway of Synthesis (continued)

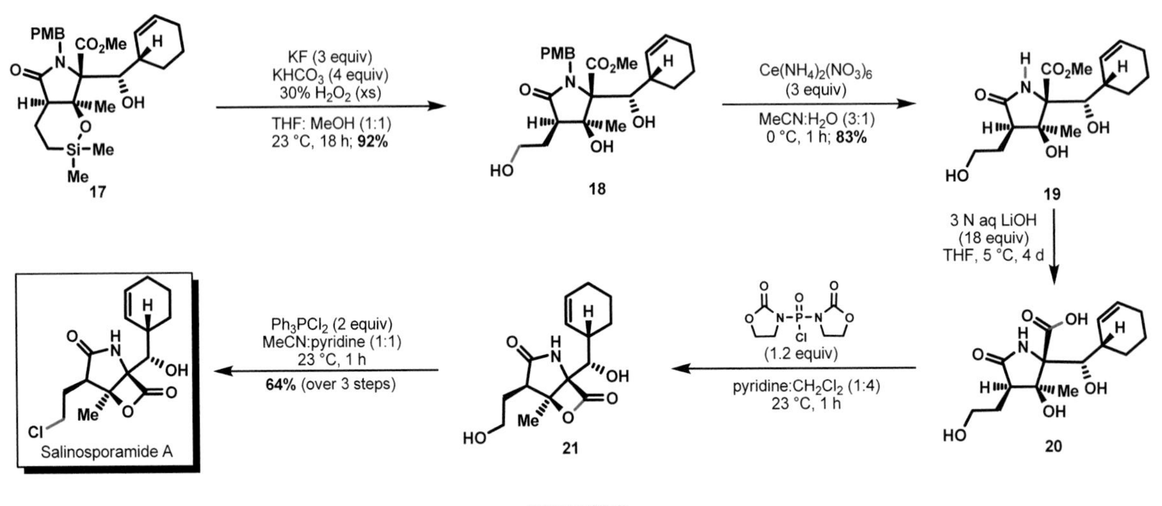

Scheme 4

Subsequent Developments – Alternative Cyclization Strategy

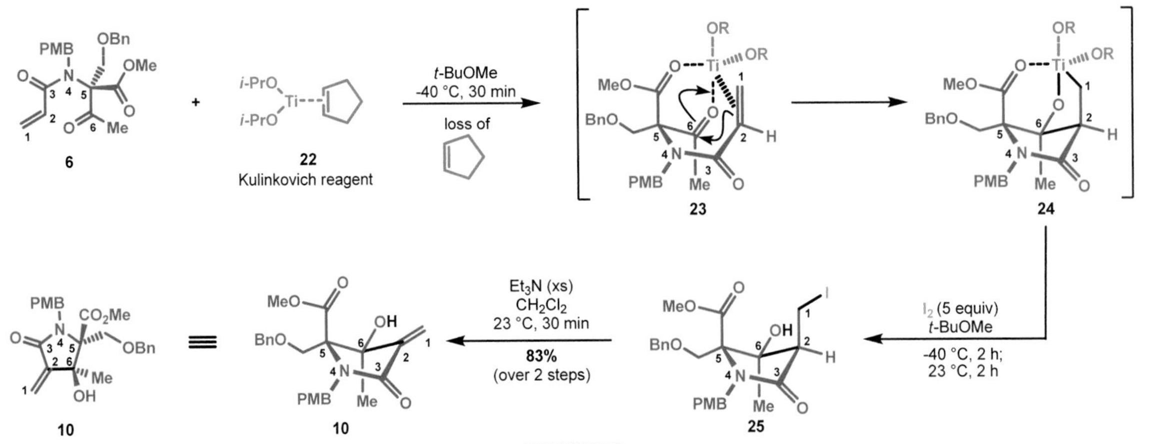

Scheme 5

Key steps in Scheme 5

The novel cyclization step that converts **6** to **25** is considered to proceed from the Kulinkovich reagent [prepared from cyclo-C_5H_{11}MgBr and Ti(Oi-Pr)$_4$] via intermediates **23** and **24**. Organotitanium intermediate **24** was quenched with iodine (I_2) and the resulting primary alkyl iodide **25** was dehydrohalogenated to afford the exocyclic enone **10**.

SALINOSPORAMIDES

Preparation of a Salinosporamide A – Omuralide Hybrid

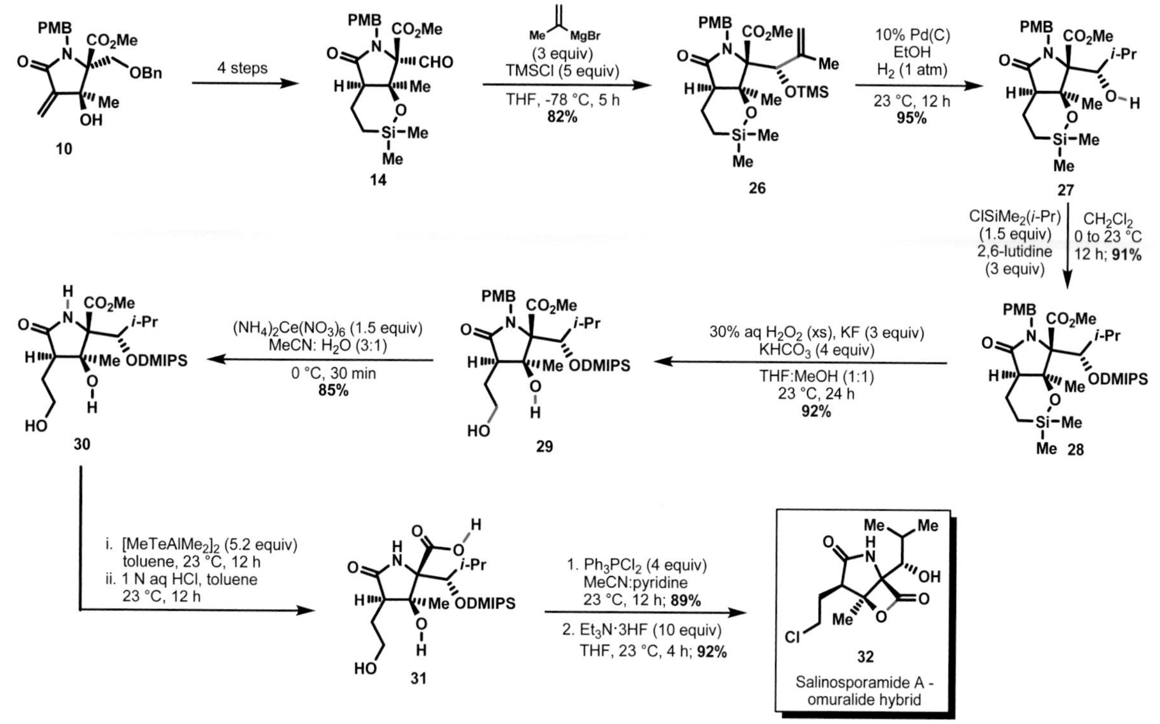

Scheme 6

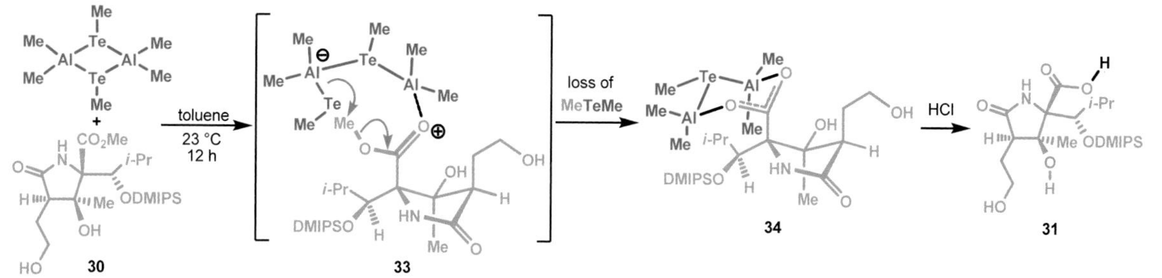

Scheme 7

Key steps in Schemes 6 and 7

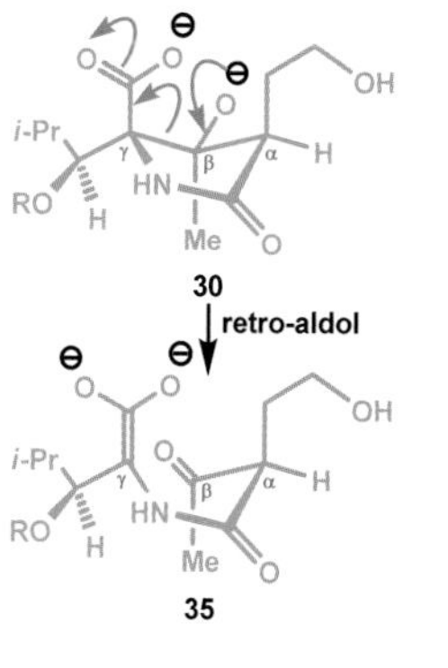

(1) The saponification of the methyl ester **30** by aqueous base proceeds in only modest yield due to competing decomposition, possibly by a retro-aldol pathway involving cleavage of the β,γ-bond of the lactam ring (**30→35**).

(2) [Me₂AlTeMe]₂, prepared by reaction of Te$^{(0)}$ and Me₃Al, is highly effective for the cleavage of sterically hindered methyl esters, as shown by a separate study with a wide variety of substrates.[4] The reaction is considered to proceed via **33** as depicted in Scheme 7. The methyl group of the methyl ester is attacked from the back side (in an S_N2 type process).

SALINOSPORAMIDES

Preparation of a β-Lactam Related to Salinosporamide A

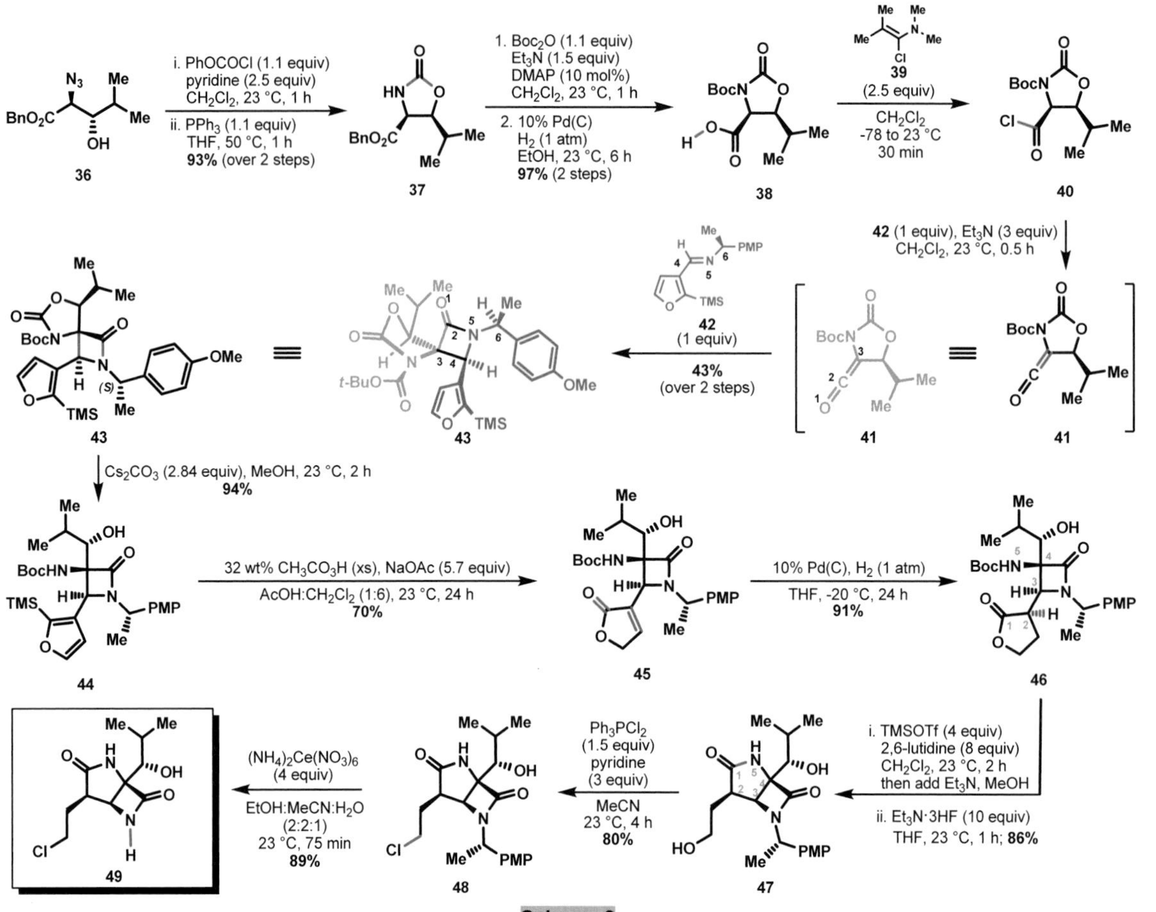

Scheme 8

Key steps in Scheme 8

(1) The rationale for the synthesis of a β-lactam **49** as a version of salinosporamide was the objective of having a proteasome inhibitor with an extended lifetime *in vivo*. The hydrolytic stability of a β-lactam is significantly greater than that for the corresponding β-lactone.

(2) The [2+2]-cycloaddition reaction of ketene **41** with the imine **42** afforded **43** stereoselectively.

(3) The catalytic hydrogenation of **45** to **46** was stereoselective at -20 °C, but less so at 23 °C.

(4) The β-lactam **49**, which is stable at pH 7 for 24 hours, was shown to inactivate the proteasome in a concentration-dependent manner.

References

1. Feling, R. H., Buchanan, G. O., Mincer, T. J., Kauffman, C. A., Jensen, P. R. & Fenical, W. Salinosporamide A: a highly cytotoxic proteasome inhibitor from a novel microbial source, a marine bacterium of the new genus Salinospora. *Angew. Chem., Int. Ed.* **2003**, 42, 355-357.

2. (a) Reddy, L. R., Saravanan, P. & Corey, E. J. A Simple Stereocontrolled Synthesis of Salinosporamide A. *J. Am. Chem. Soc.* **2004**, 126, 6230-6231; (b) Reddy, L. R., Fournier, J.-F., Reddy, B. V. S. & Corey, E. J. New Synthetic Route for the Enantioselective Total Synthesis of Salinosporamide A and Biologically Active Analogues. *Org. Lett.* **2005**, 7, 2699-2701; (c) Hogan, P. C. & Corey, E. J. Proteasome Inhibition by a Totally Synthetic beta -Lactam Related to Salinosporamide A and Omuralide. *J. Am. Chem. Soc.* **2005**, 127, 15386-15387.

3. For recent work on salinosporamide A, see: (a) Fukuda, T., Sugiyama, K., Arima, S., Harigaya, Y., Nagamitsu, T., Omura, S. Total Synthesis of Salinosporamide A. *Org. Lett.* **2008**, 10, 4239-4242; (b) Takahashi, K., Midori, M., Kawano, K., Ishihara, J., Hatakeyama, S. Entry to heterocycles based on indium-catalyzed Conia-ene reactions: asymmetric synthesis of (-)-salinosporamide A. *Angew. Chem., Int. Ed.* **2008**, 47, 6244-6246 and references therein.

4. Reddy, B. V. S., Reddy, L. R. & Corey, E. J. Dimethylaluminum methyltellurate, a new reagent for the cleavage of hindered methyl esters under exceptionally mild conditions by a novel mechanism. *Tetrahedron Lett.* **2005**, 46, 4589-4593.

LACTACYSTIN & OMURALIDE

Background

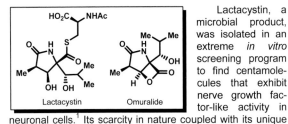

Lactacystin   Omuralide

Lactacystin, a microbial product, was isolated in an extreme *in vitro* screening program to find centamolecules that exhibit nerve growth factor-like activity in neuronal cells.[1] Its scarcity in nature coupled with its unique

biological activity stimulated many synthetic studies. The first effective synthetic approach[2] is discussed in this section.[3] During the course of this work, a related β-lactone, subsequently called omuralide, was synthesized and found to possess equivalent activity.[4] The availability of synthetic lactacystin and omuralide enabled biochemical studies which revealed that the mode of action of these compounds is actually inhibition of the proteasome, a multi-protein machine that catalyzes the hydrolysis of proteins.

Abbreviated Retrosynthetic Plan

Scheme 1

Key steps in Scheme 1

(1) The synthetic plan was devised by a bidirectional search that was guided by the S-goal of using a known chiral serine derivative (**XII**)[5] as an early intermediate and platform for control of absolute configuration. This simplifying S-subgoal allows the application of the aldol transform to clearing the quaternary stereocenter on the five-membered ring which also disconnects the 1-hydroxy-isobutyl appendage (**XI** ⇒ **XII** + **XIII**).

(2) The remaining stereocenters of the target could be cleared by application of the Mukaiyama aldol transform on intermediate **V** which could be generated by the sequence shown from lactacystin.

(3) The selection of protecting groups and reagents followed from this analysis.

LACTACYSTIN & OMURALIDE

Pathway of Synthesis

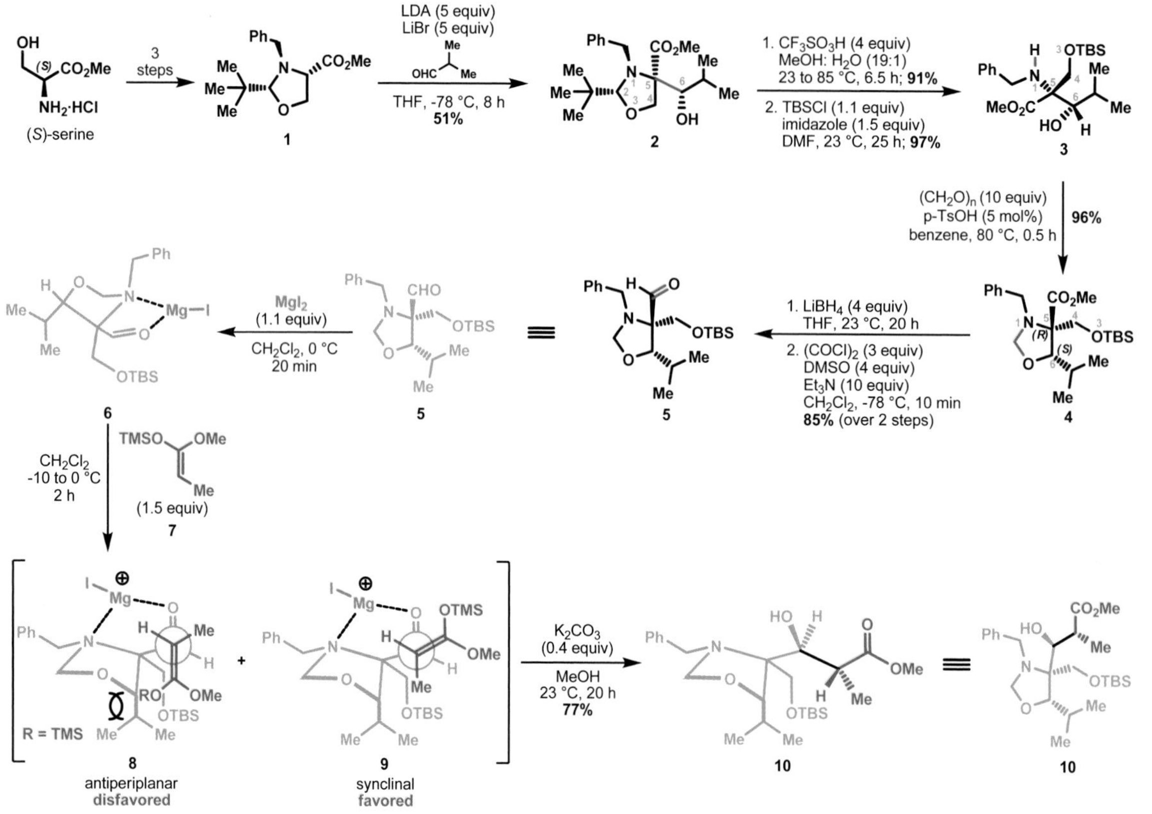

Scheme 2

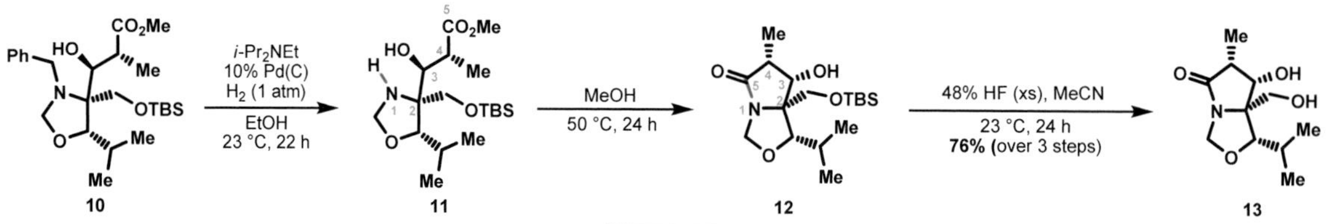

Scheme 3

Key steps in Schemes 2 and 3

(1) The doubly diastereoselective aldol reaction of **1** and isobutyraldehyde provided the readily purified crystalline intermediate **2**.

(2) The MgI$_2$-catalyzed Mukaiyama aldol reaction **5→10** was also doubly diastereoselective because of a favorable pathway via **9**.

(3) The removal of the Bn group in **10** was achieved using catalytic hydrogenation. The resulting secondary amine **11** readily furnished 5-membered lactam **12** upon heating in MeOH.

LACTACYSTIN & OMURALIDE

Pathway of Synthesis (continued)

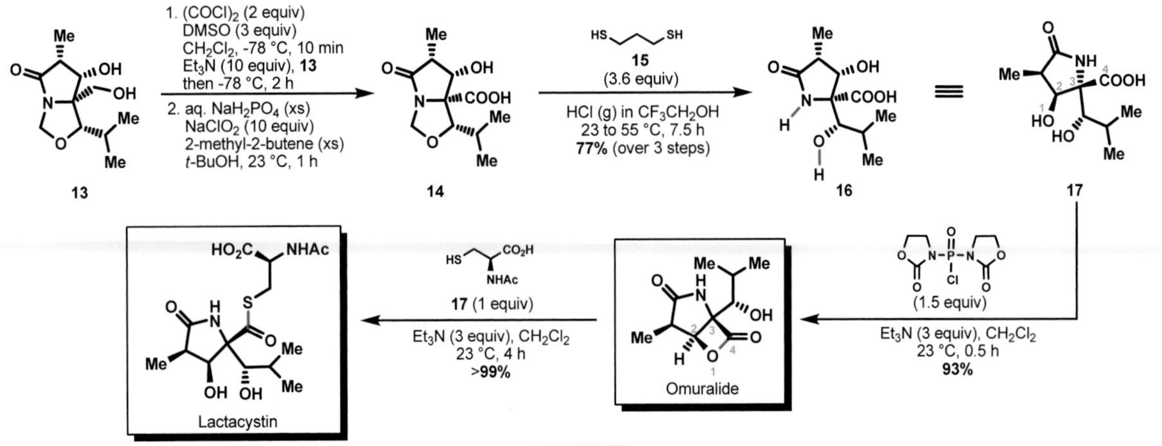

Scheme 4

Key steps in Scheme 4

In the two-step conversion of primary alcohol **13** to carboxylic acid **14**, a modified Swern oxidation procedure was employed which gave much better results than the conventional Swern procedure.

It entailed adding the solution of **13** and Et₃N in CH₂Cl₂ to the pre-formed Swern reagent (oxalyl chloride + dimethylsulfoxide) at -78 °C. Under these conditions the oxidation was completely selective for the primary alcohol and also proceeded very cleanly.

A Short and Efficient Synthesis of (–)-7-Methylomuralide[6]

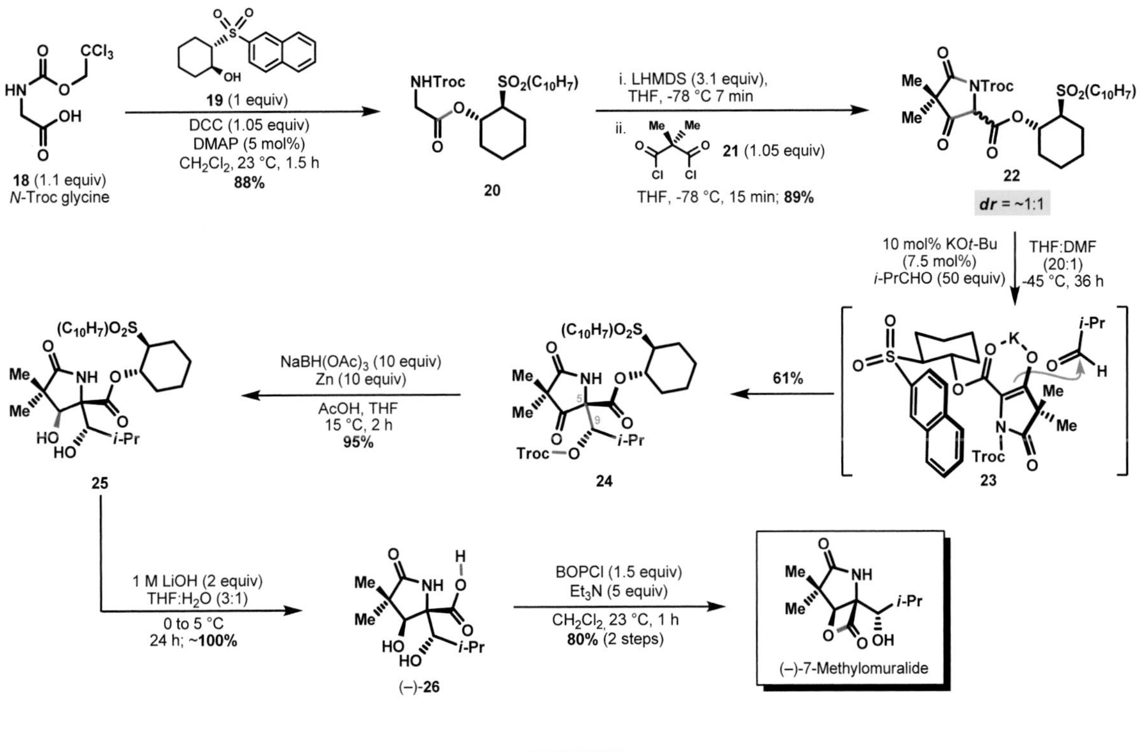

Scheme 5

LACTACYSTIN & OMURALIDE

A Short and Efficient Synthesis of (−)-7-Methylomuralide (continued)

Explanation of Scheme 5

(1) The use of the chiral controller **19** provided access to the β-keto ester **22**, which underwent face-selective aldol reaction with isobutyraldehyde to provide the crystalline coupling product **24** as a result of a subsequent N→O acyl migration.

(2) Acyl migration renders the aldol process irreversible. It is critical because without it the retroaldol process is favored.

(3) The trichloroethoxycarbonyl (Troc) group is also important for the control of diastereoselectivity.

Subsequent Developments

(1) Numerous other syntheses of lactacystin and omuralide have been developed and described,[7] confirming the extraordinary interest in any new natural product which is both scarce and biologically interesting or useful.

(2) Experiments in which tritiated lactacystin or omuralide were added to neuronal cell cultures revealed efficient uptake and remarkably selective labeling of the proteasome, a catalytic assembly of multiple protein subunits which recognizes proteins that are tagged with ubiquitin and then cleaves these to octa- or nona-peptides. The core structure of the proteasome (side-on and top views, respectively) is shown to the right.

 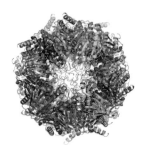

(3) The original suggestion that lactacystin is a mimic of nerve growth factor (NGF) is not correct, the apparent activity being due simply to the inhibition of degradation of NGF in neuronal cells.

(4) A covalent linkage is formed by attachment of omuralide to a catalytic threonine residue in the β-subunit of the proteasome, as was confirmed by an X-ray crystallographic study.[8]

(5) Synthetic lactacystin has become a common reagent in biological laboratories and is now commercially available.

(6) The availability of several alternative synthetic routes to compounds in the omuralide family enabled the synthesis of a range of structural analogs. Biochemical evaluation of these by measurement of the relative rates of proteasome inactivation led to the general correlation of structure and biological activity which can be summarized as follows (Scheme 6):[9]

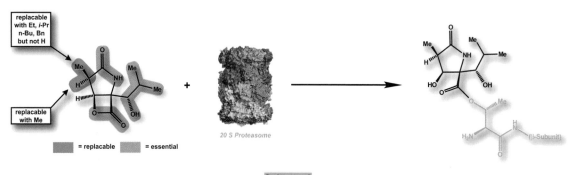

Scheme 6

LACTACYSTIN & OMURALIDE

Subsequent Developments (continued)

(7) More recently a marine natural product, related to omuralide, salinosporamide A has been isolated from the microbe *Salinospora*.[10] The synthesis of solinosporamide A is described in the previous section (pages 218-222).

Salinosporamide A

(8) Bortezomib (Velcade™), the first member of a new class of drugs called proteasome inhibitors, is mainly used for the treatment of multiple myeloma. In 2006, over 16,000 people in the US were diagnosed with this disease, which is generally fatal within a year if not treated. Multiple myeloma is a malignancy of bone marrow plasma cells (myeloma cells) that leads to disrupted bone marrow function, reduction of red blood cell levels (and also white blood cells), suppression of immune function, and severe bone erosion. The unique boronic acid moiety of bortezomib is critical for the high affinity of the drug at the active site of the enzyme. The picture below shows a mesh representation of the 20S core of proteasome with 6 bortezomib molecules (shown in red) bound inside.

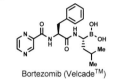

Bortezomib (Velcade™)

Bortezomib molecules

References

1. For the isolation of lactacystin, see: (a) Omura, S., Fujimoto, T., Otoguro, K., Matsuzaki, K., Moriguchi, R., Tanaka, H., Sasaki, Y. Lactacystin, a novel microbial metabolite, induces neuritogenesis of neuroblastoma cells. *J. Antibiot.* **1991**, 44, 113-116; (b) Omura, S., Matsuzaki, K., Fujimoto, T., Kosuge, K., Furuya, T., Fujita, S., Nakagawa, A. Structure of lactacystin, a new microbial metabolite which induces differentiation of neuroblastoma cells. *J. Antibiot.* **1991**, 44, 117-118.

2. Corey, E. J., Li, W. & Reichard, G. A. A New Magnesium-Catalyzed Doubly Diastereoselective anti-Aldol Reaction Leads to a Highly Efficient Process for the Total Synthesis of Lactacystin in Quantity. *J. Am. Chem. Soc.* **1998**, 120, 2330-2336.

3. For various synthetic approaches to lactacystin and derivatives by the Corey group, see: (a) Corey, E. J., Reichard, G. A. Total synthesis of lactacystin. *J. Am. Chem. Soc.* **1992**, 114, 10677-10678; (b) Corey, E. J., Choi, S. An enantioselective synthesis of (6R)-lactacystin. *Tetrahedron Lett.* **1993**, 34, 6969-6972; (c) Corey, E. J., Reichard, G. A. Synthesis of (6R,7S)-lactacystin and 6-deoxylactacystin from a common intermediate. *Tetrahedron Lett.* **1993**, 34, 6973-6976; (d) Corey, E. J., Reichard, G. A., Kania, R. Studies on the total synthesis of lactacystin. An improved aldol coupling reaction and a beta -lactone intermediate in thiol ester formation. *Tetrahedron Lett.* **1993**, 34, 6977-6980; (e) Corey, E. J., Reichard, G. A., Corey, E. J., Choi, S., Sunazuka, T., Nagamitsu, T., Matsuzaki, K., Tanaka, H., Omura, S., et al. Total synthesis of lactacystin. An enantioselective synthesis of (5R)-lactacystin. Total synthesis of (+)-lactacystin, the first non-protein neurotrophic factor. Total synthesis of (+)-lactacystin from (R)-glutamate. *Chemtracts: Org. Chem.* **1994**, 7, 266-272; (f) Corey, E. J., Li, W., Reichard, G. A. A New Magnesium-Catalyzed Doubly Diastereoselective anti-Aldol Reaction Leads to a Highly Efficient Process for the Total Synthesis of Lactacystin in Quantity. *J. Am. Chem. Soc.* **1998**, 120, 2330-2336; (g) Corey, E. J., Li, W., Nagamitsu, T. An efficient and concise enantioselective total synthesis of lactacystin. *Angew. Chem., Int. Ed.* **1998**, 37, 1676-1679; (h) Corey, E. J., Li, W.-D. Z. An efficient total synthesis of a new and highly active analog of lactacystin. *Tetrahedron Lett.* **1998**, 39, 7475-7478; (i) Corey, E. J., Li, W.-D. Z. Enantioselective synthesis of the (5S,6R,9R) and (5S,6R,9S) analogs of lactacystin β-lactone. *Tetrahedron Lett.* **1998**, 39, 8043-8046.

4. (a) Fenteany, G., Standaert, R. F., Lane, W. S., Choi, S., Corey, E. J., Schreiber, S. L. Inhibition of proteasome activities and subunit-specific amino-terminal threonine modification by lactacystin. *Science (Washington, D. C.)* **1995**, 268, 726-731; (b) Fenteany, G., Standaert, R. F., Reichard, G. A., Corey, E. J., Schreiber, S. L. A β-lactone related to lactacystin induces neurite outgrowth in a neuroblastoma cell line and inhibits cell cycle progression in an osteosarcoma cell line. *Proc. Natl. Acad. Sci. U. S. A.* **1994**, 91, 3358-3362.

5. Seebach, D., Aebi, J. D., Gander-Coquoz, M. & Naef, R. Stereoselective alkylation at C(α) of serine, glyceric acid, threonine, and tartaric acid involving heterocyclic enolates with exocyclic double bonds. *Helv. Chim. Acta* **1987**, 70, 1194-1216.

6. Shenvi, R. A. & Corey, E. J. A Short and Efficient Synthesis of (–)-7-Methylomuralide, a Potent Proteasome Inhibitor. *J. Am. Chem. Soc.* **2009**, 131, 5746-5747.

7. For other total syntheses of lactacystin and omuralide, see: (a) Sunazuka, T., Nagamitsu, T., Matsuzaki, K., Tanaka, H., Omura, S., Smith, A. B., III. Total synthesis of (+)-lactacystin, the first non-protein neurotrophic factor. *J. Am. Chem. Soc.* **1993**, 115, 5302; (b) Nagamitsu, T., Sunazuka, T., Tanaka, H., Omura, S., Sprengeler, P. A., Smith, A. B., III. Total Synthesis of (+)-Lactacystin. *J. Am. Chem. Soc.* **1996**, 118, 3584-3590. (c) Chida, N., Takeoka, J., Ando, K., Tsutsumi, N., Ogawa, S. Stereoselective total synthesis of (+)-lactacystin from D-glucose. *Tetrahedron* **1997**, 53, 16287-16298; (d) Panek, J. S., Masse, C. E. Total synthesis of (+)-lactacystin. *Angew. Chem., Int. Ed.* **1999**, 38, 1093-1095; (e) Ooi, H., Ishibashi, N., Iwabuchi, Y., Ishihara, J., Hatakeyama, S. A Concise Route to (+)-Lactacystin. *J. Org. Chem.* **2004**, 69, 7765-7768; (f) Fukuda, N., Sasaki, K., Sastry, T. V. R. S., Kanai, M., Shibasaki, M. Catalytic Asymmetric Total Synthesis of (+)-Lactacystin. *J. Org. Chem.* **2006**, 71, 1220-1225; (g) Balskus, E. P., Jacobsen, E. N. α,β-Unsaturated β-silyl imide substrates for catalytic, enantioselective conjugate additions: A total synthesis of (+)-lactacystin and the discovery of a new proteasome inhibitor. *J. Am. Chem. Soc.* **2006**, 128, 6810-6812; (h) Hayes, C. J., Sherlock, A. E., Green, M. P., Wilson, C., Blake, A. J., Selby, M. D., Prodger, J. C. Enantioselective Total Syntheses of Omuralide, 7-epi-Omuralide, and (+)-Lactacystin. *J. Org. Chem.* **2008**, 73, 2041-2051.

8. (a) Groll, M., Ditzel, L., Lowe, J., Stock, D., Bochtler, M., Bartunik, H. D. & Huber, R. Structure of 20S proteasome from yeast at 2.4 A resolution. *Nature* **1997**, 386, 463-471; (b) Groll, M., Balskus, E. P. & Jacobsen, E. N. Structural Analysis of Spiro β-Lactone Proteasome Inhibitors. *J. Am. Chem. Soc.* **2008**, 130, 14981-14983.

9. Corey, E. J. & Li, W.-D. Total synthesis and biological activity of lactacystin, omuralide and analogs. *Chem. Pharm. Bull.* **1999**, 47, 1-10.

10. Feling, R. H., Buchanan, G. O., Mincer, T. J., Kauffman, C. A., Jensen, P. R. & Fenical, W. Salinosporamide A: a highly cytotoxic proteasome inhibitor from a novel microbial source, a marine bacterium of the new genus Salinospora. *Angew. Chem., Int. Ed.* **2003**, 42, 355-357.

(−)-ANTHELIOLIDE A

Background

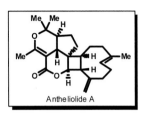

Antheliolide A

The pentacyclic natural product antheliolide A presents a unique challenge for synthesis because of its unprecedented collection of 4-, 5-, 6- and 9-membered rings and multiple stereocenters.[1] Although the originally proposed structure required subsequent correction, the absolute configuration of antheliolide A remained uncertain.[2] The bicyclic 4.9-fused subunit of antheliolide A presents an unusual problem because of the limited knowledge of how such a core can be constructed with stereochemical control and with the various appendages attached to it. That subunit also occurs in the terrestrial plant sesquiterpenoid β-caryophyllene.[3] The successful synthesis of antheliolide A summarized herein required a different approach from that used for the synthesis of β-caryophyllene and included a number of unusual steps.[4] It also clarified the absolute configuration of this natural product.

Abbreviated Retrosynthetic Plan

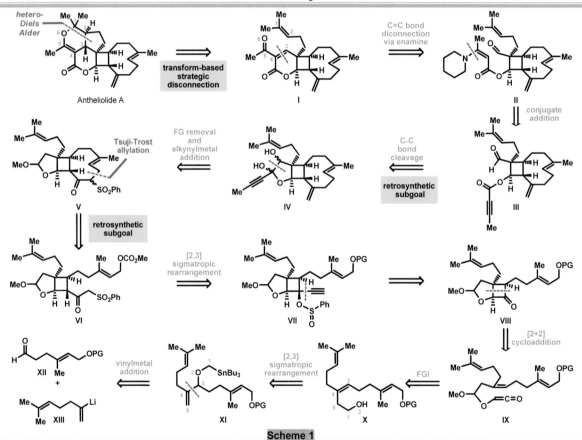

Scheme 1

Key steps in Scheme 1

(1) Several considerations suggested the early disconnection of the two oxacyclic rings of antheliolide A, including:

    (a) the chemical sensitivity of their vinyl ether and lactone units;

    (b) the possibility of applying the [4+2]-hetero-cycloaddition transform for the simultaneous disconnection of one those rings and the 5-membered carbocyclic subunit first to form **I** and then **II** and **III**;

    (c) the fact that their removal would simplify the target structure so as to facilitate the subsequent disconnection of the four-membered ring by an intramolecular [2+2]-cycloaddition transform (**VIII** ⇒ **IX**).

(2) The retrosynthetic connection between **III** and **VIII** involved subgoals **IV** and **V**. Structure **V** provided an opportunity for the key disconnection of the 9-membered ring (**V** ⇒ **VI**).

(3) The remaining key disconnection of **VIII** led to an acyclic and achiral ketene **IX** which was accessed from **XII** and **XIII** by the sequence shown.

(−)-ANTHELIOLIDE A

Pathway of Synthesis

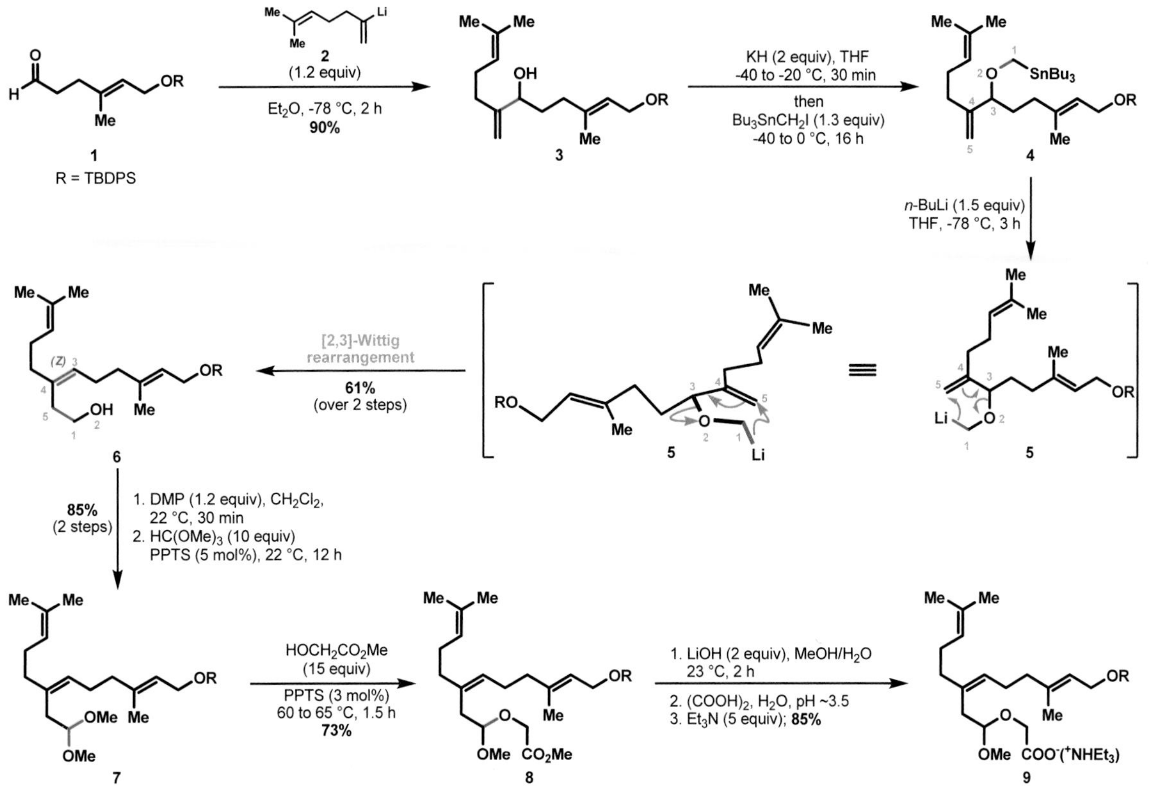

1
R = TBDPS

2
(1.2 equiv)

Et$_2$O, -78 °C, 2 h
90%

3

KH (2 equiv), THF
-40 to -20 °C, 30 min
then
Bu$_3$SnCH$_2$I (1.3 equiv)
-40 to 0 °C, 16 h

4

n-BuLi (1.5 equiv)
THF, -78 °C, 3 h

[2,3]-Wittig
rearrangement
61%
(over 2 steps)

6

5

5

85%
(2 steps)

1. DMP (1.2 equiv), CH$_2$Cl$_2$,
   22 °C, 30 min
2. HC(OMe)$_3$ (10 equiv)
   PPTS (5 mol%), 22 °C, 12 h

7

HOCH$_2$CO$_2$Me
(15 equiv)

PPTS (3 mol%)
60 to 65 °C, 1.5 h
73%

8

1. LiOH (2 equiv), MeOH/H$_2$O
   23 °C, 2 h
2. (COOH)$_2$, H$_2$O, pH ~3.5
3. Et$_3$N (5 equiv); 85%

9

Scheme 2

9

p-TsCl
(3 equiv)
Et$_3$N
(6 equiv)

toluene
reflux
7 h

10

R = TBDPS

10

[2+2]
66%

11

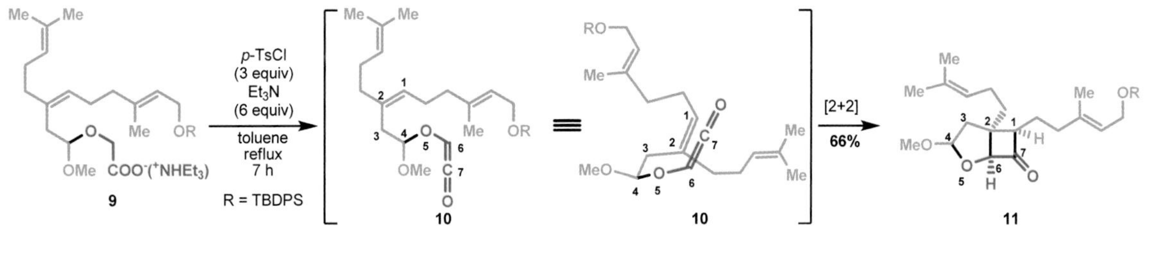

Scheme 3

Key steps in Schemes 2 and 3

(1) Still's version of the Wittig rearrangement[5] was used to effect the stereocontrolled conversion of allylic alcohol **3** to the (Z)-homoallylic alcohol **6**.

(2) The [2+2]-cycloaddition of ketene **10** to the (Z)-1,2-double bond generates all three stereocenters on the 4-membered ring in **11** with stereocontrol.

This cycloaddition is driven by the electrophilicity of C(7) of the ketene carbonyl group in **10**, the nucleophilicity of C(1) and the proximity of C(1) and C(7) in the conformation of **10** which is shown. The reaction though strictly suprafacial, is probably asynchronous.

(−)-ANTHELIOLIDE A

Pathway of Synthesis (continued)

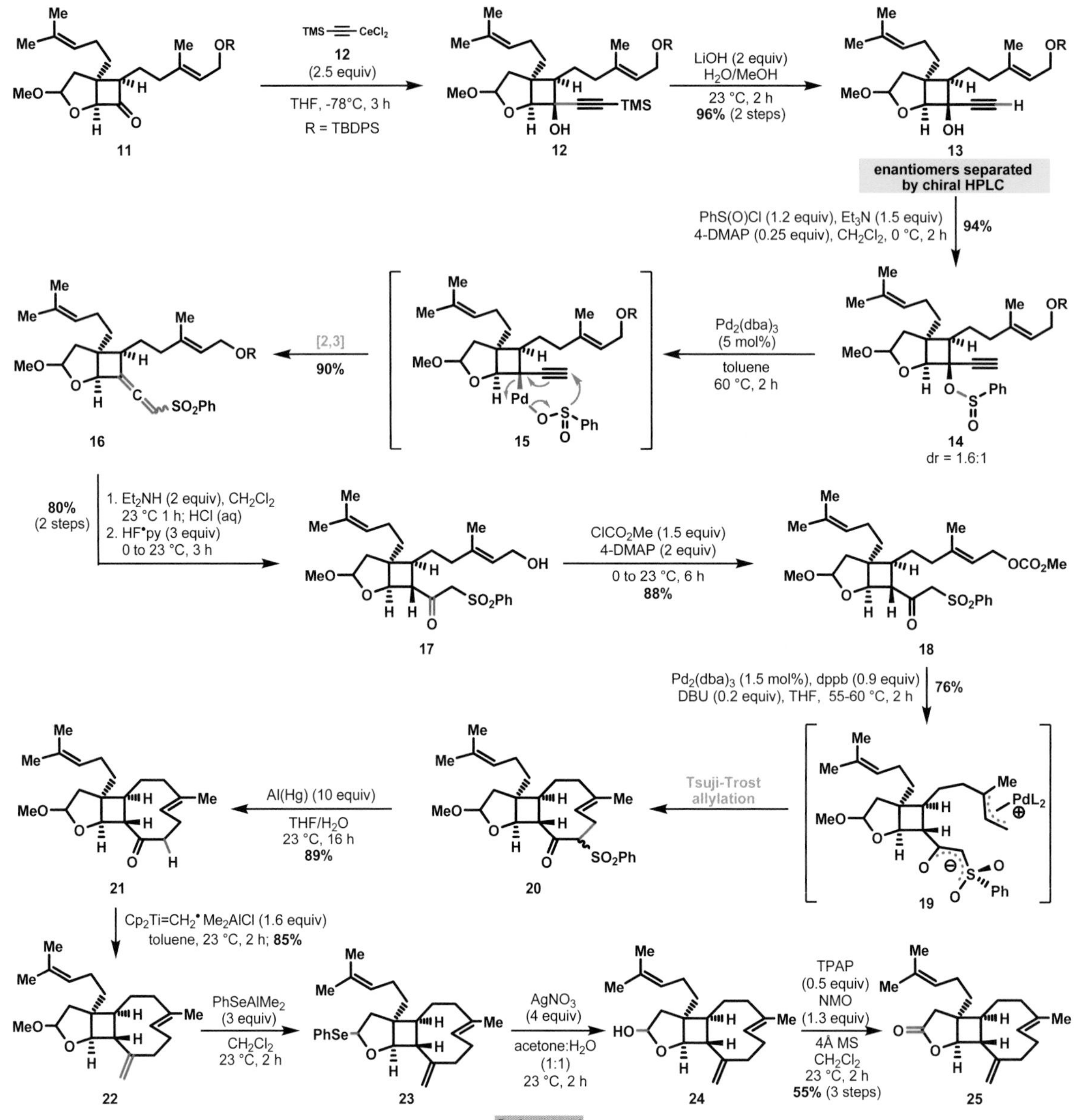

Scheme 4

Key steps in Scheme 4

(1) The racemic intermediate **13** was effectively resolved into the enantiomers by chromatography with a chiral adsorbant.

(2) The multistep conversion of ketone **17** involved a novel sequence including Pd-catalyzed [2,3]-sigmatropic rearrangement (**14**→**16**) and hydration via an enamine to **17**.

(3) The cyclization of **18** to **20** had previously been used for such ring closures.[6]

(4) Aluminum amalgam is an especially effective reagent for the reductive desulfonation of **20** to **21**.[7]

(5) The high acid sensitivity of the double bonds in the 9-membered ring of **22** necessitated a new procedure for the transformation of the hemiacetal **24**.

(−)-ANTHELIOLIDE A

Pathway of Synthesis (continued)

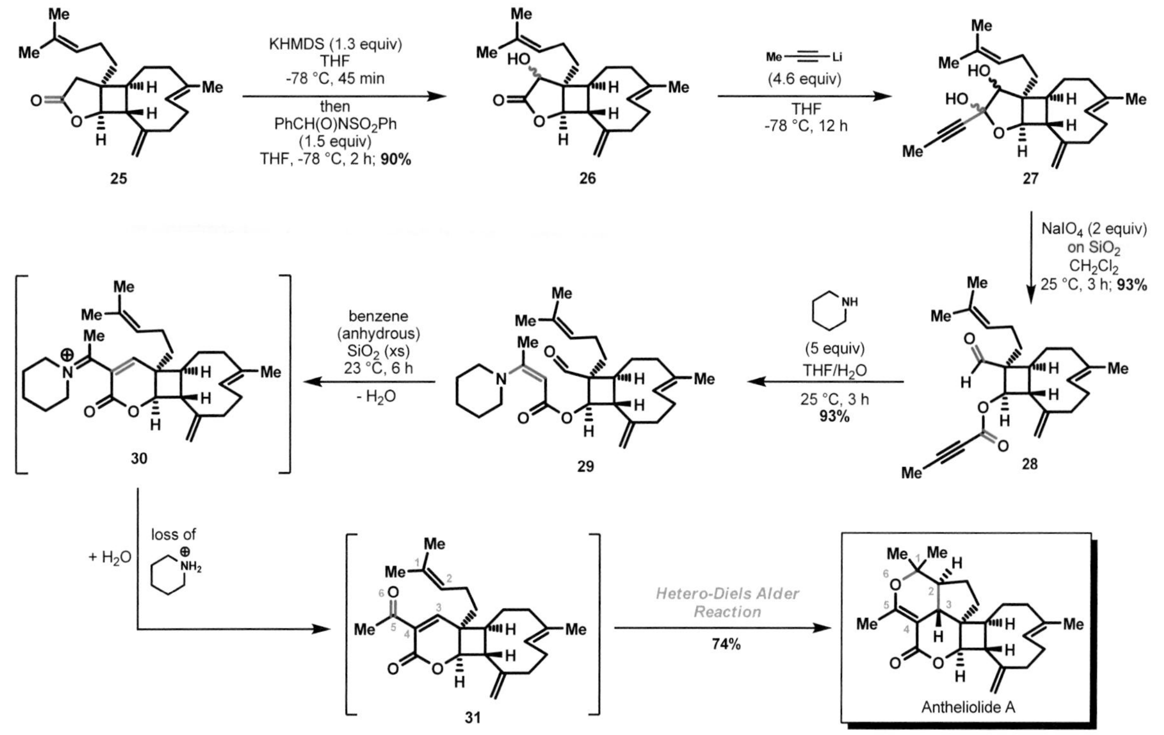

Scheme 5

Key steps in Scheme 5

(1) The conversion of the enamine-aldehyde **29** to antheliolide A was effected in a single step by stirring at ambient temperature with a suspension of silica gel in dry benzene followed by further reaction with water. The key final [4+2]-cycloaddition in **31** is stereoselective.

(2) The [4+2]-cycloaddition reaction which converts **31** to antheliolide A is a general and facile process, several examples of which have been demonstrated, including **32**→**33**→**34**.[8]

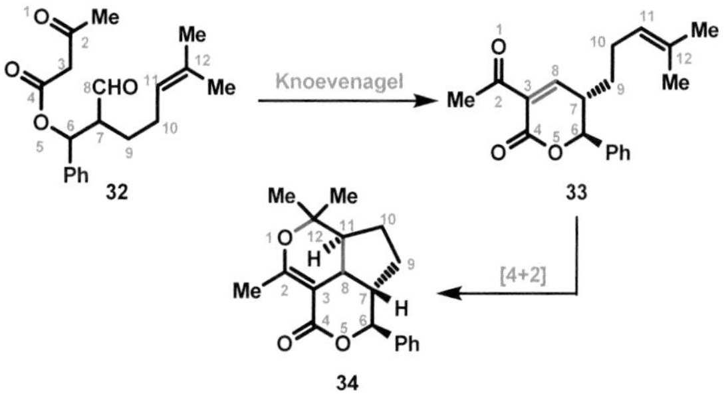

References

1. Green, D., Carmely, S., Benayahu, Y. & Kashman, Y. Antheliolide A and B: two new C24-acetoacetylated diterpenoids of the soft coral Anthelia glauca. *Tetrahedron Lett.* **1988**, 29, 1605-1608.

2. Smith, A. B., III, Carroll, P. J., Kashman, Y. & Green, D. Revised structures of antheliolides A and B. *Tetrahedron Lett.* **1989**, 30, 3363-3364.

3. Larionov, O. V. & Corey, E. J. An unconventional approach to the enantioselective synthesis of caryophylloids. *J. Am. Chem. Soc.* **2008**, 130, 2954-2955.

4. Mushti, C. S., Kim, J.-H. & Corey, E. J. Total Synthesis of Antheliolide A. *J. Am. Chem. Soc.* **2006**, 128, 14050-14052.

5. Still, W. C. & Mitra, A. A highly stereoselective synthesis of Z-trisubstituted olefins via [2,3]-sigmatropic rearrangement. Preference for a pseudoaxially substituted transition state. *J. Am. Chem. Soc.* **1978**, 100, 1927-1928.

6. Hu, T. & Corey, E. J. Short Syntheses of (±)-δ-Araneosene and Humulene Utilizing a Combination of Four-Component Assembly and Palladium-Mediated Cyclization. *Org. Lett.* **2002**, 4, 2441-2443.

7. Corey, E. J. & Chaykovsky, M. Methylsulfinyl carbanion. Formation and application to organic synthesis. *J. Am. Chem. Soc.* **1965**, 87, 1345-1353.

8. Kim, J.-H.; Corey, E.J. Unpublished results from this laboratory.

CARYOPHYLLOIDS

Background

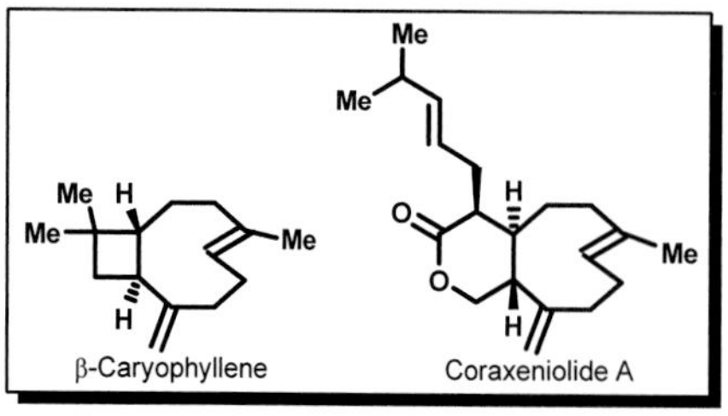

β-Caryophyllene   Coraxeniolide A

The discovery in 1960 that the long-known sesquiterpenoid β-caryophyllene[1] possesses a strikingly novel 4,9-fused ring structure was the harbinger of a new era in which countless other such unprecedented polyterpenoid structures emerged.[2]

The paucity of methods for the synthesis of such unusual acyclic structures provided a major stimulus for the development of new strategies and methods for the stereocontrolled construction of such molecules. The first synthesis of β-caryophyllene was reported in 1963.[3] The first synthesis of the most complex member of this family (caryophylloids), antheliolide A (pages 228-231) was achieved in 2006.[4]

An entirely new strategy for accessing the caryophyllide family was described in 2008 and applied not only to β-caryophyllene but also to the more complex coraxeniolide A.[5]

Abbreviated Retrosynthetic Plan

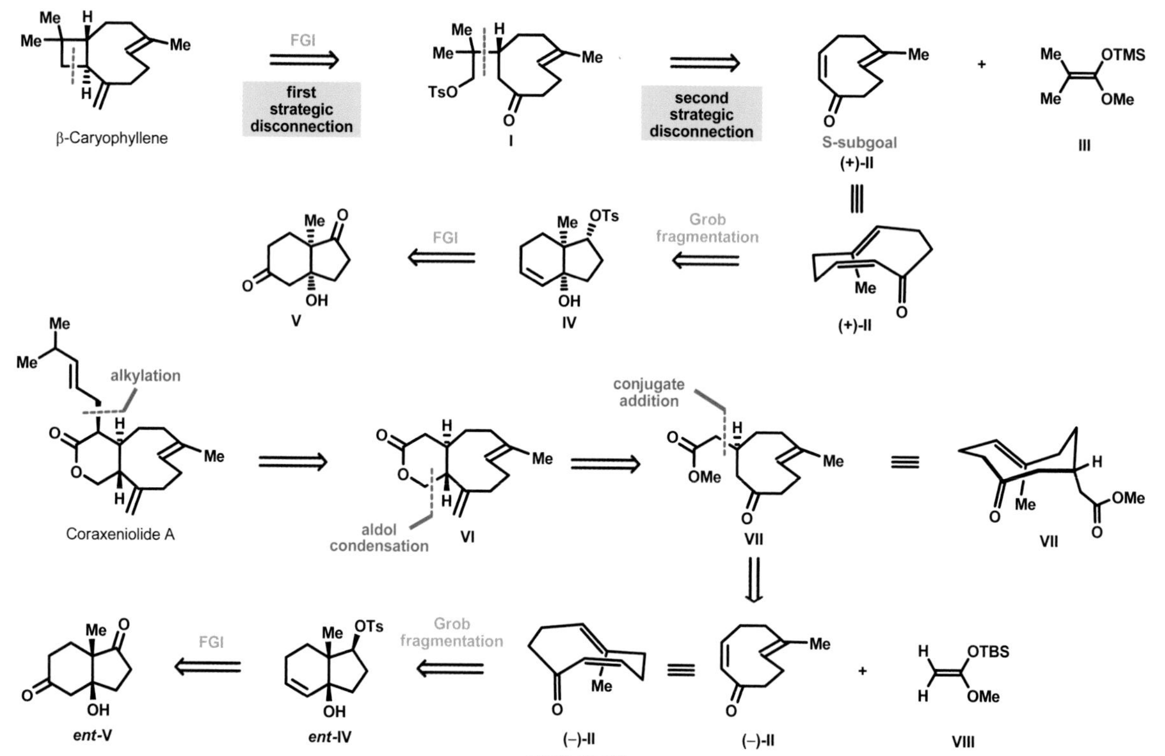

Scheme 1

Key steps in Scheme 1

(1) The retrosynthetic analysis was dependent on the stereochemical objective of establishing absolute stereochemistry in the cyclononadienone **II** which was perceived to be accessible as a key chiral intermediate and well-suited to further elaboration, e.g. to β-caryophyllene or coraxeniolide A.

(2) Using (+)-**II** as a structural subgoal the 4-membered ring of β-caryophyllene was disconnected retrosynthetically as shown. Similarly, the six-membered ring of coraxeniolide A was disconnected in stages to (−)-**II**.

(3) The enantioselective synthesis of either (+)- or (−)-**II** depended in part on the strategy developed earlier for the synthesis of β-caryophyllene.[3]

CARYOPHYLLOIDS

Pathway of Synthesis

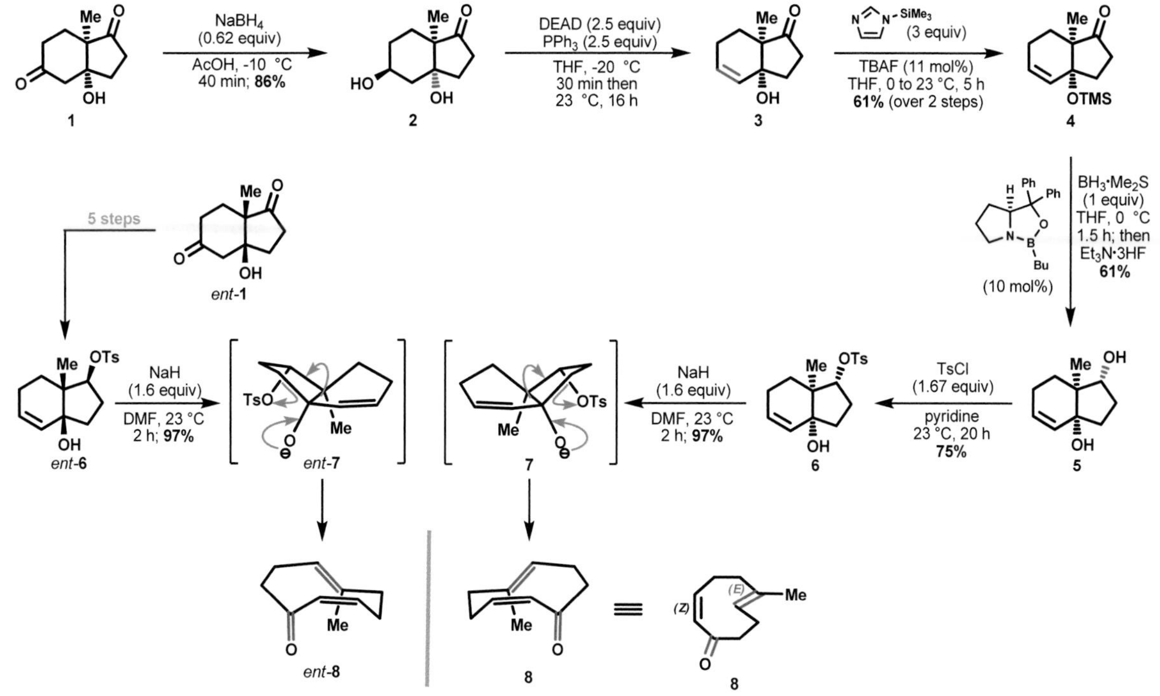

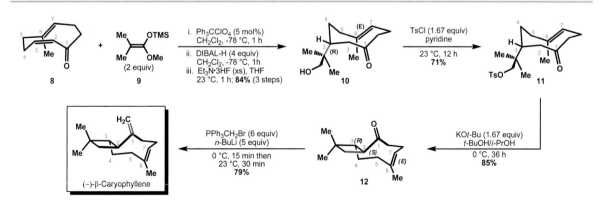

Scheme 2

Scheme 3

Key steps in Schemes 2 and 3

(1) Compounds **1** and *ent*-**1** are available by the Hajos-Parrish-modification of the Robinson annulation reaction using (*R*)- and (*S*)-proline, respectively.[6]

(2) The conversion of **6** to **8** via **7** follows the 9-membered ring construction originally applied to the synthesis of β-caryophyllene.[3]

(3) The conversion of **8** to **10** was face-selective.

CARYOPHYLLOIDS

Pathway of Synthesis (continued)

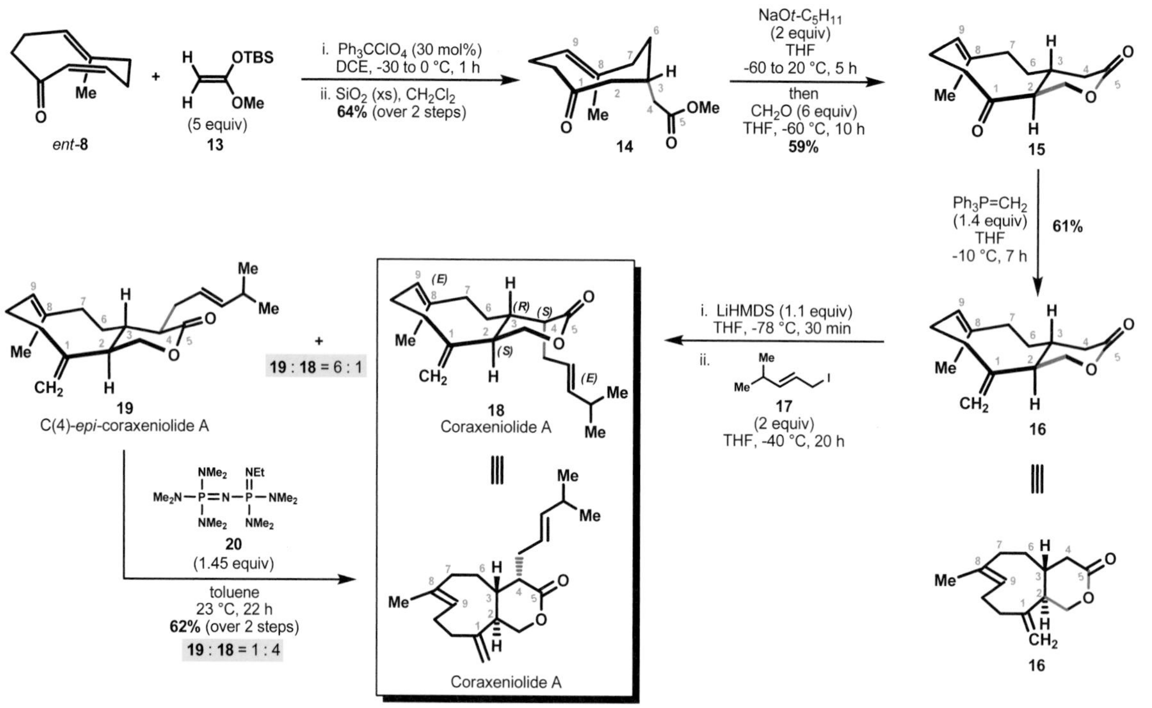

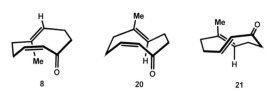

Scheme 4

Key steps in Scheme 4

(1) The conversion of *ent*-**8** to **14** was face-selective, as was the subsequent hydroxymethylation (with dry, monomeric $H_2C=O$ and $NaOt\text{-}C_5H_{11}$ as base.)

(2) The key chiral enantiomeric intermediates, **8** and *ent*-**8**, do not interconvert at 0 °C, but do so upon heating to 60 °C. They are also stable relative to isomerization to the diastereomer in which the C-CH₃ and C=O subunits are *trans*-oriented (**20**).

(3) Hartree-Fock calculations at the 6-31 G* level indicate that **20** is higher in energy than **8** by about 3.1 kcal/mol. In addition, the energy of the (*E,E*)-diastereomer **21** is 15.4 kcal/mol higher in energy than **8**.

References

1. (a) Dawson, T. L. & Ramage, G. R. Caryophyllenes. IX. Homocaryophyllenic acid. *J. Chem. Soc.* **1951**, 3382-3386; (b) Clunie, J. S. & Robertson, J. M. Structure of isoclovene. *Proceedings of the Chemical Society, London* **1960**, 82-83.

2. For reviews on some other naturally occurring caryophylloids, see: (a) Hooper, G. J., Davies-Coleman, M. T., Schleyer, M. New diterpenes from the South African soft coral Eleutherobia aurea. *J. Nat. Prod.* **1997**, 60, 889-893; (b) Miyaoka, H., Nakano, M., Iguchi, K., Yamada, Y. Two new xenicane diterpenoids from Okinawan soft coral of the genus, Xenia. *Heterocycles* **2003**, 61, 189-196.

3. (a) Corey, E. J., Mitra, R. B., Uda, H. Total synthesis of dl-caryophyllene and dl-isocaryophyllene. *J. Am. Chem. Soc.* **1963**, 85, 362-363; (b)

Corey, E. J., Mita, R. N., Uda, H. Total synthesis of dl-caryophyllene and dl-isocaryophyllene. *J. Am. Chem. Soc.* **1964**, 86, 485-492.

4. Mushti, C. S., Kim, J.-H. & Corey, E. J. Total Synthesis of Antheliolide A. *J. Am. Chem. Soc.* **2006**, 128, 14050-14052.

5. (a) Renneberg, D., Pfander, H. & Leumann, C. J. Total Synthesis of Coraxeniolide-A. *J. Org. Chem.* **2000**, 65, 9069-9079; (b) Larionov, O. V. & Corey, E. J. An unconventional approach to the enantioselective synthesis of caryophylloids. *J. Am. Chem. Soc.* **2008**, 130, 2954-2955.

6. Kürti, L. & Czakó, B. Hajos-Parrish Reaction in *Strategic Applications of Named Reactions in Organic Synthesis* 192-193 (Academic Press/Elsevier, San Diego, **2005**).

(–)-OSELTAMIVIR (TAMIFLU™)

Background

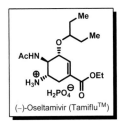

(–)-Oseltamivir (Tamiflu™)

Oseltamivir is an orally active anti-influenza medicine invented by the chemist Chung U. Kim at Gilead Sciences Co.[1] It inhibits the entry of various strains of influenza virus into cells by blocking the enzyme neuraminidase. Neuramidinase is one of the proteins on the viral coat which acts to allow release of active virus inside an infected cell after endocytosis and is important to infection. The need for a safe, enantioselective and efficient synthetic route to oseltamivir has stimulated the development of improved synthetic routes to this molecule, one of which is discussed here.[2] This route is enantio- and stereocontrolled, efficient and devoid of hazardous intermediates (e.g., azide-containing) or reagents.

Abbreviated Retrosynthetic Plan

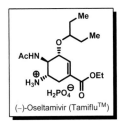

Scheme 1

Key steps in Scheme 1

(1) The retrosynthetic plan was guided by the simultaneous application of the following strategies:

  (a) clearance of stereocenters and functional groups using a mixture of mechanism-based and structure-based stereocontrolled transforms; including halolactamization to clear the 5-amino group (**V** ⇒

VI) and haloamidation to clear the 4-acetamido group (**II** ⇒ **III**);

  (b) the enantioselective Diels-Alder transform to clear the final stereocenter (**VII** ⇒ **VIII** ⇒ **IX**).

(2) The initiating stereocenter [C(1) of **VII**] serves to control the elaboration of the other stereocenters and the functionality of oseltamivir. The intermediate **VII** contains the full carbon skeleton of oseltamivir.

Pathway of Synthesis

Scheme 2

(−)-OSELTAMIVIR (TAMIFLU™)

Pathway of Synthesis

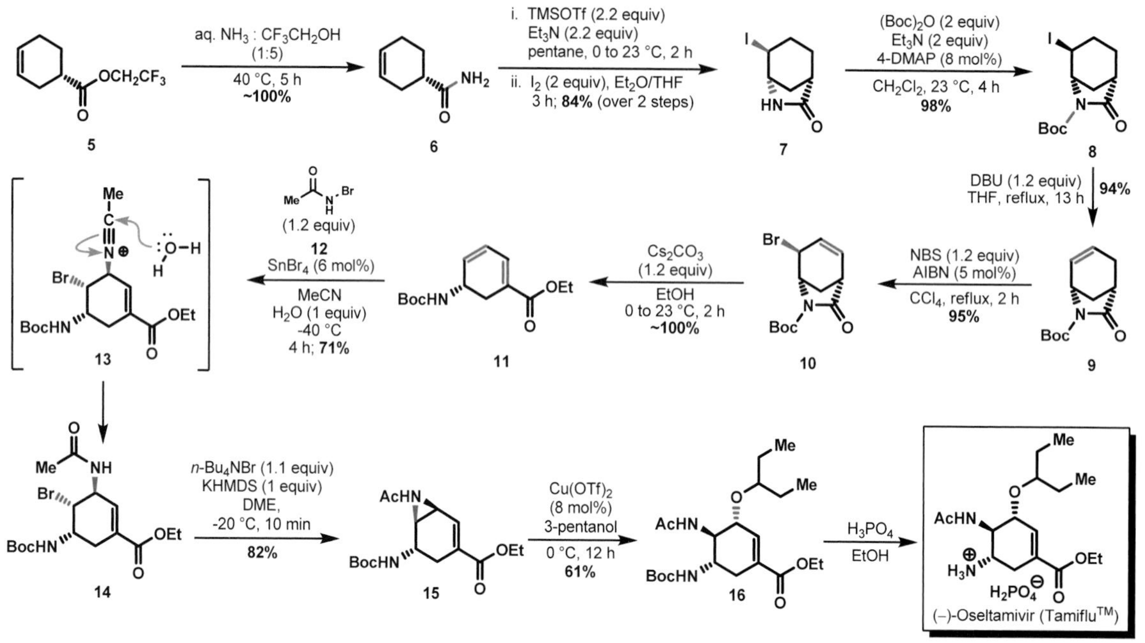

Scheme 3

Key steps in Schemes 2 and 3

(1) The enantioselective Diels-Alder reaction **1+2 → 5** using catalyst **3** is considered to proceed via the pre-transition state assembly **4** on the basis of extensive studies.[3]

(2) The conversion of **6** to **9** has also been carried out efficiently using the steps: **6** → iodolactam **7** → N-Boc derivative **8** → elimination product **9**.[4]

(3) The radical bromination **9 → 10** is position- and stereo-selective.

(4) The conversion **11→14** by bromoamidation is a novel and general method which requires CH$_3$CN as solvent and 1 equiv of H$_2$O.[5] It is considered to proceed by attack of CH$_3$CN on a bromonium ion to give intermediate **13** which reacts with water to afford **14**.[6]

References

1. (a) Moscona, A. Neuraminidase inhibitors for influenza. *N. Engl. J. Med.* **2005**, 353, 1363-1373; (b) Corey, E. J., Czakó, B. & Kürti, L. Oseltamivir (Tamiflu™) in *Molecules and Medicine* p 150 (John Wiley & Sons, Inc., Hoboken, New Jersey, **2007**).

2. Yeung, Y.-Y., Hong, S. & Corey, E. J. A Short Enantioselective Pathway for the Synthesis of the Anti-Influenza Neuramidase Inhibitor Oseltamivir from 1,3-Butadiene and Acrylic Acid. *J. Am. Chem. Soc.* **2006**, 128, 6310-6311.

3. (a) Ryu, D. H. & Corey, E. J. Triflimide Activation of a Chiral Oxazaborolidine Leads to a More General Catalytic System for Enantioselective Diels-Alder Addition. *J. Am. Chem. Soc.* **2003**, 125, 6388-6390; (b) For a review see: Corey E.J. Enantioselective Catalysis Based on Cationic Oxazaborolidines. *Angew. Chem. Intl. Ed*, **2009**, 48, 2100-2117.

4. Yeung, Y.-Y. & Corey, E. J. An efficient process for the bromolactamization of unsaturated acids. *Tetrahedron Lett.* **2007**, 48, 7567-7570.

5. Yeung, Y.-Y., Gao, X. & Corey, E. J. A General Process for the Haloamidation of Olefins. Scope and Mechanism. *J. Am. Chem. Soc.* **2006**, 128, 9644-9645.

6. For other synthetic work on oseltamivir, see: (a) Ishikawa, H., Suzuki, T., Hayashi, Y. High-yielding synthesis of the anti-influenza neuramidase inhibitor (−)-oseltamivir by three "one-pot" operations. *Angew. Chem., Int. Ed.* **2009**, 48, 1304-1307; (b) Matveenko, M., Willis, A. C., Banwell, M. G. A chemoenzymic synthesis of the anti-influenza agent Tamiflu. *Tetrahedron Lett.* **2008**, 49, 7018-7020; (c) Shie, J.-J., Fang, J.-m.,

Wong, C.-H. A concise and flexible synthesis of the potent anti-influenza agents tamiflu and tamiphosphor. *Angew. Chem., Int. Ed.* **2008**, 47, 5788-5791; (d) Zutter, U., Iding, H., Spurr, P., Wirz, B. New, Efficient Synthesis of Oseltamivir Phosphate (Tamiflu) via Enzymatic Desymmetrization of a meso-1,3-Cyclohexanedicarboxylic Acid Diester. *J. Org. Chem.* **2008**, 73, 4895-4902; (e) Trost, B. M., Zhang, T. A concise synthesis of (−)-oseltamivir. *Angew. Chem., Int. Ed.* **2008**, 47, 3759-3761; (f) Shibasaki, M., Kanai, M. Synthetic strategies for oseltamivir phosphate. *Eur. J. Org. Chem.* **2008**, 1839-1850; (g) Abrecht, S., Federspiel, M. C., Estermann, H., Fischer, R., Karpf, M., Mair, H.-J., Oberhauser, T., Rimmler, G., Trussardi, R., Zutter, U. The synthetic-technical development of oseltamivir phosphate, Tamiflu: a race against time. *Chimia* **2007**, 61, 93-99; (h) Satoh, N., Akiba, T., Yokoshima, S., Fukuyama, T. A practical synthesis of (−)-oseltamivir. *Angew. Chem., Int. Ed.* **2007**, 46, 5734-5736; (i) Yamatsugu, K., Kamijo, S., Suto, Y., Kanai, M., Shibasaki, M. A concise synthesis of Tamiflu: third generation route via the Diels-Alder reaction and the Curtius rearrangement. *Tetrahedron Lett.* **2007**, 48, 1403-1406; (j) Mita, T., Fukuda, N., Roca, F. X., Kanai, M., Shibasaki, M. Second Generation Catalytic Asymmetric Synthesis of Tamiflu: Allylic Substitution Route. *Org. Lett.* **2007**, 9, 259-262; (k) Shie, J.-J., Fang, J.-M., Wang, S.-Y., Tsai, K.-C., Cheng, Y.-S. E., Yang, A.-S., Hsiao, S.-C., Su, C.-Y., Wong, C.-H. Synthesis of Tamiflu and its Phosphonate Congeners Possessing Potent Anti-Influenza Activity. *J. Am. Chem. Soc.* **2007**, 129, 11892-11893.

(+)-PENTACYCLOANAMMOXIC ACID

Background

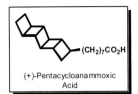

(+)-Pentacycloanammoxic
Acid

Pentacycloanammoxic acid is a membrane component of the anaerobic microbe *Candidatus Brocadia anammoxidans* which obtains energy from the conversion of NH_3 and NO_2^- to N_2 and H_2O.[1]

The microbe is shielded from toxic metabolites that are generated in this process by a particularly dense membrane formed from esters of the very rigid pentacycloanammoxic structure. The biosynthesis remains a mystery.[2] The enantiomers[3] (and the racemate[4]) of pentacycloanammoxic acid were first synthesized by the routes outlined herein.

Abbreviated Retrosynthetic Plan

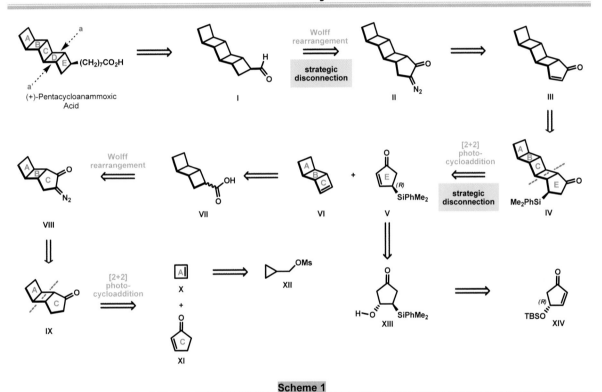

Scheme 1

Key steps in Scheme 1

(1) A major retrosynthetic problem was the identification of transforms that would permit the clearing of all nine stereocenters of pentacycloanammoxic acid.

(2) The disconnection of the **a** and **a'** bonds by the use of a cycloaddition transform was selected as a possible strategic approach since it would give the achiral tricyclic olefin **VI** (containing rings A, B and C) and a cyclic unit corresponding to ring E.

(3) The chiral 2-cyclopentenone **V** or some chiral cyclic α,β-enone appeared to be a suitable cycloaddition partner for the achiral tricyclic olefin **VI**. The silyl group in **V** can serve as a sterically large control element for cycloaddition, making the disconnection **IV** ⇒ **V** + **VI** valid stereochemically.

(4) The 2-cyclopentenone **V** seemed to offer numerous advantages over a 2-cyclobutenone, including:

   (a) easier synthetic access and

   (b) previous knowledge that 2-cyclopentenones are excellent substrates in [2+2]-photocycloadditions to olefins.

(5) The bulky silyl control element (SiPhMe$_2$) in **V** can serve to provide diastereocontrol as well as enantiocontrol.

(6) The pentacycle **IV** served as a structural subgoal to guide the retrosynthetic steps from pentacycloanammoxic acid along the path **I** ⇒ **II** ⇒ **III** ⇒ **IV**.

(+)-PENTACYCLOANAMMOXIC ACID

Pathway of Synthesis

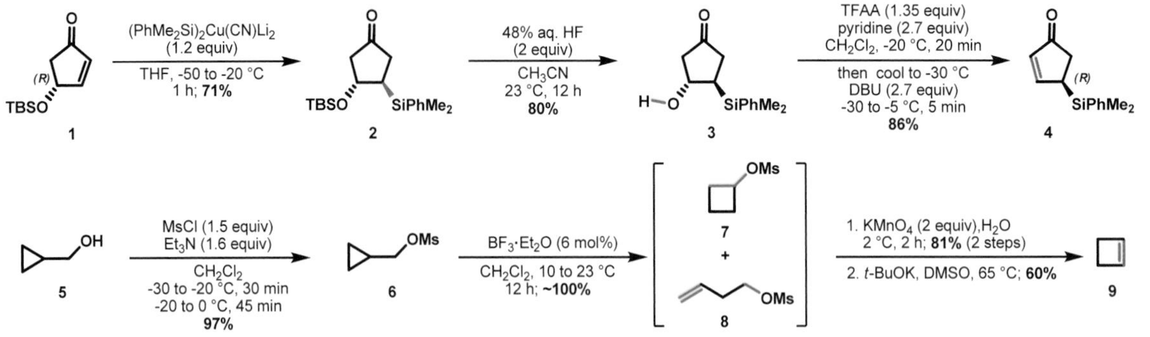

Scheme 2

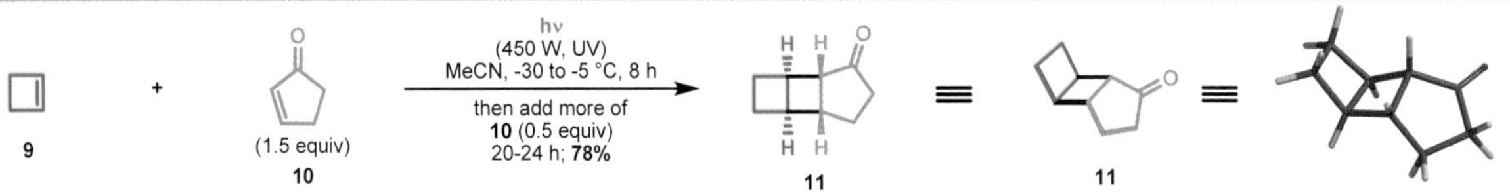

Scheme 3

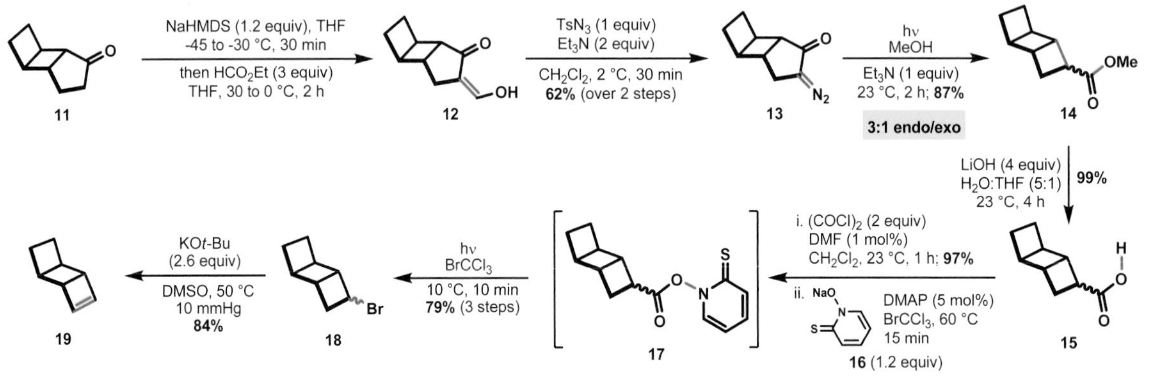

Scheme 4

Key steps in Schemes 2, 3 and 4

(1) The chiral (3*R*)-silyl-2-cyclopentanone **4** was prepared from the known α,β-enone **1**.

(2) Cyclobutene was conveniently synthesized from cyclopropylmethanol **5** on a molar scale via **6** and **7**.

(3) The [2+2]-photocycloaddition of cyclobutene **9** with 2-cyclopentenone **10** was *anti*-selective (Scheme 3).

(4) The ring-contraction of the diazo ketone **13** by Wolff rearrangement afforded a mixture of *exo* and *endo* methyl esters **14** via a ketene intermediate **13b** (see pathway at right).

(5) Brominative decarboxylation of the Barton ester **17** occurs via a radical pathway that produces bromide **18** efficiently.

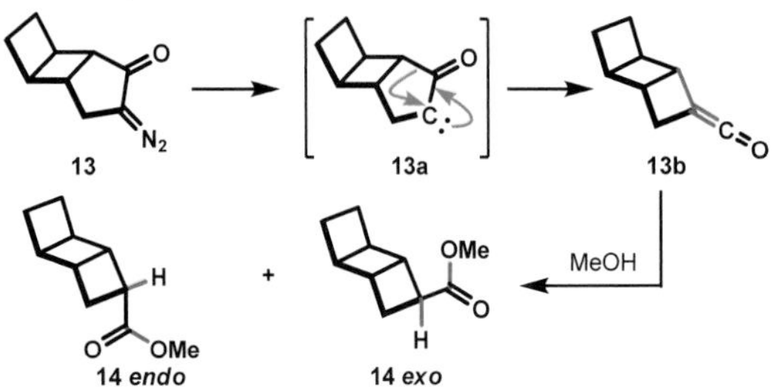

(6) The preparation of **19** by a simpler route remains challenging.

(+)-PENTACYCLOANAMMOXIC ACID

Pathway of Synthesis (continued)

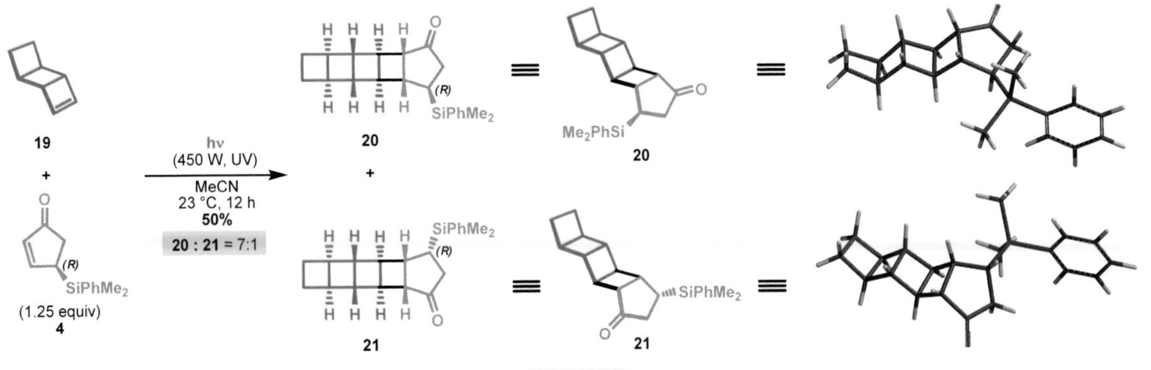

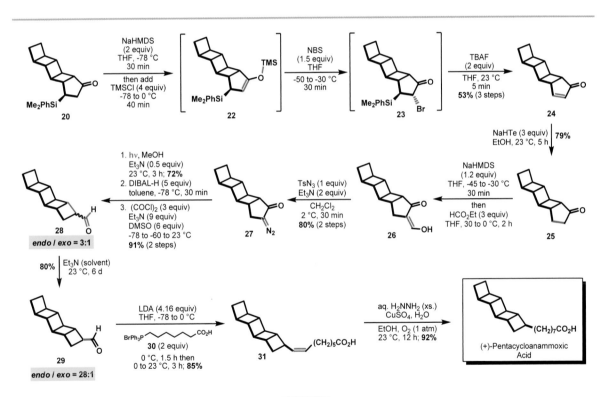

Scheme 5

Scheme 6

Key steps in Schemes 5 and 6

(1) The key photocycloaddition of **4** and **19** was 7:1 selective for **20** because addition of the olefin took place predominantly to the face of α,β-enone opposite the bulky Me$_2$PhSi substituent. It was also completely *anti* diastereoselective to form only the *anti* tetracyclic adducts **20** and **21**.

(2) The removal of the Me$_2$PhSi group from **20** was accomplished by the one-flask conversion to **24** via **22** and **23** (Scheme 6).

(3) The more stable *exo* aldehyde **29** could be obtained by equilibration of the *exo / endo* mixture **28**.

(±)-PENTACYCLOANAMMOXIC ACID

Other Developments

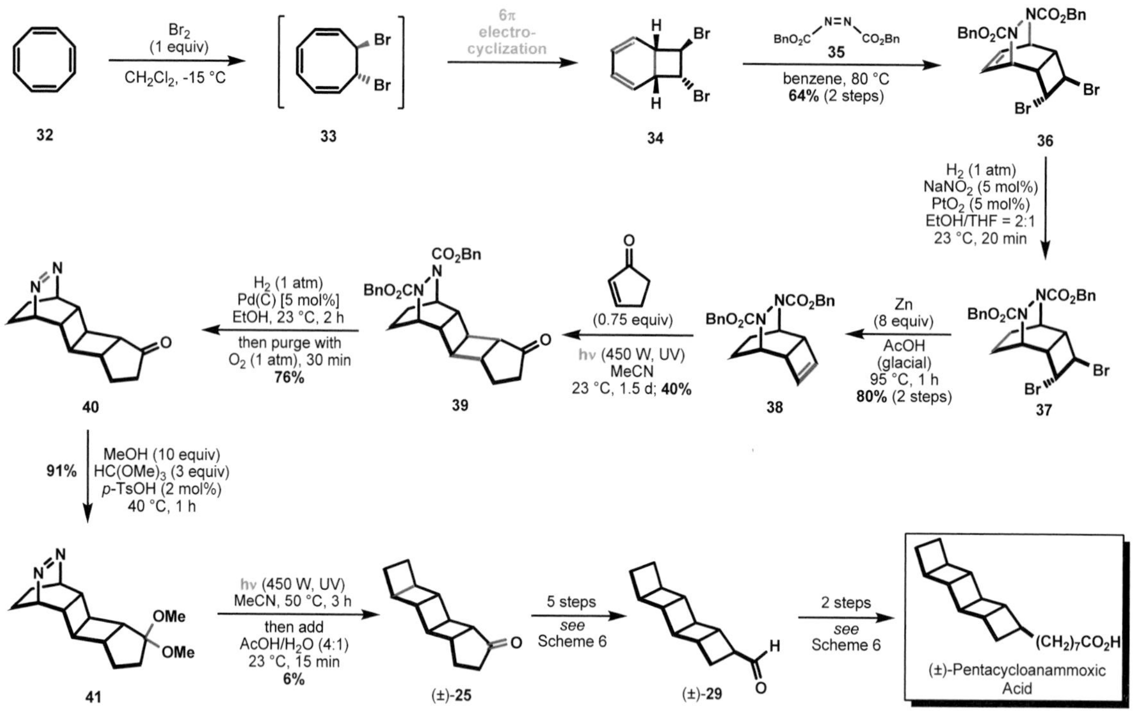

Scheme 7

Key steps in Scheme 7

(1) This synthesis of racemic pentacycloanammoxic acid started with cyclooctatetraene **32**, which contains 8 of the 12 carbon atoms in the pentacyclic nucleus, and the known dibromide **34**.

(2) The photocycloaddition of 2-cyclopentenone to **38** is *anti*-selective.

(3) The extrusion of nitrogen from **41** was complex and inefficient under all conditions tried and the yields of (±)-**25** were poor because of accompanying ring-fragmentation **41** → **42** → **43** + **44**.

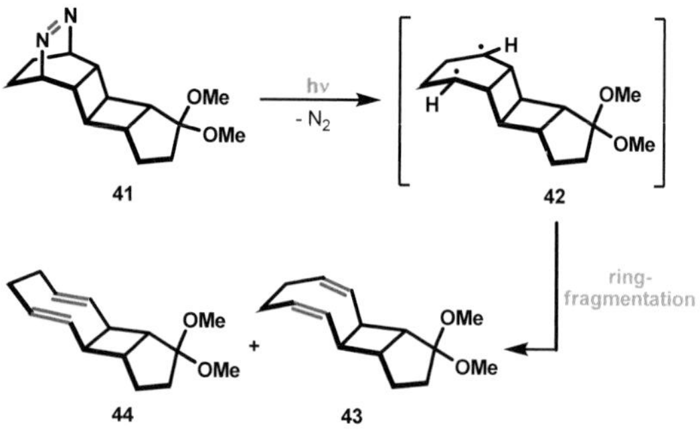

References

1. DeLong, E. F. Microbiology: All in the packaging. *Nature (London, U. K.)* **2002**, 419, 676-677.
2. (a) Jetten, M. S. M., van Niftrik, L., Strous, M., Kartal, B., Keltjens, J. T. & Op den Camp, H. J. M. Biochemistry and molecular biology of anammox bacteria. *Crit. Rev. Biochem. Mol. Biol.* **2009**, 44, 65-84; (b) Rattray, J. E., Geenevasen, J. A. J., van Niftrik, L., Rijpstra, W. I. C., Hopmans, E. C., Strous, M., Schouten, S., Jetten, M. S. M. & Damste, J. S. S. Carbon

isotope-labelling experiments indicate that ladderane lipids of anammox bacteria are synthesized by a previously undescribed, novel pathway. *FEMS Microbiol. Lett.* **2009**, 292, 115-122.
3. Mascitti, V. & Corey, E. J. Enantioselective Synthesis of Pentacycloanammoxic Acid. *J. Am. Chem. Soc.* **2006**, 128, 3118-3119.
4. Mascitti, V. & Corey, E. J. Total Synthesis of (±)-Pentacycloanammoxic Acid. *J. Am. Chem. Soc.* **2004**, 126, 15664-15665.

ATRACTYLIGENIN

Background

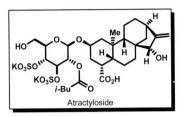

The sulfated glucose conjugate of atractyloside is a highly toxic naturally occurring compound which occurs in various plants, including *Coffea arabica* and *robustica*, the beans of which are roasted for coffee.[1] The toxicity of atractyloside, which has been known since antiquity, is due

to its blocking of ADP translocation into mitochondria and ATP biosynthesis.[2]

The first synthesis of (±)-atractyligenin[3] provided a platform for the later development of an enantioselective route.[4] The challenge in the synthesis of atractyligenin arises from a combination of several different types of complexity and the density of that complexity.[5]

Abbreviated Retrosynthetic Plan

Scheme 1

Key steps in Scheme 1

(1) To meet the challenges of topological complexity, the selected strategy utilized a cation-initiated cyclization to establish the B-ring. This, and the recognition of the C(1)-C(6) strategic disconnection (see **VII**) led to the need first to clear functionality and stereocenters retrosynthetically.

(2) That need for clearance guided the sequence to intermediate **VII** via the intervening sequence **I** ⇒ **VI**.

(3) Further analysis of **VII** suggested the pathway **VII** ⇒ **VIII** ⇒ **IX** ⇒ **XII**.

(4) Finally, the last remaining stereocenter in **XII** had to be cleared with stereocontrol, a task that was difficult, even though it may appear simple. The difficulty arises partly from the quaternary nature of the last stereocenter on an almost symmetrical ring.

ATRACTYLIGENIN

Pathway of Synthesis

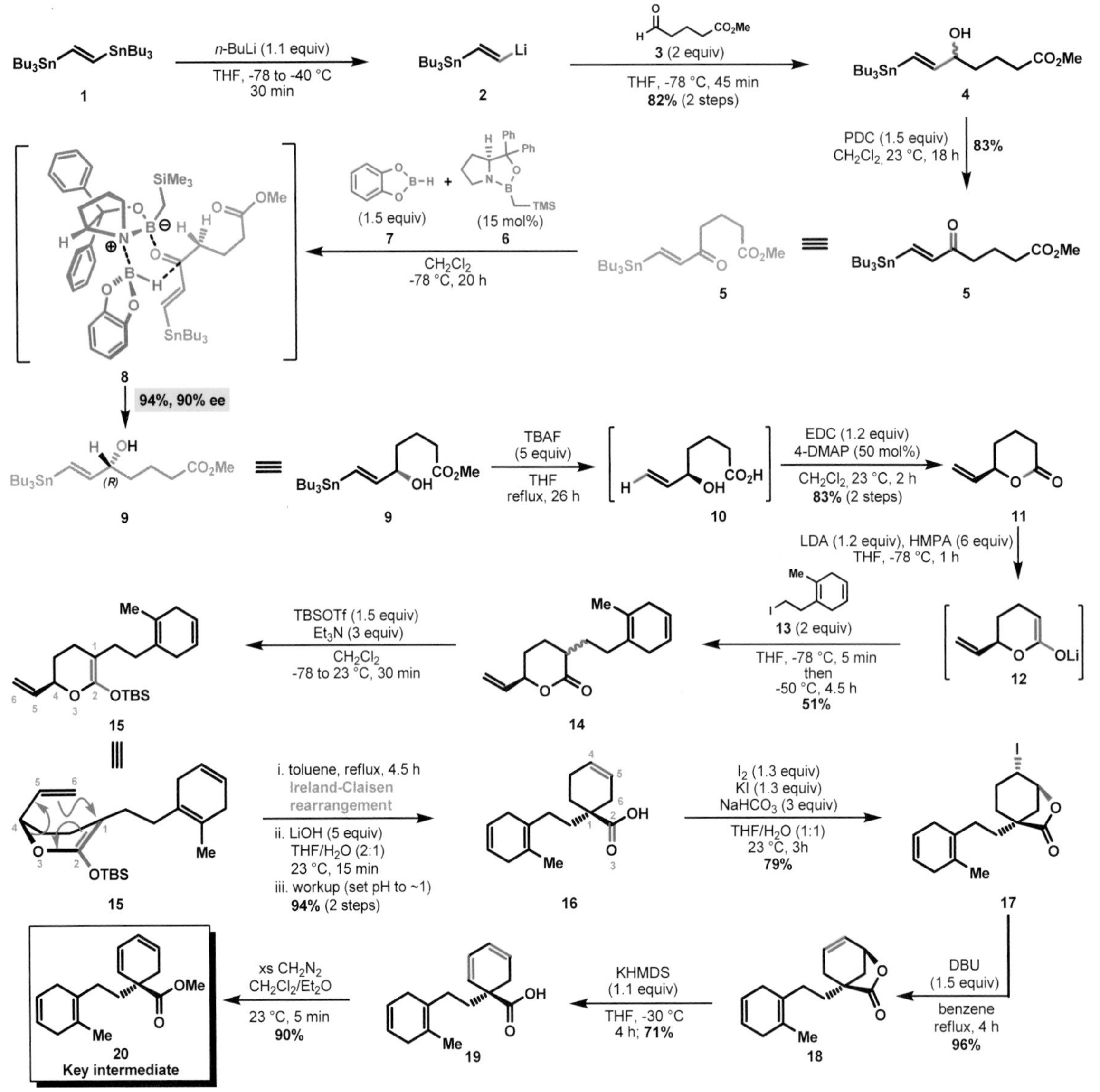

Scheme 2

Key steps in Scheme 2

(1) The enantioselective synthesis of **20**, the first chiral intermediate for the construction of the naturally occurring form of atractyligenin, depended on the establishment of absolute configuration by asymmetric reduction of the ketone carbonyl group in **5**.

(2) Ketone **5** was reduced by catecholborane **7** in the presence of chiral oxazaborolidine catalyst **6** to form chiral allylic alcohol **9**.

(3) After destannylation (**9→10**), lactonization to **11** and alkylation to **14**, Claisen rearrangement of the silyl ether **15** generated the required quaternary stereocenter in **16**.

(4) Iodolactonization of **16**, in a series of two olefin-forming eliminations and esterification gave **20**.

ATRACTYLIGENIN

Pathway of Synthesis (continued)

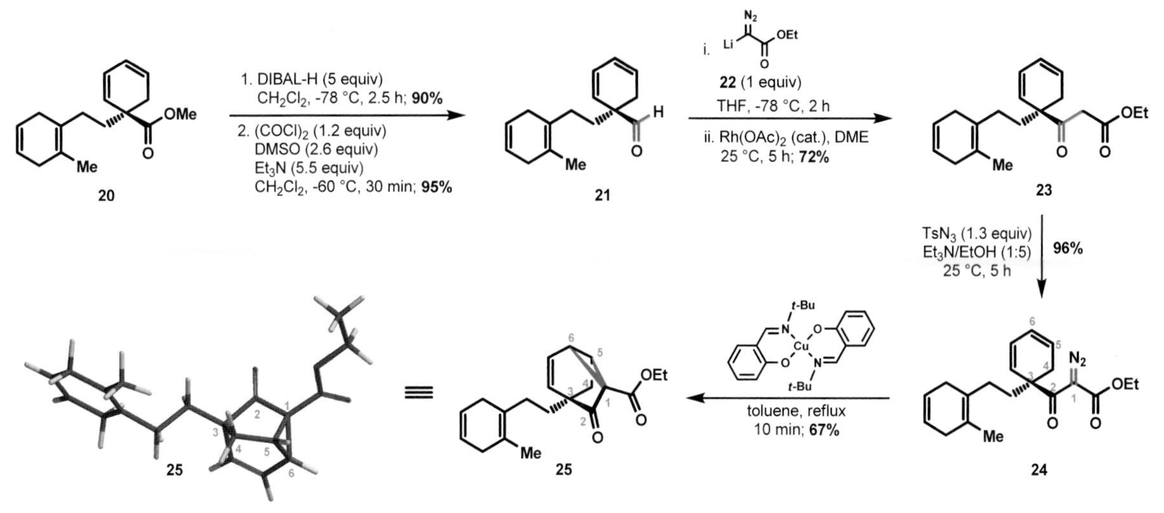

Scheme 3

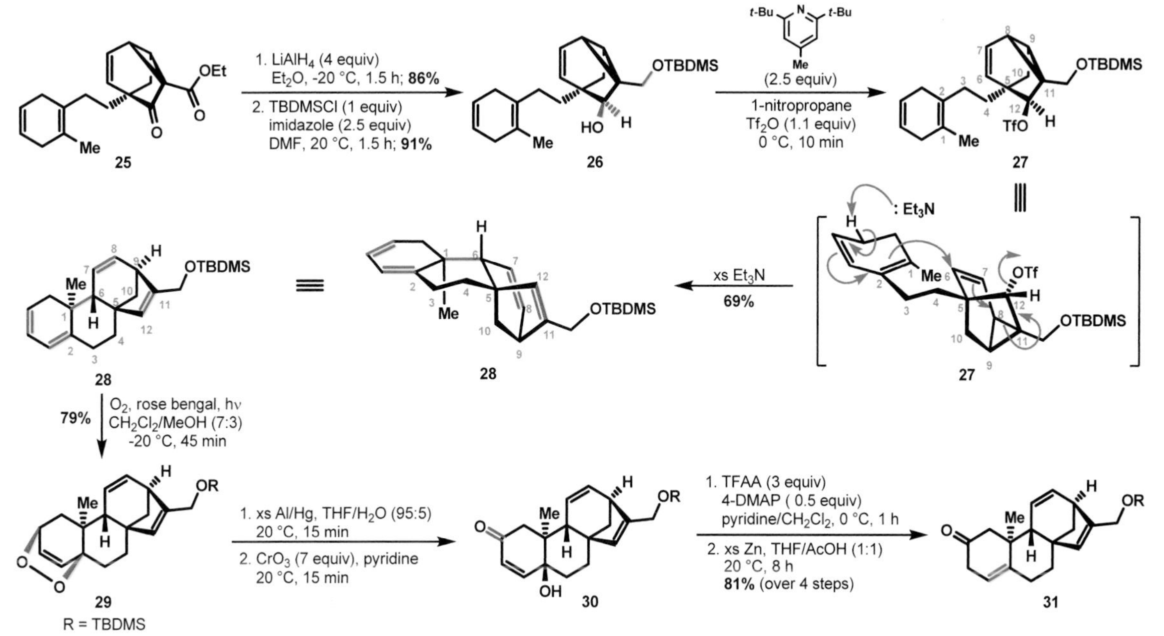

Scheme 4

Key steps in Schemes 3 and 4

(1) The initiating cation required for the cyclization to form ring B (see **28**) was generated by ionization of triflate **27** to form an allylic cation by cyclopropyl ring cleavage.

(2) The conversion **30**→**31** involved selective α-protonation of a dienol intermediate (**30a**, see at right).

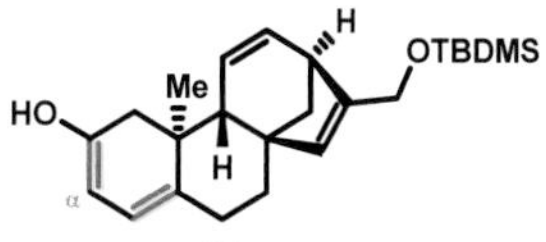

ATRACTYLIGENIN

Pathway of Synthesis (continued)

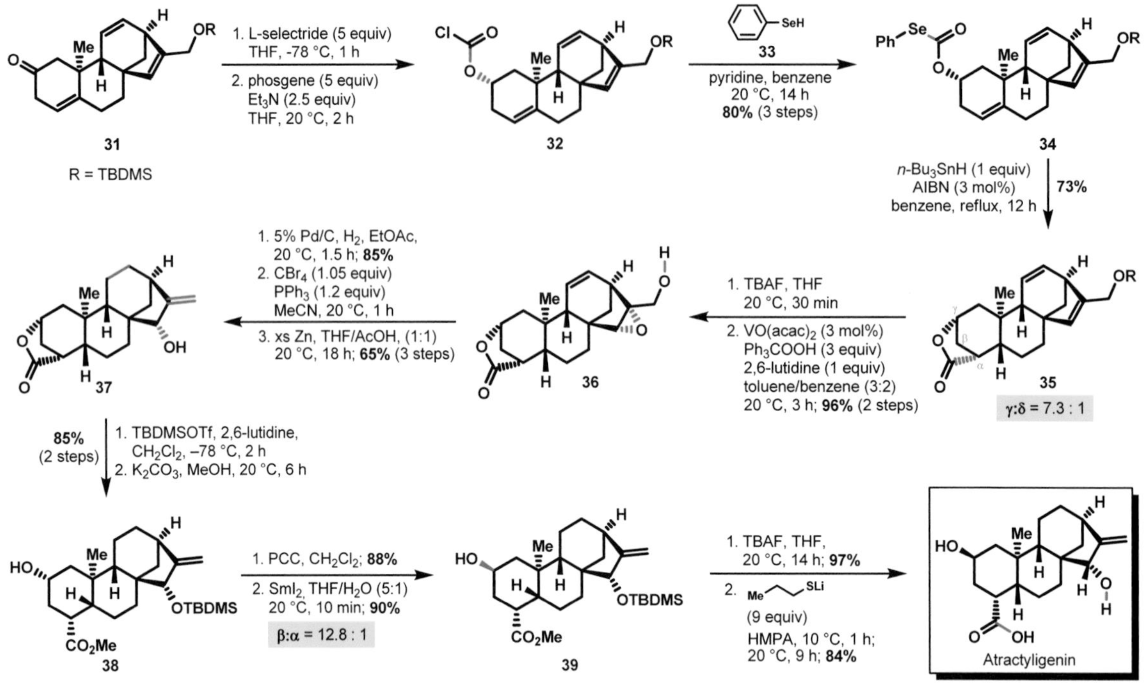

Scheme 5

Key steps in Scheme 5

(1) The conversion of selenocarbonate intermediate **34** to the lactone **35** occurs via carbonyl radical formation (**34a**) and addition to the nearby double bond (**34b**, Markovnikov orientation).[6]

(2) The endocyclic double bond of **35** was transposed to the exocyclic location in **37** with concurrent introduction of a hydroxyl group via the epoxy alcohol **36**.

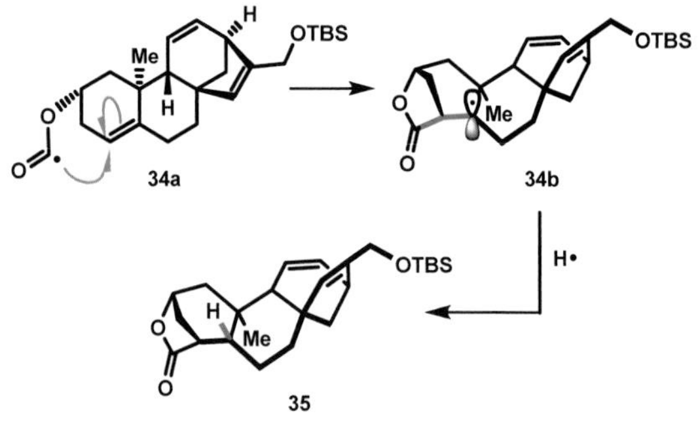

References

1. (a) Ludwig, H., Obermann, H. & Spiteller, G. Atractyligenin, an essential component of roasted coffee beans. *Chem. Ber.* **1974**, 107, 2409-2411; (b) Obermann, H. & Spiteller, G. The structures of the "coffee atractylosides". *Chem. Ber.* **1976**, 109, 3450-3461; (c) Richter, H. & Spiteller, G. A new atractyligenin glycoside from green coffee beans. *Chem. Ber.* **1978**, 111, 3506-3509.

2. *Atractyloside: Chemistry, Biochemistry and Toxicology* (eds. Santi, R. & Luciani, S.) (Piccin Medical Books, Padova (Italy), **1978**).

3. Singh, A. K., Bakshi, R. K. & Corey, E. J. Total synthesis of (±)-atractyligenin. *J. Am. Chem. Soc.* **1987**, 109, 6187-6189.

4. Corey, E. J., Guzman-Perez, A. & Lazerwith, S. E. An Enantioselective Synthetic Route to Atractyligenin Using the Oxazaborolidine-Catalyzed

Reduction of β-Silyl- or β-Stannyl-Substituted α,β-Enones as a Key Step. *J. Am. Chem. Soc.* **1997**, 119, 11769-11776.

5. Piozzi, F., Quilico, A., Fuganti, C., Ajello, T. & Sprio, V. Structure of atractyloside. *Gazz. Chim. Ital.* **1967**, 97, 935-954.

6. For reviews on the use of selenium precursors in radical reactions see: (a) Chatgilialoglu, C., Crich, D., Komatsu, M., Ryu, I. Chemistry of Acyl Radicals. *Chem. Rev.* **1999**, 99, 1991-2069; (b) Renaud, P. Radical reactions using selenium precursors. *Top. Curr. Chem.* **2000**, 208, 81-112; (c) Ogawa, A. in Main Group Metals in Organic Synthesis 813-866 (**2004**).

PSEUDOPTEROSINS

Background

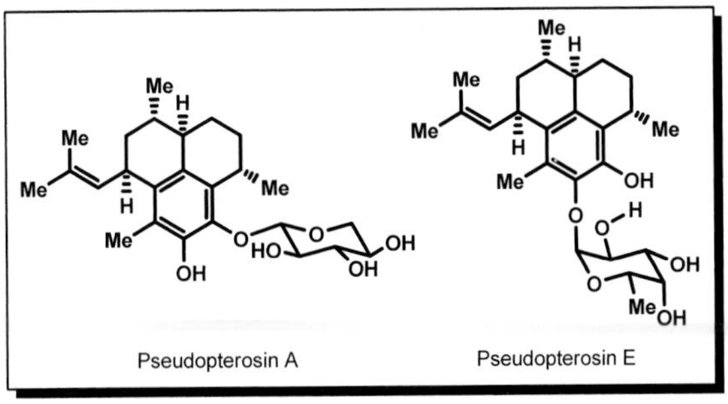

Pseudopterosin A   Pseudopterosin E

The pseudopterosins are a family of marine natural products isolated from the sea plume *Pseudopterosia elizabethae* which show strong anti-inflammatory activity,[1] apparently because they are potent agonists of adenosine. They are now being studied as accelerators of wound healing. In this section three synthetic paths to the pseudopterosins are summarized.[2]

Abbreviated Retrosynthetic Plan

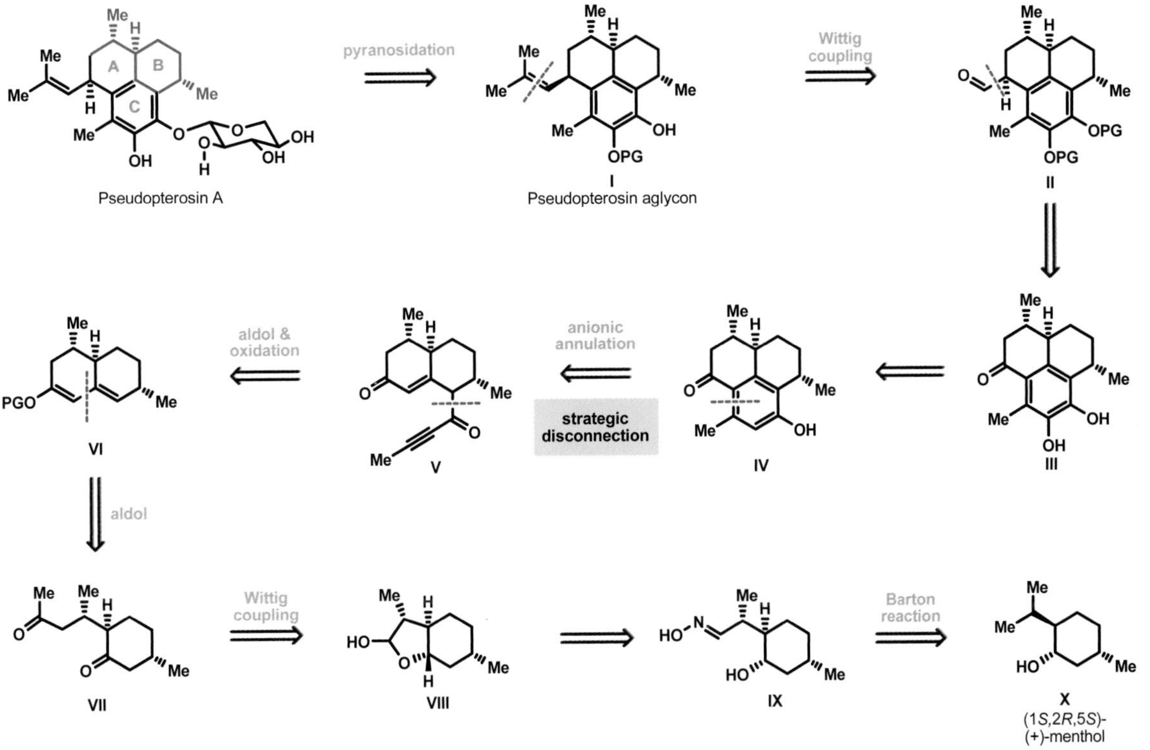

Scheme 1

Key steps in Scheme 1

(1) The first synthetic approach was devised using the inexpensive chiral molecule (+)-menthol (**X**) as a structure goal because it maps to ring B and the methyl-substituted benzylic stereocenter. The menthyl subunit is highlighted in red.

(2) The above choice guided the examination of the retrosynthetic disconnection of ring C along the path **IV** ⇒ **V** ⇒ **VI**. A novel way of implementing this disconnection was devised by the use of a mechanistically based sequence keyed by the phenolic and carbonyl functional groups.

(3) The disconnection **V** ⇒ **VI** depended on a γ-Mukaiyama aldol coupling and secondary alcohol oxidation sequence.

(4) The removal of the oxime function in **IX** results from application of the Barton nitrite ester photolysis transform.[3] The alkoxy radical corresponding to **X** selectively abstracts H from the nearest carbon that allows 1,5-H transfer.

PSEUDOPTEROSINS

Pathway of Synthesis

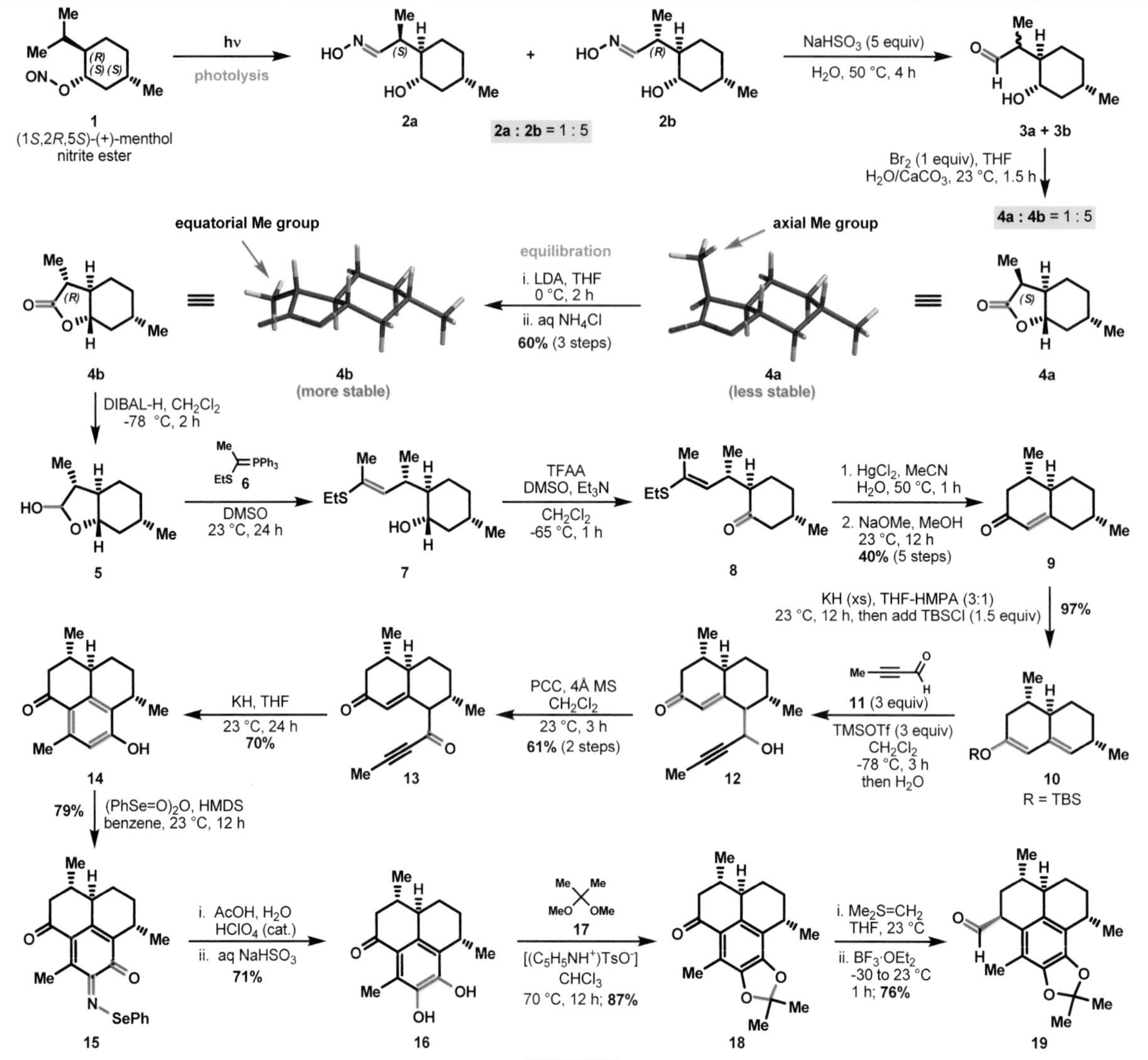

Scheme 2

Key steps in Scheme 2

(1) The conversion **1→2** occurs by homolytic cleavage of the nitrite ester function and hydrogen atom abstraction from the proximate methyl group via a six-membered H atom transfer cycle.

(2) The annulation **13→14** is thought to occur by electrocyclization of enolate **13a** to enolate **13b** (see below):

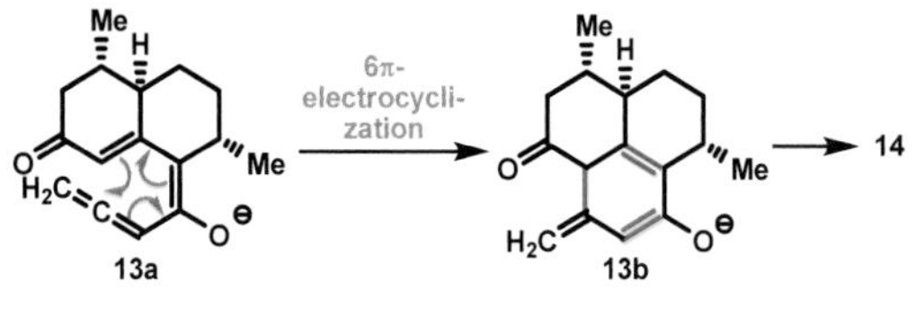

(3) The oxidative phenol *ortho*-amidation **14→15** is thought to take place via **14a→14b** (see below):[4]

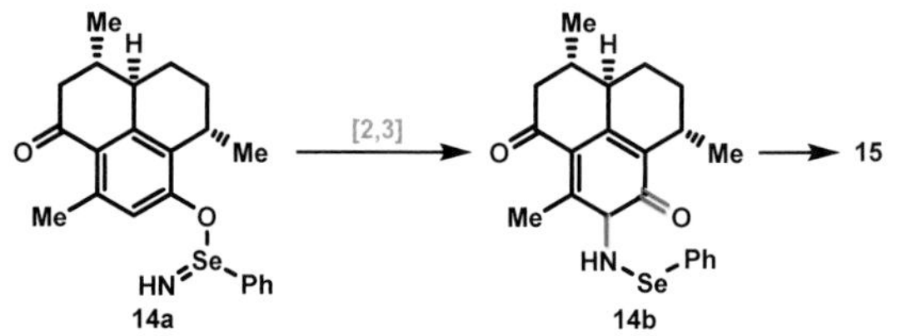

(4) The conversion of **18→19** occurs via BF₃-catalyzed rearrangement of the epoxide formed by the carbonyl methylenation of **18**.[5]

PSEUDOPTEROSINS

Pathway of Synthesis (continued)

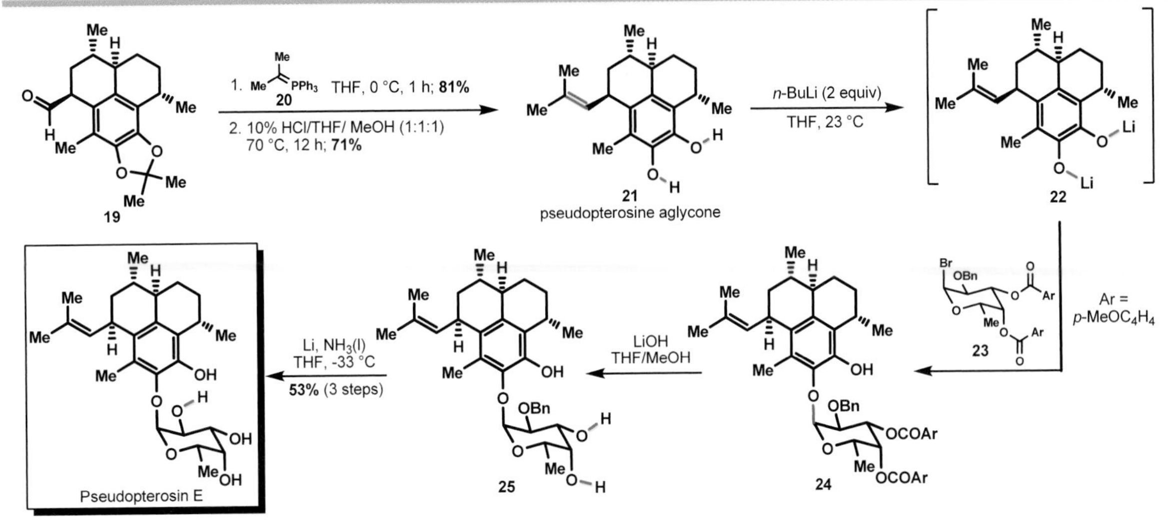

Scheme 3

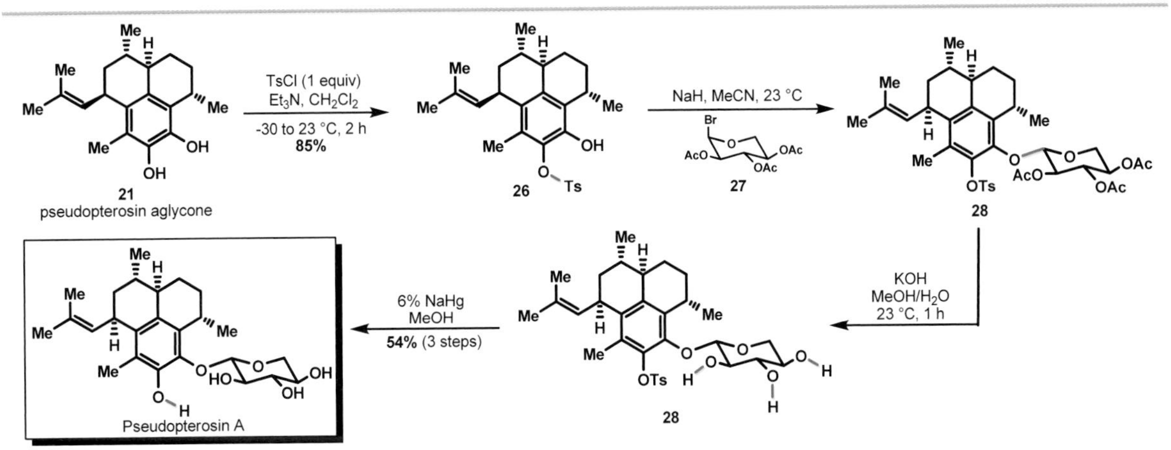

Scheme 4

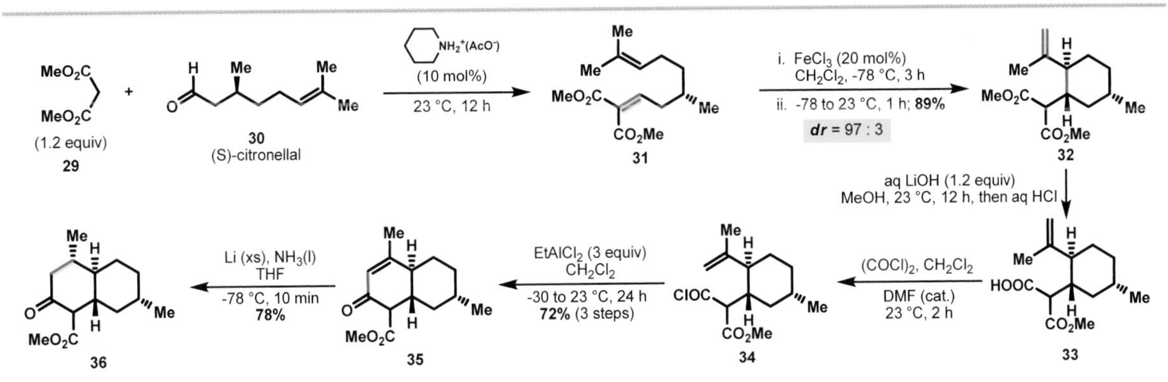

Scheme 5

PSEUDOPTEROSINS

Alternative Synthesis of Bicyclic Enone 9 and Tricyclic Ketone 14

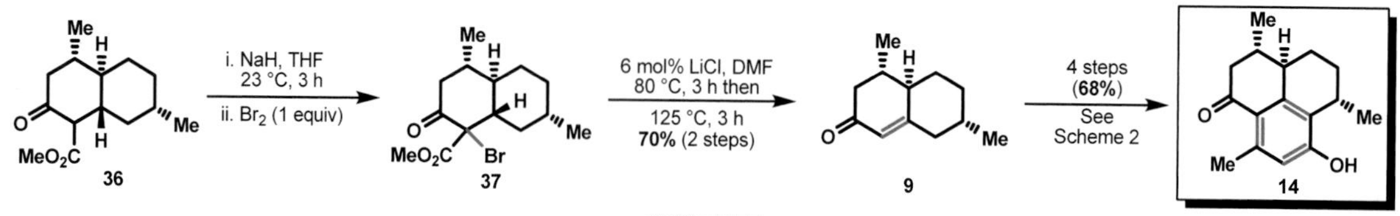

Scheme 6

Key steps in Schemes 5 and 6

The cyclization **31**→**32** is considered to be an intramolecular ene reaction of an intermediate bidentate complex **31a**.

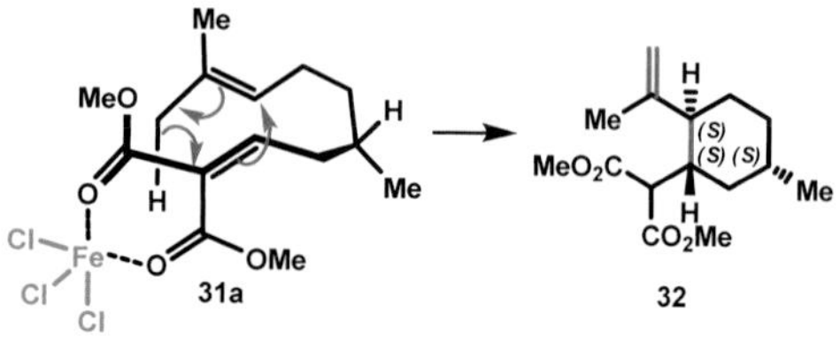

Another Approach to Pseudopterosins

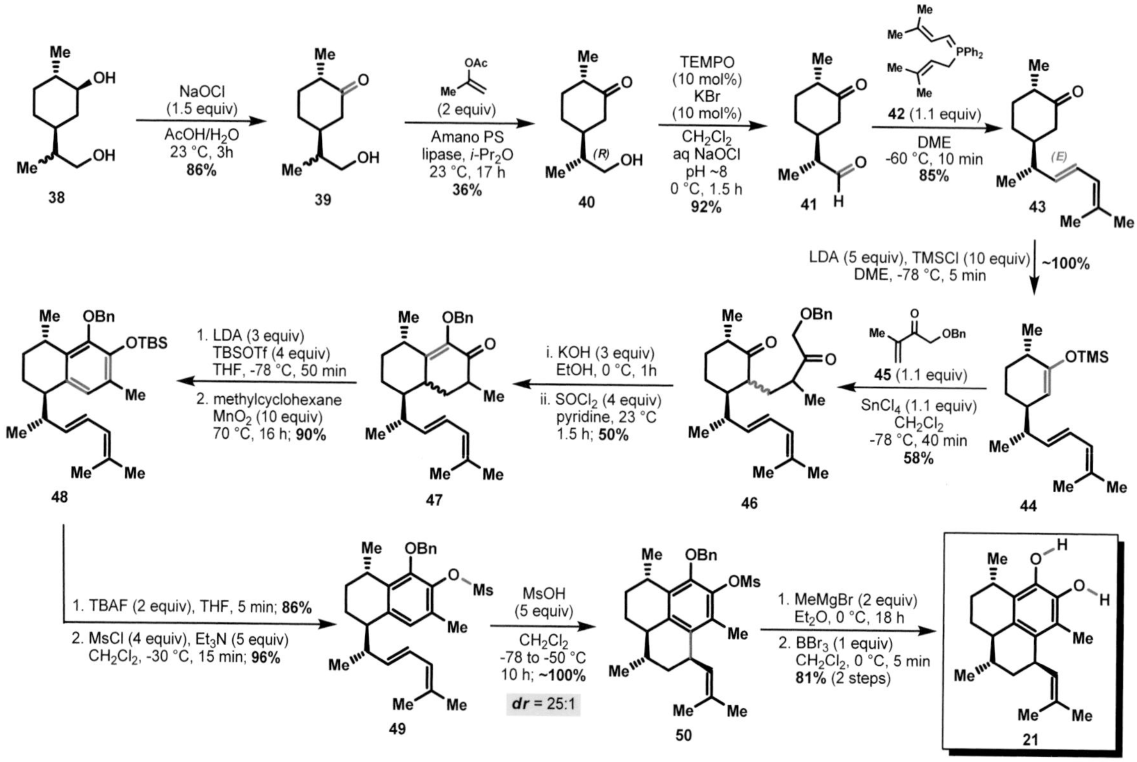

Scheme 7

Key steps in Scheme 7

(1) A 1:1 diastereomeric mixture of diol **38** is available in quantity by hydroboration-oxidation of the inexpensive limonene. One diastereomer of the derived ketone **39** is preferentially acetylated by Amano PS lipase which allows facile chromatographic purification of **40**.

(2) In the cyclization **49**→**50**, an intermediate allylic cation is formed which is directly attacked by the aromatic ring *para* to the OBn group rather than the less electron-donating OMs group (see detailed analysis of the possible transition states in Schemes 8 and 9).

PSEUDOPTEROSINS

Stereochemistry of the Cyclization 49a→50 + 53 – Analysis of Transition States

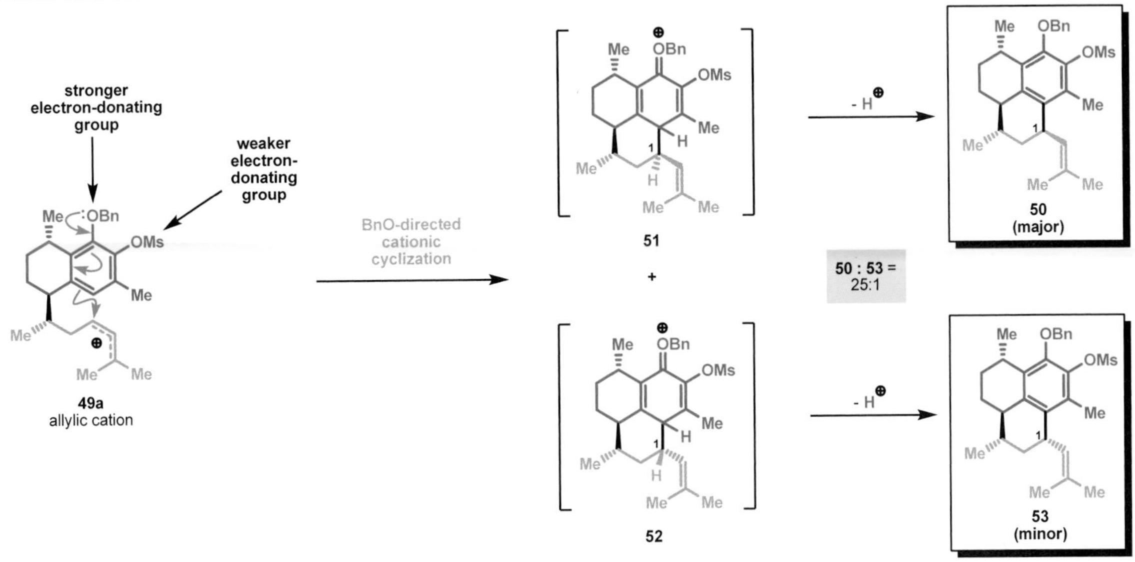

Scheme 8

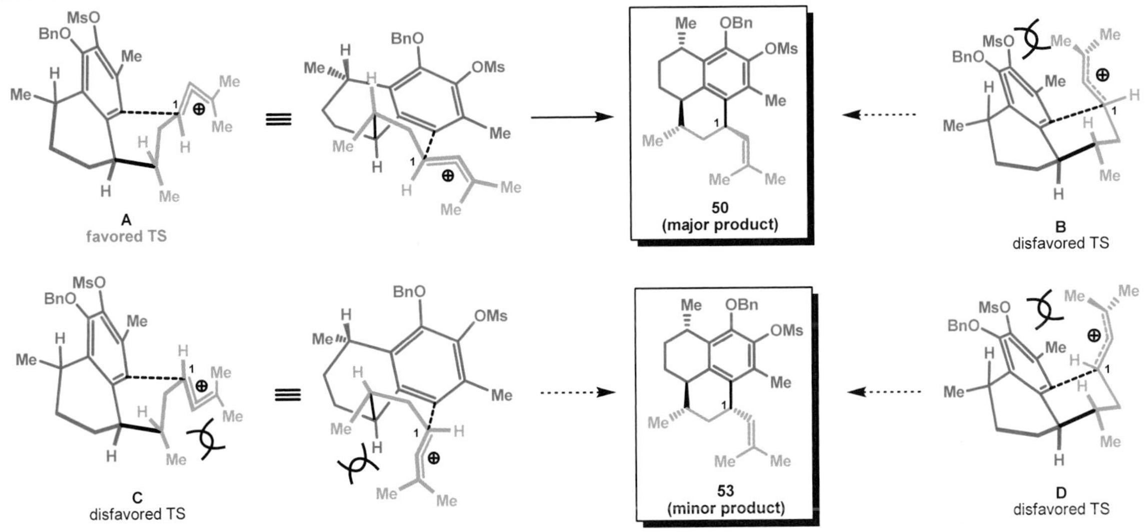

Scheme 9

Explanation of Schemes 8 and 9

(1) The cyclization of cation **49a** is considered to take place by direct attack of the secondary terminus of the cationic appendage at the benzenoid carbon *para* to the more electron-donating BnO group, for which there are two possible diastereomeric cationic intermediates **51** and **52** (Scheme 8).

(2) Of these, **52** is considerably higher in energy (than **51**) because of greater steric repulsion in the transition state (see disfavored transition states **C** and **D** in Scheme 9). Thus, the pathway **49a** → **51** → **50** predominates over **49a** → **52** → **53** by 25:1.

PSEUDOPTEROSINS

Other Developments: Synthesis of Helioporin E

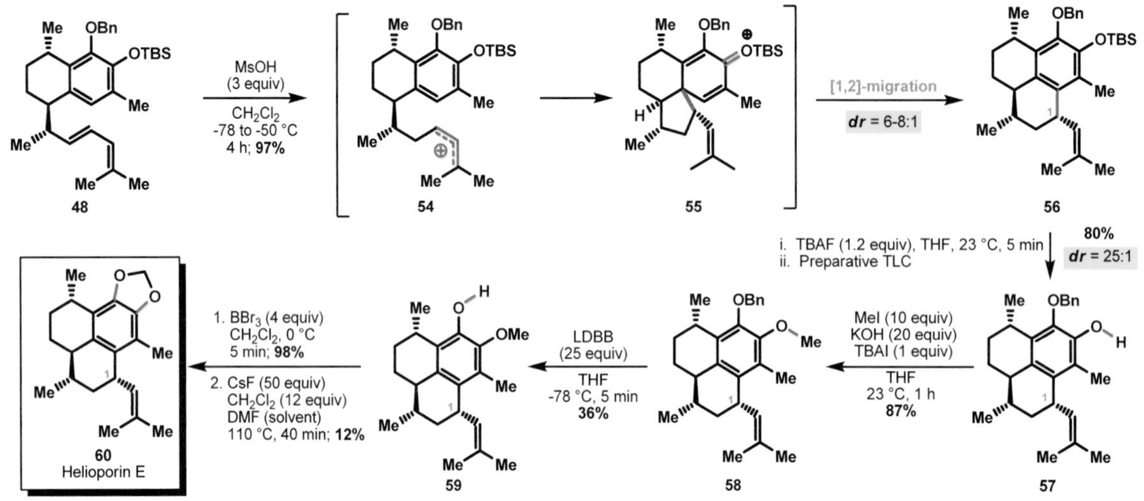

Scheme 10

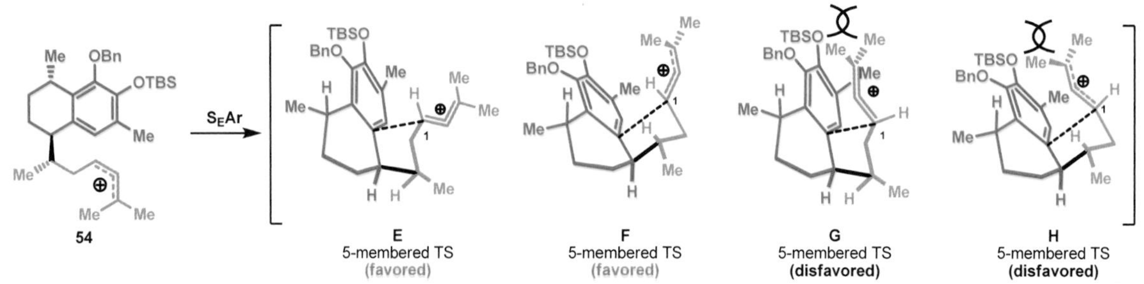

Scheme 11

Key steps in Schemes 10 and 11

(1) In the cyclization of **48**→**56**, the allylic cation **54** attacks the benzenoid ring to form the spiro intermediate **55** which then rearranges to **56** with the same configuration at C(1) as Helioporin E. The stereochemical course of this cyclization can be rationalized by comparing the 5-membered transition states, **E** to **H** (Scheme 11).

(2) The preference for the pathway **48**→**56** is a consequence of steric repulsions that destabilize transition states **G** and **H** relative to transitions states **E** and **F**.

References

1. (a) Look, S. A., Fenical, W., Matsumoto, G. K. & Clardy, J. The pseudopterosins: a new class of antiinflammatory and analgesic diterpene pentosides from the marine sea whip Pseudopterogorgia elisabethae (Octocorallia). *J. Org. Chem.* **1986**, 51, 5140-5145; (b) Look, S. A., Fenical, W., Jacobs, R. S. & Clardy, J. The pseudopterosins: anti-inflammatory and analgesic natural products from the sea whip Pseudopterogorgia elisabethae. *Proc. Natl. Acad. Sci. U. S. A.* **1986**, 83, 6238-6240; (c) Look, S. A. & Fenical, W. The seco-pseudopterosins, new anti-inflammatory diterpene-glycosides from a Caribbean gorgonian octocoral of the genus Pseudopterogorgia. *Tetrahedron* **1987**, 43, 3363-3370.

2. (a) Corey, E. J., Carpino, P. Enantiospecific total synthesis of pseudopterosins A and E. *J. Am. Chem. Soc.* **1989**, 111, 5472-5474; (b) Corey, E. J., Lazerwith, S. E. A Direct and Efficient Stereocontrolled Synthetic Route to the Pseudopterosins, Potent Marine Antiinflammatory Agents. *J. Am. Chem. Soc.* **1998**, 120, 12777-12782; (c) Lazerwith, S.

E., Johnson, T. W., Corey, E. J. Syntheses and Stereochemical Revision of Pseudopterosin G-J Aglycon and Helioporin E. *Org. Lett.* **2000**, 2, 2389-2392.

3. Kürti, L. & Czakó, B. Barton Nitrite Ester Reaction in *Strategic Applications of Named Reactions in Organic Synthesis* 42-43 (Academic Press/Elsevier, San Diego, **2005**).

4. Barton, D. H. R., Brewster, A. G., Ley, S. V. & Rosenfeld, M. N. Preparation of phenylselenoimines from phenols using diphenylseleninic anhydride and hexamethyldisilazane. *J. Chem. Soc., Chem. Commun.* **1977**, 147-148.

5. Kürti, L. & Czakó, B. Corey-Chaykovsky Epoxidation and Cyclopropanation in *Strategic Applications of Named Reactions in Organic Synthesis* 102-103 (Academic Press/Elsevier, San Diego, **2005**).

(+)-PSEUDOPTEROXAZOLE

Background

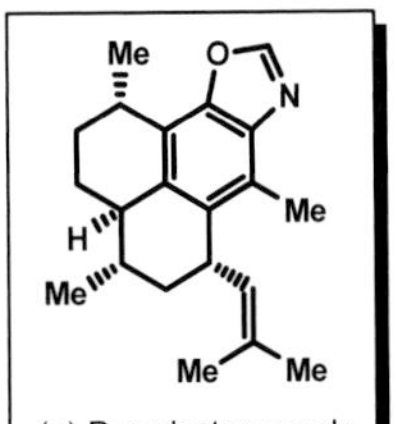

(+)-Pseudopteroxazole

Pseudopteroxazole is a potent antitubercular agent isolated from the West Indian sea whip *Pseudopterageorgia elisabethae* where it is a minor constituent.[1] The synthesis of pseudopteroxazole,[2] which was complicated by the fact that the stereochemistry originally assigned to it was incorrect, drew upon the knowledge gained in earlier syntheses of the structurally related marine natural products pseudopterosins (pages 245-250).[3] The synthesis of two diastereomers is also described in this section, including that originally proposed.

Abbreviated Retrosynthetic Plan

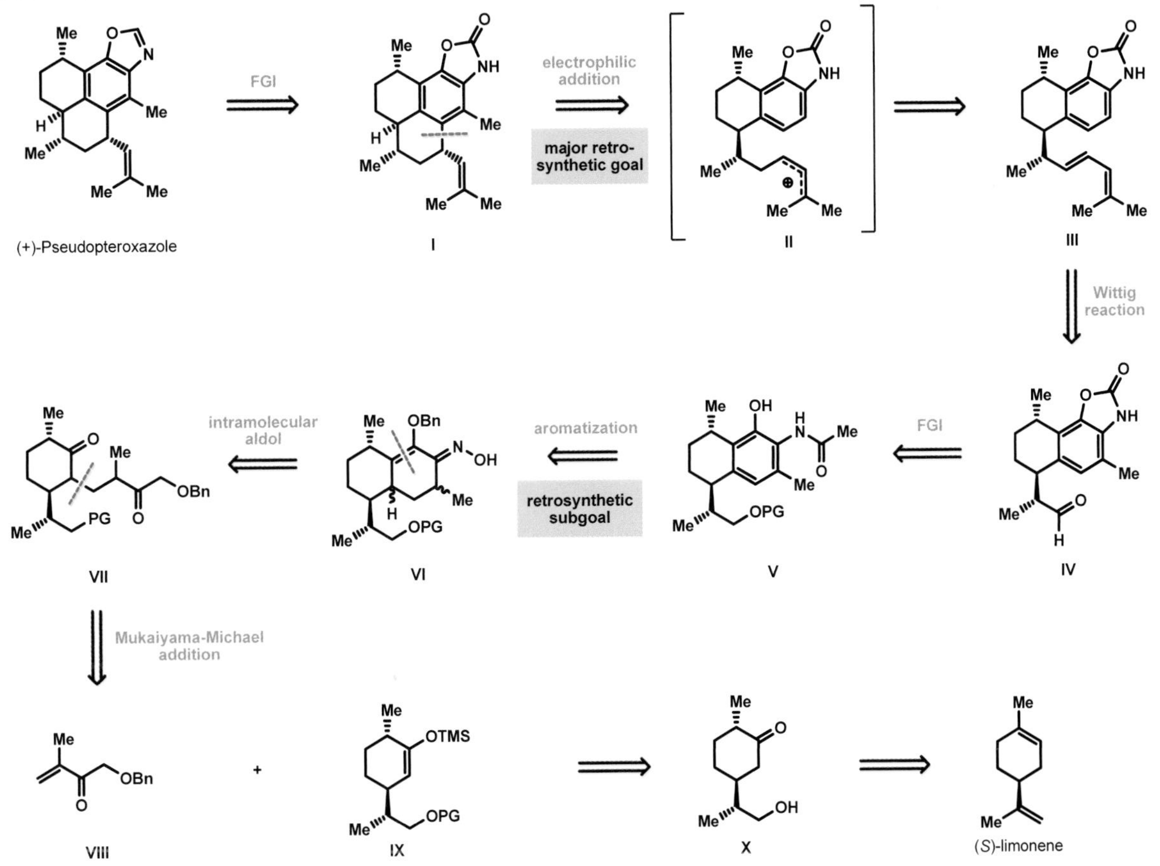

Scheme 1

Key steps in Scheme 1

(1) The disconnection of pseudopteroxazole to **III** and **IV** followed the pathway that was applied earlier for the synthesis of pseudopterosins (page 245).[3]

(2) The chiral intermediate **X** had earlier been prepared as a diastereomeric mixture from (S)-limonene.

(+)-PSEUDOPTEROXAZOLE

Pathway of Synthesis

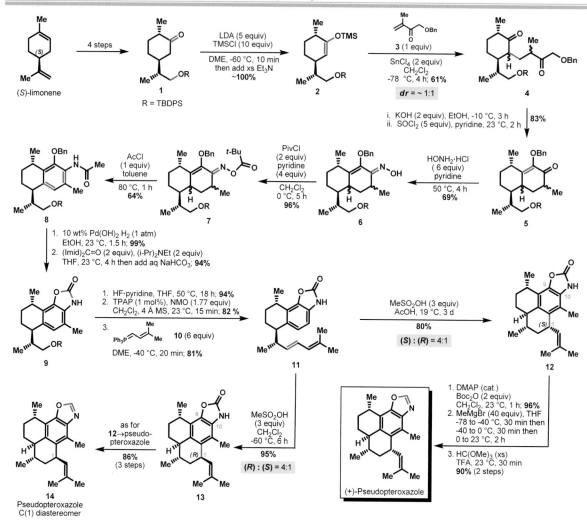

Scheme 2

Key steps in Scheme 2

(1) The ketone **1** was conveniently prepared from the inexpensive chiral starting material (S)-limonene by the novel procedure that had earlier been developed for the synthesis of pseudopterosins.[3]

(2) It is likely that the aromatization of **7** to form **8** occurs by the sequence:

   (a) N-acetylation of **7** to form the cyclohexa-dienamide **7a** and

   (b) acid-catalyzed elimination of t-BuCO$_2$H to form N-acetyl imine **7b** and

   (c) tautomerization to **8**.

(3) The acid-catalyzed cyclization of **11** gave **12** as the major product when acetic acid was used as solvent, but mainly the diastereomer **13** in CH$_2$Cl$_2$ solution.

(+)-PSEUDOPTEROXAZOLE

Mechanistic Studies on the Cyclizations 11→13 and 11→12

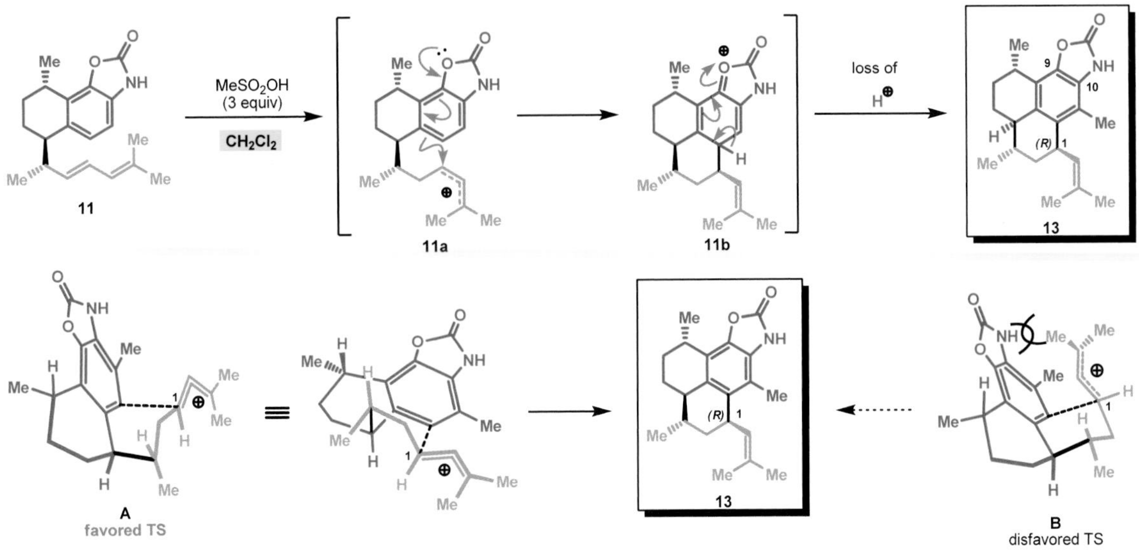

Scheme 3

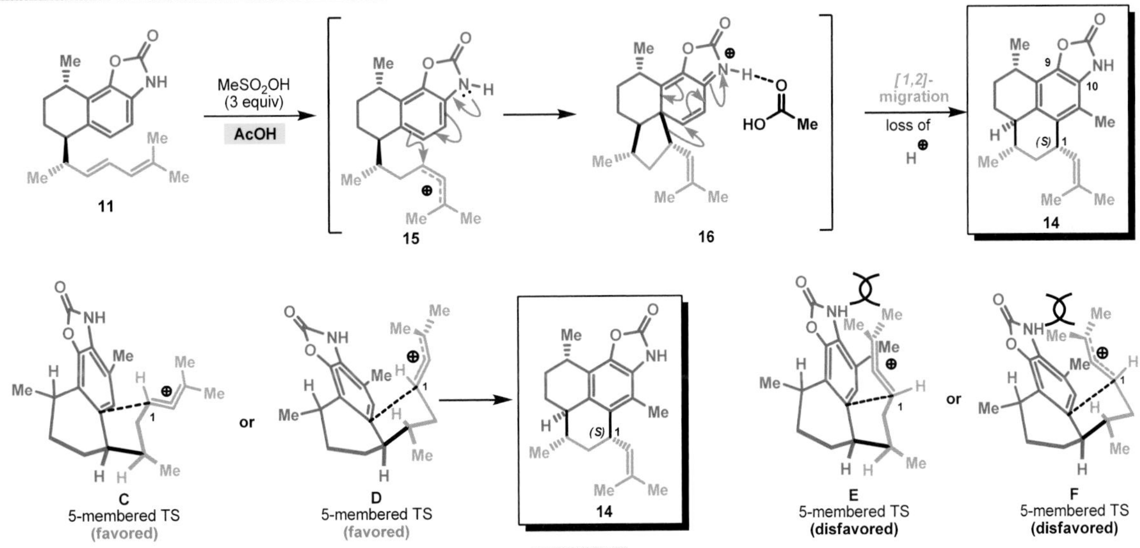

Scheme 4

Explanation of Schemes 3, 4, 5 and 6

(1) The stereochemical course of the cyclizations to form **13** and **14** in different solvents has been explained using mechanistic reasoning which parallels that presented on page 250 for the synthesis of pseudopterosin diastereomers.

(2) The target structure **29** (Scheme 6), originally assigned to pseudopteroxazole, was synthesized as shown in Schemes 5 and 6 along with the C(1) diastereomer (**30**).

(3) The *endo* quinone-Diels-Alder pathway mitigated by the oxidative step **19**→**20** (Scheme 5) predominated leading mainly to adduct **21**. The synthesis of **29** was accomplished from **5** via **27a**.[4]

(+)-PSEUDOPTEROXAZOLE

Preparation of Other Pseudopteroxazole Isomers

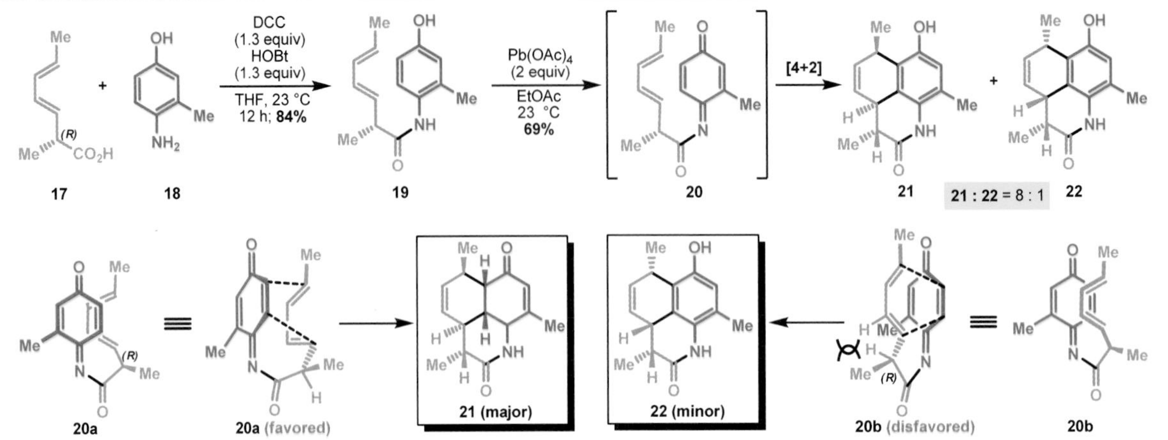

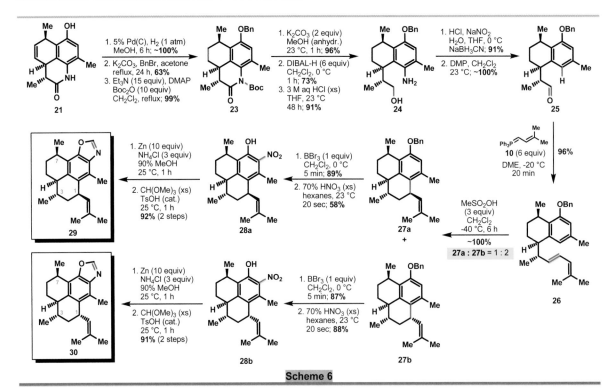

Scheme 5

Scheme 6

References

1. Rodriguez, A. D., Ramirez, C., Rodriguez, I. I. & Gonzalez, E. Novel antimycobacterial benzoxazole alkaloids, from the West Indian sea whip Pseudopterogorgia elisabethae. *Org. Lett.* **1999**, 1, 527-530.

2. (a) Johnson, T. W., Corey, E. J. Enantiospecific Synthesis of the Proposed Structure of the Antitubercular Marine Diterpenoid Pseudopteroxazole: Revision of Stereochemistry. *J. Am. Chem. Soc.* **2001**, 123, 4475-4479; (b) Davidson, J. P., Corey, E. J. First Enantiospecific Total Synthesis of the Antitubercular Marine Natural Product Pseudopteroxazole. Revision of Assigned Stereochemistry. *J. Am. Chem. Soc.* **2003**, 125, 13486-13489.

3. Corey, E. J. & Lazerwith, S. E. A Direct and Efficient Stereocontrolled Synthetic Route to the Pseudopterosins, Potent Marine Antiinflammatory Agents. *J. Am. Chem. Soc.* **1998**, 120, 12777-12782.

4. For another approach to pseudopteroxazole, see: Harmata, M., Hong, X. Benzothiazines in Synthesis. A Total Synthesis of Pseudopteroxazole. *Org. Lett.* **2005**, 7, 3581-3583.

(+)-DESOGESTREL

Background

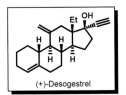

(+)-Desogestrel

Desogestrel is an unnatural synthetic steroid which is the most prescribed third-generation contraceptive agent worldwide. It has been produced commercially from the naturally occurring steroid diosgenin by a long partial synthesis.[1] A relatively short enantioselective synthesis[2] was developed based on the use of a catalytic Diels-Alder reaction to establish three rings with the required absolute configuration.[3]

Abbreviated Retrosynthetic Plan

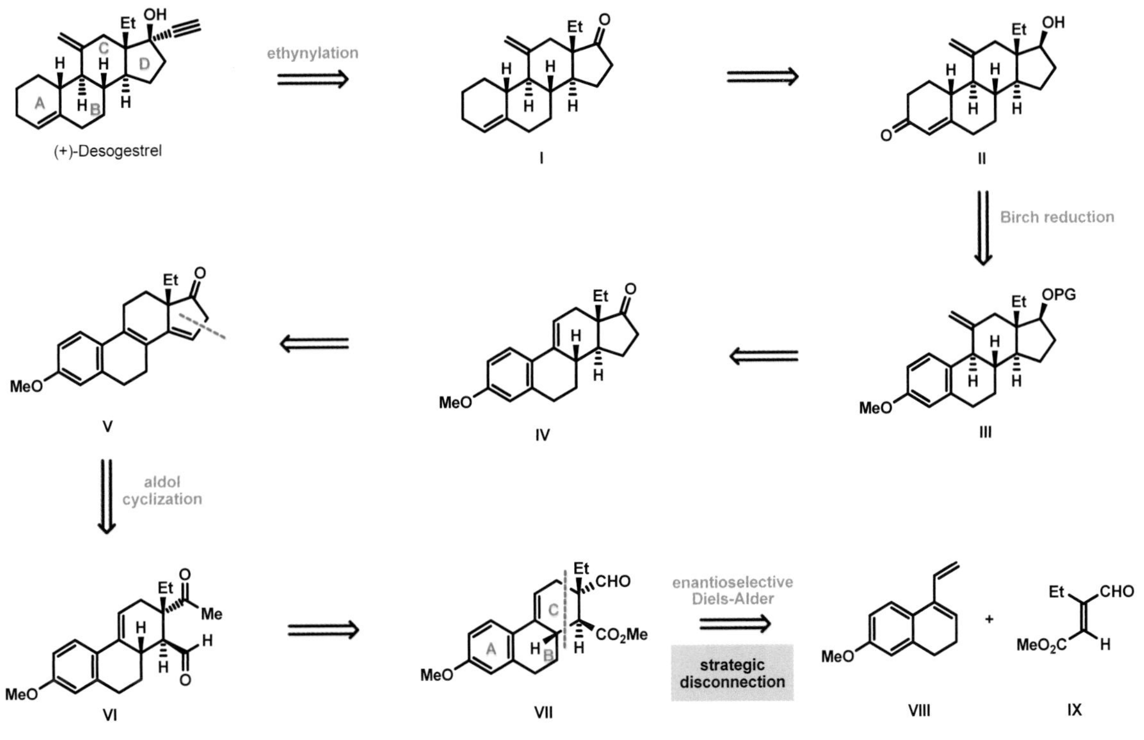

(+)-Desogestrel ⟹ ethynylation ⟹ I ⟹ II

II ⟹ Birch reduction ⟹ III

V ⟸ IV ⟸ III

V ⟹ aldol cyclization ⟹ VI

VI ⟹ VII ⟹ enantioselective Diels-Alder / strategic disconnection ⟹ VIII + IX

Scheme 1

Key steps in Scheme 1

(1) The design of the synthesis was driven by the long-range objective of applying the enantioselective Diels-Alder transform to the disconnection of ring C (**VII** ⟹ **VIII** + **IX**).

(2) The use of the benzenoid A-ring precursor (**VIII**) followed from its ready accessibility and suitable reactivity.

(3) The dienophile **IX** served to satisfy the twin requirements of an efficient catalytic Diels-Alder reaction and suitability as D-ring precursor.

(4) The intermediate **VII** was used as an S-goal to guide the retrosynthetic steps from desogestrel.

(5) The validity of the disconnection **VII** ⟹ **VIII** + **IX** and the clearance of the three stereocenters in **VII** was suggested from prior research on enantioselective Diels-Alder reactions catalyzed by chiral oxazaborolidinium ions of the type shown below.

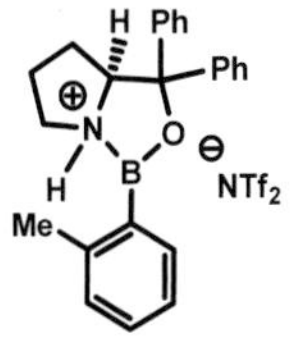

(+)-DESOGESTREL

Pathway of Synthesis

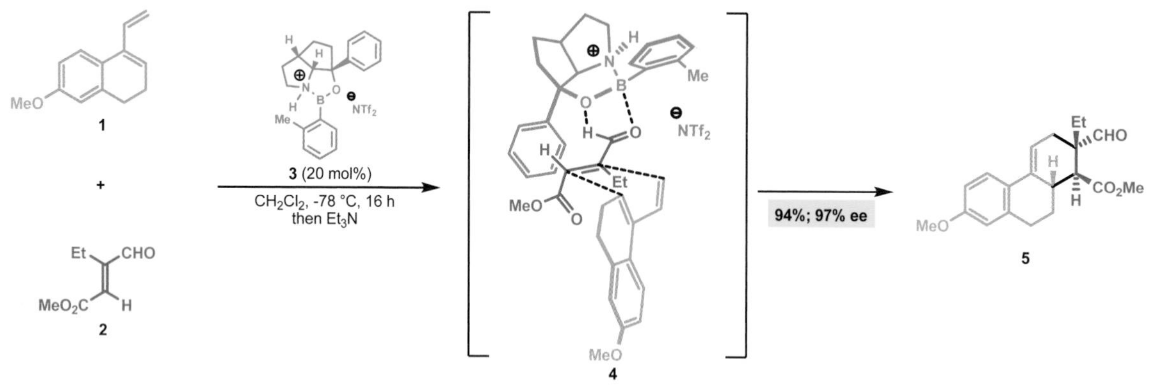

Scheme 2

Key steps in Scheme 2

The key enantioselective Diels-Alder reaction is considered to occur via the pre-transition state assembly **4**, in which the productive pathway involves coordination of the formyl group with catalyst **3** (an equivalent of the (S)-proline derived catalyst **A**).[4] The conversion of tricyclic intermediate **5** to (+)-desogestrel is shown in Schemes 3 and 4.

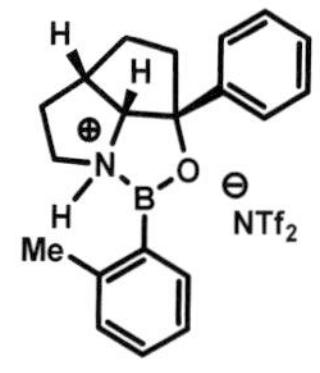

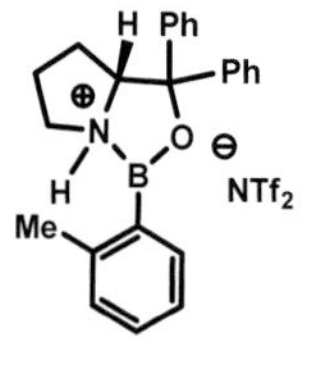

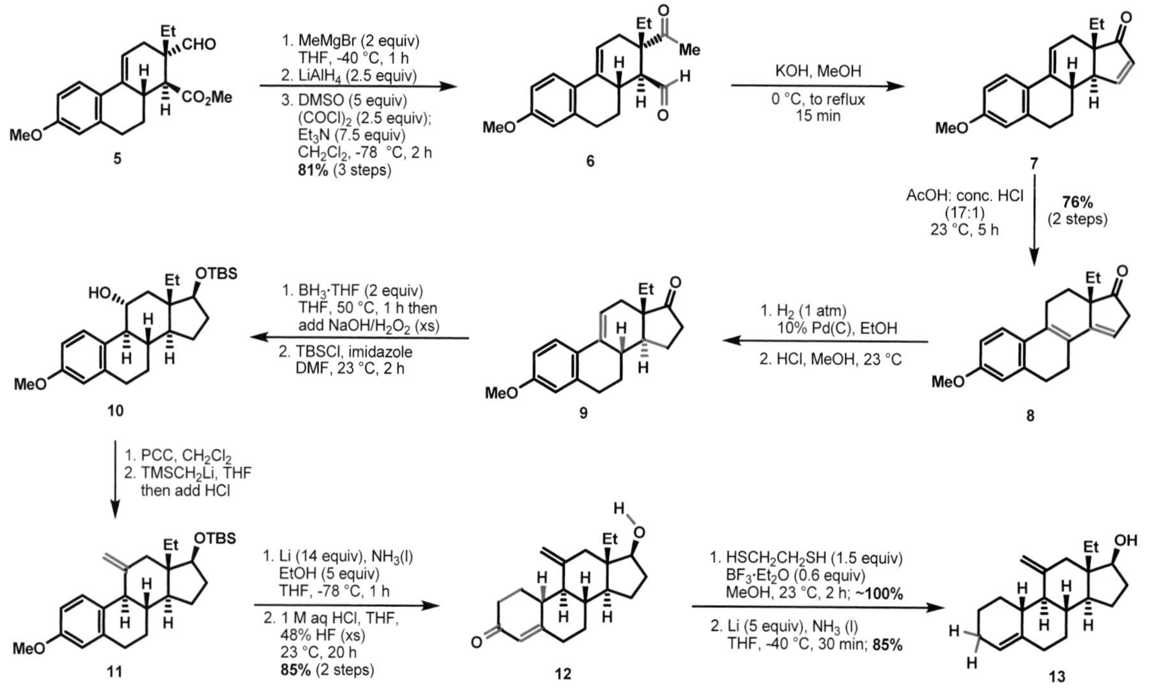

Scheme 3

(+)-DESOGESTREL

Pathway of Synthesis (continued)

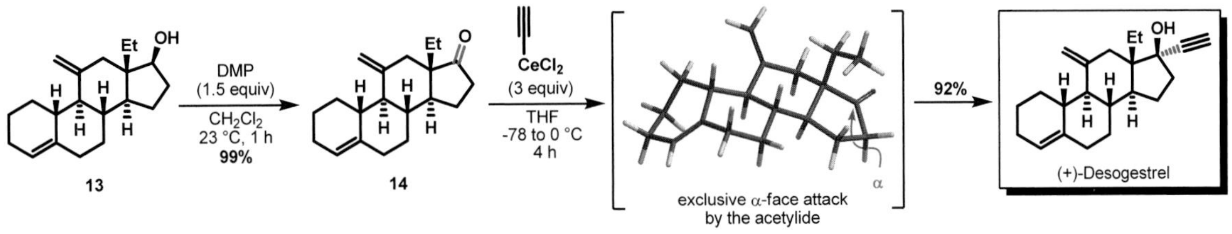

Scheme 4

Another Enantioselective Approach to (+)-Desogestrel

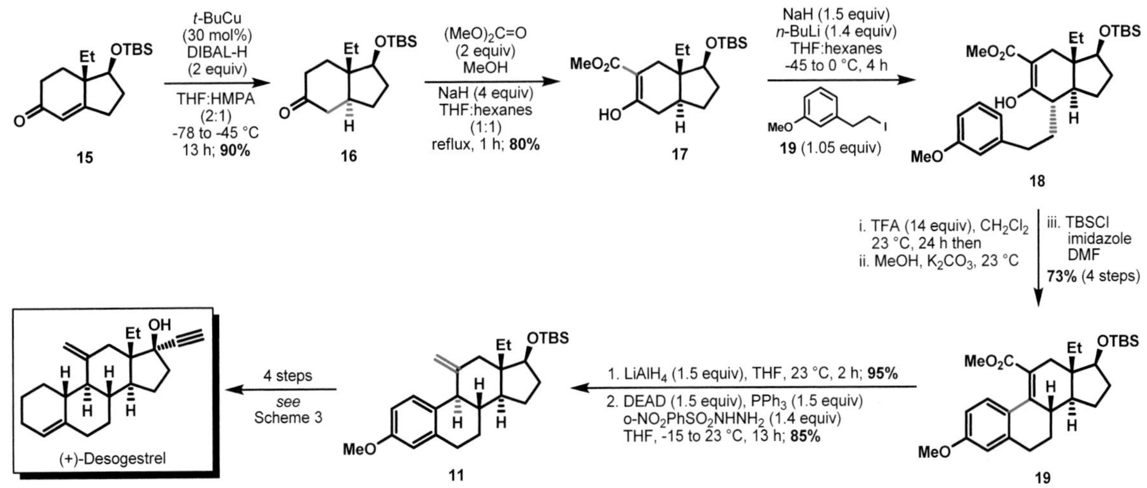

Scheme 5

Key Steps in Scheme 5

(1) This alternative route to desogestrel utilizes the chiral enone **15** which is available via (S)-proline-catalyzed Robinson annulations of methyl vinyl ketone and 2-ethylcyclopentane-1,3-dione.[5]

(2) The acid-catalyzed cyclization **18→19** is considered to occur by protonation of the carbonyl oxygen of the COOMe group in **18** and subsequent attack of the cationic C(9) on the benzenoid ring at the MeO-activated para position.

References

1. Van den Heuvel, M. J., Van Bokhoven, C. W., De Jongh, H. p. & Zeelen, F. J. A partial synthesis of 13-ethyl-11-methylene-18,19-dinor-17α-pregn-4-en-20-yn-17-ol (desogestrel) based upon intramolecular oxidation of an 11β-hydroxy-19-norsteroid to the 18 → 11β-lactone. *Recl. Trav. Chim. Pays-Bas* **1988**, 107, 331-334.

2. Hu, Q.-Y., Rege, P. D. & Corey, E. J. Simple, Catalytic Enantioselective Syntheses of Estrone and Desogestrel. *J. Am. Chem. Soc.* **2004**, 126, 5984-5986.

3. (a) For a comprehensive review, see: Chapelon, A.-S., Moraleda, D., Rodriguez, R., Ollivier, C., Santelli, M. Enantioselective synthesis of steroids. *Tetrahedron* **2007**, 63, 11511-11616; (b) For a recent total synthesis of desogestrel, see: Tietze, L. F., Krimmelbein, I. K.

Enantioselective total synthesis of the oral contraceptive desogestrel by a double Heck reaction. *Chem.--Eur. J.* **2008**, 14, 1541-1551; (c) Corey, E. J. & Huang, A. X. A Short Enantioselective Total Synthesis of the Third-Generation Oral Contraceptive Desogestrel. *J. Am. Chem. Soc.* **1999**, 121, 710-714.

4. For a review see: Corey E.J. Enantioselective Catalysis Based on Cationic Oxazaborolidines. *Angew. Chem. Intl. Ed*, **2009**, 48, 2100-2117.

5. Hajos, Z. G. & Parrish, D. R. Synthesis and conversion of 2-methyl-2-(3-oxobutyl)-1,3-cyclopentanedione to the isomeric racemic ketols of the [3.2.1]bicyclooctane and of the perhydroindane series. *J. Org. Chem.* **1974**, 39, 1612-1615.

(+)-ESTRONE

Background

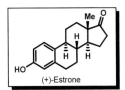

(+)-Estrone

The synthesis of the natural steroid estrogen (+)-estrone has posed a chemical challenge[1] for several decades.[2] In this section three different simple and enantioselective syntheses are outlined.[3]

Abbreviated Retrosynthetic Plan

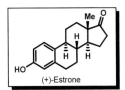

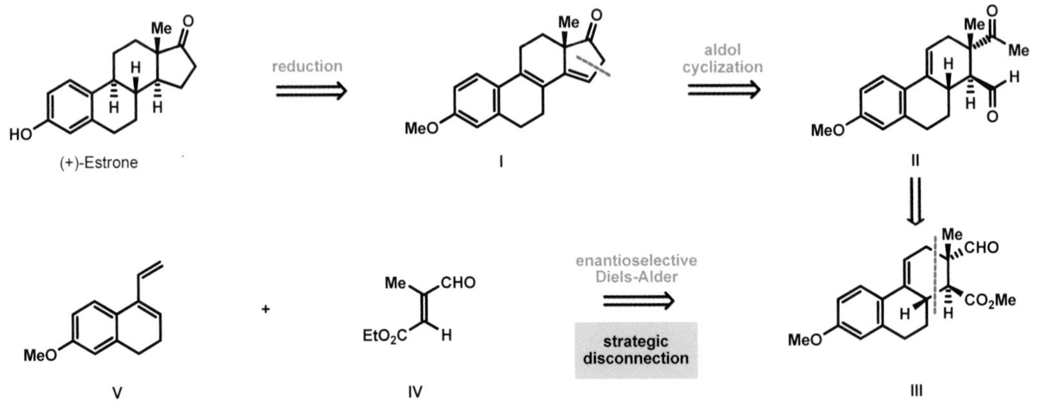

Scheme 1

Key steps in Scheme 1

The synthetic route to estrone was devised by an analysis which paralleled that used for desogestrel (see page 255).

Pathway of Synthesis

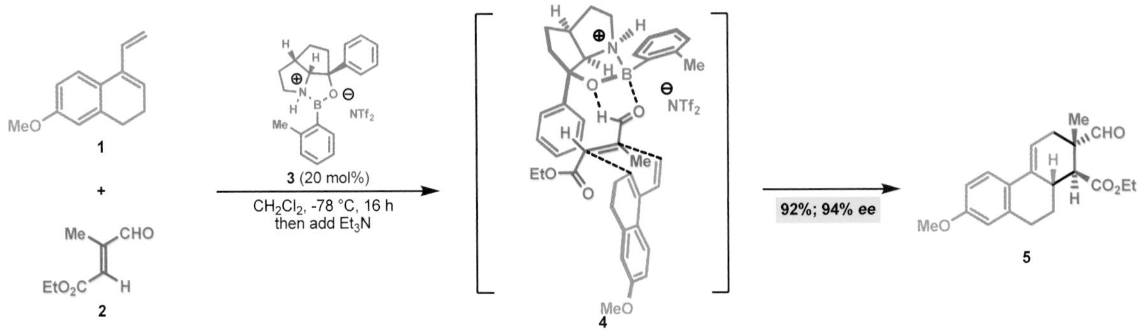

Scheme 2

Explanation of Scheme 2

The key Diels-Alder reaction between the Dane diene **1** and the α,β-enal **2** is thought to proceed via the pre-tran-

sition state assembly **4** on the basis of much supporting evidence.[4] The cycloadduct **5** was then converted to (+)-estrone using known chemistry (Scheme 3).

(+)-ESTRONE

Pathway of Synthesis (continued)

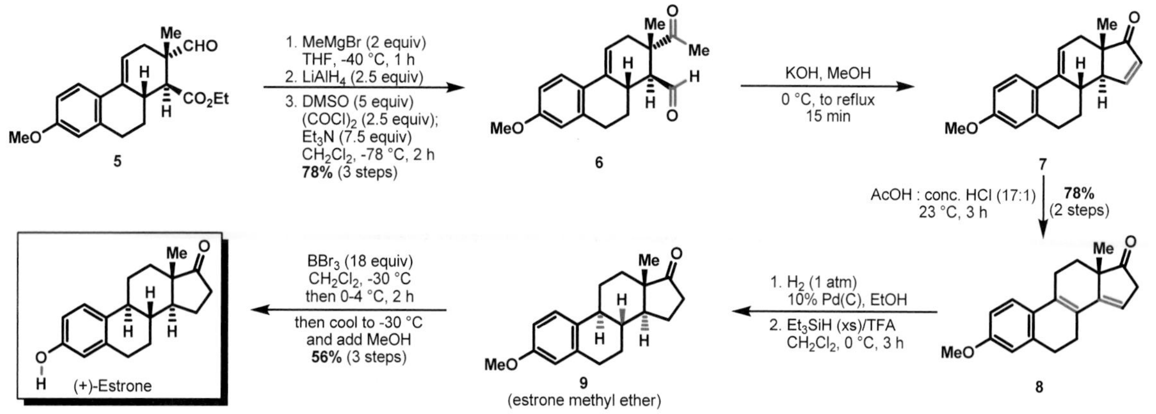

Scheme 3

Another Enantioselective Diels-Alder Cycloaddition Approach to (+)-Estrone

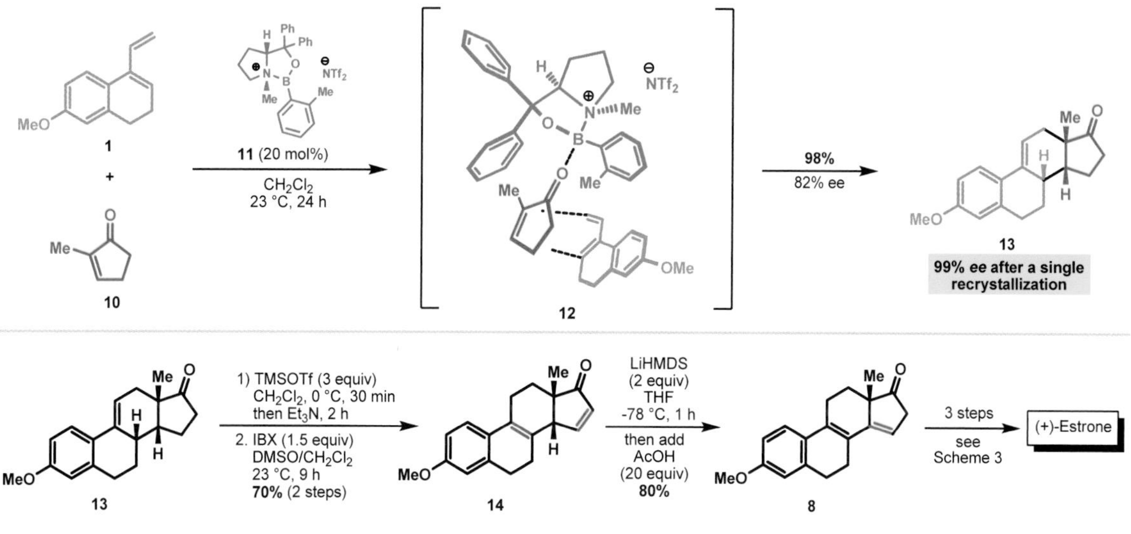

Scheme 4

Key Steps in Scheme 4

The catalyst for the initial enantioselective Diels-Alder cycloaddition was the *N*-methyl-activated oxazaborolidinium ion **11** [derived from (*R*)-diphenylpyrrolidine-methanol]. Catalyst **11**, prepared by a novel route, was more effective than the corresponding *N*-H or *N*-AlBr₂-activated oxazaborolidinium ions.[3c]

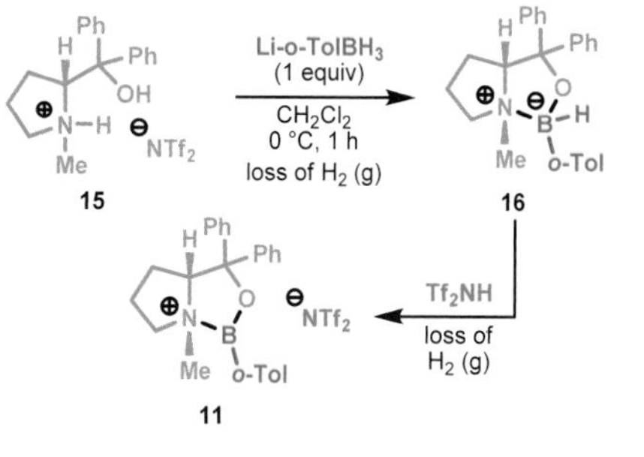

(+)-ESTRONE

Development of an Enantioselective Version of Torgov's Synthesis of Estrone

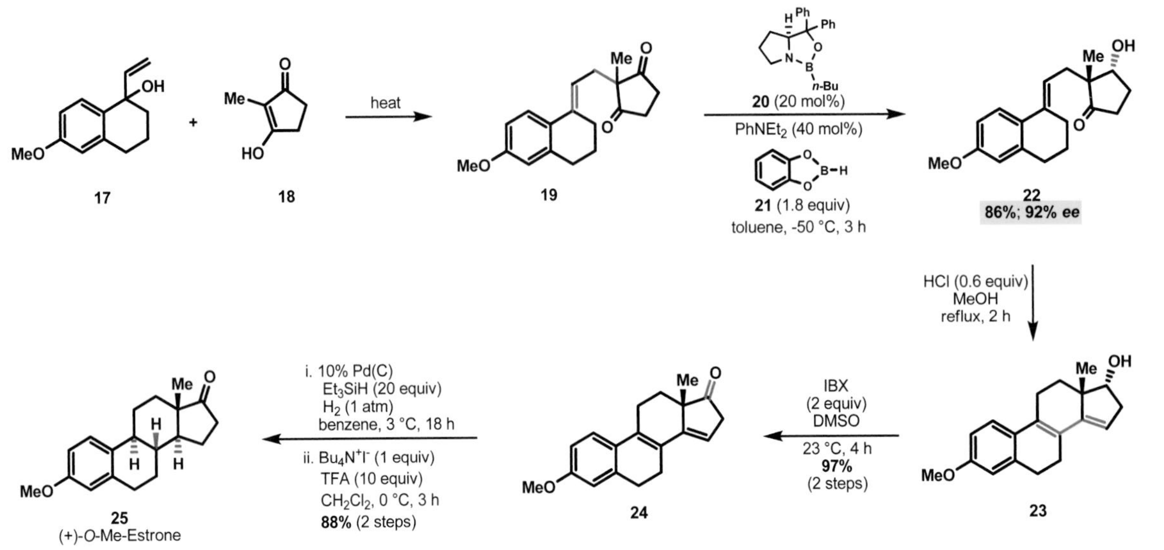

Scheme 5

Key Steps in Scheme 5

(1) The enantioselective and diastereoselective reduction of the achiral Torgov diketone **19** to form **22** required the use of carefully chosen reagents and conditions.[3b] Catecholborane **21** was the best stoichiometric reductant. Superior results were obtained by the use of N,N-diethylaniline as an additive.[3b]

(2) The beneficial effect of N,N-diethylaniline was shown in later work to be due to the fact that it catalyzed the conversion of a small amount of the deleterious impurity **26** to catalyst **20**.[5]

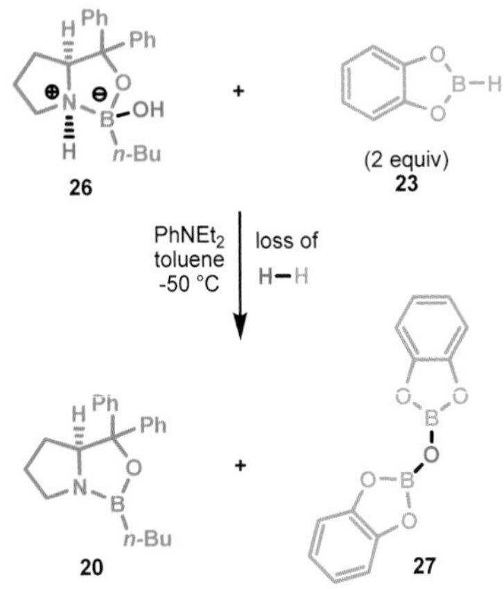

References

1. (a) Ananchenko, S. N. & Torgov, I. V. New syntheses of estrone, dl-8-isoestrone, and dl-19-nortestosterone. *Tetrahedron Lett.* **1963**, 1553-1558; (b) Enev, V. S., Mohr, J., Harre, M. & Nickisch, K. The first Lewis acid mediated asymmetric Torgov cyclization. *Tetrahedron: Asymmetry* **1998**, 9, 2693-2699.

2. For recent other syntheses of estrone, see the following publications and references therein: (a) Herrmann, P., Budesinsky, M., Kotora, M. Formal Total Synthesis of (±)-Estrone and Zirconocene-Promoted Cyclization of 2-Fluoro-1,7-octadienes and Ru-Catalyzed Ring Closing Metathesis. *J. Org. Chem.* **2008**, 73, 6202-6206; (b) Soorukram, D., Knochel, P. Formal Enantioselective Synthesis of (+)-Estrone. *Org. Lett.* **2007**, 9, 1021-1023; (c) Pattenden, G., Gonzalez, M. A., McCulloch, S., Walter, A., Woodhead, S. J. A total synthesis of estrone based on a novel cascade of radical cyclizations. *Proc. Natl. Acad. Sci. U. S. A.* **2004**, 101, 12024-12029.

3. (a) Hu, Q.-Y., Rege, P. D. & Corey, E. J. Simple, Catalytic Enantioselective Syntheses of Estrone and Desogestrel. *J. Am. Chem. Soc.* **2004**, 126, 5984-5986; (b) Yeung, Y.-Y., Chein, R.-J. & Corey, E. J. Conversion of Torgov's Synthesis of Estrone into a Highly Enantioselective and Efficient Process. *J. Am. Chem. Soc.* **2007**, 129, 10346-10347; (c) Canales, E. & Corey, E. J. Highly Enantioselective [4 + 2] Cycloaddition Reactions Catalyzed by a Chiral N-Methyl-oxazaborolidinium Cation. *Org. Lett.* **2008**, 10, 3271-3273.

4. For a review see: Corey E.J. Enantioselective Catalysis Based on Cationic Oxazaborolidines. *Angew. Chem. Intl. Ed*, **2009**, 48, 2100-2117.

5. Chein, R.-J., Yeung, Y.-Y. & Corey, E. J. Highly Enantioselective Oxazaborolidine-Catalyzed Reduction of 1,3-Dicarbonyl Compounds: Role of the Additive Diethylaniline. *Org. Lett.* **2009**, 11, 1611-1614.

(+) & (−)-EPIBATIDINE

Background

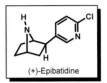

(+)-Epibatidine

Epibatidine is a toxic alkaloid secreted from the skin of the poisonous frog *Epibedobates tricolor* (ca. 1.3 mg from 1000 frogs).[1] It is ca. 300 times more potent than morphine, although binds only weakly to opioid receptor preparations. A stereocontrolled synthesis of epibatidine[2] is outlined below.[3]

Abbreviated Retrosynthetic Plan

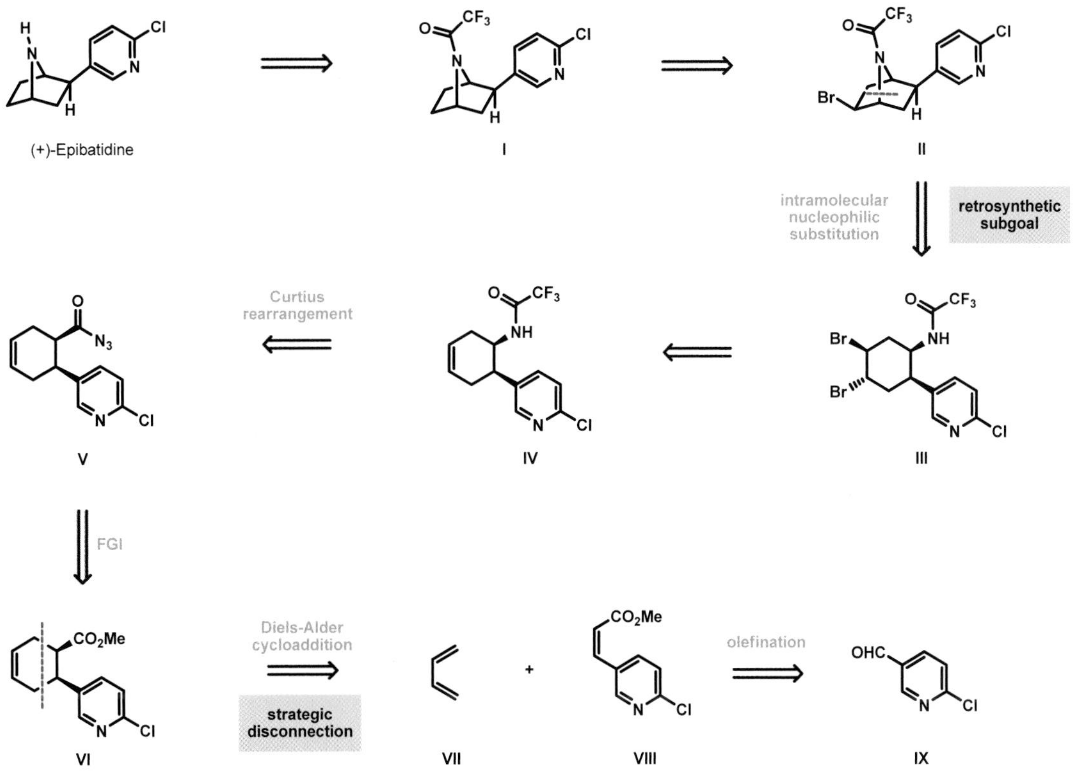

(+)-Epibatidine

I

II

intramolecular nucleophilic substitution

retrosynthetic subgoal

V

Curtius rearrangement

IV

III

FGI

VI

Diels-Alder cycloaddition

strategic disconnection

VII

+

VIII

olefination

IX

Scheme 1

Key steps in Scheme 1

(1) The retrosynthetic analysis was guided by the transform-based goal of disconnecting the entire carbon framework by means of the Diels-Alder transform. Retrosynthetic search revealed the disconnection **VI** ⇒ **VII** + **VIII** to be an attractive option because the components are readily available and the Diels-Alder reaction between them is well-precedented.

(2) Establishment of the Diels-Alder reaction in **VII** was then analyzed retrosynthetically, using **VI** as an S-goal.

(3) The disconnection of the nitrogen bridge via **II** led to the dibromide **III**, access to which would depend on the control of stereochemistry, specifically for the synthetic conversion of **IV** to **III**.

(4) The absolute configuration of epibatidine was unknown at the time when this synthesis was undertaken. Consequently, the initial objective was the synthesis of both enantiomers. In principle, the use of a chiral controller in the dienophile for the initial Diels-Alder reaction could lead to an enantioselective synthesis of the naturally occurring (−)-form.

(+) & (−)-EPIBATIDINE

Pathway of Synthesis

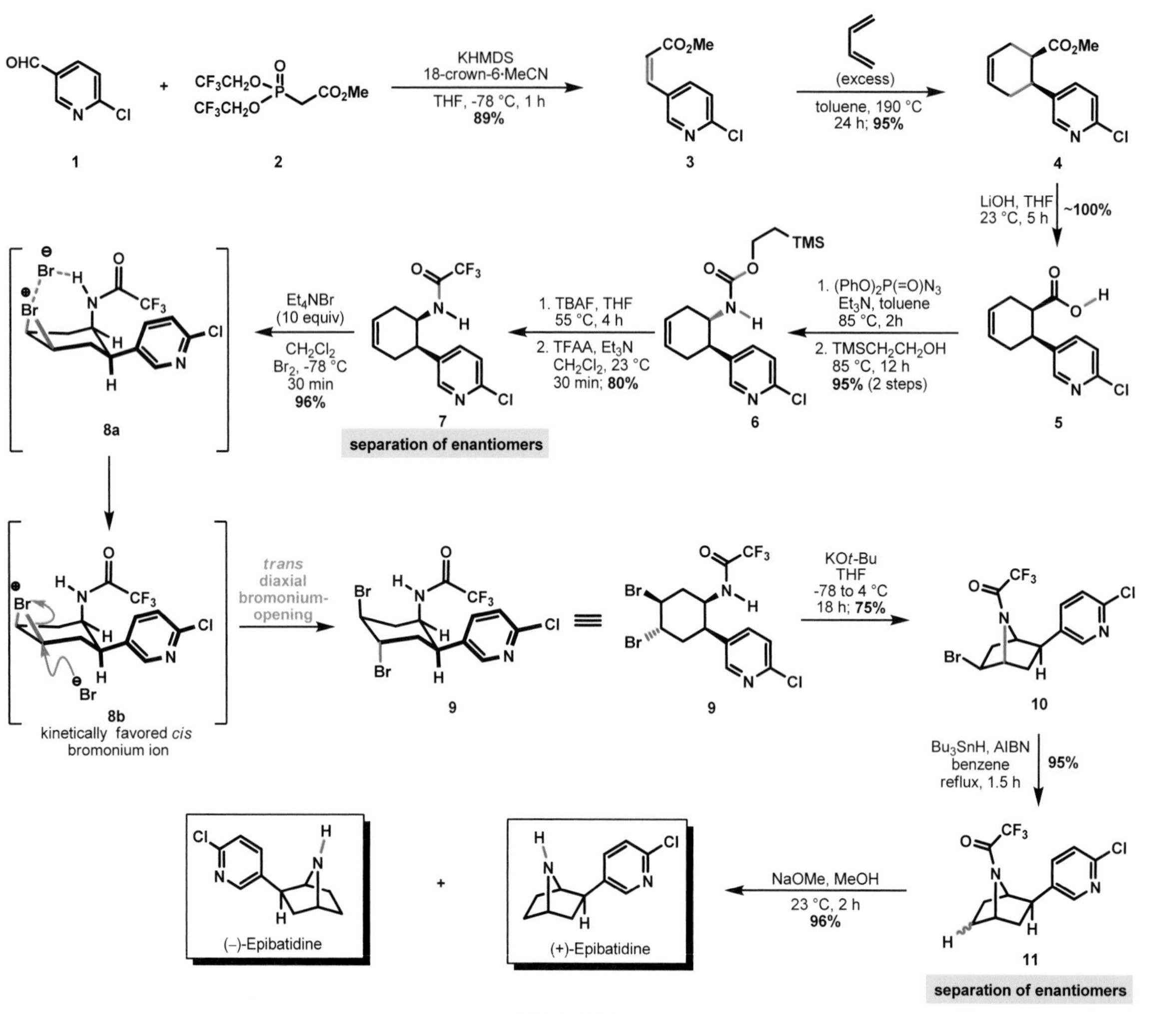

Scheme 2

Key Steps in Scheme 2

(1) The formation of the Z-ester **3** from **1** utilized Still's reagent **2**.

(2) The diastereoselectivity of the conversion of **7** to **9** is thought to be the consequence of acceleration of bromonium **8b** formation because of H-bonding as shown in pre-transition state assembly **8a**.

References

1. (a) Spande, T. F., Garraffo, H. M., Edwards, M. W., Yeh, H. J. C., Pannell, L. & Daly, J. W. Epibatidine: a novel (chloropyridyl)azabicycloheptane with potent analgesic activity from an Ecuadoran poison frog. *J. Am. Chem. Soc.* **1992**, 114, 3475-3478; (b) For a review see: Yogeeswari, P., Sriram, D., Bal, T. R., Thirumurugan, R. Epibatidine and its analogues as nicotinic acetylcholine receptor agonist: an update. *Natural Product Research, Part A: Structure and Synthesis* **2006**, 20, 497-505.

2. Corey, E. J., Loh, T. P., AchyuthaRao, S., Daley, D. C. & Sarshar, S. Stereocontrolled total synthesis of (+)- and (-)-epibatidine. *J. Org. Chem.* **1993**, 58, 5600-5602.

3. For recent approaches to epibatidine, see: (a) Kimura, H., Fujiwara, T., Katoh, T., Nishide, K., Kajimoto, T., Node, M. Synthesis of (-)-epibatidine and its derivatives from chiral allene-1,3-dicarboxylate esters. *Chem. Pharm. Bull.* **2006**, 54, 399-402; (b) Lee, C.-L. K., Loh, T.-P. Gram-Scale Synthesis of (-)-Epibatidine. *Org. Lett.* **2005**, 7, 2965-2967; (c) Evans, D. A., Scheidt, K. A., Downey, C. W. Synthesis of (-)-epibatidine. *Org. Lett.* **2001**, 3, 3009-3012; (d) Roy, B., Watanabe, H., Kitahara, T. Simple synthesis of (±)-epibatidine. *Heterocycles* **2001**, 55, 861-871; (e) Aoyagi, S., Tanaka, R., Naruse, M., Kibayashi, C. Total Synthesis of (-)-Epibatidine Using an Asymmetric Diels-Alder Reaction with a Chiral N-Acylnitroso Dienophile. *J. Org. Chem.* **1998**, 63, 8397-8406.

(+)-EUNICENONE A

Background

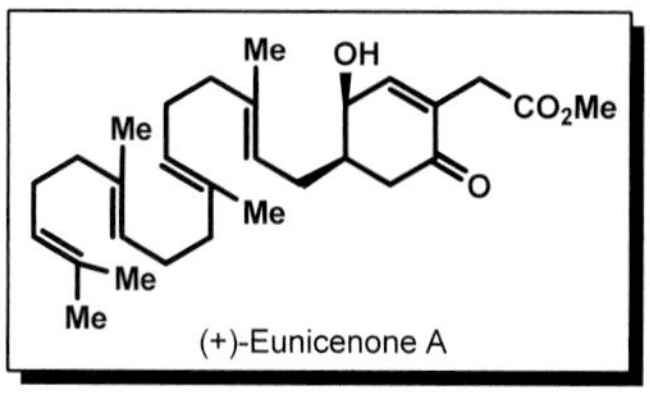

(+)-Eunicenone A

Eunicenone A is a structurally unique naturally occurring tetraprenylated 2-cyclohexenone derived from a Caribbean coral in the *Eunicea* species.[1] Tri- and tetraprenylation of proteins (e.g. the oncoprotein RAS) serves to allow attachment to cell membranes and can enable them to function.[2] In eunicenone A it is possible that the biological action is similarly dependent on membrane anchoring of the cyclic moiety. The first synthesis of eunicenone A is summarized in this section.[3]

Abbreviated Retrosynthetic Plan

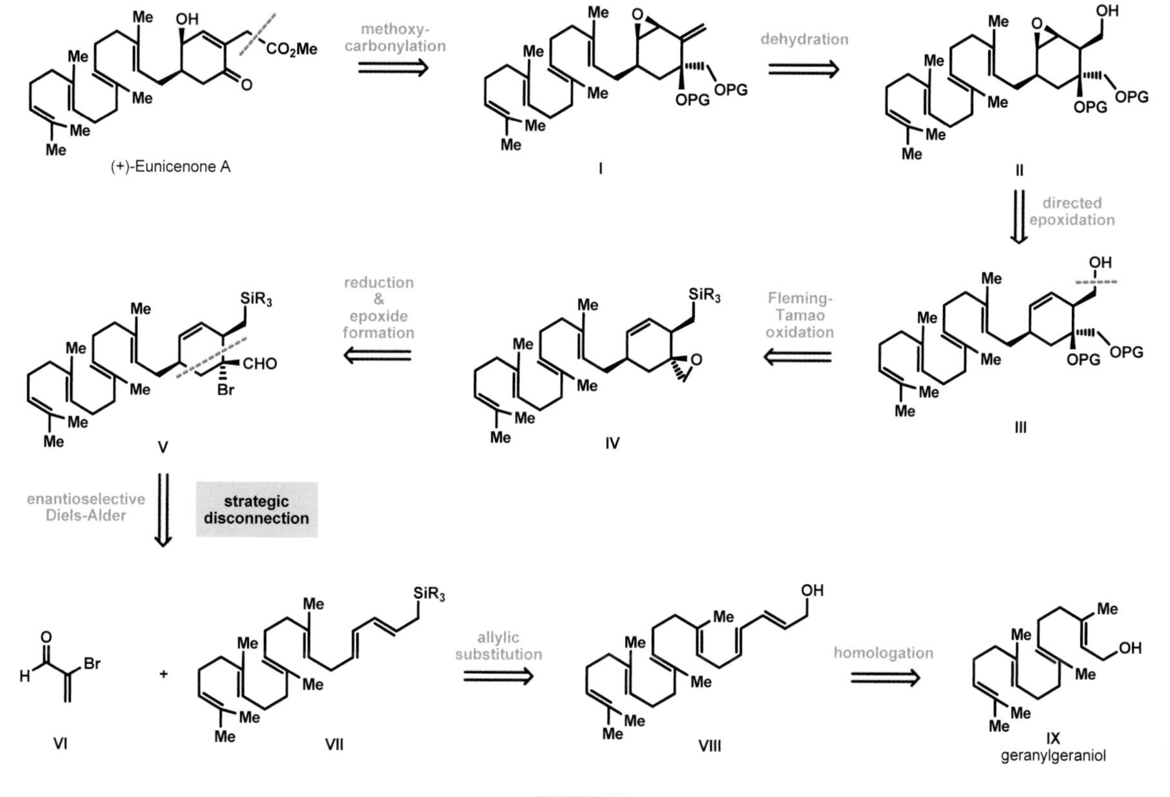

Scheme 1

Key steps in Scheme 1

(1) A key to the retrosynthetic design was the application of the enantioselective Diels-Alder transform to disconnect the six-membered ring and to clear stereocenters.

(2) Structure **V** was selected for the above disconnection for the following reasons:

  (a) It contains most of the carbon atoms of the target molecule;

(b) It can be disconnected to **VI** and **VII** which are accessible by known methods and which have suitable reactivity and;

(c) 2-bromoacrolein **VI** had previously been shown to provide excellent enantioselectivity in Diels-Alder reactions with chiral catalysts.

(3) The retrosynthetic connection between eunicenone A and **V** could be established using stereocontrolled transforms via **I** to **IV**.

(+)-EUNICENONE A

Pathway of Synthesis

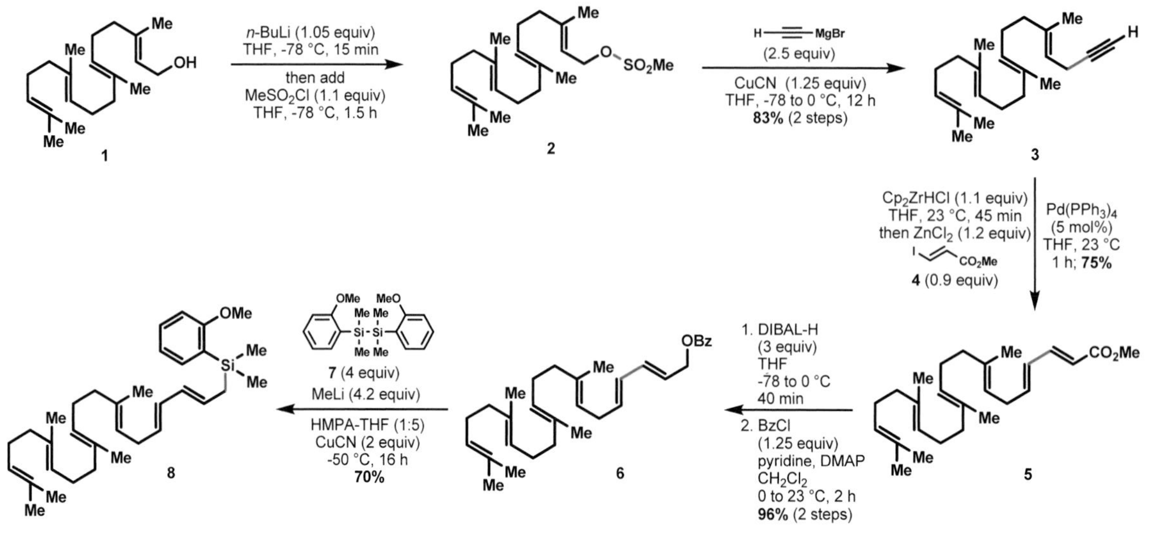

Scheme 2

Key Steps in Scheme 2

(1) The silyl-lithium reagent derived from the disilane **7** was chosen because it provides silylated intermediates which can be desilylated under very mild conditions.[4] Interestingly, serious problems were encountered during the removal (Tamao-Fleming oxidation) of the conventional phenyldimethylsilyl (PhMe$_2$Si) group.

(2) Disilane **7** is readily prepared by reaction of 2-methoxyphenyllithium with commercially available dichlorotetramethyldisilane.

(3) The reactive species during the transformation of **6** to **8** is initially (2-methoxyphenyl)dimethylsilyllithium **7a** which is then converted to the corresponding cyanocuprate **7b** under the reaction conditions and serves as the nucleophile (see below).

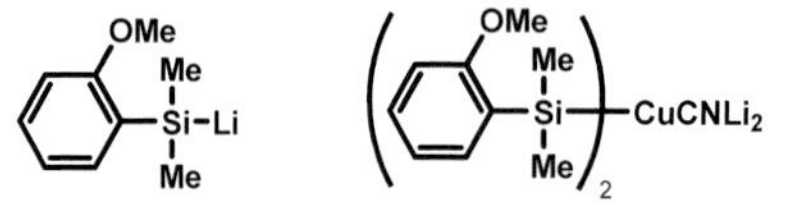

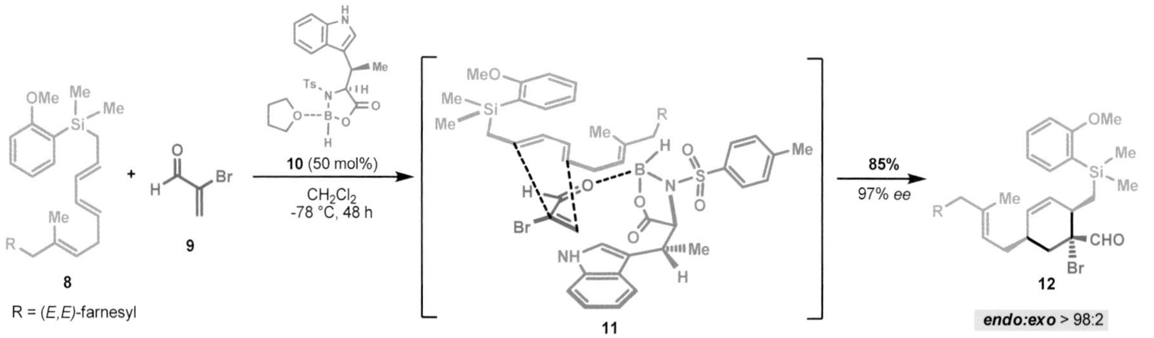

Scheme 3

Explanation of Scheme 3

(1) For a review and discussion of the mechanistic pathway of the Diels-Alder reaction of **8** and **9** with catalyst **10**, see: *Angew. Chem., Int. Ed.* **2002**, 41, 1650-1667.[5]

(2) See also the use of catalyst **10** during the synthesis of (+)-cassiol on page 267.

(+)-EUNICENONE A

Pathway of Synthesis (continued)

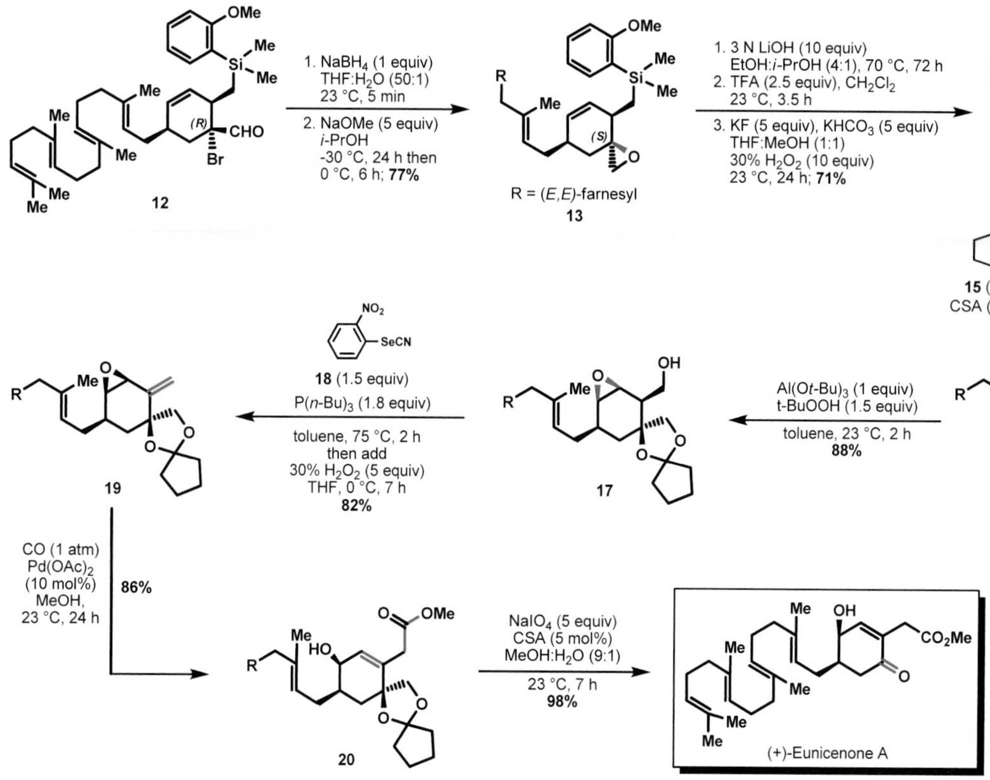

Scheme 4

Explanation of Scheme 4

(1) The conversion of **13** to **14** proceeds by LiOH-mediated epoxide ring-opening to the corresponding 1,2-diol, then Ar-Si cleavage by CF_3CO_2H and finally Tamao-Fleming C-Si oxidative cleavage.[6]

(2) The stereochemistry of epoxidation of **16**→**17** is directed by the CH_2OH group (*cis*-epoxidation).

(3) The carbonylation of the allylic epoxide **19** occurs via an intermediate π-allyl Pd(II) intermediate **19a** which then undergoes CO insertion to form an acylpalladium intermediate **19b** (see below).

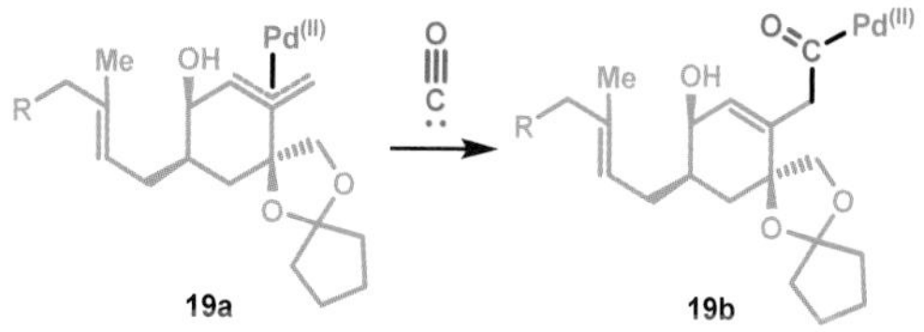

References

1. Shin, J. & Fenical, W. Eunicenones A and B: diterpenoid cyclohexenones of a rare skeletal class from a new species of the Caribbean gorgonian Eunicea. *Tetrahedron* **1993**, 49, 9277-9284.
2. Leonard, D. M. Ras Farnesyltransferase: A New Therapeutic Target. *J. Med. Chem.* **1997**, 40, 2971-2990.
3. Lee, T. W. & Corey, E. J. Enantioselective Total Synthesis of Eunicenone A. *J. Am. Chem. Soc.* **2001**, 123, 1872-1877.
4. Lee, T. W. & Corey, E. J. (2-Methoxyphenyl)dimethylsilyl Lithium and Cuprate Reagents Offer Unique Advantages in Multistep Synthesis. *Org. Lett.* **2001**, 3, 3337-3339.
5. Corey, E. J. Catalytic enantioselective Diels-Alder reactions: Methods, mechanistic fundamentals, pathways, and applications. *Angew. Chem., Int. Ed.* **2002**, 41, 1650-1667.
6. Kürti, L. & Czakó, B. Tamao-Fleming Oxidation in *Strategic Applications of Named Reactions in Organic Synthesis* 174-175 (Academic Press/Elsevier, San Diego, **2005**).

(+)-CASSIOL

Background

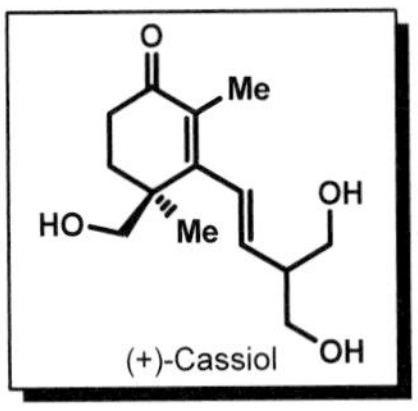

(+)-Cassiol

Cassiol occurs in trace quantities as a glycoside in the bark of Chinese cinnamon (*Cinnamum cassia*). It is notable for its potent antiulcer activity.[1] Two early syntheses involved long reaction sequences from chiral starting materials.[2] The first enantioselective route[3] is discussed in this section.[4]

Abbreviated Retrosynthetic Plan

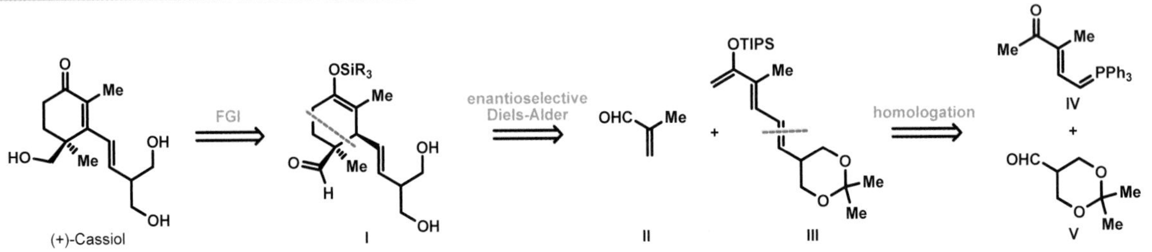

Scheme 1

Key steps in Scheme 1

(1) The synthetic plan derived from a transform-based search to find an enantioselective Diels-Alder disconnection and resulted in **I** ⇒ **II** + **III**.

(2) The diene component **III** for the cycloaddition is envisioned to arise from the Wittig coupling of stabilized ylide **IV** and aldehyde **V**.

Pathway of Synthesis

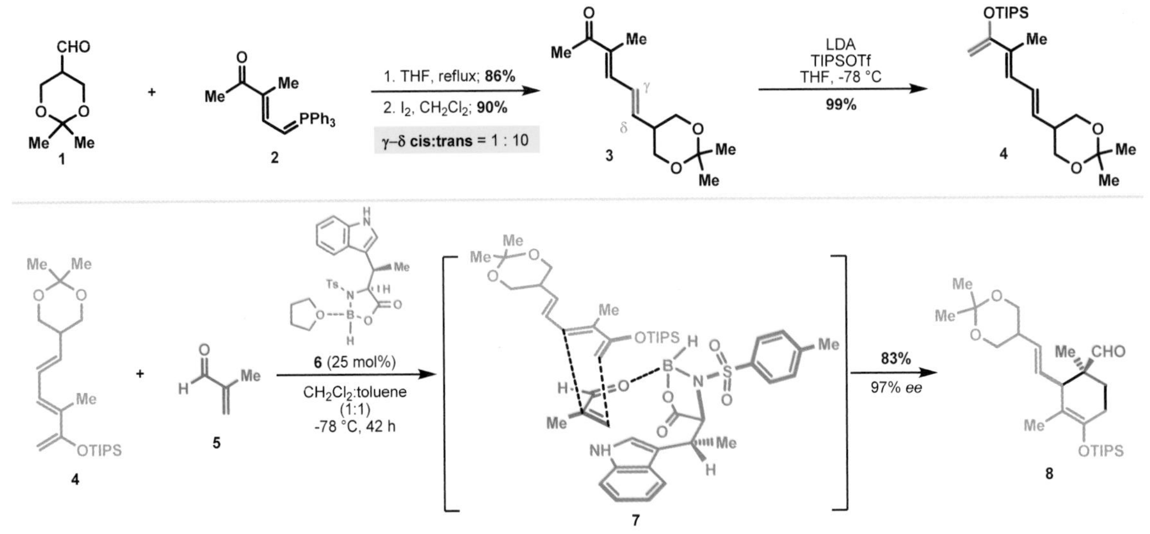

Scheme 2

(+)-CASSIOL

Pathway of Synthesis (continued)

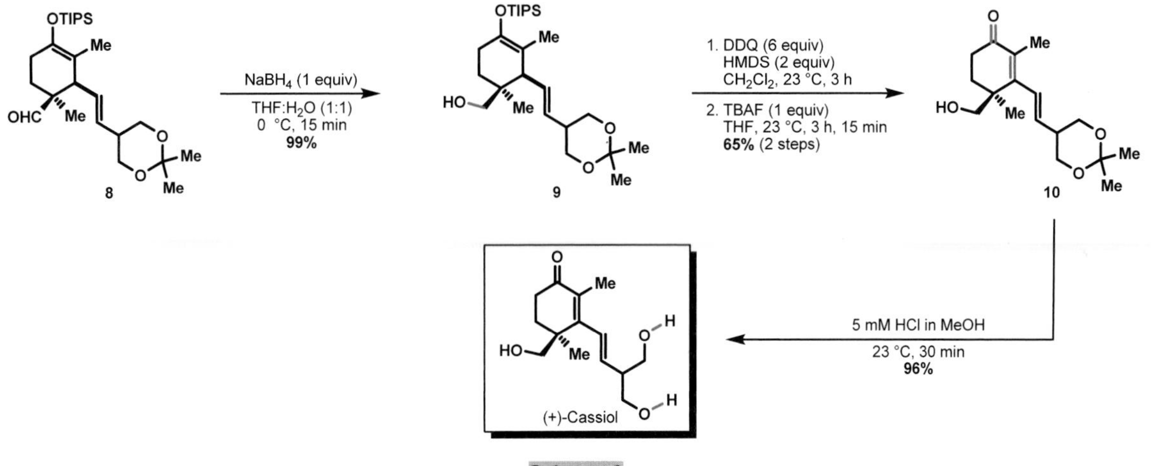

Scheme 3

Key steps in Schemes 2 and 3

(1) The crucial Diels-Alder reaction (**4** + **5** → **8**), which is highly enantioselective and diastereocontrolled, is considered to proceed via the *endo* pre-transition state assembly **7** involving catalyst **6** (Scheme 2).[5]

(2) To achieve high enantioselectivity in the Diels-Alder reaction, a number of critical observations had to be made regarding the structures of the substrates and catalyst:

  (a) the use of the TBSO and MeO analogs of diene **4** diminishes the enantiocontrol by catalyst **11**, because these groups may favor an asynchronous or a two-step pathway;

  (b) in contrast, the TIPSO substituted diene affords the cycloadduct **8** with excellent enantioselectivity;

  (c) the (R)-β-methyl catalyst **12** affords higher enantioselectivity than catalyst **11**;

  (d) the mechanistic model of the pre-transition state assembly of the catalyst, diene and dienophile (see **13**) suggests that there is likely to be serious steric repulsion between an alkyl substituent on boron and the terminal substituent on that diene [C(1)]. The model thus provides an explanation for the

superiority of the B-H catalyst[5] over the B-n-Bu analog.[3]

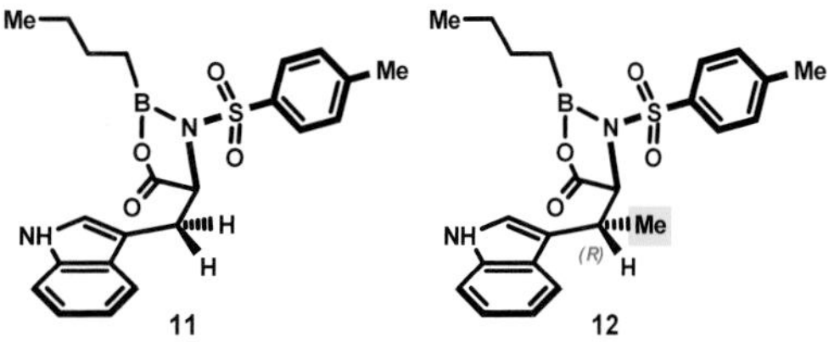

References

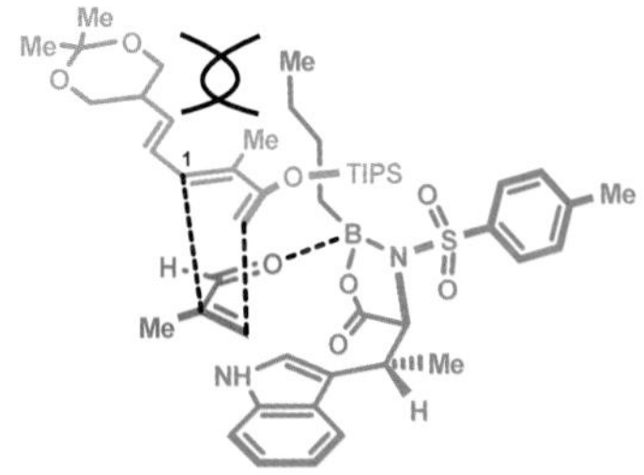

1. Shiraga, Y., Okano, K., Akira, T., Fukaya, C., Yokoyama, K., Tanaka, S., Fukui, H. & Tabata, M. Structures of potent antiulcerogenic compounds from Cinnamomum cassia. *Tetrahedron* **1988**, 44, 4703-4711.

2. (a) Takemoto, T., Fukaya, C. & Yokoyama, K. First total synthesis of (+)-cassiol. A potent antiulcerogenic compound. *Tetrahedron Lett.* **1989**, 30, 723-724; (b) Uno, T., Watanabe, H. & Mori, K. Carotenoids and degraded carotenoids. 7. Synthesis of (+)-cassiol, a potent antiulcerogenic compound. *Tetrahedron* **1990**, 46, 5563-5566.

3. Corey, E. J., Guzman-Perez, A. & Loh, T.-P. Demonstration of the Synthetic Power of Oxazaborolidine-Catalyzed Enantioselective Diels-Alder Reactions by Very Efficient Routes to Cassiol and Gibberellic Acid. *J. Am. Chem. Soc.* **1994**, 116, 3611-3612.

4. For other syntheses of cassiol, see: (a) Trost, B. M., Li, Y. A New Catalyst for a Pd Catalyzed Alder Ene Reaction. A Total Synthesis of

(+)-Cassiol. *J. Am. Chem. Soc.* **1996**, 118, 6625-6633; (b) Irie, O., Fujiwara, Y., Nemoto, H., Shishido, K. An enantioselective total synthesis of (+)-cassiol. *Tetrahedron Lett.* **1996**, 37, 9229-9232; (c) Maiti, S., Achari, B., Banerjee, A. K. A short enantioselective synthetic route to (+)-cassiol. *Synlett* **1998**, 129-130; (d) Colombo, M. I., Zinczuk, J., Mischne, M. P., Ruveda, E. A. A concise synthesis of (+)-cassiol. *Tetrahedron: Asymmetry* **2001**, 12, 1251-1253; (e) Petrova, K. V., Mohr, J. T., Stoltz, B. M. Enantioselective Total Synthesis of (+)-Cassiol. *Org. Lett.* **2009**, 11, 293-295.

5. Corey, E. J. Catalytic enantioselective Diels-Alder reactions: Methods, mechanistic fundamentals, pathways, and applications. *Angew. Chem., Int. Ed.* **2002**, 41, 1650-1667.

(−)-OVALICIN

Background

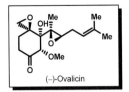

(−)-Ovalicin

Ovalicin and the microbial natural product fumagillin exhibit a range of bioactivities including antibiotic, cytotoxic and antiangiogenic (inhibition of blood vessel development) action.[1] It is thought that this occurs through inhibition of the enzyme methionine aminopeptidase via a signaling pathway called "non-canonical Wnt signaling".[2] The first enantioselective synthetic route to ovalicin[3] took advantage of steps that had been employed earlier[4] for the synthesis of (±)-ovalicin.[5]

Abbreviated Retrosynthetic Plan

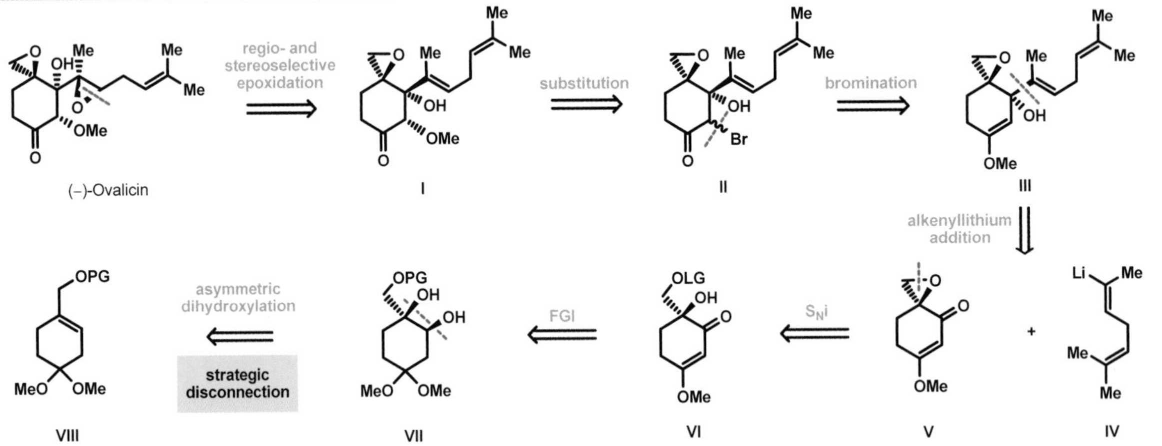

(−)-Ovalicin

regio- and stereoselective epoxidation

I

substitution

II

bromination

III

alkenyllithium addition

VIII

asymmetric dihydroxylation

strategic disconnection

VII

FGI

VI

$S_{N}i$

V

+

IV

Scheme 1

Key steps in Scheme 1

(1) The plan of synthesis was based on a retrosynthetic search to apply the enantioselective cinchona alkaloid-mediated vicinal cis-hydroxylation of a cyclohexene to clear the stereocenter in structure V, a compound that had earlier been utilized for the synthesis of (±)-ovalicin.

Retrosynthetic analysis of V led to the trisubstituted olefin VIII via VI and VII.

(2) The pathway from ovalicin to V was generated by the application of stereocontrolled transforms for the clearance of stereocenters, functionality and the 8-carbon appendage on the 6-membered ring.

Pathway of Synthesis

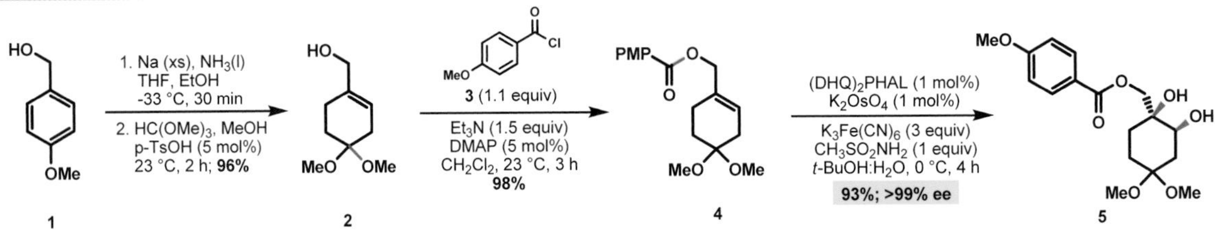

1

1. Na (xs), NH₃(l)
THF, EtOH
-33 °C, 30 min

2. HC(OMe)₃, MeOH
p-TsOH (5 mol%)
23 °C, 2 h; **96%**

2

3 (1.1 equiv)

Et₃N (1.5 equiv)
DMAP (5 mol%)
CH₂Cl₂, 23 °C, 3 h
98%

4

(DHQ)₂PHAL (1 mol%)
K₂OsO₄ (1 mol%)
K₃Fe(CN)₆ (3 equiv)
CH₃SO₂NH₂ (1 equiv)
t-BuOH:H₂O, 0 °C, 4 h

93%; >99% ee

5

Scheme 2

Key steps in Schemes 2 and 3

(1) The key enantioselective dihydroxylation (**4→5**) was guided by the model for the pre-transition state assembly that is shown as structure **7** in Scheme 3.

(2) There is much supporting evidence for the mechanistic model that is exemplified by **7**. That model correctly predicted that the 4-methoxybenzoate ester **4** should be a particularly favorable substrate since the 4-anisyl group helps to organize the pre-transition state assembly by fitting into the U-shaped binding pocket (see **7**).

(−)-OVALICIN

Pathway of Synthesis (continued)

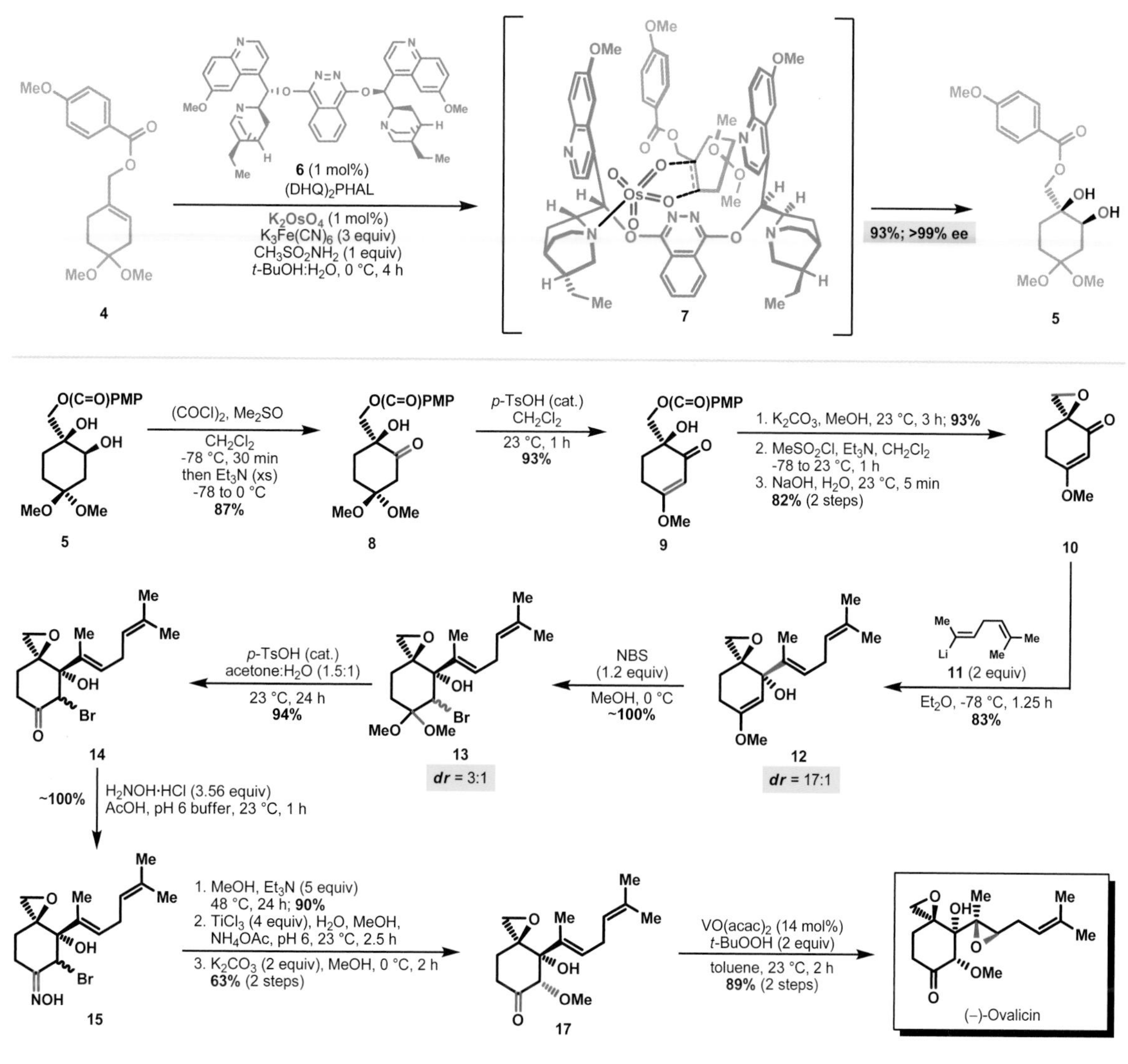

Scheme 3

References

1. (a) Sigg, H. P. & Weber, H. P. Isolation and structural elucidation of ovalicin. *Helv. Chim. Acta* **1968**, 51, 1395-1408; (b) Bollinger, P., Sigg, H. P. & Weber, H. P. Structure of ovalicin. *Helv. Chim. Acta* **1973**, 56, 819-830; (c) Ingber, D., Fujita, T., Kishimoto, S., Sudo, K., Kanamaru, T., Brem, H. & Folkman, J. Synthetic analogs of fumagillin that inhibit angiogenesis and suppress tumor growth. *Nature (London)* **1990**, 348, 555-557.

2. Zhang, Y., Yeh, J. R., Mara, A., Ju, R., Hines, J. F., Cirone, P., Griesbach, H. L., Schneider, I., Slusarski, D. C., Holley, S. A. & Crews, C. M. A chemical and genetic approach to the mode of action of fumagillin. *Chem. Biol.* **2006**, 13, 1001-1009.

3. Corey, E. J., Guzman-Perez, A. & Noe, M. C. Short Enantioselective Synthesis of (−)-Ovalicin, a Potent Inhibitor of Angiogenesis, Using

Substrate-Enhanced Catalytic Asymmetric Dihydroxylation. *J. Am. Chem. Soc.* **1994**, 116, 12109-12110.

4. Corey, E. J. & Dittami, J. P. Total synthesis of (±)-ovalicin. *J. Am. Chem. Soc.* **1985**, 107, 256-257.

5. For recent synthetic approaches toward ovalicin, see: (a) Hua, D. H., Zhao, H., Battina, S. K., Lou, K., Jimenez, A. L., Desper, J., Perchellet, E. M., Perchellet, J.-P. H., Chiang, P. K. Total syntheses of (±)-ovalicin, C4(S*)-isomer, and its C5-analogs and anti-trypanosomal activities. *Bioorg. Med. Chem.* **2008**, 16, 5232-5246; (b) Tiefenbacher, K., Arion, V. B., Mulzer, J. A Diels-Alder approach to (-)-ovalicin. *Angew. Chem., Int. Ed.* **2007**, 46, 2690-2693; (c) Brummond, K. M., McCabe, J. M. The allenic Alder-ene reaction: Constitutional group selectivity and its application to the synthesis of ovalicin. *Tetrahedron* **2006**, 62, 10541-10554 and references therein.

(−)-DYSIDIOLIDE

Background

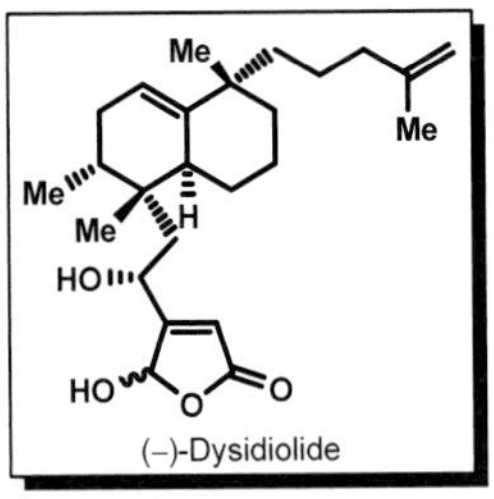

(−)-Dysidiolide

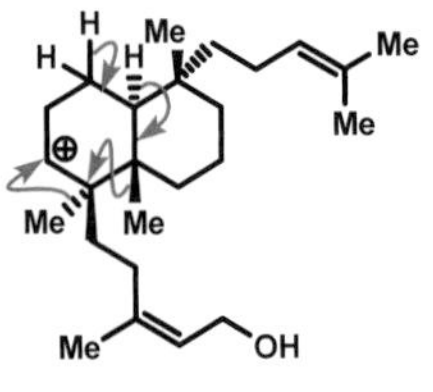

A

The sesterterpene dysidiolide, which was isolated from a Caribbean sponge, is unusual because it is one of the few centamolecules which can inhibit the phosphorylase-catalyzed P-O cleavage of phosphorylated proteins.[1] It is also of interest from a structural viewpoint, since its biosynthesis appears to involve novel cationic carbon rearrangements of a precursor such as **A**. The plan of synthesis outlined below[2] incorporated a different rearrangement.[3]

Abbreviated Retrosynthetic Plan

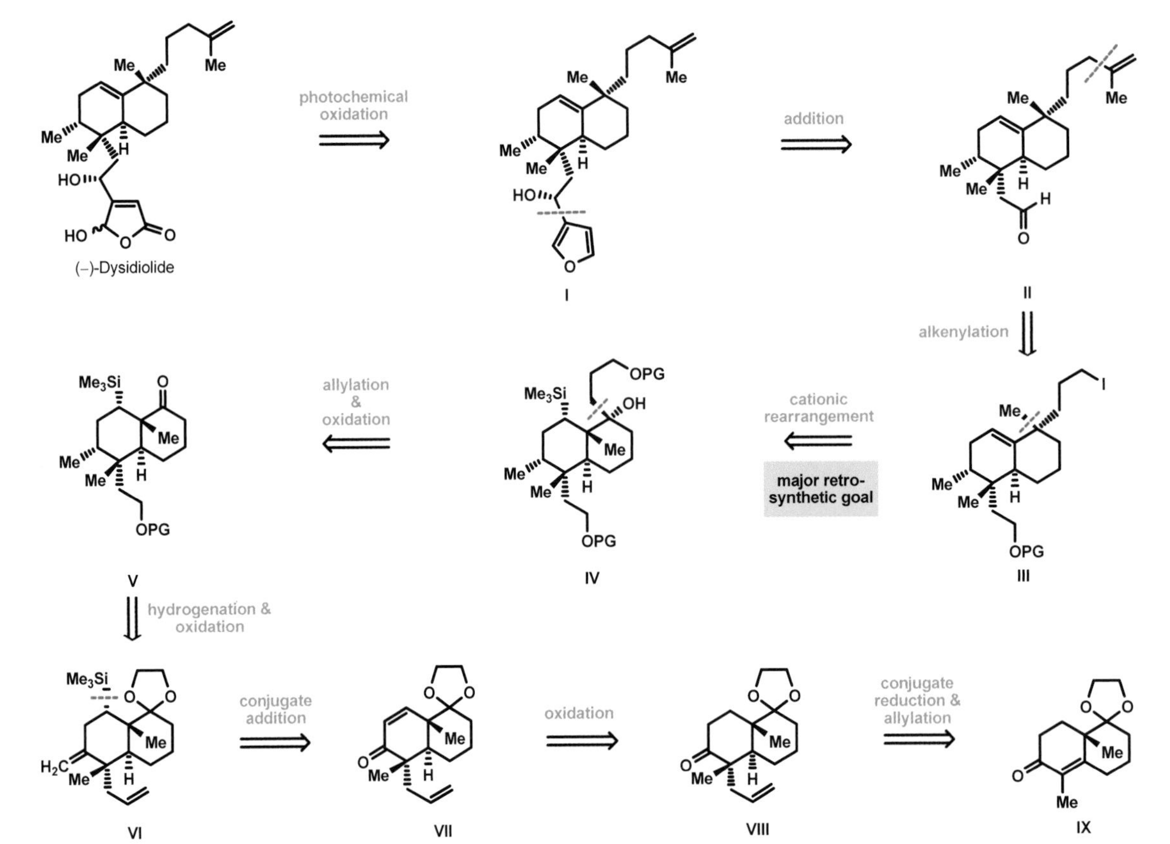

Scheme 1

Key steps in Scheme 1

(1) The design of the synthesis of dysidiolide was based on the recognition of a known chiral starting material (**IX**) that contains both rings of the target as well as useful stereocenters and functionality.[4] The use of **IX** as a retrosynthetic structure goal led to the C-C disconnections of the sequence dysidiolide ⇒ **I** ⇒ **II** ⇒ **III**.

(2) The retrosynthetic conversion **III** ⇒ **IV** utilized the cationic 1,2-methyl rearrangement transform. A trimethylsilyl group was added to **IV** both to facilitate the rearrangement and to ensure the correct location of the C-C double bond in **III**.

(−)-DYSIDIOLIDE

Pathway of Synthesis

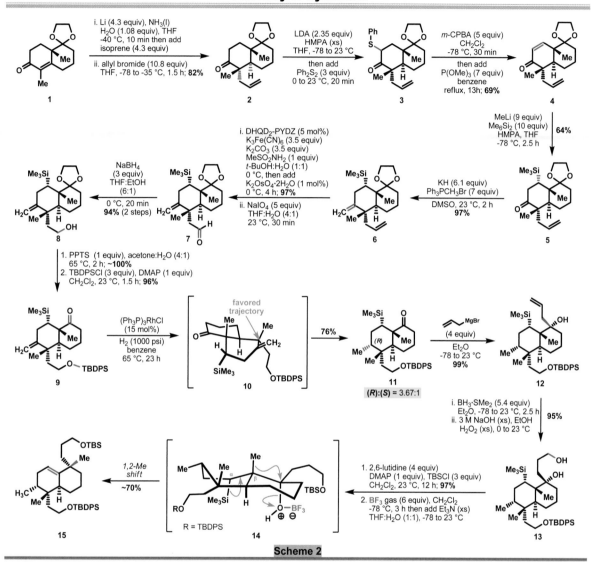

Scheme 2

Key steps in Scheme 2

(1) The chiral ketal **1** was synthesized enantioselectively by proline-catalyzed Robinson annulation (Hajos-Parrish method).[5]

(2) The use of the chiral catalyst DHQD-PYDZ allowed greater selectivity for dihydroxylation of the vinyl group of diene **6**.

(3) The rearrangement of **13** to **15** can be rationalized by the bond changes shown in **14**. The rearrangement of methyl is accelerated by orders of magnitude by the Me_3Si substituent which greatly stabilizes the tertiary cation **14a** formed initially by a 1,2-methyl shift (β-Me_3Si stabilization of a carbocation).

(−)-DYSIDIOLIDE

Pathway of Synthesis (continued)

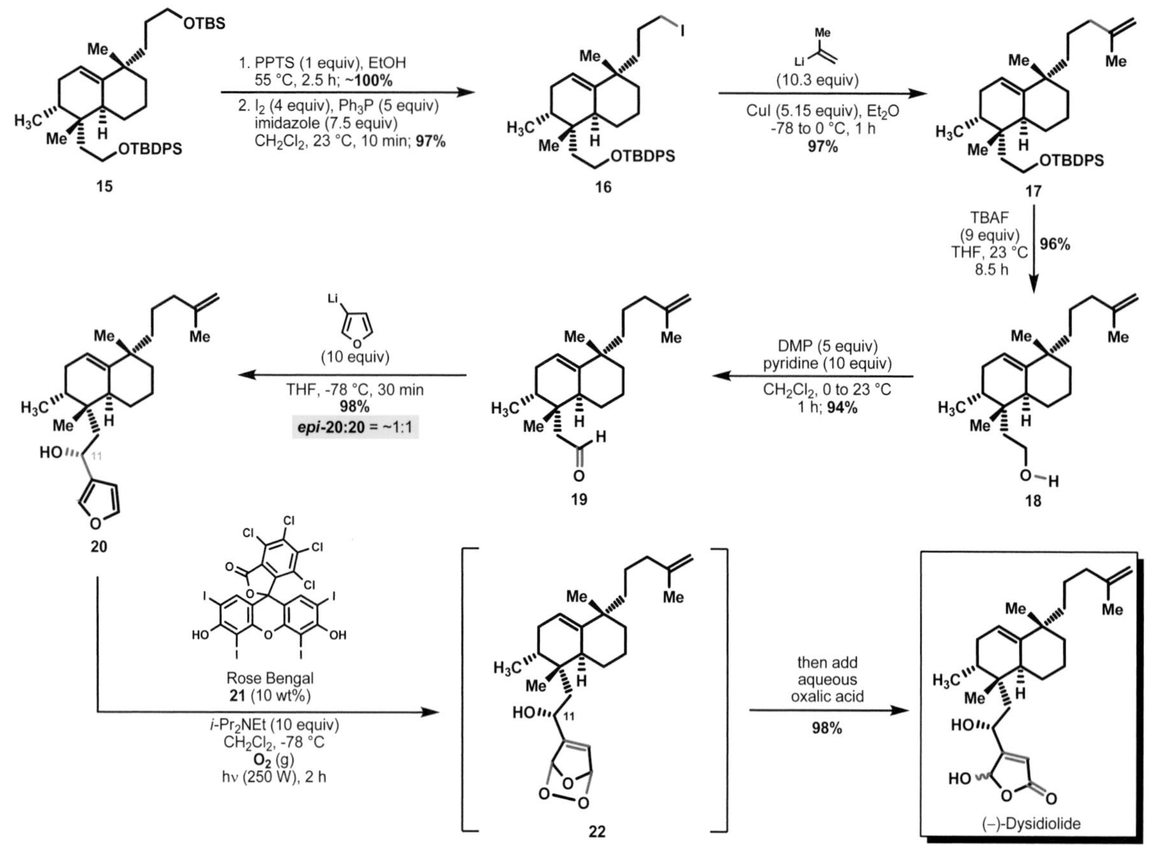

Scheme 3

Key steps in Scheme 3

The oxidation of the furan ring of **20** occurs by initial [4+2] addition of electronically excited singlet O_2 ($^1\Delta_g$ O_2, 23 kcal/mol above the ground state triplet).[6]

The resulting cycloadduct (**22**) was then treated with aqueous oxalic acid to afford (−)-dysidiolide in excellent yield.

References

1. (a) Millar, J. B. & Russell, P. The cdc25 M-phase inducer: an unconventional protein phosphatase. *Cell* **1992**, 68, 407-410; (b) Gunasekera, S. P., McCarthy, P. J., Kelly-Borges, M., Lobkovsky, E. & Clardy, J. Dysidiolide: a novel protein phosphatase inhibitor from the Caribbean sponge Dysidea etheria de Laubenfels. *J. Am. Chem. Soc.* **1996**, 118, 8759-8760.

2. Corey, E. J. & Roberts, B. E. Total Synthesis of Dysidiolide. *J. Am. Chem. Soc.* **1997**, 119, 12425-12431.

3. For other synthetic approaches to dysidiolide, see: (a) Kaliappan, K. P., Gowrisankar, P. An expedient enyne metathesis approach to dysidiolide. *Tetrahedron Lett.* **2004**, 45, 8207-8209; (b) Demeke, D., Forsyth, C. J. Total synthesis of (±)-dysidiolide. *Tetrahedron* **2002**, 58, 6531-6544; (c) Jung, M. E., Nishimura, N. Enantioselective Formal Total Synthesis of (−)-Dysidiolide. *Org. Lett.* **2001**, 3, 2113-2115; (d) Miyaoka, H., Kajiwara, Y., Hara, Y., Yamada, Y. Total Synthesis of Natural Dysidiolide. *J. Org. Chem.* **2001**, 66, 1429-1435; (e) Demeke, D., Forsyth, C. J. Novel Total Synthesis of the Anticancer Natural Product Dysidiolide. *Org. Lett.* **2000**, 2, 3177-3179; (f) Magnuson, S. R., Sepp-Lorenzino, L., Rosen, N.,

Danishefsky, S. J. A Concise Total Synthesis of Dysidiolide through Application of a Dioxolenium-Mediated Diels Alder Reaction. *J. Am. Chem. Soc.* **1998**, 120, 1615-1616; (g) Boukouvalas, J., Cheng, Y.-X., Robichaud, J. Total Synthesis of (+)-Dysidiolide. *J. Org. Chem.* **1998**, 63, 228-229.

4. Hagiwara, H. & Uda, H. Optically pure (4aS)-(+)- or (4aR)-(−)-1,4a-dimethyl-4,4a,7,8-tetrahydronaphthalene-2,5(3H,6H)-dione and its use in the synthesis of an inhibitor of steroid biosynthesis. *J. Org. Chem.* **1988**, 53, 2308-2311.

5. (a) Hajos, Z. G. & Parrish, D. R. Synthesis and conversion of 2-methyl-2-(3-oxobutyl)-1,3-cyclopentanedione to the isomeric racemic ketols of the [3.2.1]bicyclooctane and of the perhydroindane series. *J. Org. Chem.* **1974**, 39, 1612-1615; (b) Kürti, L. & Czakó, B. Hajos-Parrish Reaction in *Strategic Applications of Named Reactions in Organic Synthesis* 192-193 (Academic Press/Elsevier, San Diego, **2005**).

6. Kernan, M. R. & Faulkner, D. J. Regioselective oxidation of 3-alkylfurans to 3-alkyl-4-hydroxybutenolides. *J. Org. Chem.* **1988**, 53, 2773-2776.

ANTHERIDIC ACID

Background

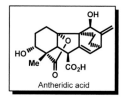

Antheridic acid

Antheridic acid is a plant regulator that was isolated from various species of fern in very small amounts.[1] Structurally related to gibberellic acid,[2] it presented a comparable challenge to synthesis. A strategy that was initially applied to the synthesis of (±)-antheridic acid[3] was modified later to allow access to the natural form.[4]

Abbreviated Retrosynthetic Plan

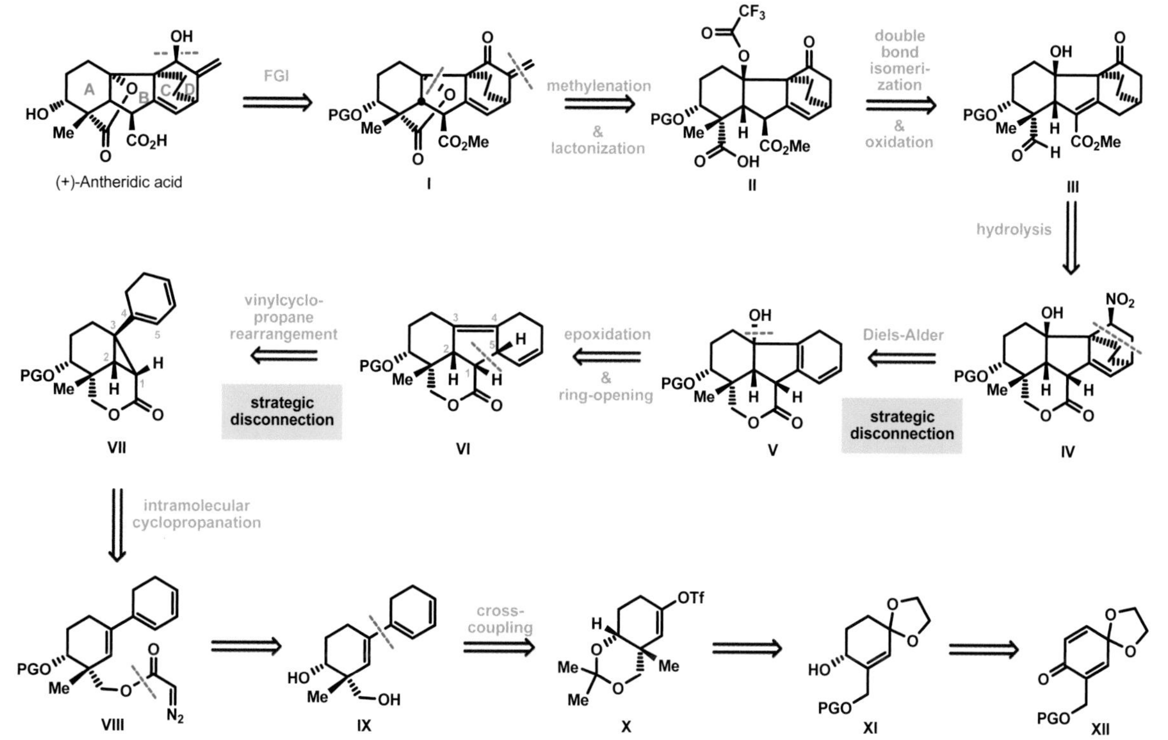

Scheme 1

Key steps in Scheme 1

(1) The initial goal of the retrosynthetic analysis was the disconnection of one of the two-carbon bridges of the bicyclo[2.2.2] octene moiety in antheridic acid by application of the Diels-Alder transform. Note that the basic retron is already present in that part (rings C/D).

(2) In order to facilitate the Diels-Alder construction of the target structure, it was simplified via the subgoals **I-IV**. The functional group interchange **III** ⇒ **IV** serves the purpose of actuating the Diels-Alder disconnection **IV** ⇒ **V**.

(3) The retrosynthetic analysis from **V** to **VIII** was guided by the goal of disconnecting the B ring. This would be carried out by the successive application of two transforms:

(a) vinylcyclopropane to cyclopentene rearrangement and;

(b) intramolecular [2+1]-cycloaddition.

(4) Finally, the sequence **VIII** ⇒ **XII** was devised to clear stereocenters and the two appendages.

273

ANTHERIDIC ACID

Pathway of Synthesis

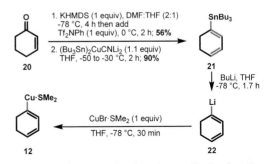

Scheme 2

Key steps in Scheme 2

(1) The initiating stereocenter for the synthesis of antheridic acid was established in the oxazaborolidine-catalyzed reduction of **6** to **9**, which is considered to occur via the assembly **8**.

(2) The introduction of the next stereocenter (**10→11**) by conjugate addition (methylation using dimethyl-lithium-cuprate) is controlled by steric screening at the β-face of the α,β-enone. The resulting lithium enolate was trapped in situ with phenyltriflimide (PhNTf$_2$) as the corresponding vinyl triflate **11**.

(3) The organocopper reagent **12** used in the coupling of **11** was prepared from 1,3-cyclohexadienyllithium **22** and an equivalent amount of cuprous bromide-methyl sulfide complex (see scheme to the right).

(4) Reagent **15** was developed specifically for the transformation of **14→16**.[3]

ANTHERIDIC ACID

Pathway of Synthesis (continued)

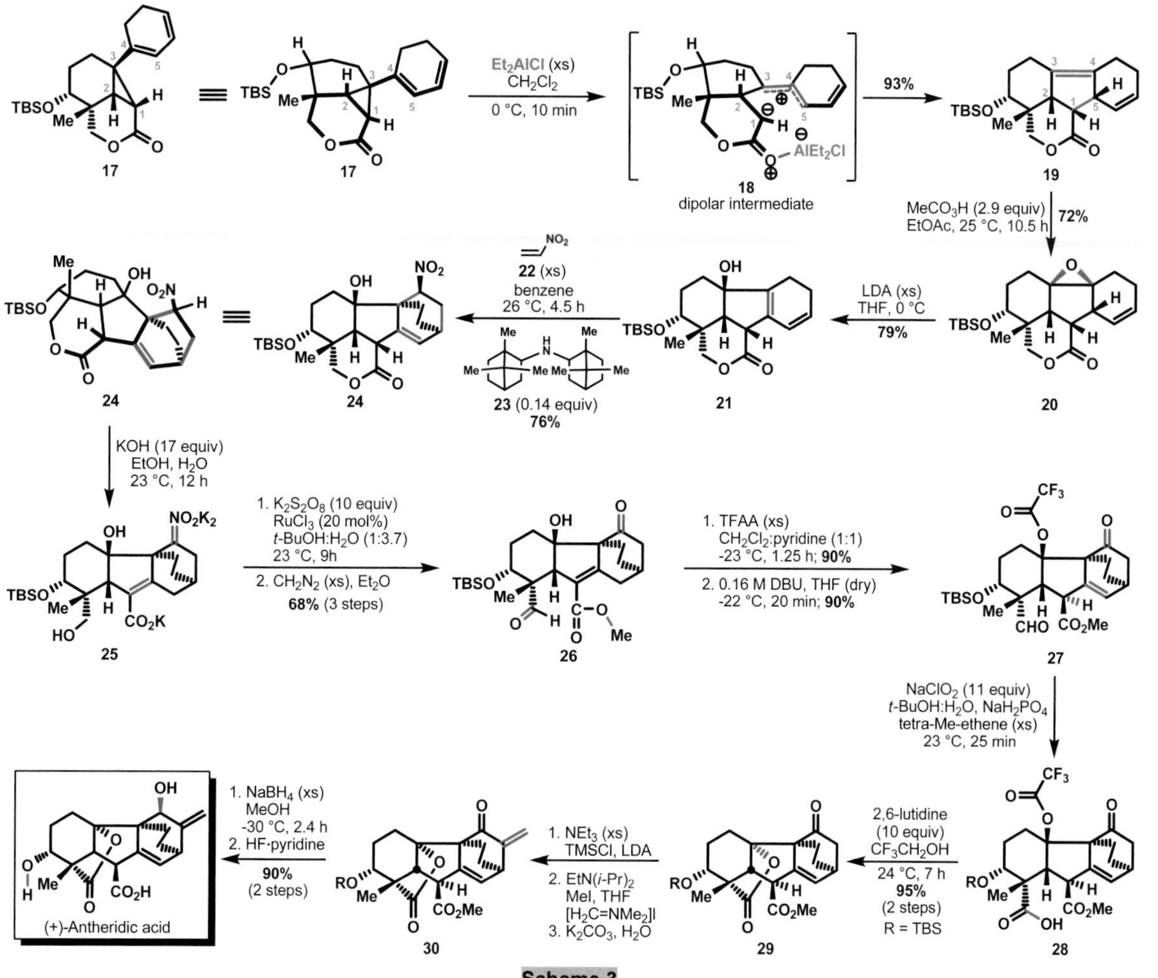

Scheme 3

Key steps in Scheme 3

(1) The novel Lewis acid-catalyzed vinylcyclopropane → cyclopentene ring-expansion **17**→**19** may be concerted.

(2) The elimination **20**→**21** is controlled by the acidity of the allylic C-H subunit.

(3) The weak base **23** is used in the Diels-Alder reaction because **21** is very acid sensitive.

(4) The oxidation **27**→**28** generates chlorinating byproducts which are removed by trapping with $Me_2C=CMe_2$.

References

1. (a) Naf, U., Nakanishi, K. & Endo, M. On the physiology and chemistry of fern antheridiogens. *Bot. Rev.* **1975**, 41, 315-359; (b) Corey, E. J., Myers, A. G., Takahashi, N., Yamane, H. & Schraudolf, H. Constitution of antheridium-inducing factor of Anemia phyllitidis. *Tetrahedron Lett.* **1986**, 27, 5083-5084; (c) Takeno, K., Yamane, H., Nohara, K., Takahashi, N., Corey, E. J., Myers, A. G. & Schraudolf, H. Biological activity of antheridic acid, an antheridiogen of Anemia phyllitidis. *Phytochemistry* **1987**, 26, 1855-1857.

2. Mander, L. N. Twenty years of gibberellin research. *Nat. Prod. Rep.* **2003**, 20, 49-69.

3. Corey, E. J. & Myers, A. G. Total synthesis of (±)-antheridium-inducing factor (A_{An}) of the Fern Anemia phyllitidis. Clarification of stereochemistry. *J. Am. Chem. Soc.* **1985**, 107, 5574-5576.

4. Corey, E. J. & Kigoshi, H. A route for the enantioselective total synthesis of antheridic acid, the antheridium-inducing factor from Anemia phyllitidis. *Tetrahedron Lett.* **1991**, 32, 5025-5028.

3β,20-DIHYDROXYPROTOST-24-ENE

Background

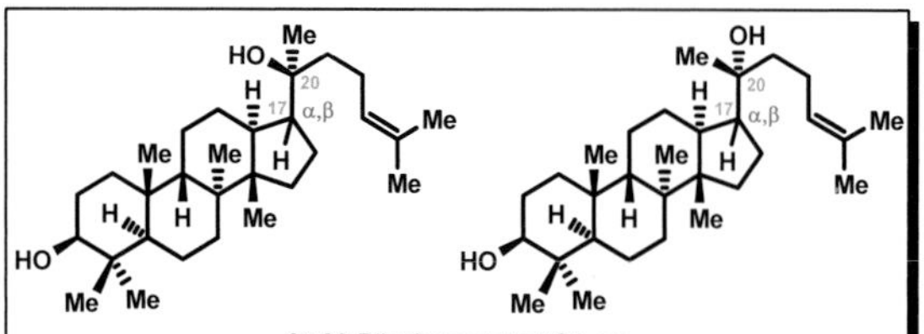

3β,20-Dihydroxyprotost-24-ene

The protostane skeleton represents a hypothetical triterpenoid which is thought to be a transient intermediate in the biosynthesis of lanosterol and cholesterol from (S)-2,3-oxidosqualene.[1] The syntheses of four C(17) and C(20) diastereomeric dihydroxyprotostenes [17α and 17β, 20(R) and 20(S)] that are described in this section were carried out to obtain these unknown compounds for biosynthetic studies as well as to address the challenge of their synthesis.[2] The last part of this section describes the biosynthetic results that depended on the availability of the four diastereomeric synthetic 3β,20-dihydroxyprotost-24-enes.

Abbreviated Retrosynthetic Plan

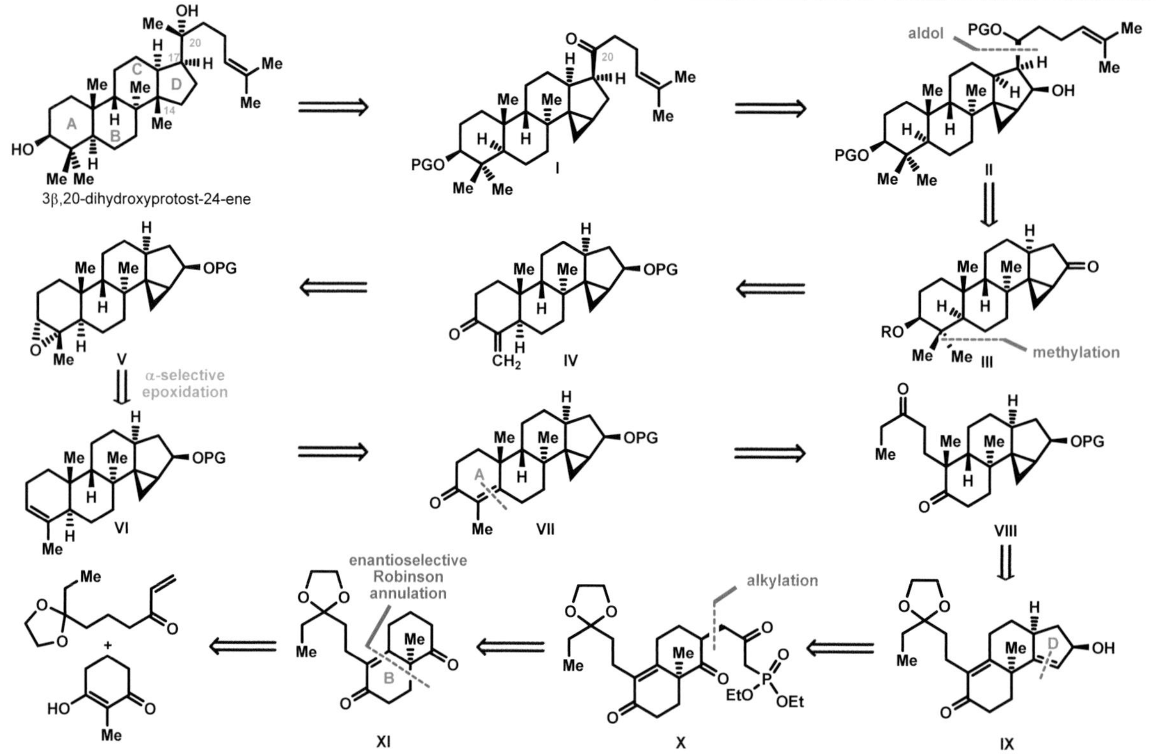

Scheme 1

Key steps in Scheme 1

(1) The unusual *trans-syn-trans* fusion of the A/B/C ring subunit of the protostane nucleus, which is crucial for cholesterol biosynthesis, had to be taken into account in the retrosynthetic design because there were no known synthetic routes to such structures and some previous failures.[3] The B-ring in the protosterol core is constrained to have a twist boat conformation rather than the more stable chair.

(2) Other elements which guided the retrosynthetic planning included:

(a) the available early intermediates containing an initiating stereocenter which could serve as long-range S-goals;

(b) the need to incorporate stereochemically valid trans-forms for disconnecting rings and appendages and;

(c) the selection of powerfully simplifying sterically controlled transforms for strategic bond disconnections.

(3) The bicyclic structure **XI** was selected as a logical chiral subgoal since it is accessible by a known enantioselective method, contains the carbons of the A/B/C ring system and is well-suited for attachment of the D ring.

(4) An early synthetic objective was the disconnection of the appendage at C(17) and the methyl at C(14) with stereochemical control (**I** ⇒ **II** ⇒ **III** ⇒ **IV** ⇒⇒ **IX**). In addition, the steps **IV** ⇒ **V** ⇒ **VI** ⇒ **VII** serve to disconnect ring A.

3β,20-DIHYDROXYPROTOST-24-ENE

Pathway of Synthesis

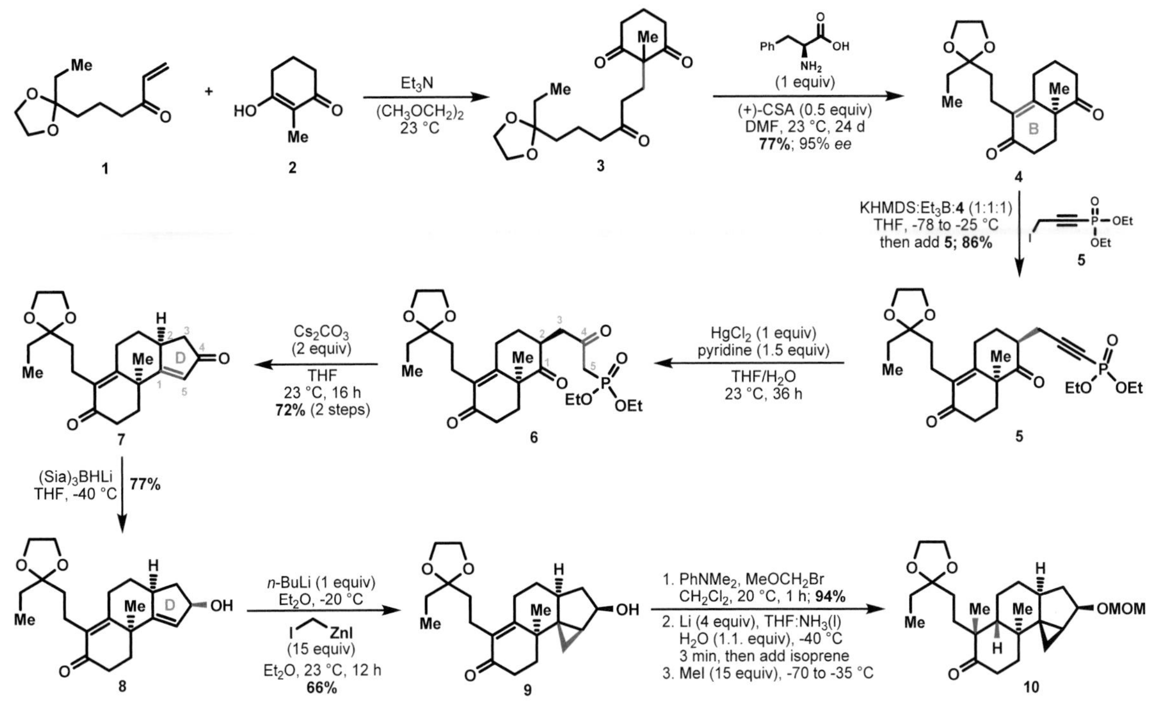

Scheme 2

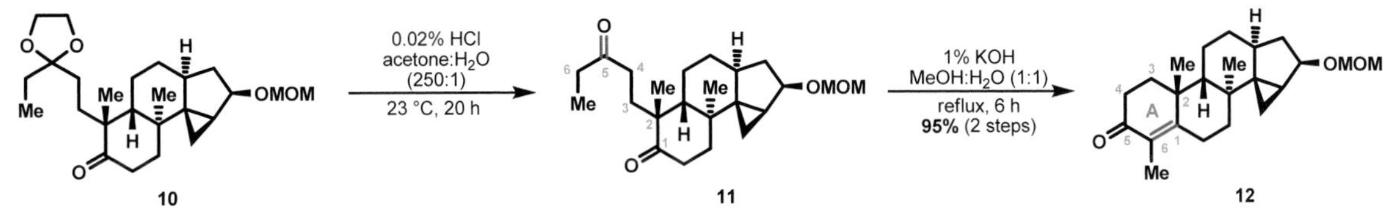

Scheme 3

Key steps in Schemes 2 and 3

(1) (S)-phenylalanine promoted the enantioselective enamine-mediated aldol cyclization of **3** to the bicyclic Robinson annulation product **4**, setting the stage for the attachment of ring D by the sequence **4→5→6→7**. The initiating stereocenter at C(2) of **4** controls the configuration at the next-formed stereocenter of **6** and **7** and also the third stereocenter in **8**.

(2) The allylic hydroxyl group on ring D of **8** directs the syn-methylenation of the nearby double bond. In addition, the α-methylenation (**8→10**) is even more favored because of steric screening by the proximate angular methyl group (see the figure at right).

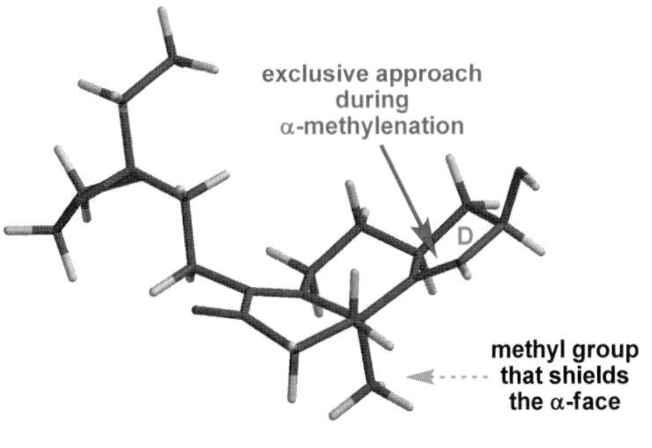

exclusive approach during α-methylenation

methyl group that shields the α-face

8

3β,20-DIHYDROXYPROTOST-24-ENE

Pathway of Synthesis (continued)

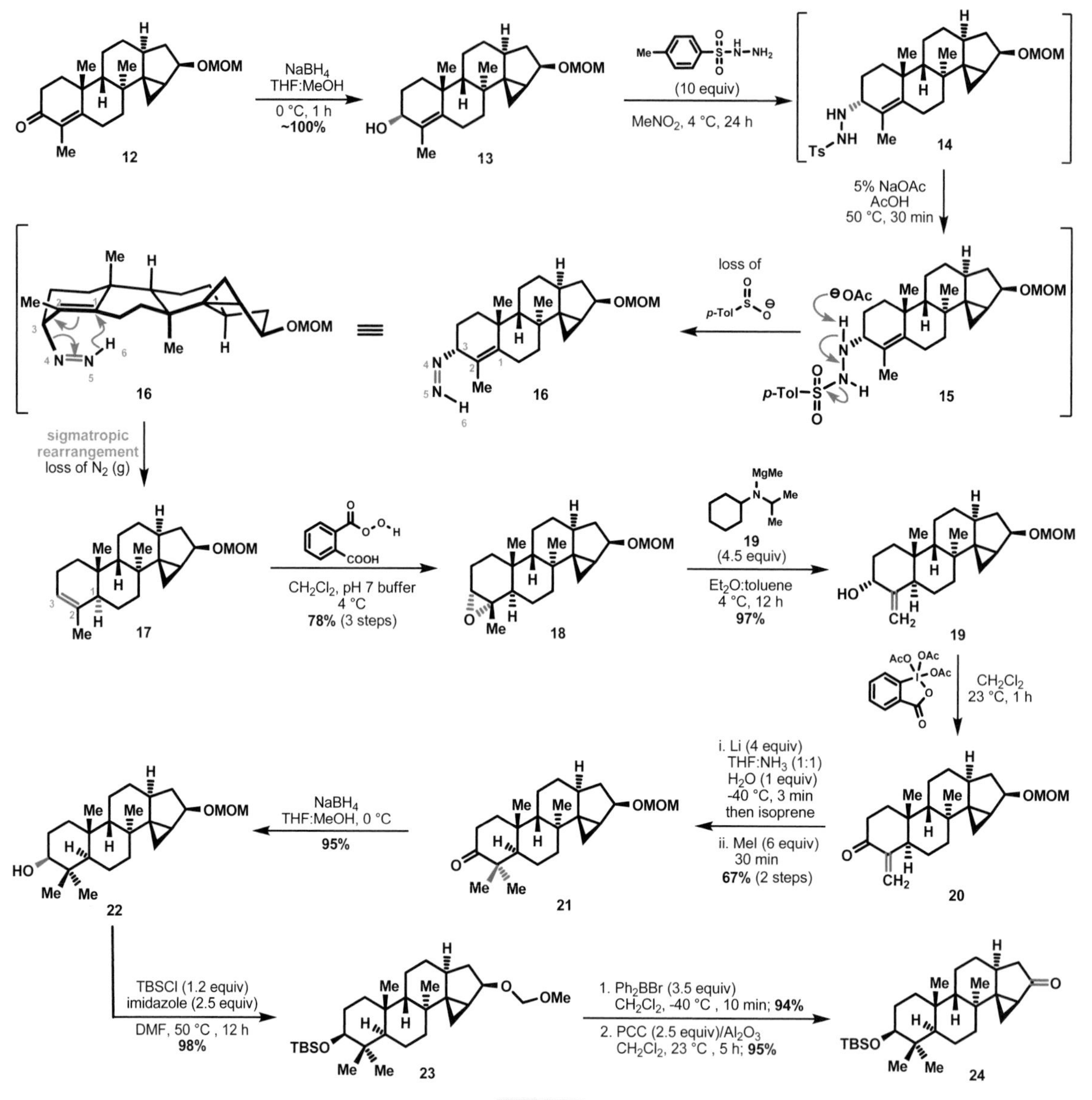

Scheme 4

Key steps in Scheme 4

(1) The displacement **13**→**14** takes place with inversion directly with the need for hydroxyl activation in CH_3NO_2 as solvent.

(2) The reductive transposition of C=C, **14**→**17**, occurs suprafacially to form the A/B *trans* product **17** (see the transition state representation at right).

TS* of the suprafacial sigmatropic rearrangement of **16** to **17**

3β,20-DIHYDROXYPROTOST-24-ENE

Pathway of Synthesis to 17α Protosterols

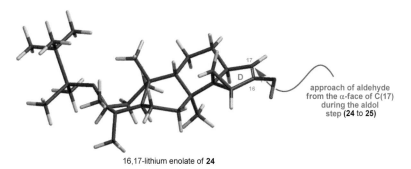

Scheme 5

Key steps in Scheme 5

(1) The cyclopropyl methylene group of **24** serves not only as a precursor of the 14α-methyl group of **28**, but also as a control element that ensures selective formation of the 16,17-enolate that is required for the aldol step **24→25**.

(2) The (20R)- and (20S)-diastereomers of **30** can be separated chromatographically.

approach of aldehyde from the α-face of C(17) during the aldol step **(24 to 25)**

16,17-lithium enolate of **24**

3β,20-DIHYDROXYPROTOST-24-ENE

Pathway of Synthesis to 17β Protosterols

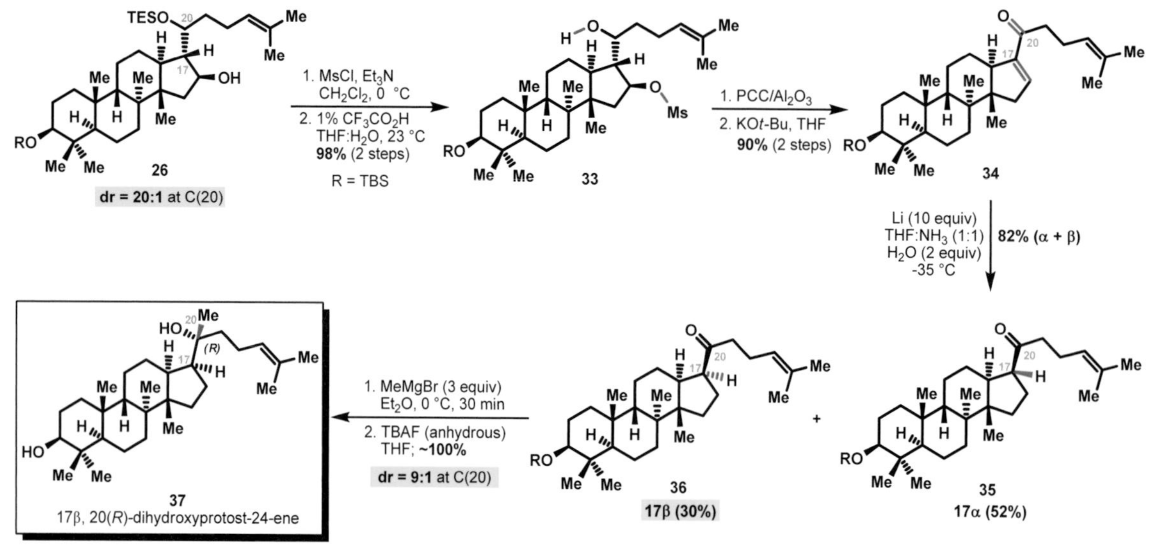

26
dr = 20:1 at C(20)

1. MsCl, Et$_3$N
   CH$_2$Cl$_2$, 0 °C
2. 1% CF$_3$CO$_2$H
   THF:H$_2$O, 23 °C
98% (2 steps)
R = TBS

33

1. PCC/Al$_2$O$_3$
2. KOt-Bu, THF
90% (2 steps)

34

Li (10 equiv)
THF:NH$_3$ (1:1) | 82% (α + β)
H$_2$O (2 equiv)
-35 °C

37
17β, 20(R)-dihydroxyprotost-24-ene

1. MeMgBr (3 equiv)
   Et$_2$O, 0 °C, 30 min
2. TBAF (anhydrous)
   THF; ~100%
dr = 9:1 at C(20)

36
17β (30%)

+

35
17α (52%)

Scheme 6

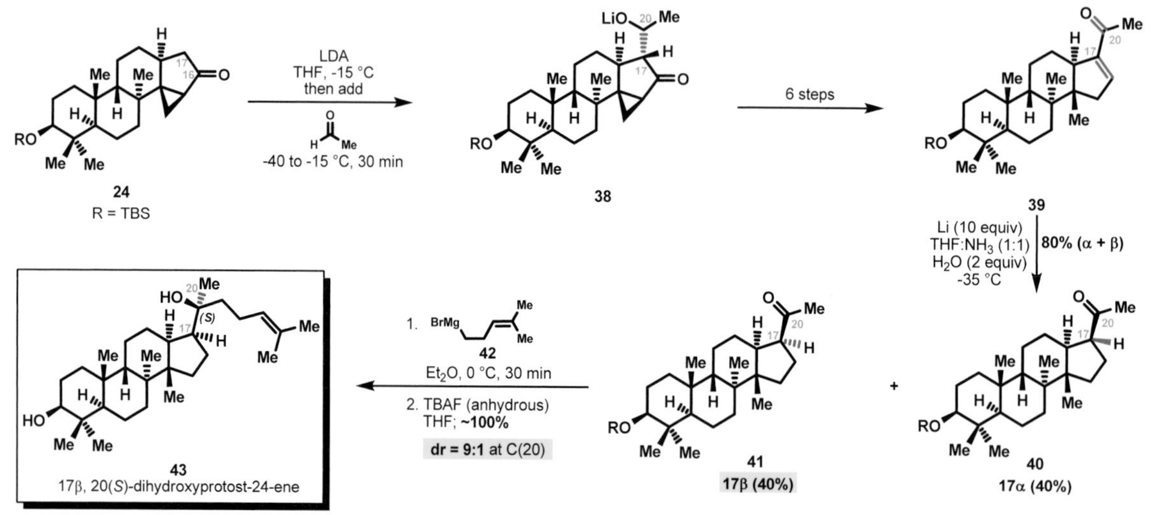

24
R = TBS

LDA
THF, -15 °C
then add

-40 to -15 °C, 30 min

38

6 steps

39

Li (10 equiv)
THF:NH$_3$ (1:1) | 80% (α + β)
H$_2$O (2 equiv)
-35 °C

43
17β, 20(S)-dihydroxyprotost-24-ene

1. BrMg
   42
   Et$_2$O, 0 °C, 30 min
2. TBAF (anhydrous)
   THF; ~100%
dr = 9:1 at C(20)

41
17β (40%)

+

40
17α (40%)

Scheme 7

Key steps in Schemes 6 and 7

The Grignard addition reactions to the 17β-ketones **36** and **41** to form **37** and **43**, respectively, are both highly diastereoselective (ca. 9:1). One reason for this preference is the proximity of the ketone 17β-sidechain to the 14β-methyl group which restricts rotation and orientation of the sidechain (see figure at right).

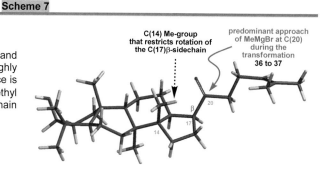

C(14) Me-group that restricts rotation of the C(17)β-sidechain

predominant approach of MeMgBr at C(20) during the transformation 36 to 37

3β,20-DIHYDROXYPROTOST-24-ENE

Subsequent Developments

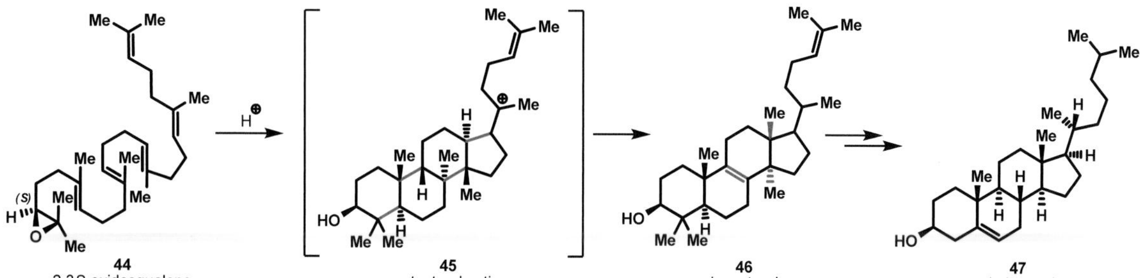

44 2,3*S*-oxidosqualene; **45** protosterol cation; **46** lanosterol; **47** cholesterol

Scheme 8

Description of Scheme 8

(1) The process for the synthesis of the four C(17) and C(20) 3β, 20-dihydroxyprotost-24-enes that is described above has been used to obtain insights on the stereochemical and mechanistic details of the biosynthesis of cholesterol (**47**) and its precursor lanosterol (**46**) which are biosynthesized from 2,3(*S*)-oxidosqualene (**44**). The overall biosynthesis of cholesterol is summarized in Scheme 8. A single enzyme converts 2,3(*S*)-oxidosqualene to lanosterol in a single step via highly reactive, transient carbocationic intermediates, including the protosterol cation.[1] Suprafacial 1,2-migration of H and Me lead from the protosterol cation (**45**) to lanosterol (**46**) on the same enzyme by an energetically downhill (exothermic) pathway.[4]

(2) The synthetic efficiency of the construction of lanosterol from **44** has permitted this step to endure through eons of evolution as the initial phase of cholesterol biosynthesis, even though some 18 biochemical steps and numerous enzymes are required to transform lanosterol to cholesterol. It remains beyond the capabilities of modern chemical synthesis to duplicate the remarkable one-step cyclization of 2,3(*S*)-oxidosqualene to lanosterol; indeed, it looms as one of the greatest challenges for the future.

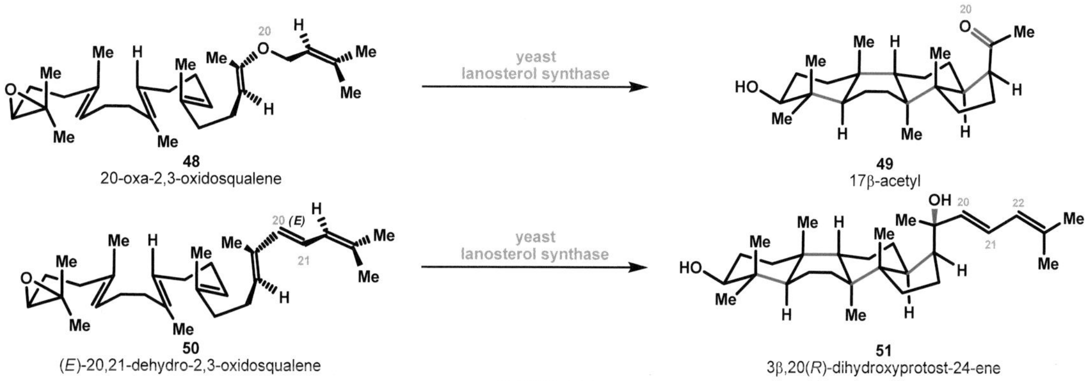

48 20-oxa-2,3-oxidosqualene; **49** 17β-acetyl; **50** (*E*)-20,21-dehydro-2,3-oxidosqualene; **51** 3β,20(*R*)-dihydroxyprotost-24-ene

Scheme 9

Description of Scheme 9

(1) The first definitive result on the stereochemistry of the biosynthetic pathway to protosterol was obtained in an experiment on the cyclization of 20-oxa-2,3-oxidosqualene by yeast lanosterol synthase, which gave as major product the tetracyclic methyl ketone **49**, identical with a synthetic sample that was prepared from the 17-ketone **24** (Scheme 5).[5]

(2) The second definitive result, which confirmed the first, emerged from the transformation by yeast lanosterol synthesis of (*E*)-20,21-dehydro-2,3-oxidosqualene **50** to the dehydro derivative of 3β,20(*R*)-dihydroxyprotost-24-ene **51**.[6] The stereochemistry of the biosynthetic product **51** was confirmed by selective hydrogenation of the 22,23-double bond to give a compound that was identical with totally synthetic **32** (22*R* diastereomer, see Scheme 5 on page 279).[6]

3β,20-DIHYDROXYPROTOST-24-ENE

Biosynthesis of Lanosterol from 2,3-Oxidosqualene

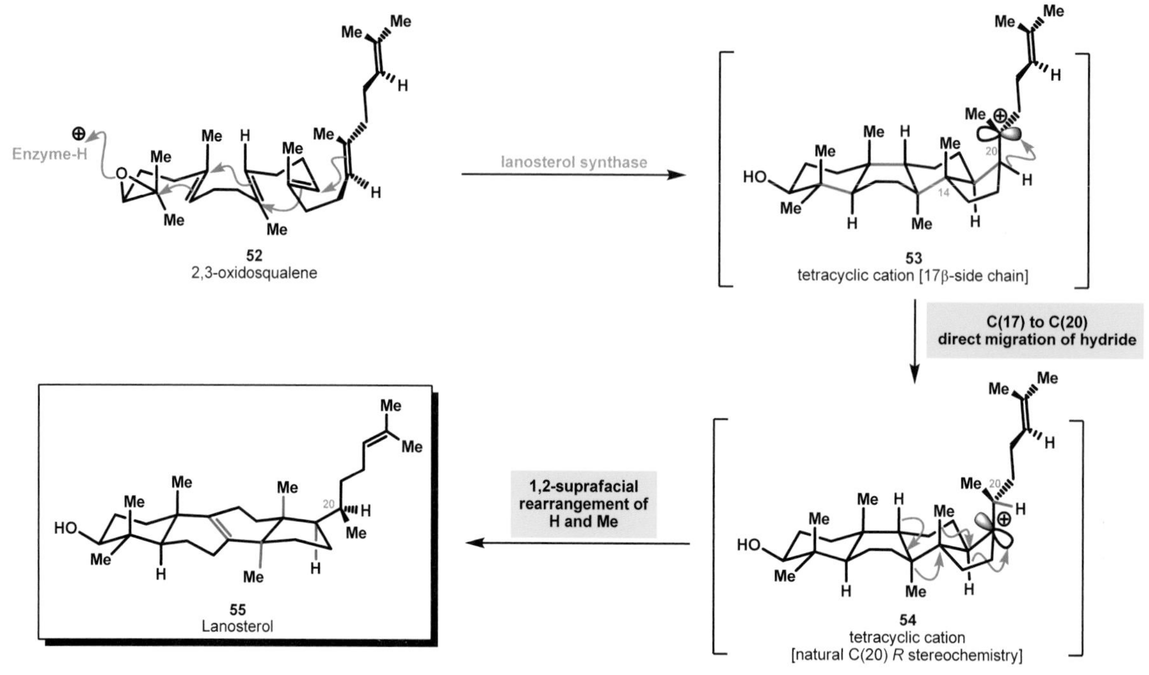

Scheme 10

BF₃-Initiated Rearrangements of 17β-Protosterols

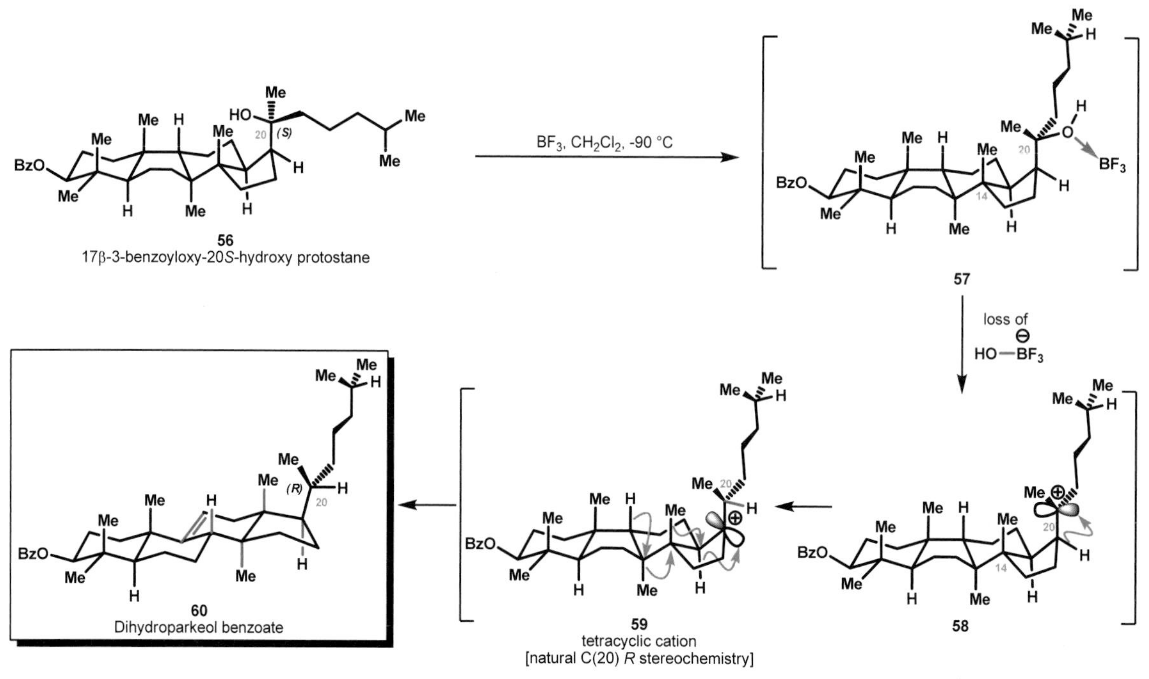

Scheme 11

3β,20-DIHYDROXYPROTOST-24-ENE

BF₃-Initiated Rearrangements of 17β-Protosterols (continued)

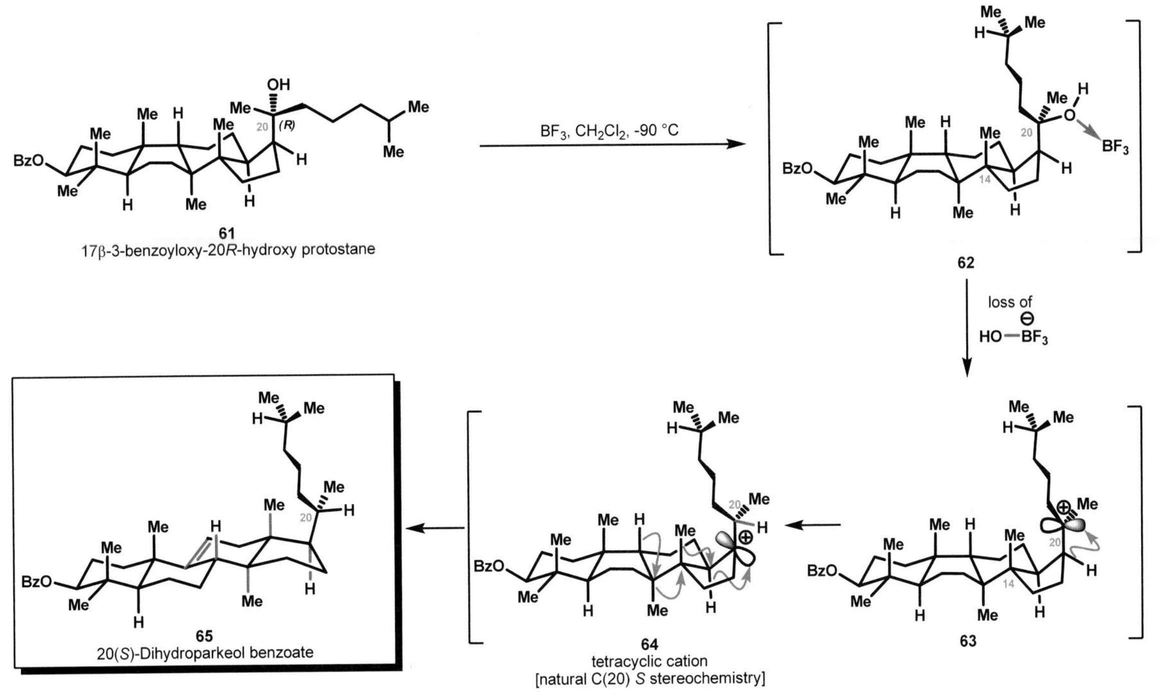

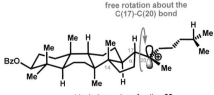

61
17β-3-benzoyloxy-20R-hydroxy protostane

BF₃, CH₂Cl₂, -90 °C

62

loss of
HO—BF₃⁻

63

64
tetracyclic cation
[natural C(20) S stereochemistry]

65
20(S)-Dihydroparkeol benzoate

Scheme 12

Description of Schemes 10-13

(1) The directly formed conformation of the 17β-cation **53** allows a 1,2-hydride shift from C(17) to C(20) after a least motion rotation about the C(17)-C(20) bond of <60° to give the lanosterol 20R configuration in **54**, whereas the formation of the unnatural 20S configuration requires a rotation about the C(17)-C(20) bond of ca. 120°.

(2) Furthermore, there is a large barrier to that 120° rotation resulting from the spatial proximity of the 14β-methyl group and 17β-sidechain (see tetracyclic cation **53**). Thus, both stereoelectronic and steric factors favor the natural 20R stereochemistry in the biosynthetic product.

(3) In contrast, it is clear that lanosterol synthase would be severely challenged to control the configuration at C(20) of lanosterol if the tetracyclization were to produce the 17α-sidechain as a metastable intermediate.

(4) Chemical experiments gave results which are fully in accordance with the above analysis. As outlined in Scheme 11, treatment of the 17β-3-benzyloxy-20(S)-hydroxy protostane **56** with BF₃ in CH₂Cl₂ at -90 °C for 10 minutes afforded the benzoate of the natural product dihydroparkeol (natural 20R configuration) **60**.

(5) As outlined in Scheme 12, the corresponding 20R-hydroxy diastereomer **61** rearranged under the same conditions to unnatural 20S dihydroparkeol benzoate **65**.

(6) In contrast, as shown in Scheme 13, the rearrangement of either 17α-sidechain oriented benzoates (i.e., 20R- or

20S-hydroxy 17a-protostannyl 3-benzoates) took place without any stereoselection whatsoever at C(20) to give a 1:1 mixture of 20R and 20S-dihydroparkeol benzoates.

(7) These results support the assignment of the 17β-sidechain orientation to the tetracyclic protostanyl cation **53** in lanosterol biosynthesis and the steps that are outlined in Scheme 10. Previously it had been thought that the 8-carbon sidechain at C(17) in the intermediate tetracyclic protostanyl cation was α-oriented.[7]

(8) As a result of the availability of synthetic 17α- and 17β-protosterol derivatives (**66** and **67**, respectively), the reasons why evolution settled on the 17β-pathway for lanosterol biosynthesis rather than the alternative 17α-arrangement of the cationic sidechain are apparent from the experimental results in Schemes 12 and 13.

(9) In contrast, the α-sidechain cation corresponding to cation **63**, has free rotation about the C(17)-C(20) bond (see figure below).

free rotation about the
C(17)-C(20) bond

α-sidechain version of cation **63**

3β,20-DIHYDROXYPROTOST-24-ENE

Biosynthesis of Lanosterol from 2,3-Oxidosqualene

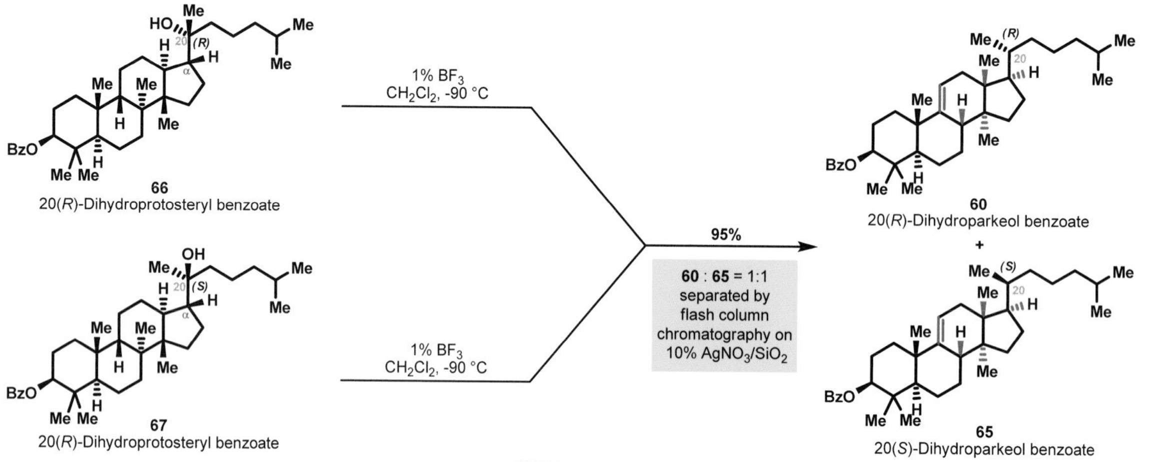

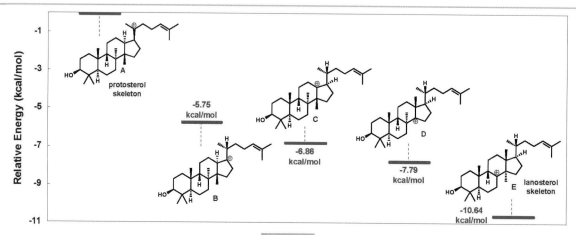

Scheme 13

The rearrangement of the protosterol core to the lanosterol/parkeol core, which is clearly thermodynamically favorable, is substantially driven by the conformationally strained twist-boat B-ring of the protosterol structure. The barrier to the rearrangement is also low, a point that is supported by Hartree-Fock calculations at the DFT6-31G* level (Scheme 14).[4]

Scheme 14

References

1. Wendt, K. U., Schulz, G. E., Corey, E. J. & Liu, D. R. Enzyme mechanisms for polycyclic triterpene formation. *Angew. Chem., Int. Ed.* **2000**, 39, 2812-2833.
2. Corey, E. J. & Virgil, S. C. Enantioselective total synthesis of a protosterol, 3β,20-dihydroxyprotost-24-ene. *J. Am. Chem. Soc.* **1990**, 112, 6429-6431.
3. Ireland, R. E. & Hengartner, U. Total synthesis of steroidal antibiotics. I. Efficient, stereoselective method for the formation of *trans-syn-trans*-perhydrophenanthrene derivatives. *J. Am. Chem. Soc.* **1972**, 94, 3652-3653.
4. Kürti, L., Chein, R.-J. & Corey, E. J. Conformational Energetics of Cationic Backbone Rearrangements in Triterpenoid Biosynthesis Provide an Insight into Enzymatic Control of Product. *J. Am. Chem. Soc.* **2008**, 130, 9031-9036.
5. Corey, E. J. & Virgil, S. C. An experimental demonstration of the stereochemistry of enzymic cyclization of 2,3-oxidosqualene to the protosterol system, forerunner of lanosterol and cholesterol. *J. Am. Chem. Soc.* **1991**, 113, 4025-4026.
6. Corey, E. J., Virgil, S. C. & Sarshar, S. New mechanistic and stereochemical insights on the biosynthesis of sterols from 2,3-oxidosqualene. *J. Am. Chem. Soc.* **1991**, 113, 8171-8172.
7. (a) Eschenmoser, A., Ruzicka, L., Jeger, O. & Arigoni, D. Triterpenes. CXC. A stereochemical interpretation of the biogenetic isoprene rule of the triterpenes. *Helv. Chim. Acta* **1955**, 38, 1890-1904; (b) Cornforth, J. W. Olefin alkylation in biosynthesis. *Angew. Chem., Int. Ed. Engl.* **1968**, 7, 903-911.

(+)-DAMMARENEDIOL II

Background

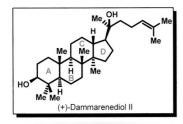

(+)-Dammarenediol II

Dammarenediol is the primary product of tetracyclization of 2,3-(S)-oxidosqualene in many plants.[1] It differs from 3β,20-dihydroxy-protost-24-ene (pages 276-284), the prede-

cessor of lanosterol and other sterols, in respect to backbone stereochemistry and the conformation of 2,3(S)-oxidosqualene from which it is formed enzymically. Specifically, dammarenediol possesses a *trans-anti-trans-anti-trans* arrangement of the A/B/C/D ring system. There are numerous tetracyclic plant triterpenes that share the nucleus of dammarenediol. This section outlines the first synthesis of dammarenediol.[2]

Abbreviated Retrosynthetic Plan

Scheme 1

Key steps in Scheme 1

(1) The retrosynthetic plan depended on the application of the cation-olefin cyclization transform to disconnect rings A, B and C.

(2) Since the direct disconnection of the tetracyclic network of dammarenediol is not feasible, two subgoals had to be utilized:

(a) first, disconnection of the C(20)-C(21) bond to form I and

(b) second, the disconnection of ring D at the C(13)-C(17) bond using the aldol transform via the sequence I ⇒ II ⇒ III.

(3) The disconnection V ⇒ VI + VII + VIII depended on the development of new methodology.

Pathway of Synthesis

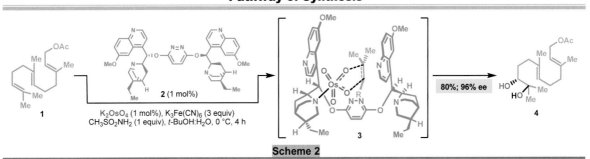

2 (1 mol%)

K_2OsO_4 (1 mol%), $K_3Fe(CN)_6$ (3 equiv)
$CH_3SO_2NH_2$ (1 equiv), t-BuOH:H_2O, 0 °C, 4 h

80%; 96% ee

Scheme 2

(+)-DAMMARENEDIOL II

Pathway of Synthesis (continued)

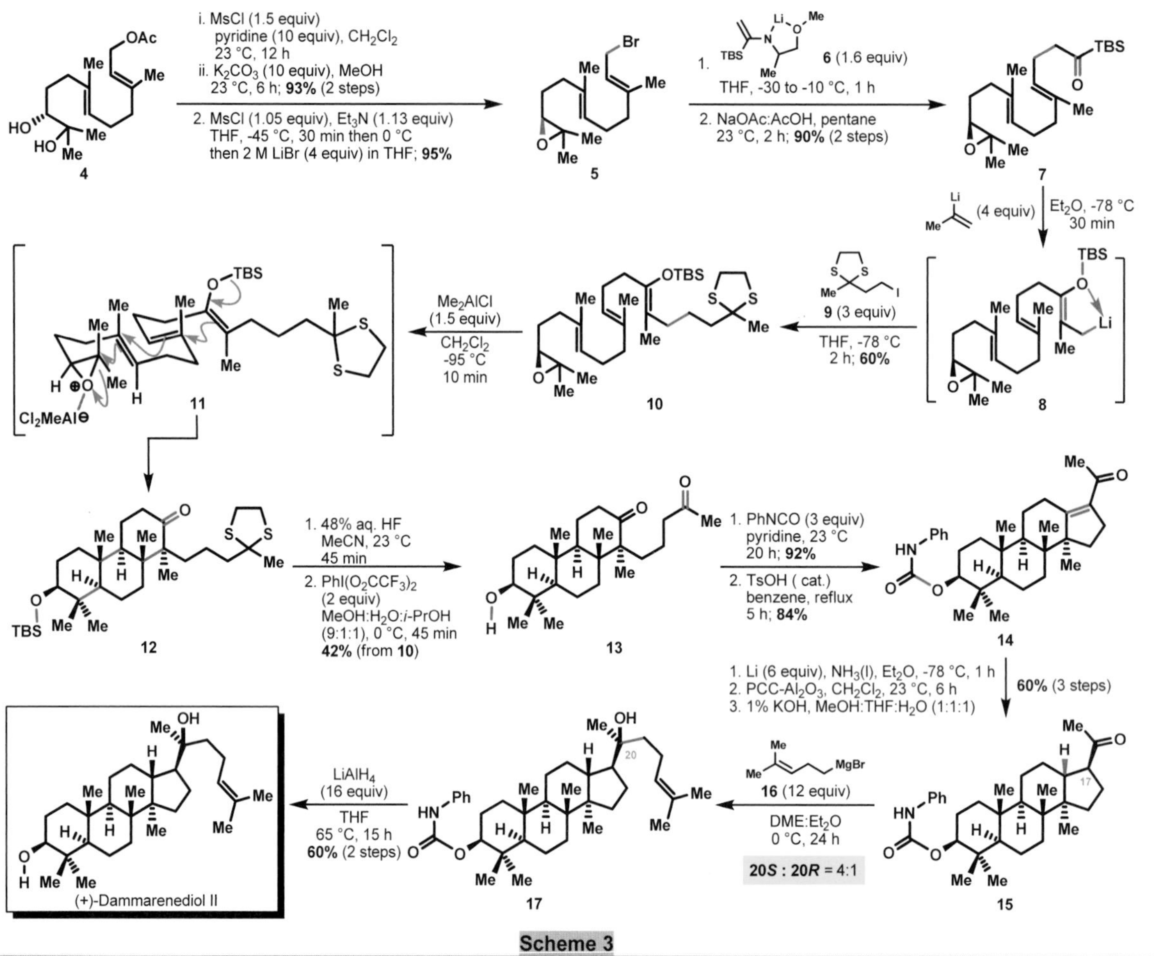

Scheme 3

Explanation of Schemes 2 and 3

(1) The introduction of the first stereocenter for the synthesis was carried out by catalytic enantioselective dihydroxylation of the isopropylidene subunit using **2** as catalyst (Scheme 2).

(2) The three-component coupling that was developed for the synthesis outlined here, specifically **7→8→10**, is a very powerful and general construction that forms an effective tactical combination with cation-olefin polycyclization.[3] The process occurs by the sequence:

(a) carbonyl addition of 2-propenyllithium to **7**;

(b) Brook rearrangement (**7a→7b**, see below) and

(c) nucleophilic coupling of the resulting chelated lithium derivative **8** with the iodide **9**.

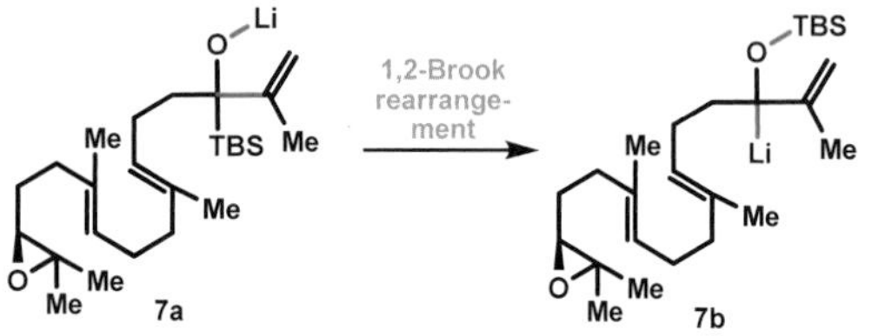

References

1. Mills, J. S. The constitution of the neutral tetracyclic triterpenes of dammar resin. *Chem. Ind. (London, U. K.)* **1956**, 189-190.

2. (a) Corey, E. J. & Lin, S. A Short Enantioselective Total Synthesis of Dammarenediol II. *J. Am. Chem. Soc.* **1996**, 118, 8765-8766; (b) Johnson, W. S., Bartlett, W. R., Czeskis, B. A., Gautier, A., Lee, C. H., Lemoine, R., Leopold, E. J., Luedtke, G. R. & Bancroft, K. J. The fluorine atom as a cation-stabilizing auxiliary in biomimetic polyene cyclizations: total synthesis of dl-dammarenediol. *J. Org. Chem.* **1999**, 64, 9587-9595.

3. Corey, E. J., Lin, S. & Luo, G. Stereospecific synthesis of tetrasubstituted Z-enol silyl ethers by a three component coupling process. *Tetrahedron Lett.* **1997**, 38, 5771-5774.

(+)-LANOSTENOL

Background

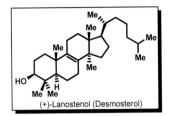

(+)-Lanostenol (Desmosterol)

Lanostenol is biosynthesized from (S)-oxidosqualene by a remarkable one-step, non-stop process which is mediated by lanosterol synthase.[1] It is the precursor of cholesterol and the other biologically important steroids. A short, enantioselective synthesis of lanostenol is outlined here.[2] Two earlier approaches were very lengthy and not enantiocontrolled.[3]

Abbreviated Retrosynthetic Plan

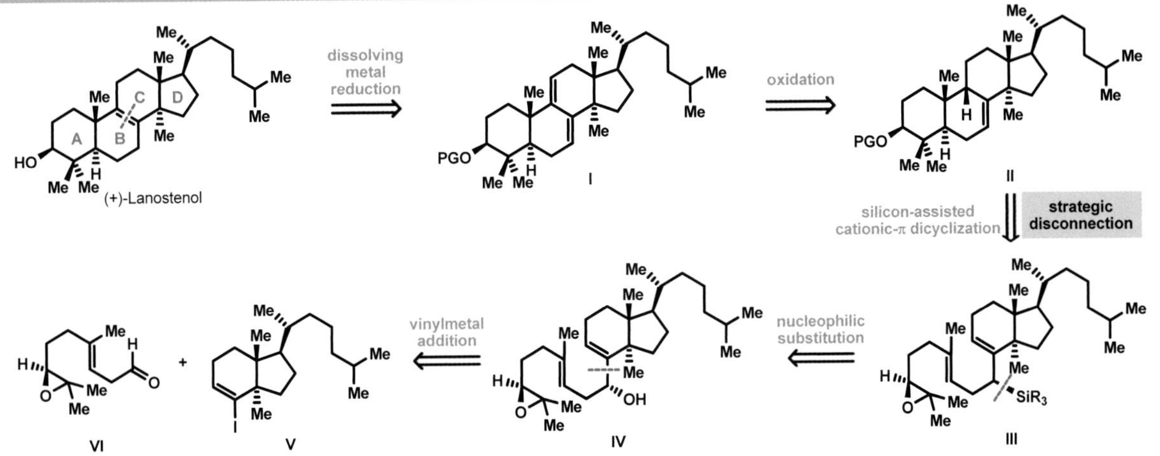

Scheme 1

Key steps in Scheme 1

(1) The plan of synthesis took advantage of the availability of two chiral precursors:

   (a) the epoxide of geraniol (S-enantiomer)[4] and

   (b) a commercially available ketone (**1**, Grundemann ketone, see Scheme 2) containing the C and D rings of lanostenol. Ketone **1** can be synthesized enantioselectively.[5]

(2) The disconnection of the A and B rings by application of cation-olefin cyclization transform was not feasible directly, but became possible after retrosynthetic conversion via **I** ⇒ **II**.

(3) A silyl group was introduced retrosynthetically (**II** ⇒ **III**) to facilitate the disconnection of rings A and B.[6]

Pathway of Synthesis

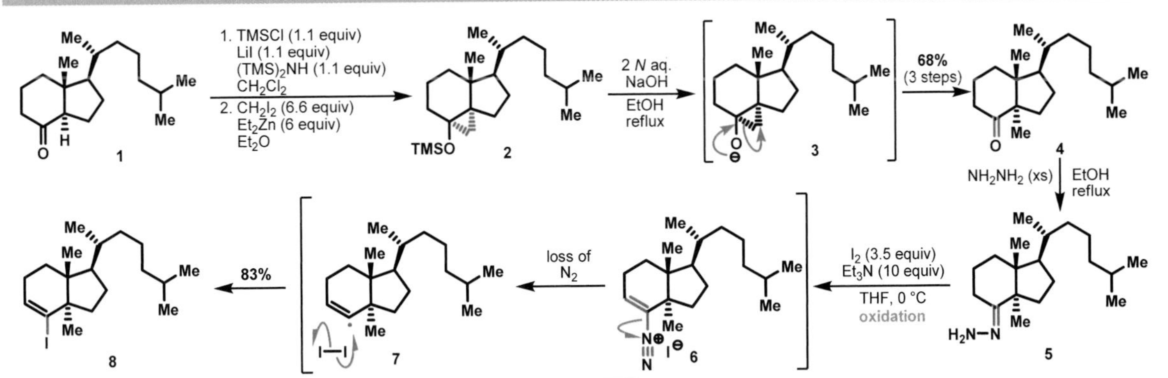

Scheme 2

(+)-LANOSTENOL

Pathway of Synthesis (continued)

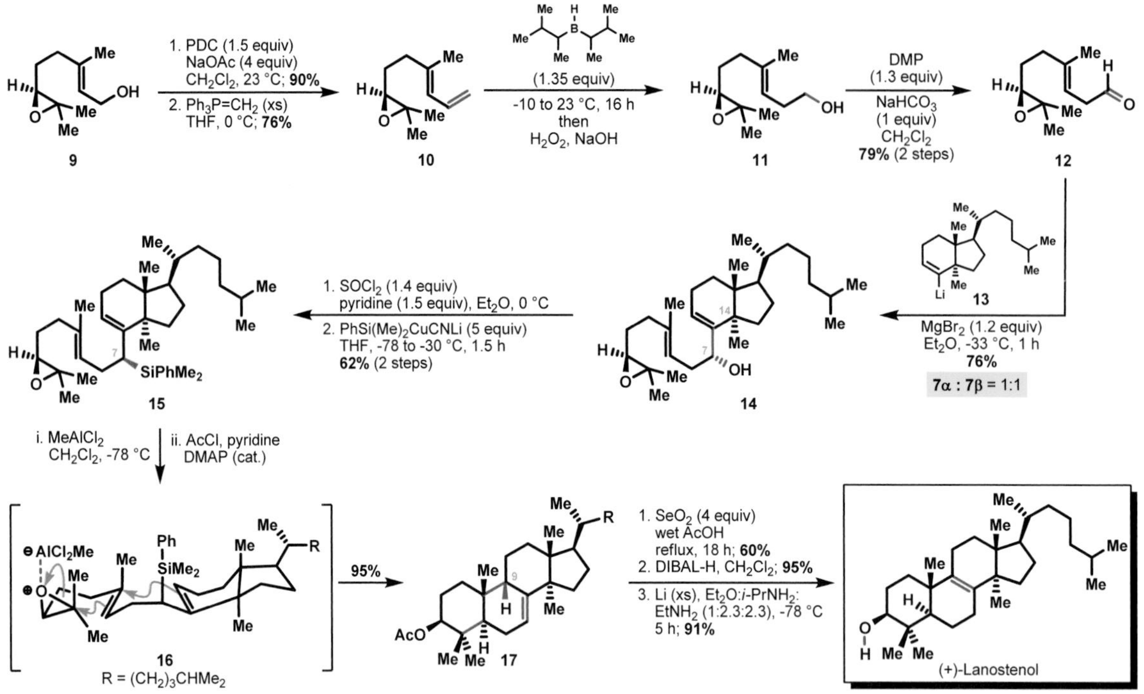

Scheme 3

Explanation of Scheme 2

(1) The angular methylation of the Grundemann ketone **1** was not possible directly, but could be effected by the unusual TMS-enol ether cyclopropanation / base-mediated fragmentation sequence (**1→2→4**).

(2) The vinyl iodide **8** was prepared by the two-step Barton procedure[7] from **4** via hydrazone **5**. The excess iodine (I$_2$) oxidizes the hydrazone to the corresponding vinyldiazonium iodide **6** which forms a vinyl radical **7** upon losing a molecule of nitrogen.

Explanation of Scheme 3

(3) The cyclization **15→17** proceeds via the unusual boat-like conformer **16**, as shown by comparison of

the 9β-H product **17** with the known 3α-H diastereomer **18** (see below).

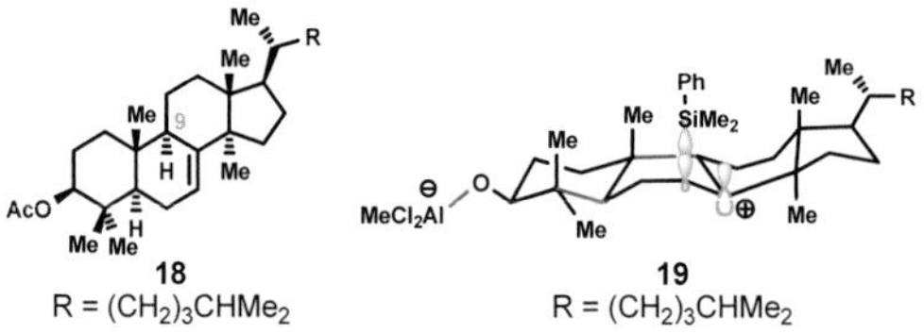

(4) The stereochemistry of the cyclization corresponds to that expected for an axial-like orientation of the PhMe$_2$Si group in **16**. That orientation of Si provides maximum stereoelectronic stabilization of the β-silyl cation intermediate **19** (see above).

References

1. Wendt, K. U., Schulz, G. E., Corey, E. J. & Liu, D. R. Enzyme mechanisms for polycyclic triterpene formation. *Angew. Chem., Int. Ed.* **2000**, 39, 2812-2833.

2. Corey, E. J., Lee, J. & Liu, D. R. First demonstration of a carbocation-olefin cyclization route to the lanosterol series. *Tetrahedron Lett.* **1994**, 35, 9149-9152.

3. (a) Woodward, R. B., Sondheimer, F. & Taub, D. The total synthesis of cholesterol. *J. Am. Chem. Soc.* **1951**, 73, 3548; (b) Woodward, R. B., Patchett, A. A., Barton, D. H. R., Ives, D. A. J. & Kelly, R. B. The synthesis of lanostenol. *J. Am. Chem. Soc.* **1954**, 76, 2852-2853; (c) Johnson, W. S. Fifty years of research. A tribute to my co-workers. *Tetrahedron* **1991**, 47, xi-l.

4. Corey, E. J., Noe, M. C. & Shieh, W. C. A short and convergent enantioselective synthesis of (3S)-2,3-oxidosqualene. *Tetrahedron Lett.* **1993**, 34, 5995-5998.

5. Wovkulich, P. M., Barcelos, F., Batcho, A. D., Sereno, J. F., Baggiolini, E. G., Hennessy, B. M. & Uskokovic, M. R. Stereoselective total synthesis of 1α,25S,26-trihydroxycholecalciferol. *Tetrahedron* **1984**, 40, 2283-2296.

6. Fleming, I., Barbero, A. & Walter, D. Stereochemical Control in Organic Synthesis Using Silicon-Containing Compounds. *Chem. Rev.* **1997**, 97, 2063-2192.

7. Barton, D. H. R., Bashiardes, G. & Fourrey, J. L. An improved preparation of vinyl iodides. *Tetrahedron Lett.* **1983**, 24, 1605-1608.

(+)-PROTOLIMONOID

Background

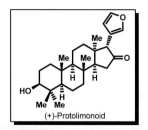

(+)-Protolimonoid

The limonoid family of natural products, now numbering in the hundreds, consists of tetracyclic triterpenes that may be regarded as further oxidation products of a core protolimonoid member.[1] The first recognized member of this series, limonin, which was identified a half-century ago, occurs abundantly in citrus (especially in the seeds). In this section an enantioselective synthetic route to the core protolimonoid is summarized.[2]

Abbreviated Retrosynthetic Plan

(+)-Protolimonoid

$\Longrightarrow$ hydroboration & oxidation $\Longrightarrow$ **I**

$\Longrightarrow$ Pd(0)-catalyzed cross-coupling $\Longrightarrow$ **II**

$\Downarrow$ angular alkylation & triflation

V $\Longleftarrow$ 5-exo-trig radical cyclization **strategic disconnection** $\Longleftarrow$ **IV** $\Longleftarrow$ C=C bond cleavage $\Longleftarrow$ **III**

$\Downarrow$ reduction & xanthate formation

VI $\Longrightarrow$ cationic cascade cyclization **strategic disconnection** $\Longrightarrow$ **VII** $\Longrightarrow$ nucleophilic addition & Brook rearrangement $\Longrightarrow$ **VIII**

Scheme 1

Key steps in Scheme 1

(1) A key element of the retrosynthetic strategy was the application of the powerfully simplifying cation-olefin polycyclization transform.

(2) Because this transform could not be directly applied to the target molecule, it was necessary to disconnect the furan and methyl appendages, which led to intermediates **I**, **II** and **III**.

(3) Because the disconnection of the ring system of **III** was also not possible by application of the cation-olefin polycyclization transform, the D-ring was disconnected next, leading to the sequence **III** ⇒ **IV** ⇒ **V** ⇒ **VI**.

(4) The application of the cation-olefin cyclization transform to **VI**, which was well-precedented, led to the generation of the acyclic intermediate **VII** with the clearance of five stereocenters.

(+)-PROTOLIMONOID

Pathway of Synthesis

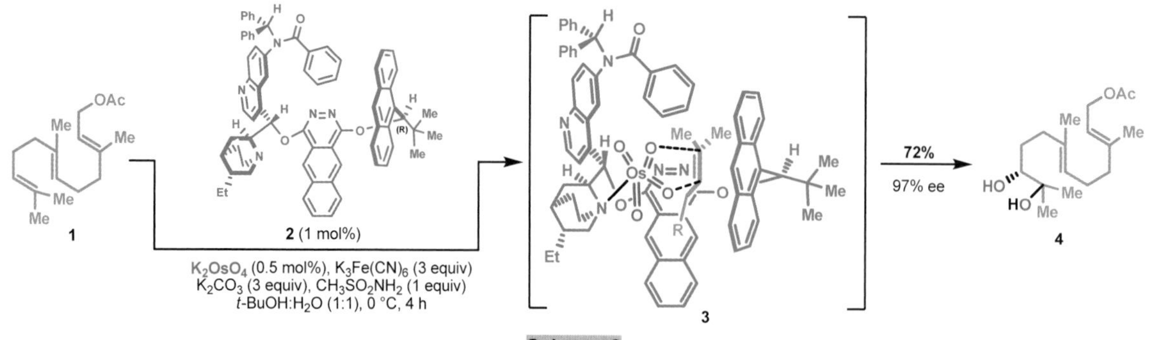

Scheme 2

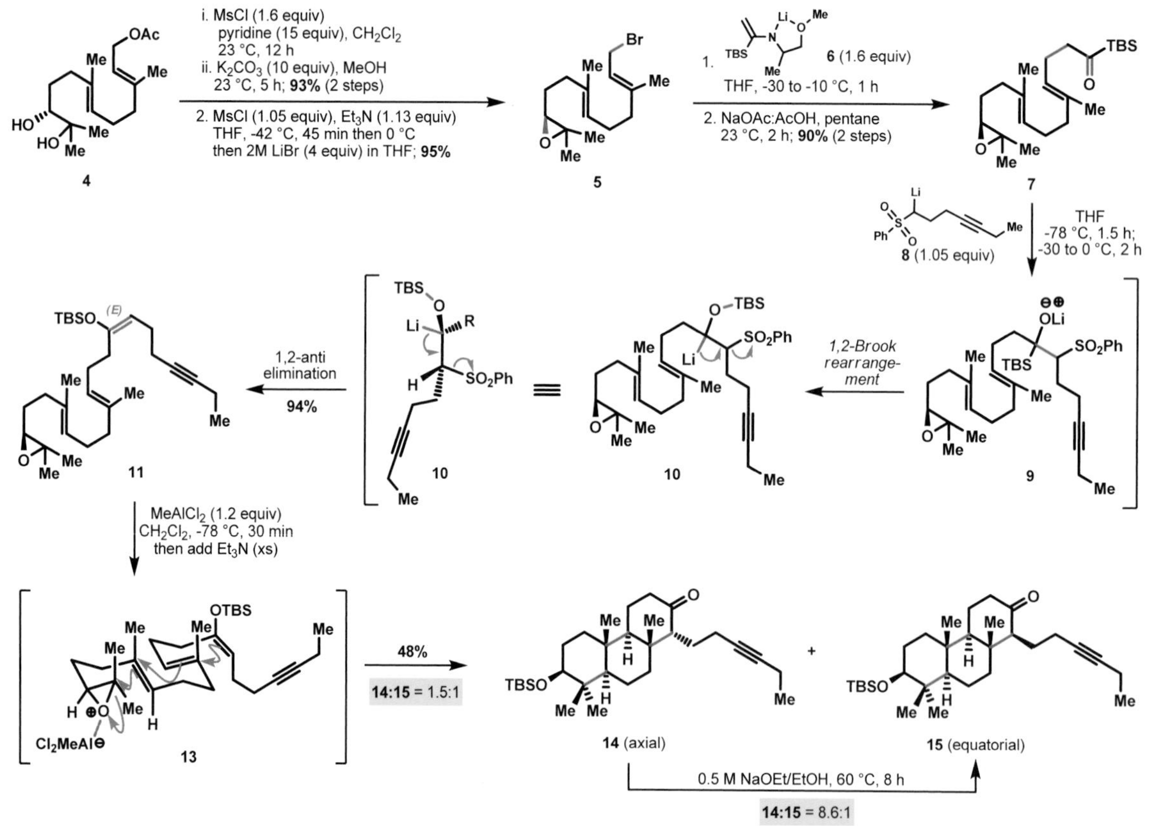

Scheme 3

Explanation of Schemes 2 and 3

(1) The initiating stereocenter for the synthesis was introduced by site- and enantioselective dihydroxylation of farnesyl acetate (1) using the specially designed catalyst 2.[3] The formation of the required 1,2-diol 4 is considered to result from pre-transition state assembly 3.

(2) The Brook-rearrangement/elimination methodology for the conversion of 7 to 11 had been developed previously for the stereoselective C-C coupling.[4] It follows the pathway 7→9→10→11.

(+)-PROTOLIMONOID

Pathway of Synthesis (continued)

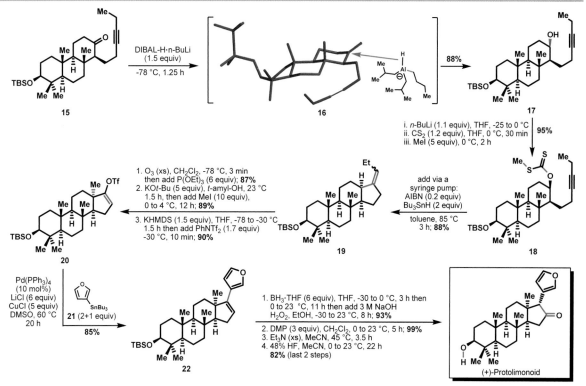

Scheme 4

Key steps in Scheme 4

(1) The cyclization **18→19** proceeds via the 2° radical **18b** which results from abstraction of the MeSCSO fragment by Bu₃Sn-radical (see mechanism on the right).

(2) A number of other routes to limonoids had been developed earlier.[5]

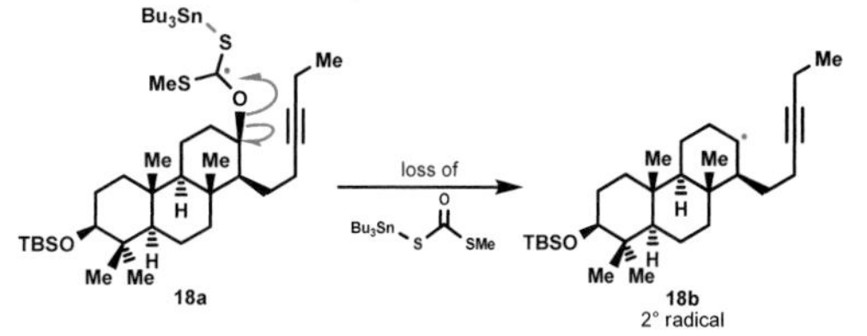

Other Developments: Synthesis of Azadiradione[5b]

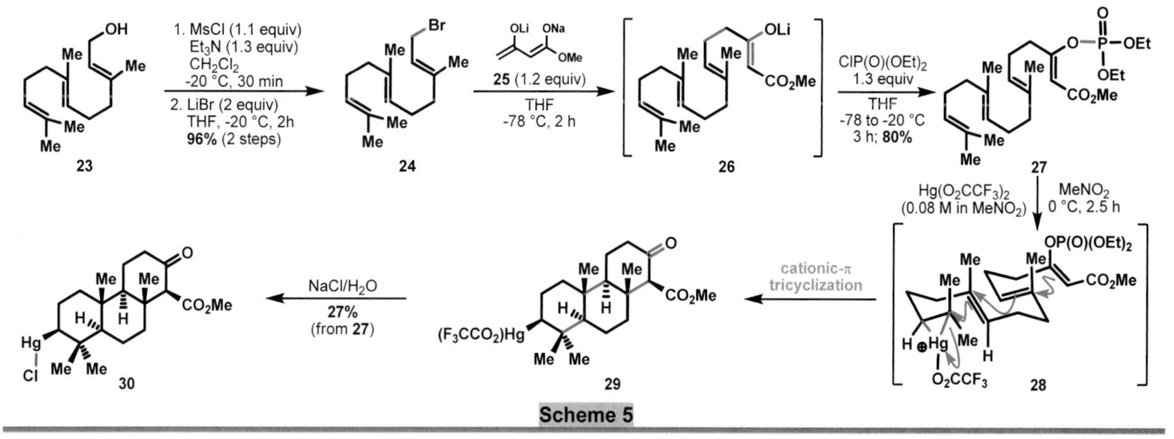

Scheme 5

(+)-PROTOLIMONOID

Other Developments: Synthesis of Azadiradione (continued)

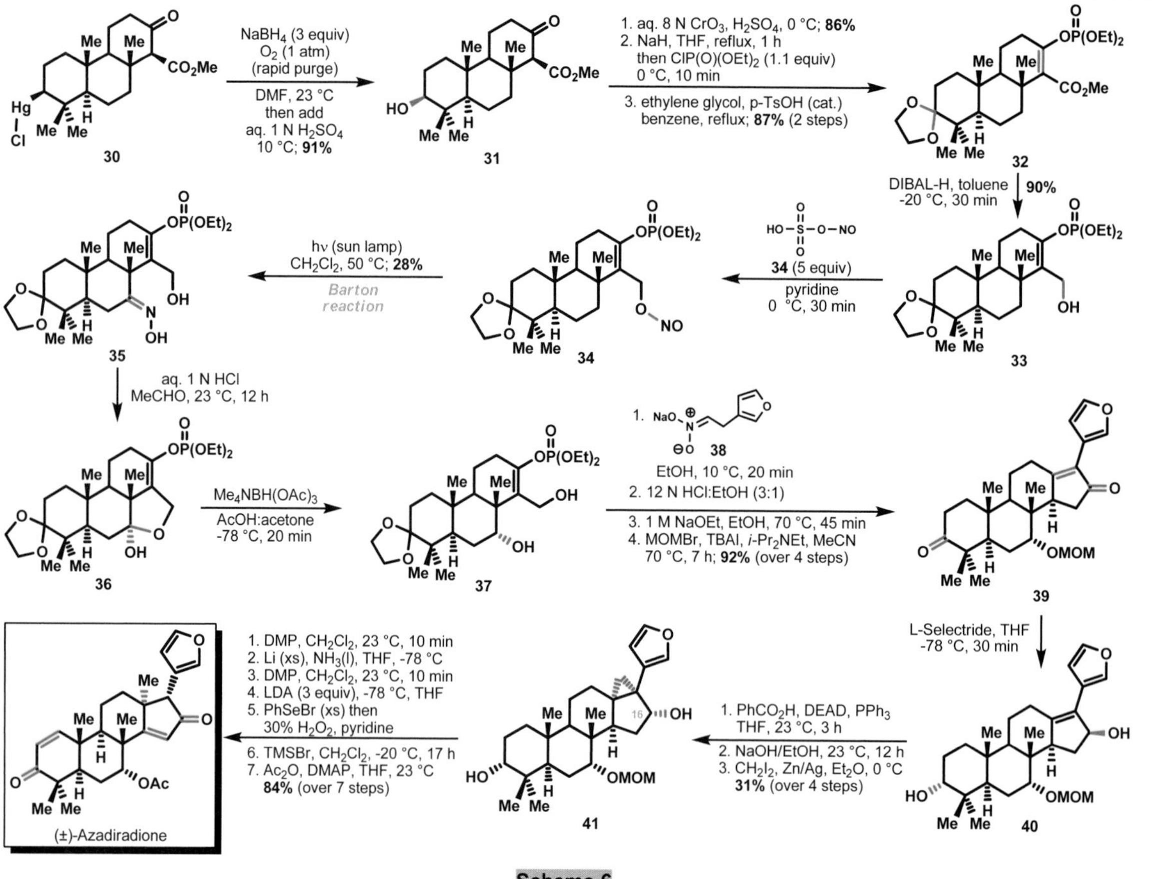

Scheme 6

Key steps in Schemes 5 and 6

The transformation **37→39** occurred by the following novel highly efficient sequence: (a) elimination of phosphate from **37** to form an α-methylene ketone **37b**; (2) Michael addition of the nitronate anion **38** to give **37d**; (c) Nef reaction[6] to convert the nitronate function in the Michael adduct **37d** to ketone **37e** and (4) aldol cyclization to afford **39** (see scheme at right).[5b]

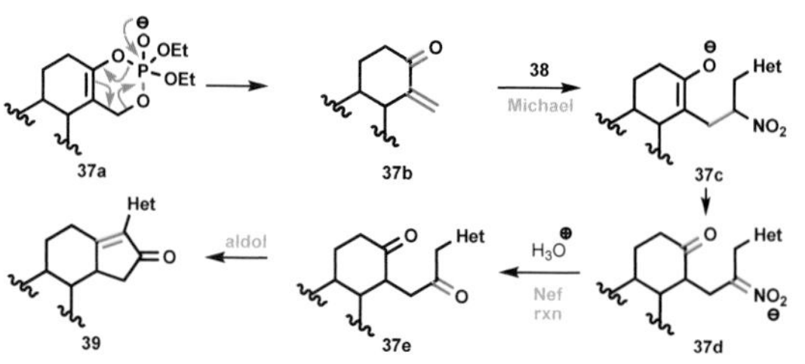

References

1. (a) Arigoni, D., Barton, D. H. R., Corey, E. J., Jeger, O., Caglioti, L., Dev, S., Ferrini, P. G., Glazier, E. R., Melera, A., Pradhan, S. K., Schaffner, K., Sternhell, S., Templeton, J. F. & Tobinaga, S. Constitution of limonin. *Experientia* **1960**, 16, 41-49; (b) Arnott, S., Davie, A. W., Robertson, J. M., Sim, G. A. & Watson, D. G. Structure of limonin. *Experientia* **1960**, 16, 49-52; (c) Dreyer, D. L. Limonoid bitter principles. *Fortschr. Chem. Org. Naturst.* **1968**, 26, 190-244.
2. Behenna, D. C. & Corey, E. J. Simple Enantioselective Approach to Synthetic Limonoids. *J. Am. Chem. Soc.* **2008**, 130, 6720-6721.
3. Corey, E. J. & Zhang, J. Highly Effective Transition Structure Designed Catalyst for the Enantio- and Position-Selective Dihydroxylation of Polyisoprenoids. *Org. Lett.* **2001**, 3, 3211-3214.
4. For previous examples of the Brook rearrangement/olefination sequence, see: (a) Corey, E. J., Luo, G., Lin, L. S. A Simple Enantioselective Synthesis of the Biologically Active Tetracyclic Marine Sesterterpene Scalarenedial. *J. Am. Chem. Soc.* **1997**, 119, 9927-9928; (b) Corey, E.

J., Luo, G., Lin, L. S. Exceptionally simple enantioselective syntheses of chiral hexa- and tetracyclic polyprenoids of sedimentary origin. *Angew. Chem., Int. Ed.* **1998**, 37, 1126-1128; (c) Mi, Y., Schreiber, J. V., Corey, E. J. Total Synthesis of (+)-α-Onocerin in Four Steps via Four-Component Coupling and Tetracyclization Steps. *J. Am. Chem. Soc.* **2002**, 124, 11290-11291.
5. (a) Corey, E. J., Reid, J. G., Myers, A. G. & Hahl, R. W. Simple synthetic route to the limonoid system. *J. Am. Chem. Soc.* **1987**, 109, 918-919; (b) Corey, E. J. & Hahl, R. W. Synthesis of a limonoid, azadiradione. *Tetrahedron Lett.* **1989**, 30, 3023-3026.
6. Kürti, L. & Czakó, B. Nef Reaction in *Strategic Applications of Named Reactions in Organic Synthesis* 308-309 (Academic Press/Elsevier, San Diego, **2005**).

(+)-α-ONOCERIN

Background

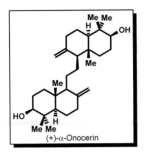

(+)-α-Onocerin

Onocerin is a structurally unique C_2-symmetric tetracyclic triterpene, the constitution of which was deduced by Barton in 1955. The corresponding hydrocarbon, onoceradiene, which is also a natural product, was synthesized in 1957 by cation-olefin cyclization.[1] This approach was applied shortly afterward to a synthesis of α-onocerin itself.[2] The first enantioselective synthesis of α-onocerin is summarized in this section.[3]

Abbreviated Retrosynthetic Plan

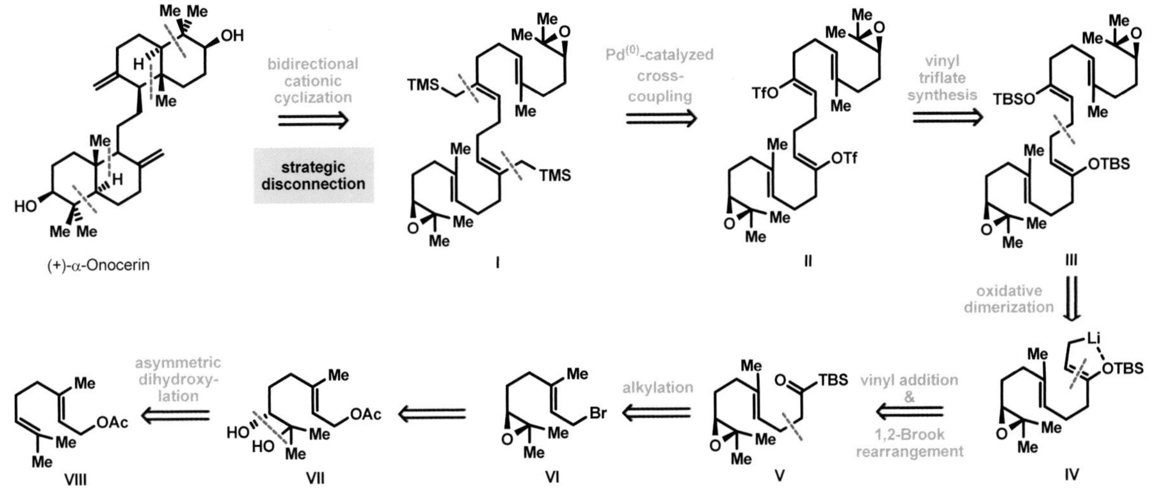

(+)-α-Onocerin ⟹ (bidirectional cationic cyclization) (strategic disconnection) I ⟹ (Pd(0)-catalyzed cross-coupling) II ⟹ (vinyl triflate synthesis) III

III ⟹ (oxidative dimerization) IV ⟸ (vinyl addition & 1,2-Brook rearrangement) V ⟸ (alkylation) VI ⟸ VII ⟸ (asymmetric dihydroxylation) VIII

Scheme 1

Key steps in Scheme 1

(1) The disconnection of the ring system and clearance of stereocenters by the application of the epoxide-initiated cation-olefin cyclization transform formed the basis for the plan and led to the intermediates **I**, **II** and **III**.

(2) The simplification of **II** was driven by the selection of geranyl acetate as a starting goal (S-goal).

Pathway of Synthesis

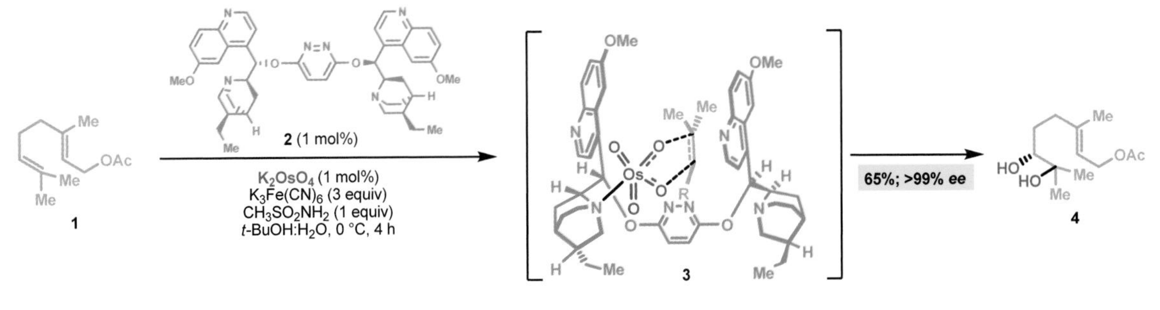

1 + **2** (1 mol%), K₂OsO₄ (1 mol%), K₃Fe(CN)₆ (3 equiv), CH₃SO₂NH₂ (1 equiv), t-BuOH:H₂O, 0 °C, 4 h → [**3**] → 65%; >99% ee → **4**

Scheme 2

(+)-α-ONOCERIN

Pathway of Synthesis (continued)

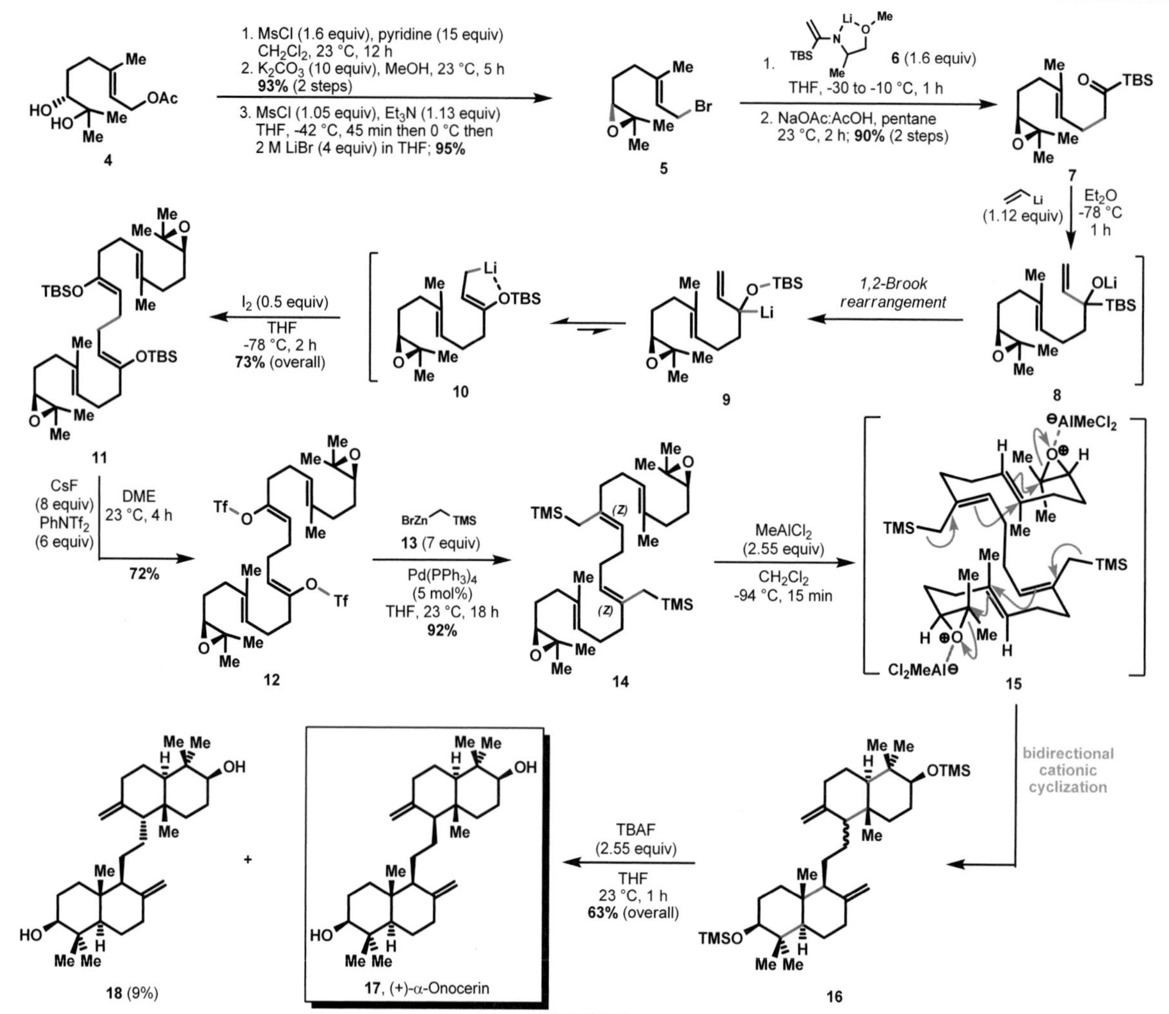

4

1. MsCl (1.6 equiv), pyridine (15 equiv) CH$_2$Cl$_2$, 23 °C, 12 h
2. K$_2$CO$_3$ (10 equiv), MeOH, 23 °C, 5 h **93% (2 steps)**
3. MsCl (1.05 equiv), Et$_3$N (1.13 equiv) THF, -42 °C, 45 min then 0 °C then 2 M LiBr (4 equiv) in THF; **95%**

5

1. **6** (1.6 equiv) THF, -30 to -10 °C, 1 h
2. NaOAc:AcOH, pentane 23 °C, 2 h; **90% (2 steps)**

7

Li (1.12 equiv), Et$_2$O, -78 °C, 1 h

8

1,2-Brook rearrangement

9

10

I$_2$ (0.5 equiv) THF -78 °C, 2 h **73% (overall)**

11

CsF (8 equiv), PhNTf$_2$ (6 equiv), DME 23 °C, 4 h **72%**

12

BrZn—TMS **13** (7 equiv) Pd(PPh$_3$)$_4$ (5 mol%) THF, 23 °C, 18 h **92%**

14

MeAlCl$_2$ (2.55 equiv) CH$_2$Cl$_2$ -94 °C, 15 min

15

bidirectional cationic cyclization

16

TBAF (2.55 equiv) THF 23 °C, 1 h **63% (overall)**

18 (9%)

17, (+)-α-Onocerin

Scheme 3

Explanation of Schemes 2 and 3

(1) A four-component coupling allowed the conversion of **9** to **11** in a single step using the powerful methodology developed earlier,[4] and homocoupling of intermediate **10**.

(2) The conversion of the vinyloxysilane **11** to the corresponding vinyl triflate **12** was carried out by a novel procedure which is also generally useful. The experimental protocol calls for the stirring of **11** with an excess of vacuum-dried (300 °C) CsF and dry *N*-phenyltrifluoromethanesulfonimide (**A**) in dry dimethoxyethane in a sealed flask at 23 °C. The active triflating agent may be *in situ*-generated trifluoromethanesulfonyl fluoride (**B**, see equation at right, bp -21 °C) which must be contained by using a sealed reaction vessel. The formation of the vinyl triflate **12** from vinyloxysilane **11** requires all the reactants, since it does not occur with **B** alone or with

B + CsF mixtures in DME. It is clear that reagent **A** plays a role in solubilizing CsF and promoting the fluoride-induced silyl ether cleavage (**11→11a**).

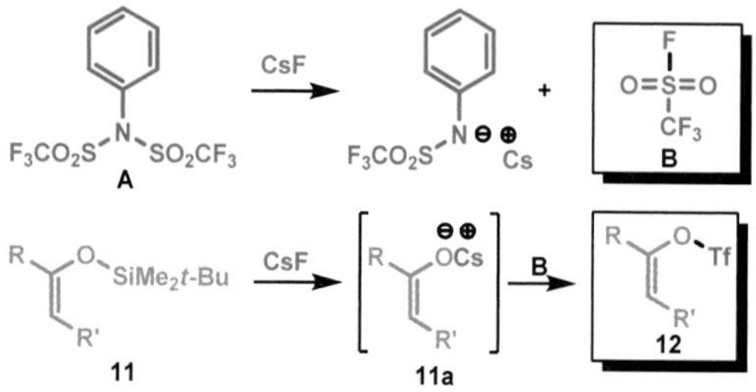

(3) The cyclization of the diepoxide **14** is not completely diastereoselective but favors the required disatereomer **17** over the non-C$_2$-symmetric **18**.

(+)-α-ONOCERIN

The Impact of Olefin Stereochemistry on the Course of Tetracyclization

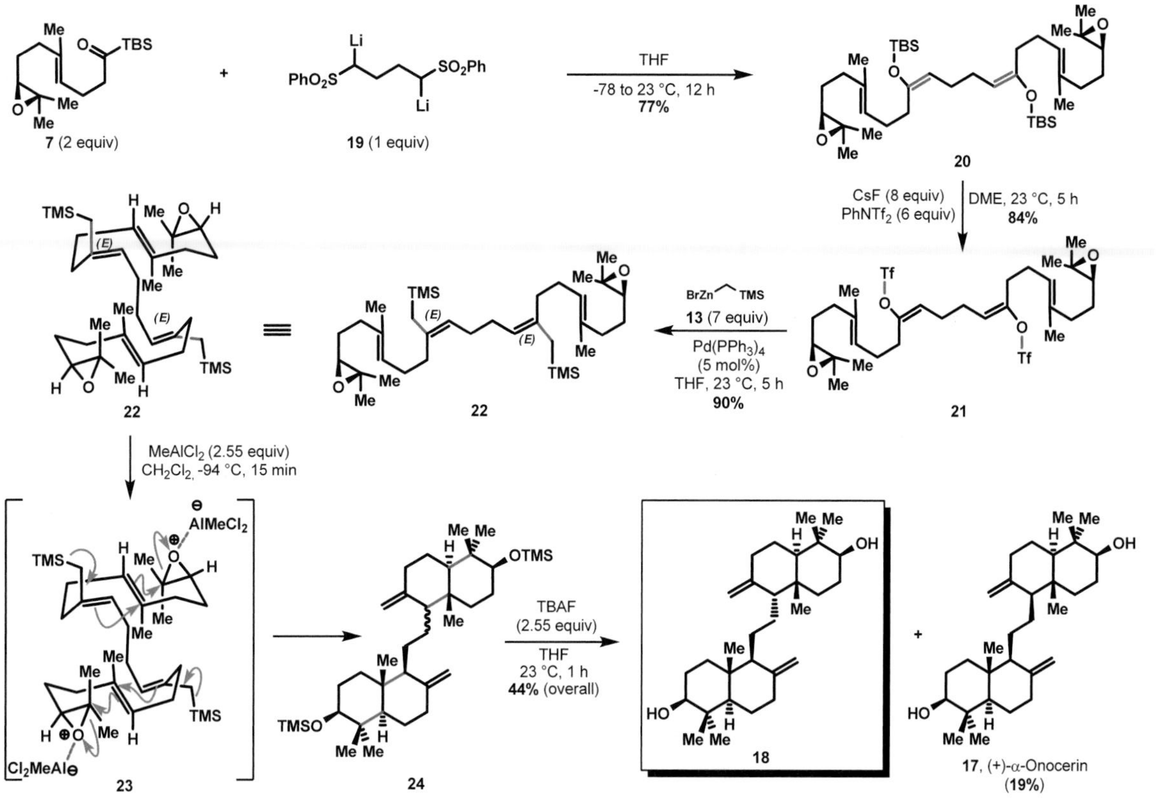

Scheme 4

Key steps in Scheme 4

(1) The synthesis of (E,E)-vinyloxysilane **20**, which is isomeric with the (Z,Z)-vinyloxysilane **11** that was utilized for the synthesis of α-onocerin (see Scheme 3), was accomplished in a novel way by coupling of **7** and **19** via a double Brook rearrangement / β-elimination sequence.

(2) The cyclization of **22** produced mainly the non-C$_2$-symmetric diastereomer of onocerin (**18**), as anticipated from a concerted cyclization and from the cyclization of **15** to α-onocerin (**17**) as major product (Scheme 2).

(3) Based on the results of cyclization studies, it seems likely that the MeAlCl$_2$-induced cation-olefin cyclizations of **14** and **22** (especially **22**) proceed to a significant extent through chair-boat A/B transition states (steroid A/B folding[5]).

(4) In conclusion, the results also show that the design of substrates that selectively favor one cyclization pathway over the other may be possible in chemical systems as well as for enzymes.[6]

References

1. (a) Corey, E. J., Sauers, R. R. Total synthesis of pentacyclosqualene. *J. Am. Chem. Soc.* **1957**, 79, 3925-3926; (b) Corey, E. J., Sauers, R. R. The synthesis of pentacyclosqualene (8,8'-cyclöonocerene) and the α- and β-onoceradienes. *J. Am. Chem. Soc.* **1959**, 81, 1739-1743.
2. Stork, G., Meisels, A. & Davies, J. E. Total synthesis of polycyclic triterpenes. Total synthesis of (+)-α-onocerin. *J. Am. Chem. Soc.* **1963**, 85, 3419-3425.
3. (a) Mi, Y., Schreiber, J. V. & Corey, E. J. Total Synthesis of (+)-α-Onocerin in Four Steps via Four-Component Coupling and Tetracyclization Steps. *J. Am. Chem. Soc.* **2002**, 124, 11290-11291; (b) Domingo, V., Arteaga, J. F., Quilez del Moral, J. F. & Barrero, A. F.

Unusually cyclized triterpenes: occurrence, biosynthesis and chemical synthesis. *Nat. Prod. Rep.* **2009**, 26, 115-134.
4. (a) Corey, E. J. & Lin, S. A Short Enantioselective Total Synthesis of Dammarenediol II. *J. Am. Chem. Soc.* **1996**, 118, 8765-8766; (b) Corey, E. J., Lin, S. & Luo, G. Stereospecific synthesis of tetrasubstituted Z-enol silyl ethers by a three component coupling process. *Tetrahedron Lett.* **1997**, 38, 5771-5774.
5. Wendt, K. U., Schulz, G. E., Corey, E. J. & Liu, D. R. Enzyme mechanisms for polycyclic triterpene formation. *Angew. Chem., Int. Ed.* **2000**, 39, 2812-2833.
6. Yee, N. K. N. & Coates, R. M. Total synthesis of (+)-9,10-syn- and (+)-9,10-anti-copalol via epoxy trienylsilane cyclizations. *J. Org. Chem.* **1992**, 57, 4598-4608.

(−)-SERRATENEDIOL

Background

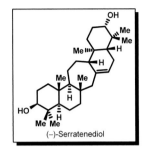

(−)-Serratenediol

Serratenediol, the parent pentacyclic triterpene of a family of >30 natural products, possesses an unusual central seven-membered ring (ring C).[1] It appears to be formed by enzymic cationic cyclization of bis-oxidosqualene, in common with α-onocerin.[2] A cationic cyclization route to serratenediol is outlined here.[3]

Abbreviated Retrosynthetic Plan

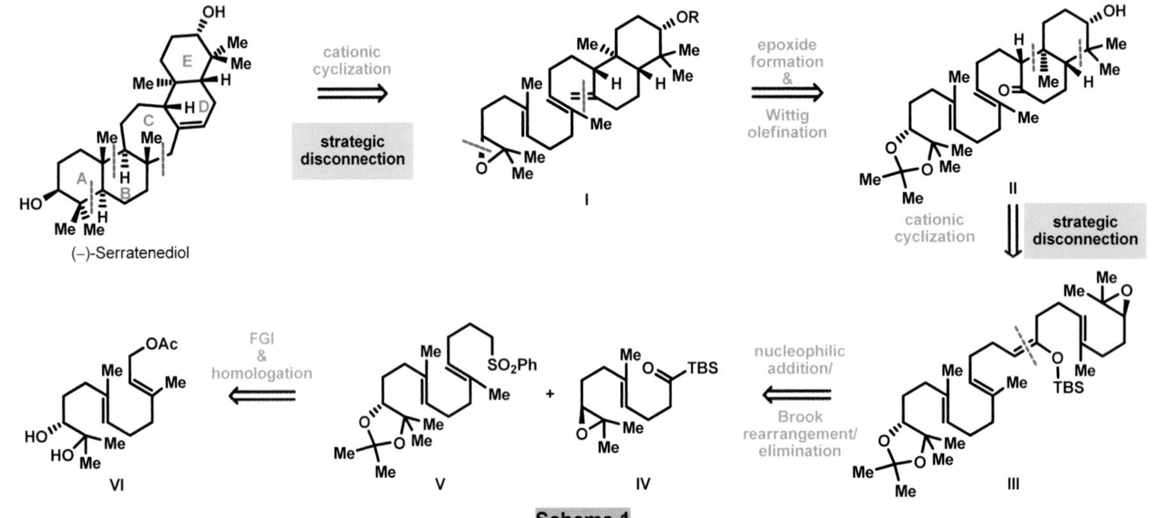

(−)-Serratenediol

cationic cyclization

strategic disconnection

I

epoxide formation & Wittig olefination

II

cationic cyclization

strategic disconnection

FGI & homologation

VI

V

+

IV

nucleophilic addition/

Brook rearrangement/ elimination

III

Scheme 1

Key steps in Scheme 1

Direct application of the epoxide-initiated cation-olefin cyclization to serratenediol generates the bicyclic intermediate **I**, which could be further disconnected via **II** to the acyclic intermediate **III**.

Disconnection of **III** at the indicated double bond generates **IV** and **V** which are available from geraniol and farnesol, respectively.

Pathway of Synthesis

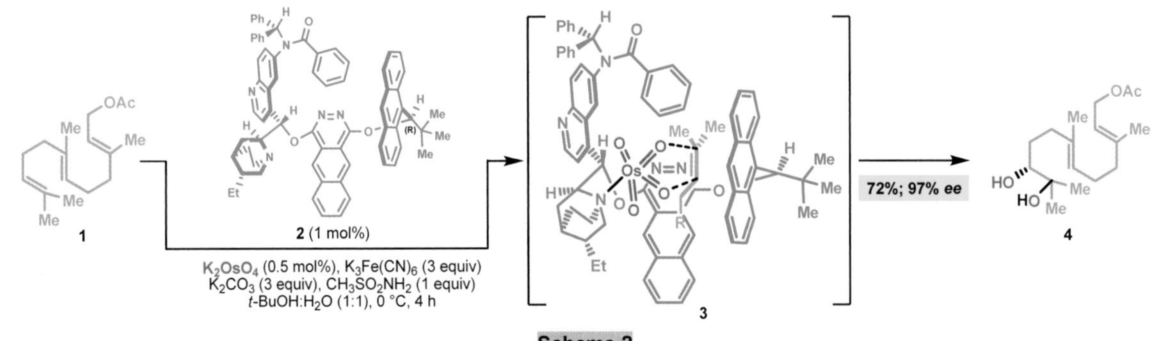

1

2 (1 mol%)

K₂OsO₄ (0.5 mol%), K₃Fe(CN)₆ (3 equiv)
K₂CO₃ (3 equiv), CH₃SO₂NH₂ (1 equiv)
t-BuOH:H₂O (1:1), 0 °C, 4 h

3

72%; 97% ee

4

Scheme 2

(−)-SERRATENEDIOL

Pathway of Synthesis (continued)

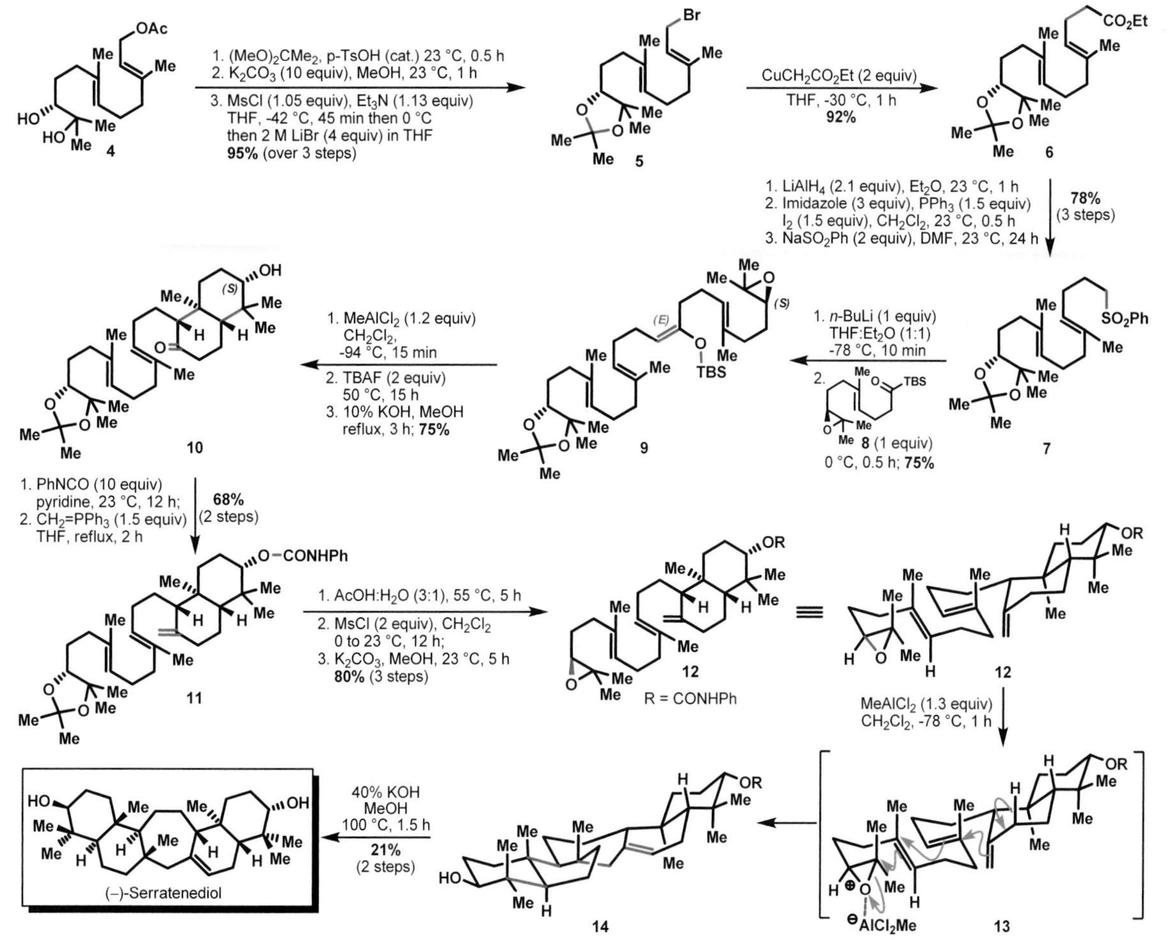

Scheme 3

Explanation of Schemes 2 and 3

(1) The reaction of the sulfone-stabilized lithio derivative of **7** with the acylsilane **8** occurs via a sequence:[4]

    (a) carbonyl addition;

    (b) Brook rearrangement[5] of Si from C to O and

    (c) β-elimination of $PhSO_2^-$ to form the (E)-coupling product **9**.

(2) The conversion of **12** via **13** to **14** represents a rare case of seven-membered ring formation by cation-olefin cyclization.

References

1. For articles on the isolation and the elucidation of the structure, see: (a) Inubushi, Y., Sano, T., Tsuda, Y. Serratenediol-a new skeletal triterpenoid containing a seven-membered ring. *Tetrahedron Lett.* **1964**, 1303-1310; (b) Rowe, J. W. Triterpenes of pine barks. Identity of pinusenediol and serratenediol. *Tetrahedron Lett.* **1964**, 2347-2353; (c) Rowe, J. W., Bower, C. L. Triterpenes of pine barks. Naturally occurring derivatives of serratenediol. *Tetrahedron Lett.* **1965**, 2745-2750; (d) Dev, S., Gupta, A. S. CRC Handbook of Terpenoids. Triterpenoids. **1987**, Vol II., pp 565-584 (CRC Press, Boca Raton, Florida).

2. Barton, D. H. R. & Overton, K. H. Triterpenoids. XX. The constitution and stereochemistry of a novel tetracyclic triterpenoid. *J. Chem. Soc.* **1955**, 2639-2652.

3. Zhang, J. & Corey, E. J. A simple enantioselective synthesis of serratenediol. *Org. Lett.* **2001**, 3, 3215-3216.

4. For other examples of the Brook rearrangement/olefination sequence from these laboratories, see: (a) Corey, E. J., Luo, G., Lin, L. S. A Simple Enantioselective Synthesis of the Biologically Active Tetracyclic Marine Sesterterpene Scalarenedial. *J. Am. Chem. Soc.* **1997**, 119, 9927-9928; (b) Corey, E. J., Luo, G., Lin, L. S. Exceptionally simple enantioselective syntheses of chiral hexa- and tetracyclic polyprenoids of sedimentary origin. *Angew. Chem., Int. Ed.* **1998**, 37, 1126-1128; (c) Mi, Y., Schreiber, J. V., Corey, E. J. Total Synthesis of (+)-α-Onocerin in Four Steps via Four-Component Coupling and Tetracyclization Steps. *J. Am. Chem. Soc.* **2002**, 124, 11290-11291.

5. Kürti, L. & Czakó, B. Brook Rearrangement in *Strategic Applications of Named Reactions in Organic Synthesis* 64-65 (Academic Press/Elsevier, San Diego, **2005**).

(−)-SCALARENEDIAL

Background

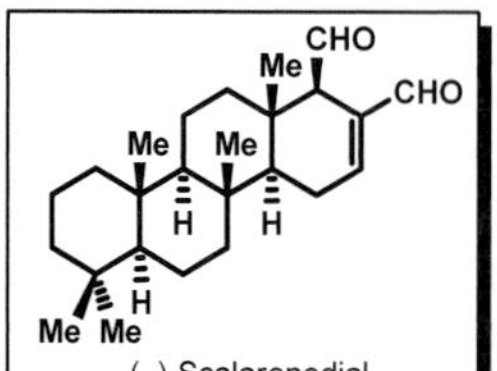

(−)-Scalarenedial

The sesterterpenoid scalarenedial is a potent antitumor agent with multiple other biological activities. Sesterterpenoids are also very ancient marine natural products because derivatives occur in sea sediments and in petroleum.[1] A direct synthetic pathway[2] is described here that illustrates a powerful tetracyclization reaction.[3]

Abbreviated Retrosynthetic Plan

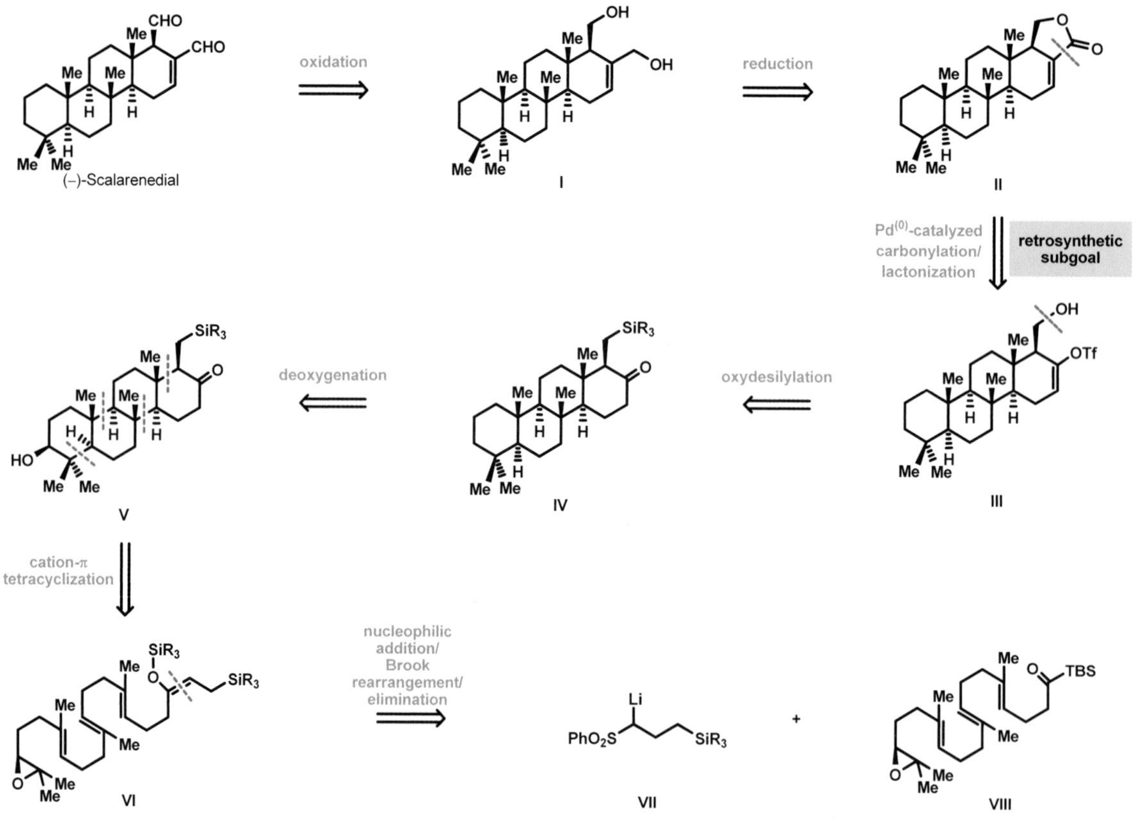

Scheme 1

Key steps in Scheme 1

(1) At the heart of the retrosynthetic design was the conversion of the target structure to an intermediate suitable for disconnection by application of the cation-olefin cyclization transform, in this case the step **V** ⇒ **VI**.

(2) The intermediate **V** was accessed by stepwise disconnection of scalarenedial to **I** ⇒ **II** ⇒ **III** and **IV**.

(3) The silyl group (SiR₃) in **IV** served as a replacement for a hydroxyl group (OH), which would have been a liability in the cyclization step and the subsequent ketone to vinyl triflate conversion

(−)-SCALARENEDIAL

Pathway of Synthesis

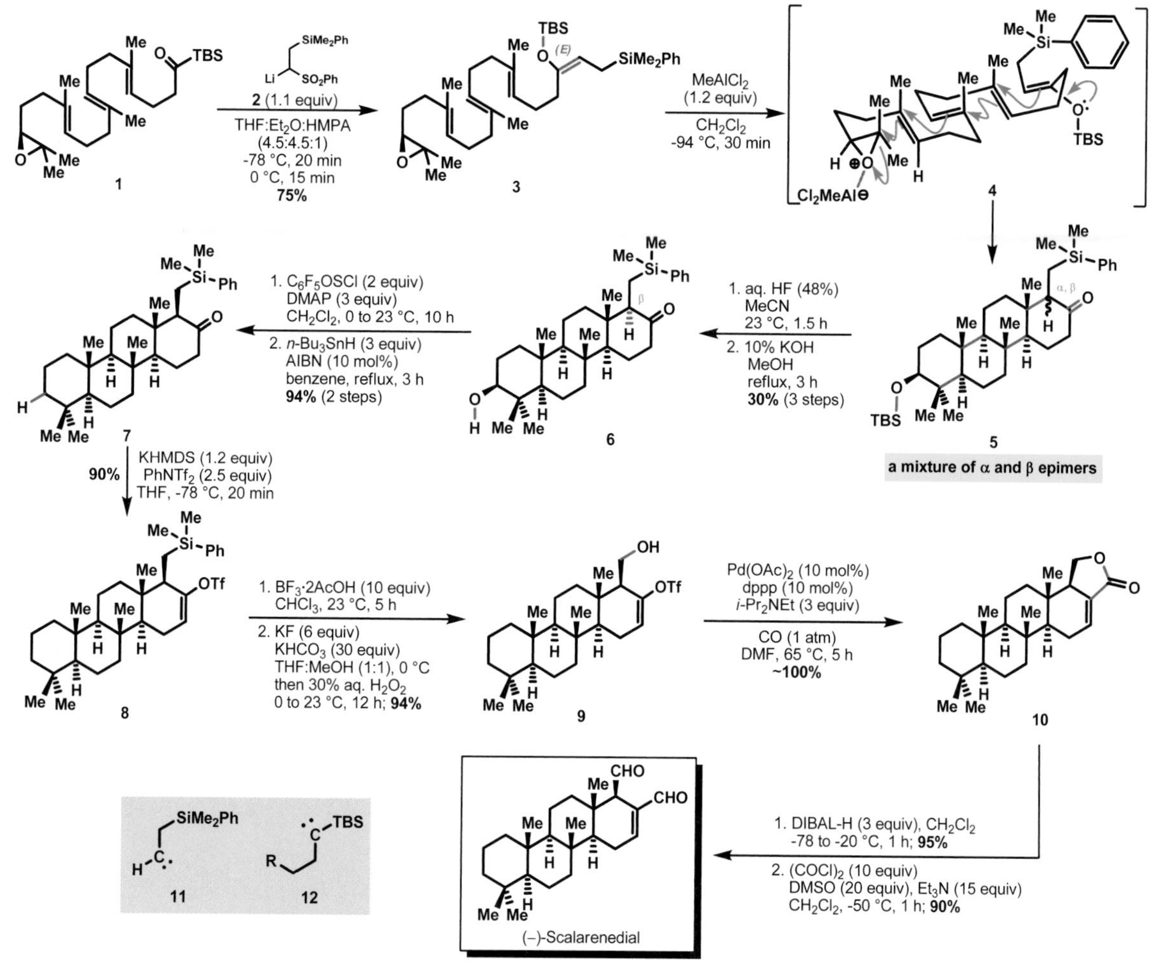

Scheme 2

Explanation of Scheme 2

The conversion of **3** to **5** is a dramatic example of the power of cation-olefin cyclizations, which in this case generates 4 rings and 8 new stererocenters.

The sulfone-stabilized nucleophile **2** serves as the equivalent of carbenoid **11** and the acylsilane **1** as an equivalent of a carbenoid **12** (see bottom of Scheme 2). The newly created double bond in **3** can be formed conceptually by the joining of **11** and **12**.

References

1. For recent reviews on the structure elucidation and syntheses of various sesterterpenoids, see: (a) Ungur, N., Kulcitki, V. Synthesis of scalarane sesterterpenoids. *Rec. Res. Dev. Org. Chem.* **2003**, 7, 241-258; (b) De

2. Corey, E. J., Luo, G. & Lin, L. S. A Simple Enantioselective Synthesis of the Biologically Active Tetracyclic Marine Sesterterpene Scalarenedial. *J. Am. Chem. Soc.* **1997**, 119, 9927-9928.

3. For other synthetic approaches to scalarenedial, see: (a) Soetjipto, H., Furuichi, N., Hata, T., Katsumura, S. Stereocontrolled synthesis of a tetracyclic sesterterpene, (+)-scalarenedial. *Chem. Lett.* **2000**, 1302-

Rosa, S., Mitova, M. Bioactive marine sesterterpenoids. *Stud. Nat. Prod. Chem.* **2005**, 32, 109-168; (c) Liu, Y., Wang, L., Jung, J. H., Zhang, S. Sesterterpenoids. *Nat. Prod. Rep.* **2007**, 24, 1401-1429. 1303; (b) Furuichi, N., Hata, T., Soetjipto, H., Kato, M., Katsumura, S. Common synthetic strategy for optically active cyclic terpenoids having a 1,1,5-trimethyl-trans-decalin nucleus: syntheses of (+)-acuminolide, (-)-spongianolide A, and (+)-scalarenedial. *Tetrahedron* **2001**, 57, 8425-8442.

POLYPRENOIDS OF SEDIMENTARY ORIGIN

Background

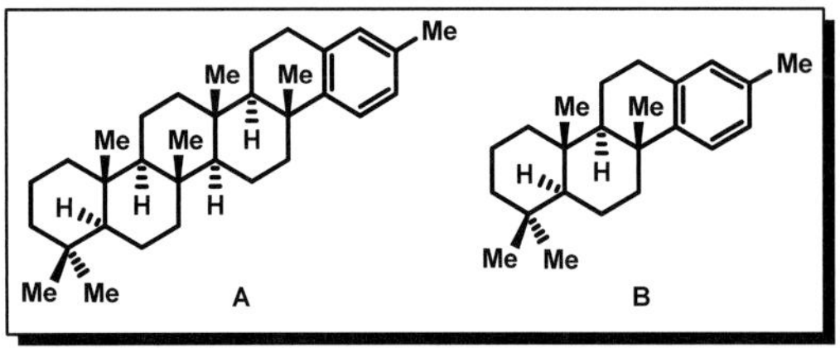

Shale deposits contain several chiral polycyclic compounds which are thought to arise by the decomposition over a long time span of an ancient family of cyclopolyprenoids.[1] A short synthesis of two of these unusual molecules is outlined here.[2]

Abbreviated Retrosynthetic Plan

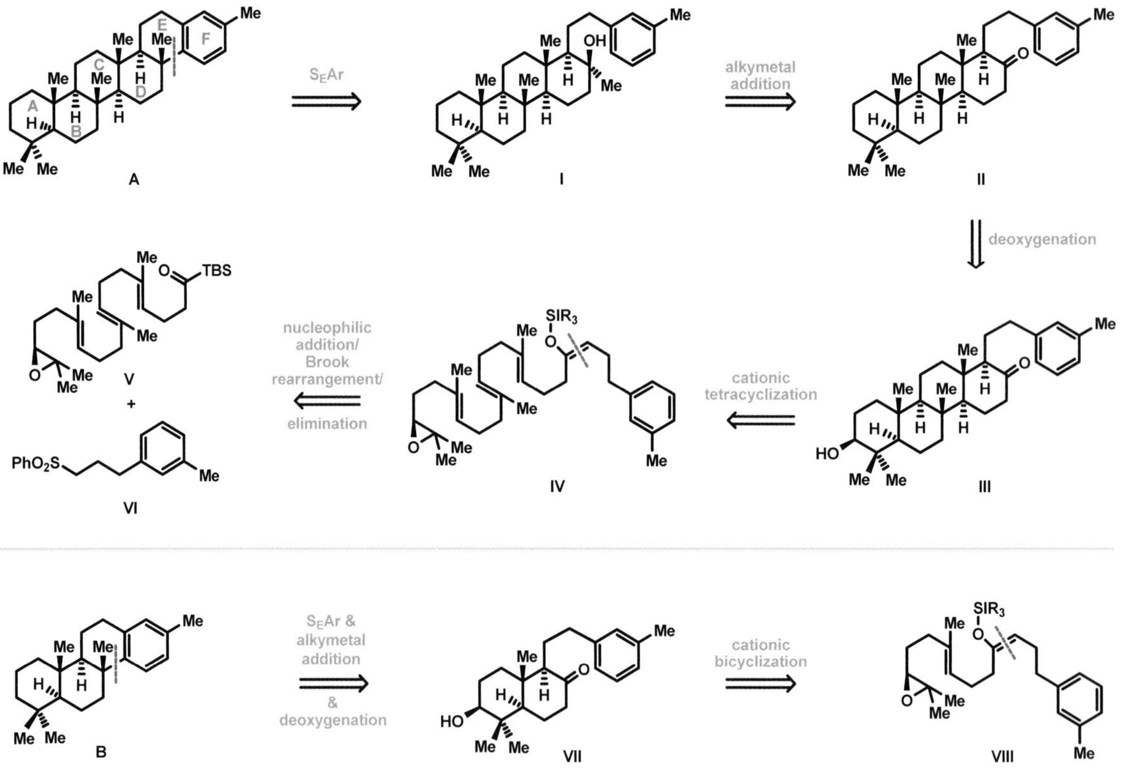

Scheme 1

Key steps in Scheme 1

(1) The retrosynthetic analysis was guided by the goal of applying the cation-olefin cyclization transform to disconnect the polycyclic network with stereochemical control. The disconnection of target **A** was carried out in two stages:

(a) disconnection of the E ring to **I** and

(b) further disconnection of **III** to **IV**.

(2) The chiral epoxide **IV** was used as the initiator of polycyclization to control absolute configuration.

(3) The chiral epoxide **IV** was disconnected to ketone **V** and sulfone **VI** using the tandem nucleophilic addition/Brook rearrangement/elimination transform.

POLYPRENOIDS OF SEDIMENTARY ORIGIN

Pathway of Synthesis

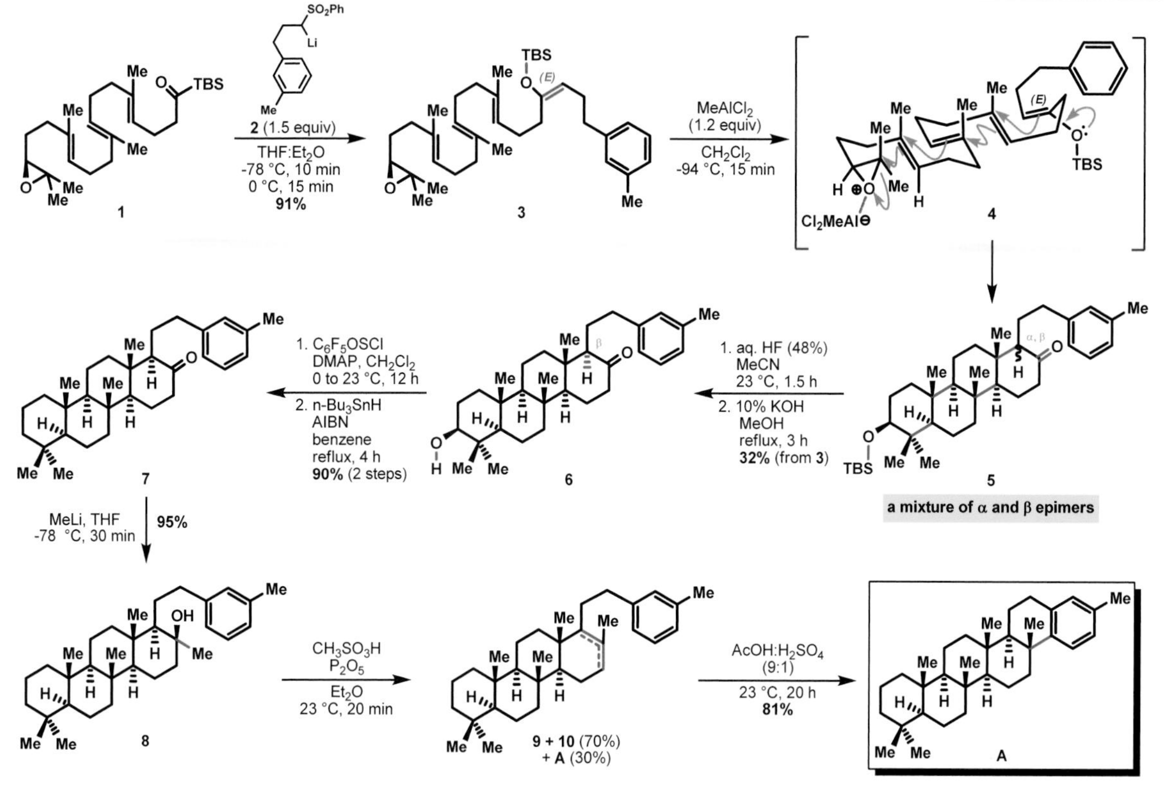

Scheme 2

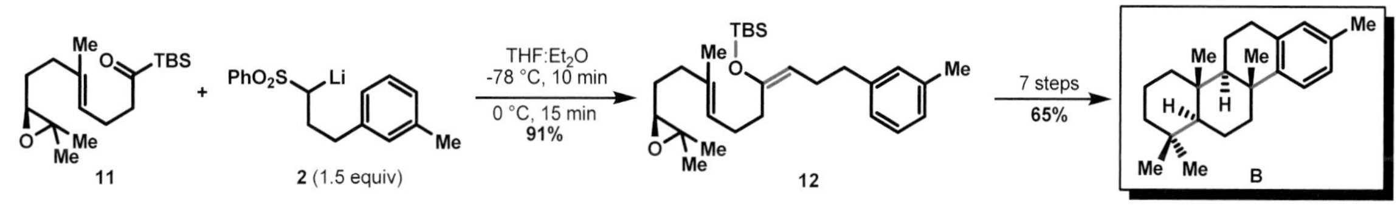

Scheme 3

Explanation of Schemes 2 and 3

(1) The rapid synthesis of **5** from **1** results from the constructive power of the coupling process **1 + 2 → 3** and the tetracyclization **3 → 5**.

(2) The success of the synthesis also depends on the availability of methodology[3] for the enantioselective formation of the key chiral epoxide **1**.

References

1. Schaeffer, P., Poinsot, J., Hauke, V., Adam, P., Wehrung, P., Trendel, J.-M., Albrecht, P., Dessort, D. & Connan, J. New optically active hydrocarbon in sediments: indication of extensive biological cyclization of higher regular polyprenols. *Angew. Chem.* **1994**, *106*, 1235-1238 (See also *Angew. Chem., Int. Ed. Engl.*, **1994**, 1166-1169).
2. Corey, E. J., Luo, G. & Lin, L. S. Exceptionally simple enantioselective syntheses of chiral hexa- and tetracyclic polyprenoids of sedimentary origin. *Angew. Chem., Int. Ed.* **1998**, *37*, 1126-1128.
3. (a) Corey, E. J., Noe, M. C. & Lin, S. A mechanistically designed bis-cinchona alkaloid ligand allows position- and enantioselective dihydroxylation of farnesol and other oligoprenyl derivatives at the terminal isopropylidene unit. *Tetrahedron Lett.* **1995**, *36*, 8741-8744; (b) Corey, E. J. & Noe, M. C. A Critical Analysis of the Mechanistic Basis of Enantioselectivity in the Bis-Cinchona Alkaloid Catalyzed Dihydroxylation of Olefins. *J. Am. Chem. Soc.* **1996**, *118*, 11038-11053.

TRITERPENES OF THE β-AMYRIN FAMILY

Background

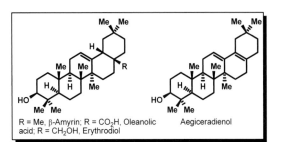

R = Me, β-Amyrin; R = CO_2H, Oleanolic acid; R = CH_2OH, Erythrodiol   Aegiceradienol

There are several hundred natural products that are derived from the precursor β-amyrin as parent, including those molecules shown at left.[1] This section describes the first enantioselective syntheses of β-amyrin, erythrodiol, and oleanolic acid via the intermediate aegiceradienol.[2]

Abbreviated Retrosynthetic Plan

(+)-β-Amyrin

reduction

I

dissolving metal reduction

II

oxidation & cyclopropane cleavage

V

Pd$^{(0)}$-catalyzed cross-coupling

retrosynthetic subgoal

IV [(+)-Aegiceradienol]

cyclopropanation

III

enolate triflation

VI

cation-π tricyclization

strategic disconnection

cation-π tricyclization

strategic disconnection

XI

CBS reduction & SNi

XII

FGI & alkylation

XIII +

XIV

VII

VIII + IX + X

Scheme 1

Key steps in Scheme 1

(1) Although β-amyrin is biosynthesized by enzymic cation-olefin cyclization of (S)-2,3-oxidosqualene, the chemical retrosynthetic disconnection to this precursor is not feasible. Consequently, a retrosynthetic route from β-amyrin to the tetracycle **VI** was devised via intermediates **I** to **V**.

(2) Ketone **VI** was then disconnected to **VII** by application of the epoxide-initiated cation-olefin cyclization to **VII**.

(3) Intermediate **VII** was then disconnected to three reactants, the chiral epoxide **VIII**, dibromide **IX** and 2-propenyllithium **X**.

TRITERPENES OF THE β-AMYRIN FAMILY

Pathway of Synthesis

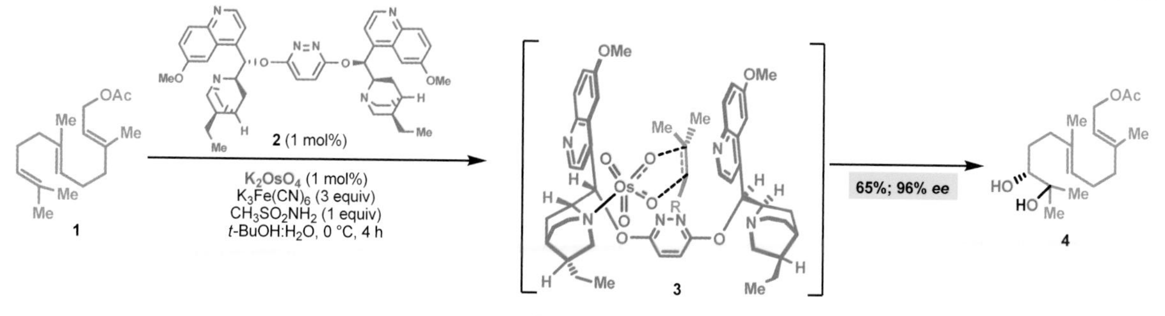

Scheme 2

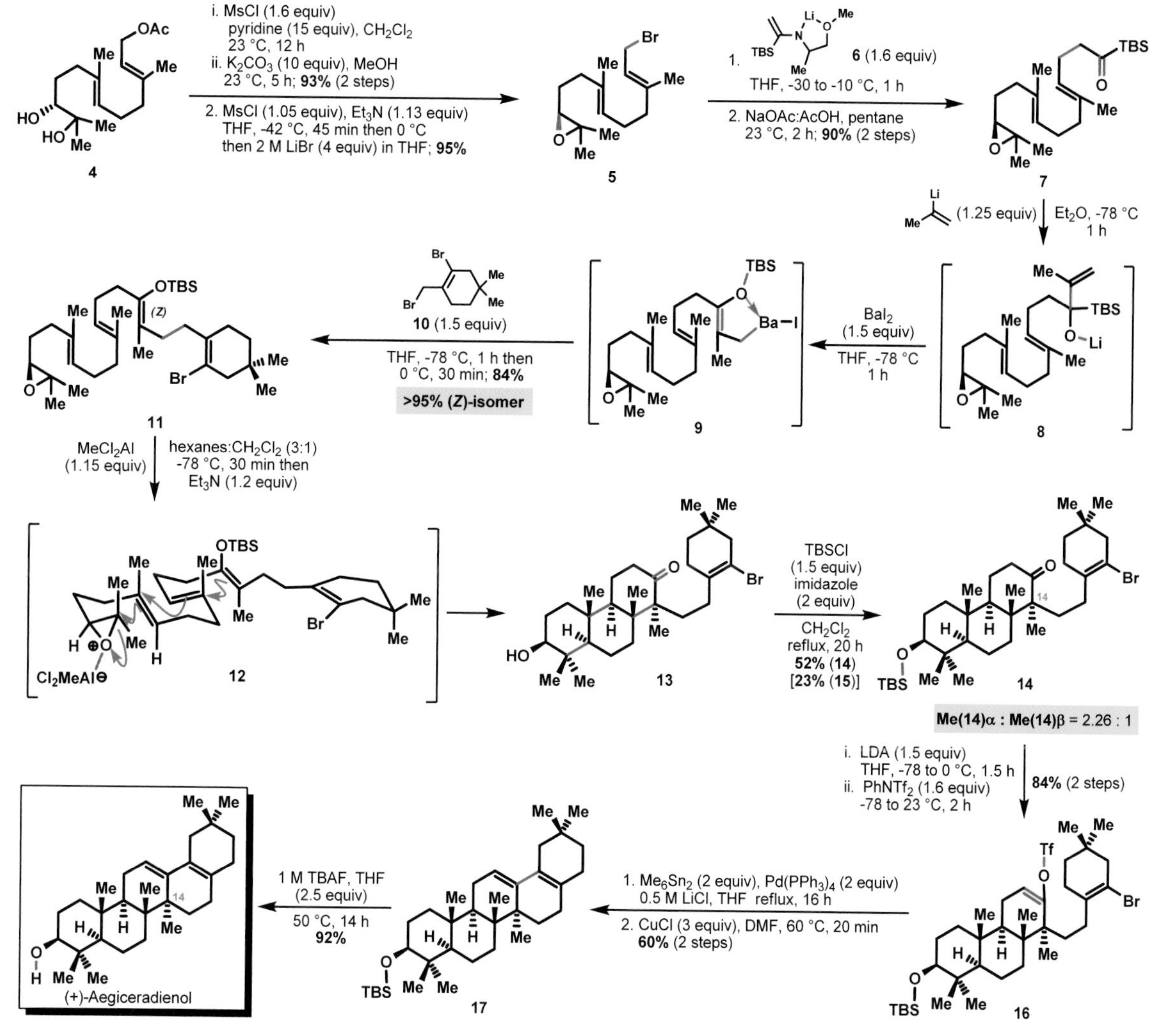

Scheme 3

TRITERPENES OF THE β-AMYRIN FAMILY

An Alternative Pathway of Synthesis

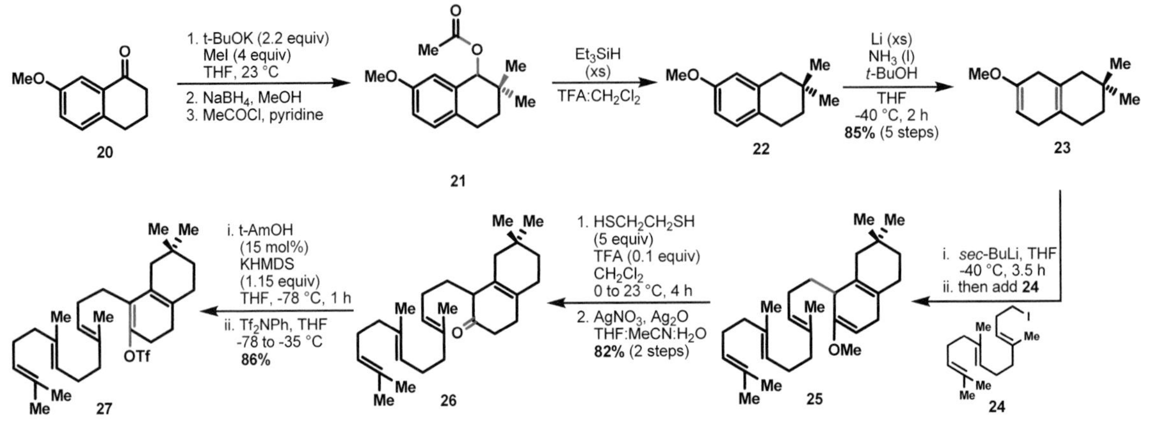

Scheme 4

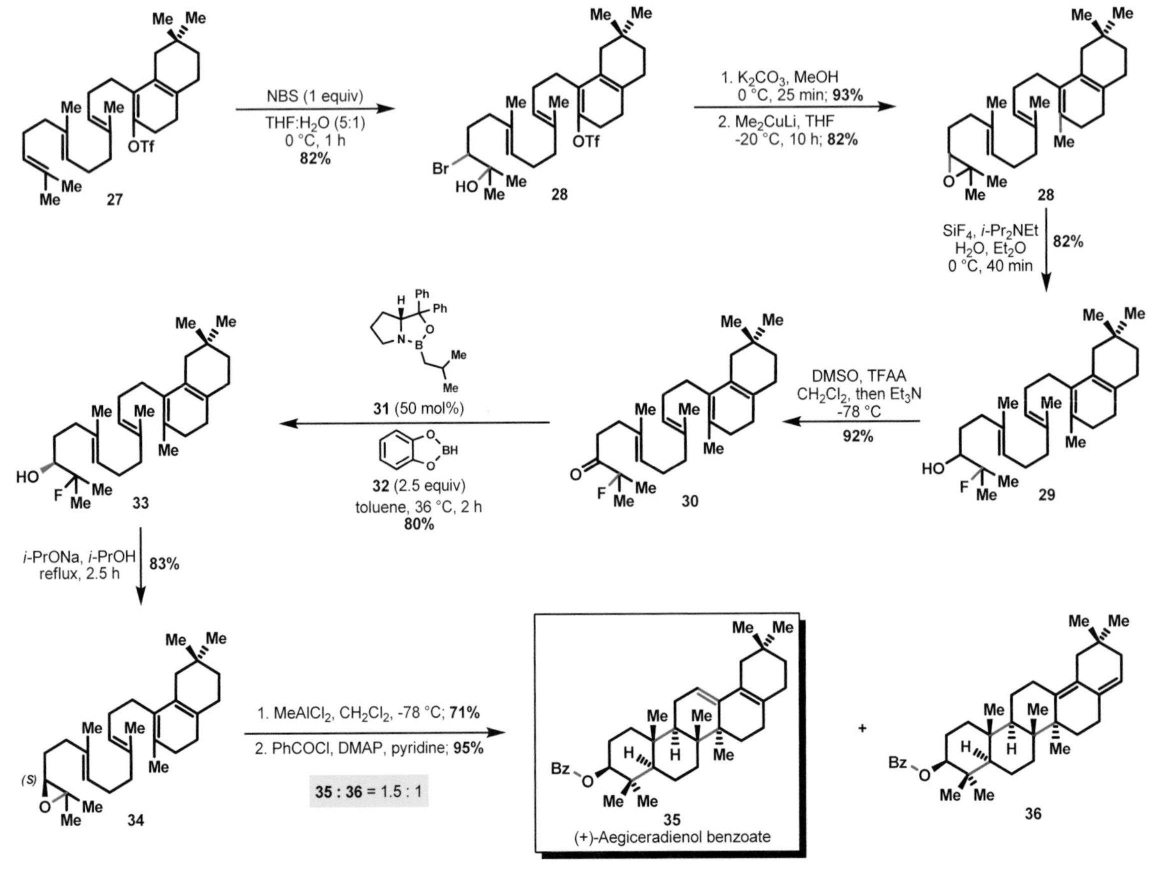

Scheme 5

TRITERPENES OF THE β-AMYRIN FAMILY

An Alternative Pathway of Synthesis (Continued)

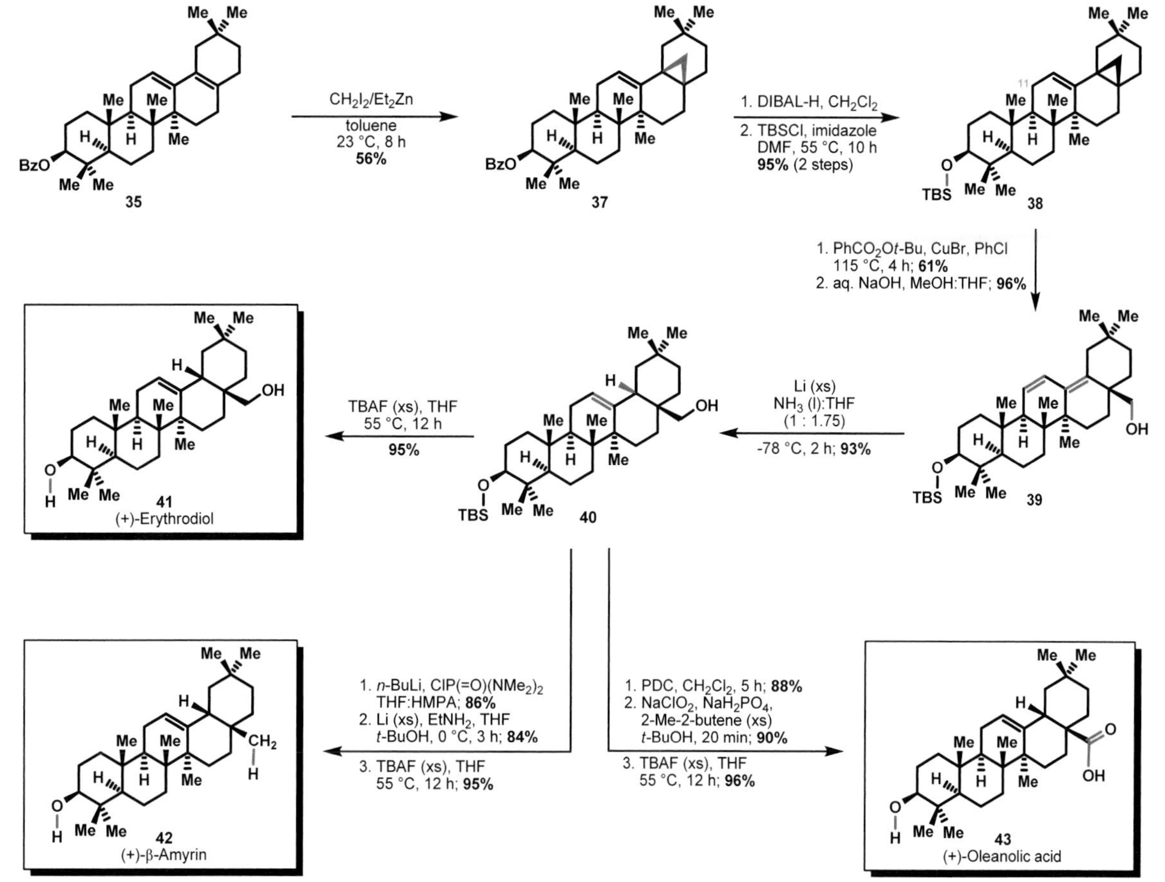

Scheme 6

Explanation of Schemes 4, 5 and 6

(1) The coupling of **23** and **24** (Scheme 4) occurs via the lithio derivative of **23** (a pentadienyl anion, **23a**).

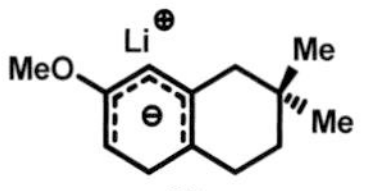

(2) The conversion of **25** to **26** (Scheme 4) was carried out in this way to prevent formation of the α,β-unsaturated isomer of **26** (see below, **26a**).

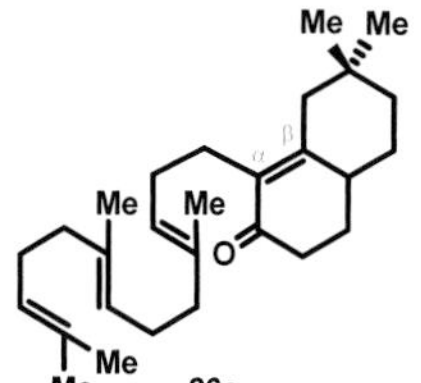

(3) The epoxide-fluorohydrin conversion (**28**→**29**, Scheme 5) illustrates a novel use of SiF₄.[3]

(4) The transformation[4] **38**→**39** (Scheme 6) takes place by way of an initiating C(11) radical, ring-opening, capture of PhCO₂ radical and benzoate hydrolysis (see below).[5]

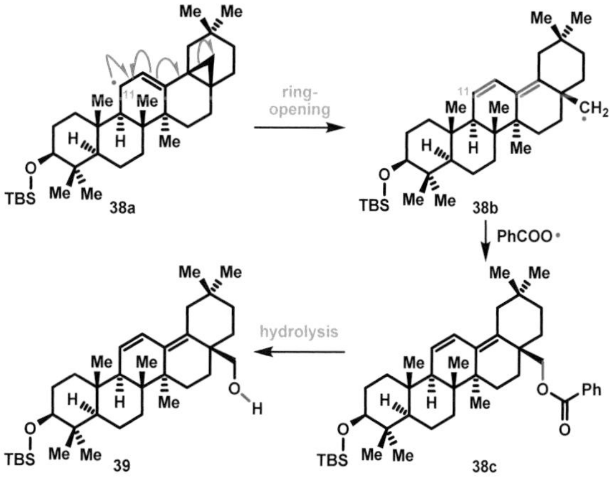

TRITERPENES OF THE β-AMYRIN FAMILY

Recent Developments: Germanicol and Other Pentacyclic Triterpenes

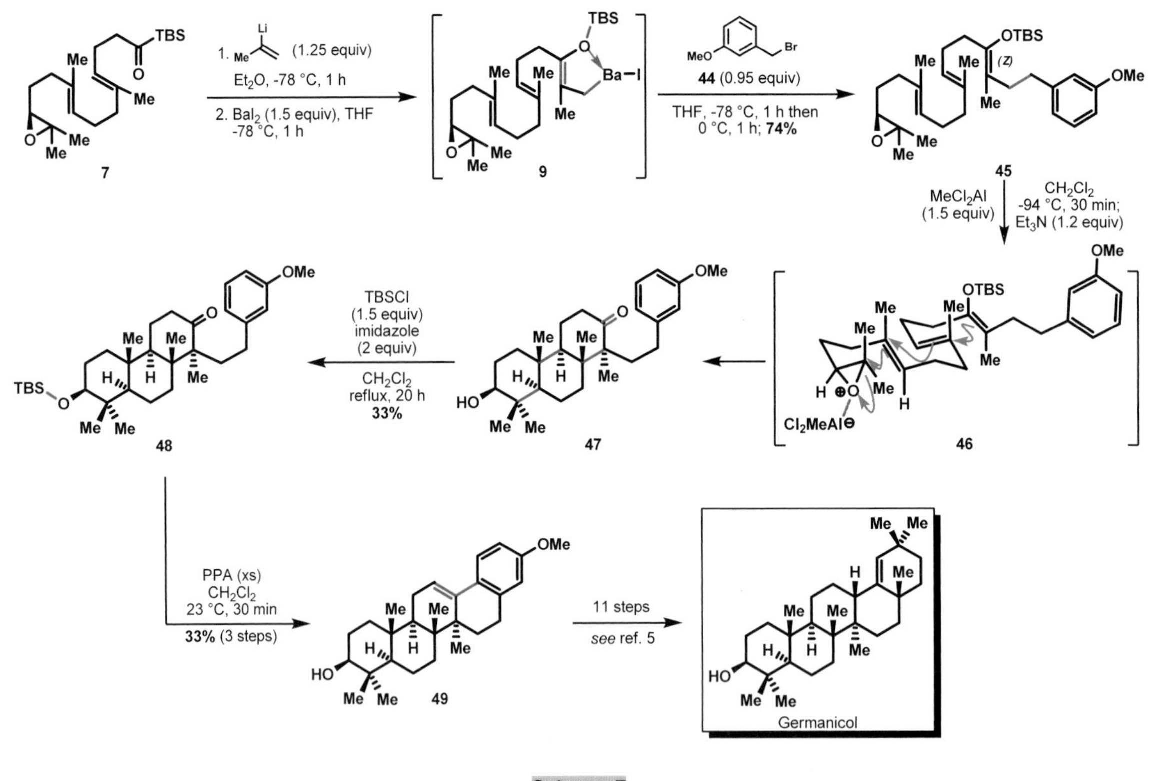

Scheme 7

Explanation of Scheme 7

(1) The synthesis of the racemic triterpene germanicol originally required 32 steps and gave an overall yield of only 0.1%.[6] The shorter and more efficient route shown utilized the chiral epoxide **7** to generate **49** in 4 steps.[7]

(2) The organobarium intermediate **9** was more suitable for coupling to form **45** than the corresponding lithio derivative.

A Short Enantioselective Total Synthesis of Lupeol

The direct, one-step pentacyclization of (S)-2,3-oxidosqualene to the pentacyclic triterpene lupeol is a remarkable example of biosynthetic efficiency. The chemical emulation of this elegant biosynthetic conversion remains an elusive challenge in modern synthesis. In fact, there has only recently been a successful multistep enantioselective synthesis of lupeol.[8] An earlier synthesis of racemic lupeol involved a relay sequence via an oxidation product of natural lupeol.[9]

Shown in Scheme 8 is a simple and very short enantioselective solution to the longstanding challenge posed by the lupeol molecule. The enantioselective synthesis of lupeol depended on the careful choice of substrate for stereocontrolled cation olefin polycyclization,

which was facilitated by earlier studies outlined in Scheme 7.[7] In addition, the further one-step conversion of (+)-lupeol to the naturally occurring pentacyclic triterpenes germanicol, δ-amyrin, 18-epi-β-amyrin, taraxasterol, ψ-taraxasterol and α-amyrin via cationic intermediates was demonstrated.

Although the synthesis of lupeol described in Scheme 8 is very different from the process used in biosynthesis, it shares the use of cationic cyclization to simplify and shorten the synthesis. It is important to note, however, that in a chemical setting, careful choice of the substrates for cyclization is absolutely crucial for success.

TRITERPENES OF THE β-AMYRIN FAMILY

A Short Enantioselective Total Synthesis of (+)-Lupeol

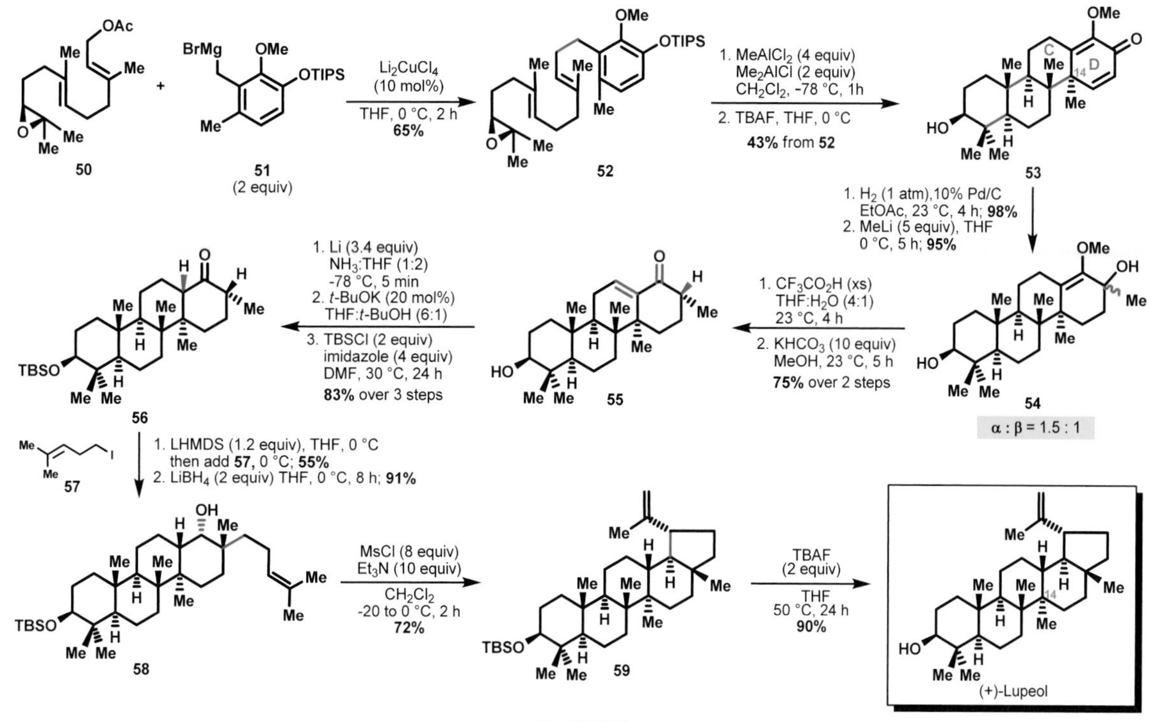

Scheme 8

Explanation of Scheme 8

The enantioselective synthetic pathway described in Scheme 8 includes two unusually interesting ring-forming steps. First, the conversion of **52** to tetracycle **53**, which depends on careful choice of the substituents on the aromatic ring, in order to (1) activate that ring for a sterically difficult cyclization, (2) channel the cyclization to a single tetracyclic product, (3) establish functionality in **53** that allows rapid execution of the final steps of the synthesis and (4) minimize the possibility of Lewis acid coordination to the electron supplying groups (which would deactivate the aromatic ring). Second, the *unusually facile* cationic conversion of **58** to **59** under essentially non-acidic conditions allows the clean formation of the acid-sensitive lupeol molecule.

References

1. Dev, S., Gupta, A. S. & Patwardhan, S. A. CRC Handbook of Terpenoids. Triterpenoids. 321-499 (CRC Press, Boca Raton, Florida, **1989**).

2. (a) Corey, E. J. & Lee, J. Enantioselective total synthesis of oleanolic acid, erythrodiol, β-amyrin, and other pentacyclic triterpenes from a common intermediate. *J. Am. Chem. Soc.* **1993**, 115, 8873-8874; (b) Huang, A. X., Xiong, Z. & Corey, E. J. An Exceptionally Short and Simple Enantioselective Total Synthesis of Pentacyclic Triterpenes of the β-Amyrin Family. *J. Am. Chem. Soc.* **1999**, 121, 9999-10003.

3. For the ring-opening of epoxides and oxetanes with other fluoride sources, see: (a) Shimizu, M., Nakahara, Y. Ring-opening fluorination of epoxides using hydrofluoric acid and additives. *J. Fluorine Chem.* **1999**, 99, 95-97; (b) Shimizu, M., Kanemoto, S., Nakahara, Y. Regioselective ring-opening fluorination of oxetanes with silicon tetrafluoride. *Heterocycles* **2000**, 52, 117-120; (c) Yoshino, H., Nomura, K., Matsubara, S., Oshima, K., Matsumoto, K., Hagiwara, R., Ito, Y. A mild ring opening fluorination of epoxide with ionic liquid 1-ethyl-3-methylimidazolium oligo hydrogenfluoride (EMIMF(HF)2.3). *J. Fluorine Chem.* **2004**, 125, 1127-1129; (d) Park, S. A., Lim, C. H., Chung, K.-H. Epoxide opening with tetrabutylammonium fluoride (TBAF). *Bull. Korean Chem. Soc.* **2007**, 28, 1834-1836.

4. Kharasch, M. S., Sosnovsky, G. & Yang, N. C. Reactions of tert-butyl peresters. I. The reaction of peresters with olefins. *J. Am. Chem. Soc.* **1959**, 81, 5819-5824.

5. For recent reviews on the Kharasch-Sosnovsky free-radical chain oxidation, see: (a) Eames, J., Watkinson, M. Catalytic allylic oxidation of alkenes using an asymmetric Kharasch-Sosnovsky reaction. *Angew. Chem., Int. Ed.* **2001**, 40, 3567-3571; (b) Andrus, M. B., Lashley, J. C. Copper catalyzed allylic oxidation with peresters. *Tetrahedron* **2002**, 58, 845-866; (c) Le Paih, J., Schlingloff, G., Bolm, C. Kharasch-Sosnovsky type allylic oxidations in *Transition Metals for Organic Synthesis (2nd Edition)* **2004**, Vol 2, pp 256-265; (d) Frison, J.-C., Legros, J., Bolm, C. Copper- and palladium-catalyzed allylic acyloxylations in *Handbook of C-H Transformations* **2005**, 2, pp 445-454, 492-494.

6. Ireland, R. E., Baldwin, S. W., Dawson, D. D., Dawson, M. I., Dolfini, J. E. & Newbold, J. Total synthesis of an unsymmetrical pentacyclic triterpene. DL-germanicol. *J. Am. Chem. Soc.* **1970**, 92, 5743-5746.

7. Surendra, K. & Corey, E. J. Rapid and Enantioselective Synthetic Approaches to Germanicol and Other Pentacyclic Triterpenes. *J. Am. Chem. Soc.* **2008**, 130, 8865-8869.

8. Surendra, K. & Corey, E. J. A Short Enantioselective Total Synthesis of the Fundamental Pentacyclic Triterpene Lupeol. *J. Am. Chem. Soc.* **2009**, 131, 13928-13929.

9. Stork, G., Uyeo, S., Wakamatsu, T., Grieco, P. & Labovitz, J. Total synthesis of lupeol. *J. Am. Chem. Soc.* **1971**, 93, 4945-4947.

(−)-GLABRESCOL & ISOMERS

Background

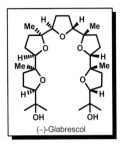

(−)-Glabrescol

Glabrescol is a chiral, C_2-symmetric oxygenated squalene derivative which occurs in the Caribbean plant *Spathelia glabrescens* and is a member of a family of oxasqualenoids. It was originally assigned an achiral C_S-symmetric (meso) structure in part because it appeared to have an optical rotation of zero.[1] The correct structure was established by the synthesis outlined in this section.[2] A description of the synthesis of four C_S-symmetric isomers[3] is also included at the end of this section.[4]

Abbreviated Retrosynthetic Plan

(−)-Glabrescol

$\Longrightarrow$ bidirectional cascade pentacyclization

I

$\Longrightarrow$ enantioselective epoxidation

II

$\Downarrow$ dimerization

III

$\Longleftarrow$ FGI

farnesyl acetate

(−)-Glabrescol
(hydrogens have been removed for clarity)

Scheme 1

Key steps in Scheme 1

(1) The synthetic simplification of the glabrescol structure followed from its relationship to the biosynthetic precursor squalene and the clear possibility of disconnection of all five tetrahydrofuran rings by C-O cleavage. Repeated C-O cleavage of glabrescol using the hydroxyl-initiated oxygen displacement transform generates I.

(2) The disconnection of glabrescol to I should be valid since it had previously been established that in the synthetic direction tetrahydrofuran ring-formation is preferred over tetrahydropyran ring-formation.

(3) An energetically favorable conformation for (−)-glabrescol is shown above. The hydrogen atoms have been removed for clarity.

(−)-GLABRESCOL & ISOMERS

Pathway of Synthesis

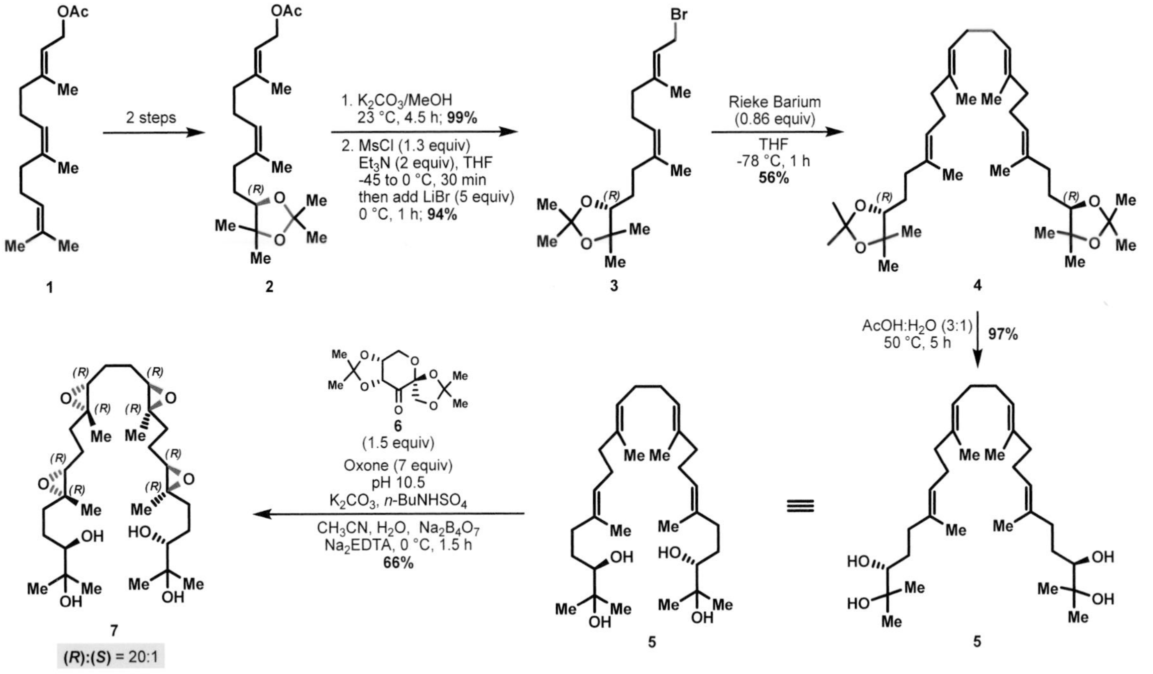

Scheme 2

Key steps in Scheme 2

(1) The use of the Noe-Lin catalyst[5] with OsO$_4$ as oxidant allowed the enantioselective Sharpless dihydroxylation of the terminal double bond of **1** and the two-step synthesis of **2**.

(2) The barium metal-mediated coupling (**3**→**4**) occurred without E/Z isomerization.[6]

(3) The catalyst **6**, originated by Shi[7] and co-workers, effects epoxidation via a dioxirane intermediate.[8]

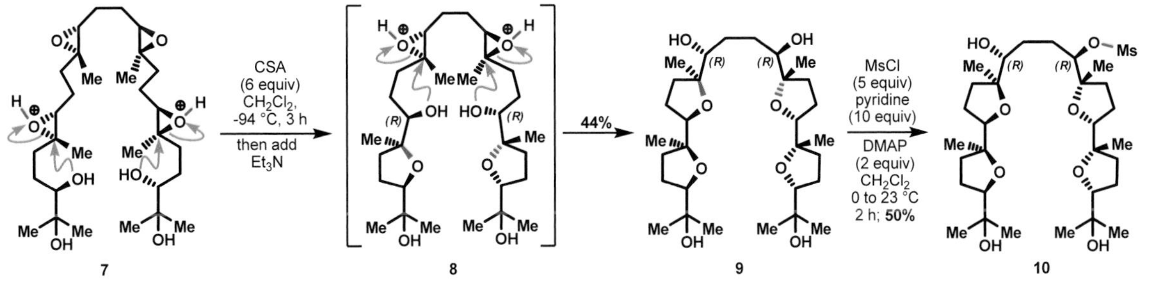

Scheme 3

Key steps in Schemes 3 and 4

(1) The formation of the first of the four tetrahydrofuran rings occurs in one step with each epoxide-displacement taking place with inversion of configuration at the carbon which is attacked by oxygen, thereby converting **7** via **8** to **9**.

(2) Diol **9** was selectively converted to the corresponding monomesylate **10**. Under conditions of solvolysis (NaOAc in AcOH, see Scheme 4), the vicinal nucleophilic tetrahydrofuran oxygen participates to form the intermediate oxonium ion **11** with inversion. This then undergoes a second displacement by the free hydroxyl group (see **12**) to afford (−)-glabrescol.[9]

(–)-GLABRESCOL & ISOMERS

Pathway of Synthesis (continued)

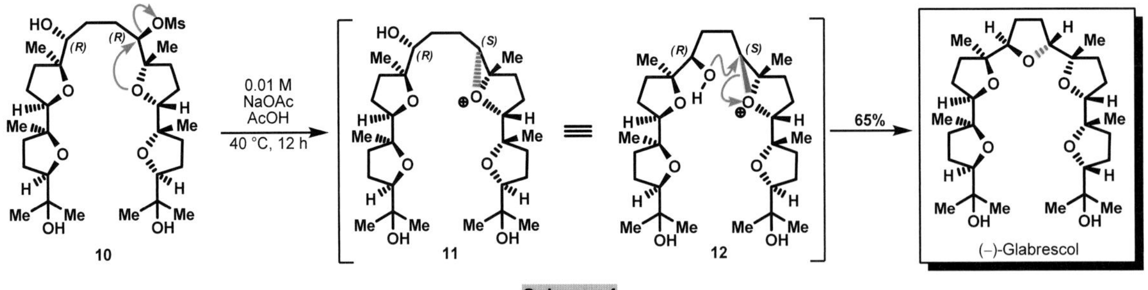

Scheme 4

Synthesis of the Proposed C_S-Symmetric Structure of Glabrescol (A) and an Isomer (B)

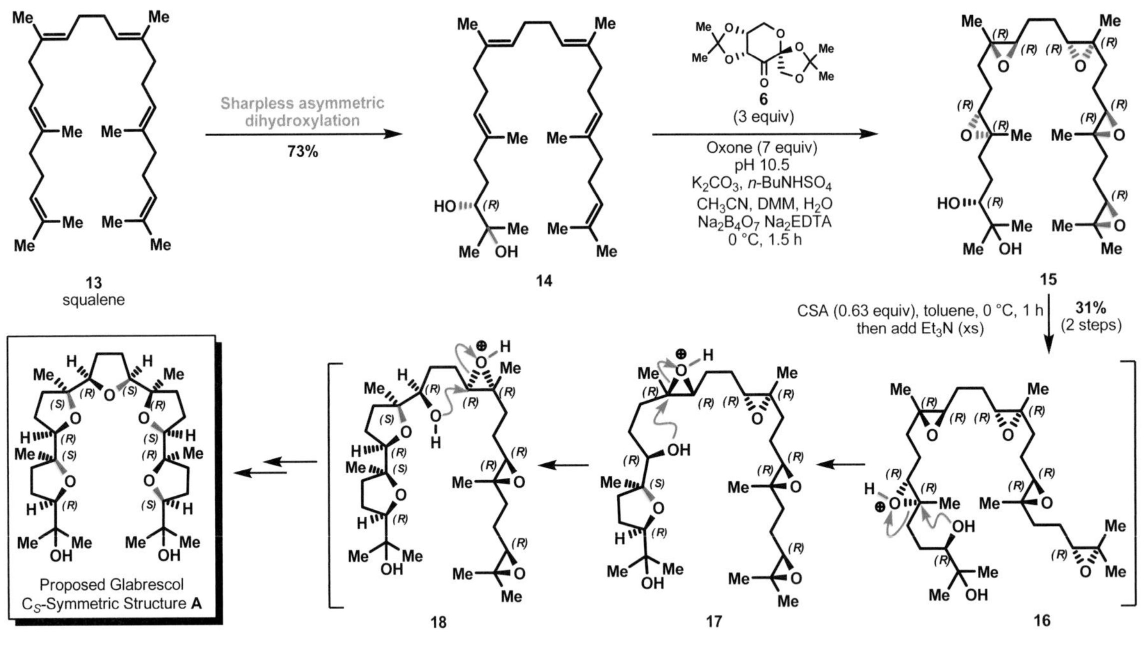

Scheme 5

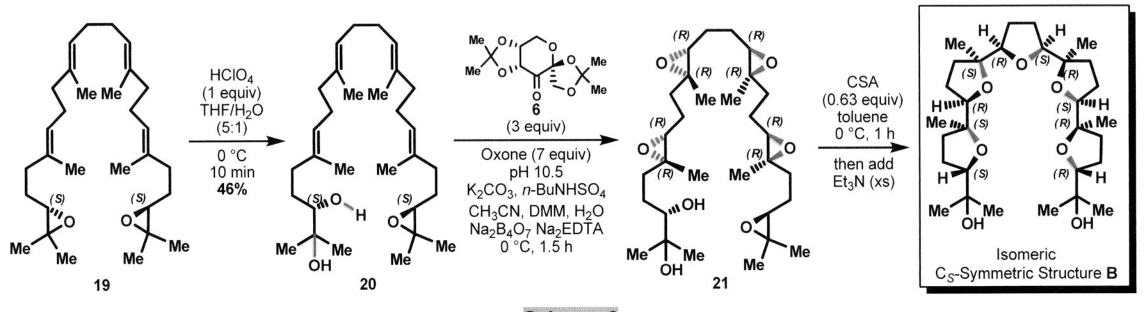

Scheme 6

(–)-GLABRESCOL & ISOMERS

Synthesis of C_S-Symmetric Glabrescol Isomers C and D

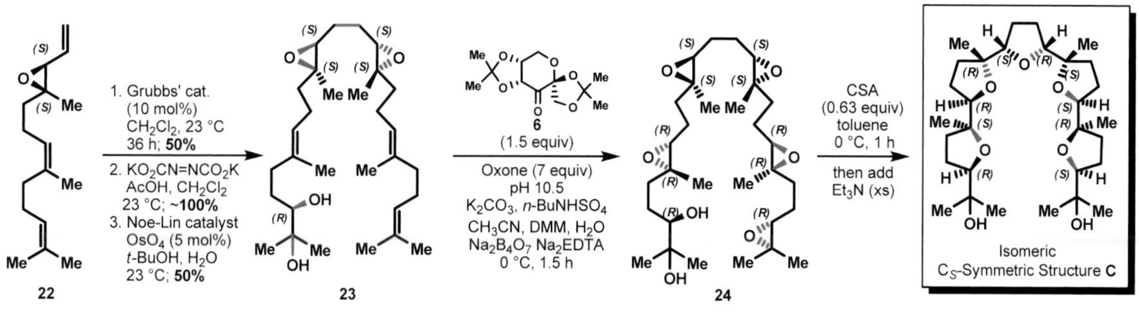

Scheme 7

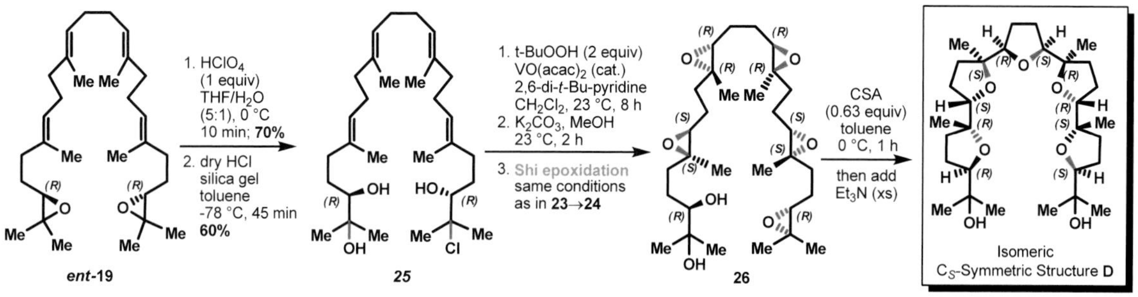

Scheme 8

References

1. Harding, W. W., Lewis, P. A., Jacobs, H., McLean, S., Reynolds, W. F., Tay, L.-L. & Yang, J.-P. Glabrescol. A unique squalene-derived penta-THF diol from Spathelia glabrescens (Rutaceae). *Tetrahedron Lett.* **1995**, 36, 9137-9140.

2. Xiong, Z. & Corey, E. J. Simple Enantioselective Total Synthesis of Glabrescol, a Chiral C2-Symmetric Pentacyclic Oxasqualenoid. *J. Am. Chem. Soc.* **2000**, 122, 9328-9329.

3. Xiong, Z. & Corey, E. J. Simple Total Synthesis of the Pentacyclic Cs-Symmetric Structure Attributed to the Squalenoid Glabrescol and Three Cs-Symmetric Diastereomers Compel Structural Revision. *J. Am. Chem. Soc.* **2000**, 122, 4831-4832.

4. For another synthesis of two C_S-symmetric glabrescol isomers, see: (a) Morimoto, Y., Iwai, T., Kinoshita, T. Revised Structure of Squalene-Derived PentaTHF Polyether, Glabrescol, through Its Enantioselective Total Synthesis: Biogenetically Intriguing Cs vs. C$_2$ Symmetric Relationships. *J. Am. Chem. Soc.* **2000**, 122, 7124-7125; (b) Hioki, H., Kanehara, C., Ohnishi, Y., Umemori, Y., Sakai, H., Yoshio, S., Matsushita, M., Kodama, M. What is the Structure of glabrescol? stereoselective synthesis of reported glabrescol. *Angew. Chem., Int. Ed.* **2000**, 39, 2552-2554.

5. Corey, E. J., Noe, M. C. & Lin, S. A mechanistically designed bis-cinchona alkaloid ligand allows position- and enantioselective dihydroxylation of farnesol and other oligoprenyl derivatives at the terminal isopropylidene unit. *Tetrahedron Lett.* **1995**, 36, 8741-8744.

6. (a) Corey, E. J. & Shieh, W. C. A simple synthetic process for the elaboration of oligoprenols by stereospecific coupling of di-, tri-, or oligoisoprenoid units. *Tetrahedron Lett.* **1992**, 33, 6435-6438; (b) Corey, E. J., Noe, M. C. & Shieh, W. C. A short and convergent enantioselective

synthesis of (3S)-2,3-oxidosqualene. *Tetrahedron Lett.* **1993**, 34, 5995-5998.

7. Wang, Z.-X., Tu, Y., Frohn, M., Zhang, J.-R. & Shi, Y. An Efficient Catalytic Asymmetric Epoxidation Method. *J. Am. Chem. Soc.* **1997**, 119, 11224-11235.

8. For reviews on the Shi Asymmetric Epoxidation and other related organocatalytic epoxidations of olefins, see: (a) Wong, O. A., Shi, Y. Organocatalytic Oxidation. Asymmetric Epoxidation of Olefins Catalyzed by Chiral Ketones and Iminium Salts. *Chem. Rev.* **2008**, 108, 3958-3987; (b) Shi, Y. in Handbook of Chiral Chemicals (2nd Edition) **2005** pp147-163 (CRC Press); (c) Kürti, L., Czakó, B. Shi Asymmetric Epoxidation in *Strategic Applications of Named Reactions in Organic Synthesis* **2005**, pp 410-411 (Academic Press/Elsevier, San Diego); (d) Shi, Y. Organocatalytic Asymmetric Epoxidation of Olefins by Chiral Ketones. *Acc. Chem. Res.* **2004**, 37, 488-496.

9. For other members of the oxasqualenoid family, see: (a) Suzuki, T., Suzuki, M., Furusaki, A., Matsumoto, T., Kato, A., Imanaka, Y., Kurosawa, E. Constituents of marine plants. 62. Teurilene and thyrsiferyl 23-acetate, meso and remarkably cytotoxic compounds from the marine red alga Laurencia obtusa (Hudson) Lamouroux. *Tetrahedron Lett.* **1985**, 26, 1329-1332; (b) Sakemi, S., Higa, T., Jefford, C. W., Bernardinelli, G. Venustatriol: a new antiviral triterpene tetracyclic ether from Laurencia venusta. *Tetrahedron Lett.* **1986**, 27, 4287-4290; (c) Itokawa, H., Kishi, E., Morita, H., Takeya, K., Iitaka, Y. A new squalene-type triterpene from the woods of Eurycoma longifolia. *Chem. Lett.* **1991**, 2221-2222; (d) Tinto, W. F., McLean, S., Reynolds, W. F., Carter, C. A. G. Quassiol a, a novel squalene triterpene from Quassia multiflora. *Tetrahedron Lett.* **1993**, 34, 1705-1708.

(+)-DAFACHRONIC ACID A

Background

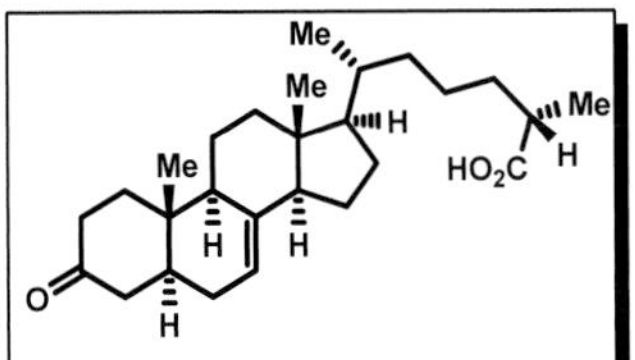

(+)-Dafachronic acid A

The lifetime of the nematode *C. elegans* can be extended from about 2 weeks to 12 weeks by loss of function of two genes which code for the formation of the small molecule dafachronic acid A, which is a natural ligand for a nuclear hormone receptor DAF-12.[1] Thus, dafachronic acid A is part of a control mechanism that links metabolism, reproduction and lifespan to the availability of nutrients in the environment of *C. elegans*.[2] A synthesis of this potent regulator[3] is described.[4]

Abbreviated Retrosynthetic Plan

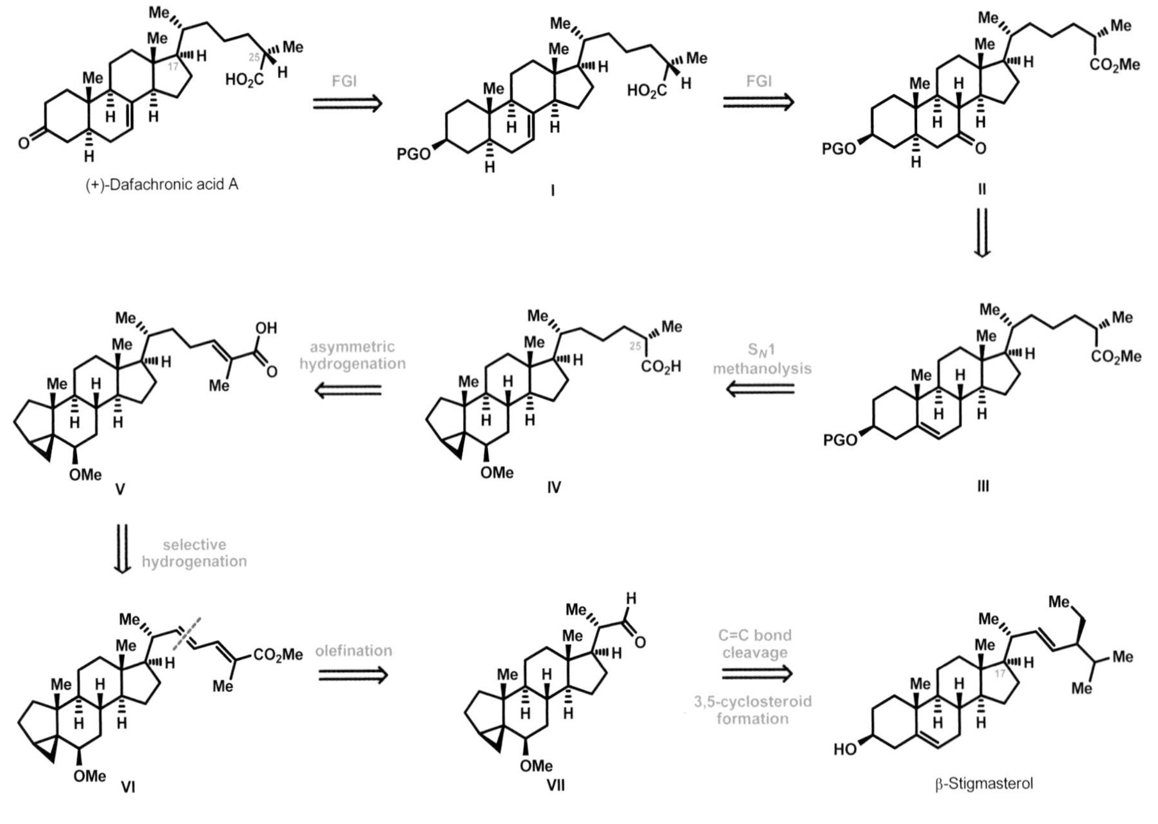

Scheme 1

Key steps in Scheme 1

(1) The design of the synthetic route was based on the use of an available sterol (β-stigmasterol) as a starting material.

(2) The choice of starting material then required the replacement of the sidechain at C(17) with one containing the C(25) stereocenter and carboxylic acid function.

(3) The C(25) stereocenter was cleared by application of the catalytic enantioselective hydrogenation transform **IV** ⇒ **V**.

(4) Disconnection of the double bond in the β-stigmasterol side chain necessitated the protection of the more reactive $\Delta^{5,6}$-olefinic linkage. This was accomplished via the 3,5-cyclosteroids **IV** ⇒ **V**.

(+)-DAFACHRONIC ACID A

Pathway of Synthesis

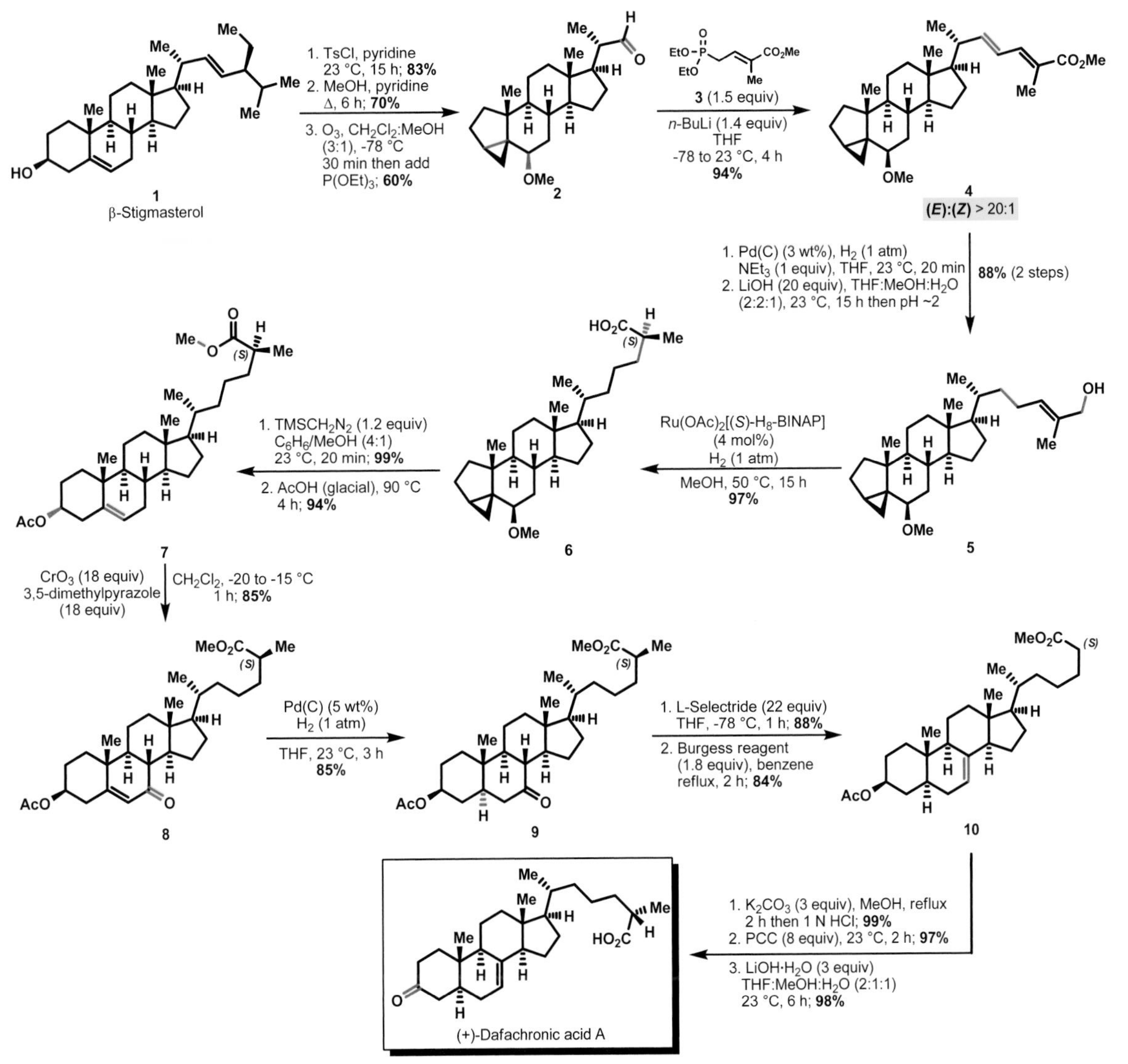

Scheme 2

Key steps in Scheme 2

(1) The stereoselective Ru-catalyzed reduction **5→6** was chosen on the basis of previous work with simple α,β-unsaturated acids.[5]

(2) The transposition of the Δ^5-olefinic linkage in **7** to the Δ^7-position was best accomplished via the sequence **7→10**.

Key steps in Scheme 3

(3) The less bioactive 25(R)-diastereomer of dafachronic acid A was synthesized using the Ireland-Claisen rearrangement **15→17** as a key step.[6]

(4) Another key operation was the highly position-selective reduction of the $\Delta^{5,7}$-diene **11** and benzoylation to form the Δ^7-benzoate **12**.

(5) Steric screening of the Δ^7-double bond in **12** allowed selective oxidative cleavage of the side chain double bond to form **13**.

(6) The intermediate aldehyde **13** can also serve as a precursor of 25(S)-dafachronic acid by use of the chain elongation-enantioselective hydrogenation process shown in Scheme 2.

313

(+)-DAFACHRONIC ACID A

Synthesis of the 25-(*R*)-Diastereomer from β-Ergosterol

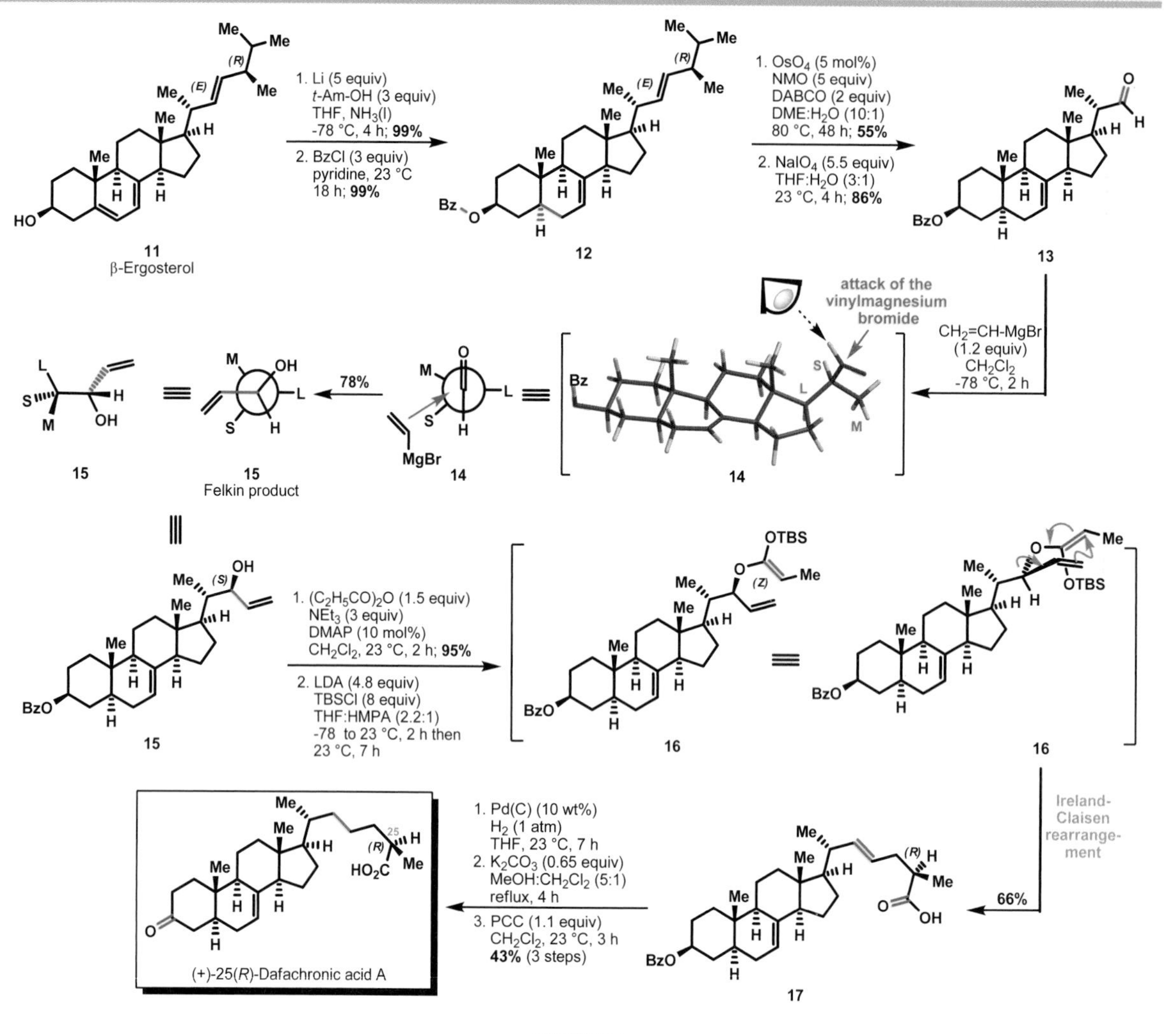

11
β-Ergosterol

1. Li (5 equiv)
   t-Am-OH (3 equiv)
   THF, NH₃(l)
   -78 °C, 4 h; **99%**
2. BzCl (3 equiv)
   pyridine, 23 °C
   18 h; **99%**

12

1. OsO₄ (5 mol%)
   NMO (5 equiv)
   DABCO (2 equiv)
   DME:H₂O (10:1)
   80 °C, 48 h; **55%**
2. NaIO₄ (5.5 equiv)
   THF:H₂O (3:1)
   23 °C, 4 h; **86%**

13

attack of the
vinylmagnesium
bromide

CH₂=CH-MgBr
(1.2 equiv)
CH₂Cl₂
-78 °C, 2 h

15

15
Felkin product

78%

14

MgBr

14

15

1. (C₂H₅CO)₂O (1.5 equiv)
   NEt₃ (3 equiv)
   DMAP (10 mol%)
   CH₂Cl₂, 23 °C, 2 h; **95%**
2. LDA (4.8 equiv)
   TBSCl (8 equiv)
   THF:HMPA (2.2:1)
   -78 to 23 °C, 2 h then
   23 °C, 7 h

16

16

Ireland-
Claisen
rearrange-
ment

(+)-25(*R*)-Dafachronic acid A

1. Pd(C) (10 wt%)
   H₂ (1 atm)
   THF, 23 °C, 7 h
2. K₂CO₃ (0.65 equiv)
   MeOH:CH₂Cl₂ (5:1)
   reflux, 4 h
3. PCC (1.1 equiv)
   CH₂Cl₂, 23 °C, 3 h
   43% (3 steps)

17

66%

Scheme 3

References

1. (a) Gerisch, B. & Antebi, A. Hormonal signals produced by DAF-9/cytochrome P450 regulate C. elegans dauer diapause in response to environmental cues. *Development (Cambridge, United Kingdom)* **2004**, 131, 1765-1776; (b) Antebi, A. Physiology: the tick-tock of aging? *Science* **2005**, 310, 1911-1913; (c) Motola, D. L., Cummins, C. L., Rottiers, V., Sharma, K. K., Li, T., Li, Y., Suino-Powell, K., Xu, H. E., Auchus, R. J., Antebi, A. & Mangelsdorf, D. J. Identification of ligands for DAF-12 that govern dauer formation and reproduction in C. elegans. *Cell (Cambridge, MA, United States)* **2006**, 124, 1209-1223.

2. Brenner, S. *Nobel Lecture*, http://nobelprize.org/nobel-prizes/medicine/laureates/2002/brenner-lecture.pdf.

3. Giroux, S. & Corey, E. J. Stereocontrolled Synthesis of Dafachronic Acid A, the Ligand for the DAF-12 Nuclear Receptor of Caenorhabditis elegans. *J. Am. Chem. Soc.* **2007**, 129, 9866-9867.

4. For other dafachronic acid analogs and their biological evaluation, see: (a) Martin, R., Schmidt, A. W., Theumer, G., Krause, T., Entchev, E. V., Kurzchalia, T. V., Knoelker, H.-J. Synthesis and biological activity of the (25R)-cholesten-26-oic acids-ligands for the hormonal receptor DAF-12 in Caenorhabditis elegans. *Org. Biomol. Chem.* **2009**, 7, 909-920; (b) Martin, R., Daebritz, F., Entchev, E. V., Kurzchalia, T. V., Knoelker, H.-

J. Stereoselective synthesis of the hormonally active (25S)-Δ⁷-dafachronic acid, (25S)-Δ⁴-dafachronic acid, (25S)-dafachronic acid, and (25S)-cholestenoic acid. *Org. Biomol. Chem.* **2008**, 6, 4293-4295; (c) Martin, R., Schmidt, A. W., Theumer, G., Kurzchalia, T. V., Knoelker, H.-J. Stereoselective synthesis of (25R)-dafachronic acids and (25R)-cholestenoic acid as potential ligands for the DAF-12 receptor in Caenorhabditis elegans. *Synlett* **2008**, 1965-1968.

5. (a) Ohta, T., Takaya, H., Kitamura, M., Nagai, K. & Noyori, R. Asymmetric hydrogenation of unsaturated carboxylic acids catalyzed by BINAP-ruthenium(II) complexes. *J. Org. Chem.* **1987**, 52, 3174-3176; (b) Uemura, T., Zhang, X., Matsumura, K., Sayo, N., Kumobayashi, H., Ohta, T., Nozaki, K. & Takaya, H. Highly Efficient Enantioselective Synthesis of Optically Active Carboxylic Acids by Ru(OCOCH₃)₂[(S)-H8-BINAP]. *J. Org. Chem.* **1996**, 61, 5510-5516.

6. (a) Ziegler, F. E. The thermal, aliphatic Claisen rearrangement. *Chem. Rev.* **1988**, 88, 1423-1452; (b) Kürti, L. & Czakó, B. Ireland-Claisen Rearrangement in *Strategic Applications of Named Reactions in Organic Synthesis* 90-91 (Academic Press/Elsevier, San Diego, **2005**).

(+)-BREFELDIN A

Background

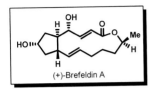

(+)-Brefeldin A

The antibiotic brefeldin A[1] interferes with intracellular protein trafficking, specifically intracellular membrane transport involving the endoplasmatic reticulum and the Golgi apparatus.[2]

This effect is the result of binding of brefeldin A to a family of GTP-exchange factors that play a role in membrane association. Brefeldin A is a valuable reagent for studies of intracellular protein trafficking. Since the first synthesis of (±)-brefeldin A,[3] there have been several others.[4] A recent enantioselective synthetic route is outlined in this section.[5]

Abbreviated Retrosynthetic Plan

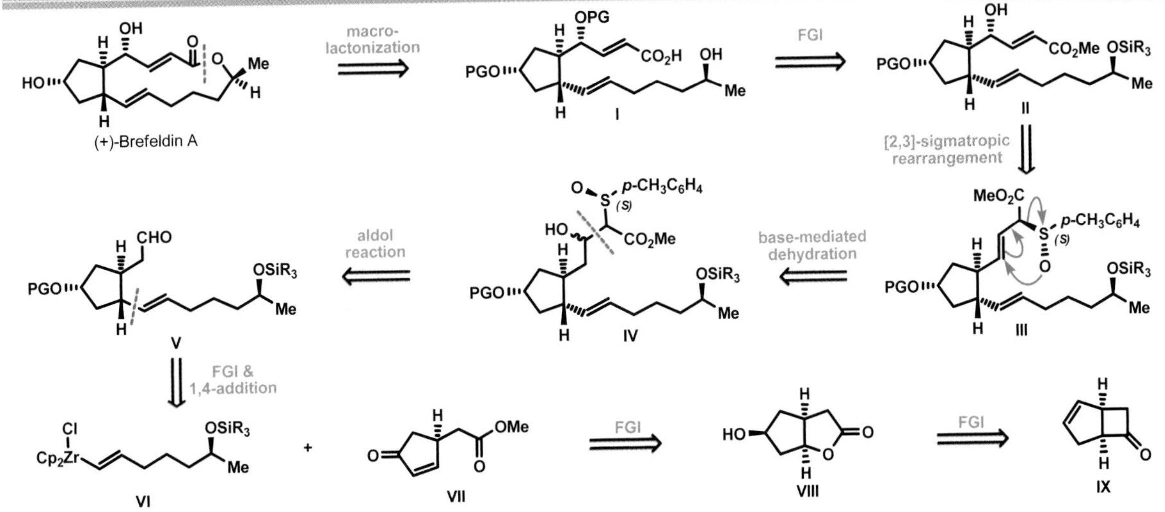

(+)-Brefeldin A ⟹ (macro-lactonization) **I** ⟹ (FGI) **II** ⟹ ([2,3]-sigmatropic rearrangement) **III** ⟹ (base-mediated dehydration) **IV** ⟹ (aldol reaction) **V** ⟹ (FGI & 1,4-addition) **VI** + **VII** ⟹ (FGI) **VIII** ⟹ (FGI) **IX**

Scheme 1

Key steps in Scheme 1

(1) The retrosynthetic plan was guided by the recognition of an available chiral starting point (**IX**) that contained appropriate functionality for introducing the two appendages and the hydroxyl group on the cyclopentane subunit.

(2) Using **IX** as an S-goal the hydroxy acid **I** was transformed via **II**, **III** and **IV** to the aldehyde **V**, which was further simplified to **IX** as shown.

Pathway of Synthesis

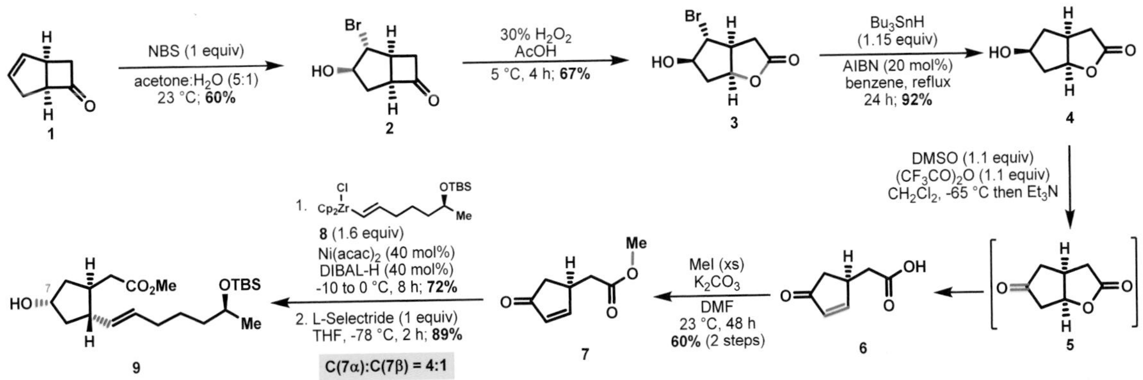

1 → NBS (1 equiv), acetone:H$_2$O (5:1), 23 °C; **60%** → **2** → 30% H$_2$O$_2$, AcOH, 5 °C, 4 h; **67%** → **3** → Bu$_3$SnH (1.15 equiv), AIBN (20 mol%), benzene, reflux, 24 h; **92%** → **4**

4 → DMSO (1.1 equiv), (CF$_3$CO)$_2$O (1.1 equiv), CH$_2$Cl$_2$, -65 °C then Et$_3$N → **5** → MeI (xs), K$_2$CO$_3$, DMF, 23 °C, 48 h, **60% (2 steps)** → **6** → **7**

7 → 1. Cp$_2$Zr(Cl) ... OTBS ... Me, **8** (1.6 equiv), Ni(acac)$_2$ (40 mol%), DIBAL-H (40 mol%), -10 to 0 °C, 8 h; **72%**; 2. L-Selectride (1 equiv), THF, -78 °C, 2 h; **89%**, C(7α):C(7β) = 4:1 → **9**

Scheme 2

(+)-BREFELDIN A

Pathway of Synthesis (continued)

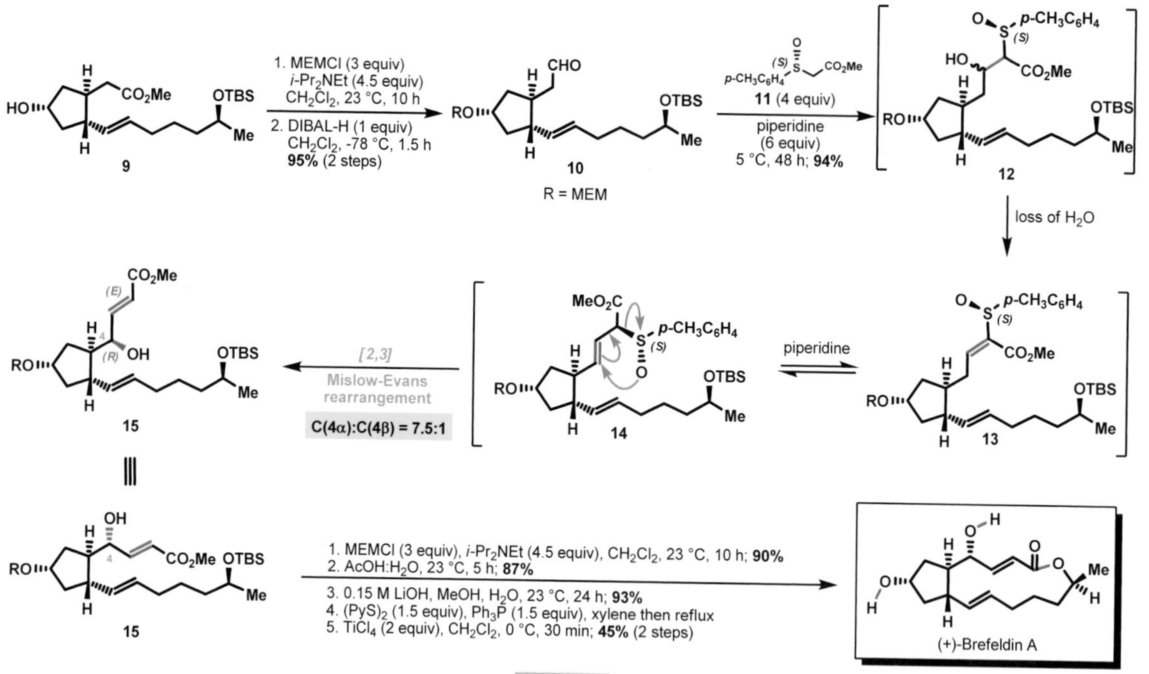

Scheme 3

Explanation of Scheme 2

(1) The position-selective conversion of **1**→**2** may occur via the corresponding hydrate, geminal diol **16**.

(2) The alkenyl zirconium reagent **8** used in the nickel-catalyzed conjugate addition[6] (**7**→**9**) was obtained by hydrozirconation of the terminal alkyne **17** with Cp$_2$ZrHCl (Schwartz's reagent).[7]

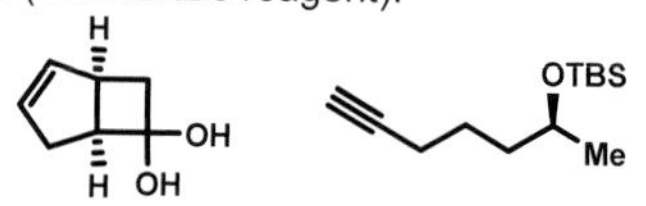

Explanation of Scheme 3

(3) The formation of **15** in one-step from **10** is considered to occur via intermediates **12-14**. In the presence of piperidine vinyl sulfoxide **13** isomerizes to the corresponding allylic sulfoxide **14**, which undergoes a facile [2,3]-sigmatropic rearrangement (Mislow-Evans rearrangement[8]).

(4) The macrolactonization[9] of the dihydroxy acid corresponding to **15** had previously been applied to the synthesis of (±)-brefeldin A.[3b]

References

1. (a) Sigg, H. P. The constitution of brefeldin A. *Helv. Chim. Acta* **1964**, 47, 1401-1415; (b) Weber, H. P., Hauser, D. & Sigg, H. P. Structure of brefeldin A. *Helv. Chim. Acta* **1971**, 54, 2763-2766.
2. Nebenfuhr, A., Ritzenthaler, C. & Robinson, D. G. Brefeldin A: deciphering an enigmatic inhibitor of secretion. *Plant Physiol.* **2002**, 130, 1102-1108.
3. (a) Corey, E. J. & Wollenberg, R. H. Total synthesis of (±)-brefeldin A. *Tetrahedron Lett.* **1976**, 4705-4708; (b) Corey, E. J., Wollenberg, R. H. & Williams, D. R. Total synthesis of (±)-brefeldin A. (Part IV). *Tetrahedron Lett.* **1977**, 2243-2246.
4. For recent total syntheses of Brefeldin A, see: (a) Wu, Y., Gao, J. Total Synthesis of (+)-Brefeldin A. *Org. Lett.* **2008**, 10, 1533-1536; (b) Lin, W., Zercher, C. K. Formal Synthesis of (+)-Brefeldin A. Application of a Zinc-Mediated Ring Expansion Reaction. *J. Org. Chem.* **2007**, 72, 4390-4395; (c) Seo, S.-Y., Jung, J.-K., Paek, S.-M., Lee, Y.-S., Kim, S.-H., Suh, Y.-G. An olefin disconnection strategy for the practical synthesis of (+)-brefeldin A: olefin cross metathesis and intramolecular Horner-Wadsworth-Emmons olefination. *Tetrahedron Lett.* **2006**, 47, 6527-6530;

(d) Trost, B. M., Crawley, M. L. 4-Aryloxybutenolides As "Chiral Aldehyde" Equivalents: An Efficient Enantioselective Synthesis of (+)-Brefeldin A. *J. Am. Chem. Soc.* **2002**, 124, 9328-9329.
5. Corey, E. J. & Carpino, P. A simplified synthesis of (+)-brefeldin A. *Tetrahedron Lett.* **1990**, 31, 7555-7558.
6. Schwartz, J., Loots, M. J. & Kosugi, H. Nickel-catalyzed conjugate addition of alkenylzirconium species to α,β-unsaturated ketones. *J. Am. Chem. Soc.* **1980**, 102, 1333-1340.
7. Kürti, L. & Czakó, B. Schwartz Hydrozirconation in *Strategic Applications of Named Reactions in Organic Synthesis* 400-401 (Academic Press/Elsevier, San Diego, **2005**).
8. Kürti, L. & Czakó, B. Mislow-Evans Rearrangement in *Strategic Applications of Named Reactions in Organic Synthesis* 292-293 (Academic Press/Elsevier, San Diego, **2005**).
9. Kürti, L. & Czakó, B. Corey-Nicolaou Macrolactonization in *Strategic Applications of Named Reactions in Organic Synthesis* 108-109 (Academic Press/Elsevier, San Diego, **2005**).

COLNELEIC ACID

Background

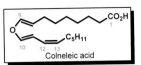

Colneleic acid

Colneleic acid, an unusual natural product that was isolated from potato (*Solanium tuberosum*),[1] remained a curiosity until it was discovered that an enzyme in potato is capable of converting arachidonic acid into the 5(*S*)-hydroperoxyeicosa- 6(*E*), 8, 11, 14(*Z*)-tetraenoic acid (15-HPETE),[2] the precursor of the leukotriene family of mammalian immune regulators.[3] The potato enzyme that forms 5-HPETE also converts linoleic or linolenic acid to the corresponding 9(*S*)- and 13(*S*)-hydroperoxides. That lipoxygenase activity[4] is inhibited by colneleic acid. A stereocontrolled synthesis of colneleic acid is set forth below.[5]

Abbreviated Retrosynthetic Plan

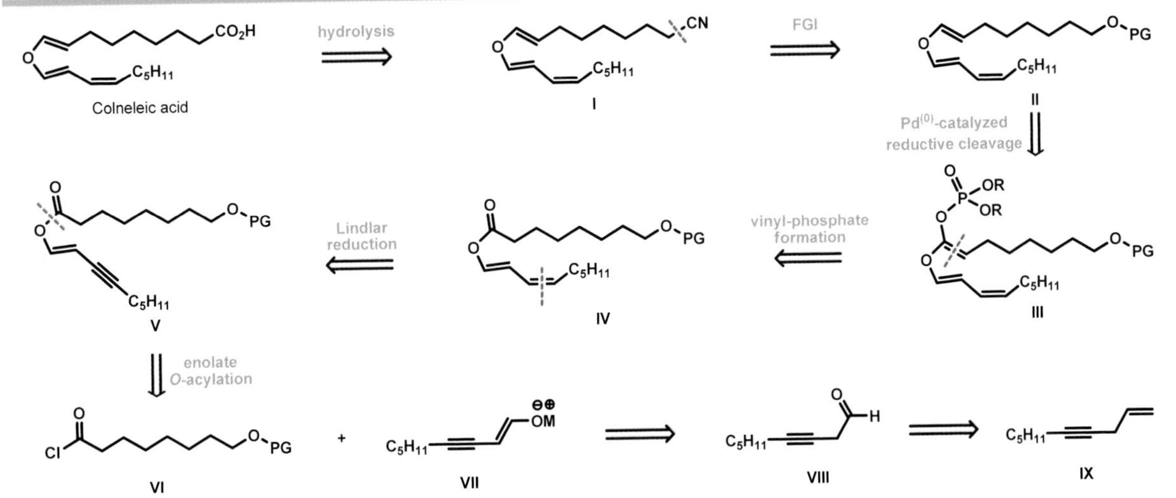

Scheme 1

Pathway of Synthesis

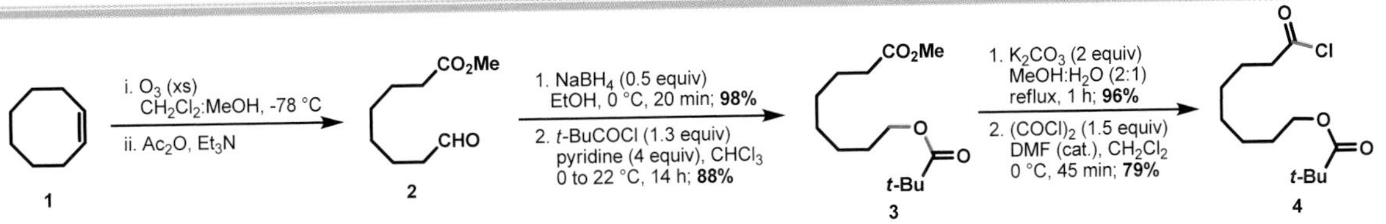

Scheme 2

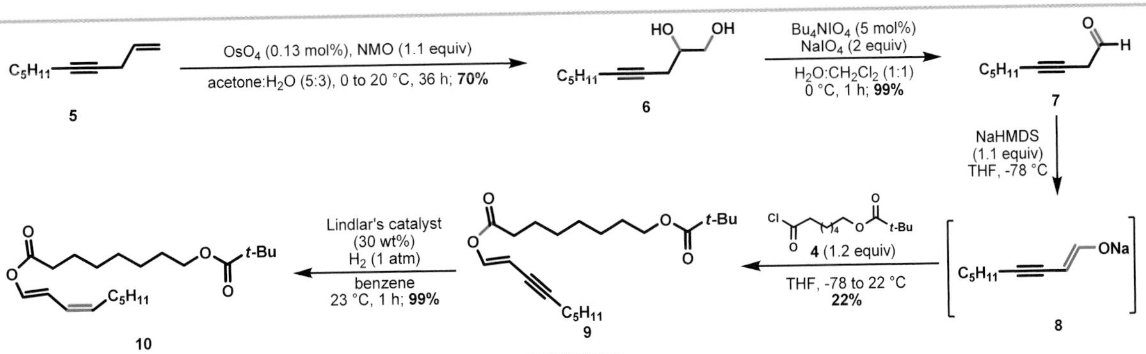

Scheme 3

COLNELEIC ACID

Pathway of Synthesis (continued)

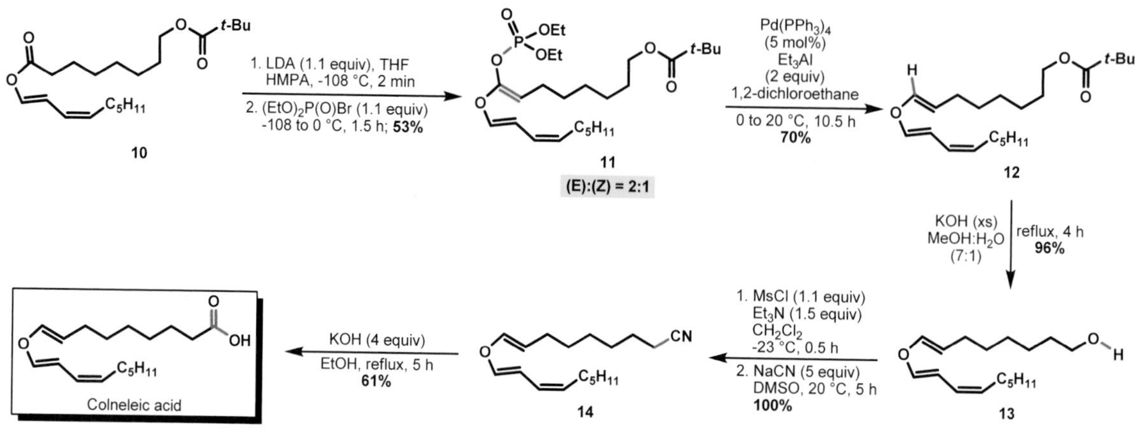

Scheme 4

Explanation of Scheme 3

(1) The conversion of **5** via **6** to the very sensitive aldehyde **7** represents a superior method for the synthesis of β,γ-unsaturated aldehydes such as **7**.

(2) Ynenol ester **9** is highly acid-sensitive and easily polymerized, therefore its preparation was accomplished by the O-acylation of sodium enolate **8** with acid chloride **4** at -78 °C. The partial hydrogenation of **9** was carried out under Lindlar's conditions to furnish **10**.

Explanation of Scheme 4

(3) Bis-enol phosphate **11** was prepared from **10** by quenching the lithium enolate derived from **10** with diethylphosphorobromidate. Compound **11** was isolated as a 2:1 mixture of C=C isomers. These isomers were separated only at the very end of the synthetic sequence (HPLC separation of methyl esters).

(4) The phosphate group in **11** was exchanged for hydrogen under Pd(0)-catalyzed conditions using Et$_3$Al as the hydride source.

(5) It was conjectured that colneleic acid might be biosynthesized from 9(S)-hydroperoxyoctadeca-10(E), 12(Z)-dienoic acid **15** by a Baeyer-Villiger-like C→O rearrangement. A biomimetic synthesis along these lines was demonstrated experimentally by the sequence shown below.[6]

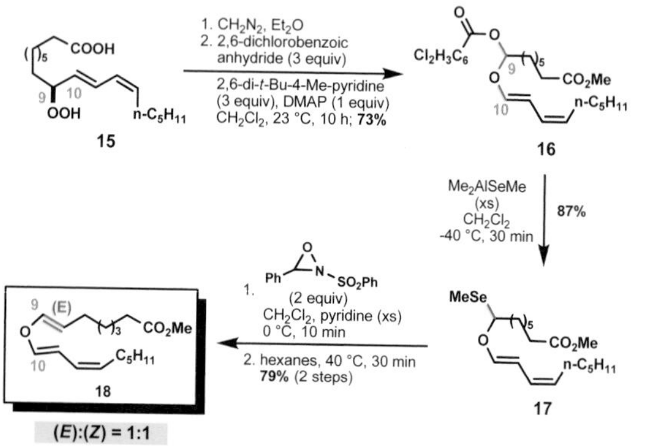

References

1. (a) Galliard, T. & Phillips, D. R. Enzymic conversion of linoleic acid into 9-(1',3'-nonadienoxy)-8-nonenoic acid, a novel unsaturated ether derivative isolated from homogenates of Solanum tuberosum tubers. *Biochem. J.* **1972**, 129, 743-753; (b) Galliard, T., Philips, D. R. & Frost, D. J. Novel divinyl ether fatty acids in extracts of Solanum tuberosum. *Chem. Phys. Lipids* **1973**, 11, 173-180.
2. (a) Corey, E. J., Albright, J. O., Barton, A. E., Hashimoto, S. Chemical and enzymic syntheses of 5-HPETE, a key biological precursor of slow-reacting substance of anaphylaxis (SRS), and 5-HETE. *J. Am. Chem. Soc.* **1980**, 102, 1435-1436; (b) Corey, E. J., Lansbury, P. T., Jr. Stereochemical course of 5-lipoxygenation of arachidonate by rat basophil leukemic cell (RBL-1) and potato enzymes. *J. Am. Chem. Soc.* **1983**, 105, 4093-4094; (c) Corey, E. J., Cheng, X. M. *The Logic of Chemical Synthesis* (Wiley, **1989**).
3. Corey, E. J., Clark, D. A., Goto, G., Marfat, A., Mioskowski, C., Samuelsson, B. & Hammarstroem, S. Stereospecific total synthesis of a

"slow reacting substance" of anaphylaxis, leukotriene C-1. *J. Am. Chem. Soc.* **1980**, 102, 1436-1439.
4. (a) Grechkin, A. Recent developments in biochemistry of the plant lipoxygenase pathway. *Prog. Lipid Res.* **1998**, 37, 317-352; (b) Grechkin, A. N. & Tarchevsky, I. A. The lipoxygenase signaling system. *Russian Journal of Plant Physiology (Translation of Fiziologiya Rastenii (Moscow))* **1999**, 46, 114-123; (c) Grechkin, A. N. & Gardner, H. W. Biocatalysis by the plant lipoxygenase pathway: Hydroperoxide-metabolizing enzymes in *Lipid Biotechnology* 183-201 (**2002**).
5. Corey, E. J. & Wright, S. W. Total synthesis of colneleic acid. *J. Org. Chem.* **1990**, 55, 1670-1673.
6. Corey, E. J., Nagata, R. & Wright, S. W. Biomimetic total synthesis of colneleic acid and its function as a lipoxygenase inhibitor. *Tetrahedron Lett.* **1987**, 28, 4917-4920.

(+)-ARBORONE & (−)-GEORGYONE

Background

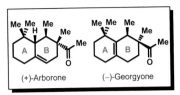

(+)-Arborone   (−)-Georgyone

Georgyone is an artificial odorant that is used, in racemic form, in many popular fragrances because of its pleasant woody odor. Arborone is an even stronger woody scent.[1] It is also sold in racemic form in admixture with several other related compounds that constitute the main components of "Iso E Super".[2] Recent advances in enantioselective synthetic methodology[3] allowed the design and demonstration of the first enantioselective syntheses[4] of these valuable molecules and the assignment of absolute configuration of the active odorant.[5] Remarkable achievements in the understanding of the mechanism of olfaction[6] made it possible to propose possible molecular binding modes of georgyone and arborone on the responsive olfactory G-protein-coupled receptors.[7]

Abbreviated Retrosynthetic Plan

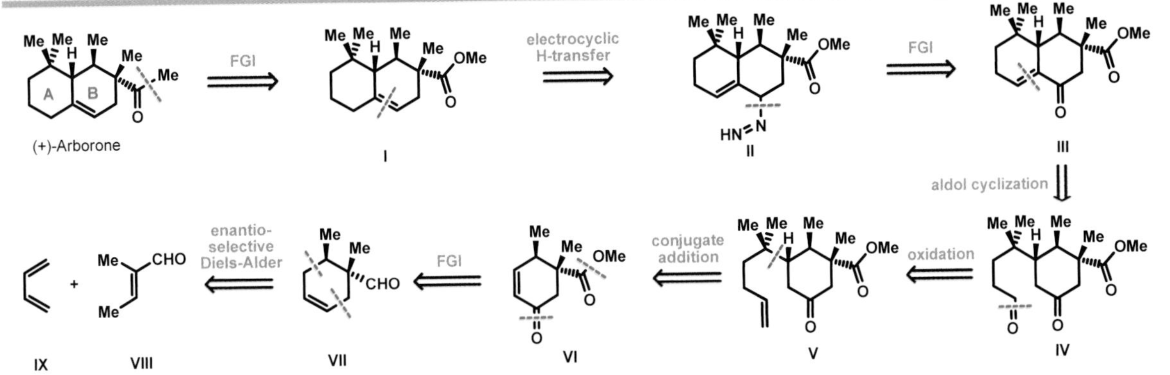

Scheme 1

Key Steps in Scheme 1

(1) The retrosynthetic plan for the synthesis of each of the enantiomers of arborone was derived by use of an enantioselective transform-based strategy, specifically based on catalysis of Diels-Alder reactions by chiral oxazaborolidinium ions. Even though arborone possesses the retron for direct Diels-Alder disconnection of ring B, the corresponding Diels-Alder reaction cannot be effected. Thus, it was necessary to disconnect ring A first. The sequence: arborone ⇒ I ⇒ ⇒ VII led to the realizable disconnection VII ⇒ VIII + IX.

(2) A similar strategy was used for the target georgyone:

  (a) disconnection of ring A and

  (b) disconnection of ring B by application of the enantioselective Diels-Alder transform.

Pathway of Synthesis

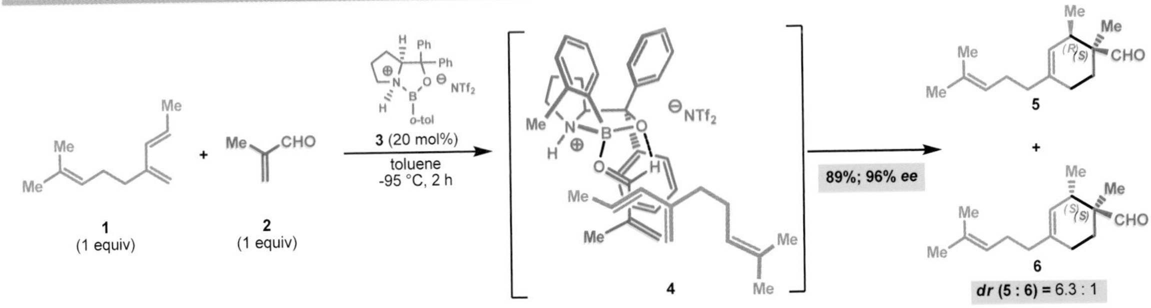

Scheme 2

(+)-ARBORONE & (−)-GEORGYONE

Pathway of Synthesis (continued)

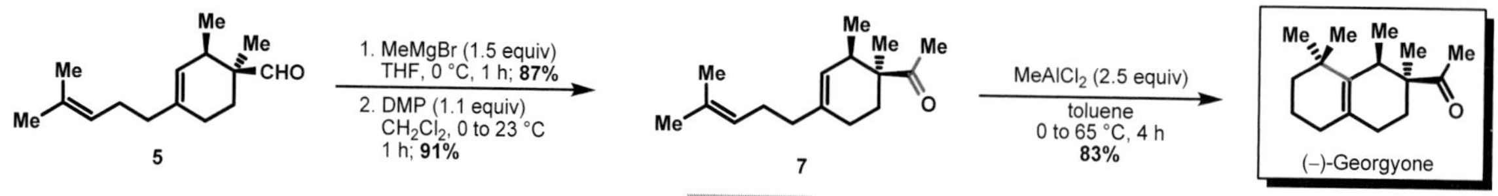

Scheme 3

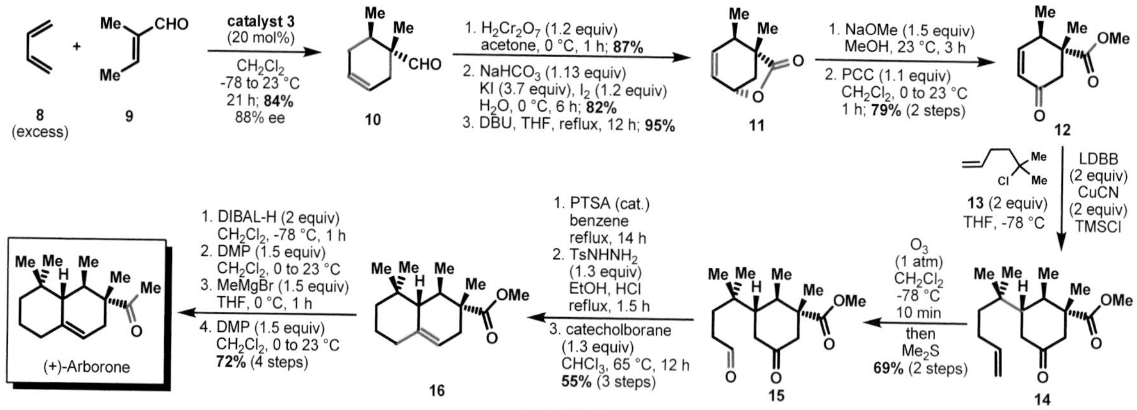

Scheme 4

Key Steps in Schemes 2, 3 and 4

(1) The principal adduct in the enantioselective Diels-Alder reaction for the synthesis of georgyone was the *endo* adduct **5** (*endo:exo* ratio = 6:3:1, Scheme 2).

(2) Cation-olefin cyclization of **7** led cleanly to (−)-georgyone which possesses an intense and pleasing woody-odor in contrast to the enantiomer which lacks the woody scent entirely and has an unpleasant musty odor (Scheme 3).

(3) The conjugate addition **12**→**14** was diastereoselective due to steric screening of the CH₃ group on the γ-carbon of the α,β-enone **12** (Scheme 4).

(4) In contrast to (+)-arborone which is a very powerful woody odorant, (−)-arborone has very little odor.

References

1. Satyanarayana, P., Rao, P. K., Ward, R. S. & Pelter, A. Arborone and 7-oxodihydrogmelinol: two new ketolignans from Gmelina arborea. *J. Nat. Prod.* **1986**, 49, 1061-1064.

2. Kraft, P., Bajgrowicz, J. A., Denis, C. & Frater, G. Odds and trends: recent developments in the chemistry of odorants. *Angew. Chem., Int. Ed.* **2000**, 39, 2980-3010.

3. For a review see: Corey E.J. Enantioselective Catalysis Based on Cationic Oxazaborolidines. *Angew. Chem. Intl. Ed*, **2009**, 48, 2100-2117.

4. For related synthetic work, see: (a) Takahashi, M., Takada, K., Matsuura, D., Takabe, K., Yoda, H. First total syntheses of new phenylpropanoid lignans, (±)-aglacin K stereoisomer and analogue of (±)-arborone. *Heterocycles* **2007**, 71, 2113-2118; (b) Hicken, E. J., Corey, E. J. Stereoselective Synthesis of Woody Fragrances Related to Georgyone and Arborone. *Org. Lett.* **2008**, 10, 1135-1138.

5. Hong, S. & Corey, E. J. Enantioselective Syntheses of Georgyone, Arborone, and Structural Relatives. Relevance to the Molecular-Level Understanding of Olfaction. *J. Am. Chem. Soc.* **2006**, 128, 1346-1352.

6. Serizawa, S., Miyamichi, K., Nakatani, H., Suzuki, M., Saito, M., Yoshihara, Y. & Sakano, H. Negative Feedback Regulation Ensures the One Receptor-One Olfactory Neuron Rule in Mouse. *Science* **2003**, 302, 2088-2094.

7. (a) Frater, G., Bajgrowicz, J. A. & Kraft, P. Fragrance Chemistry. *Tetrahedron* **1998**, 54, 7633-7703; (b) Ronnett, G. V. & Moon, C. G proteins and olfactory signal transduction. *Annu. Rev. Physiol.* **2002**, 64, 189-222; (c) Brenna, E., Fuganti, C. & Serra, S. Enantioselective perception of chiral odorants. *Tetrahedron: Asymmetry* **2003**, 14, 1-42; (d) Axel, R. Scents and sensibility: A molecular logic of olfactory perception (nobel lecture). *Angew. Chem., Int. Ed.* **2005**, 44, 6111-6127; (e) Buck, L. B. Unraveling the sense of smell (nobel lecture). *Angew. Chem., Int. Ed.* **2005**, 44, 6128-6140; (f) Bentley, R. The nose as a stereochemist. Enantiomers and odor. *Chem. Rev.* **2006**, 106, 4099-4112.

(+)-(11R,12S)-OXIDOARACHIDONIC ACID

Background

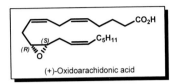

(+)-Oxidoarachidonic acid

11,12-Oxidoarachidonic acid, which is produced by endothelial cells lining the human vasculature, possesses strongly vasodilating, anti-inflammatory and anti-atherosclerotic properties leading to a cardioprotective function.[1] It also influences angiogenesis, blood pressure and renal function by the activation of certain ion channels. It was first synthesized in racemic form considerably before its biological activity was recognized.[2]

The first enantioselective synthesis is summarized below.[3]

Abbreviated Retrosynthetic Plan

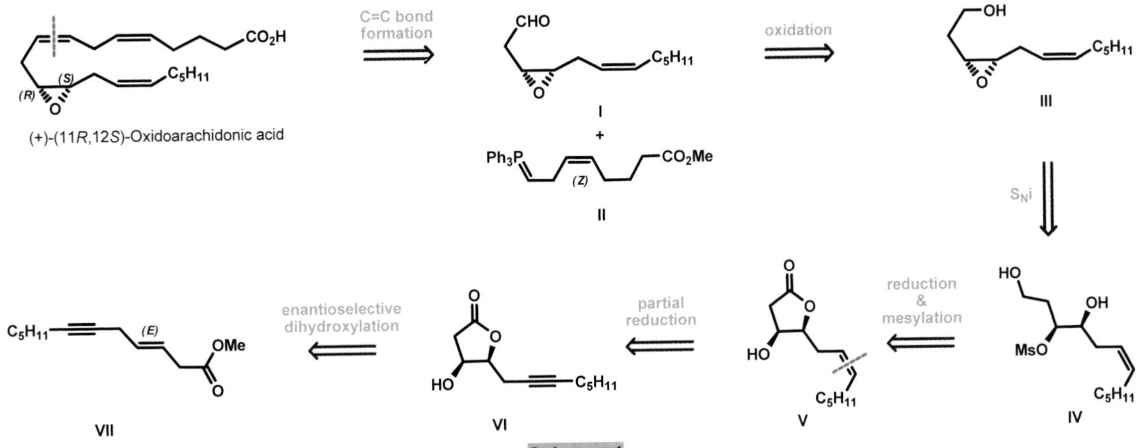

Scheme 1

Key Steps in Scheme 1

(1) A key driver of the retrosynthetic planning was the goal of using catalytic asymmetric dihydroxylation of a mono-olefinic substrate to establish the C(11) and C(12) stereocenters.

(2) The strategy of disconnecting the target near the center dictated the disconnection to **I** and **II**.

(3) Intermediate **I** was converted retrosynthetically via **III** to **VI** to the acetylenic olefin **VII** as an appropriate substrate for the asymmetric dihydroxylation.

(4) The Z alkene stereochemistry in **V** would arise by the partial hydrogenation of disubstituted alkyne **VI**.

Pathway of Synthesis

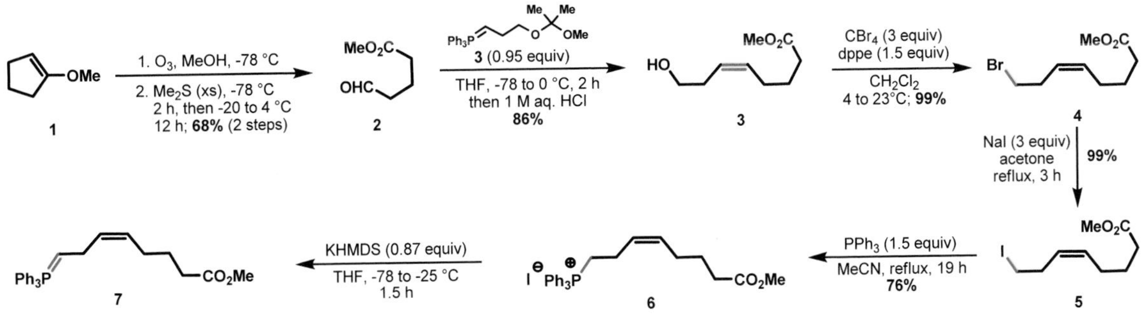

Scheme 2

(+)-(11R,12S)-OXIDOARACHIDONIC ACID

Pathway of Synthesis (continued)

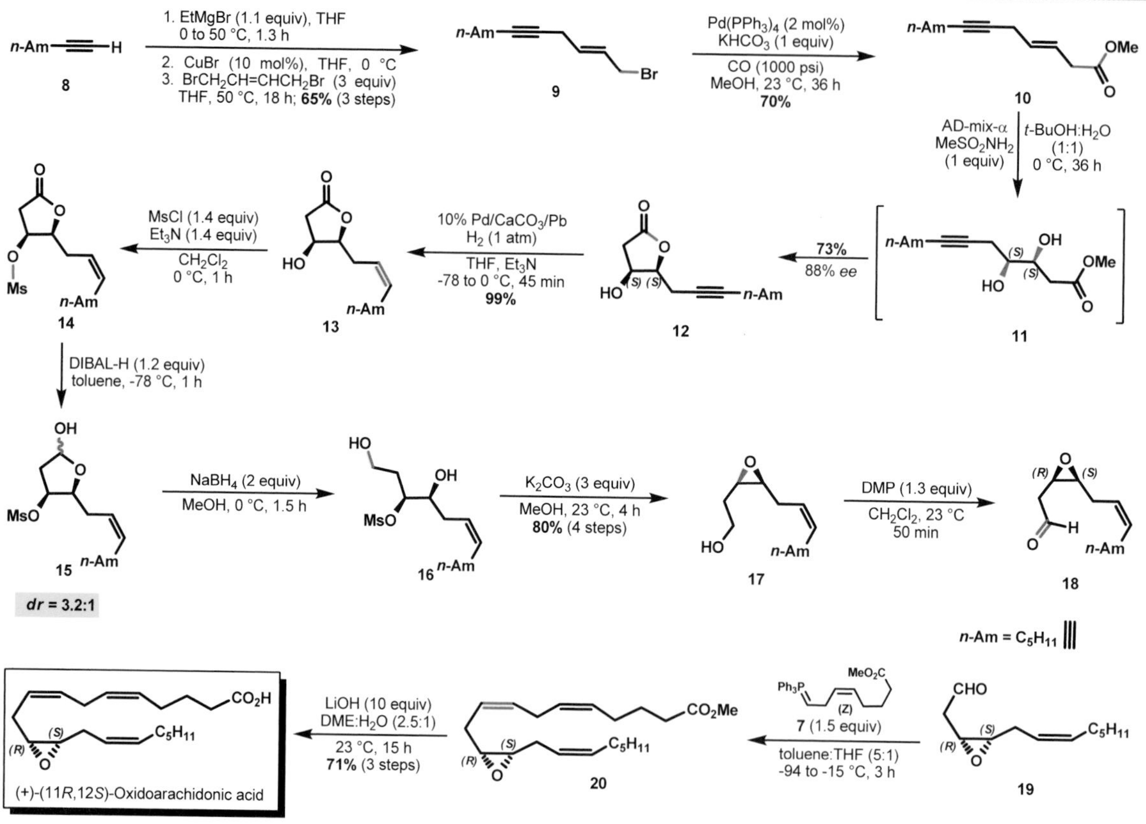

Scheme 4

n-Am = C_5H_{11}

Explanation of Scheme 3

(1) The choice of **10** for the asymmetric dihydroxylation[4] step was based on the following considerations:

  (a) the acetylenic function is resistant to dihydroxylation under conditions that suffice for olefins;

  (b) the C-C triple bond subunit is an excellent predecessor of the (Z)-CH=CH subunit and

  (c) **10** is readily made from 1-heptyne **8**.

(2) The γ-lactone function in **12** allowed facile discrimination between the two hydroxyl groups in **11** that were introduced by asymmetric dihydroxylation. The remaining hydroxyl group could readily be activated by mesylation of **13**, setting the stage for the epoxide closure step **16**→**17**.

References

1. (a) Ross, R. Atherosclerosis--an inflammatory disease. *N. Engl. J. Med.* **1999**, 340, 115-126; (b) Fisslthaler, B., Popp, R., Kiss, L., Potente, M., Harder, D. R., Fleming, I. & Busse, R. Cytochrome P450 2C is an EDHF synthase in coronary arteries. *Nature* **1999**, 401, 493-497; (c) Node, K., Huo, Y., Ruan, Z. X., Yang, B., Spiecker, M., Ley, K., Zeldin, D. C. & Liao, J. K. Anti-inflammatory properties of cytochrome p450 epoxygenase-derived eicosanoids. *Science (Washington, D. C.)* **1999**, 285, 1276-1279; (d) Campbell, W. B. New role for epoxyeicosatrienoic acids as anti-inflammatory mediators. *Trends Pharmacol. Sci.* **2000**, 21, 125-127; (e) Larsen, B. T., Campbell, W. B. & Gutterman, D. D. Beyond vasodilatation: non-vasomotor roles of epoxyeicosatrienoic acids in the cardiovascular system. *Trends Pharmacol. Sci.* **2007**, 28, 32-38.

2. Corey, E. J., Marfat, A., Falck, J. R. & Albright, J. O. Controlled chemical synthesis of the enzymically produced eicosanoids 11-, 12-, and 15-HETE from arachidonic acid and conversion into the corresponding hydroperoxides (HPETE). *J. Am. Chem. Soc.* **1980**, 102, 1433-1435.

3. Han, X., Crane, S. N. & Corey, E. J. A Short Catalytic Enantioselective Synthesis of the Vascular Antiinflammatory Eicosanoid (11R,12S)-Oxidoarachidonic Acid. *Org. Lett.* **2000**, 2, 3437-3438.

4. Kürti, L. & Czakó, B. Sharpless Asymmetric Dihydroxylation in *Strategic Applications of Named Reactions in Organic Synthesis* 406-407 (Academic Press/Elsevier, San Diego, **2005**).

(−)-WODESHIOL

Background

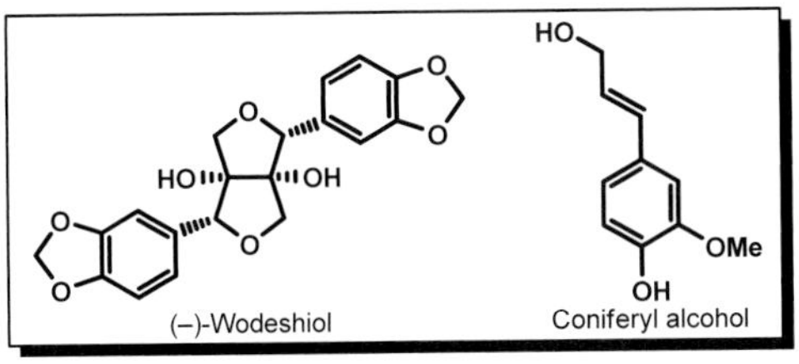

(−)-Wodeshiol

Coniferyl alcohol

Wodeshiol[1] is a member of the large lignan class of natural products[2] and is uniquely symmetric about the central 5,5-ring fusion bond. The lignans are a large and structurally diverse group of molecules which are biosynthesized in chiral form from the achiral precursor coniferyl alcohol, by various sequences of oxidation, dimerization and cyclization events. A short and simple synthesis of wodeshiol using two different catalytic enantioselective processes is presented below[3]

Abbreviated Retrosynthetic Plan

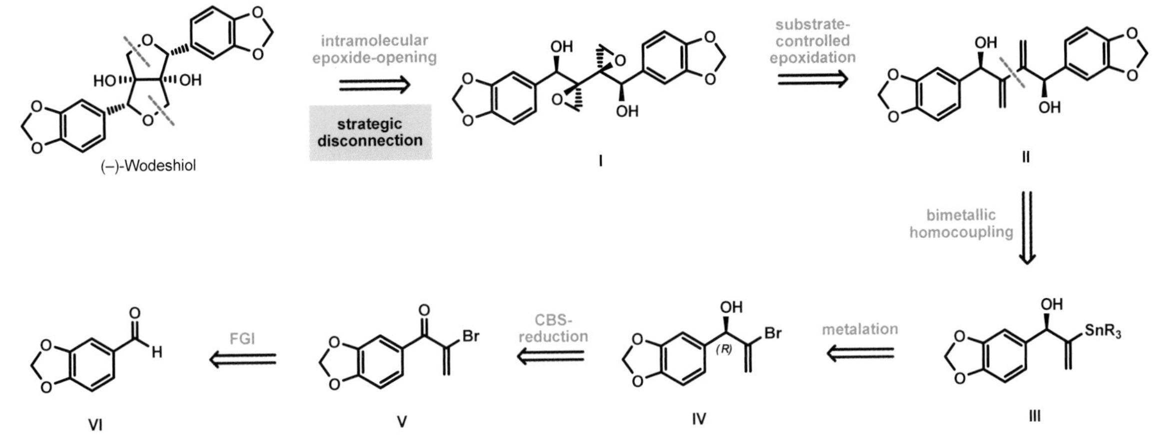

(−)-Wodeshiol

intramolecular epoxide-opening

strategic disconnection

I

substrate-controlled epoxidation

II

bimetallic homocoupling

VI

FGI

V

CBS-reduction

IV

metalation

III

Scheme 1

Key steps in Scheme 1

(1) The fused tetrahydrofuran core was disconnected by application of the alcohol-epoxide displacement transform to form I. The symmetry of wodeshiol allowed a two-fold disconnection.

(2) Application of the substrate-controlled Sharpless epoxidation transform to the symmetrical bis-epoxide I allowed clearance of two stereocenters with formation of II, which could be further simplified to the allylic alcohol III.

(3) A novel C-C coupling transform employing bimetallic catalysis was applied for the simplification of II.[4]

(4) The initiating stereocenter in IV was cleared using the CBS reduction transform to form V, which was ultimately derived from aromatic aldehyde VI.

Pathway of Synthesis

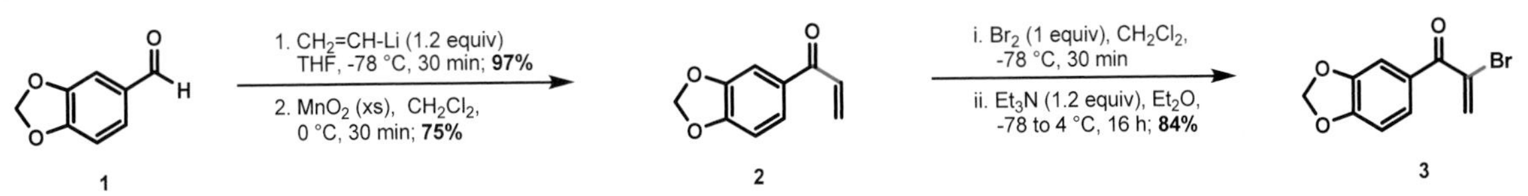

1

1. CH$_2$=CH-Li (1.2 equiv) THF, -78 °C, 30 min; 97%

2. MnO$_2$ (xs), CH$_2$Cl$_2$, 0 °C, 30 min; 75%

2

i. Br$_2$ (1 equiv), CH$_2$Cl$_2$, -78 °C, 30 min

ii. Et$_3$N (1.2 equiv), Et$_2$O, -78 to 4 °C, 16 h; 84%

3

Scheme 2

(−)-WODESHIOL

Pathway of Synthesis (continued)

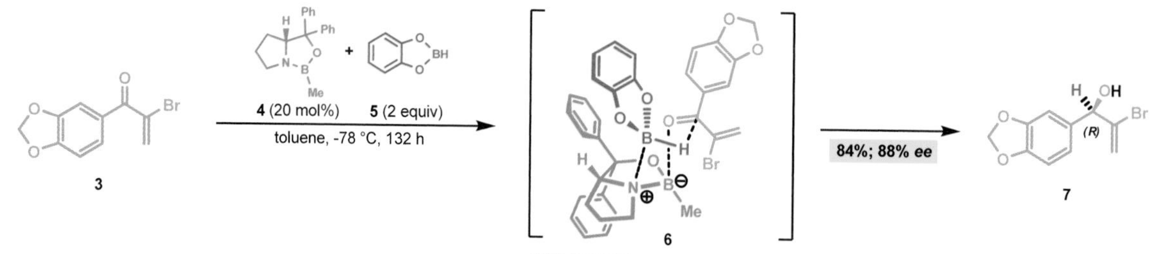

Scheme 3

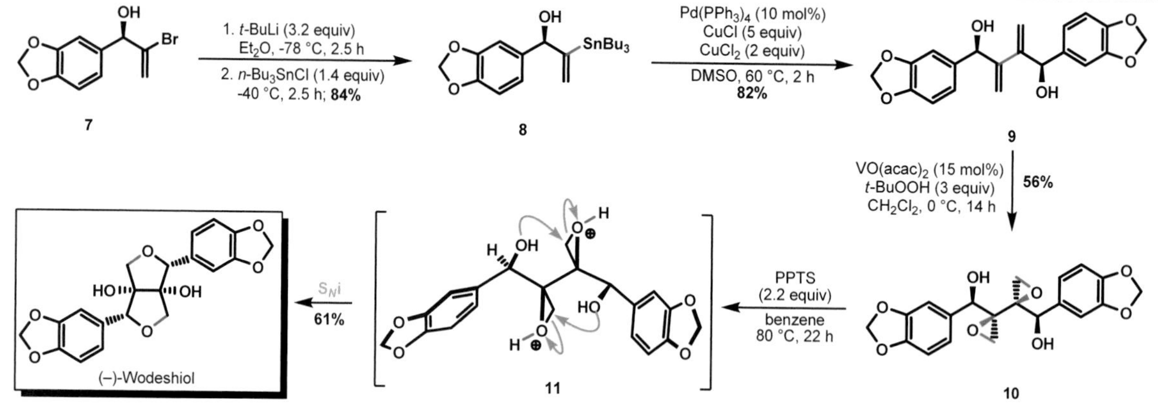

Scheme 4

Explanation of Scheme 3

(1) The introduction of the initiating stereocenter by the enantioselective reduction[5] of **3** was effected using the catalyst **4** likely by way of the pre-transition state assembly **6**.

Explanation of Scheme 4

(2) The coupling of two molecules of **8** is considered to occur by the sequence:

a. transmetallation of **8** to a vinylcopper species **12**

b. trasmetallation of vinylpalladium **13** with **12**

c. reductive elimination of **14** to give **9**.

(3) The coupling conditions indicated for **8**→**9** have been shown to be generally applicable and very effective.

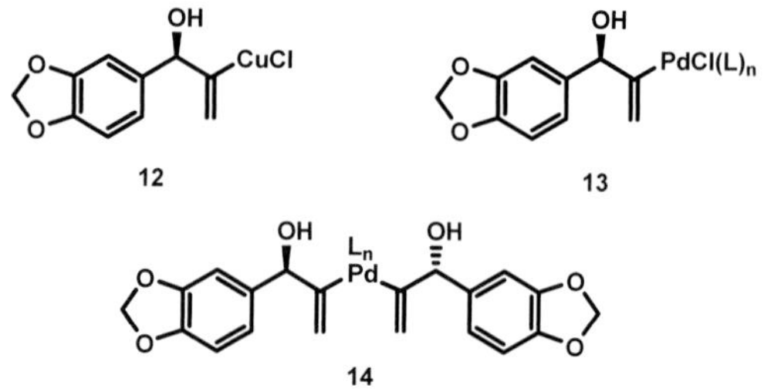

References

1. Anjaneyulu, A. S. R., Ramaiah, P. A., Row, L. R., Pelter, A. & Ward, R. S. Structure of wodeshiol. First of a new series of lignans. *Tetrahedron Lett.* **1975**, 2961-2964.
2. For recent reviews on the synthesis of lignans, see: (a) Bouyssi, D., Monteiro, N., Balme, G. Transition metal-mediated strategies toward lignans and related natural compounds. *Curr. Org. Chem.* **2008**, 12, 1570-1587; (b) Del Signore, G., Berner, O. M. Recent developments in the asymmetric synthesis of lignans. *Stud. Nat. Prod. Chem.* **2006**, 33, 541-600; (c) Brown, R. C. D., Swain, N. A. Synthesis of furofuran lignans. *Synthesis* **2004**, 811-827.
3. (a) Han, X. & Corey, E. J. A Catalytic Enantioselective Total Synthesis of (-)-Wodeshiol. *Org. Lett.* **1999**, 1, 1871-1872; (b) Cho, B. T. Recent

advances in the synthetic applications of the oxazaborolidine-mediated asymmetric reduction. *Tetrahedron* **2006**, 62, 7621-7643.
4. Piers, E., Gladstone, P. L., Yee, J. G. K. & McEachern, E. J. Intermolecular homocoupling of alkenyltrimethylstannane functions mediated by CuCl: preparation of functionalized conjugated diene and tetraene systems. *Tetrahedron* **1998**, 54, 10609-10626.
5. Kürti, L. & Czakó, B. Corey-Bakshi-Shibata Reduction (CBS Reduction) in *Strategic Applications of Named Reactions in Organic Synthesis* 100-101 (Academic Press/Elsevier, San Diego, **2005**).

(–)-SIRENIN

Background

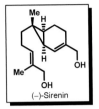

(–)-Sirenin

Sirenin is an extremely powerful pheromone of the water mold *Allomyces*.[1] An early synthesis led to racemic sirenin.[2] In this section a modern enantioselective version of that route is outlined.[3] A number of other routes to (±)-sirenin have been described.[4]

Abbreviated Retrosynthetic Plan

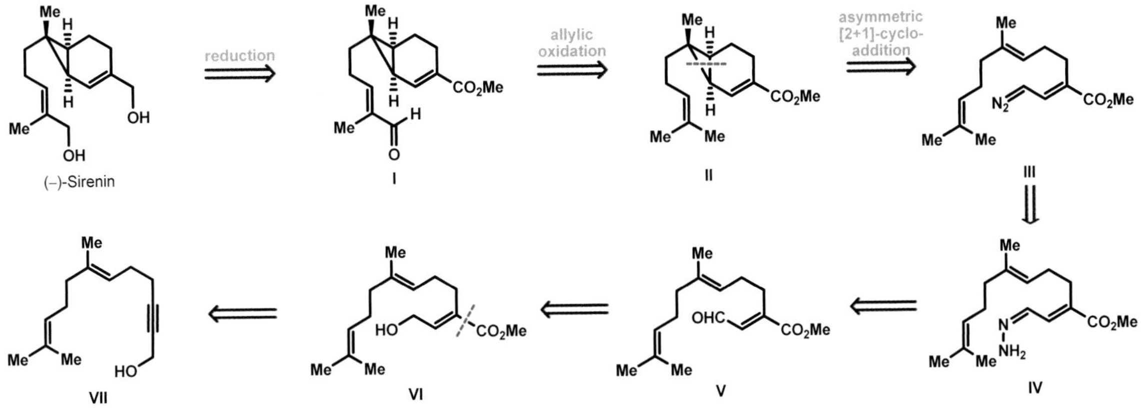

Scheme 1

Key Steps in Scheme 1

The plan was based on the selection of the enantioselective, intramolecular [2+1]-cycloaddition

transform as a powerfully simplifying disconnection of both rings of sirenin, specifically **II ⇒ III**.

Pathway of Synthesis

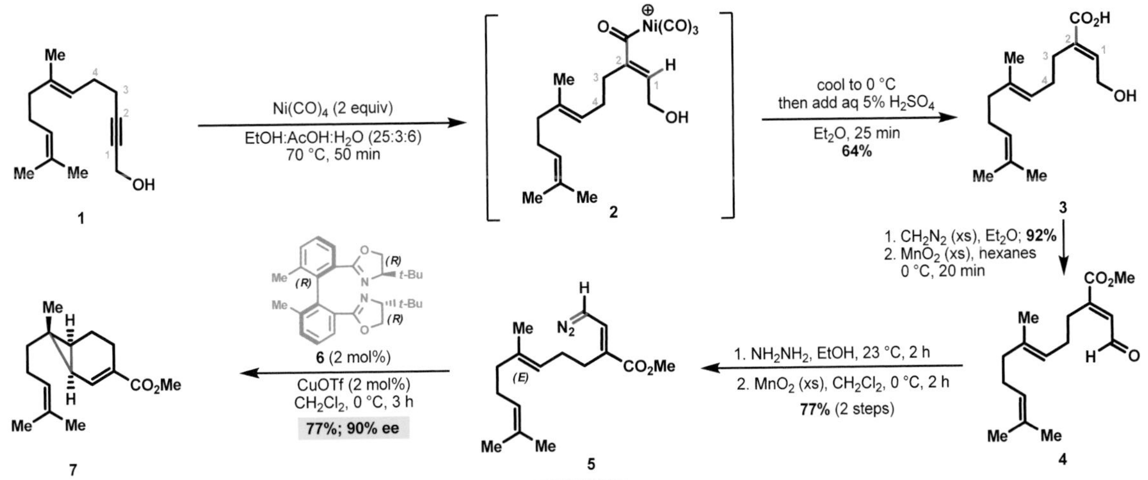

Scheme 2

(−)-SIRENIN

Pathway of Synthesis (continued)

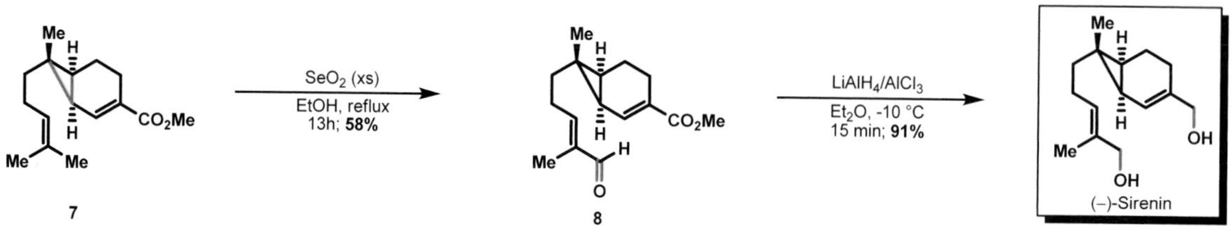

SeO$_2$ (xs)
EtOH, reflux
13h; **58%**

LiAlH$_4$/AlCl$_3$
Et$_2$O, -10 °C
15 min; **91%**

(−)-Sirenin

7   **8**

Scheme 3

9   **10**   **11**   **12**   **13**

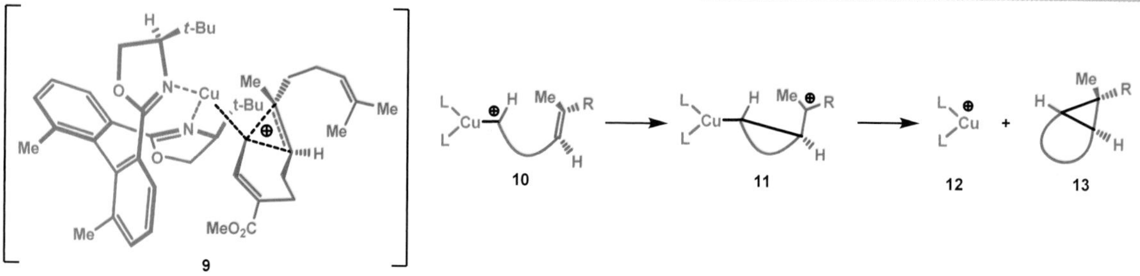

Scheme 4

Key Steps in Schemes 2, 3 and 4

(1) The hydrocarboxylation of the acetylenic alcohol **1** was based on J.W. Reppe's synthesis of acrylic acid from acetylene using Ni(CO)$_4$ as a catalyst.[5] The reaction is position- and stereoselective. The pathway is considered to involve an addition of H$^+$ and Ni(CO)$_3$ to the C-C triple bond, followed by the insertion of CO and hydrolysis of the intermediate shown (**2**, Scheme 2).

(2) The Cu(I)-complex of **6** was the most effective catalyst for the enantioselective [2+1]-cycloaddition **5→7**, the pathway of which is thought to be that expressed **9→10→11→13** (Scheme 4).

(3) A number of other chiral Cu(I) complexes were studied with the ligands (**14** through **17**) shown on the right. The observed ee's of **7** are indicated below each complex.

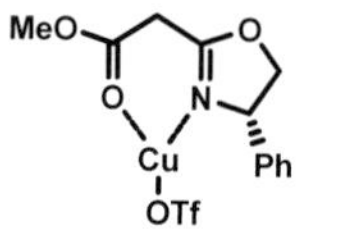

14
0% ee

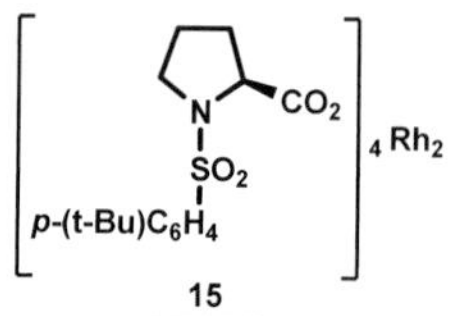

15
4% ee

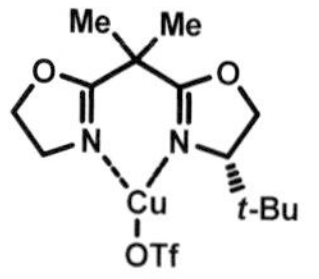

16
26% ee

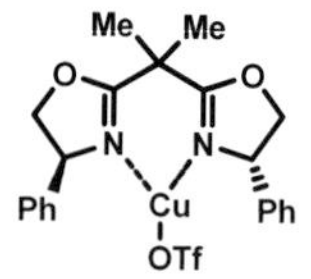

17
68% ee

References

1. Machlis, L., Nutting, W. H., Williams, M. W. & Rapoport, H. Production, isolation, and characterization of sirenin. *Biochemistry (Mosc).* **1966**, 5, 2147-2152.

2. For an early approach to racemic sirenin by the Corey group, see: (a) Corey, E. J., Achiwa, K., Katzenellenbogen, J. A. Total synthesis of dl-sirenin. *J. Am. Chem. Soc.* **1969**, 91, 4318-4320; (b) Corey, E. J., Achiwa, K. Simple synthetic route to dl-sirenin. *Tetrahedron Lett.* **1970**, 2245-2246.

3. Gant, T. G., Noe, M. C. & Corey, E. J. The first enantioselective synthesis of the chemotactic factor sirenin by an intramolecular [2 + 1] cyclization using a new chiral catalyst. *Tetrahedron Lett.* **1995**, 36, 8745-8748.

4. For recent syntheses of sirenin, see: (a) Kitahara, T., Horiguchi, A., Mori, K. Synthetic microbial chemistry. Part 19. The synthesis of (-)-sirenin, sperm attractant of the water mold Allomyces macrogynus. *Tetrahedron*

1988, 44, 4713-4720; (b) Harding, K. E., Strickland, J. B., Pommerville, J. New synthesis of (±)-sirenin and a physiologically active analog. *J. Org. Chem.* **1988**, 53, 4877-4883; (c) Mandai, T., Hara, K., Kawada, M., Nokami, J. A new total synthesis of dl-sirenin. *Tetrahedron Lett.* **1983**, 24, 1517-1518.

5. Reppe, W., Magin, A., Schuster, C., Keller, R., Kroper, H., Klein, T., Kerckow, F. W., v. Blank, G., Merkel, K., Scheller, H., Weschky, L., Wolff, K., Schweckendiek, W., Hecht, O., Gassenmeier, E. & Simon, A. Carbonylization I. Interaction of acetylene with carbon monoxide and compounds having a reactive hydrogen atom; synthesis of α,β-unsaturated carboxylic acids and their derivatives. *Liebigs Ann. Chem.* **1953**, 582, 1-37.

INDEX

INDEX